PROCEEDINGS
SIXTH INTERNATIONAL CONGRESS
INTERNATIONAL ASSOCIATION OF ENGINEERING GEOLOGY
VOLUME 1

COMPTES-RENDUS
SIXIEME CONGRES INTERNATIONAL
ASSOCIATION INTERNATIONALE DE GEOLOGIE DE L'INGENIEUR
VOLUME 1

T0166245

Comptes-rendus
sixième congrès international
association internationale de
géologie de l'ingénieur

6–10 AOUT 1990 / AMSTERDAM / PAYS-BAS

Rédacteur
D.G. PRICE
Université de Technique de Delft, Pays-Bas

VOLUME 1
Premier thème: Cartographie et études géotechniques du terrain

A.A. BALKEMA / ROTTERDAM / BROOKFIELD / 1990

Proceedings
Sixth International Congress
International Association
of Engineering Geology

6–10 AUGUST 1990 / AMSTERDAM / NETHERLANDS

Editor
D.G. PRICE
Delft University of Technology, Delft, Netherlands

VOLUME 1
Theme one: Engineering geological mapping and site investigation

A.A.BALKEMA / ROTTERDAM / BROOKFIELD / 1990

The texts of the various papers in this volume were set individually by typists under the supervision of each of the authors concerned.

Les textes des divers articles dans ce volume ont été dactylographiés sous la supervision de chacun des auteurs concernés.

Complete set of four volumes / Collection complète de quatre volumes: ISBN 90 6191 130 3
Volume 1: ISBN 90 6191 131 1
Volume 2: ISBN 90 6191 132 X
Volume 3: ISBN 90 6191 133 8
Volume 4: ISBN 90 6191 134 6
Volume 5: ISBN 90 6191 135 4
Symposia / Colloques: ISBN 90 6191 136 2

Published by:
© 1990 A.A. Balkema, Postbus 1675, 3000 BR Rotterdam, Netherlands
Distributed in the USA & Canada by: A.A. Balkema Publishers, Old Post Road, Brookfield, VT 05036, USA
Printed in the Netherlands

Publié par:
© 1990 A.A. Balkema, Postbus 1675, 3000 BR Rotterdam, Pays-Bas
Distribué aux USA & Canada par: A.A. Balkema Publishers, Old Post Road, Brookfield, VT 05036, USA
Imprimé aux Pays-Bas

Foreword

These volumes comprise the Proceedings of the Sixth International Congress of the International Association of Engineering Geology (IAEG). The discipline of Engineering Geology is dedicated to the recognition and solution of problems that may arise at the interface between Geology and the Activities of Man and is thus one of the broadest fields of endeavour within Science and Engineering. It is truly multi-disciplinary and plays a role in Civil, Mining and Environmental Engineering.

This Congress celebrates the twenty-fifth anniversary of the IAEG and the themes for the Congress were chosen to offer the greatest opportunity to display the wide range of interest and scientific activity falling within the discipline of Engineering Geology.

The themes were as follows:

Theme 1: Engineering Geology Mapping and Site Investigation
1.1 Engineering geological and environmental maps and plans
1.2 Sampling and testing in boreholes
1.3 Laboratory and in situ testing and instrumentation
1.4 Procedures, classification and interpretation for engineering design.

Theme 2: Remote sensing and geophysical techniques
2.1 Aerial and terrestrial photography and photogrammetry
2.2 Multispectral sensing and radar
2.3 Geophysics on land
2.4 Geophysics overwater

Theme 3: Hydro-Engineering Geology
3.1 Groundwater flow in dams, reservoirs, tunnels and excavations
3.2 Reduction of groundwater flow by dewatering, injection and other methods
3.3 Groundwater supply – Methods and environmental consequences
3.4 Groundwater pollution and waste disposal – Environmental impact

Theme 4: Surface Engineering Geology
4.1 Natural geotechnical hazards and the environment
4.2 Foundations
4.3 Stability of excavated slopes and embankments
4.4 Infrastructure (roads, railways, pipelines etc.)

Theme 5: Underground Engineering Geology
5.1 Tunnels and shafts
5.2 Large permanent underground openings
5.3 Engineering and environmental problems caused by subsidence brought about by underground extraction of minerals, oil and gas
5.4 Underground storage of energy, liquids and waste

Theme 6: Engineering Geology of Land and Marine Hydraulic Structures
6.1 Flood control and erosion protection; environmental impact
6.2 Offshore structures and seabed stability
6.3 Coastal protection and land reclamation
6.4 Harbours, causeways and breakwaters

Theme 7: Construction materials
7.1 Exploration
7.2 Exploitation – Methods and environmental impact
7.3 Testing and classification
7.4 Problem materials

As well as these general themes the Congress included four symposia dealing with topics considered to be of particular importance today. These were

1. Computer use in Engineering Geology
2. Environmental protection, pollution and waste disposal
3. Coastal protection and erosion, including the engineering and environmental consequences of rises in sea level
4. Engineering Geology in the oil industry

The reader will observe that many of the sub-theme and symposia titles include the word 'environment'. This is not merely a response to the current fashion to link many forms of scientific work to environmental issues, but rather the confirmation that engineering geologists have been for many years concerned with environmental problems. IAEG conferences, commissions and publications have dealt with such topics as mass movements, subsidence and seismicity and in 1979 an IAEG conference in Poland dealt with the topic of 'Changes of the geologic environment under the influence of man's activity'. Because geology related environmental problems are multi-facetted engineering geologists are particularly suited not only to warn of their approach but also to aid in their solution. It seems clear that Engineering Geology must and will become committed to this activity.

David G. Price

6th International IAEG Congress / 6ème Congrès International de AIGI, © 1990 Balkema, Rotterdam. ISBN 90 6191 130 3

Avant-propos

Ces volumes constituent les Actes du Sixième Congrès International de l'Association Internationale de Géologie de l'Ingénieur (AIGI). La matière de la Géologie de l'Ingénieur est consacrée à l'identification et à la solution des problèmes pouvent survenir entre la Géologie et les Activités de l'Homme et elle représente de ce fait l'un des terrains d'applications le plus vaste dans le cadre de la Science et de l'Ingénierie. C'est une branche véritablement multi-disciplinaire qui a son rôle à jouer dans l'Ingénierie Civile, de l'Exploitation minière et de l'Environnement.

Ce Congrès commémore le vingt-cinquième anniversaire de l'AIGI et le choix des thèmes du Congrès permet d'offrir le plus d'occasions possibles de présenter le large registre d'intérêts et d'activités scientifiques appartenant au domaine de la Géologie de l'Ingénieur.

Les thèmes sont les suivants:

Premier thème: Cartographie et études géotechniques du terrain
1.1 Cartes et plans géotechniques et de l'environnement
1.2 Echantillonnage et essais dans les trous de sonde
1.3 Expérimentation et instrumentation en laboratoire et sur le terrain
1.4 Procédures, classification et interprétation pour des plans techniques

Deuxième thème: Télédétection et techniques géophysiques
2.1 Photographie et photogrammétrie aériennes et terrestres
2.2 Détections multispectrales et radar
2.3 Géophysique terrestre
2.4 Géophysique marine

Troisième thème: Géologie Hydrotechnique
3.1 Courant phréatique dans les digues, réservoirs, tunnels et excavations
3.2 Réduction du courant phréatique par drainage, par injection et par d'autres méthodes
3.3 Approvisionnement phréatique – Méthodes et conséquences pour l'environnement
3.4 Pollution de l'eua souterraine et traitement des déchets: Influence sur l'environnement

Quatrième thème: Géologie de l'ingénieur de surface
4.1 Dangers géotechniques naturels et l'environnement
4.2 Fondations
4.3 Stabilité des pentes excavées et des digues
4.4 Infrastructure (routes, chemins de fer, pipelines, etc.)

Cinquième thème: Géologie de l'ingénieur souterraine
5.1 Tunnels et fosses
5.2 Grands vides souterrains
5.3 Problèmes techniques et de l'environnement causés par la subsidence à la suite de l'extraction souterraine de minéraux, de pétrole et de gaz
5.4 Stockage souterrain de l'énergie, des liquides et des déchets

Sixième thème: Géologie de l'ingénieur des structures hydrauliques sur terre et sur mer
6.1 Contrôle de la marée et protection contre l'érosion; influences sur l'environnement
6.2 Structures offshore et stabilité du fond de la mer
6.3 Protection littorale et accroissement territoriale
6.4 Ports, levées et bries-lames

Septième thème: Materiaux de construction
7.1 Exploration
7.2 Methodes de l'exploitation et influences sur l'environnement
7.3 Essais et classification
7.4 Matériaux problématiques

En même temps que ces thèmes d'intérêt général le Congrès compte également quatre symposiums traitant des sujets considérés aujourd'hui comme étant d'une importance toute particulière. Ceux-ci étaient:

1. L'ordinateur dans la géologie de l'ingénieur.
2. Protection de l'environnement, pollution et traitement des déchets.
3. Protection littorale et érosion, y compris les conséquences techniques et les conséquences pour l'environnement de la hausse du niveau de la mer.
4. Géologie de l'ingénieur dans l'industrie pétrolière: Colloque organisé en association avec la Société de technologie sousmarine.

Le lecteur notera que le terme 'environnement' figure dans un grand nombre de titres du thème subsidiaire et des symposiums. Il ne s'agit pas simplement d'une réponse à la mode actuelle qui cherche à relier de nombreuses formes d'activités scientifiques à des questions écologiques, mais bien d'une confirmation du fait que les géologues ingénieurs sont depuis longtemps concernés par les problèmes de l'environnement. Les conférences, les commissions et les publications de l'AIGI ont traité de sujets tels que les mouvements de masse, les affaissements de terrain et les mouvements sismiques et, en 1979, lors d'une conférence de l'AIGI en Pologne, a été traité le thème des 'Changements de l'environnement géologique sous l'influence de l'action humaine'.

Les problèmes de l'environnement relatif à la géologie pouvant présenter des aspects très variés les géologues ingénieurs sont particulièrement compétents pour prévenir de leur apparition mais aussi pour aider à les résoudre. Il est clair que la Géologie de l'Ingénieur sera amenée à être confrontée à ces activités.

David G. Price

INTERNATIONAL ASSOCIATION OF ENGINEERING GEOLOGY
ASSOCIATION INTERNATIONALE DE GEOLOGIE DE L'INGENIEUR

EXECUTIVE COMMITTEE / COMITE EXECUTIF

President/Président	Prof. Dr O.White	Canada
Secretary General/Secrétaire Général	Dr L.Primel	France
Treasurer/Trésorier	M.P.Masure p.i.	France
Past Presidents/Anciens Présidents	Prof.A.Shadmon	Israel
	Acad.Prof.Dr Qu Zaruba	Czechoslovakia/Tchécoslovaquie
	Prof.Dr M.Arnould	France
	Acad.Ye. M.Sergeev	USSR/URSS
	Prof. Dr M.Langer	FR Germany/Allemagne, RF
Honorary President/Président d'Honneur	Prof. Dr M.Arnould	France
Vice-Presidents/Vice-Présidents (On a regional basis/Selon la région)		
Africa/Afrique	Dr R.Brancart	Ivory Coast/Côte d'Ivoire
North America/Amérique du Nord	Prof. Dr G.Kiersch	USA/EU
South America/Amérique du Sud	Prof. N.Chiossi	Brazil/Brésil
Asia/Asie	Dr A.S.Balasubramaniam	S.E.Asia/Asie
Australasia/Australasie	M.J.P.Trudinger	Australia/Australie
Eastern Europe/Europe de l'Est	Prof. V.Osipov	USSR/URSS
Western Europe/Europe occidentale	Prof. Dr R.Oliveira	Portugal

ORGANISING COMMITTEE / COMITE D'ORGANISATION

E.Oele, Chairman	Geological Survey of The Netherlands
H.Beijer, Secretary General	Geological Survey of The Netherlands
J.G.Bakker, Treasurer	Delft Geotechnics
J.E.Hageman	4-G Consult
D.G.Price	Delft University of Technology

SCIENTIFIC COMMITTEE / COMITE SCIENTIFIQUE

D.G.Price, Chairman	Delft University of Technology
N.Rengers, Secretary	International Institute for Aerosurvey and Earth Sciences (ITC)
F.B.J.Barends	Delft Geotechnics
F.C.Dufour	Netherlands Organisation for Applied Scientific Research
C.D.Green	Shell
J.Harteveldt	Fugro-McClelland
H.P.S.van Lohuizen	Delft University of Technology
P.M.Maurenbrecher	Delft University of Technology
E.F.J.de Mulder	Geological Survey of the Netherlands
J.D.Nieuwenhuis	Delft Geotechnics
H.E.Rondeel	Free University, Amsterdam
P.N.W.Verhoef	Delft University of Technology
J.van der Weide	Delft Hydraulics

COMMITTEE OF RECOMMENDATION / COMITE DE RECOMMANDATION

K.J.Beek	Rector of the International Institute of Aerospace Survey and Earth Sciences (ITC)
G.Blom	Director-General, Rijkswaterstaat
F.A.G.Collot d'Escury	President, AKZO Salt Division
C.W.M.Dessens	Director-General for Energy, Ministry of Economic Affairs
J.C.van der Lippe	Chairman, Netherlands Association of Highway Engineers
J.Th.Meyers	Chairman, Association Central Dredging Contractors
J.M.Ossewaarde	President, Royal Institute of Engineers
H.Speelman	Director, TNO-DGV Institute of Applied Geoscience
E.van Spiegel	Director-General for Science Policy, Ministry of Education and Science
Chr.Staudt	Director, Geological Survey of The Netherlands
W.Stevelink	Director, Delft Geotechnics
C.van Veen	Former Chairman, Industrial Council for Oceanology
J.van Veen	Exploration manager, Netherlands Petroleum Company
F.A.M.de Vilder	Chairman, General Association of Building Contractors
H.J.Zwart	Chairman, Royal Netherlands Geological and Mining Society

LIST OF SPONSORS / LISTE DES SPONSORS

Akzo Zout Chemie BV
Ballast Nedam Beton en Waterbouw BV
Dienst Grondwaterverkenning TNO
Eerste Nederlandse Cement Industrie (ENCI)
Fugro McClelland Engineers BV
Grind Verkoopkantoor BV
Grondmechanica Delft
Grontmij NV
Investeringsmaatschappij Nederland Energy BV
Koninklijk Nederlands Geologisch Mijnbouwkundig Genootschap
Koninklijke Luchtvaart Maatschappij KLM
Koninklijke Nederlandse Academie van Wetenschappen
Ministerie van Buitenlandse Zaken
Ministerie van Economische Zaken
MOS Beheersmaatschappij BV
Nederlandse Aardolie Maatschappij
Nordmeyer KG
PRC Holding BV
Rijks Geologische Dienst
Rijkswaterstaat
Waterloopkundig Laboratorium

6th International IAEG Congress / 6ème Congrès International de AIGI, © 1990 Balkema, Rotterdam. ISBN 90 6191 130 3

Table of contents
Table des matières

Opening Lecture
Conférence d' Ouverture

E. F.J. de Mulder   Engineering geology in the Netherlands   3
Géologie de l'ingénieur aux Pays-Bas

1 *Engineering geological mapping and site investigation*
Cartographie et études géotechniques du terrain

1.1 *Engineering geological and environmental maps and plans*
Cartes et plans géotechniques et de l' environnement

H.Albrecht   Mapping of the mechanical properties of a salt dome   23
U.Hunsche   Cartographie des propriétés méchaniques d'une voûte salifère
I.Plischke
O.Schulze

E.Alonso   Environmental-geological mapping and evaluation in the Cantabrian   31
E.Francés   Mountains, Spain
A.Cendrero   Cartographie et évaluation de la géologie et de l'environnement dans la région
cantabrique, Espagne

Ilia Broutchev   Compiling of special engineering-geological maps   39
Sur le dressement de cartes géotechniques spéciales

F.Brunori   G.I.S. in geology: an application for engineering geology maps   45
S.Moretti   G.I.S. en géologie: une application pour les cartes de géologie technique

J.M.Buachidze   Construction of small-scale sketch maps for mountain-folded areas   51
E.A.Djavakhishvili   with the example of Georgia
K.I.Djandjgava   Composition de cartes schematiques à petite échelle pour les régions de
montagnes à plissement à l'exemple de la Géorgie

Chen Jinren   Engineering geological mapping in the northern shelf of South China   55
Pang Gaocun   Sea
Cartographie géotechnique du plateau au nord de la mer de Chine méridionale

Vishnu D.Choubey   Engineering geological mapping for tunnels in the Himalayas: A rock   57
Gopal Dhawan   mass classification approach
                    Cartographie géotechnique pour les tunnels dans l'Himalaya: Une approche par
                    la classification des masses rocheuses

Vishnu D.Choubey   Landslide hazard zonation in the Garhwal Himalaya: A terrain   65
Pradeep K.Litoria   evaluation approach
                    Determination des zones à risques de glissement de terrain dans l'Himalaya: Une
                    approche par l'évaluation du terrain

E.Conte   G.I.S. aid in regional planning of quarrying activity   73
T.Farenga   La contribution du G.I.S. dans la planification régionale des carrières
P.Loiacono
M.Semeraro

T.Cserny   Some results of engineering geological mapping in the Lake Balaton   79
                    region
                    Résultats dans la cartographie géotechnique de la région du Lac Balaton

M.G.Culshaw   Applied geology maps for land-use planning in Great Britain   85
A.Forster   Cartes de géologie appliquée pour le plan d'occupation des sols en
J.C.Cripps   Grande-Bretagne
F.G.Bell

E.Francés   Environmental mapping applied to the planning of urban and natural   95
J.R.Díaz de Terán   park areas in the north coast of Spain
A.Cendrero   Cartes géoscientifiques pour la planification de parcs naturels et zones urbaines
D.Gómez Orea   sur la côte nord de l'Espagne
T.Villarino
J.Leonardo
L.Saiz

E.H.G.Goelen   Enseignements tirés de la comparaison des données des cartes   103
J.P.Dam   géotechniques et d'essais postérieurs à l'exécution des cartes
S.Ghiste   Comparison of geotechnical maps with more recent data
Cl.Polo Chiapolini
I.Bolle

T.P.Gostelow   An investigation into the origin and engineering geology of periglacial   111
                    slope deposits in Wenlock Shale, Ironbridge, UK using interactive
                    digital terrain modelling
                    Une étude sur l'origine et la géologie de l'ingénieur des dépôts de pentes
                    périglaciaires dans le Wenlock Shale (Silurien), Ironbridge, UK, utilisant la
                    modélisation de terrain par interaction digitale

H.L.Gunaratna   The engineering geology of the Samanalawewa dam site   119
                    La géologie de l'ingénieur du site du barrage de Samanalawewa

I.Hadzinakos   Engineering geological mapping and related geotechnical problems in   127
D.Rozos   the wider industrial area of Thessaloniki, Greece
E.Apostolidis   Cartographie géotechnique et autres problèmes géotechniques de la région
                    industrielle des alentours de la ville de Thessalonique, Grèce

R. Holzer
J. Vlčko
Special zoning for pumped-storage plant on Plešivská planina karst 135
plateau
Le zonage spécial pour la centrale à accumulation sur le plateau de Plešivec

T.Y. Irfan
A. Cipullo
Application of engineering geological investigation and mapping 141
techniques for stability assessment of an urban slope in weathered
rocks in Hong Kong
L'application des techniques de recherches et de cartographie géotechniques
pour l'évaluation de la stabilité d'un versant urbain de rochers dégradés à Hong
Kong

Alarico A.C. Jácomo
Antonio Carlos
Oliva Ribeiro
Etude des sols naturels pour inventariser les possibilités 151
d'implantation de centrales hydroélectriques dans la région
Amazonienne
Study of natural soils of borrow areas in the inventory and feasibility stages of
hydroelectric power plants in the Amazon region

J. Kalterherberg
The engineering geological mapping 1:25 000 of 159
Nordrhein-Westfalen, FRG A regional engineering geological map
Cartographie géotechnique à l'èchelle 1:25 000 de Nordrhein-Westfalen, RFA
Une carte géotechnique régionale

E. Krauter
J. Feuerbach
M. Witzel
The engineering geological map of Mainz/Rhine, FRG 163
La carte géotechnique de Mayence/Rhin, RFA

A. Lakov
K. Anguelov
Engineering geological mapping of natural slopes 169
Classification géotechnique de versants naturels

Li Shenglin
Shi Bin
Ren Runhu
The principles and practice on drawing the maps of urban 175
environmental engineering geology
Théories et pratiques dans l'établissement de cartes géotechniques de
l'environnement urbain

Petar Lokin
Milica
Bajić-Brković
Destruction and protection of geological environment in urban areas 181
Discussion on Belgrade example
Destruction et protection de l'environnement géologique dans des zones urbaines
– discussion sur Belgrade

Reinaldo Lorandi
Adail Ricardo
Leister Gonçalves
Maria Lúcia Calijuri
Surface geotechnical mapping of São Carlos, Brazil 189
Carte geotechnique de surface de São Carlos, Brésil

M. Masson
P. Buquet
C. Allet
Application d'une cartographie géoenvironnementale au littoral du 195
Roussillon
Application of a geoenvironmental cartography to the Roussillon's littoral

O. Mimouni
I. Saidoun
Carte géotechnique de la région d'Alger – Cas d'étude: La zone de 203
l'Aéroport Houari Boumédiène et de l'Université de Bab Ezzouar
Geotechnic map of Algiers region – Case study: The Houari Boumédiène airport
area and the University of Bab Ezzouar area

K. Ourabia
K. Benallal
Alger: Cartographie et problèmes géotechniques 211
Algiers: Geotechnical mapping and problems

Russell L.Owens   Collapsible soil hazard mapping along the Wasatch Range, Utah,   221
Kyle M.Rollins   USA
                         Cartographie des sols à danger d'affaissement le long des Montagnes Wasatch,
                         dans l'état d'Utah, aux USA

M.A.Philippart   Results of the pilot study of Ingeokaart Amsterdam   229
                         Les résultats de l'étude épreuve d'Ingeokaart Amsterdam

Niek Rengers   Large-scale engineering geological mapping in the Spanish Pyrenees   235
Robert Soeters   Cartographie géotechnique à grande échelle dans les Pyrénées espagnoles
Paul A.L.M.van
Riet
Edwin Vlasblom

Niek Rengers   Analytical photogrammetry for measurement and plotting of steep   245
Michael Weir   rock slopes
Jan Willem Rösingh   Utilisation de la photogrammétrie analytique pour la mesure et la représentation
                         de versants rocheux escarpés

G.G.Strizhelchik   Evaluation of suitability of the North Armenia territories for major   253
                         construction works
                         L'estimation de l'aptitude des terrains de l'Arménie du nord pour les grands
                         travaux de construction

A.van Schalkwyk   Engineering geological mapping for urban planning in developing   257
G.V.Price   countries
                         Cartes géotechniques pour l'urbanisme dans des pays en voie de développement

C.J.van Westen   Mountain hazard analysis using a PC-based GIS   265
J.B.Alzate Bonilla   Analyse des risques naturels en terrain montagneux, un domaine d'application
                         pour SIG sur PC

L.V.Zuquette   Geotechnical mapping: A basic document to urban planning   273
N.Gandolfi   La cartographie géotechnique: Un document fondamental pour la planification
                         urbaine

1.2 *Sampling and testing in boreholes*
Echantillonnage et essais dans les trous de sonde

M.E.Barton   The geotechnical investigation of geologically aged, uncemented   281
S.N.Palmer   sands by block sampling
                        L'investigation géotechnique, par le moyen d'échantillonnage en bloc, des sables
                        sans ciment géologiquement mûris

Giulio C.Borgia   Considerations on subsoil exploration by means of instantaneous   289
Giovanni Brighenti   drilling parameters recording
Ezio Mesini   Considérations sur la reconnaissance du sous-sol par enregistrement instantané
                        des paramètres de forage

A.J.Coerts   Determining the spatial variation of CPT properties   297
                        Détermination de la variation latérale des propriétés CPT

G. Greeuw F. Schokking	Piezocone and other measurements in an overconsolidated glacial clay Piezocone et autres mesures dans une argile glaciaire préconsolidée	303
G. P. Huijzer	Automated stratigraphic classification of CPT data Classification automatique et stratigraphique de données pénétrométriques	309
R. A. N. Mackean M. S. Rosenbaum	Geostatistical characterisation of the SPT Caractérisation géostatistique du SPT	317
O. Mohamed Ahmed O. Shire Yusuf A. Dahir Hassan R. Mortari	Standard penetration tests on eolian sands of Somalia Essais standard de pénétration sur les sables éoliens de la Somalie	323
F. Schokking R. Hoogendoorn H. C. van de Graaf	A new method for deep static cone penetration testing Une nouvelle méthode de pénétrométrie statique profonde	329

1.3 *Laboratory and in-situ testing and instrumentation*
Expérimentation et instrumentation en laboratoire et sur le
terrain

Namir K. S. AL-Saoudi Fahmi S. Jabbar	Prediction of swelling pressure of soil from suction measurements Prédiction de la pression du gonflement des sols par la mesure de la suction	339
L. Dobereiner A. L. L. Nunes C. G. Dyke	Developments in measuring the deformability of sandstones Développements dans la mesure des déformabilités des grès	345
E. J. Ebuk S. R. Hencher A. C. Lumsden	Determination of residual bond strength by the pulling test method La détermination de la force de liens résiduels par la méthode de l'essai de traction	357
M. K. El-Rayes F. A. K. Hassona M. A. Hassan A. M. Hassan	Stress-history dependent behaviour of soft clay L'influence de l'histoire de l'effort sur le comportement d'argile mou	363
H. Giannaros J. Christodoulias	Pressuremeter, laboratory and in situ test correlations for soft cohesive soils Corrélations entre essais pressiométriques, in situ et en laboratoire sur des sols mous cohérants	371
Jiří Herštus	The effects of geological history on the behaviour of Miocene claystones on a regional scale Les effets de l'histoire géologique sur le comportement d'argilolithes du Miocène à une échelle régionale	377
Jing-Wen Chen	Development and application of a hollow cylinder triaxial cell Développement et application d'une cellule cylindrique triaxiale creuse	385

Jun Fang Chen	The novel rock fracture toughness testing method: Cracked-Chevron-Notched Brazilian Disc method Une nouvelle méthode d'essai de la ténacité de fracture de roche: Méthode de disque Brésilien de l'Encoche à Chevron craqué	389
W. Kamp M. J. Cockram	Possibilities of true triaxial experiments in the laboratory for rock mechanics of the Delft University of Technology Possibilités des expérimentations traxiales authentiques dans le laboratoire de mécanique des roches à l'Université de Technologie de Delft	393
U. Kołodziejczyk	On the shear strength parameters of some silty clays Caractéristique de la résistance au cisaillement des argiles silteuses	399
G. Koukis D. Rozos	Geotechnical properties of the Neogene sediments in the NW Peloponnesus, Greece Caractéristiques géotechniques des sédiments du Neogene en Péloponnèse du N-O, Grèce	405
J. Krajewska-Pininska Z. Karska	Application of acoustic emission parameters as a criterion of stress-strain behaviour of rocks Les paramètres de l'émission acoustique comme un critère du comportement des roches du point de vue de la résistance et de la déformation	413
Li Rongqiang Kong Defang Nie Dexing	Study on mechanical property of fault gouge under ground stress environment Etude des propriétés méchaniques des failles dans la roche broyée, dans l'environnement de la contrainte du sol	419
R. Mortari T. Gerardi L. Budassi	Observations on different procedures for the oedometer test Observations sur différentes exécutions de l'essai oedométrique	425
S. F. Oni	Characterisation of tropical weathering profile derived from Precambrian gneiss-migmatite complex in parts of South Western Nigeria Caractéristiques du profil de l'altération tropicale, provenant de complexes gneiss-migmatite précambriens, dans quelques régions du Sud-Ouest nigérien	431
P. Previatello G. Rossato	On oedometric and shear moduli of sands Sur les modules oedométriques et de cisaillement des sables	441
M. M. Reyad	Study of the effect of some factors on shear strength of expansive soils in Egypt Etude des effets de quelques facteurs sur la résistance au cisaillement de sols expansifs en Egypte	447
Rakesh Sarman Donald F. Palmer	Dielectric behavior of rocks under uniaxial compressive stress Le changement de la constante diélectrique de rochers sous une compression uniaxiale	451
Rakesh Sarman Abdul Shakoor	Prediction of volumetric increase of selected mudrocks La prédiction de l'augmentation volumétrique des roches argileuses	459
R. K. Srivastava A. V. Jalota Ahmad A. A. Amir	Shear behaviour and strength prediction studies on an Indian quartzite and sandstone Une étude qui prédit l'allure et la résistance de cissaillement sur le grès et le quartzite indiens	467

E. I. Stavridakis K. A. Demiris T. N. Hatzigogos	Correlation of slaking with the unconfined compressive strength of cement stabilized clayey admixtures Corrélation du pourcentage de l'érosion avec la résistance à la compression simple des mixtures d'argile stabilisés avec le ciment	473
Tianbin Li Lansheng Wang	An approach to the determination of geostress using the Kaiser effect Recherche en détermination de géostress au moyen de l'effet de Kaiser	481
L. E. Torres	Les caractéristiques géotechniques des terrains résiduels The geotechnical characteristics of the residual soils	487
P. N. W. Verhoef H. J. van den Bold Th. W. M. Vermeer	Influence of microscopic structure on the abrasivity of rock as determined by the pin-on-disc test L'influence de la texture microscopique sur l'abrasivité des roches, déterminée avec l'essai 'pin-on-disc'	495
Y. X. Wu Z. H. Zhang Z. M. Ling	Quantitative approach on micro-structure of engineering clay Etude quantitative de la microstructure de sols cohérents liée aux travaux	505
S. Xu, P. Grasso A. Mahtab	Use of Schmidt hammer for estimating mechanical properties of weak rock Utilisation du Schmidt hammer pour l'estimation des propriétés méchaniques des roches faibles	511
H. Yamaguchi	Physico-chemical and mechanical properties of peats and peaty ground Propriétés physico-chimique et mécanique de tourbes et terres tourbeuses	521
H. Yamaguchi Y. Mori I. Kuroshima M. Fukuda	Geotechnical properties of tertiary mudstone ground Propriétés géotechniques de terre pélite de l'ère tertiaire	527
N. F. Zorn P. E. L. Schaminée	Triaxial testing of rock/soil under cryogenic conditions Test triaxial de roches/sols sous conditions cryogéniques	535

1.4 *Procedures, classification and interpretation for engineering design*
Procédures, classification et interprétation pour des plans techniques

Joseph Olusola Akinyede Keith Turner Niek Rengers	Use of remote sensing to estimate earthwork volumes for road construction Utilisation de la télédétection pour estimer le volume de travail lors de la construction d'une route	543
M. Arenillas I. Cantarino R. Martinez A. Pedrero	Hydric resources generated by snow melting on the Spanish mountain Ressources hydriques produites par la fonte de la neige dans les montagnes espagnoles	551

Oliver B. Barker	A holistic approach to detailed geotechnical data gathering in southern Africa Une approche holistique pour détailler des données géotechniques recueillies dans le sud de l'Afrique	559
V. Bräuer A. Pahl	Special purpose engineering geological methods for mapping and interpretation of rock mechanical phenomena Méthodes de géotechnique specialisée pour la cartographie et l'interprétation de phénomènes de mécanique des roches	567
Vishnu D. Choubey Shailendra Chaudhari	Engineering performance evaluation of the foundation rocks of Narmada Sagar Dam Project, Central India Evaluation des performances techniques des fondations rocheuses du projet de barrage de Narmada Sagar, Inde Centrale	575
J. H. de Beer	Stages of geotechnical investigation for townships Etapes d'investigations géotechniques pour communes	581
Ed F. J. de Mulder	Recent developments in Urban Geology Développements récents dans la géologie urbaine	585
Y. R. Dhar V. D. Choubey H. S. Pandalai	Site characterisation for tunnels in jointed rock masses Caractéristiques de site, pour tunnels, dans des masses rocheuses compactes	593
A. Eusebio E. Rabbi P. Grasso	Geological and geotechnical characterization of the morainic 'Amphitheater of Rivoli' in NW Italy Caractéristiques géologique et géotechnique de 'l'Amphithéâtre de Rivoli' dans le N-O de l'Italie	601
M. Favaretti P. Previatello	Recent soils of the Po River: Statistical analysis of geotechnical properties Les terrains récents du Pô: Analyse statistique des propriétés géotechniques	609
S. R. M. Ferreira	Geology and pedology related to collapsible soils in Pernambuco-Brazil Géologie et pédologie rapportées aux sols à structure ouverte de Pernambouc-Brésil	617
P. G. Froldi S. Mantovani	An experimental-statistical approach to study the shear strength of rock joints Une approche statistique-expérimentale à l'étude du comportement au cisaillement des discontinuités naturelles	623
L. K. Ginzburg	Characteristic features of engineering geological survey on landslide prone slopes Les particularités de la reconnaissance du sol de fondation des pentes, subissant des glissements	627
H. R. G. K. Hack C. S. Kleinman G. J. de Koo	A case history of the problems of determining the ground parameters of 300 sites in 6 months in Indonesia L'histoire des problèmes entraînés en déterminant les paramètres du sol sur 300 endroits, en 6 mois, en Indonésie	635

Hu Ruilin	Evaluation and microcomputer mapping on construction foundation suitability in Ningbo	643
	Evaluation de l'aptitude de la fondation à la construction à Ningbo et cartographie par micro-ordinateur	
J.M.Kate	Anisotrophy assessment of different rocks	649
	Estimation des anisotropies de différents rochers	
Kehe Wei Dianxuan Liu	The zoning system for assessment of weathering states of granites	657
	Le système de classification pour l'estimation de l'état d'altération de granites	
Ibrahim Komoo Jasni Yaakub	Engineering properties of weathered metamorphic rocks in Peninsular Malaysia	665
	Les propriétés techniques de roches métamorphiques altérées en Malaisie	
R.R.Kronieger	Geotechnical uniformity of the Weichselien Loess sequence in South Limburg, the Netherlands	673
	L'uniformité géotechnique dans la séquence du loess weichselien dans le sud du Limbourg, Pays-Bas	
Liang Jinhuo Sun Guangzhong	Geological prediction in the Jundushan Tunnel	683
	Prédiction géologique dans le tunnel de Jundushan	
C.E.Ligtenberg-Mak P.V.F.S.Krajicek C.Kuiter	Geological study of flow slide sensitive sediments	691
	Etude géologique sur la sensibilité des sédiments pour la rupture par écoulement	
P.L.Narula Y.P.Sharda	Dispersive soils – A constraint for embankment dams in outer Himalaya	697
	Sols dispersés-une contrainte pour les barrages en terre dans l'Himalaya extérieur	
H.J.Olivier	Some aspects of the engineering-geological properties of swelling and slaking mudrocks	707
	Considérations sur les propriétés des argilites gonflantes dégradables appliquées à la géologie de l'ingénieur	
V.I.Osipov	Physico-chemical fundamentals of soil microrheology	713
	Les bases physico-chimiques de la microrhéologie de sols meubles	
S.J.Plasman R.R.Kronieger	An engineering geological classification of the Loess deposits in the Netherlands	725
	Une classification pour la géologie de l'ingénieur des dépôts de Loess aux Pays-Bas	
M.Šamalíková	Regional engineering geological evaluation of weak zones	733
	Evaluation géotechnique régionale des zones affaiblies	
Shufang Xiao Renjiu Pang Minxun Ding	The fractal analysis of strength of intercalated clay layer in dam foundation	739
	Etude d'analyse fractionnelle de la résistance de la couche d'argile intercalée dans la fondation d'un barrage	

Soedibjo Geotechnical classification and determination of the rock materials 745
properties used for the embankment of Wadaslintang Dam, Central
Java, Indonesia
Classification et détermination géotechnique des propriétés des matières
rocheuses utilisées au remblai du barrage de Wadaslintang, Java centrale,
Indonesie

V.N.Sokolov Engineering-geological classification of clay microstructures 753
Classification des microstructures des roches argileuses en géologie de
l'ingenieur

O.Tesař The classification of rocks for underground structures and results of 761
engineering – geological observations
La classification des roches pour les constructions souterraines et les résultats de
l'observation de la géologie de l'ingénieur

D.G.Toll Use of an expert system in site investigation 767
P.B.Attewell Utilisation d'un système expert dans l'investigation des sites

Martin Th.van Another approach to discontinuity shear strength assessments, based 775
Staveren on investigations of bore cores
Une autre approximation des estimations de la force de coup d'incohérences
basée sur des investigations de carottes de forage

A.A.M.Venmans Classification of Dutch peats 783
E.J.den Haan Classification des tourbes néerlandaises

S.T.Wang A numerical simulation study of core disking 789
R.Q.Huang Recherche sur la simulation numérique de carotte discoïde

Xu Jixian Simulation tridimensionnelle de la blocométrie naturelle de massifs 797
R.Cojean rocheux
Three-dimensional simulation of natural rock mass granulometry

Zhou Chuangbing Application of surface spline function to engineering geology 803
Application de fonction 'spline' de surface à la géologie de l'ingénieur

Opening Lecture
Conférence d'Ouverture

6th International IAEG Congress / 6ème Congrès International de AIGI, © 1990 Balkema, Rotterdam. ISBN 90 6191 130 3

Engineering geology in the Netherlands
Géologie de l'ingénieur aux Pays-Bas

E. F. J. de Mulder
Geological Survey of the Netherlands, Haarlem, Netherlands

ABSTRACT: Windmills, cows and wooden shoes: that is the traditional image foreigners have of the Netherlands. These three elements reflect the problems this country has with its soil conditions and its permanent struggle against water encroaching from the sea and upward percolation. Holland, the western part of the Netherlands, where the major activities are concentrated, lies almost entirely below sea level. Without the protection of the dunes and sea dikes in the west, without the river dikes in the centre and the east and without constant pumping Holland would be drowned immediately. Beacuse of the thick succession of unconsolidated Holocene deposits almost all constructions have to be built on a foundation of piles that are supported by firm Pleistocene sands. These particular geographical, hydrological and geological conditions have contributed to extensive studies in soil-mechanical engineering, to the development of several special surveying techniques, such as Cone Penetrometer Testing, and to the development of special geological mapping techniques in areas where no natural outcrops occur. The location of the Netherlands on a delta in the subsiding North Sea Basin downstream of the main European rivers Rhine and Meuse causes scarcity of coarse construction material, like gravel, an almost complete lack of dimension stone but an abundance of clays and silts for the ceramic industry. In a densely populated and low-lying country with soft deposits, man-induced land subsidence is a major problem. Large engineering works such as the reclamation of part of the central Lake IJssel cause subsidence of the adjacent land areas resulting in costly damage to constructions. Recent research in engineering geology in the Netherlands focusses on mathematical representations of geological parameters and numerical modelling. This kind of research is expected to contribute to future progress in engineering geology.
The development of computer expert systems, of Three-Dimensional Geographic Information Systems, of time-dependent geoscientific simulation models and the improvement of existing surveying methods, both for direct and indirect parameter assessment on a micro and a macro scale are foreseen for the near future.

RESUME: Des vaches, des sabots de bois et des moulins à vent: voilà les trois éléments essentiels de l'image traditionelle évoquée par les Pays-Bas. En même temps, ces éléments sont l'expression des problèmes de ce pays vis à vis de l'état du sol et de sa lutte perpétuelle contre l'eau de mer et l'eau du sol qui menace de l'envahir. La Hollande, c'est à dire la partie occidentale des Pays-Bas, où se trouvent les centres à population dense et les régions industrielles les plus importantes, est située presque entièrement en dessous du niveau de la mer. Sans la protection des dunes, des digues de mer à l'ouest, des digues fluviales au centre et à l'est, et sans le pompage incessant de l'eau du sol, la Hollande serait inondée. A cause des paquets denses de sédiments holocènes non consolidés, presque tous les bâtiments en Hollande sont construits sur pilotis dans des couches de sable pleistocènes bien stables. Ces circonstances géographiques et géologiques exceptionelles ont sans doute influencé la pratique considérable de la mécanique du sol en tant que science et profession et le développement de plusieurs techniques spécifiques de l'exploration du sol telles que la "Cone Penetrometer". Ces conditions ont également donné lieu au développement de techniques cartographiques spécialement prévues pour la géologie dans une région où les affleurements naturels du sol font défaut. Les cartes en question montrent, de façon tridimensionnelle, la composition des couches dans les sédiments Holocènes. La situation du pays en aval de grands fleuves européèns comme le Rhin et la Meuse, avait pour conséquence le manque de gravier en tant que matériel de construction et entraînait la surabondance d'argile pour l'industrie de céramique grossière. Dans un pays de basses terres, situé au bord de la mer, qui doit faire face à une pression démographique considérable et à un sol facilement compressible, l'affaissement du sol constitue une problème essentiel. En cas d'interventions à grande échelle dans l'équilibre actuel, telle que l'aménagement d'un nouveau polder, les bâtiments risquent de subir des dégâts considérables par suite d'un affaissement du sol. A présent, deux éléments jouent un rôle important dans le domaine de la recherche fondamentale de la géologie de l'ingénieur, à savoir l'approche mathématique concernant de la correlation des données du sol et la recherche concernant la variabilité à l'intérieur des couches. Pareille recherche contribue aux évolutions futures prévues dans le domaine de la géologie de l'ingénieur aux Pays-Bas, telles que le développement de systèmes d'expertise, de systèmes d'informations géographiques tridimensionnels, de modèles de simulation géo-scientifiques temporels et de techniques directes et indirectes améliorées d'exploration du sol, aussi bien à l'échelle micro- qu'à l'échelle macro-géologique.

1. INTRODUCTION

The Dutch have a blasphemic expression: "God created the world but the Dutch created the Netherlands". When we compare old maps with recent ones there seems to be some truth in this expression. Large parts of the country have been reclaimed from the sea. The Netherlands are protected by the natural coastal defence system of the dunes and by impressive sea dikes.
If the dikes failed; 60% of the Netherlands would be flooded by the sea. The continuous struggle against the invading sea is aptly described in the motto of the coat of arms of the Province of Zeeland which states: "Luctor et emergo": "I struggle and I emerge". However, this saying became macabre reality when the dikes failed indeed and the whole Province was flooded in the catastrophic flood of February 1953. This struggle against the sea effectively started some 1000 years ago, when the former inhabitants of this swampy area constructed artificial mounds to protect their homes and properties from flooding. Later, these hills were connected and thus the first primitive dikes were constructed. Under the pressure of repeated marine transgressions this technology has been improved ever since. Now, there are no major problems in dike construction anymore.

This development explains why Holland (which is the western part of the Kingdom of the Netherlands) is, and has always been, a paradise for civil engineers: in the first place, it is they who have built our artificial coastal-defence system, but engineers also were needed to design special foundations in Holland, where soil conditions are so poor that no building can survive that is not built on a foundation of piles. Without exageration, we can say that Holland in total is built on piles: every year some 900,000 piles are driven through the weak and soft top layers (Heijnen, 1985).

Therefore, it is obvious that foundation engineering and soil mechanical engineering became very important professions in this country. The Netherlands produced many great soil mechanical engineers, such as Terzaghi, Koppejan, and Keverling Buisman. The latter founded the soil mechanics department at the Technical University in Delft in 1934. "Delft Geotechnics" was established in the same year (Visser et al., 1987).

Then, the implementation of the "Delta Project" which entailed the construction of a sophisticated system of dikes, locks and barriers, after the disastrous flood of 1953, was a new challenge for hydraulic and soil mechanical engineers and generated the development of new techniques and methods. This, moreover, resulted in an expansion of the Dutch geotechnical industry, in which about 700 engineers, engineering geologists, technicians, laboratory analysts and field operators are employed now. Furthermore, the Dutch geotechnical industry is very active abroad, about 35% of their annual income is earned there.

Until quite recently, geologists were not much involved in engineering practice in the Netherlands. The traditional opinion among geologists in this country was that geology was not practised in such a muddy and flat country, in which the highest "mountain" rises only 321 metres (above mean sea level). This view changed only some 30 years ago when geotechnical engineers started to realize that constructions can be designed much cheaper and safer when geological

Figure 1: Depth contour map (in m below sea level) of the top of the Pleistocene deposits. The low areas in the west formed the main corridors for the Holocene transgressions.

information on the spatial distribution of geotechnically relevant beds in the subsurface would be available. At about the same time the Geological Survey started a new programme of systematic geological mapping in the southwest of this country, where the earlier-mentioned Delta Project became implemented. A succession of young geologists became involved in the particular engineering aspects and problems of Quaternary Geology. Co-operation between geologists and engineers became self evident in such projects. The previous director of the Geological Survey, Bob Hageman, and the Survey's mapping geologist in this area, Frans van Rummelen started this co-operation from the geological side. Engineering geology soon became more popular and since 1972 courses in engineering-geological mapping were given in the International Institute for Aerospace Survey and Earth Sciences (ITC) in Enschede. Since 1976 this discipline has been taught in Delft Technical University by David Price, the Chairman of the Scientific Committee of today's Congress. More than 30 Dutch engineering geologists completed their studies at this University since. Almost all Dutch engineering geologists are united in the "Ingenieursgeologische Kring" ("Engineering-Geological Cycle") of the Royal Geological and Mining Society of the Netherlands (KNGMG), which Cycle now has 197 members.

During the past 15 years, the importance of (applied) Quaternary geological know-how was exported to various countries, mainly in SE Asia, where Quaternary geological mapping programmes were introduced and many courses, seminars and workshops were organised. In this way the use of applied Quaternary geological information was promoted both to fellow earth scientists but especially to engineers, planners, and decision makers.

Before discussing a.o. some specific case histories of engineering geology in the Netherlands, a brief outline of the subsurface conditions in this country will be given.

2. GEOLOGY OF THE NETHERLANDS

At the end of the Mesozoic, some 70 million years ago, the Netherlands formed part of an extensive subsiding basin, the North Sea Basin. This basin was bordered to the south and east by the Palaeozoic Massifs of the Ardennes and the Eifel respectively. Due to the differences in rate of subsidence, the thickest sediment series accumulated in the central part of the basin located in the northern section of the present North Sea. Consequently, the base of the relatively soft Tertiary and very soft Quaternary deposits slopes from east to west, lying at about 1,000 m in the west, whereas it is outcropping in the southeast and extreme east of the country. The Tertiary and Quaternary deposits have been affected by vertical movements only, which created rising horsts and subsiding grabens in a NW-SE trending pattern. No significant folding has been documented for the Dutch post-Mesozoic deposits.

The sedimentary sequence of the Quaternary (which started about 2.5 million years ago) is characterised by changing sedimentary environments resulting from strong climatic fluctuations, glacio-isostacy and -eustacy. This was expressed by shifting coastlines and river systems and by the great variety in the Quaternary deposits. During the Pleistocene, the earlier part of the Quaternary, very cold, glacial periods alternated with relatively warm periods. Beause the North Sea Basin was still subsiding thick successions of sand, gravel, clay, and peat were deposited. The major part of these Pleistocene sediments is of fluviatile origin, but, especially in the west, thick successions of marine Pleistocene deposits occur in the subsurface as well. The northern parts of the Netherlands were covered by inland ice during two successive glacial periods. This resulted in the sedimentation of glacial deposits, such as boulder clays and periglacial deposits, mainly consisting of medium to coarse-grained sands. Deep glacial troughs (tunnel valleys) were created during the first period of glaciation (Elsterian), while relatively broad glacial valleys flanked by ice-pushed ridges were formed during the second (Saalian). The remnants of these ridges in the central and eastern parts of the Netherlands give the Dutch landscape a fairly hilly appearance (Figure 3). During the various glacial periods the southern part of the shallow North Sea was dry, while during the last ice age (Weichselian) the river Thames was even a tributary of the river Rhine.

As a consequence of the tectonic setting at the margin of the subsiding North Sea basin, the Pleistocene deposits increase in thickness in a northwesterly direction to about 400 m in the western part of the Netherlands. They consist mainly of fluviatile sands that constitute the main aquifers from which fresh groundwater is exploited in the areas above sea level. Very coarse-grained sands and gravels are restricted to the upstream parts of the delta of the rivers Rhine and Meuse, in the southeast and of the older river systems of German origin in the east.

The youngest Pleistocene sediments generally consist of fine-grained, wind-blown sands, laid down during the last glacial period, the Weichselian.
Site investigations show that these cover sands have high bearing capacities and, as a consequence, these are preferentially used as support for pile foundation in the areas covered by soft younger deposits.

At the end of the last ice age, some 18,000 years ago, the sea level began to rise again. Via topographic depressions inherited from the Saalian glaciation, the western part of the Netherlands was flooded by the sea once more and soft clays and sands of Holocene age were deposited in a tidal-flat environment (Figure 1).
Later, a coastal-barrier system began to develop in the present coastal area. Protected from the sea by these barriers, large peatswamps could develop. Since the twelfth century AD breaching of the coastal barriers by the sea and flooding of the peat areas occurred. From then on human intervention in the natural coastal environment became ever more pronounced as evidenced by the construction of dikes and the excavation of the thick peat bogs behind the coastal barriers. A more systematic coastal defence system including the construction of impressive sea dikes (Figure 2) became increasingly urgent during late medieval times when a number of new inlets were formed by the encroaching sea. The natural process of dune formation was stimulated by constructing low dikes in the dune area. From the sixteenth century onwards beaches suffering from severe erosion were proctected by breakwaters. In the same period our ancestors started to reclaim the many large lakes that resulted from the extensive peat diggings in Holland. Reclamation was done initially with the help of windmills. After the lakes had been reclaimed the former lake floor was cultivated, thus enlarging the national territory of a country that continues to search restlessly for expansion within its own frontiers.

The very poor geotechnical condition of the subsoil of the Quaternary deposits in the west of the country, combined with one of the highest population densities on earth gives a very special dimension to engineering geology and applied Quaternary geology in the Netherlands, as can be seen in the cross section of Figure 3. The ice-pushed ridges of some 100 m high that have good foundation conditions in the east are sparsely populated. However, the very soft clays (indicated by vertical hatching in the section) and peat beds (indicated in dark grey) of Holocene age that are intersected by tidal channels, filled with somewhat stronger Holocene sands, in Holland are among the most densely populated areas. A striking feature of the western part of the Netherlands is the diversity in land-use. Here we find bulb fields next to urbanised areas, historic buildings with a foundation of shallow but numerous wooden foundation piles next to modern buildings with deep, concrete foundation piles. In and beyond the polders many different water levels exist which result in the pictures of ships sailing above the rooftops of houses. Windmills and electrically powered water pumping stations serve to maintain the correct water level. The Amsterdam Congress Centre is situated at 0.5 m below sea level and 11.5 m of Holocene clay and peat beds separate this lecture theatre from firm Pleistocene sand beds that support this building via numerous concrete piles. Because of the typical and -on an international scale- somewhat unusual geograhical, hydrological and geological conditions, in this paper specific attention will be given to the western and central part of the Netherlands, although very interesting studies have been performed in the sandy areas above sea level in the north, east and south as well as in the Mesozoic limestones in the extreme southeast of the country.

3. SURVEYING METHODS

The I.A.E.G. Commission on Site Investigation, chaired by David Price, has created some order in the jungle of site investigation and laboratory testing techniques and procedures, their advantages and disadvantages (Price, 1981). All offshore and onshore surveying methods that are currently being used in deltaic areas with thick successions of soft sediments are applied in the Netherlands. The development of several of these methods however, is strongly related to this country and some of these will be reviewed here.

3.1 Hand drilling equipment
In spite of recent technological developments in drilling equipment traditional hand drilling equipment is used as much as ever. This is due to the fact that this type of equipment is easy to handle, light-weight, has a high durability, is cheap to purchase and maintain, while permission for making boreholes using these devices is practically always given since it does not interfere with the daily work in the countryside. A set of hand drilling equipment costs about 500 US$ and generally consists of an Edelman auger for sampling sand and clay above the phreatic level, gouges for sampling clay and peat below the level, a T-shaped device for manual operation (Figure 4), and several extension rods (Oele et al., 1983). In the mid 1970s a very practical and cheap device for sampling sand below the groundwater table was developed. It is named after the inventor: the Van der Staay suction corer. It consists of two p.v.c. tubes, one fitting inside the other. Once the groundwater level has been reached by means of, for example, an Edelman auger, the set of tubes is installed in the borehole. Sampling is done by pushing the outer tube into the ground while the inner tube is lifted. This device has been and still is used extensively in the Netherlands and in SE Asia (Van de Meene et al., 1979).

3.2 Begemann Continuous Sampler
A spectacular development in drilling was the invention of the Begemann continuous sampler (Begemann, 1971). This device envelops the sample in a sleeve-like stocking which is stored between the inner and the outer barrel, while a special mud supports the sample and eliminates friction between the sample and the inner barrel of the sampler (Figure 5). This equipment provides undisturbed samples of superb quality for laboratory analysis and testing. If the Begemann sampler is pushed down by a Cone Penetration Test (CPT) rig, up to 30 m long, continuous samples with diameters of 29 or 66 mm can be collected, both onshore and offshore, in soft to moderately compacted sediments.

3.3 Cone Penetrometer
Most geotechnical engineers from abroad associate the Dutch geotechnical expertise with the Cone Penetration Test (CPT), often called the Dutch cone penetrometer. It

Figure 2: Aerial photograph of the Hondsbossche Zeewering, located along the Northwest coast of the Netherlands, between the village of Camperduin and Petten. This sea dike standing 11.50 m above sea level was constructed in 1792 at a place where the natural coastal defence system of the dunes was eroded. Note the strong coastal erosion both north and south of the dike (Photograph by KLM/Aerocarto).

was invented in 1931 by Pieter Barendsen and is used to determine the bearing capacity of foundation beds. A cone-shaped probe is pushed into the ground at a constant speed of 2 cm/sec. In the cone the resistance of the subsoil strata is recorded which is read in the measuring cabin. Various types of cones have been developed, such as the Begemann adhesion jacket cone which can additionally record the local frictional resistance and by means of which the lithology of the penetrated beds can roughly be determined. The first electrical cone was developed in 1948; mechanical cones are hardly used anymore. Many other types of cones have been developed since, such as the piezocone, the electrical resistivity cone, the nuclear in-situ density-measuring cone, the seismic cone, the thermal resistivity cone, the heatflow cone, etc. (Figure 6).

Site investigation for foundation purposes is done mainly by CPTs in the Netherlands. The performance of a CPT can be considered as a mini pile test, since all relevant soil parameters are directly or indirectly derived from the CPT. The development and extensive use of CPTs in the Netherlands is most probably the reason that hardly any Standard Penetration Tests (SPTs) are performed in this country. CPTs are increasingly applied abroad and international scientific interest for this technique has been focussed in a series of CPT conferences. CPTs are fast and cheap to perform, accurate to read and easy to interpret but can only be used in unconsolidated to poorly consolidated deposits, both onshore and offshore. The average depth is about 15 metres, but occasionally depths of more than 70 metres have been reached. Occasionally, CPTs are even used for geological mapping activities (de Mulder & Westerhoff, 1985).

3.4 Georadar
A most promising technique for surveying the shallow subsurface above the groundwater level is Georadar. It is applied in many countries now but this technique is mainly used for the detection of karstic solution cavities, the determination of the thickness of the ice cover or overburden, groundwater depth, and for the detection of artifacts in the subsurface, such as cables and drums. Although the records are still quite difficult to read this method can be applied successfully for geological purposes. From a recently performed Georadar survey over an ice-pushed ridge in the central part of the Netherlands the dip and tectonic configuration of the various beds in this ridge could be determined accurately (Figure 7). It is standard procedure to store the data digitally.

3.5 High Resolution Seismic Reflection method
This method is based on the same principles as the standard seismic reflection used for the exploration of hydrocarbons in the deep subsurface. The main difference is that in the High Resolution Seismic Reflection method, specific attention is paid to data acquisition parameters, such as: much shorter distances between the geophones and use of smaller dynamite loads. This generally results in very clear images of the reflectors in the shallow subsurface, starting from a depth of about 30 - 50 m (Figure 8).
From such images it is even possible to interpret the depositional environment in quite some detail. The availability of such records is of particular interest for groundwater surveys in areas with thick successions of unconsolida-

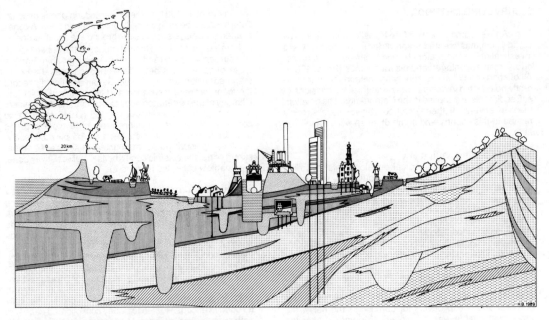

Figure 3 : Artist impression of the relationship between land-use and the geological conditions of the subsurface in the western and central part of the Netherlands. Note the various water levels, the predominantly soft Holocene layers and the significance of the Pleistocene foundation beds.

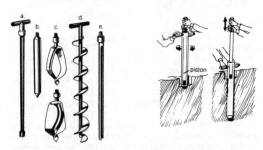

Figure 4:
Left: Hand drilling equipment:a) T-shaped device for manual operation with an extension rod; b) Dachnowsky sampler for clay and peat; c) two Edelman augers, above for clay and below for coarse sand; d) flight auger; e) gouge for sampling clay and peat below the groundwater table.
Right: Hand drilling equipment: the Van der Staay suction corer, consisting of two p.v.c. tubes of slightly different diameter and a piston. The inner tube functions as a suction barrel and the outer one as a suction tube.

ted deposits, such as the Netherlands. During the past decade, Dutch earth scientists made significant progress in developing and improving of this method.
The TNO Institute for Applied Geoscience has contributed greatly to the improvement of this method and they now have a field party for High Resolution Seismic Reflection permanently in action (Meekes et al., 1989).
Further progress in this field is to be expected from the development of a portable high frequency vibrator for shallow seismic reflection by the State University of Utrecht.

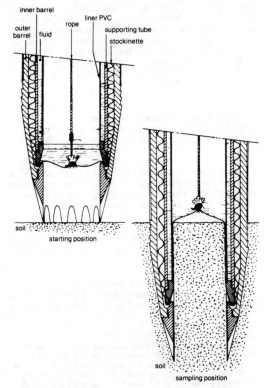

Figure 5: The Begemann 66 mm sampler for continuous, undisturbed samples.

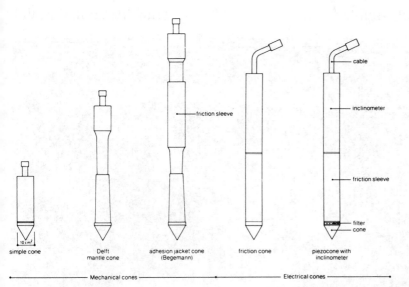

simple cone

Delft
mantle cone

adhesion jacket cone
(Begemann)

friction cone

piezocone with
inclinometer

cable

inclinometer

friction sleeve

filter
cone

friction sleeve

10 cm²

Mechanical cones

Electrical cones

Figure 6: Various types of mechanical and electrical cones.

4. MAPPING

Maps that provide information on the subsoil conditions can be produced ad-hoc or systematically. The first type of subsurface map is usually prepared for a special problem such as the relation of soil conditions to a particular construction or, for example, to show the distribution of high-quality sands for the glass industry. As in most other countries, mapping of the subsurface is done in a systematic manner in the Netherlands. In the rest of the world geological mapping deals mainly with surface outcrops, but in the Netherlands geological mapping focusses on the succession of strata in the subsurface. Systematic mapping programmes include:
* geological mapping of the onshore subsurface down to about 50 metres, at a scale of 1:50.000;
* geological mapping of the North Sea floor, at a scale of 1:250,000 and 1:100,000 in the coastal zone;
* geological mapping of the deep subsurface, at a scale of 1:250.000;
* pedological (soil) mapping down to about 1,20 metres, at a scale of 1:50.000;
* geomorphological mapping, at a scale of 1:50,000;
* groundwater mapping, at a scale of 1:50,000.

The first three geological mapping programmes are carried out by the Geological Survey. The shallow-onshore mapping programme is about half completed. Some 25 map sheets accompanied by extensive explanatory notes have been published. For the Holocene areas a special legend has been developed by the Geological Survey (Hageman, 1963). By means of this so-called profile-type legend a three-dimensional image of the succession of Holocene strata is represented on the map. The development of this legend was meant to provide engineers with relevant data on the succession and composition of the uppermost beds. The geological map at a scale of 1:50,000 is accompanied

by a number of additional maps at a scale of 1:100,000 which show the distribution, depth, thickness and/or facies of specific beds. In Holocene areas a depth contour map of the top of the firm Pleistocene deposits, relevant for both geotechnical and geohydrological purposes is added as a standard. The geological maps of the North Sea at a scale of 1:250,000 are a joint production of the British Geological Survey and the Geological Survey of the Netherlands. Three out of a total of eleven mapsheets have been published. Apart from the regular geological data, these maps contain substantial engineering-geological information. A mapping programme on the deep subsurface (in excess of 500 metres depth), of the country which started recently, will encompass 15 mapsheets, four of which are in preparation now.

The pedological map (soil map) of the Netherlands is a product of the Winand Staring Centre (formerly: STIBOKA) in Wageningen. Fieldwork for the pedological map, which covers the entire country, has just been completed (in 1990) and the last map sheet will be published before 1993. All pubished and draft maps have been digitised and the Geographical Information System of the Winand Staring Centre contains 104 map sheets (van der Pouw, 1990). The geomorphological map of the Netherlands is a joint production of the Winand Staring Centre and the Geological Survey. Just over half the total number of map sheets has been published now. Like the soil map, the groundwater mapping programme of the Netherlands has been completed recently. These maps were produced by the Institute of Applied Geoscience in Delft. Unlike the other maps, these maps have not been printed but are available as manuscript maps in black and white.

No systematic engineering geological mapping programme exists in the Netherlands. An inquiry carried out in the twelve provinces and various ministries showed that planners and decision-makers are only marginally interested in a stan-

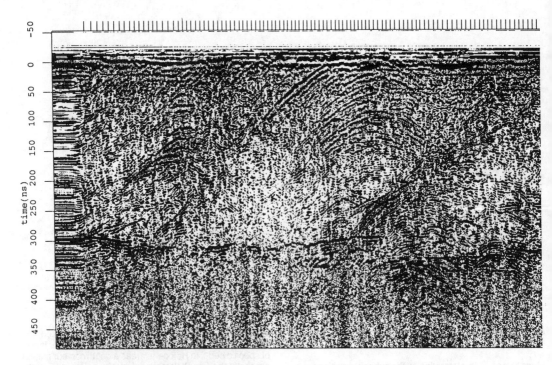

Figure 7: Georadar record of a survey over an ice-pushed ridge in the central part of the Netherlands, the Veluwe, near Apeldoorn. This survey was performed by Delft Geotechnics in cooperation with TNO Technical Physical Institute in Delft and was commissioned by the Geological Survey of the Netherlands. Note the position of the groundwater level at about T=300, and the folding and faulting of the sand beds in the ridge; horizontal scale: 1 cm = approximately 5 m.

dard set of engineering geological maps on a national and/or a regional scale. What they need are specific thematic maps dealing with areas and subjects that are of special (political) interest at that particular moment. Maps on items such as the distribution of construction materials, the potential settling of the soil, and the impact of digging or dredging activities on the landscape are currently made by the Geological Survey and are based on existing geological maps, on a re-evaluation of the field data and on the addition of other (e.g. geotechnical, geohydrological, etc.) parameters (de Mulder, 1984; 1987; Schokking & Hoogendoorn, 1989; Hillen & van der Zwan, 1990). In order to be able to make periodical up-dates of thematic maps and to match such maps and other data, for instance the locations of sand/gravel pits in combination with a thematic map on construction materials or the location of groundwater extraction points in combination with settlement maps, an increasing number of these maps is produced by means of Geographical Information Systems.

Another request put forward by planners and decision-makers concerned integration of the results of the systematic earth-science mapping activities above mentioned. To meet this need, the Geological Survey arranged for co-

operation between the earth-science mapping institutes, which led to a the publication of a brochure (Geological Survey et al., 1986). This brochure contains a large number of thematic maps on scales ranging from 1:50,000 to 1:250,000 for one particular area in the centre of the country.

Occasionally, engineering geological maps have been produced for several towns and parts of larger urbanised areas in the Netherlands. Such maps have been made for the town of Venray and its surroundings by the Technical University in Delft (van Winkoop, 1981) and jointly by the Geological Survey and Delft Geotechnics for the towns of Leiden (see Figure 9), Lisse, Sliedrecht, Koudekerke a/d Rijn and parts of the city of Amsterdam (de Mulder & Hillen in press).

A systematic geochemical mapping programme does not yet exist in the Netherlands. Plans for the implementation of a more or less systematic geochemical mapping programmes for specific areas and on specific targets are in preparation.

With the introduction of Geographical Information Systems

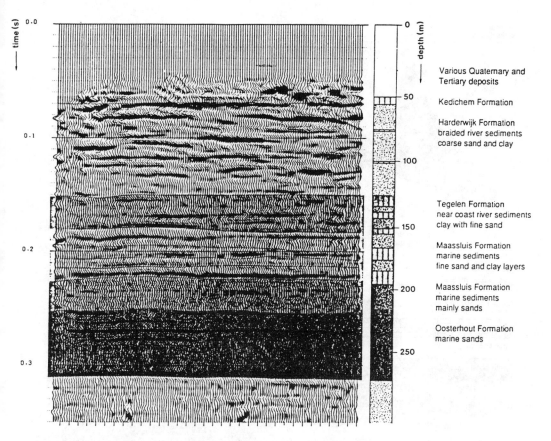

time (s)

0.0
0.1
0.2
0.3

depth (m)

0
50
100
150
200
250

Various Quaternary and
Tertiary deposits

Kedichem Formation

Harderwijk Formation
braided river sediments
coarse sand and clay

Tegelen Formation
near coast river sediments
clay with fine sand

Maassluis Formation
marine sediments
fine sand and clay layers

Maassluis Formation
marine sediments
mainly sands

Oosterhout Formation
marine sands

Figure 8: Geologically interpreted High-Resolution Seismic record. Result of a survey in the SW of the Netherlands by the TNO Institute for Applied Geoscience in Delft. Note the differences in continuity of reflectors in the various lithostratigraphic units, corresponding to differences in sedimentary environment; horizontal scale: 1 cm = approximately 25 m.

modern earth scientists have abandoned the traditional view that maps have "eternal" value. Via the GIS new maps can be produced rapidly when new data become available or when decision-makers raise slightly different questions that have to be answered by engineering geologists. Therefore, maps can be considered as products of a temporary value, nowadays.

5. CONSTRUCTION MATERIALS

As a result of the specific position of the Netherlands on a delta in the subsiding North Sea Basin and to the almost total absence of hardrock, natural construction materials can only be found in the unconsolidated Quaternary sediments. Exceptions are the occurrences of limestones and dolomites in the extreme southeast and east of the Netherlands providing cement and calcareous additives for fertilizers.

Most gravel deposits are found in upstream deposits of the Dutch sections of the Rhine and Meuse Pleistocene river courses. Near-surface deposits of gravel occur in the extreme SE part of the Netherlands, in the Province of Limburg. In the south of this province gravel is found in river terraces formed by the repeated incision of the river Meuse. However, almost all exploited gravel in the Netherlands comes from in the central part of the Province of Limburg, where in the subsiding Central Graben Rhine and Meuse material was continuously deposited. The total annual demand for gravel in the Netherlands is about 17 million tons, of which 10 million tons is produced in this country. Although gravel resources in the subsurface of the Province of Limburg are sufficient to cover the demand expected in the next 100 years, gravel exploitation is restricted now and will be terminated in the coming decade for reasons of landscape preservation (van Montfrans et al., 1988). On the North Sea only one gravel occurrence, related to glacial sedimentation, has been found so far. Its extension is limited. This means that gravel production in the Netherlands will almost

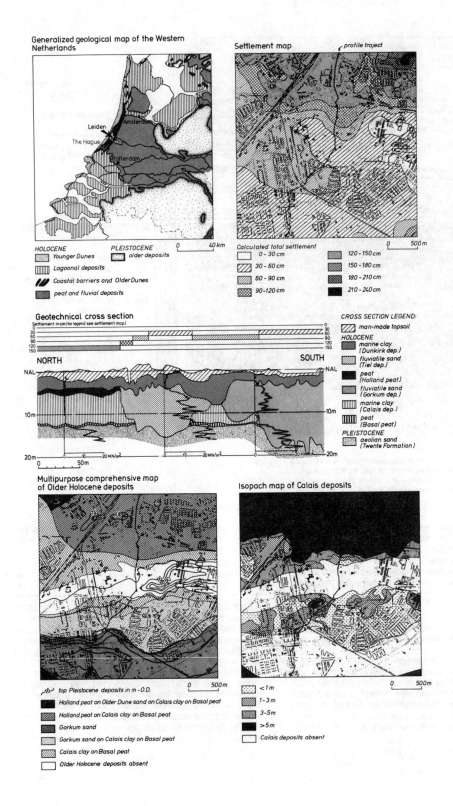

Generalized geological map of the Western Netherlands

HOLOCENE
- Younger Dunes
- Lagoonal deposits
- Coastal barriers and Older Dunes
- peat and fluvial deposits

PLEISTOCENE
- older deposits

Leiden
The Hague
Amsterdam
Rotterdam

0   40 km

Settlement map

profile traject

Calculated total settlement
- 0 – 30 cm
- 30 – 60 cm
- 60 – 90 cm
- 90 – 120 cm
- 120 – 150 cm
- 150 – 180 cm
- 180 – 210 cm
- 210 – 240 cm

0   500 m

Geotechnical cross section

Settlement in cm (for legend see settlement map)

NORTH   SOUTH

NAL   NAL

10m   10m

20m   20m

0   50m

0   10   20 MN/m²

CROSS SECTION LEGEND:
- man-made topsoil

HOLOCENE
- marine clay (Dunkirk dep.)
- fluviatile sand (Tiel dep.)
- peat (Holland peat)
- fluviatile sand (Gorkum dep.)
- marine clay (Calais dep.)
- peat (Basal peat)

PLEISTOCENE
- aeolian sand (Twente Formation)

Multipurpose comprehensive map of Older Holocene deposits

- top Pleistocene deposits in m – O.D.
- Holland peat on Older Dune sand on Calais clay on Basal peat
- Holland peat on Calais clay on Basal peat
- Gorkum sand
- Gorkum sand on Calais clay on Basal peat
- Calais clay on Basal peat
- Older Holocene deposits absent

0   500m

Isopach map of Calais deposits

- <1m
- 1–3 m
- 3–5 m
- >5m
- Calais deposits absent

0   500m

12

come to a complete stop in the near future. It is the government's policy to increase imports of gravel and to encourage the use of alternative aggregates made out of recycled materials.

Exploitable coarse sand that can be used for concrete and mortar is spread over a considerably larger area in the Netherlands than gravel. It covers the central and eastern parts of the country, along the upstream parts of the Pleistocene rivers. The annual demand for coarse sand is about 15 million tons, which amount can just be covered by our national production.

Calcareous sands for sand-brick production contain a high percentage of quartz grains and are almost devoid of clay and organic matter. At or near the surface such sands occur in the central and northern parts of the country. Although only present in a very small amount, the Dutch quartz sands of Miocene age, which occur locally at surface in the south of the Province of Limburg, are well known in Europe for their extremely high quality (>99.8% quartz particles after cleaning; de Mulder, 1984). After quarrying the quartz sands are washed, dried and classified. The present annual production in two quarries is about 350,000 tons.

Sand for backfilling and surface raising is found at the surface almost everywhere except in the west. In the west of the Netherlands such sands are found in the dunes and in the buried Holocene tidal channels or at relatively great depth in the Pleistocene fluviatile deposits. The annual demand is about 80 million tons which amount can be produced from local sources.

Clay and silt is found along the Holocene river courses, mainly in the central part of the country. A recent investigation (Hillen & van de Zwan, 1990) showed that good-quality clay for brick/roofing tile production and for reinforcement of dikes is present in large quantities. The resources were calculated at $1.8 \times 10^9 m^3$, while the annual demand for such clays is about 5 million tons only. For the production of fine ceramic products certain clays of Tertiary age are very suitable.
The Dutch government stimulates production of sand and gravel from the Dutch sector of the North Sea and the Waddenzee, between the chain of islands in the north and the Dutch mainland because of the limited onshore reserves of some construction materials and their concern about the environmental impact of the exploitation of construction materials. Sand for backfilling, which is to be used in the western parts of the Netherlands is currently mainly obtained offshore now. As indicated above, the offshore occurrences of gravel are very limited.

6. CASE HISTORIES

In this chapter some case histories of engineering and environmental- geological studies that have been carried out over the past 10 years both onshore and offshore will be presented. For more detailed reviews and additional case histories one is referred to Oele et al., 1983; de Leeuw (ed.), 1985; de Mulder, 1986; de Mulder & Bakker, 1989. Such

Figure 9: Engineering geological map- set showing the susceptibility to settlement in the centre of the city of Leiden.

case histories demonstrate that the application of engineering geological information is essential for feasibility studies, even in a country where it is commonly thought that anything can be constructed everywhere. Moreover, the use of (engineering) geological data may lead to substantial cost savings, as has been demonstrated recently for engineering geological maps (de Mulder, 1989).
Apart from one (6.5) the presented case histories all refer to engineering problems in Quaternary, poorly consolidated deposits. However, some very interesting engineering geological investigations have been carried out in hard rock, both at or near the surface and in the deep subsurface of the Netherlands. Several studies have been performed on the subsurface storage of energy, both in rock salt (for the storage of natural gas), in aquifers (for the storage of hot water) and in Carboniferous rock (for the generation of subsurface hydroelectric energy, the OPAC-project: Braat & de Haan, 1983).

6.1 Land reclamation
Land reclamation of fresh-water lakes and parts of the sea has been common practice since the 13th century in the Netherlands. Until recently, land reclamation was considered to reveal only positive effects, such as provision of new, often fertile land for agriculture or for other purposes or for improvement of the water management. However, man became aware of the fact that land reclamation was often accompanied by land subsidence in the immediate vicinity or -in some instances- even at large distance from new polders. Moreover, it is only since the mid nineteen-seventies that these negative effects can be predicted reliably by means of groundwater simulation models.

In the early nineteen-eighties a major study has been carried out to calculate the potential costs of the adverse effects that would result from the construction of such a new polder ("Markerwaard") in Lake IJssel, in the central part of the Netherlands (Figure 10). As a result of this reclamation project, the groundwater conditions would change dramatically: the present groundwater flow is from the lake to the mainland; if the new polder would be constructed the groundwater flow would reverse: from the mainland into the new polder. The piezometric level of the groundwater in the first aquifer would be reduced considerably (Figure 11). The soft clay and peat beds that constitute the subsurface in the neighbouring mainland would become drained and the land surface might subside. This, in turn, would cause an increased negative skin friction on the foundation piles under constructions. In particular the foundations of the historic houses in the small cities bordering Lake IJssel would be adversely affected by this phenomenon. Calculation of the potential damage to constructions, carried out by a team of geologists, geohydrologists and geotechnicians from four different research institutes, showed that damage claims resulting from the creation of this new polder could amount to more than 300 million US dollars.

Moreover, damage to agriculture and to ecology would result from creating the new polder. This is caused by changes in the moisture/air distribution in the soil, which would affect plant life, e.g., in nature reserves. These effects would generate a shift in species composition of the vegetation and so -indirectly- in birdlife.

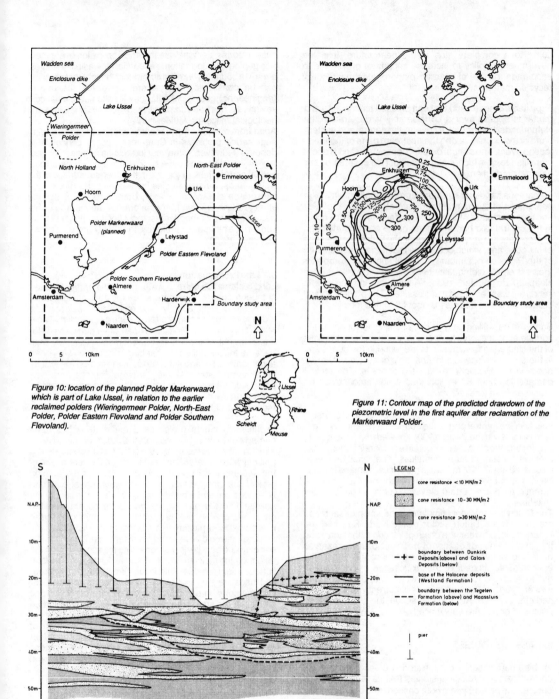

Figure 10: location of the planned Polder Markerwaard, which is part of Lake IJssel, in relation to the earlier reclaimed polders (Wieringermeer Polder, North-East Polder, Polder Eastern Flevoland and Polder Southern Flevoland).

Figure 11: Contour map of the predicted drawdown of the piezometric level in the first aquifer after reclamation of the Markerwaard Polder.

Figure 12: Section showing the distribution pattern of cone resistance zones in the subsurface of one of the three tidal inlets in the Oosterschelde mouth.

14

However, these negative effects of polder construction can be mitigated by taking countermeasures. The most promising approaches all advocate compensation of the lowering of the piezometric levels by artificial recharge of the topmost aquifers. Three possible countermeasures were studied in detail: Injection wells/recirculation systems, infiltration grooves and infiltration wells. The effects of these countermeasures on the geohydrological system and the costs of the still expected damage to buildings were calculated using a combination of models. These calculations proved that injection wells and recirculation systems are the best option (Claessen et al., 1989).

6.2 Inland structures
As a result of both a very high population density (442 inhabitants/km²) and a high degree of industrialisation, only very few countries on this planet can compete with the high density of infrastructural networks in the Netherlands.
Such networks include 4,360 km of navigable waterways and 2,500 km of railway systems for passenger traffic. Consequently, numerous bridges and viaducts have been

constructed on the junctions of these infrastructural elements. Because of the poor subsoil conditions a great variety of new techniques for the foundation and construction of infrastructural elements, such as roads, railways, bridges, tunnels, and sluices have been developed in the Netherlands (Heijnen, 1986; Brons, 1986).

The main foundation problem for the construction of linear elements in the western and most densely populated part of the Netherlands is differential settlement. To assess the most effective and most economic foundation method a thorough programme of site investigation is performed. As discussed above, performing Cone Penetration Tests constitutes the major part of such a programme. Subsequently, the CPT graphs are interpreted geologically and geotechnically followed by the construction of a "geotechnical profile". Verification of the cross-section is mostly done by means of a limited number of Begemann continuous samples. The results of CPT, i.e., cone resistance q_c, local friction f_s and the friction ratio f_s/q_c x 100%, determine the stratification and soil type. By using the piezocone, which enables assessment of in-situ water pressures, an even

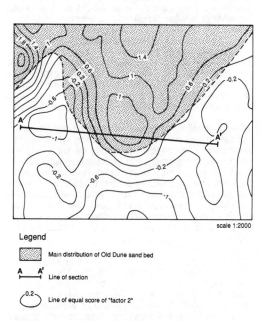

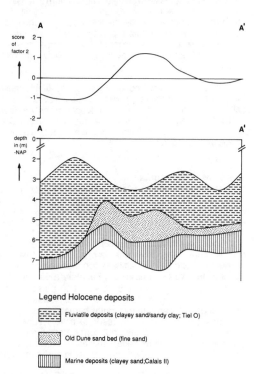

Legend

Main distribution of Old Dune sand bed

A ⊢——⊣ A' Line of section

0.2 Line of equal score of "factor 2"

scale 1:2000

Legend Holocene deposits

Fluviatile deposits (clayey sand/sandy clay; Tiel O)

Old Dune sand bed (fine sand)

Marine deposits (clayey sand;Calais II)

Figure 13: The geotechnical characteristics of a particular sand bed (Old Dune Sands), derived from the part 3-7 m below ordnance level of numerous CPT graphs in a small area, are quantified as "factor 2". The distribution and geotechnical variability of this sand bed can be estimated both horizontally and vertically. In the cross section, the correlation between lithostratigraphic units and "factor 2" is shown. The detailed map (scale 1:2,000) shows the scores of "factor 2" in relation to the main distribution of this sand bed (inferred from the boreholes) in the southeastern part of the city of Leiden.

more refined representation of the subsurface conditions can be obtained.

An example of the applicability of such a "geotechnical profile" is the one made at the junction of the motorway encircling Amsterdam and the waterway, east of this city. The interpreted results of CPTs and borings showed that the original geological situation, consisting of a sand layer with high bearing capacity at 12 to 25 m below mean sea-level, has been altered greatly. Archive searches proved that these discontinuities in bearing capacity followed from subrecent sand dredging. The man-made channels were backfilled with silty material displaying very low cone resistances.

These deviations from the normal geological configuration had a major impact on the subsequent design of the crossing, from north to south consisting of a submerged tunnel (the "Zeeburger Tunnel"), a dam, and a bridge. Considerably deeper piling was required for the tunnel foundation (50 m instead of 20 m) at the location of the sand dredging channels. In total, three different pile systems with varying pile lengths were applied. The southern extension of the dam was restricted because of the soft backfill of the man-made channel. Finally, the first supports of the bridge over the lake required more expensive, tubular, open-ended steel piles, reaching down to much deeper strata, instead of the normally applied pile type (Krajicek & Kruizinga, 1982; Kruizinga & Tan, 1989).

Despite the substantial foundation problems for this causeway as a result of the very poor local subsoil conditions, no alternative routing for this set of constructions was ever considered seriously. Unfortunately, for such constructions the quality of the subsoil has not played a decisive role in decision making yet.

6.3 Coastal structures
The major part of the flood-prone low-lying western part of the Netherlands is protected by the natural coastal defence system of dunes. Where dunes are absent or where this natural defence system is too narrow to provide reliable protection against flooding, sea dikes up to some 12 m high have been constructed (Figure 2). In the southwestern part of the country, the delta consists of a series of east-west trending estuaries, in which tides can reach great heights. During the severe storms in early February 1953, the dikes failed in several places and about 2,000 persons were drowned in this region. As a consequence, it was decided that in most of these estuary mouths barriers would have to be constructed. This would result in a very substantial shortening of the coastline and, consequently, reinforcement and costly maintenance works of the existing weakened dikes surrounding the estuaries would not be required.

In the mouth of one of these estuaries, the Oosterschelde, a storm-surge barrier has been constructed. This barrier consists of 65 giant piers between which 62 gates are suspended. The gates are closed about once in a year, during severe storms. The construction of this unique and technologically very advanced storm-surge barrier was completed in 1987. The foundation of this barrier caused many problems since the seabed conditions vary greatly over small distances. Depending on the depth of the sea-

floor (up to 45 metres) and geological conditions, the foundations of the piers had to be put on firm Pleistocene sands and strong, consolidated sandy clays or on weak and loose sands of Late Holocene age. These latter sands proved to be very susceptible to liquefaction. For stabilising the Holocene sands vibro-compaction methods were employed using giant water-jet needles. However, only the unconsolidated young sands could be compacted by this method whereas the bearing capacity of the firm Pleistocene sands was affected negatively by the penetration of the needles. This made it imperative to carefully map the Holocene/Pleistocene interface, to avoid loss of bearing capacity of the Pleistocene beds intended to carry the enormous piers of the barrier. Geological maps could be made by using the extremely large quantity of subsurface data produced by the extensive programme of site investigations carried out in the mouth of the Oosterschelde, which comprised continuously sampled borings (Begemann-type) and thousands Cone Penetration Tests (CPTs). All these data were used to construct the "geotechnical profiles" described above or cone-resistance cross-sections of the type shown in Figure 12 (de Mulder, 1979).

6.4 Offshore structures
The discovery of major oil and gas occurrences in the northern part of the North Sea in the late 1960s led to the installation of jack-up rigs, fixed platforms, and pipelines in water depths in excess of 100 m. The structures needed were exceptionally large and the deep and heavy foundations required presented problems that could not be solved with the techniques available at the time. The two main problems were inadequate geotechnical data and insufficiently powerful hammers to drive the deep, high-capacity piles. There were also problems connected with the heavily overconsolidated clays and dense sands typical for the northern North Sea, which are quite different from the soft clays in the Gulf of Mexico, the only area for which extensive offshore experience was available at the time.

Site-investigation techniques which could provide reliable soil parameters were improved. Based on extensive experience with cone penetration testing on shore, offshore penetrometer equipment was introduced, which could obtain in-situ data at great depths below the seabed. At about the same time, the Dutch introduced the concept of dedicated geotechnical drilling vessels, which were capable of deploying this new equipment successfully. Such vessels are equiped with a heave-compensation system, a laboratory and a seabed frame with a clamp to immobilise the drillstring during in-situ testing.

The routine geotechnical investigation for jack-up rig foundations consists of one shallow (25-30 m) borehole with various CPTs and sampling. The basic programme for a fixed platform with 4, 6 or 8 legs comprises one or two boreholes with CPTs and sampling to a depth of about 100 m. The foundation of these platforms consists of open ended pipe piles with diametres of 42 to 54 inches driven to depths of 50 to 80 m below seabed. The conventional site investigation for pipelines consists of vibrocore sampling at regular distances. Recent projects for pipelines include shallow CPTs. Gravity Base Structures (GBS) have not been installed in the Dutch Sector of the North Sea yet. This alternative for the conventional steel jackets has been considered now for Block F3 in this Sector. The proposed structure for oil production and storage has a rectangular concrete base. The total site investigation programme for this piled platform consisted of three deep and twenty shallow borings, various piezocone tests and three near-continuous sampled borings. The major foundation aspects of the GBS are the horizontal sliding resistance and the degradation of shear resistance due to cyclic loading. Therefore, an extensive laboratory testing programme of static and cyclic triaxial tests and simple shear tests has been scheduled. The results of these tests and the in-situ test data will be used to derive the design parameters for finite element models of the GBS foundation.

The quality of offshore site-investigation techniques now rivals those used on land and this technology has become a world standard for major platforms and other large constructions in offshore areas (de Ruiter & Richards, 1983).

6.5 Nuclear waste disposal
In environmental impact studies geoscientific information is applied only incidentally in the Netherlands at present. One of the studies in which such information has been and is being used extensively, concerns the selection of sites for the permanent disposal of radioactive waste in deep-seated salt bodies. Since 1985, geological, geohydrological and safety studies have been performed on this subject. Geoscientific literature and model studies, which focussed on the possible effects of (peri-)glacial processes on the geological stability of salt bodies during the next 100,000 years, were completed recently. Ground freezing, differential loading by an ice mass and glacial erosion may have had marked effects on the stability of salt bodies in the subsurface of the Netherlands. This study showed that -disregarding the emission of greenhouse gases- an occurrence of a major glaciation in northwestern Europe can reasonably be assumed in the second half of the next 100,000 years. Such a glaciation might also affect the Netherlands. Furthermore, this study showed that -analogous to the events during the Elsterian glaciation- the effects of glacial erosion will most probably have the greatest impact on the stability of the salt bodies of all the processes studied. The worst-case glaciation scenario for the next 100,000 years predicts that the effects of (peri-)glacial and other geological processes may result in a 540 m reduction in thickness of the overburden above a salt body. This amount is the sum total of fluviatile erosion (50 m), diapiric erosion of salt (40 m), and glacial erosion (450 m) (Wildenborg et al., 1990).

7. RECENT RESEARCH

Some recent research topics in the Netherlands are: firstly, the development of numerical models for simulating the effects of human interference in the natural geological environment, and secondly, the development of expert systems for performing routine geological interpretation activities by means of a computer, so that geologists can devote their time more effectively to solving problems at higher analytical levels. Two examples of such research activities presently carried out in the Netherlands are discussed briefly here.

As mentioned above the method most widely used for engineering-geological site investigation in the soft Quater-nary deposits in the Netherlands is the Cone Penetration Test (CPT). One of the main advantages of the resulting CPT graphs is the objective recording of some of the physical characteristics of the subsurface beds. Such graphs can be interpreted in two different manners:
1) purely geological, i.e., deducing the depositional environment from the sediments penetrated, and
2) engineering-geological, i.e., predicting the present and future behaviour of the soil layers.

In fact, knowledge of sedimentary processes and environments is a prerequisite for a reliable prediction of soil behaviour. This type of interpretation work has always been done and is still being done visually and manually. However, two Dutch engineering geologists (G.J. Huyzer and A.J. Coerts) are currently working on the development of methods for decribing CPT- graphs in mathematical terms, and for differentiating units that have a geological and/or an engineering relevance. This might especially be useful for the correlation of the bulk of CPT graphs in a small area. An example of such a mathematical approach is given in Figure 13. This figure shows the approximate horizontal and vertical distribution of a particular sand bed (Old Dune Sands), the geotechnical characteristics (quantified as "factor 2") are derived from the CPT graphs. A special Congress poster is devoted to this study (A.J. Coerts). In this poster, special attention is given to the derivation of depositional environments of selected deposits from the horizontal and vertical variation in CPT properties of individual beds. This type of fundamental research study may result in the development of expert systems for engineering-geological correlation in areas that are largely made up of poorly consolidated deposits, like the Netherlands.

Other recent research is closely related to government policies in the Netherlands. The Central Government provides annual financial support to municipalities located in areas that have weak subsurface conditions, such as soft clays and peat. This is because such municipalities have significantly higher expenses for construction and maintenance of public infrastructural works than towns built on predominantly firm soils. However, the current subsidy system is subject to much criticism, primarily because it is based on a soil-quality criterion that is only marginally related to the extra costs involved.

The Government therefore requested three leading Dutch geological and geotechnical institutes/companies (Geological Survey of the Netherlands in Haarlem, Delft Geotechnics and Heidemij Adviesbureau BV in Arnhem) to develop a new criterion that may serve as a sound basis for a new subsidy system. This new criterion emphasises the correlation between the physical quality of the subsurface and the extra costs of construction and maintenance. A special simulation model has been developed for this purpose. In this model the cost of realising and maintaining many types of land-use is related to several soil-quality elements in the entire range of lithological sequences present in the Netherlands. This model showed that the major part of the extra costs for construction and maintenance can be expressed as a function of two soil quality elements: the susceptibility to settlement, and the phreatic groundwater conditions (Figures 14 and 15). Before the new subsidy system can be implemented the susceptibility to settlement and the groundwater levels have to be assessed in detail in the build-up areas of all towns involved. As indicated earlier, the suscep-

A

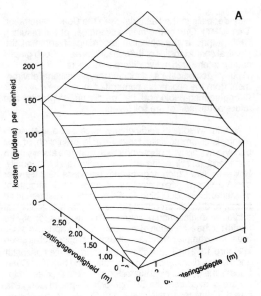

B

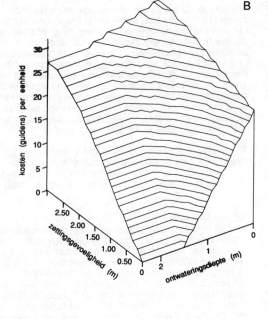

Figure 14: Examples of the relationship between extra costs and quality of the soil (on the two horizontal axes), expressed in susceptibility to settlement ("zettingsgevoeligheid") and groundwater depth ("ontwateringsdiepte") for two types of land-use: sewerage (a), and cemeteries (b). The extra cost is indicated on the vertical axis expressed in m² (a) or in metre length units (b).

tibility to settlement can be derived from the distribution and thickness of compressible beds in the subsurface, the geotechnical properties of these beds and the groundwater levels.

It is evident that the cost involved in assessing the susceptibility to settlement in each individal town is closely related to the quantity of additional site investigations that have to be performed. Therefore, it is of major importance to evaluate all existing geological and geotechnical data. Fortunately, large numbers of borehole logs and CPT-data exist and are accessible for most towns in the western part of the Netherlands. The Central Government now has to decide whether this sophisticated new subsidy criterion will be implemented or not. If the decision is positive, a large database will be set up containing geological, geotechnical and hydrological data for many hundreds of urban areas and an equal amount of detailed (scale 1:5,000) geological and engineering-geological maps will become available for almost all major cities in the Netherlands.

This is an example of the potential relevance of geological and geotechnical data to management purposes in urbanised areas. Moreover, it demonstrates the ability of geoscientists to provide support to government bodies in solving their problems. It is obvious that the results of the subsurface inventories in these urban areas will become a powerful tool for engineering and environmental planning and management in urban areas.

8. FUTURE DEVELOPMENTS IN THE NETHERLANDS

Out of all the items presently covered by research studies in engineering geology in several government, semi-government and private institutes and companies in the Netherlands, significant progress can be expected in the development and improvement of expert systems, relational databases, and 3-D Geographical Information Systems. In the near future, all borehole data and groundwater data will be stored in large computer databases. These data will be provided with a quality label in order to identify their relevance for, for instance, engineering-geological modelling purposes. Because of the large quantity of data and the fact that CPT files are held in many private companies in the Netherlands, it is not expected that CPT data will become centrally accessible in the near future. Relating databases to earth-scientific and climatic information from various national institutes has begun. This will lead to an integrated approach of regional modelling and mapping of groundwater flow patterns. The presentation of geological data in a traditional, qualitative manner will be replaced increasingly by quantitative data. The reliability of contour lines and other lines will be indicated on maps, and most of the printed (engineering) geological maps will be replaced by (colour) computer prints. Development and application of simulation models is expected to expand greatly, both to understand earth processes better, as to predict soil and groundwater behaviour as a result of planned activities. Since the real value of certain geological imput parameters in simulation models cannot be determined precisely, more emphasis will have to be given to the development of stochastic

18

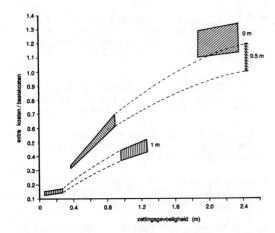

Figure 15: This figure shows the relationship between simulated extra costs and basic costs (unrelated to the quality of the soil) for all public infrastructural works and the quality of the soil, expressed as the susceptibility to settlement (horizontal axis: "zettingsgevoeligheid") ranging from 0 to 2.5 metres by a hypothetical load of 50 kN/m2. The various groundwater levels (0 m, 0.5 m and 1 m) are indicated by hatching (vertical: 1 m; to the right: 0.5 m; to the left: 0 m). The costs have been calculated for 10 municipalities (large and small ones).

simulation models. To approach the combined effects of all relevant geological processes in the course of (geologic) time, further development of time-dependent stochastic models is foreseen. This type of research is already being performed by a number of internationally operating teams of earth scientists from various disciplines.

Earth-science models are becoming ever more important in decision making. The approach adopted in the programme for selecting sites for the permanent disposal of high-level nuclear waste, suggests that local geological conditions will play a more decisive role in environmental planning and management in future. Existing surveying techniques are expected to improve, whereas no fundamental new technological breakthroughs are foreseen in the very near future. For the shallow subsurface (0-10 m) promising developments are foreseen in the Georadar technique in the near future. At Delft Geotechnics a programme for improving the Georadar signal transmission and data processing is in progress. This programme aims at developing an improved antenna system by means of which the radar beam can be steered in the direction chosen and at developing algorithms that are analogous to those currently used in seismic processing. The programme is intended to be completed by 1996. For the longer term, expert systems for automatic data interpretation have to be developed (van Deen, 1989).

The same kind of improvement is expected in high-resolution seismic-reflection techniques leading to a still better resolution of the seismograms in the upper part of the record (30-60 m). With respect to the in-situ measurements and sampling methods, the number of cone-shaped probes for special purposes, to be used in conjunction with CPT-equipment, will undoubtly increase. Since the variation of

lithology is of utmost relevance as an imput parameter in many of the geotechnical and geohydrological simulation models, probes for determining the variation in lithology in-situ on a micro scale and techniques for undisturbed sampling at greater depths should be developed. With the increasing availability of such site investigation tools geologists, planners, and decision makers will probably become more and more interested in the subsurface conditions between 50 m and 500 m depth.

Acknowledgements

The author is very grateful for the contributions to this paper of specialists working in different fields of engineering geology in the Netherlands. Many thanks are due to Gerrit van der Zwaag of FUGRO-McClelland Engineers B.V. in Leidschendam, to Jan Kruizinga of Delft Geotechnics, to Alfred Coerts of the State University in Utrecht, to David Price and Niek Rengers of the Technical University in Delft, to Taeke Stavenga, Roeland Hillen, and Erno Oele in the Geological Survey in the Netherlands. Mrs. Thea van de Graaff-Trouwborst revised the English text and Mrs. Nynke Boorsma took care of the French translation of the abstract.

REFERENCES

Begemann, H.K.S.Ph. (1971): Soil sampler for taking an undisturbed sample 66 mm in diameter and with a maximum length of 17 metres. Proc. IV Asian Regional Conf. SMFE, Bangkok.

Braat, K.B. & J.F. de Haan (1983): Ondergrondse Pompaccumulatiecentrale:perspectievenvoor Nederland (in Dutch). Energiespectrum 7, p. 74-83.

Brons, K.F., (1986): Construction methods - in: Leeuw, E.H. de: The Netherlands Commemorative Volume XI, I.C.S.M.F.E., San Fransisco p. 71-82.

Claessen, F.A.M., A.J. van Bruchem, E. Hannink, J.G. Hulsbergen, E.F.J. de Mulder (1987): Secondary effect of the reclamation of the Markerwaard Polder. Geologie & Mijnbouw 67.

Deen, J.K. van (1989): Niet destructieve technieken: grondradar (in Dutch) - in: LGM-Mededelingen 97, p. 67-71, Delft.

Geological Survey of the Netherlands, Institute for Applied Geosciences & National Institute for Soil Mapping (1986): Subsoil Uncovered Haarlem, Delft, Wageningen.

Graaf, H.C. van de & J. Vermeiden (1988): Half a century of Static Cone Penetration Techniques. LGM-Mededelingen 95, 36 p., Delft.

Hageman, B.P. (1963): A new method of representation in mapping alluvial areas. Verh. Kon. Ned. Geol. Mijnbouwk. Gen, Geol. Ser., 21-1, Jub. Conv., pt. 2, p. 211-219.

Heijnen, W.J. (1985): Design of Foundations and Earthworks - in: Leeuw, E.H. de: The Netherlands Commemorative Volume XI, I.C.S.M.F.E., San Fransisco, p. 53-70.

Hillen, R. & J.T. van der Zwan (1990): Clay resources in the Netherlands, an inventory. Proc. 6th I.A.E.G. Congress Amsterdam.

Krajicek, P.V.F.S. & J. Kruizinga (1982): CPTs, an excellent aid to determine soil parameters and deviations in the soil profile. E.S.O.P.T. II, Amsterdam.

Kruizinga, J. & G.L. Tan (1989): Geotechnical aspects of the Zeeburger Tunnel in Amsterdam. Conf. on Immersed Tunnel Techniques, Manchester.

Kruse, G.A.M. (1989): Verkenningen en geologisch modelleren (in Dutch) - in: LGM-Mededelingen 97, p. 62-66, Delft.

Leeuw E.H. de (1985): the Netherlands Commemorative Volume XI, I.C.S.M.F.E., San Fransisco.

Meekes, J.A.C., van Overmeeren, R.A., van Will, M.F.P. & W.D. Langeraar (1989): Evaluation of geophysical methods for the reconnaissance of aquifers for thermal energy storage; Phase 2. TNO-report OS 89-72.

Meene, E.A. van de, van der Staay, J. & T.L. Hock (1979): The Van der Staay suction corer, a simple apparatus for drilling in sand below the groundwater table. Rijks Geologische Dienst, Haarlem.

Montfrans, H.M., Graaff, L.W.S. de, Mourik, J.M. van & W.H. Zagwijn (1988): Delfstoffen en samenleving (in Dutch) - in: Geologie van Nederland, 2, Rijks Geologische Dienst, Haarlem.

Mulder, E.F.J. de (1979): Engineering-geological Investigations in the Mouth of the Eastern Scheldt (SW Netherlands). Geol. & Mijnbouw, 58 (4): 471-476.

Mulder, E.F.J. de (1984): A geological approach to traditional and alternative aggregates in the Netherlands - Bull. I.A.E.G., 29, pp. 49-57.

Mulder, E.F.J. de (1986): Applied and Engineering Geological Mapping in the Netherlands- Proc. V Int. Conar. I.A.E.G Buenos Aires, 6, p. 1755-1759.

Mulder, E.F.J. de (1987): Recent developments in Environmental Geology in the Netherlands - in: P. Arndt & G.W. Lüttig (eds.): Mineral Resources extraction, environmental protection and land-use planning in the Verlagsbuchhandlung, Stuttgart.

Mulder, E.F.J. de (1988): Thematic geological maps for urban management and planning. Proc. Int. Symp. on Urban Geology. Shanghai, ESCAP.

Mulder, E.F.J. de (1989): Thematic Applied Quaternary maps: a profitable Investment or expensive wallpaper? In: E.F.J. de Mulder & B.P. Hageman (Editors), Applied Quaternary Research. Balkema, Rotterdam, pp. 105-117.

Mulder, E.F.J. de & W.E. Westerhoff (1985): Geology - in: Leeuw, E.H. de: The Netherlands Commemorative Volume XI, I.C.S.M.F.E., San Fransisco, p. 19-28.

Mulder, E.F.J. de & J.G. Bakker (1989): Constructions - in: McCall & Marker (eds.): Earth Science Mapping related to planning and development.

Mulder, E.F.J. de & R. Hillen, in press: Preparation and Application of Thematic Engineering and Environmental geological Maps in the Netherlands

Oele, E., W. Apon, M.M. Fischer, R. Hoogendoorn, C.S. Mesdag, E.F.J. de Mulder, B. Overzee, A. Sessören & sampling techniques, maps and their application. In: M.W. van den Berg & R. Felix (Editors), Special issue in the honour of J.D. de Jong. Geol. & Mijnbouw, 62, pp. 355-372.

Pouw, B.J.A. van der (1990): Soil mapping in the Netherlands. Proc. Meeting Heads of Soil Surveys in EC countries. Silsoe Campus, Great Brittain.

Price, D.G. (Chairman) (1981): Report of the I.A.E.G. Commission on Site Investigation. I.A.E.G. Bull., 24, p. 185-226.

Ruiter, J. de & A.F. Richards, 1983: Marine geotechnical investigations, a mature technology- Geotechnical Practice in Offshore Engineering, American Society of Civil Engineers, New York.

Schokking, F. & A. Hoogendoorn (1989): Computer-aided compilation of settlement maps for the Provinces of Friesland and Gelderland, the Netherlands. Preprints 25th Conf. Eng. Group of the Geol. Soc. of London.

Visser, W.A., Zonneveld, J.I.S. & A.J. van Loon (eds.) (1987): Seventy-five years of Geology and Mining in the Netherlands (1912-1987).

Wildenborg, A.F.B., J.H.A. Bosch, E.F.J. de Mulder, R. Hillen, F. Schokking & K. van Gijssel (1990): A review: Effects of (peri-)glacial processes on the stability of rock salt. Proc. VIth Int. Congress I.A.E.G., Amsterdam, Balkema.

Winkoop A.A. van, 1981: Ingenieursgeologische atlas van Venray (in Dutch, unpubl. report)- Technical University of Delft, Dept. of Mining Engineering, Section of

1 Engineering geological mapping and site investigation
Cartographie et études géotechniques du terrain
1.1 Engineering geological and environmental maps and plans
Cartes et plans géotechniques et de l'environnement

6th International IAEG Congress / 6ème Congrès International de AIGI, © 1990 Balkema, Rotterdam. ISBN 90 6191 130 3

Mapping of the mechanical properties of a salt dome
Cartographie des propriétés méchaniques d'une voûte salifère

H. Albrecht, U. Hunsche, I. Plischke & O. Schulze
Federal Institute for Geosciences and Natural Resources, Hannover, FR Germany

ABSTRACT: For reliable model calculations on underground openings in rock salt formations, the mechanical properties of the rock salt must be investigated, because its mechanical behavior varies considerably, particularly its creep behavior. Therefore, the rock has to be classified and characterized and the same material laws and parameters have to be used for all parts of the salt dome with mechanically similar behavior. These homogeneous parts are not always correlatable with stratigraphic layers, however.

A mapping and characterization procedure based on laboratory tests is outlined. It includes creep and failure tests under uniaxial and triaxial conditions. The results demonstrate the need for classification, as well as the suitability of the method.

This mapping method will be used for the in situ investigation of the Gorleben salt dome, a candidate site for a radioactive waste repository in the FRG.

RESUME: Pour la réalisation de calculs modèles sûrs portant sur des excavations souterraines dans des formations salifères il faut que les caractères mécaniques du sel soient étudiés parce que son comportement mécanique varie considérablement - avant tout son comportement de fluage. C'est pourquoi la roche doit être classifiée et caractérisé et les mêmes règles et paramètres physiques doivent être appliqués pour toutes les parties du diapirs qui montrent un comportement mécanique semblable. Néanmoins ces parties homogènes ne peuvent pas toujours être corrélées avec des unités stratigraphiques.

Un procédé pour le levé et pour la caractérisation basé sur des tests de laboratoire est présenté. Ce procédé comprend des tests de fluage et de rupture sous des conditions de stress uniaxial et triaxial. Les résultats montrent la nécessité d'une classification et l'utilité de la méthode.

Cette méthode de levé cartographique sera utilisée lors de l'étude in-situ du diapir salifère de Gorleben, qui représente l'un des sites possibles pour le stockage des déchets radioactifs en RFA.

1 INTRODUCTION

Knowledge of the mechanical behavior of rock salt is important for the dimensioning and the long-term safety analysis of a permanent repository for radioactive wastes in a salt dome, as well as for the design of mines and caverns. The structure of the salt dome has to be taken into account in the model calculations. This will be demonstrated on an example, the candidate site for a permanent repository in the FRG, the Gorleben salt dome.

The rather complicated geology of the Gorleben salt dome is shown in Figure 1. It must be stressed that our knowledge is limited because information has been obtained only by exploration from ground level, mainly by drilling. Without underground exploration the tectonic, structural, and stratigraphic details are only tentative.

The order of the stratigraphic units of the rock salt in the Gorleben salt dome was established from the data from 44 boreholes drilled to the top of the salt dome, four boreholes to about 2000 m at the margin of

23

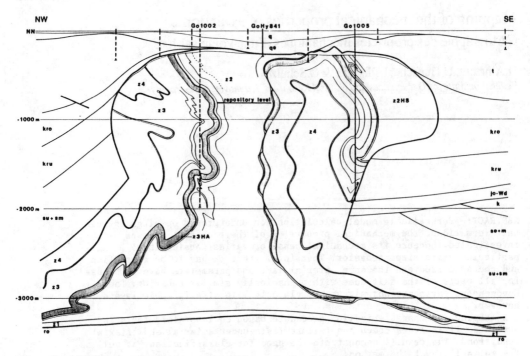

NW   Go1002   GoHy841   Go1005   SE

Figure 1:   Gorleben salt dome, cross section derived from borehole data
(after Bornemann 1987). Depth planned for the underground
investigation is about 840 m.

the salt dome, and two boreholes in preparation for the shafts (about 1000 m). They
form the basis for the detailed analysis
leading to Figure 1.

Halitic, carnallitic and anhydritic rocks
differ considerably in their mechanical behavior. This can be a problem for the construction of a mine or cavern in a salt
dome, although such construction is done
primarily in the halitic rock, i.e. rock
salt. In the case of the Gorleben salt
dome, rock salt forms nearly 90 % of the
Gorleben salt dome (Jaritz et al. 1986).
However, the mechanical behavior of rock
salt can also vary considerably, depending
on its chemical and mineralogical composition and the petrography. Therefore, the
mechanical properties of the rock salt in
all parts of the planned mine have to be
examined. It cannot be assumed that the
mechanical behavior is correlatable with
the stratigraphic layers in the halite.

1.1 Observations

Examples are shown in Figure 2 of creep
tests carried out on rock salt samples from
the Gorleben salt dome taken from different
locations and stratigraphic layers of the
Leine cycle (z3) and Stassfurt cycle (z2)
(Bornemann 1987).

Although samples 1 and 2 were loaded with
a differential stress of $\Delta\sigma$ = 12 MPa, they
exhibit a larger transient creep deformation and much higher steady-state creep
rates than samples 3 and 4, loaded with
$\Delta\sigma$ = 14 MPa. Generally the steady-state
creep rates can differ by a factor of up to
about 30 under similar conditions. These
differences are also observed in uniaxial
creep tests at temperatures up to 180 °C.
Our experiments also demonstrate that even
samples from similar stratigraphic layers
can show large differences (compare samples
2 and 4 in Fig. 2).

Comparable results are provided by true
triaxial strength tests on cubic samples.
Different rock salt types yield a difference in strength of up to 20 % (Hunsche &
Albrecht 1990; Hunsche 1990a & 1990b). Comparison of these failure tests with creep
tests show that the more ductile rock salt
types have a lower strength than the less
ductile ones (see Fig. 3). It is generally
observed that high strength materials fail
in a brittle manner, whereas ductile materials have a lower strength. This is also
true for rock salt.

1.2 Investigation of mechanical properties

The stress, strain, and temperature distributions in a model rock formation can be calculated by the finite element method. The geology and structure of a salt dome can be taken into account with this method (Nipp 1988), some simplifications will normally be made, however. In cases where differences in the mechanical behavior have to be considered in detail, as shown for rock salt in Figure 2, a more complex model is needed and the mechanical properties of the rocks have to be specified in detail. The model calculations are even more complicated if the differences in the mechanical properties of the rock salt also have to be taken into account and thus it must be checked whether this is necessary. Criteria for the decision can be derived only from numerical sensitivity studies of the various material parameters. For a site-specific study of this kind we must know the differences in the mechanical properties of the rock salt, in addition to the structure and geology.

In conclusion, the rock salt has to be characterized so that all parts of the salt dome with mechanically similar behavior will be represented by the same constitutive laws or empirical equations and by the same set of parameters. Therefore, the salt dome has to be mapped, i.e. parts that are homogeneous with respect to their mechanical behavior are to be determined.

Two procedures are possible for identifying and mapping the homogeneous parts:
1. "First principles analysis": The chemical and mineralogical composition, distribution of impurities, structure, texture, grain boundary, etc. determine mechanical behavior and can, in principle, be modelled by theoretical material laws. In specific cases this procedure is quite successful. In the work of Skrotzki & Haasen (1981) and Vogler & Blum (1990), for instance, the influence of impurities and second-phase particles on the dominant deformation mechanisms (mainly dislocation movement) has been described using models based on solid-solution hardening or parti-

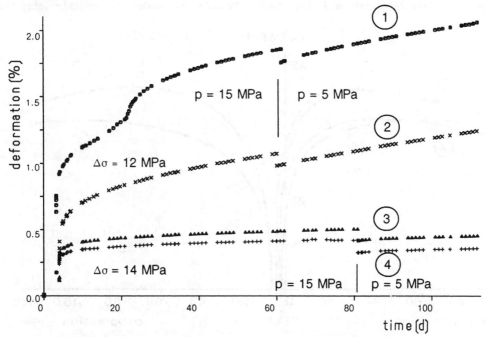

Figure 2: Creep tests (triaxial compression tests) on rock salt samples from different stratigraphic layers and locations in the Gorleben salt dome. T = 50 °C.

(1) Go1002, z2HS2,   501 m   (3) Go5001, z3BK/BD, 734 m
(2) Go5002, z30SU,   481 m   (4) Go5001, z30SU,   881 m
p: confining pressure,
Δσ: differential stress
sample size: 100 mm in diameter and 250 mm long

25

cle hardening. However, the theoretical basis and the methods for determining the parameters are not sufficient to be able to predict the mechanical behavior of salt rocks completely.

2. "Mapping": In addition to geologic mapping and complicated geomechanical in situ measurements, laboratory experiments on samples from the site can supply most of the information required for defining the homogeneous parts. This leads to a more applicable method: Mapping of the mechanical properties.

2 MAPPING METHOD

For mapping during in situ exploration, a laboratory test program has been elaborated on the basis of our present knowledge (see Table 1). It yields the information on the mechanical properties of the rock salt required to determine the homogeneous parts of a salt dome.

This testing program includes short-term tests used mainly to investigate failure behavior. A failure criterion derived from

numerous tests at the BGR is given by

$$\tau_{0B} = \tau_0 (\sigma_0, m, T).$$

The octahedral shear stress at failure, τ_{0B}, depends on the octahedral normal stress, σ_0, on the load geometry as expressed by the Lode parameter, m, and on temperature, T. A detailed discussion is given by Hunsche & Albrecht (1990) and Hunsche (1990a, 1990b).

In addition to short-term tests, the testing program includes long-term tests to investigate creep behavior. Steady-state creep rate can be described by a creep law as follows (Hunsche 1984):

$$\dot{\varepsilon}_s = A \cdot \sigma^n \cdot \exp(-Q/RT).$$

The first aim of the testing program is to identify the homogeneous parts. For this purpose the tests during the first part of the testing program will be run on samples taken during the in situ exploration of the Gorleben salt dome. About 25 km of drifts are planned and samples are to be taken about every 200 m. Thus, there will be specimens from about 125 sampling locations. Results are needed very quickly after the

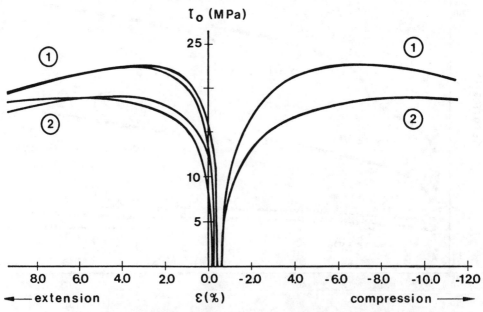

Figure 3: True triaxial strength tests on cubic samples with different hardening and failure behavior.
Test conditions: room temperature, compression, octahedral normal stress σ_0 = 20 MPa, loading rate $\dot{\tau}_0$ = 21.35 MPa/min, edge length of cubes a = 57.5 mm.

sample:	(1) Go5001, z3BK/BD (hard)	(2) Go5002, z3OSU (ductile)
failure strength, τ_{0B}:	22.6 MPa	19.1 MPa
strain at failure, ε_B:	6.5 %	10.1 %

26

Table 1: Laboratory test program for identification and
characterization of homogeneous parts

Type of test	information	
Part one of the program:		
tests to identify the homogeneous parts of the rock salt		
1. uniaxial strength tests in compression		
– at room temperature	failure strength	σ_B
– strain-rate controlled, 3 diff. rates	strain at failure	ε_B
– with loading and unloading cycles	elastic modulus	E(stat)
– from ultrasonic measurement	Young's modulus	E(dyn)
2. triaxial creep tests		
– with confining pressure, p = 20 MPa	creep rates	$\dot{\varepsilon}_s$
– at room temperature	stress exponent	n
– stress changes at 20 MPa		
Part two of the program:		
tests to characterize the homogeneous parts of the rock salt		
3. true triaxial strength tests on cubes	failure strength	σ_B
– at room and at elevated temperatures	strain at failure	ε_B
– stress control (deviatoric tests)	residual strength	σ_R
4. triaxial strength tests in compression		
– with confining pressure, p: 2, 5, 20 MPa	failure strength	σ_B
– at room temperature	strain at failure	ε_B
– strain rate controlled, 3 diff. rates	elastic modulus	E(stat)
– with loading and unloading cycles		
5. uniaxial creep tests	creep rates	$\dot{\varepsilon}_s$
– at room temperature and at 50 & 150 °C	activation energy	Q
– stress changes at 15,10, & 4 MPa, resp.	stress exponent	n
6. triaxial creep tests		
– with confining pressure, p = 20 MPa	creep rates	$\dot{\varepsilon}_s$
– elevated temperatures at 50 & 150 °C	activation energy	Q
– stress changes at 12 & 6 MPa, resp.	stress exponent	n

samples have been taken. Therefore, testing procedures were chosen which will not take much time but which nevertheless give clear information on the mechanical properties.

The second aim of the testing program is to characterize each homogeneous part by constitutive laws or empirical equations with the values for their parameters. The number of specimens that must be analyzed for mapping and characterizing the homogeneous parts of the Gorleben salt dome will depend on the number of homogeneous parts present. If there are four homogeneous parts, about 1000 specimens will be required.

3 CONCLUSIONS

The results of the laboratory tests demonstrate that the mechanical behavior of rock salt can vary greatly and that mapping of homogeneous parts in a rock salt formation is the prerequisite for valid model calculations for the safety analysis. A testing

program has been developed for this purpose (Table 1). The validity of the listed tests has been checked by using them on different types of rock salt. The efficiency of our laboratory program as a mapping method, as well as the capacity of our laboratory, has been demonstrated.

An example is given in Figure 3 for the differences in strength and failure behavior (test 3 in Table 1). Cubic rock salt samples taken from the boreholes for the shafts at the Gorleben site were compression loaded until failure. The tested specimens clearly exhibit different failure strengths and strain at failure.

An example is given in Figure 4 for differences in creep (test 5 in Table 1). Samples 1 and 2, taken from the "Hauptsalz" of the Stassfurt cycle (z2), showed faster creep than the other three samples, as was the case for specimens from the "Hauptsalz" under different conditions shown in Figure 2. Sample 5, taken from the "Bank/Bänder" Salt of the Leine cycle (z3) also has the lowest creep rate, as was the case for sam-

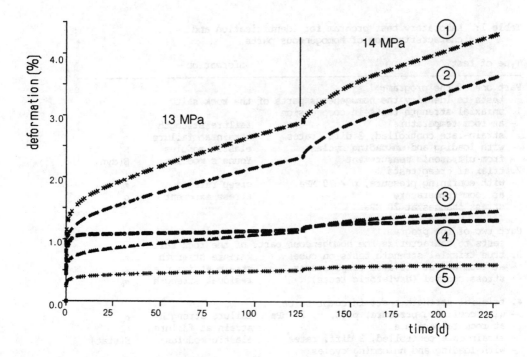

Figure 4: Uniaxial creep tests on rock salt at room temperature

(1) Go1002, z2HS2,   508 m   (3) Go5001, z3OSU,   712 m
(2) Go1002, z2HS3,   546 m   (4) Go5001, z3LSO,   701 m
                                 (5) Go5001, z3BK/BD, 733 m

ple 3 from the "Bank/Bänder" Salt shown in
Figure 2. In these examples the rock salt
samples, taken from different stratigraphic
layers as well as from the same stratigraphic layer, clearly show differences in
their mechanical behavior and obviously
belong to different homogeneous parts.

Although our method is a very useful tool
for identifying the homogeneous parts of a
salt dome, research still has to be done on
the first method, the "first principles
analysis". For instance, humidity has a
large influence on creep rate in uniaxial
tests (by a factor of 20) and variation of
the confining pressure between 0 and 3 MPa
considerably influences creep rate. Moreover, impurities can dissolve as temperature rises resulting in solid-solution
hardening. These effects are not understood
sufficiently at present to take them into
account in the constitutive laws.

Financial support by the Federal Ministry
for Research and Technology (BMFT) is
gratefully acknowledged.

REFERENCES

BORNEMANN, O. (1987): Die geologische Er-
kundung des Salzstocks Gorleben. Kern-
technik, 50: 138 - 142.

HUNSCHE, U. & H. ALBRECHT (1990): Results
of true triaxial strength tests on rock
salt. Engineering Fracture Mechanics,
35: 867–877.

HUNSCHE. U. (1990a): On the fracture behavior of rock salt. In: Constitutive Laws
of Plastic Deformation and Fracture, Proceedings of the Nineteenth Canadian Fracture Conference, Ottawa, Canada 1989;
Eds.: Krausz, A.S., J.I. Dickenson,
J.-P.A. Immanigeon, & W. Wallace. Kluwer
Acad. Publ., Dordrecht (in press).

HUNSCHE, U. (1990b): A failure criterion
for natural polycrystalline rock salt.
Proc. Int. Conf. on Constitutive Laws for
Engineering Materials (ICCLEM), 11 - 13
Aug. 1989, Chongquing, China. Int.
Academic Publ., Beijing (in press).

JARITZ, W., O. BORNEMANN, W. GIESEL & H.
VIERHUFF (1986): Geoscientific investi-
gation of the Gorleben site. Siting,
Design and Constuction of Underground
Repositories for Radioactive Wastes,
Proc. of an International Symp., Hannover
(FRG), March 1986; IAEA-SM-289/44, p. 371
- 384. International Atomic Energy
Agency, Vienna.

NIPP, H.-K. (1988): Numerische Modellierung
zur Integrität der Barriere Salzstock.
BGR report, October 1988, Archives No.
103 819.

SKROTZKI, W. & P. HAASEN (1981): Hardening
mechanisms of ionic crystals on {110} and
{100} slip planes. J. Phys., 42 (C3):
119-148.

VOGLER, S. & W. BLUM (1990): Micromechani-
cal modelling of creep in terms of the
composite model. Fourth International
Conference on Creep and Fracture of Engineering Materials and Structures,
Swansea, April 1990 (in press).

6th International IAEG Congress / 6ème Congrès International de AIGI, © 1990 Balkema, Rotterdam. ISBN 90 6191 130 3

Environmental-geological mapping and evaluation in the Cantabrian Mountains, Spain

Cartographie et évaluation de la géologie et de l'environnement dans la région cantabrique, Espagne

E. Alonso
Departamento de Ecología, Universidad de León, Spain

E. Francés & A. Cendrero
DCITTYM (División de Ciencias de la Tierra), Universidad de Cantabria, Spain

ABSTRACT: A two-level system of mapping and assessment of environmental units and its application to two areas of the Cantabrian region are presented. The methodology used includes, first, the definition of first-order units, on the basis of physiographic, climatic and morphostructural features and their evaluation for generic land-use activities, through the use of a suitability index and, secondly, the definition of more detailed units, on the basis of bedrock, regolith, landform, soils and vegetation. These units are then described in terms of their engineering geological, natural hazards and ecological features and used as the basis for the derivation of other maps of interest for planning, such as natural hazards, quality for conservation, soils and soil capability, potential vegetation or land-use recommendations and limitations. The method provides a means for rapid and flexible evaluation and planning at different scales and levels of detail.

RÉSUMÉ: On présente une méthode de cartographie et évaluation de l'énvironnement à deux niveaux et son application dans deux zones de la Cordillère Cantabrique. La méthodologie utilisée comprend, d'ábord, la définition d'unités de premier ordre, sur la bàse des caractéristiques physiographiques, climatiques et morpho-structurelles. La "vocation" de ces unités pour différentes activités génériques est evaluée en utilisant un indice d'aptitude. Deuxiémement, des unités plus detaillées sont définies sur la base de la lithologie, matériaux de surface, morphologie, sols et végétation. Ces unités sont décrites du point de vue de la géologie de l'ingénieur, les risques naturels et l'écologie, et employées pour la dérivation d'autres cartes d'intérêt pour la planification et l'aménagément (cartes de risques naturels, qualité pour la conservation, sols et aptitude des sols, végétation potentielle ou recommandations et limitations pour l'utilisation du terrain. La méthode décrite permet l'évaluation rapide et flexible pour la planification á échelles et niveaux de détails différents.

INTRODUCTION

The use of geoscientific maps of various types provides a useful basis for the flexible combination of engineering geological information with other environmental data, in order to obtain an integrated assessment of land capability for different kinds of activities.

Two procedures of environmental-geological mapping and assessment are presented applied to two different areas of the Cantabrian Mountains and at two different scales. The areas studied (Fig. 1) cover the basins of two rivers, the Nansa (500 km², altitudes 0-2000 m) and the Esla (600 km², altitudes 1.000-2.200 m). The

31

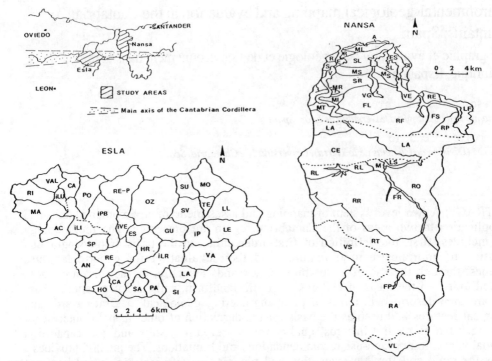

Fig. 1. Location map (top) and first-order subdivisions in the Esla and Nansa valleys. The initials in the Esla correspond to the "small valleys" of Table 1. The initials in the Nansa indicate different "morphodynamic systems" (A=coastal cliffs; R= raised marine terraces; RT=cuestas; CE=karstic massif; FR=alluvial terraces; VS=glacial valleys... etc.).

lower part of the Nansa valley (Francés, 1987) is constituted by moderately folded and faulted Mesozoic sedimentary rocks, with well developed soils. The upper part of this valley and the one of the Esla (Alonso, 1987) are formed by strongly folded and faulted Paleozoic sedimentary sequences, with a few small plutonic intrusions. The soils are not so well developed. There is a fairly marked contrast between the mild and densely populated coastal areas in the Nansa and the high, thinly populated mountain zones. This contrast also exists between the northern and the southern part of the cordillera. The differences in the natural environment and in the degree and type of land occupation, between the coastal and mountain areas, make it adivisable to use different methods and levels of detail in the mapping and assessment of each type of area.

In the high, thinly populated zones, semi-detailed analysis and planning at the regional level are normally sufficient. In the coastal, densely populated areas, characterization and evaluation of land units must be more detailed, for planning at the local level (Lüttig, 1.987).

The aims of this work were:
-To define and map homogeneous, integrated environmental units that could be evaluated at different levels of detail, for different types of planning situations.
- To establish the relationships between geological engineering-geological and other abiotic parameters, on the one hand, and soils and vegetation on the other to define criteria for the derivation of soils and potential vegetation maps and maps of land-use recommendations.

METHODOLOGY AND RESULTS

The methodology used included two main steps. In the first one, larger subdivisions of the territory were defined on the basis of physiographic, climatic, morphological and lithostructural features. An initial assessment of the general carrying capacity or suitability of such larger subdivisions for a series of generic activities was made.

In the case of the Nansa (Fig. 1), the valley was first divided into "morphodynamic environments" (MDE) on the basis of climatic and physiographic features (Christian, 1957; Cendrero and Díaz de Terán, 1987). These in turn were divided into "morphodynamic systems" (MDS) on the basis of lithostructure and morphology. A total of 42 MDS's were obtained and mapped. In the case of the Esla (Alonso and Cendrero, 1988) the approach was different and the first subdivison was made using "small valleys" (SV) that is, the secondary basins of the different tributaries to the main river. In this case 25 subdivisions were obtained (Fig. 1).

The general assessment of suitability, to determine those generic activities for which each larger subdivision has a better defined "vocation" was made using two approaches. First, an analysis of principal components (Sneath and Sokal, 1973) was carried out, in order to obtain groupings of subdivisions according to the similitude of their abiotic parameters. For instance, the SV's of the Esla were grouped into eight categories, using as components: average altitude, average slope, gradient of the river channels, main orientation, calcareous bedrock, silicieous bedrock, absence of soil, subalpine climatic conditions (Fig. 2).

The SV's thus grouped were, in turn, assessed in terms of their carrying capacity for the following activities: agriculture, ranching, wood production, reforestation, tourism-residential, conservation, ski stations. This assessment was based on the distribution, in each valley, of the different "types" (units in a thematic map) of the "elements" considered. These "elements" were: slope, altitude, orientation, bedrock,

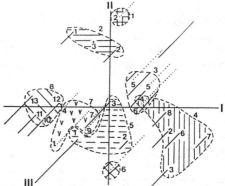

Fig. 2. Analysis of principal components for SV's on the basis of their abiotic parameters, showing their grouping into 8 categories.

surficial deposits, vegetation.

A suitability index was obtained:

$$C_{ai} = \sum_{j=1}^{22} R_{iej}$$

$$R_{iej} = ej \times V_{iej}$$

where:

C_{ai} = suitability of the valley for activity $\underline{i}$.

ej = % of the valley occupied by type $\underline{e}$ of element $\underline{j}$.

V_{iej} = value of type $\underline{e}$ of element $\underline{j}$ for activity $\underline{i}$.

The results are represented in Table 1. This procedure enables the rapid identification of those activities which should be given priority in each larger subdivision and it can be applied to general assessments at scales 1:100,000 or smaller.

The determination of the precise locations which are most advisable for different land-use activities requires a more detailed analysis. The procedure followed consisted in the definition and mapping of smaller units within the larger subdivisions (MDS's or SV's). These smaller subdivisions were defined on the basis of bedrock, landform, regolith, and vegetation and they were named "morphodynamic units" (MDU) and "morphodynamic elements" (MDE), the latter being divisions of the former. Fig. 3 shows the MDU's and MDE's present in the coastal area of the Nansa.

Table 1

Valley/Activity	Agriculture	Ranching	Wood Prod.	Refores.	Tour.Urb.	Conser.	Ski Stat.
San Pelayo(SP)	1430,34	1347,84	1374,84	1175,62	1211,93	1065,59	901,15
Acebedo(AC)	1247,02	1328,56	1207,69	1184,69	1021,09	905,18	1022,67
Maraña(MA)	1408,91	1327,72	1139,35	1081,34	1133,87	997,77	973,28
Riosol(RI)	1532,38	1499,13	1447,49	1163,67	1281,00	1167,83	961,58
Valdosin(VAL)	1116,48	1395,15	1239,89	1236,78	1109,99	914,67	783,64
Carcedo(CA)	1137,11	1473,50	1178,52	1396,81	1108,45	855,25	980,65
Polvoredo(PO)	1284,38	1402,87	1378,44	1228,63	1250,73	1060,91	894,59
Ret-Pont(REP).	1417,39	1530,11	1643,03	1244,75	1283,33	1190,85	1000,58
Oza(OZ)	1085,07	1438,86	1323,29	1345,19	1027,18	884,55	1054,09
Hormas(HR)	1583,28	1534,69	1597,34	1338,69	1419,54	1258,71	981,48
Guspiada(GU)	862,40	1326,32	1027,13	1434,63	834,51	624,96	1070,04
Salc-Vall.(SV)	933,54	1339,15	1200,63	1593,20	892,56	724,58	904,47
Susiella(SU)	1024,93	1596,07	1273,04	1201,13	855,33	764,06	1346,97
Mostajal(MO)	880,77	1478,61	986,31	1212,44	842,53	632,36	1290,56
Llanaves(LL)	722,58	1288,95	857,28	1355,94	814,58	559,23	1099,92
Lechada(LE)	672,59	1295,92	800,36	1328,80	753,82	502,13	1281,98
Valponguero(VA)	870,28	1367,66	997,05	1292,24	843,84	654,61	1174,30
La Rasa(LA)	1425,48	1647,05	1560,52	1427,94	1240,97	1046,00	1063,97
Siero(SI)	1663,73	1532,32	1661,02	1394,51	1358,84	1190,33	968,19
Pando(PA)	1699,21	1487,99	1681,74	1338,78	1471,88	1267,62	948,49
Salio(SA)	1593,62	1455,36	1639,85	1315,59	1363,99	1209,99	935,87
Carande(CAR)	1826,02	1402,13	1476,30	1288,38	1416,53	1236,86	892,83
Escaro(ES)	1377,00	1632,27	1499,10	1571,29	1286,71	1013,41	1053,09
Anciles(AN)	1175,84	1250,90	1044,05	1158,00	1083,67	837,03	826,90
Horcadas(HO)	1351,75	1390,70	1412,81	1397,56	1145,21	923,57	964,16
Total	1242,26	1362,84	1237,03	1177,51	1097,95	915,54	967,00

Each MDU and MDE was then described with detail on the basis of different abiotic and biotic parameters. At this level of the analysis engineering geological and natural hazards features are of particular relevance. The main items used for the characterization and description of MDU's and MDE's at this, more detailed, level were: climate, landform bedrock, thickness of the regolith, bearing capacity, corrosivity, ease of excavation, permeability, slope stability, soil type, soil capacity, slope, altitude, collapse hazard, flood hazard, slope instability hazard, groundwater vulnerability, present vegetation, potential vegetation, fauna, landscape quality, landscape fragility, sites of scientific or cultural interest. Information on land-use recommendations and limitations was also added.

MDU's were also used to derive other types of environmental maps which are relevant for planning purposes, like the synthetic soils map, the map of potential vegetation or the map of quality for conservation (Cendrero et al., 1986). As MDU's can be defined in terms of their climate, morphology, bedrock, regolith and vegetation, they include the factors that, in principle, determine soil genesis, so that MDU's with equivalent features from the point of view of soil genesis will also represent soil units. Using this principle, a soils map (Francés et al., 1988) can be elaborated taking a limited number of samples and extrapolating the data to other. equivalent MDU's.

Similarly, MDU's contain a wealth of information on abiotic parameters of biogeographical significance. Therefore, if correlations are established between the

Fig. 3. M.D.U.'s and M.D.E.'s (dashed limits) in the coastal area of the Nansa

different types of natural vegetation formations and the relevant abiotic parameters, it will be possible to determine, which areas would be covered by each type of vegetation if no human interference had taken place. Thus, the map of potential vegetation would indicate what type of vegetation would be adequate in each place for the restoration of natural conditions. The correlations between vegetation and abiotic parameters were made by means of a square grid with a cell of 6.52 ha. Different types of profiles were obtained on the basis of the frequently distributions observed (Fig. 4). To obtain "indexed ecological profiles" (Gauthier et al., 1976) for each vegetation formation a contingency table was made, relating it with all abiotic parameters (Table 2).

A probability test was applied to each observed frequency, a, (Daget and Godron, 1982) in order to define the level of statistical significance (three levels, positive and negative) (Fig. 5). Using the results of these indexed profiles, together with field observations on the natural vegetation formations actually present, the MDU's were used to derive a map of potential vegetation.

The map of quality for conservation was obtained by means of a weighing/scaling method, like the one discribed by Cendrero and Díaz de Terán (1987).

Table 2

CLASSES OF VARIABLE L

	type K	other types	Total
presence	a	b	m=a+b
Veg.A			
absence	c	d	n=c+d
Total	r=a+c	s=b+d	N=m+n+r+s

a= number of cells in which variable L presents type k and , simultaneously, formation A is present.
m= total number of cells where formation A is present.
r= total numer of cells in which variable L presents type K.
N= total number of cells.

Finally, a map of land-use recommendations was obtained, also on the basis of MDU's. To obtain this map, a series of land-use activities were first defined. For the Nansa valley they included: intensive agriculture, dairy farming, extensive stock raising, conservation, productive reforestation, protective reforestation, exploitation of existing forests. These activities were assigned to each MDU according to the criteria expressed in the following, successive steps:

1.- Conservation is recommended for MUD's of class 5 (maximum) of quality for conservation.

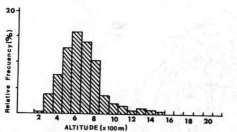

Fig. 4. Frequency distribution of "oak forest" in relation to altitude.

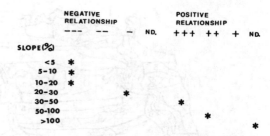

Fig. 5. Indexed profile showing the negative relationship between "oak forest" and gentle slopes and the positive relationship with steep slopes. N.D.: insuficient data.

2.- Among the remaining MDU's, intensive agriculture is recommended for those with good quality soils (classes A, B and C in the synthetic soils map).

3.- Dairy farming is recommended for MDU's with lower quality soils (classes C and D) covered by meadows.

4.- Extensive stock raising is recommended for the remaining MDU's within the "glacial valleys" system.

5.- Among the remaining MDU's, those with very low quality soils (classes D and E), slopes greater than 50% and evidence of erosion were recommended for protective reforestation.

6.- The rest of MDU's with soils of class D and slopes lower than 50%, covered by pastureland and with Jurassic limestone as bedrock were recommended for extensive stock raising.

7.- MDU's with soils of class E and slopes lower than 50% were recommended for productive reforestation.

8.- In other MDU's, already covered by forests, the sustained exploitation of such forests was recommended.

9.- No specific recomendation was made for the remaining MDU's. Obviously, these are the units in which no importat limitations are present, on the basis of ecological or productivity considerations.

However, land-use activities which require construction, such as residential, industrial or recreational ones, might suffer from limitations derived from engineering geological factors such as geological or geomorphological hazards, or foundation conditions. To take this into account, the information refering to such factors in each

MDU was extracted and represented on the final map in the form of land-use limitations. Fig. 6 is a sample of the final map of land-use limitations and recommendations.

This kind of mapping and evaluation is suitable for semidetailed or detailed assessments, at scales 1:50,000 or greater.

CONCLUSIONS

- The method described makes it possible to proceed, in the mapping and assessment of integrated environmental units, from smaller scale, large units and general assessments to larger scale, small units and detailed assessments.

- Morphodynamic units are an extremely useful basis for the storage of information on the environment. They provide the framework for the combination of data on engineering geology, natural hazards, geomorphology, soils and ecology, in order to obtain an integrated evaluation of land units for different planning purposes.

- Morphodynamic units also constitute a basic representation of the territory from which different kinds of thematic, purpose-oriented maps can be obtained, such as synthetic soils maps, maps of natural hazards and risks, potential vegetation maps, maps of quality for conservation or maps of land-use recommendations and limitations.

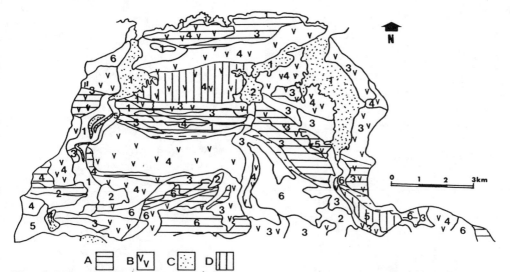

Fig. 6. Map of land-use recommendations and limitations of the lower part of the Nansa valley. 1: conservation, first priority 2: conservation, second priority; 3: intensive agriculture; 4: dairy farming; 5: protective reforestation; 6: productive reforestation. A: aquifer vulnerability, B: soil vulnerability, C: flood hazard, D: slope instability hazard.

REFERENCES

Alonso, A. (1987). Inventariación, análisis y evaluación integrada del medio natural en la comarca de Riaño, León. Ph. D. Thesis, Univ. of León. 618 pp.

Alonso, A. and Cendrero, A. (1988). Valoración territorial de unidades-valle para diferentes actividades, a partir de parámetros abióticos, en la montaña de Riaño (León). Actas II Congr. Mundial Vasco, Biol. Ambiental, T. 1. 235-253.

Cendrero, A. and Díaz de Terán (1987). The environmental map system of the University of Cantabria, Spain. In: P. Arndt and G. Lüttig (Eds.) Mineral Resources Extraction, Environmental Protection and Land-Use Planning in the developing and industrial countries. E. Scheizerbart Verlag, Stuttgart. 149-181.

Cendrero, A., Nieto, M., Robles, F., Sánchez, J. et al. (1986). Mapa Geocientífico de la Provincia de Valencia a escala 1:200.000. Diputación Provincial de Valencia, Valencia, 2 T. 71+350 pp.

Christian, C.S. (1957). The concept of land units and land systems. Proc. 9 th Pacific Sci. Congr., 20. 74-81.

Daget, P. et Godron, M. (1982). Analyse frequentielle de l'écologie des espèces dans les communautés. Masson Ed., París, 163 pp.

Francés, E. (1987). Cartografía geocientífica integrada del Valle del Nansa; su relación con la cobertera vegetal y con la vocación de uso del territorio. Ph. D. Thesis, Univ. of Oviedo. 1174 pp.

Francés, E., Martínez, V. and Cendrero, A. (1982). Un método de cartografía edafológico-sintética sobre unidades morfodinámicas y su aplicación en la vertiente cantábrica. Actas II Congr. Geol. España, Com.,2, Granada. 471-474.

Gauthier, B., Godron, M. et Lepart, J. (1976). Un type complementaire de profil ecologique: le profil écologique "indice". Can. J. Bot., 55: 2859-2865.

Lüttig, G. (1987). Large scale maps for detailed planning. In F.Ch. Wolff, ed. Geology for environmental planning. Geol. Survey of Norway, Special Publ. 2, Trondheim, 71-76.

Sneath, P.H.A. and Sokal, R.R. (1973) Numerical taxonomy: the principles and practice of numerical classification. W.H. Freeman, S. Francisco. 318pp.

Compiling of special engineering-geological maps

Sur le dressement de cartes géotechniques spéciales

Ilia Broutchev
Geotechnical Laboratory at the Bulgarian Academy of Sciences, Sofia, Bulgaria

ABSTRACT: Ideas and proposals for directing engineering geology to some spheres, having not been sufficiently dealt with so far, are discussed. The following necessity is reasoned: a) compiling of review engineering-geological maps of vast territories - regions, continents and the globe; b) widening the scope of engineering-geological maps of different scales by compiling new types of maps; c) unification of engineering-geological maps by applying unanimously acknowledged principles, methods and legends. All this will allow to come to new estimations and regularities concerning the engineering-geological conditions, destructive processes in the earth crust and the risks they create for human society.

RESUME: Dans le but d'orienter la géologie d'ingénieur dans quelques domaines qui sont peu considérés jusqu'à présent, sont discutées des idées et des propositions. La nécessité est argumentée par: a) dressement de cartes géotechniques d'aperçu de grands territoires - régions, continents et le globe terrestre; b) - élargissement du spectre des cartes géotechniques en différentes échelles au moyen de dressement de nouvelles sortes de cartes; c) unification des cartes géotechniques par application de principes, méthodologies et légendes générales. Tout cela donnera la possibilité d'atteindre de nouvelles estimations et régularités concernant les conditions géotechniques, les processus de destruction de la croûte terrestre et les risques du'elles portent pour la société humaine.

INTRODUCTION

A review on the problem of place and importance of engineering geology for science and society leads to the conclusion that its role is constantly increasing and that it considerably contributes to the development of modern civilization. This is a historical tendency which has started since the age of primitive man who lived in caves and used stone weapons to survive. Nowadays the relation man-nature is more complicated and the task of science is to regulate this relation in such a way that no harm may be done to either party. Engineering geology has its own part in solving this problem as far as it studies the features of the geological environment, in which modern society lives. Reclamation of vast territories and their constant widening necessitates the acquaintance with the natural conditions, engineering-geological inclusive, in these territories. This has originated the idea that engineering geology should be directed towards regional and global estimations of these conditions. Of course this does not eliminate the studies performed on small areas for

concrete objectives. The necessity of qualitative and quantitative estimates of destructive processes in the earth's crust, the risk they create, what part of the total natural factors' risk it is, is obvious. The degree to which geological risks endanger mankind is unknown. For example what is the vulnerability and resistance of this thin soil layer which supports the entire flora and fauna on land? These questions and their answers place engineering geology on a new quality level and gives it higher and relative importance among the other sciences. This became obvious also at the last international geological congresses, where even scientists from other branches of geology stressed on this, taking into consideration in particular its role for protection of the geologic environment. Facing engineering geology with regional and global tasks determines new directions of its future development. One of these tasks is the compilation of review engineering-geological maps of large areas, as well as widening of the spectrum of special maps of different scales. The reasoning of this necessity, certain estimations, ideas and proposals are considered in the present paper.

The research and practical activities of compiling engineering-geological maps have developed intensively during the past decades. Theoretical and methodological concepts have been improved. The nomenclature and content of the maps has been diversified, their use for different purposes has been widened. All this deserves a positive estimation. However, during the past years, with the development of urbanization, industry, energetics, with the increased impact on the environment, also geologic environment inclusive, new problems have emerged, which cannot be solved by the traditional engineering-geological maps and more. This is due to their space restrictions with respect to area and depth as well as with respect to their content satisfying the traditional requirements of standard construction works. This necessitates the development of new types of engineering-geological maps, which should comprise larger areas and depths of the earth surface and which should include new elements in its content aimed at meeting the requirements of the newly emerged problems.

1. NECESSITY OF REVIEW ENGINEERING-GEOLOGICAL MAPS OF LARGE TERRITORIES

Review engineering-geologic maps of larger territories - groups of countries, geographic regions, continents, and of the globe are not known to have been compiled so far. The lack of interest in this category of maps up to now is explained by their negligible practical use for concrete purposes which is due to their small scales. However, a number of changes, having occurred in science and in the world enforce the reconsideration of the problem. More and more arguments in favour of such maps are appearing.

Here are some of these arguments (Tabl. 1):

a) The creation of economic units comprising large groups of contries, which is the tendency and prospect in the world, will require a good knowledge also of the natural, engineering-geological inclusive, conditions of the respective territories. Initial, tentative data on these conditions may be of good use for managing and planning institutes, organizations and firms.

b) A great number of activities in industry and building go beyond the scope of some countries. Construction of long and complicited linear equipments is planned and realized - transcontinental highways, high-speed railways, gas-mains, oil pipelines, electric transmission networks. Hydrothermal systems along large rivers lie on the territory of several countries. Joint construction is being realized in frontier regions of neighbour countries. The availability even of small-scale engineering-geological maps may assist with the initial planning and choice of versions.

c) The problems of protection of the environment have become more and more acute. Some of them have global importance. The tendency of a close engagement of the engineering geology with these problems was clearly expressed at the last world congresses in geology (Moscow, 1984 and Washington, 1989). To solve these problems engineering-geological maps should be compiled.

d) The engineering geology, like all other sciences, requires its own reproduction, development and the outlining of forthcoming and more distant prospect. The existing pragmatism of solving concrete tasks for given projects will be kept and developed in future as well, but new ideas and new philosophy are necessary, which could meet the requirements of the 21st century. In this respect the review engineering-geological maps may be helpful.

e) A number of regularities concerning space distribution and development of different destructive geological processes and phenomena may arise when analysis larger territories. This concerns seismic zones, volcanoes, deserts and their widening, the zones of eternal freezing, weathering and denudation processes, movement of masses along the slopes, changes of sea and ocean coast lines, regional fluctutations of the ground water level, etc. Some of these are related to the climate, hydrographic network, technogenic activity of man. Qualitative results on the balance of destructed, transported and deposited earth and rock masses during different kinds of processes, on the tendencies of their development and impact on man's environment, may be achieved in future. This will allow to develop better founded strategy of counteraction against the processes of destruction. Such kind of results could be of importance also when solving more general geological problems.

Besides the arguments stated for the necessity of review engineering-geologic maps, there is another substantial argument of methodological nature. The countries which have engineering geological zonation, based on a regional principle, had as a starting point the territories within their national boundaries and these territories were divided into engineering-geological units. Thus the maps of the different countries have different style of the content and are incomparable. If the globe is accepted as a starting point for the engineering-

Table 1. Reasons for the necessity of review engineering-geological maps

Arguments	Scientific value and practical application
Creation of regional economic units	For obtaining tentative information on the engineering-geological conditions of the respective territories
Planning and construction of transcontinental linear equipments	A preliminary estimate of the engineering-geological conditions, choice of prospective variants of layouts, planning of research
Protection and utilization of the geological medium	Development of conceptions, models, estimations for protection of the earth area against destructive processes
Development of engineering geology as a science	Outlining of forthcoming and more distant prospects and trends of future development
Finding possibilities for the sake of establishing regularities	The regularities of spreading and development of a number of risks may be established at studying large territories
Unity of methods, principles and categories	Possibilities of comparing, joining on and comfortable use of the engineering-geological maps will be created

geological zonation and it is divided into differing in rank units on the basis of generally accepted principles, features and criteria, then each country and its corresponding map will take its proper predetermined place. Only in this case is possible to compare the maps and to join them on along the national boundaries of the countries. Maps and monographs of the national territories (Kamenov and Iliev, 1964 , Matula and Pašek, 1986) will serve as a starting point.

The compilation of review maps of large territories, realized by international working groups with the participation of many countries will open up possibilities for new scientific contacts, for summarizing a considerable amount of information and for general elevation of the level of engineering geology in the whole world. Even only this suffices as a motive provid-

ing reasons for the realization of this activity. The cognitive importance of these maps should not be under-estimated as well. Such maps will ensure the access of larger public circles to the engineering-geological information and will elevate the prestige of engineering geology among these circles.

The development of new territories with unfavourable climatic and engineering-geological conditions will increase in future and the source of an initial information will be these maps. Their further progress as methods, principles, legends and their practical realization are possible only under the conditions of international co-operation. Convenient time for this is the decade of UN dedicated to natural calamities. The International Association of Engineering Geology may play a positive role with its managing, coordinating and organizing functions. All this would not be possible without the active participation of national groups and organizations in engineering geology.

Airborn and space methods, applied through national and international programmes will be also useful for the engineering-geological mapping of regions, continents and of the entire globe.

As far as in the near future is envisaged construction of a permanent inhabited station at the Moon and landing of a piloted space-ship on Mars, the necessary conditions also include the engineering-geological characteristics of the areas chosen for landing. The maps of zonation of these space bodies with respect to their suitability for landing of space-ships would be valuable and useful. Even only the land slides, known to exist on Mars, are quite interesting in this respect (Lucchitta, 1978).

In other branches of geology analogous activities have been developing for a long time now. A great number of geologic, tectonic and other kinds of maps of a number of large regions have been compiled. Even more narrowly specialized maps have been produced. A working group from 7 countries compiled a series of maps of the Carpathian-Balkan region in a scale of 1:1 000 000 (geologic, tectonic, of the magmatic formations, of the minerals, hydrogeologic). This was performed by the countries-members of the Carpathian-Balkan Geological Association. Useful results were obtained of the scientific projects for study of the seismicity on the Balkan Peninsula, developed with the help of UNESCO. The hydrogeological map of Europe in a scale of 1: 1 500 000 (Struckmeier, 1989) will be soon ready. The experience of research and compiling work from all over the world and from separate regions may be useful when compiling engineering-geological maps.

2. NEW TYPES OF SPECIAL ENGINEERING-GEOLOGICAL MAPS

A considerable widening of the scope of problems of engineering geology has been observed during the past decades. This is caused by the requirements of the advance of science and technique, by the appearance and development of new spheres of human activity such as the intensive urbanization, nuclear energy, development of deeper ground areas, penetration into shelf zones, creation of mega polices, construction of large but complicated, sensitive to the processes in the geologic medium, equipments. In a great number of the cases it is necessary to come into territories of complicated and unfavourable engineering-geologic conditions. Nature and the requirements of mordern economy and construction come into conflicts. Thus the task of engineering-geology becomes more difficult but at the same time it acquires greater importance. May be Ter-Stepanyan is right suggesting that the stage after Quaternary will be a stage of engineering geology (Ter-Stepanyan, 1985). The problem of protecting the environment, facing now mankind, has its engineering-geological aspects. On these grounds engineering geology orientates more fully towards solving the problem at the same time including in its subject protection and rational utilization of geological environment (Sergeev, 1984). This deep penetration of engineering geology into human life, although being a natural science, has its impact on man's spiritual sphere. Examples for this may be given also by ancient mythology. Even philosophic generalizations about man and its environment were made. It is paradox but life on the Earth appeared thanks to the destructive processes in the earth's crust, to which engineering geology is trying to counteract.

All these circumstances are forcing engineering geology to estimate its present state, its results and faults, theoretical base and methodology. This also concerns engineering-geological maps. They have been serving well so far, but it is obvious that a new necessity is appearing of widening and deepening of this sphere on the part of engineering geology. This is a very vast theme, and in the present case only some proposals, concerning the compilation of new types of engineering-geological maps, meeting present-day needs, are discussed. Depending on their purpose they may be of different scales - from plans of individual sites up to review ones about whole countries. Each map presents the state of the site at the moment of its compiling. Maps, presenting slightly changing with time engineering-geological elements may be used for a long period of time. However, when the changes of some elements occur more rapidly, a remapping would be necessary at intervals depending on the intensity of the processes and the importance of the given territory. This concerns mainly large-scale maps of regions of active physicogeological processes and a more considerable technogenic activity. In some cases it is necessary to know and illustrate on the maps the geologic structure of the greater depths up to a basement rock, which may lie at several hundred meters, and even deeper, below the surface. Although this makes the task more complicated, it is possible to solve the problem by using an adequate legend.

The nominations of the maps, discussed in the present work, are tentative so far. More

Table 2. Kinds of engineering-geological maps and their destination

Kinds of maps	Destination
Engineering-seismogeological	For estimate of seismic danger
Ecogeological	For protection and rational use of earth environment
Depth engineering-geological	For development and use of underground areas; extraction of minerals
Archaeo-geological	To help the protection of archeological monuments and places of historical interest
Radioengineering-geological	For zonation of territories according to the degree of suitability for storing radioactive wastes
Marine engineering-geological	Estimate of the engineering-geological conditions of the shelf area and the possibilities of using it
Engineering-geological maps of artificial soils	To estimate the state and quality of surface equipments, cultivated layer, industry wastes, artificially improved soils
Engineering-geological maps for terrain regulation	To improve landscape and characteristics of terrains demaged by human activities
Engineering-geological maps for building materials	For the types and location of natural building materials and raw materials
Paleoengineering-geological	For estimate of engineering-geological conditions in the past
Map of exotic phenomena	For registration of valuable rock phenomena, caves, protected geological sites, etc.

43

suitable names are likeable to be given in future discussions and final versions to be accepted. Here are some examples of special engineering-geological maps which are necessary: engineering-seismogeologcal map, ecological maps, marine engineering-geological map, depth engineering-geologic map, archaelogical-geological map, radio-engineering-geological map (terrain zonation according to the degree of suitability for storing radioactive wastes), marine engineering-geological map, etc. (Table 2). This list may be made longer adding considerable number of new special engineering-geological maps and this obviously widens the scope of engineering geology adding different new scientific and practical activities. It is natural for each country to have its dominating problems and the world as a whole is a complex mosaic consisting of countries and groups with their specific engineering-geological conditions. It is expected as a future prospect that all the countries will have their own atlases of engineering-geological maps, comprising both traditional and new types of maps. These will be valuable sources of synthesysed information, useful for scientific generalizations and conclusions, for planning research works, for estimates of territories for different purposes, etc. If these maps are compiled using one and the same methods and legends, it will be possible to join on the maps along the boundaries and in this way to obtain maps of larger regions.

3. UNIFICATION OF ENGINEERING-GEOLOGICAL MAPS

Compilation of engineering-geological maps is intensively developing now which should be positively estimated. However, at the same time different principles, approaches, methods and legends are used. This creates difficulties for comparing and using these maps. It is even more difficult to summarize these maps and to join them on so that maps of larger territories may be obtained. The unification valid for the geological maps, using equal approaches and legends accepted all over the world, has not been achieved for the engineering-geological maps.

The necessity of unification of the engineering-geological maps, of the compiling methods and legends, has become urgent. This will be connected with a number of difficulties due to the considerable differences of the engineering-geological conditions of the separate countries, the different degree of development of the engineering-geological investigations in these countries. Nevertheless, directing research works into this sphere will allow to compare the future unified engineering-geological maps to use them as a source of scientific conclusions and regularities, to combine them into maps of larger territories. These reasons are quite sufficient to undertake unification of the engineering-geological maps within the frames of their different types.

CONCLUSIONS

The formulated ideas and the proposals made are considered to be a base of an international discussion, where the problems stated may be further developed.

REFERENCES

Kamenov, B., Iliev, I. (1963). Engineering-Geological Subdivision of the PR Bulgaria. Works on the Geology of Bulgaria, series Eng. Geol. and Hydrogeol., vol. II, 5-123 (in Bulgarian).

Lucchitta, B.K. (1978). A Large Landslides on Mars. Geological Society of America Bulletin, vol. 89, 11, 1601-1609.

Matula M., Pašek, J. (1986). Regionálna inzinierska geológia CSSR, 296 pp. ALFA - Bratislava, SNTL - Praha.

Sergeev, E.M. (1984). Scientific-Technical Progress and Environment Protection. 27 Int. Geol. Congress, General volume, 54-59. Moskow (in Russian).

Struckmeier, W.F. (1989). La carte hydrogéologique international de l'Europe. Hydrogéologie 2, BRGM, France, 115-118.

Ter-Stepanian, G.I. (1985). Beginning of the Quinari or the Technogene. An Engineering Geological Analysis. Communication Lab. of Geomechanics IGES AS Arm. SSR, No 5, 100 pp. Publ. House of the Arm. Acad. of Sciences, Yerevan. (in Russian).

6th International IAEG Congress / 6ème Congrès International de AIGI, © 1990 Balkema, Rotterdam. ISBN 90 6191 130 3

G.I.S. in geology: an application for engineering geology maps
G.I.S. en géologie: une application pour les cartes de géologie technique

F. Brunori & S. Moretti
Earth Sciences Department, University of Florence, Italy

ABSTRACT: In the modern geological cartographic representation the acquisition and treatment of the territorial data is very difficult to manage without a Geographic Information System based on automatic procedure. This paper wants to give an example of application of one of these system, treating the data for an erosion hazard map in an area located in central Italy (Atri, Abruzzo).

RESUME: Dans la moderne cartographie géologique la représentation, l'acquisition et l'élaboration des données territoriales sont trés difficile à exploiter sans un systéme geographique d'information basé sur des pratiques automatiques. Ce rapport veut donner un exemple d'application d'un de ces systèmes, qui s'occupe des donnèes avec le but de réaliser une carte sur le risque d'érosion dans une aire de l'Italie centrale (Atri, Abruzzo).

1 INTRODUCTION

Due to the fast technological development and the new socio-economic requirements, a radical change in approaching the different steps of the basic and thematic classical cartography, is necessary.

The landscape characteristics survey, linked with earth sciences, needs automatic systems to carry out most difficult and time wastig elaborations, due to the inter-actions of the different factors acting on the landscape itself.

With such systems, it is possible to store quickly the data and then elaborate such information by means of quite complex criteria and mathematical methodologies.

The link between G.I.S. (Geographical Information Systems) and modern cartography gives an important contribution to a rapid development for representing and elaborating the information obtained by means of basic studies, such as field work (Burrough 1988).

This fact is especially important in studies dealing with the knowledge of the landscape evolution (Kirkby at Al. 1987).

The G.I.S. used by the Applied Geology Research Group of the Earth Sciences Department has allowed the realization and the utilization of a data bank containig information about land unit identification, which is also integrated with historical data. Such interpolation, between surveying and automation, allows us to obtain well detailed maps, improving both their reding and their potential use.

The result of such a procedure, will be a better knowledge of the land potentiality, from a physical point of view; furthermore introducing the right parameters (e.g. economical, social etc.), the procedure will allow the user to estrapolate some point data all over a previously defined land unit.

Such a kind of cartographic representation has the advantage of giving immediate information which are also typologically and geographically referred to.

The sample area is on the Adriatic part of the central Apennine and more precisely, in the Atri council between the Vomano and the Piomba torrents (Fig. 1).

2 G.I.S. GENERAL FEATURES

The used software (GEOSYS) is applied, and is especially useful for the management of the territorial elements. It is formed by different integrated packages. The functions, that such a programme performs, are orga-nized as follow:

1. Acquisition of data. Digitizing, editing and file creation. This part of the

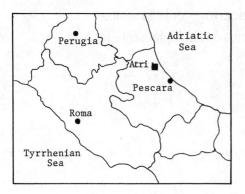

Figure 1. - Location of the sample area.

programme acquires and storages the graphic, cartographic and descriptive (attribute) data.

2. Management and data analysis. This part organizes and search the territorial information, then it processes and produces data for well defined conditions.

3. Graphic representation. This part creates a graphical output depending on the device which is employed (plotter, printer, atc.).

4. Conversion. It changes and writes the data in standard files (ASCII code).

Phisically the programme is formed by eleven integrated packages. Most of these packages have their own function that gives the possibility to run the programme even without the others, with the exception of the two fundamental ones that activate the programme (EDICAR and VIGRAF).

This programme has been written in FORTRAN 77 language with graphical standard for Calcomp and Tektronix. The software packages work on operative system VMS-VAX, Digital and it takes a memory of about 20 Mbytes. It runs also under PC/MS-DOS as shown in Figure 2.

3 TYPOLOGY AND APPLICATION OF CARTOGRAPHIC DATA

The used software is especially intended for acquisition of referred territorial data, with which it is possible to treat geographical information of enhancing elements, that can be virtually intended as points, lines, polygons or grid, attributes, drawings and images associated to them.

Furthermore it is possible, not only to

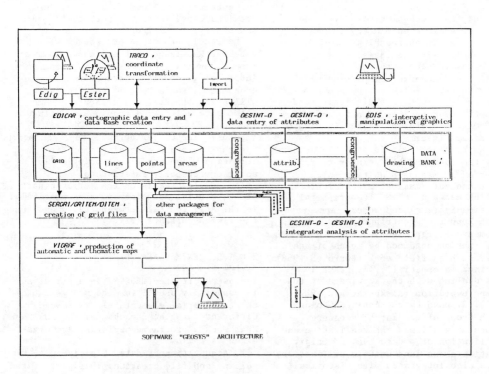

Figure 2. - Software GEOSYS architecture.

manage the basic cartography, but also to overlay all the geographic and descriptive information for the production of derived cartography, tables and statistical graphs.

The acquisition of the basic cartography, has been organized on the basis of thematic planes in such a way as to eventually produce the different themes both separately and logic overlying as shown in Figure 3.

The original cartographic documents utilized for this work were topographic maps 1:25.000 scale, a geological map of the area with 1:100.000 scale and a geomorphological map with 1:25.000 scale.

The information taken from the geomorphological map was essentially related to the badland areas in such a way as to allow the identification and the cartographic representation of this area.

Particular attention was given to the digitizing of the badland ·boundary. In fact the scale of the map is not the most suitable for precise work, however the obtained results, compared with the field checks, resulted to be rather precise.

The final and general task, is of course, the realization of basic maps on the following subjects: geology, hydrogeology, land use and thematic maps as : slope stability, erosion susceptibility etc.

The example of this paper is referred at the first phase, consisting on the realization of the erosion hazard map of a badland area in Atri.

4 GEOLOGICAL OUTLINES

The sample area chosen for this research is located in the Teramo Province and includes the whole Atri municipality (Fig. 1).

The landscape is characteristic of the badland areas that occur on the marine pliocenic deposits with a prevalently fine texture. It is characterized by a series of sub-parallel rivers, mainly west-est oriented. These rivers, with their erosive action cut down the deposits; this fact makes it possible to identify quite well the stratigraphy of the area.

Beginning from the older deposits, the geology, shown in Figure 4, can be resumed as follows:

Middle pliocenic deposits consisting of clay and marls; on these materials the most part of the badlands is formed. Such deposits pass in concordance into sand and sandy-clay of the upper Pliocene.

Going up the succession, the lower Calabrian deposits are found (Molinari 1984); these materials are prevalently sandy clay interbedded with marls and fine sand. The following component of this succession is formed by sands and clay with some coastal conglomerate levels dated to the lower Calabrian. To these deposits, is associated

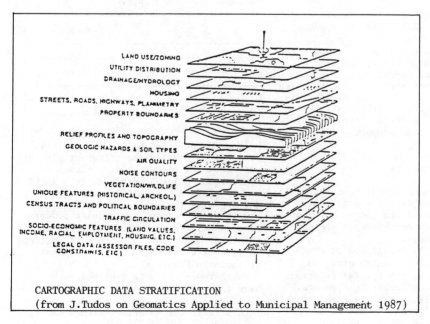

CARTOGRAPHIC DATA STRATIFICATION
(from J.Tudos on Geomatics Applied to Municipal Management 1987)

Figure 3. - Cartographic data stratification.

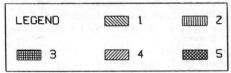

LEGEND   1   2
   3   4   5

Figure 4. - Geological map of the sample
area. 1 Conglomerates; 2 sands and clay ;
3 clay with marls and fine sand; 4 sandy
clay; 5 clay and marls.

lagoonar clay (Parea and Velloni 1984). The
last deposit of this succession is a very
resistent continental conglomerate, forming
very steep escarpments bordering remnent
platou on which are most villages and town
of the area (e.g. Atri, Mutignano, Silvi
etc.).

The morphology of this area is strongly
influenced by the lithologic succession.
In fact there are some more or less
extensive surfaces which indicate the more
resistent materials covering the more
erodible underlying deposits.

In the areas where such a coveer has been
removed by the erosive processes, there is
a completely different morphology in which
badlands and rolling areas alternate, of
course not in the same proportion.

5 RESULTS AND DISCUSSION

The first clear result comes from the real
possibility to drow from the data bank,
information regarding the geological and
morphological characteristics in the area
which is being studied.

The geology can be overlapped on the 3-
dimensional morphology, which gives an
immediate picture of the spatial relations
between these two cartographic elements.
In fact, there exists very clear relatioship
between the stratigraphic position of
various deposits and their topographic

location.

As previously described, the harder
deposits (conglomerates) are found on the
top of the main reliefs; on the other hand
the ones on which the badland areas (calan-
chi) are developed , are mostly located
on the middle slopes (unit 2).

Table 1 shows the relative percentage
of each geological unit with respect to
the sample area and the distribution of
the surface covered with badlands (Fig. 5),
which results to be 15% of the whole area.

Figure 5. - Distribution of badland areas
in the sample area.

Table 1 - Distribution of the geological
units in the area and percentage of
badlands for each unit.

unit	%	badlands %
1	28.5	5
2	34.5	82
3	0.8	/
4	16.6	5
5	19.6	8

After completing the data base, it is
possible to overlay the different elements
e.g. geology, geomorphology, slopes and
the corresponding attributes. This enables
one to produce a thematic cartography
dealing with the subject being studied.

By this procedure, the erosion process
is evidenciated in detail, by enhancing
the areas where such a process is particularly active and where it is incipient
as shown in Figure 6.

Furthermore the data bank allows us to
have information on the previous landscape
(Pliocenic surface) in such a way that it
reconstructs the amount of eroded materials,
measured in volume of earth removed.

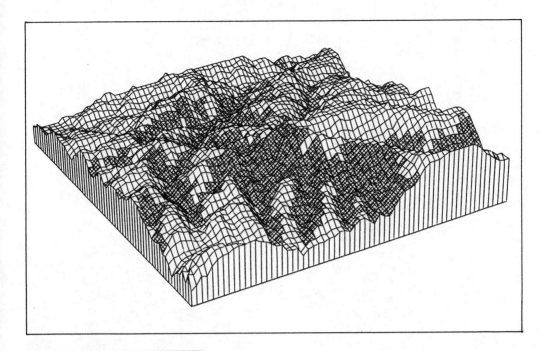

HORIZZONTAL SCALE 1·50000
VERTICAL SCALE 1·15000

ZENITH ANGLE 60   (IN SEXAGESIMAL DEGREES)
AZIMUT ANGLE 30

VIEW DISTANCE 18 KM.

Figure 6. – Distribution of the badland area. 3-dimensional view of the sample area.

The volumetric calculation, of course is only referred to as the total amount of material eroded from the Pliocene and up to now it is not possible to have its distribution over the time.

REFERENCES

Burrough, P.A. 1988. Principles of Geographical Information Systems for Land Resources Assessement. Ed. Oxford.
Kirkby, M.J., Naden, P.S., Burt, T.P., Butcher, D.P. 1987. Computer Simulation in Physical Geography. John Wiley & Sons.
Molinari, E. 1984. La dispersione dei materiali nei depositi terrazzati pleistocenici della fascia costiera abruzzese nella zona di Atri-Silvi. Boll. Soc. Geol. It. 103: 529-537.
Velloni, R. and Parea, G.C. 1984. Late Quaternary marine terraces in central Italy. 5th European Congress of Sedimentology.

Construction of small-scale sketch maps for mountain-folded areas with the example of Georgia

Composition de cartes schematiques à petite échelle pour les régions de montagnes à plissement à l'exemple de la Géorgie

J.M.Buachidze, E.A.Djavakhishvili & K.I.Djandjgava
Sector of Hydrogeology & Engineering Geology, Georgia, USSR

ABSTRACT: Engineering-geological maps scaled down to 1:600.000 and 1:1.000.000 reflect engineering-geological structure of large regions zoning them to major taxonomic units. Herein areas are distinguished according to rock strength indices, tectogenetic morphological peculiarities and formation principle. Regions, subregions and districts are outlined based on other various features.

RESUME: Une carte ingénieur-géologique à l'échelle 1:600.000 et 1:1.000.000 représente la structure ingénieur-géologique de grandes régions s'est subdivisées en unités taxonomiques des différentes degrés. Des régions se sont mise en relief d'après leur indice de la solidité, de la particularité de tectonique morphologique et du principe de la formation. Régions, sous-régions et arrondissements se sont distingués selon leur indice de différente forme.

The object of engineering-geological maps scaled down to 1:600.000 and 1:1.000.000 involves reflection of engineering-geological environment of large regions and their zoning to major taxonomic units to a degree sufficient for substantiation of preliminary projects of planning structures and proposals for further engineering-geological researches.

Georgia presents a part of the Alpine Orogen and is peculiar for rather involved tectonic structure, variability of lithological composition of rocks and the degree of their lithification. Herein we come across nearly all the possible stratigraphical rock complexes from Paleozoic up to the Quaternary deposits, represented by magmatic, terrigenous-clayey and carbonated formations. The region is characterized by rather intricate geomorphology, vertical and horizontal zonality of climatic conditions and abundance of various exogenetic geological processes. Therefore construction of sketch engineering-geological map of Georgia as a mountain-folded area is a rather complicated problem.

Small-scale maps should bear obvious, easily read and rather vast information.

There were two primary principles assumed as the basis at mapping. On maps scaled at 1:600.000 strength properties of rocks acquire the paramount importance, i.e. especial prominence is given to engineering-geological rock groups and their strength indices are represented by appropriate dense colours. Map scaled at 1:1.000.000 is based on rock-association principles with stratigraphic and lithological background. Herein rocks are mapped in accordance with adopted in geology colouring.

On the map scaled at 1:600.000 (Buachidze et al. 1968) we distinguish the below engineering-geological rock groups: rocks – given in red colour; rocks (r) and semi-rocks (sr) – in different tints of green just depending upon prevalence of rocks or semi-rocks (r sr or sr r); semi-rocks – in orange; semi-rocks and plastic rocks – in yellow; plastic rocks – in pale blue, and loose rocks – in grey colour. Lithology and stratigraphy are given by coloured striation, complying with geological standards. Mode of striation indicates the lithological composition. Rock strength indices are listed in a zoning chart applied to the map. This chart comprises data on recent geological processes: e.g. on the pattern of weathering – rock debris, en bloc or grussy one, etc., with indication of foremost exogenetic processes peculiar for the present rock type. Table 1 exemplifies a pattern of rock specification in accordance with the proposed chart.

Table 1

Rock			Gneisses, granitoids
Age index			Pz
Engineering-geological rock groups	Rocks	Red	~~~~~
	Semi-rocks and rocks	sr>r> r>sr> Green	
	Semi-rocks	Orange	
	Semi-rocks and plastic rocks	Yellow	
	Plastic rocks	Pale blue	
	Loose rocks	Grey	
Strength indices $R_{comp.}$ 10^5 Pa	Fresh		1500–1800
	Weathered		10–50
Recent geological processes	Mode of weathering		En bloc
	Key process		Collapses, avalanches

Superficial deposits are mapped in the terms of genetical types (alluvial-boggy, alluvial, proluvial, eluvial and other types) which are marked by coloured specks. Two-layered composition of the mantle rocks is also envisaged. Thus the upper layer is depicted by hollow specks, and the lower layer – by solid specks. Various combinations of specks indicate composition and thickness of these soils (table 2).

Recent geological processes are plotted by coloured vertical and tilted striation. Local major landslides, mudflows, etc. are marked by special conventional designations.

Strength indices for each specific engineering-geological rock complex are presented in a legend along with evidence on recent geological processes: i.e. mode of weathering – rock-debris, en-bloc, grussy one and so on, indicating, which process is most peculiar for a certain engineering-geological rock group.

Each engineering-geological rock group corresponds to a certain mapped district. Altogether 29 districts are distinguished on the map, which are united in areas according to the morphological-tectonical features. Thus, the territory of Georgia is subdivided to 8 areas: I – that of the crystalline core of the Great Caucasus anticlinorium axis; II – that of the highland part of the fold system of the Great Caucasus Southern Slope; III – that of upland west part of the fold system of the Great Caucasus Southern Slope; IY – that of west submersion zone of the Georgian Block; Y – that of the Dzirula crystalline mass; YI – that of the east submersion zone of the Georgian Block; YII – that of Adjara-Trialety fold system; YIII – that of Artvina-Bolnisi Block.

On the map scaled at 1:1.000.000 rocks are given at their age succession through standard colours; rock strength indices are depicted by various oblique striation complying in colour with stratigraphical rock complex, which is represented by means of somewhat denser colouring.

Herein we may recognise the below engineerinh-geological rock groups: 1. rocks with rigid bondings – mapped through vertical and partially tilted striation; these

Table 2

Genetical type of rocks and weathering subzone	Soil	Layer and its thickness, m						
		I				II		
		0-5	5-10	10-20	20	0-5	5-10	10-20
Alluvial-proluvial	Pebbles	o	o o	o o / o o	o o / o o / o	●	● ●	● ● / ●

rocks may be subdivided to rocks - unilinear striation; anizotropic rocks - two-linear striation; rocks and semi-rocks - striation tilted to the left. 2. Another group involves rocks with rigid bonding or lacking it altogether and comprises rocks and loose soils which are marked by thin vertical striation; semi-rocks - horizontal striation; semi-rocks and plastic rocks - bilinear horizontal striation. 3. The third group involves rocks having no rigid bondings; these are plastic rocks plotted with rightward tilt of the black striation, while the black specks inside the striation indicate rock composition; laminated and loose rocks which are depicted by varying black specks. Strength indices of each group are presented in the legend, lithological composition being listed in a column.

Exogenetic geological processes are mapped through coloured specks or corresponding conventional designations, while size of specks accounts for intensity of the process.

The above combined stratigraphical and engineering geological rocks groups are united in formations, subformations and facia, in all 26 units. The latters make up 12 groups of formations. These are as follows: geosynclinal groups: I - Early Hercynic one; II - Hercynic one; III - Early Alpine one; IY - Middle Alpine one; Y - Late Alpine one; geosynclinal (with retarded development): YI - Middle and Late Alpine; subplatform groups: YII - Early Alpine one; YIII - Middle Alpine one; orogenic groups (those of internal troughs): IX - Early Orogenic one; X and XI - Late Orogenic ones (those of superficial deposits), and XII - marine

formations (those of the Black Sea).

The both maps show the seismic zone with estimated seismicity of 8 points.

REFERENCES

Buachidze, J.M., Areshidze, G.M., Djavakhishvili, E.A., Djandjgava,K.I. et al. 1968. Opit sostavleniya obzornikh inzhenerno-geologicheskikh kart masshtaba 1:600.000 v gorno-skladchatikh oblastyakh. Tezisi dokladov 3-go regionalnogo covetschania po inzhenernoi geologii. Leningrad.

Buachidze, J.M., Areshidze, G.M., Djavakhishvili, E.A., Djandjgava, K.I. et al. Inzhenerno-geologicheskaya karta Gruzinskoi SSR masshtaba 1:1.000.000. Trudi 25-oi nautchno-tekhnicheskoi konpherentsii GPI. 1979. Tbilisi.

Engineering geological mapping in the northern shelf of South China Sea
Cartographie géotechnique du plateau au nord de la mer de Chine méridionale

Chen Jinren & Pang Gaocun
Guangzhou Marine Geological Survey, People's Republic of China

ABSTRACT : This paper expounds the principles, contents and methods of marine enginering geological mapping. Proper arrangement and colouring make the maps clean in patterns and outstanding in key points.

RESUME: Nous présentous ici le principe, le contenu et la façon de faire le plan de géologie constructive de l'océan. Les arrangement et les couleurs raisonnables font le plan clair et le mettre en relief.

Marine engineering geological survey in the Pearl River Mouth Basin of South China Sea is a project with technical assistance from the United Nations Development Program(CPR/85/044/01/1). Four international sheets have been surveyed so far, and Chinese and foreign experts have high opinion of the results acquired. The paper will discuss mapping methods of marine engineering geology based on the experience obtained during these years.

1 Purposes And Principles of The Mapping

The main purposes of the mapping of marine engineering geology are to provide (1) basic maps for ocean developments, seafloor engineering construction and environmental protection, (2) engineering geological information for exploration and extraction of offshore oil & gas resources and solid mineral resource, and (3) basis for site survey and special study(KERRY J. CAMPBELL, 1982).
The mapping follows the following principles: (1) to present engineering geological features and their change patterns within survey area, (2) to mark the stability of seafloor and potential geohazards

2 Construction of The Map

The main plan of marine engineering geo-

logical map is a plan view. Map's title and scale are on the top of the map frame, and an engineering geological section at the bottom. Borehole column and geotechnical test results are on the left side. A reproduction of typical feature of geohazard is on the left side of the bottom, and a chart of zonation of enginering geology on the right side of the bottom.

2.1. Engineering geological section

The geological section is made in the direction along which geological charactor change significantly , this section shows the geological features and their change patterns of each engineering geological unit. The boreholes, sampling locations and CPT stations must be marked on the section. The horizonal scale of the section is as same as that of plan view.

2.2. Borehole column and geotechnical test results

The expression of engineering geological sequences and their positions are on the left side of the column. The description of strata, the geotechnical test results and CPT curves are on the right side of the column.

2.3. Reproduction of data

The reproduction on large scale showes the typical features, in order to make users to pay more attention to them.

2.4. Chart of zonation

Zones and subzones should be usually divided. The chart includes the name, symbol and main characteristics of engineering geology of each zone and subzone.

2.5. Legend

Legends are arranged in following sequence: engineering geological zoning, soil types and their boundarys, hydrology and meteorology, topography and geomorphology, bottom erosion and accumulation, geological structure and active fault, earthquake, potential geohazard, the location of borehole and coring station, etc..

3 Contents of Map

The following contents should be included in the plan view(STANDARD,1989): The factors of hydrology and meteorogy, such as tropical cyclones(degrees, routes), waves, tides, bottom currents (speeds, directions).
Bathymetric lines at a certain intervals.
Geomorphic patterns such as tidal ditch, shoal, sand dam, underwater delta, fan, valley , low-lying land, ancient lagoon, underwater terrace, ancient coast etc..
Soil types and space distributions, soil structures(including their ages, causes, engineering geological character).
Geological structure units and subunits, combination and developing regularity of faults, the age, scale and depth below seafloor of faults should be marked.
Locations of earthquake centres,intensities and dates of earthquake.
Geotechnical character including mechanical and physical character.
Geohazards such as bottom slide or slump, shallow gas and mud diapir.
Zonation of engineering geology.
In addition, the locations of boreholes, coring stations, CPT positions and engineering geological section line should be also mapped on the plan view.

4. Expressive Form

The firt order of expressive mark of marine engineering geological map expresses zonation. Zones and subzones are coloured generally. Colour constribution is of little chromatism, are used to express subzones within a zone, and indicate they are of same general character. The second order of expressive mark indicates soil type and layer structure. Soil type is expressed by dark dot, line and veined sign. Layer structure is expressed by blue horizontal line(single line, dual line and triline). For example, symbol CI indicates sillty clay and one layer structure; symbol CI·SC(5--7) expresses dual layer structure, first layer is sillty clay, and second layer is clayey sand, and the depth of their interface is 5--7m symbol CI·SC(5--7)/SM(19--20) indicates trilayer structure, the third layer is clayey sand, and the depth of the relevant interface is 19--20m. the third order of expressive mark expresses environmental factor and geohazardous factor. Fault is showed by red line. Earthquake, slide, slump, shallow gas, etc. are expressed by red colour uniformly. Others are indicated by black and grey colour.
The map which is of substantial content presents bright administrative level and focal poit stressed through rational construction and colour arrangement, even if a nongeologist can also understand and use it.

REFERENCE

KERRY J. CAMPBELL - MICHAEL R. PLOESSEL (1982): Regional Engineering Geological Mapping of Offshore Areas. Proceedings. 4th Congress International Association of Engineering Geology, India. Volume VII, P. 187.
NATIONAL STANDARD BUREAU OF PEOPL'S REPUBLIC OF CHINA (1989): Map-editing Standard of Synthetical Enineering Geological Map. Science Public Press.

Engineering geological mapping for tunnels in the Himalayas
A rock mass classification approach

Cartographie géotechnique pour les tunnels dans l'Himalaya
Une approche par la classification des masses rocheuses

Vishnu D.Choubey
Department of Applied Geology, Indian School of Mines, Dhanbad, India

Gopal Dhawan
National Hydroelectric Power Corporation, India

ABSTRACT : The procedure of geological mapping and discontinuity survey incorporating basic geotechnical data for engineering classification of rocks has been discussed to forecast geoenvironment for tunnels in the Himalayas. It provides reasonably reliable data base for understanding probable modes of structural instability and evaluating the rock mass by the empirical methods prior to tunnelling operation in a geostructural unit. A case study of rock mass assessment and support estimation for a part of 4.7 Km long head race tunnel at Goriganga Hydroelectric Project, Uttar Pradesh, India is presented in brief.

1 INTRODUCTION

The behaviour of rock mass around a tunnel is influenced by a wide variety of geological, hydrological, engineering and physico-mechanical parameters while tunnelling. Their assessment along a tunnel route in the Himalayas is extremely difficult and sometimes unrealistic because of complex structural set up of geological formations. The past experience on correlation of predicted and actual rock mass conditions in Himalayan Tunnels is rather not encouraging. In addition to complex structural set up, the main restraints in understanding the rock mass, prior to tunnelling in the Himalayas, are paucity of outcrops, inaccessible terrain under dense forest cover and low priority for explorations related to tunnels in comparison to other civil structures, resulting in poorly documented geological maps for evaluating the rock mass before tunnelling. To break this stalemate partially, need for developing the Engineering Geological Map (EGM) for tunnels with detailed record of geotechnical parameters is stressed. This calls for spelling out the procedural steps to prepare an EGM containing complete data on rock units to classify the rock mass for preliminary support design by empirical methods. It is attempted by dealing a case study of rock mass evaluation and support assessment for a part of head race tunnel of Goriganga Hydroelectric Project, in eastern sector of the Kumaon Himalayas falling under Pithoragarh District, U.P., India.

2 GEOLOGICAL MAPPING AND DISCONTI-NUITY SURVEY

The EGM portrays the basic geology incorporating the lithology, the structure and the cover characteristics (Shome 1989). For large tunnelling projects, cartographic presentation of outcrops and overburden boundaries on a 1:5,000 scale is considered adequate. For detailed geotechnical evaluation and proper application of the rock mass classification approach for support design, each and every exposure/outcrop of the area has to be studied carefully to record rock mass characteristics, nature of discontinuities and the hydrological conditions influencing the geological formations. The methodology of rock mass classification based geological mapping and its outcome is discussed hereunder.

2.1 Data collection

Each exposure/outcrop is numbered on the ground and plotted on the contour plan. Different types of superficial cover are mapped with approximate estimate of depth of bed rock under the cover. The permanent and intermittent springs, seepages and streams are presented on the map. The lithological contacts, surface menifestations of faults/ thrusts and the major traces of master discontinuities are also picked by surveying techniques.

The finer details of rock mass characteristics and discontinuity parameters of an exposure are recorded on format sheets as per the checklist of items required for assessing rock mass conditions. The joint/discontinuity is described by its spatial orientation, persistance, spacing, aperture, filling, roughness and alteration, whereas the mass description includes the rock type, strength of the intact rock mass, degree of weathering, geological structure, stress reduction factor and volumetric count of joints (Jv) apart from the hydrogeological conditions. The basic geotechnical description of these parameters and procedure for their measurement is laid down by International Society of Rock Mechanics (ISRM 1978). These documentation sheets form an integral part of the EGM.

2.2 Data processing

For all the exposure of a geological formation, orientation values of discontinuities are divided into geometrical groups to assign a set number to each reading on the data sheet. For each set, general range of values for orientation, roughness, aperture, filling, persistance and spacing, are determined.

Exposures having identical rock strength Jv, stress reduction factor, degree of weathering and discontinuity characteristics are grouped together (IAEG 1976). Such a group of exposures falling under one type of structural set-up will form a geotechnical domain or unit irrespective of the lithological boundaries. A judicious extrapolation of the unit boundaries can be attempted, based on understanding of the geological structure and available exploratory control in sensitive areas. For each unit, final discontinuity characteristics are to be computed by synthesising the data on exposures, drifts and drill holes piercing through a particular unit. Wherever possible, representative variants for different discontinuity sets in a unit with depth may be identified by correlating the

surface and the subsurface data. The discontinuity readings for a unit are sieved out of data sheets to estimate mean orientations and dispersion of the principal sets by stereographic techniques or a computer programme for separating heterogeneous set of orientations into subsets (Huang and Charlesworth 1989). The mean values for a unit are marked on an overlay showing interpreted unit boundaries, selected values of foliation/bedding indicating structure, the master joints, lineaments, thrusts, fault and broad lithocharacters. The overburden characteristics barring the river borne material, are not presented on this overlay for better appreciation of disposition of geotechnical units and structural elements affecting the rock mass. Thus, an interpreted EGM for a tunnelling project emerges.

2.3 Data analysis for rock mass stability

The data generated for the preparation of the EGM and its allied documentation sheets can be useful for structural instability assessment and rock mass evaluation of the tunnelling media.

2.3.1 Major structural instability assessment

Discontinuities within a rock mass play a critical role in determining the extent of likely instability surrounding a proposed excavation. Intersection of two or more discontinuities form blocks or wedges, free to fall or slide. The mean orientations of the main discontinuity sets are studied by developing hypothetical sections along and across the tunnel face and isometric diagrams to understand geometry of the blocks. Stereographic methods are simple and accurate for determination of the shape and volume of a structurally defined wedge (Hoek and Brown 1980). Application of block theory to establish removability of blocks by Shi's theorem has been recently recognised as an extremely useful tool (Goodman and Shi 1986). This offers an effective way for optimization of design of rock bolt supports for tunnels. However, in view of the precision of procedure for describing and locating the key blocks for establishing their support requirements by block theory, it is emphasised to use Shi's theorem during excavation only. Whereas general mode of failure due to intersection amongst members of various discontinuity sets of a unit in preconstruction phase, can be fairly well analysed by simple stereographic techniques.

2.3.2 Rock mass evaluation

Many classification systems are in vogue for quantitative assessment of rock mass conditions and support requirements. Amongst these, the Geomechanic Classification (RMR-system), proposed by Bieniawski (1973) and the NGI tunnelling quality index (Q-system) proposed by Barton et al (1974), include sufficient information on factors influencing the stability of an underground opening.

Q-values for a unit which is supposed to be homogeneous are estimated by computing average values of rock quality designation (RQD), joint set number (Jn), joint roughness number (Jr), Joint alteration number (Ja), Joint water reduction factor (Jw) and stress reduction factor (SRF). The six parameters, chosen to describe the rock quality Q, are combined in following way (Barton et al 1974).

$$Q = (RQD/Jn) \times (Jr/Ja) \times (Jw/SRF) \quad ...(1)$$

The first quotient (RQD/Jn) is defined to be a crude measure of the relative block size. The second quotient (Jr/Ja) is an approximation of inter-block shear strength and Jw/SRF represents active stresses. The numerical value of Q ranges from 0.001 (for exceptionally poor quality squeezing ground) to 1000 (for exceptionally good quality rock which is practically unjointed). This range of Q-values is divided into 9 categories of physical description for rock mass quality. This is related to support requirements depending upon excavation span and intended use of excavation.

The RMR - Geomechanic classification, evolved from several earlier systems, has undergone several modifications (Bieniawski 1976) since its first introduction in 1973. The five basic classification parameters considered for modified Geomechanic classification are rock strength, RQD, spacing of joints, condition of joints and ground water conditions. A rating is allocated to each parameter and the overall rating for the rock mass is arrived at by adding the rating for each parameter. This overall rating is adjusted for accounting effect of joint orientations by applying the corrections to estimate the final rock mass rating (RMR). The interpretation of the RMR in terms of stand up time and unsupported span for an underground excavation can be used for evaluating optimum size of the opening. This also provides a crude estimate of cohesion and friction angle of the rock mass for stability analysis.

Comparison and correlation between the RMR and Q-value has been attempted by Bieniawski (1976) and Rutledge (1978). Relationship is expressed in terms of following

equations :

$$RMR = 9 \ In \ Q + 44 \ (Bieniawski \ 1976) \quad ...(2)$$

$$RMR = 13.5 \ log \ Q + 43 \ (Rutledge \ 1978) \quad ...(3)$$

2.4 Support estimation and design philosophy

Barton et al (1974) have illustrated the method of determining rock mass quality (Q) from surface exposures. Rough measurement of discontinuity shear resistance is possible by an empirical equation (Barton and Choubey 1977). This can be used for estimating supports for an imaginary tunnel at certain depth. Bieniawski (1975) has presented case studies for prediction of rock mass behaviour by the geomechanic classification. Based on these principles, the authors have compared the results of RMR and Q-systems on exposures and actual supports provided to support the rock mass when excavated in shape of tunnels in the Himalayas. In most of the cases, RMR and Q-values, estimated on exposed portions of a unit, are found to be reasonably close to their respective values computed on the basis of parameters observed during excavation. This has inspired to assess rock mass behaviour of units projected at the tunnel grade on the basis of the RMR and the Q-values estimated on exposures while conducting engineering geological mappings for a tunnelling project.

Experience with jointed rocks in the Himalayas indicate that a combination of rock mass evaluation, structural instability assessment and stress induced rock failures (when excavation is exceptionally deep) has to be taken into account rather than considering either of these factors in isolation for support design. In practice, it was found comparatively safer if the support recommendations offered for the RMR and Q-system proposed by Bieniawski and Barton respectively are coupled with results of structural instability analysis in a low to medium stress regime. This approach optimises the length and direction of the rock bolts depending on the weight, volume, shape and type of potential key blocks, formed by the intersection of pre-existing weak planes and the tunnel surface.

3. A CASE STUDY - ROCK MASS EVALUATION FOR A TUNNEL IN THE HIMALAYAS

To illustrate relevance of the rock mass classification approach for evaluating tunnellability in Himalayan terrain, a case study of the head race tunnel for Goriganga Hydro-electric Project in the Lesser Himalaya, is discussed. This is a medium sized Hydro-

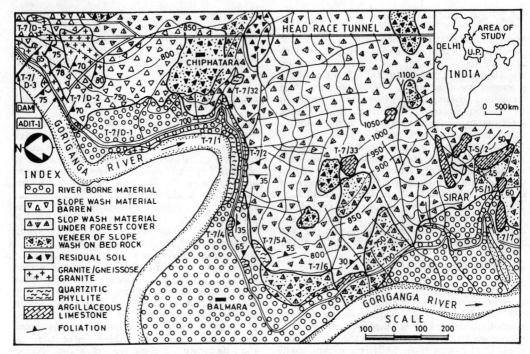

FIG.1. GEOLOGICAL MAP OF A PART OF HEAD RACE TUNNEL AREA
GORIGANGA HYDROELECTRIC PROJECT

power project under advance stage of investigations on river Goriganga in Pithoragarh District, Uttar Pradesh, India. The 4720 m long tunnel is envisaged to connect the reservoir and the surface power house. Broadly, it will pass through granitic rocks and calcareous meta-sedimentaries.

The area was geologically mapped on 1:5,000 scale. Exposures were numbered on the ground and plotted on the map. Their lithological, structural and discontinuity characteristics were recorded on format sheets discussed earlier. Main discontinuities are also drawn on the map. Various types of overburden were classified based on their geological, engineering and land-use characteristics to decipher rock-overburden ratio in critical low cover reaches. A part of this map is presented in Fig.1. Based on the lithological, structural and cover characteristics on the map and discontinuities data on format sheets, geostructural domains (units) of similar geotechnical properties are developed on the interpreted EGM (Fig.2). Wherever available, the subsurface information has also been used to differentiate geostructural domains. For each domain, mean orientations for all the sub-sets are plotted on the interpreted EGM and their average characters are tabulated separately. Notional lines along which hypothetical cross sections

for different units are developed at tunnel grade, are marked with reference to proposed excavation direction (A-A' to I-I' on Fig.2). Some of the finer details related to structural aspects and individual discontinuity orientations are deleted from Fig.2 because of multiple reductions of the original to bring it to the required size. Due to paucity of space, allocated by the publisher, it is out of scope of the paper to present details of analysis for all the ten geostructural units, interpreted on the basis of foregoing discussions. However, to demonstrate outcome of the rock mass classification based mapping procedure for rock mass assessment and support design, the unit III is chosen to be discussed in brief as a type example.

3.1 Discussions on unit III

The rocks, forming unit III, are fine grained quartzitic phyllite with subordinate bands of calcareous phyllites. These are projected at the tunnel grade from RD 335 m to 520 m. In the vicinity of its contact with granitic rocks, foliation displays erratic structural disposition in the upper half of the unit. Towards its basal part the foliation attains sympathetic attitude to the regional structures, striking WNW-ESE with moderate to steep

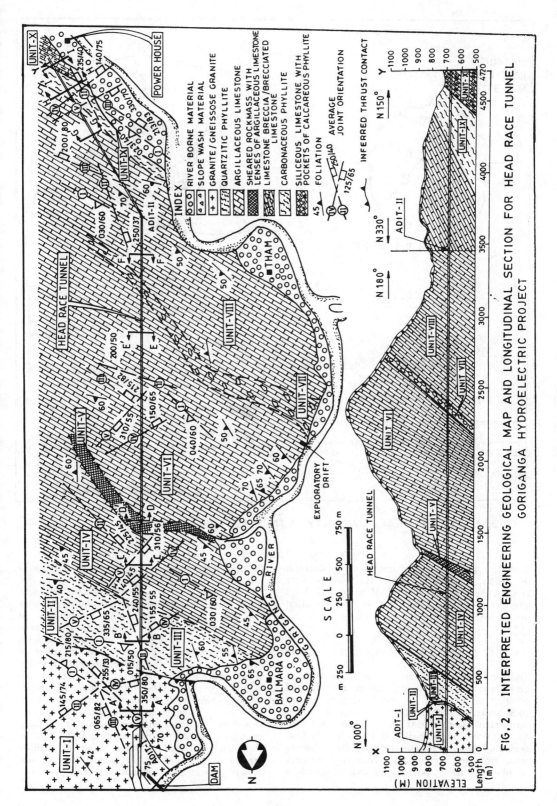

FIG. 2 . INTERPRETED ENGINEERING GEOLOGICAL MAP AND LONGITUDINAL SECTION FOR HEAD RACE TUNNEL
GORIGANGA HYDROELECTRIC PROJECT

61

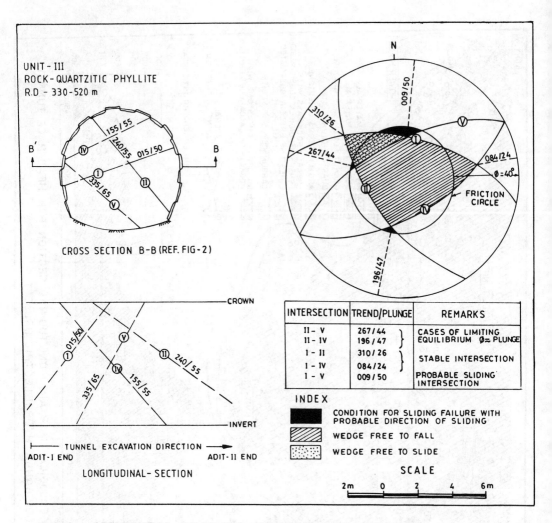

UNIT-III
ROCK-QUARTZITIC PHYLLITE
R.D – 330-520 m

CROSS SECTION B-B (REF. FIG-2)

CROWN

INVERT

TUNNEL EXCAVATION DIRECTION →
ADIT-I END   ADIT-II END

LONGITUDINAL- SECTION

FRICTION CIRCLE

Ø=40°

INTERSECTION	TREND/PLUNGE	REMARKS
II – V	267/44	CASES OF LIMITING
II – IV	196/47	EQUILIBRIUM Ø≃PLUNGE
I – II	310/26	STABLE INTERSECTION
I – IV	084/24	
I – V	009/50	PROBABLE SLIDING INTERSECTION

INDEX

■ CONDITION FOR SLIDING FAILURE WITH PROBABLE DIRECTION OF SLIDING

▨ WEDGE FREE TO FALL

▦ WEDGE FREE TO SLIDE

SCALE

2m   0   2   4   6m

FIG. 3.   STRUCTURAL STABILITY ANALYSIS FOR UNIT III

northern dips. The uniaxial compressive strength of rocks, based on point load index test results, varies between 42 MPa to 110 MPa when tested along and across the foliation planes respectively.

3.1.1 The possible modes of structural instability

The rock mass is bisected by four sets of prominent discontinuities. These are corresponding to sets I, II, IV and V of sedimentary rock units. Set III is absent in unit III, under discussion. Hypothetical face log (corresponding to section line B÷B') and a longitudinal section, showing disposition of discontinuity sets, has been drawn to postulate geometry of planes forming potentially unstable wedges

(Fig.3). Stereographic projections reveal that a group of gravity wedges may develop due to intersection of discontinuities of sets I, II and IV. However, this pattern may not be taken as a principal mechanism of failure throughout the length of the tunnel to be driven in unit III due to following reasons, analysed on the basis of average characteristics of the discontinuities forming these wedges (Table 1).

1. In most of the rock exposures either of the two sets, II or IV are present at a time. Simultaneous occurrence of set II and set IV along with set I could be noticed on very few rock exposures of unit III. Hence, this mode of failure may occur at limited stretches of the tunnel where set II and IV will appear simultaneously.

Table 1. Rock discontinuity characteristics for unit III

Joint Set	Average Orienta- tion	Persistance (m)	Aperture (mm)	Spacing (cm)	Condition	Special Characteristics
I	015/50	≥ 1, ≤ 10	Tight	5-50	Smooth undulating/ Planar	One discontinuity along Set I is > 10 m in persistance
II	240/55	≥ 1, ≤ 3	Tight	50-200	Rough Irregular Stained Surfaces	R M R - 53 Class III Fair Rock C - 150-200 KPa
IV	155/55	≥ 0.5, ≤ 2	Tight	50-200	Slightly smooth Planar Stained Surfaces	Friction Angle - 35°-40°
V	335/65	≥ 1, ≤ 2	Tight	50-200	Smooth Planar/Undula- ting Tightly Healed Joints	

2. Persistance of set IV is very low. It seldom exceeds 1 m. Therefore, potentiality of the wedge formed by the intersection of set I, II and IV is considerably reduced.

Another mode of structurally controlled failure in unit III may exist due to intersection of sets I, II and V. This is a sliding wedge which may get mobilised along planes, defined by set V or interaction of sets I and V.

3.1.2 Rock mass evaluation by geomechanic classification

The table 2 embodies average description of basic geomechanical parameters with their corresponding ratings for unit III.

Table 2 The input parameters for RMR computations

Classification parameter	Description Value	Rating
Strength	42 to 110 MPa	7
RQD	80%	17
Average spacing of closest set	120 to 320 mm	10
Condition	Slightly smooth, undulating, stained, tight joints	20
Ground Water	Moderately large inflow	4
	Total score	58

The tunnel has been oriented such that the dominent joint set strikes perpendicular to the tunnel axis with a dip of 50° against the drive direction, hence, a rating adjustment of -5 is to be accommodated for this situation. Thus, the final RMR becomes 53 which places the rock mass in class III with a description of fair.

3.1.3 Tunnelling quality index by Q-system

The parameters required for determination of

Q-values of unit III are rated in table 3.

Table 3. The input parameters for Q-values.

Item	Description	Value
Rock Quality	Good	RQD=80%
Joint sets	Either II or IV absent, hence, three sets	Jn = 9
Joint roughness	Slightly smooth	Jr = 2
Joint alteration	Surface staining	Ja = 1
Joint water	Moderately large inflow	Jw = 0.5
SRF	Low stress	SRF = 2.5

The numerical value of Q can be determined from equation (1) in following manner.

$$Q = 80/9 \times 2/1 \times 0.5/2.5 = 3.5$$

This value of Q falls under poor tunnelling quality index.

3.1.4 Correlations

Excellent correlations between the RMR and the Q-value for unit III are observed. The RMR from equations (2) and (3) is calculated as 55 and 50 respectively against its estimated value of 53, based on rating of the parameters observed in the field.

3.1.5 The support design

The rock mass constituting the unit III conforms to the class III, the fair rock of RMR system. As per the Bieniawski's guide lines, 4 m long bolts spaced 1.5 to 2 m with 50 to 100 mm thick shotcrete in the crown are recommended to be adopted for permanent support of the rock mass.

The Q-value for the same unit is 3.5 and the equivalent dimension (B/ESR) is 6.25 for

10 m dia horse shoe shaped tunnel (chosen for simplicity of comparison with Bieniawski's guide to supports). This places the rock mass in support category 22 for which 3 m long, untentioned grouted rock bolts at 1 m spacing with 25 mm to 75 mm shotcrete are suggested as suitable support measures. It can be seen that there is marginal variation in design of supports recommended by Bieniawski and Barton. The RMR-system proposes slightly longer bolts with thicker layer of shotcrete but the bolt spacing is greater than the recommended spacing by Q-system.

A review of structural instability reveals that in event of formation of wedges due to intersection of sets I, II, IV or I, II, V, the longer rock bolts will be more effective. Therefore, if set IV or V are occurring simultaneously with sets I and II, 4 m long bolts, as suggested by Bieniawski, may be preferred otherwise 3 m long rock bolts spaced at 1 to 2 m are expected to be sufficient for supporting the tunnel section. Some of the bolts in each section may be aligned perpendicular to set V for increasing frictional resistance along these planes to check possible sliding and mobilisation of new wedges.

4. CONCLUSIONS

Engineering geological maps, incorporating basic data for rock mass classification, may prove to be useful for understanding rock mass behaviour and estimating support requirements for Himalayan tunnels during feasibility stage investigations. This will enable to evaluate the tunnelling media through various methods in vogue. Application of RMR and Q-systems during excavation of tunnels in the Himalayas are encouraging Also, the limited efforts on correlation of rock quality parameters indicate close relationship between surface and subsurface conditions. This opens a possibility of support assessment for a tunnelling project by empirical methods with a backup of detailed engineering geological mapping and discontinuity survey.

ACKNOWLEDGEMENTS

The authors wish to express their sincere thanks to Shri Nanda Gopal, Chief Engineer and Shri M.R. Bandyopadhyay, Chief (Geology), National Hydroelectric Power Corporation, India for encouragements while practicing the approach in the field. The help received from Shri S.A. Zaki for data analysis and Shri P.K. Litoria during finalising the manuscript is thankfully acknowledged.

REFERENCES

Barton, N.Lien, R. and Lunde, J. (1974). Engineering classification of rock masses for the design of tunnel support. Rock. Mech., 6(4), 189-236.

Barton, N. and Choubey, V. (1977). The shear strength of rock joints in theory and practice. Rock. Mech., 10, 1-54.

Bieniawski, Z.T. (1973). Engineering classification of jointed rock masses. Transactions, South African Institution of Civil Engineers, 15(12), 335-344.

Bieniawski, Z.T. (1976). Rock mass classification in rock engineering. Proc. Symp. on exploration for rock engineering, 97-106. Balkema, Rotterdam.

Goodman, R.E., and Shi, G.H. (1986). The application of block theory to the design of rock bolt supports for tunnels. 35th Geomechanics Colloquy in Salzburg.

Hoek, E. and Brown, E.T. (1980). Underground excavations in rock, 183-189. The Institution of Mining and Metallurgy, London.

Huang, Q. and Charlesworth, H. (1989). A Fortran-77 program to separate a heterogeneous set of orientations into subsets . Computers and Geosciences, 15(1), 1-7.

International Association of Engineering Geology (1976). Engineering Geological Maps: A guide to their preparation, 21 pp. UNESCO Press, Paris.

Shome, S.K. (1989). What is an engineering geological map (EGM) and how it should be prepared. Jour. of Engineering Geology (India), 18(3&4), 51-55.

Landslide hazard zonation in the Garhwal Himalaya
A terrain evaluation approach

Determination des zones à risques de glissement de terrain dans l'Himalaya
Une approche par l'évaluation du terrain

Vishnu D. Choubey & Pradeep K. Litoria
Department of Applied Geology, Indian School of Mines, Dhanbad, India

ABSTRACT: The Himalayas have witnessed several landslides in the recent years. The associated hazards could have been reduced considerably provided the proneness of a zone to sliding could be predicted. To map such zones, a terrain classification map (TCM) has been prepared assuming slope as basic element. A combination of TCM and geological/structural map portrays that the frequency of landslides increases while approaching closer to thrusts and ridges. Also, a physiographic and land-use map (PLM) has been prepared, using medium scale aerial photographs and four hazard prone zones are categorised - (1) old stable landslide (2) old active landslide (3) recently active landslide and (4) potential landslide zones. It is noticed that most of the escarpments and barren slopes pertain to either potential landslide zones or recently active landslides. An overlay of TCM on PLM provides a fair assessment of slope, vegetation, geology/structures and land-use on each facet for the preparation of final landslide hazard zonation map.

1 INTRODUCTION

The studied area is situated in a part of Dehradun and Tehri-Garhwal districts of Uttar Pradesh, India (Fig.1). The altitudinal variations, observed in the area, range from 300 to 2350m above Mean Sea Level. River Ganges, the holy shrine for pilgrims, pass through this area together with Chandrabhaga, Hiyunl and Gulab Gad as other streams. The contact of the Siwaliks and the Lesser Himalayas is marked near Lakshman Jhula. The area constitutes a part of Garhwal Syncline of Krol Belt, excellently described by Auden (1934) which consists of Simla slates, Chandpur, Nagthat, Krol, Tal and Subathu Formations. The northern limb of the syncline is intricately faulted whereas the southern limb is intensely folded (Gaur & Dave, 1978). Several landslides and other mass movements have been occurring along Rishikesh-Deoprayag road and its surrounding high hills. These on-going hazardous processes pose serious geo-environmental problems to inhabitants, traffic and river valley projects. To reduce the damage, caused by such hazards, their prediction is a must which can be done only by detailed mapping, risk evaluation

FIG.1. LOCATION MAP OF AREA OF STUDY

Table 1. Stratigraphic sequence in the Garhwal Syncline, Rishikesh (after Gaur & Dave 1978)

	Intrusive	Metadolerite
	Subathu Formation	Pisolitic laterite, shales, limestones and sandstones
	---------------------------------------	UNCONFORMITY --
TAL GROUP	Upper Tal Formation	Variegated sandstones
	Lower Tal Formation	Calcareous siltstones, carbonaceous shales, greywackes
KROL GROUP	Krol E Formation	Cherty dolostone
	Krol D Formation	Dolostone with subordinate shales
	Krol C Formation	Bluish black limestone
	Krol A Formation	Alternate shale-limestone sequence
BLAINI GROUP	Infra Krol Formation	Thinly laminated slate-quartzite association
	Blaini Formation	Tillites, limestones, shales, siltstones and sandstones
SIMLA-JAUNSAR GROUP	Nagthat Formation	Purple sandstones, greywackes, grit
	Chandpur Formation	Phyllite-quartzite association
	Chaosa Formation	Veriegated slate/meta-siltstone, phyllites & sandstones

and zonation of such hazards in order of their severity. With this view, an attempt is made to prepare a terrain classification map (TCM) of the area, assuming slope as basic factor. Further, a detailed physiographic and land-use map (PLM) is prepared

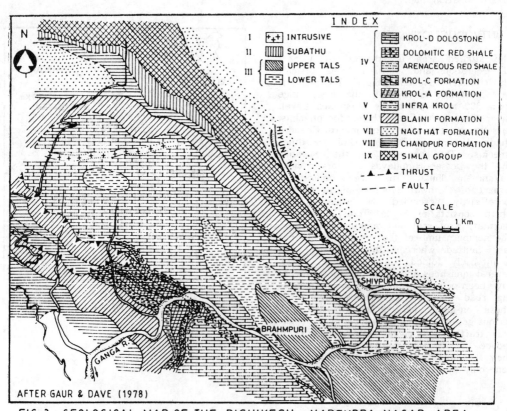

AFTER GAUR & DAVE (1978)

FIG.2. GEOLOGICAL MAP OF THE RISHIKESH - NARENDRA NAGAR AREA

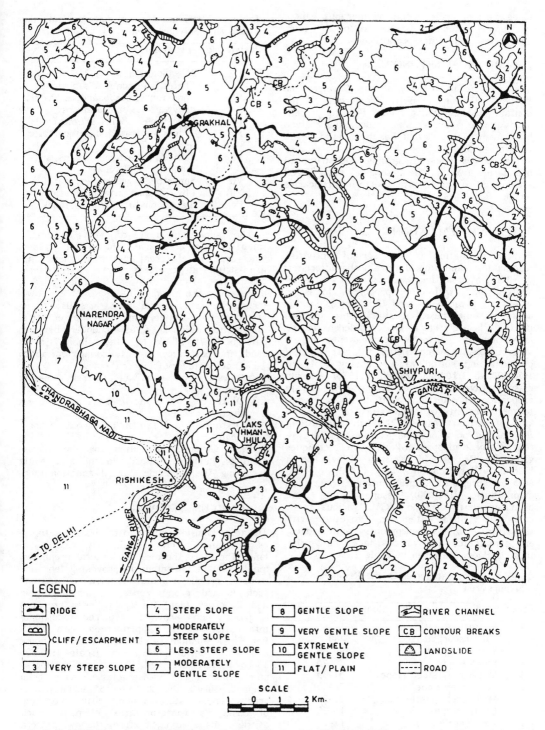

LEGEND

⬡ RIDGE		4 STEEP SLOPE		8 GENTLE SLOPE		⬡ RIVER CHANNEL	
⬡ CLIFF/ESCARPMENT		5 MODERATELY STEEP SLOPE		9 VERY GENTLE SLOPE		CB CONTOUR BREAKS	
2		6 LESS-STEEP SLOPE		10 EXTREMELY GENTLE SLOPE		⬡ LANDSLIDE	
3 VERY STEEP SLOPE		7 MODERATELY GENTLE SLOPE		11 FLAT/PLAIN		----- ROAD	

SCALE

1 0 1 2 Km.

FIG.3.TERRAIN CLASSIFICATION MAP OF RISHIKESH–SHIVPURI AREA.

using medium scale aerial photographs which is supplemented by demarcation of four categories of landslide prone zones.

2 GEOLOGICAL SETTING

The area forms a part of the Siwaliks and Lesser Himalayan Belt, demarcating the contact near Lakshman Jhula. For the present study, the stratigraphic succession given by Gaur & Dave (1978) has been used which is given in Table 1.

The various litho-stratigraphic and structural boundaries given in the geological map (Fig.2) are used to demarcate pattern boundaries on the TCM.

3 TERRAIN EVALUATION

Terrain evaluation is the set of activities, leading to compilation of terrain characteristics or terrain qualities, extracted by terrain analysis along with other properties to assign values to a piece of land (land unit). In this approach, the basic smallest unit mapped is the 'Facet' which is considered to be consisting of one or more elements and is still reasonably homogeneous for all practical purposes and suited to a mapping scale of 1:50,000 to 1:100,000 (Mitchell 1973). Terrain classification is used to predict the terrain properties at sites not visited, by reference to measurements of other sites on the same kind of terrain (Way 1973). With this notion, the terrain on a part of Rishikesh-Deoprayag road has been mapped, following the techniques given by Pachauri & Krishna (1984) and a 'Terrain Classification Map' (TCM) of the area is prepared (Fig.3), using topographic map (Scale 1:50,000), assuming the slope as basic unit. On the TCM, various facet categories (slope categories), together with landslides, cliffs/escarpments and ridges, are demarcated and various facet categories, thus obtained, are given in Table 2.

Table 2. Facet categories in the Rishikesh-Shivpuri area

Facet	Facet category	Slope in degrees
1	Ridge top	-
2	Cliff/escarpment	>60
3	Very steep slope	46-60
4	Steep slope	36-45
5	Moderately steep slope	31-35
6	Less steep slope	26-30
7	Moderately gentle slope	21-25
8	Gentle slope	16-20
9	Very gentle slope	11-15
10	Extremely gentle slope	6-10
11	Flat or plain	0-5
12	River channels and banks	-

Table 3. Various terrain patterns in the Rishikesh - Shivpuri area

Pattern No.	Litho-unit
I	Metadolerite
II	Shales, limestones & sandstones
III	Sandstones, carbonaceous shales, greywackes, calc-arenites
IV	Dolostones, shales & limestones
V	Thinly laminated slate-quartzite association
VI	Tillites, limestones, shales, silt-stones & sandstones
VII	Sandstones, grewackes, grit
VIII	Phyllite-quartzite association
IX	Slates, phyllites & sandstones

The superimposition of the geological map of the area over the TCM, divides the area in different patterns where the litho-structural contacts serve as pattern boundaries. Thus, total nine divisions are made which are given in Table 3. The two thrusts, present in the area, are also transferred on the TCM together with the geological information which acted as a boundary between pattern IV,VI,VIII & IX. This final TCM is used in obtaining the following variables:
1 Facet category (slope category)
2 Distance of each facet from nearest ridge
3 Distance of landslides from nearest thrust
4 Distance of landslides from nearest ridge
These variables are used for the assessment of landslide hazards in the area while preparing the landslide hazard zonation map.

4 LAND-USE MAPPING

The term land-use refers to the human activities on land which are directly related to land. Therefore, besides the natural processes, man's role is important in provoking landslide hazards in the Himalayan terrain. To study such hazard prone zones, a 'Physiographic and Land-use Map' (PLM) has been prepared (Fig.4) which provides required additional information to be supplemented with TCM. For this, medium scale panchromatic black and white aerial photographs (scale 1:50,000 approximately) are used. The various physiographic units and land-use classes have been mapped, during the scanning of aerial photographs under stereo-vision, with the help of basic photo-interpretation elements (such as size, shape, tone, texture, shadow, pattern etc.) following the scheme given by Anderson (1971). The physiographic units, thus identified, are as under:

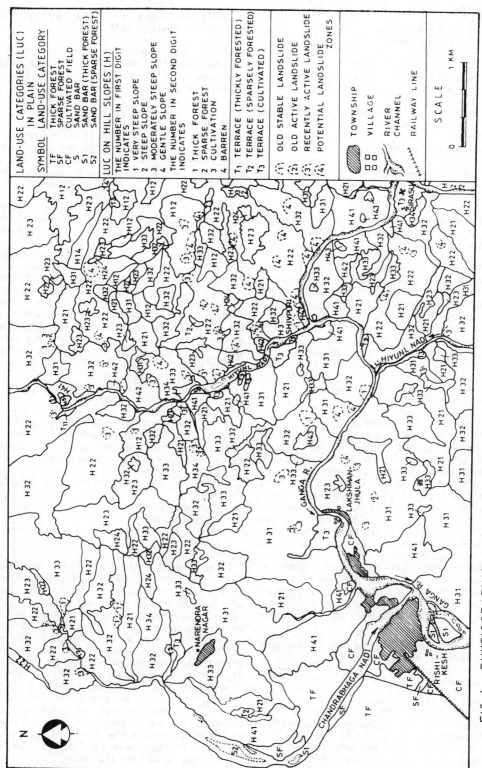

LAND-USE CATEGORIES (LUC)
IN PLAIN
SYMBOL   LAND-USE CATEGORY

TF   THICK FOREST
SF   SPARSE FOREST
CF   CULTIVATED FIELD
S   SAND BAR
S1   SAND BAR (THICK FOREST)
S2   SAND BAR (SPARSE FOREST)

LUC ON HILL SLOPES (H)

THE NUMBER IN FIRST DIGIT
INDICATES
1 VERY STEEP SLOPE
2 STEEP SLOPE
3 MODERATELY STEEP SLOPE
4 GENTLE SLOPE
THE NUMBER IN SECOND DIGIT
INDICATES
1 THICK FOREST
2 SPARSE FOREST
3 CULTIVATION
4 BARREN

T1 TERRACE (THICKLY FORESTED)
T2 TERRACE (SPARSELY FORESTED)
T3 TERRACE (CULTIVATED)

(1) OLD STABLE LANDSLIDE
(2) OLD ACTIVE LANDSLIDE
(3) RECENTLY ACTIVE LANDSLIDE
(4) POTENTIAL LANDSLIDE
     ZONES

        TOWNSHIP
        VILLAGE
        RIVER
        CHANNEL
        RAILWAY LINE

SCALE

0   1 KM

FIG. 4. PHYSIOGRAPHIC AND LAND-USE MAP OF RISHIKESH - SHIVPURI AREA

69

1 Hills
- Cliffs/escarpments
- Very steep slopes
- Steep slopes
- Gentle slopes
2 River terraces
3 Sand bars
4 Flood plains
The land-use classes which are identified under stereo-vision are:
1 Forest
- Dense/thick forest
- Sparse forest
2 Cultivated land/slope
3 Barren land/slope
4 Rivers
5 Residential areas

The physiography has a direct bearing over the land-use in any area, as there exists a distinct relation between physiography and land-use. Therefore, an attempt has been made to establish relation between the two which is given in Table 4. This relation was confirmed during field checks which corroborates the above statement.

As the aerial photographs provide a synoptic view of the area, it serves as an excellent tool for regional studies, therefore, besides the above land-use classes, an attempt is made to demarcate various landslides/landslide prone zones on the same map (PLM) and these are classified in four categories:
1. Old active landslide
2. Old stable landslide
3. Recently active landslide
4. Potential landslide zones

A scrutiny of PLM leads to some interesting observations that the landslide categories, discussed earlier, are situated either on the margins of prominent drainage lines or in the vicinity of rivers/streams and faults or near the ridges. A combination of PLM with TCM conforms to these observations, thus, corroborating the physiographic and structural control over the landslides. Also, it is noticed that most of the escarpments and barren slopes pertain to either potential landslide zones or recently active landslides, indicating physiographic control over land-use In this way, the TCM and PLM forms an integral part of the landslide hazard zonation map.

Table 4. Relation of land-use with physiography in the Rishikesh-Shivpuri area.

Physiography	Land-use
Cliffs/escarpments	Usually barren
Very steep slopes	⌈Barren or sparsely/densely
Steepslopes	⌊(occasionally forested)
Gentle slopes	Forested/cultivated
Terraces	Forested/cultivated
Sand bars	Forested/cultivated
Flood plains	Forested/cultivated

5 CONSTRUCTION OF THE LANDSLIDE HAZARD ZONATION MAP

Most of the information about the hazard situation was obtained from detailed analysis of slopes with the help of TCM (Fig.3) and PLM (Fig.4) using medium scale aerial photographs. Although, the aerial photographs have been extensively used for the land hazard mapping but in the present study there has been slight over estimation of hazards because of the two restraints: (1) In the area, with such steep slopes (especially the regions with high relief), just by using aerial photographs, it is not possible to detect all the hazards because of presence of shadow zones on the other side of the sun facing side of the hills (2) The poor quality of photographs further restricted the discernment of currently active and former processes.

Therefore, to overcome such handicaps and minimize the over estimation, help was taken from TCM which contained complete information about various slope categories, cliffs/escarpments and contour breaks. These features were confirmed on the PLM and the details were supplemented by combining the two maps. The final combined map was used for the evaluation and zonation of landslide hazards in the Rishikesh-Shivpuri area.

5.1 Hazard estimation and zonation

In order to evaluate the proneness of a particular facet to landslide hazards, the following factors (individually as well as in combination with other factors) are taken into consideration:
- Facet category
- Distance of each facet from nearest ridge
- Distance of each facet from nearest thrust/fault
- Closeness of the facet to river/stream
- Lithology
- Forest/vegetative cover

It is noticed that most of the landslides and cliffs/escarpments are situated either on the river banks or in the vicinity of rivers (Fig.3), therefore, the river valleys with steep slopes having sparse/no vegetation are found to be more susceptible to future landslips. All the facet categories marked as 2 & 3 (slope range 45-60° and above) are assessed as highly prone zones for landslides. The facet categories 4 & 5 (slope range 31-45°) are assessed having moderately high probability to sliding when it contains vegetative cover and other factors are unfavourable, otherwise, in the reverse case, it is assessed as less probable to sliding. The facet categories 6 and below are least probable to sliding if the other factors are favourable.

70

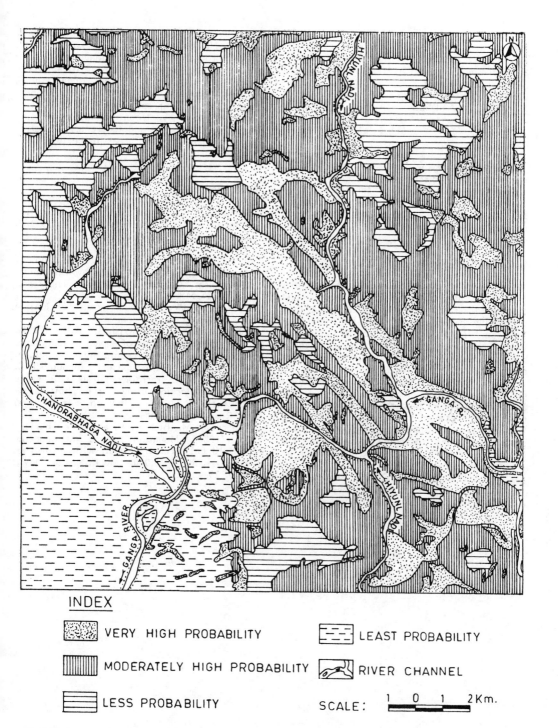

INDEX

░	VERY HIGH PROBABILITY	⊟	LEAST PROBABILITY
▓	MODERATELY HIGH PROBABILITY	≈	RIVER CHANNEL
▤	LESS PROBABILITY		SCALE: 1 0 1 2Km.

FIG.5. LANDSLIDE HAZARD ZONATION MAP OF RISHIKESH-SHIVPURI AREA

Thus, total four landslide hazard prone zones are demarcated in the order of degree of severity of hazards on the Landslide Hazard Zonation Map (LHZM) of Rishikesh-Shivpuri area (Fig.5).

ZONE NUMBER	PROBABILITY OF LANDSLIDE HAZARD
I	High
II	Moderately high
III	Less
IV	Least

While assigning the degree of hazardousness, the factors discussed earlier, together with presence/absence of forest/vegetative cover, are taken into account and due weightage is given to each factor whenever it needed (Zimmermann et al 1986).

6 CONCLUSIONS

The present study incorporates the mapping, evaluation and zonation of landslides and associated hazard prone areas along a part of Rishikesh-Deoprayag road. For this, the TCM and PLM served as sound data base in the preliminary stage. The cliffs/escarpments ridges, landslides and various slope categories are demarcated on the TCM and PLM as well. The litho-structural boundaries, exhibited on geological map, helped in dividing the area in different patterns and the interrelationship between various factors (including land-use) is established successfully on physiographic and and tectonic basis, leading to demarcation of four landslide hazard prone zones on the final LHZM.

The LHZM is, in fact, a snapshot of the hazard situation in the Rishikesh-Shivpuri area, portraying a regional picture of landslide hazards, especially while aligning new highway in the Himalayan terrain. The areas where these processes cause damage to lives, property and other public utilities, there is a necessity to carry out such studies in detail which will enable the planners to implement proper remedial/corrective measures to prevent/control these on-going hazardous processes and to prepare effective land-use plans for the areas to be developed in the future.

REFERENCES

Anderson, J.R.(1971). Landuse classification scheme. Photogramm.Engg.,37, 379-388.

Auden, J.B.(1934). Geology of the Krol Belt. Rec.GSI.,67(4), 357-444.

Gaur, G.C.S. and Dave, V.K.S.(1978). Geology and structure of a part of Garhwal Syncline, Rishikesh, Garhwal Himalaya. Him.Geol., 8(1), 524-549.

Mitchell, C.W.(1973). Terrain evaluation, 30 pp. Longman, London.

Pachauri, A.K. and Krishna, A.P. (1984). Terrain classification of a part of Himalayas using multistage sampling technique. 5th ACRS (Late paper) (unpublished).

Way, D.S.(1973). Terrain analysis : A guide to site selection using aerial photo-interpretation, 1-35. Dowden, Hutchinson & Ross Inc.,Straundsburg.

Zimmermann,M., Bichsel, M. and Kienholz, H. (1986). Mountain hazard mapping in the Khumbu Himal, Nepal, with prototype map, scale 1:50,000. Mount.Res.& Dev., 6(1), 29-40.

G.I.S. aid in regional planning of quarrying activity

La contribution du G.I.S. dans la planification régionale des carrières

E. Conte, T. Farenga, P. Loiacono & M. Semeraro
Geo s.r.l., Bari, Italy

ABSTRACT: This paper explains the methodology and the first results concerning the drawing up of thematic maps for a tool of territorial planning on regional scale. This tool is the Regional Plan of Quarries Activities (R.P.Q.A.) in Apulia (Italy). For its drawing up it has been needed to preliminarly acquire the cognitive scenario on the territory - soil usage, geolithology, regulation and other constraints, current situation of quarrying industry, etc. - Starting from this cognitive scenario, it has been possible to draw up the plan maps by a crossed and comparative reading of the several acquired information levels, taking advantage of a suitable Geographic Information System (G.I.S.).

RESUME: Cet étude illustre la méthodologie et les premiers résultats relatifs à la rédaction de cartes thématiques pour un instrument de planification du territoire à l'échelle régionale. Cet instrument est le plan regional pour les activités extractives de la Pouille, pour la rédaction duquel il a été nécessaire acquérir an préalable le tableau cognitif du territoire. En partant de ce tableau cognitif il a été possible en se servant d'un apte G.I.S., rédiger la cartographie du plan a'travers une lécture croisée et comparée des variés niveaux d'information acquis. En outre le G.I.S., se place même come un valable support pour la successive gestion du plan.

1 INTRODUCTION: R.P.Q.A. AIMS AND PURPOSES

The R.P.Q.A. main purpose is to recognize the most suitable areas for quarrying activities, where to allow the existing quarries stoping and the opening of new ones, in the period of the next decade.

This goal has the aim of exercising the razionalization of quarrying activities on regional scale.

Achieving it involves the searching for the right balance of production needs - related to quarry materials requirements for fulfilling the market demand -, resources availability - in relation to their space arrangement and their amount and properties - and restrictions imposed on special usage of the soil subject of planning. Relating to this last point, it needs to take into the right account the various characterized constraints, existing on the territory, ex-lege or ex-facto.

2 INFORMATION LEVELS FOR INPUT AND THE CHANCE FOR A G.I.S.

In order to achieve the above mentioned purposes and to produce a proper map, defining the areas subject to a quarrying activity in the next future, it needed to arrange the acquiring of an articulate and complex data set, divided into many information levels. The acquiring data bases are individuated and subdivided into two main classes: graphic and alphanumeric data bases.

In the first class we can find areal, linear and punctual information - such as several kinds of ex-lege imposed constraints, different conditions of the current soil usage, quarrying activities location. In the second class we can find data about individual quarrying activities (size, capacity and productive resource, employed staff, comsumption, product diffusion radius, etc.), data about the available materials features (chemical-physical, mechanical, mineralogical features, etc.), data about quarry materials needs (quantity, material characteristics, etc.).

As above explained, it is plain that the information is qualitatively varied in typology (geographic -territorial data, numerical data, descriptive data); it is originated by extremely varied sources often incompatible among them (national, regional and local cartography with different frames of reference and aerophotos; absolute and statistical geotechnical data; collection of information cards on size characteristics of active or abandoned quarries) and moreover not all updated. so therefore, the premises are tipical and congenial for the use of a Geographical Information System (G.I.S.), which can allow an integrated managementof all data, enabling fast operations of consultation and graphic displaying and supplying intermediate and final work tools for a simple reading (basic maps, thematic maps, tables and graphs), allowing to work on different territorial areas (regional, provincial, municipal areas, or their sub-areas).

The G.I.S. tasks, that we recognize like essential, are the following:
- data acquisition, both cartographic and descriptive elements;
- geographic and alphanumeric data bases management, for the most various operations, as modifications, insertions, cancellations, etc.;
- data processing;
- thematic mapping output in any scale, also following alphanumeric data processing operations.

Graphic and alphanumeric data bases can be input on different hardware supports and they can be managed by different software.

3 PLAN THEMATIC MAPS

It has been already mentioned the vast amount of acquired data that have allowed the drawing up of basic maps, such as the geolithological map, the active quarries map, the constraints map, etc.

Of course, this cartography can be updated at any moment, basing on new data availability.

Starting from this basic maps, we can get the plan cartography,through a suitable procedure, which allows the planner to be present during automatic processing operations for a cartographic creation. If necessary, the plan cartography can follow the drawing up of a   intermediate level maps, got by a synthesis of several basic maps or by a trasformation of alphanumeric data in graphic information.

As an example of a map, got as a synthesis

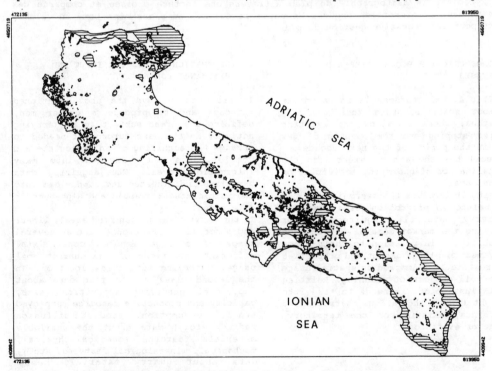

Fig. 1 Regional Territory of Apulia: Map of the Absolutely Constraint to quarrying activity (outlined areas)

of several basic maps, we report the map of those constraints which hinder the quarrying activity practice.

In this map it can be recognized those areas in which one or more than one of the following constraints are present: landscape constraint, archeological constraint, special naturalistic interest oases, woods, etc. (see fig. 1). Quarry stoping is forbidden in the constrainted areas; so the therein included active quarries will have a brief period to dispose the active cessation and to arrange a different dislocation for plants.

As an example of a transformation map, we report the map displaying the density of active quarries over a part of the regional territory (Province of Bari). This density is calculated as quarries number on municipal surface measure ratio. (see fig. 2).

It is foreseen to articulate the thematic plan maps mainly in the following two

elaborations: resources map and quarry activity constraints map.

The resources map recognizes and displays those areas in which it can be found material to be used, in one or more utilization fields, directly or by transformation received - limestone for ornamental applications, for concrete, for iron purposes, clay for bricks and concrete, etc. -. Material usage is to be related to its intrinsic nature and to its available amount such as areal size and thickness of the utilizable bank.

The resources map has been got through several working steps, always with a large man-machine interaction, that was required because considered essential.

The first working step starts from the geolithological map and from punctual stratigraphic knowledge - literature data, specific studies and investigations made in the past, considerations about active or abandoned quarries faces, etc. -. A careful

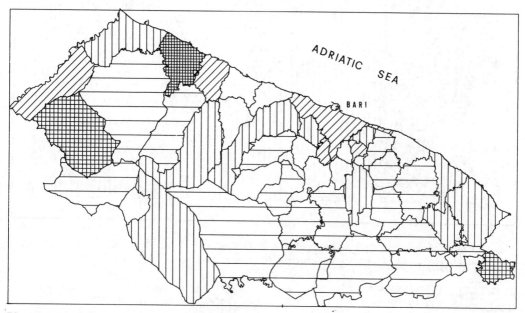

Fig. 2 Province of Bari: Map of the density (d) of active quarries over each municipal area (d=number of quarries x 1000/kmq)

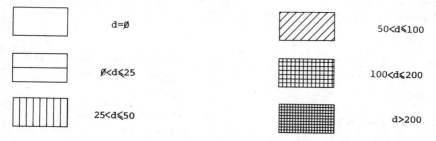

d=∅

∅<d≤25

25<d≤50

50<d≤100

100<d≤200

d>200

comparative analysis of this knowledge allows to "a priori" exclude, automatically, the outcrops with low size and/or thickness as well as outcrops marked

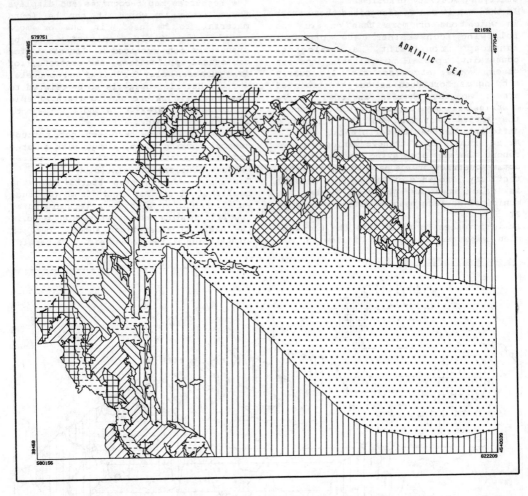

Fig. 3 Map of Resources over a part of the Bari Province (fg. 176 – Barletta – of the "Carta d'Italia", published by Geographic Military Institute – I.G.M.)

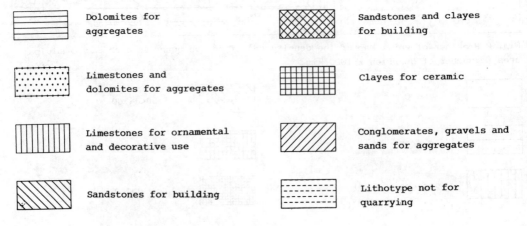

Dolomites for aggregates

Sandstones and clayes for building

Limestones and dolomites for aggregates

Clayes for ceramic

Limestones for ornamental and decorative use

Conglomerates, gravels and sands for aggregates

Sandstones for building

Lithotype not for quarrying

76

by peculiar geomorphologic conditions. This is made possible also saving the chance of controlling lithotypes settled in a stratigraphic succession and if necessary including them among resources.

A second step provides a conjunction among various existing lithotypes, in relation to the intrinsic materials nature and to the characteristics required for their usage in different productive fields.

This is possible because the availability of a sampling points close net existing on the regional territory. So, this information is memorized in the G.I.S. by a code referred to a relational data base, loaded on a personal computer, only for a handy management. An interactive comparison between the intrinsic materials characteristics and the requirements for their usage allows to assign typological resource codes to each area in the geolithological map. In this second step, planners surely have a decisional task but they can be aided by computer in knowledge acting and comparison among variable data sets. A third and last step allows the resource map automatic output (see fig. 3), through the conjunction among lithotypes marked by the same resource typological code.

The quarry activity constraints map recognizes and displays those areas which will be, in the next decade:

- submitted to the "quarry activity constraints", imposed whereas quarrying activity has an absolute priority over any other activity, because of the presence of "unique" material in quantitative, qualitative and territorial distribution characteristics;
- suitable for stoping, thanks to the presence of resources and the market demand;
- subject to a special discipline (stoping modality, reclamation modality, etc.) because restrictions on soil usage, even if they are not hindering;
- subject to an absolute prohibition for stoping because they belong to hindering constrained areas on condition that the material present there do not have the characteristics of "unicity" above mentioned.

Also in this second case, some intermediate level maps made up by a computer support were drawed up before the final one. Among those maps, there is one showing the current quarrying potentiality, for which a different degree of "covering" has been assigned to the territory. This degree means the capacity to meet the market demand only using materials actually quarried by the active quarries.

For the last mentioned map, planners cannot report an extract display before the official acceptance by the Public Administration.

4 G.I.S. AS A TOOL FOR THE NEXT PLAN MANAGEMENT

The large amount of available data represents an information source useful for plan management. Management has to be intended as a checking, coordinating and controlling action on quarrying activities to achieve the purposes expected by the planning and scheduling tool. Therefore all actions have to be checked, controlled and coordinated to apply the planning design. All the data bases in input, that are the basic information for the plan studies, can be used in all the operations needed for managing the plan itself.

During the managing, it becomes also possible the need for a modification of acquired data because they can be either incomplete right from the start or they can change in time. Therefore it is possible to need an updating of the plan, if the necessary modification is so large for number and quality to invalidate the achieved results. The updating procedure is quickly attainable using a G.I.S.

So, our aim is to insert in the G.I.S. all the next geological, geotechnical, hydrogeological studies to constantly update the actual plan. Moreover, to achieve the same goal, the possibility to get the connection with other information systems, that belong to the Public Administration, becomes estremely useful and largely necessary. This way of acting allows to update any other data, common to different fields of studies, such as new regulation constraints imposed on territory, soil usage estimated modifications, etc.

5 CONCLUSIONS

The planners' task has met the G.I.S. as a valid aid during decision making operations of the planning tool drawing up. In fact, G.I.S. allows on one side a more quickly comparison among several design hypothesis, and on the other one it enables to examine all available data, relating them one each other like user needs, in a way that can be very difficult to do manually.

Moreover the plan that will be carried out, will allow a better future management of quarrying activities on regional scale.

REFERENCES:

Biasini, A., Galletto, R., Mussio, P.,

Rigamonti, P. (1983). La cartografia e i sistemi informativi per il governo del territorio, Franco Angeli Editore, Milano.

Burrough, P.A. (1986). Principles of Geographical Information Systems for Land Resources Assessment. Monographs on soil and Resources Survey n. 12, Oxford University Press.

Durand, M. (1988). La micro-informatique comme aide à la création des cartes géologiques, Bulletin n. 38 of the International Association of Engineering Geology, Paris.

Shoor, R. (1989). Beyond G.I.S., Computer Graphics World.

Preoceedings of the 1 Conferenza /Esposizione nazionale Italiana sui sistemi Informativi Cartografici - AMFM International European Division, Roma 1989.

Some results of engineering geological mapping in the Lake Balaton region
Résultats dans la cartographie géotechnique de la région du Lac Balaton

T.Cserny
Hungarian Geological Institute, Budapest, Hungary

SUMMARY: Lake Balaton Region is the largest recreational and touristic area in Hungary. The Hungarian Geological Institute has conducted up-to-date complex geological and sedimentological investigations of the region since 1965. Two engineering-geological mappings, the first one on the scale of 1:10 000, the second one on the scale of 1:50 000, were carried out in the lakeshore area. At the same time, sedimentological and geological investigation of lacustrine sediments has also been carried out since 1981.

The principal aims of engineering-geological mapping and actuogeological investigation are: (1) to carry out a complex geological investigation of loose Quaternary and Pannonian sedimentary rocks and lacustrine sediments; (3) to give proposal for systematical development of the Balaton Recreational Area; (4) to find proper methods for the reduction of eutrophization.

The paper shows the main results of investigation and their applicability, including a brief description of some engineering-geological and sedimentological features of Lake Balaton Region.

RÉSUMÉ: Les environs du lac Balaton représentent le plus important centre de repos et de tourisme de la Hongrie. L'Institut Géologique de Hongrie (MÁFI) poursuit depuis 1965 la recherche moderne complexe géologique et sédimentologique de cette région, d'une manière continue. Dans la région contière on a exécuté deux travaux de cargoraphie ingénieurgéologique: d'abord celle en une échelle 1:10 000 puis en 1:50 000. Simultanément se poursuivent depuis l'année 1981 les recherches sédimentologiques et géologiques des sédiments lacustres aussi.

Les buts de la cartographie ingénieurgéologique et des recherches de géologie actuelle sont les suivants: (1) recherches de géologie complexe des sédiments lacustres moux et des roches sédimentaires du Quaternaire et du Pannonien; (2) reconstruction de l'histoire de développement du lac et de ses alentours; (3) propositions concernant le développement systématique de l'enceinte de repos du Balaton; (4) recherche des méthodes convenables pour la diminution de l'eutrophisation de l'eau. L'étude expose le résultats principaux et l'applicabilité des résultats de recherche et décrit brièvement quelques caractères ingénieurgéologiques et sédimentologiques du Balaton et de ses alentours.

Introduction

Lake Balaton, the largest lake in Central Europe, represents the most important recreational and touristic centre not only in Hungary but in the whole region (Fig.1). The lake oriented NE--SW stretches in a length of nearly 80 km, and has a maximum width of nearly 15 km. The water has an average depth of 3.3 m, a total water surface of 600 sq.km. and a water mass of approx. 2 cu.km. Its water catchment area and lakeshore belt have been investigated in several phases since the beginning of this century.

Due to development of settlements the Hungarian Geological Institute, have been studying the geological environs of Lake Balaton since 1966. Included in these complex geological study programmes are as follows (Fig.2):
-- between 1967 and 1979, the engineering--geological mapping of the environs of Lake Balaton, on the scale of 1:10 000,
-- between 1982 and 1990, the engineering--geological mapping of the extended recreational zone of Lake Balaton, on the scale of 1:50 000,
-- between 1981 and 1990, the complex geological, actuo-geological study of La-ke Balaton were carried out.

The aim of this paper is to give a brief description on the above-listed investigations and the applicability of results thereof, laying stress on a few, methodologically new or important maps allowing us to draw an engineering-geological--sedimentological picture of the region.

Investigations, and results

The mapping (between 1967 and 1975),

covered a 3 to 6 km wide lakeshore belt of Lake Balaton, representing an area of 780 sq.km. to be mapped.

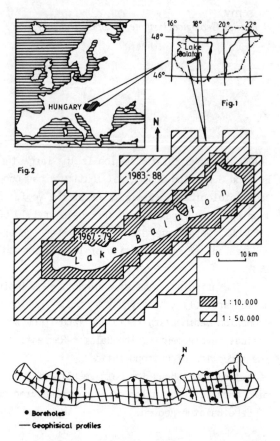

Fig.1

Fig.2

1983-88

1967-79

0   10 km

▨ 1:10.000
▨ 1:50.000

● Boreholes
— Geophisical profiles

The mapping has resulted in an atlas containing the following thematical maps showing the endowments of the area concerned:
-- observation maps (the technical conditions of the area, and the layout of geological exposures, boreholes and wells),
-- geomorphological maps (applied geomorphology, slope categorization, slope exposure)
-- geological maps (surface and subsurface versions, isopach map of Quaternary sediments, geophysical parameter maps),
-- hydrogeological maps (relative depth and altitude of ground-water table, hydro-

geochemical maps,

-- foundation maps,

-- complementary maps (economic-geological, microseismic zoning, agrogeological and environment-geological maps),

-- synoptic maps (for the evaluation of different districts for constructional suitability).

Of the above-listed ones, the geological and applied geomorphological maps, map showing the altitude of ground-water table and constructional suitability maps were published in 1982, on the scale of 1:50 000.

Atlases of scale 1:10 000 are useful in making expertises on regions, when developing plans for regional urban planning or when carrying out a preliminary allocation of linear projects. Maps, on the scales of 1:20 000 and 1:50 000 are useful when working out plans and conception for regional areal planning.

It seemed to be advisable to extend the recreational zone of Lake Balaton, in order to protect the environs of the lake, to improve water quality and to increase lakeshore public areas. A new mapping programme of scale 1:50 000 launched in 1982 was aimed at the complex study of the water catchment area (5200 sq.km.) of Lake Balaton (Fig. 2).

The mapping was carried out on the basis of a method formely approved, but in compliance with the new conditions. The engineering-geological map series includes 7 to 10 map versions that provide engineering-geological and hydrogeological fundaments for land-use plans required by the reconstruction of lakeshore areas as well as for studies needed for the development of background areas.

The eutrophization and filling-up of Lake Balaton have increased at an exponential rate in the past few decades. In order to establish scientific foundation for the environment protection interventions associated with these processes, and to improve the understanding of geological history of the lake, the actuo-geological study of Lake Balaton was also launched in 1981.

As a result of the exploration, the isopach map of loose mud in Lake Balaton, the seismo-stratigraphic--tectonic map of the basement, on the scale of 1:50 000, and the palaeoecological and palaeoclimatological reconstruction of the lake were completed.

Results are also useful in working out plans for interventions working against the eutrophization of the lake (such as mud dredging and mud traps) and designs of hydro-engineering constructions (wharfs, harbour).

Some engineering-geological and sedimentological features of Lake Balat on and its region

There are several, methodologically new maps completed on the scale of 1:50 000 within the research programmes described before, summarizing the endowments of Lake Balaton from the aspects of practice and environment protection - within the range of the opportunity allowed by the scale. These maps are as follows:

-- Map of zoning according to suitability for construction;

-- Vulnerability-to-pollution maps, and maps showing factors working against the fertility of soil;

-- Isopach map of the loose mud of the lake, and the seismostratigraphic-tectonic map of the basement of Lake Balaton.

In the map of zoning according to suitability for construction all important information a designer may need for the complex utilization of the area concerned are collected and systematized on the basis of the existing geological, geomorphological, hydrogeological and foundation maps. However, this map not only shows factors causing problems when constructing a project, but also gives a complex evaluation for each section of the area, and combining them into districts, it offers proposal for their suitability for constructional purposes.

Based on its geological structural features Lake Balaton is situated on the margin of the Transdanubian Midmountains (towards the north) and on the rim of the Hilly Region of Transdanubia (towards the south). Based on their geomorphological and geological structures both aforesaid engineering-geological regions can be divided into 8 engineering-geological divisions. Each division can be subdivided, on the basis of the microrelief, into 4 engineering-geological subdivisions which are as follows: alluvial plain; denudation surfaces with slight slope; heavily broken-up terrains and valleys; and basins in midmountains and hilly regions. On the basis of slope categorization, dynamic phenomena, ground-water situation and aggressivity of the water, lithological structure and load-bearing capacity calculated on the foundation for each formation, each subdivision are further divided into ranges. Evaluating the ranges obtained as described before, according to their suitability for construction, the following four categories have been used: very favourable to housing;

regular to housing; not recommended for housing. These areas are distinguished by different colours. For areas not recommended for housing, the unfavourable factors such as slopes steeper than 35 degrees, intensive extension of geographical processes (landslide, erosion, éboulement, slumping, moving stone flow), the ground-water position exceeding 1 m, aggressive property of ground-water (SO_4 > 400 mg/litre; Cl > 500 mg/litre), and the value of load-bearing capacity of soil related to the foundation being lower than 0.1 N/mm^2, are indicated by shade-lines.

In response to the new requirements concerning, primarily, the environment protection and arising in the course of the mapping, the vulnerability-to-pollution maps and maps showing the factors working against the fertility of soil were also completed (P. Farkas, 1987). the vulnerability-to-pollution map indicates - based mainly on the filtration properties of formations - the areas that are particularly sensitive to contamination (e.g. open karst surface, terraces built up from gravel and sand, etc.), and at the same time it gives a proposal for the optimum allocation of waste disposal sites.

The map showing the factors working against the fertility of soil separates and illustrates - in regard with agricultural standpoint included - the ares subject to problems associated with erosion (active and potential, areal and linear), deflation, or unfavourable mechanical composition of subsurface beds (high clay, sand or gravel content, lime beds, heterogeneous clasts or filling-up), unfavourable water effects (open water table, marsh), or the extreme pH conditions of soil.

Maps completed as a result of geophysical logging and underwater drilling carried out at Lake Balaton illustrate the thickness conditions for the colloidal mud and for the whole Quaternary mud, as well as the surface, the varied lithological structure and the structural and tectonical conditions of the solid basement of the lake. The pre-Quaternary basement has uneven surface and is covered by lacustrine mud which has an average thickness of 5 m. The upper part of 0.5 to 1 m (max. 1.5 m) of this mud is in colloidal state. Maximum thickness of mud (10 m) was measured in the mouth of River Zala ensuring the major part of water replacement of the lake, whereas the minimum thickness (0 m) was observed in the strait of Tihany.

Using radiocarbon method the oldest sediment was dated 12 to 14 thousand years BP. Based on it and the measured values of mud thickness an average rate of 0.4 mm pro year is obtained for silting up.

Seismograms have allowed us to distinguish 7 groups, and within 2 groups 6 beds, as well as to differentiate the horizontal, transcurrent and vertical (normal) faults. The distinguished groups and beds represent, on one hand, formations that are separated from one another lithologically, and on the other hand, sequences with different conditions of position.

References

Bodor, E. (1987): Formation of the Lake Balaton palynological aspects - Holocene environment in Hungary. Contr. of the INQUA Congress Canada, 1987, pp. 77-80.

Boros, J.-T. Cserny (1987): Engineering geological characteristics of the Quaternary in the Lake Balaton region. - Pleistocene environment in Hungary Geographical Research Institute Hungarian Academy of Sciences, Budapest.

Boros, J.-T.Cserny-G.Csillag, Á.Kurimay (1985): Engineering geological map series of the environs of Lake Balaton, scale 1:50 000, MÁFI, Budapest.

Cserny, T. (1977): Az 1:25 000-es méretarányú építésföldtani mintatérképek szerkesztésének elvi alapjai (Guidelines for engineering-geological mapping on the scale 1:25 000), Földt. Int. Évi Jel. 1975-ről, pp. 315-318, in Hungarian.

Cserny, T. (1984): Relation of clay content and some lithophysical characteristics of Upper Pannonian sedimentary rocks based on the evidence obtained in the course of mapping of the surroundings of lake Balaton. - Proc. 6th Conference on Soil Mech. and Found. Eng., Budapest.

Cserny, T. (1987): Results of recent investigations of the Lake Balaton deposits. - Holocene environment in Hungary Geographical Research Institute Hungarian Academy of Sciences, Budapest.

Cserny, T.-R.Corrada (1989): Complex geological investigation of Lake Balaton (Hungary) and its results. - Acta Geol. Hung. 32/1-2.

Farkas, P. (1987): A talajerózió új, tér-
képszerű ábráolási módszere (A new
method of map-like representation of
soil erosion). Földt. Int. Évi Jel.
1985-ről, pp. 287-294, in Hungarian.

Guóth, P. (1974): Guidelines for engineering-geological mapping on the scale
of 1:10 000. Special papers 1974/2,
Hung. Geol. Inst., Budapest.

Máté, F. (1987): A Balaton-meder recens
üledékeinek térképezése (Mapping of
the recent sediments of the bottom of
Lake Balaton). Földt. Int. Évi Jel.
1985-ről, pp. 366-379, in Hungarian.

Müller, G.-F.Wagner (1978): Holocene car-
bonats evolution in Lake Balaton
(Hungary): a response to climate and
impact of man - in modern and ancient
lake sediments. Blackwell Sci. Publ.,
pp. 57-81.

Raincsák-Kosáry, Zs.-T.Cserny (1984): A
Balaton környéki építésföldtani tér-
képezés eredményei (Results of the engineering geological mapping of the
Balaton region). Földt. Int. Évi Jel.
1982-ről, pp. 49-57, in Hungarian.

Zólyomi, B. (1987): Degree and rate of
sedimentation in Lake Balaton - Pleistocene environment in Hungary. Contr.
of the INQUA Congress Canada, 1987,
pp. 57-79.

Applied geology maps for land-use planning in Great Britain
Cartes de géologie appliquée pour le plan d'occupation des sols en Grande-Bretagne

M.G.Culshaw & A.Forster
Engineering Geology Research Group, British Geological Survey, Keyworth, Nottingham, UK

J.C.Cripps
Department of Geology, University of Sheffield, Sheffield, UK

F.G.Bell
Department of Geology and Applied Geology, University of Natal, Durban, South Africa

ABSTRACT: Land-use planning seeks to resolve the conflict between man's need to utilise land for housing, industry and the infrastructure, the extraction of minerals, the disposal of waste or the provision of recreational areas, and his need to protect the environment, whether in its unspoilt state or as sites of historic or scientific interest. In addition, people's perceptions of what is environmentally acceptable change as their knowledge, experience and expectations of life increase. Thus, the perceived balance between the advantages and disadvantages of a particular form of development alters. Therefore, planners frequently have to reassess the costs and benefits of such activities and require up to date information from many specialists, including geologists, to do this. In the past, geology sometimes has been neglected. This is not surprising because it is only as urban areas begin to expand into geologically less favourable locations that the significance of the geology becomes more apparent.

This paper draws attention to many of the geological factors that affect the planning process and illustrates how geologists have attempted to provide planners with the geological information they require. Particular reference is made to recent work on applied geological mapping in Great Britain.

RESUME: Le plan d'occupation des sols s'efforce de résoudre le conflit entre les besoins de l'homme d'utiliser le sol pour le logement, l'industrie et ses infrastructures, l'extraction de minéraux, le traitement des déchets ou la création d'aires de loisirs, et son besoin de protéger l'environnement, soit en bon état, soit en tant que sites d'intérêt historique ou scientifique. De plus, la perception qu'a une personne de ce qui est acceptable vis-à-vis de l'environnement change lorsque ses connaissances, son expérience et ce qu'elle attend de la vie augmentent. En même temps, la perception de la balance entre les avantages et les inconvénients d'une forme particulière de développement s'altère. Il en résulte que la planification doit fréquemment réajuster les coûts et les bénéfices de telles activités et recquiert pour cela des informations récentes émanant de plusieurs spécialistes dont le géologue. Dans le passé, la géologie a parfois été négligée. Ce n'est pas surprenant puisque ce n'est que lorsque les zones urbaines commencent à se répandre sur des terrains géologiquement moins favorablesque l'importance de la géologie devient plus apparente.

Cet article attire l'attention sur beaucoup de facteure géologiques qui affectent le processus de planification et illustre comment les géologues ont tenté de fournir les informations géologiques nécessaires. Une référence est particulièrement faite aux récents travaux sur la cartographie géologique appliquée en Grande-Bretagne.

1 INTRODUCTION

National Geological Surveys have existed for a long time, the British Geological Survey, for example, having been created in 1835. In the early decades of their existance, people saw the functions of such Surveys as being to provide geological maps at appropriate scales and to identify sources of minerals. For many Surveys, worldwide, these are still the primary functions. Although, in some countries, population growth and rapid urbanisation in the face of geological hazards has stimulated hazard and risk mapping. In the "developed" world, the emphasis has changed as indiginous mineral resources have been used up and as nationwide mapping has neared completion. New problems have arisen as the economic emphasis has changed from mineral exploitation, industrialisation and urbanisation to "high technology", service industries and the suburbanisation of populations.

Mineral resources, upon which many cities were founded (literally, in some cases), have become exhausted and inner city industrial areas have become disused or derelict. In recent years this process of rundown of urban centres has been reversed and the traditional industrial areas, such as docklands and their surrounding factories, have become locations for new housing and new businesses. Because these sites are often in, or close to, city centres, the land has great commercial value. In addition, modern construction techniques, which allow the building of much larger structures, use the land more efficiently. However, this newly available land carries with it the legacy of its previous industrialisation. Cities such as Glasgow are underlain, in part, by old, shallow, pillar and stall workings for coal. Deterioration of the mines and the higher ground loadings imposed by modern structures built above the old workings raise the spectre of ground collapse into old workings (Browne & Hull 1985, Browne et al. 1986a, 1986b, Forsyth et al. 1983, 1984, 1985).

In Southampton and other cities where development is taking place on land reclaimed from river estuaries, soft alluvial sediments provide poor foundation conditions (Edwards et al. 1987). Inland sites have other problems; in London, for example, a rising groundwater table has flooded basements and led to the redesign of some foundations (Simpson et al. 1989)

Therefore, it is evident that the rebuilding of many British cities is beset by a number of geologically-related problems. However, the conventional stratigraphically-based geological map is of only limited help in assessing the nature of these problems. Such maps are often difficult for non-geologists, such as planners and engineers, to understand, and do not provide the kinds of information that are needed. For example, whilst landslips may be shown on such maps, their type and degree of activity, and their angles of slope may not. The extent of mineworkings and groundwater conditions are not shown and the quality and quantity of potential mineral resources (such as aggregate) may have to be inferred. Even the lithologies of the different strata may not be clearly indicated.

Therefore, in many ways, the conventional geological map is inadequate for the needs of planners, developers and engineers. To meet these needs, a different type of geological map has been devised. This is

usually referred to as an environmental geology map (EGM), though other names such as thematic geology maps and applied geology maps (Great Britain), geoscientific maps (Spain) and land use potential maps (Germany) have been used.

The production of comprehensive suites of EGM's in Britain only began in the early 1980's following a pilot study on the Glenrothes area of Fife, Scotland, by Nickless (1982). This study produced 27 separate maps of an area of 100 km^2 at a scale of 1:25 000, covering such aspects as, stratigraphy and lithology of bedrock and superficial deposits, rockhead contours, engineering properties, mineral resources and workings, groundwater conditions and landslip potential. The maps were primarily for use by local and central planners, but were also useful sources of information for other users, including civil engineers, mineral extraction companies and developers. The main feature of the study was the presentation of each element of the geology on a separate map sheet in a way that was easy for non-geologists to understand. A further feature was the attempt to interpret these elements so as to provide data other than mere outcrop distribution and lithology. Thus, a conventional stratigraphically-based geological map might give, for example, no indication of whether sand and gravel deposits were likely to be exploitable, both in terms of quality or quantity. Such information would be indicated on an EGM.

Nickless's (1982) approach followed practice in the the USA and Europe which has been briefly reviewed by Doornkamp et al (1987). However, the methodology also built upon the techniques developed for engineering geological mapping elsewhere in Britain by Anon (1977), Cratchley et al. (1979) (S E Essex), Gostelow and Tindale (1980) (Cromarty, Scotland) and Gostelow and Browne (1981) (Firth of Forth, Scotland). The Firth of Forth study has been summarised by Gostelow and Browne (1986). Similar projects were carried out overseas (Culshaw et al. 1979, Culshaw and Sutarto 1981, Cratchley et al. 1982). The development of the EGM approach from earlier engineering geological mapping work in Great Britain has been described briefly by Culshaw et al. (1988).

2 THE COMMISSIONING OF EGM PROJECTS

Primary geological mapping, revision and remapping in the UK is carried out by the British Geological Survey (BGS) at a scale of 1:10 000. It has been going on since the Survey was founded, though mapping in the early part of the nineteenth century was usually at the smaller scale of 1:63 360 (1 inch to 1 mile). Mapping at the 1:10 000 scale will be completed early in the next century.

Almost all the environmental geological mapping projects undertaken in Britain to date have been commissioned by the DoE for the following reasons:

"a) To develop techniques for presenting geological, and related, information in forms which can be readily used for planning of land use control of development, and protection of the environment by those concerned with these matters but who do not have an earth science training -- this is aimed at trial areas which cover as wide a range of geological and land-use circumstances as possible.

b) To improve mapping of coal field areas in which problems for development and redevelopment are particularly concentrated in Great Britain" (Marker pers. comm.).

However, there are two ways in which these projects are funded. The BGS has its own priorities for the order in which mapping will take place, based on environmental and scientific factors and the state of existing mapping. In the same way, the DoE, after consultation with local authorities, identifies priority areas for which EGM's are required. Where priorities coincide, EGM's usually will be produced by the BGS, with remapping on a cost-sharing basis. However, where the BGS considers remapping to have a low priority, or where geological maps are already of an acceptable standard, the DoE usually will put contracts out to tender. The successful tenderer (in some cases the BGS iteself) then carries out the EGM contract without basic geological remapping. Non-BGS contractors are engineering or geological consultants, sometimes with a stronq academic background and input. Differences of opinion have arisen when the BGS has considered the standard of existing geological maps to be inadequate but the DOE has taken the view that existing maps, with limited revision based on site investigation reports and some field checking, are sufficient to meet the objectives of a study for land-use planning purposes.

The various projects commissioned by DoE in Britain have been listed by Culshaw et al. (1988). This list indicates that the mapping projects fall into two types. Firstly, there are basic geological mapping projects (BGM's) which include some maps additional to the traditional stratigraphically-based ones. Secondly, there are the EGM's which are considered further below. Since Culshaw et al. (1988) listed the BGM's and EGM's, a number of

further projects have commenced or will start shortly. These are listed in Table 1.

Table 1. Current environmental geology mapping projects

LOCATION	TYPE	CONTRACTOR
St Helens	EGM	Geomorphological Services Ltd.
Birmingham (West)	EGM	BGS
Wrexham	EGM	BGS
Leeds (South/central)	EGM	BGS
Livingston	EGM	BGS
Stirling	EGM	BGS

3 THE AIM OF EGM'S

As EGM's are carried out for the DoE, which also has certain responsibilities with respect to local government and to planning, the studies are aimed, primarily, at providing information related to the planning process. However, the projects also aim to provide information for other professionals, including engineers and geologists. Brook and Marker (1987) summarised these aims as follows:

1. i) detailed level; to provide information usable in the formulation of Local Plans

   ii) broader level; to provide the foundation for sound Structure Plan policies and a means of evaluating them

2. to give a context for the consideration of development proposals

3. to facilitate rapid access to the collected basic data for professional engineers and geologists.

The planning policy framework into which aims 1 and 2 fit was described briefly by Brook and Marker (1987) and in more detail by Worth (1987). Put simply, the planning policy framework exists at three levels:

1. Local policies resulting in Local Plans produced by District authorities

2. Strategic policies resulting in Sturcture Plans produced by County or Regional authorities

3. National policies produced by central government and resulting in Government Circulars and Advice Notes.

In some instances, both local and strategic policies are produced by one local authority where a two-tier local authority structure does not exist. Henry (1987) illustrated how geology was taken into account in formulating local and strategic plans for the Lothian Region of Scotland. This did not include the use of EGM's (which did not exist for the Lothian Region at that time). However, the production of EGM's should improve the planning process with respect to geology.

4 EGM METHODOLOGY

Fundamental to the production of EGM's is the traditional, stratigraphically-based geological map. Where modern, up to date geological maps do not exist it is essential to survey areas without cover or to revise existing maps. Without this base, other maps on various themes cannot be prepared or will be inaccurate. Nickless (1982), in his pilot EGM study of the Glenrothes area, divided the maps produced into three types:
1. Basic data maps
2. Derived maps
3. Environmental potential map
These three types were discussed further by Brook and Marker (1987). Basic data maps may be considered to be factual maps showing, for example, rock or soil type, rockhead contours, areas of undermining, shaft or adit locations, borehole locations, made ground or fill, landslips and slope angles, engineering properties and mineral deposits (including groundwater). The derived, or interpretative, maps might show mineral resources, foundation conditions or underground storage potential. These maps may be produced at a variety of scales, though the primary ones are 1:10 000, 1:25 000 or 1:50 000. Within one project maps may be at more than one scale depending upon the availability of data and the information required.

The "environmental potential maps" of Nickless (1982) were further discussed by Brook and Marker (1987) who emphasised their importance to planners. These maps are compiled from the factual and interpretative maps. They present, in general terms, the resources for development with respect to, for example, mineral, groundwater or agricultural potential which might be used in development or which should not be sterilised by building or contamination from landfill sites; also the constraints on development such as areas with poor foundation conditions, land susceptible to landslipping or subsidence, or land likely to be subjected to flooding. These maps summarise the information contained in the factual and interpretative maps in a way that can be more readily understood by non-geologists. The maps may be produced at a scale similar to the factual or interpretative maps (for example, 1:25 000)

or at a smaller scale (for example 1:50 000 1:80 000 or 1:100 000) if this is more appropriate. An example, showing constraints on development for part of the Torbay area of south west England, is shown in Fig. 1. This project was described by Lee et al. (1988) and in more detail by Anon. (1988).

Brook and Marker (1987) discussed the likely reaction of geologists to this "environmental potential map" because concern had been expressed that these maps were too generalised, and might give the impression that where no problems were depicted, no further investigation was required before development. Brook and Marker (1987) disagreed with this viewpoint because the basic aim was to inform planners of "relevant factors" so that expert professional advice could be sought. It should be added that EGM project reports and maps carry riders to the effect that the project results should not be considered as a substitute for on-site investigation.

A further problem concerns the use of these maps in making local planning decisions which may affect property values and other financial matters. In South Wales, the maps from a study of natural landslips (Conway et al. 1980) have been used to assist the planning authorities in deciding whether or not to grant planning permission for the development of individual plots of land. As such decisions may affect the value of land or property, it is important not only that maps used to assist the decision-making process are as accurate as possible but also that the limitations on their accuracy are clearly indicated.

5 COMMENTS ON THE COST BENEFIT OF EGM'S

In Britain, no attempt has been made to carry out a cost benefit analysis on the production of EGM's. Such an analysis would not be easy to carry out, particularly in the short term when all benefits (for example, from the exploitation of mineral resources) may not be apparent. However, such a study has been attempted in the Netherlands by de Mulder (1988, 1989) for environmental geological mapping of Amsterdam, which lies on Quaternary deposits. This study showed that the "earn-back time" was about 3.2 years if only the costs of desk studies and ground investigations were considered, but less than 1.2 years if savings due to better planning were included.

Although geological conditions for many urban areas in Britain are very different from those of Amsterdam, it is likely that such an analysis would indicate a similarly

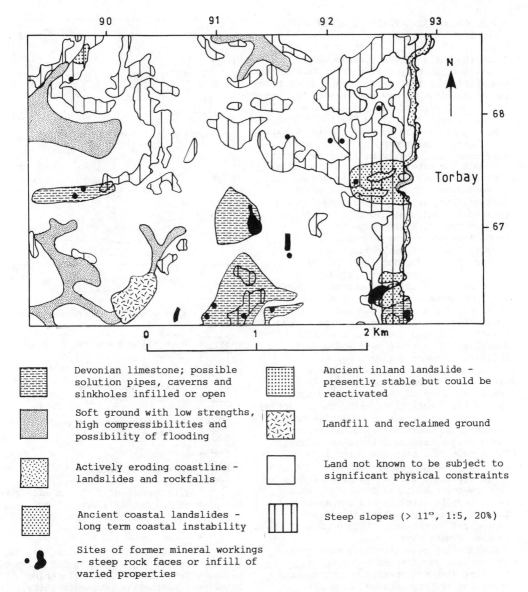

Devonian limestone; possible solution pipes, caverns and sinkholes infilled or open

Ancient inland landslide - presently stable but could be reactivated

Soft ground with low strengths, high compressibilities and possibility of flooding

Landfill and reclaimed ground

Actively eroding coastline - landslides and rockfalls

Land not known to be subject to significant physical constraints

Ancient coastal landslides - long term coastal instability

Steep slopes (> 11°, 1:5, 20%)

Sites of former mineral workings - steep rock faces or infill of varied properties

Fig. 1 Ground characteristics for planning and development for part of the Torbay area (after Anon 1988).

rapid "earn-back time". An illustration of this is the case of the Black Country spine road in the English West Midlands, northwest of Birmingham. In March 1988 the cost of the road was estimated at £50M, but by August 1989 these costs had risen to £140M. £45M of this increase was due to "inflation and unforeseen ground conditions" (Bishop 1989). The road passes over some particularly difficult ground including sites of a gas works, a power station,

railway sidings, derelict sewage works and open mine shafts, all of which needed reclaiming or treating. Had an EGM study of the area been carried out before the feasibility study for the road, it is probable that many of the problems would have been identified in advance, either specifically or in general terms, and that some of the additional costs would not have been incurred.

6 METHODOLOGY AS APPLIED TO RECENT EGM PROJECTS IN BRITAIN

Since the early 1980's considerable experience has been gained in Britain in the preparation of EGM's. The description below is not specific to any one project but draws on the work carried out for a number of projects, particularly in England and Wales.

Most of these projects initially required basic geological remapping. This work is time-consuming, requiring field geologists to "walk the ground" as well as to collect borehole records, mining information, site investigation reports and other data. However, this mapping forms a fundamental part of the national geological database, as well as providing the basis for the production of the environmental (or thematic/applied) geological maps. Ultimately the mapping leads to the production of maps of the solid (pre Quaternary) and drift (Quaternary) geology and the extent of fill and made ground (Fig. 2), and commonly diagrams of the depth to rockhead and formation thickness, and the extent of various potential mineral deposits, for example sand and gravel, building stone or fuller's earth.

Using data collected during the mapping, other environmental geology maps are compiled. Some of these, usually produced by the field geologists, show the extent and depth of known or suspected active or abandoned mineworkings (Fig. 3) and their associated entrances (shafts and adits). These maps have been discussed in more detail by Culshaw et al. (1988).

Hydrogeologists assess available data on groundwater conditions to produce maps showing wells, springs and seeps as well as the groundwater potential of different geological formations. This work may result in one or more maps.

Engineering geologists use geological information acquired during mapping, together with engineering geological and geotechnical data obtained from site investigation reports, to produce an asssssment of the engineering geology of the solid and drift deposits of the area (for example, Forster 1990). Part of this work usually involves the creation of a geotechnical database containing in situ and laboratory geotechnical test results obtained from the site investigation reports. The data are stored in a computer database and, after validation, are analysed to produce summary geotechnical parameter values for each geological formation or deposit. The summary values are tabulated for the various formations (for example, Forster et al. [1987], based on work by Forster et al. [1985] for the Bath area of

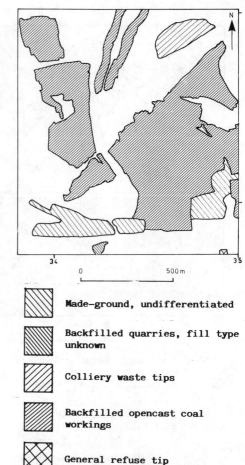

Made-ground, undifferentiated

Backfilled quarries, fill type unknown

Colliery waste tips

Backfilled opencast coal workings

General refuse tip

Fig. 2 Distribution of fill and made ground for part of the Rothwell area of West Yorkshire (after Williamson and Giles 1984).

south west England). The tables also may contain engineering comments, for example, on the effect of water on engineering behaviour, weathering, excavatability, foundation conditions and slope conditions for the formations. The data banking, validation, analysis and presentation of the geotechnical data was discussed in more detail by Forster and Culshaw (1990).

Solid and drift deposits with similar geotechnical characteristics are grouped together into various engineering geological units following the methodology of Anon. (1972) as developed by Gostelow and Browne (1986). Typical groupings for the Deeside area of North Wales were given by Campbell and Hains (1988). Engineering geological maps based on these groups are produced for the solid and drift.

Maps showing areas of similar slope angle

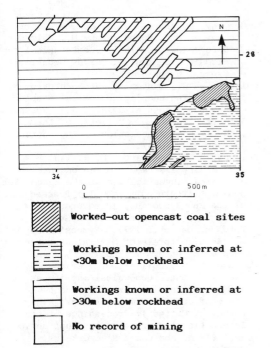

34   500m   35

0   500 m

▨ **Worked—out opencast coal sites**

Workings known or inferred at <30m below rockhead

Workings known or inferred at >30m below rockhead

No record of mining

Fig. 3 Underground and opencast mining in part of the Rothwell area of West Yorkshire (after Williamson and Giles 1984).

and extent of landslipped ground also may be presented were appropriate. Slope categories may vary, depending on local relief but typical categories might be <3°, 3°-7°, 7°-11°, 11°-15° and >15°. Landslips may be shown only in terms of their areal extent unless local circumstances require more detailed representation in terms of type and degree of activity. Where local conditions require it, other maps might be produced showing, for example sinkholes and subsidence features in areas of natural solution (for example Cooper 1989), areas susceptible to flooding or areas of contaminated ground. All projects will include a map showing the geographical location of data sources (boreholes, trial pits, etc).

The information provided in these factual and interpretative maps is summarised in additional maps showing constraints on, and resources for development, as described above.

The exact format of the different maps is not specified at the beginning of a project because at that time the geological conditions and the data available are not fully known. Work is regularly reviewed during the project by a steering committee composed of representatives of the client (DoE), the contractor (BGS or consultant), the local authorities for the area of the

project and, if appropriate, other organisations that might have a legitimate interest in the work or be able to provide constructive comments.

The environmental geology maps are accompanied by a project report which summarises the geology, engineering geology and hydrogeology of the study area, and comments on each of the thematic maps. The implications of the geology for planners, in terms of constraints and resources, is described in non-technical language. A glossary of technical terms and various appendices relating to data sources and test methodologies also is included. More technical open file reports on the geology of the constituent 1:10 000 scale maps, on the engineering geology of the area and on particular topics, such as landslipping, also may be compiled and provided as part of the project documentation.

In the early reports, maps were produced in monochrome by die-line methods to enable cheap and rapid reproduction. However, more recently, the value of producing maps in several colours has been appreciated as being a clearer way of representing information for non-specialists. The extra cost is considered to be justified in terms of easier understanding. Reports are illustrated with additional maps, sections, diagrams and graphs as appropriate. In the last three to four years special effort has been put into developing computer techniques for the production of the various maps. Some of these techniques involve the use of geographical information systems (GIS's) which are the subject of much research for a wide range of applications and which, when fully developed, will revolutionise the types of maps that can be produced, on demand, for EGM projects.

Following approval by the DoE, reports are placed on open file and made available for purchase. Marketing of the reports is being improved to make potential users more aware of their existence.

7 CONCLUSIONS

The development of EGM's in Britain has improved the ways in which geological information is presented and made available to a wide range of users. A methodology has been established which is followed in broad terms by most organisations producing this type of map. This methodology is subject to a continuing process of development so that the geological information can be appropriate, understandable and useful to planners, engineers and developers as they strive to meet the changing needs of society in a constantly changing world.

ACKNOWLEDGEMENTS. The authors thank Dr Brian Marker for his helpful comments. This paper is published with the permission of the Director of the British Geological Survey (NERC).

REFERENCES

Anon. 1972. Code of Practice for Foundations. CP 2004. British Standards Institution, London.

Anon. 1977. South Essex geological and geotechnical survey. Open File Report. British Geological Survey, Keyworth.

Anon. 1988. Planning and Development: Applied Earth Sience Background. Torbay. Doornkamp, J. C. (Ed.). Geomorphological Services (Publications and Reprographics) Ltd., Newport Pagnell.

Bishop, P. 1989. Costs triple on Black Country spine road. New Civil Engineer, 31 August 1989, 10.

Brook, D. & Marker, B. R. 1987. Thematic geological mapping as an essential tool in land-use planning. In: Culshaw, M. G., Bell F. G., Cripps, J. C. & O'Hara, M. (Eds.), Planning and Engineering Geology. Engineering Geology Special Publication No 4, Geological Society, London. 211-214.

Browne, M. A. E. & Hull, J. H. 1985. The environmental geology of Glasgow, Scotland - a legacy of urban surface and subsurface mining. In: Glaser, J. D. & Edwards, J. (eds.), Proceedings of the 20th Forum on the Geology of Industrial Minerals. Baltimore, Maryland. 141-152.

Browne, M. A. E., McMillan, A. A. & Forsyth, I. H. 1986a. Urban geology: Glasgow's hidden industrial heritage. Geology Today, 2, 74-78.

Browne, M. A. E., Forsyth, I. H. & McMillan, A. A. 1986b. Glasgow, a case study in urban geology. Journal of the Geological Society, 143, 509-520.

Campbell, S. D. G. & Hains, B. A. 1988. Deeside (North Wales) thematic geological mapping. Technical Report WA/88/2. British Geological Survey, Keyworth.

Conway, B. W., Forster, A., Northmore, K. J. & Barclay W. J. 1980. South Wales Coalfield landslip survey. Technical Report EG/80/4. British Geological Survey, Keyworth.

Cooper, A. H. 1989. Airborne multispectral scanning of subsidence caused by Permian gypsum dissolution at Ripon, North Yorkshire. Quarterly Journal of Engineering Geology, 22, 219-229.

Cratchley, C. R., Conway, B. W., Northmore, K. J. & Denness, B. 1979. Regional geological and geotechnical survey of South Essex. Bulletin of the International Association of Engineering Geology, 19, 30-40.

Cratchley, C. R., Hobbs, P. R. N., Petrides, G. & Loucaides, G. 1982. Geotechnical map of Nicosia. Geological Survey Department of Cyprus, Nicosia.

Culshaw, M. G., Duncan, S. V. & Sutarto, N. R. 1979. Engineering geological mapping of the Banda Aceh alluvial basin, Northern Sumatra, Indonesia. Bulletin of the International Association of Engineering Geology, 19, 40-47.

Culshaw, M. G. & Sutarto, N. R. 1981. A preliminary assessment of an engineering geological investigation of the Banda Aceh basin, Northern Sumatra. In: Proceedings of the 2nd Symposium, Integrated Geological Survey of Northern Sumatra, Bandung, December 1977. Symposium Report of the Directorate of Mineral Resources, Indonesia, No 3, 1, 77-85.

Culshaw, M. G., Bell, F. G. & Cripps, J. C. 1988. Thematic geological mapping as an aid to engineering hazard avoidance in areas of abandoned mineworkings. In: Forde, M. C. (ed.), Proceedings of the 2nd International Conference on Construction in Areas of Abandoned Mineworkings, Edinburgh, June 1988. Engineering Technics Press, Edinburgh. 69-76.

De Mulder, E. F. J. 1988. Engineering geological maps: a cost-benefit analysis. In: Marinos, P. G. & Koukis, G. C. (eds.), Proceedings of an International Symposium on The Engineering Geology of Ancient Works, Monuments and Historical Sites, Preservation and Protection, Athens, September 1988. A. A. Balkema, Rotterdam. 3, 1347-1357.

De Mulder, E. F. J. 1989. Thematic applied Quaternary maps: a profitable investment or expensive wallpaper? In: De Mulder, E. F. J. & Hageman, R. P. (eds.), Proceedings of the INQUA Symposium on Applied Quaternary Studies, Ottawa, August 1987. A. A. Balkema, Rotterdam. 105-117.

Doornkamp, J. C., Brunsden, D., Cooke, R. U., Jones, D. K. C. & Griffiths, J. S. 1987. Environmental geology mapping: an international review. In: Culshaw, M. G., Bell, F. G., Cripps, J. C. & O'Hara, M. (eds.), Planning and Engineering Geology. Engineering Geology Special Publication No 4, Geological Society, London. 215-219.

Edwards, R. A., Scrivener, R. C. & Forster, A. 1987. Applied geological mapping: Southampton area. Research Report ICSO/87/2, British Geological Survey, Keyworth.

Forster, A. 1990. A method of producing engineering geology maps of Nottingham, England, for use by planners and engineers (this volume).

Forster, A. & Culshaw, M. G. 1990. The use of site investigation data for the preparation of engineering geological maps and reports for use by planners and civil engineers. Engineering Geology (in press).

Forster, A., Hobbs, P. R. N., Monkhouse, R. A. & Wyatt, R. J. 1985. An environmental geology study of parts of west Wiltshire and south east Avon. Technical Report, British Geological Survey, Keyworth.

Forster, A., Hobbs, P. R. N., Wyatt, R. J. & Entwisle, D. C. 1987. Environmental geology maps of Bath and the surrounding area for engineers and planners. In: Culshaw, M. G., Bell, F. G., Cripps, J. C. & O'Hara, M. (eds.), Planning and Engineering Geology. Engineering Geology Special Publication No 4, Geological Society, London. 221-235.

Forsyth, I. H., McMillan, A. A., Browne, M. A. E. & Ball, D. F. 1983. Account accompanying environmental geology maps of Glasgow (National Grid sheet NS66). Technical Report NL/83/1. British Geological Survey, Edinburgh.

Forsyth, I. H., McMillan, A. A., Browne, M. A. E. & Ball, D. F. 1984. Account accompanying environmental geology maps of Glasgow (National Grid sheet NS56). Technical Report NL/84/3. British Geological Survey, Edinburgh.

Forsyth, I. H., Paterson, I. B. & Hall, I. H. S. 1985. Account accompanying environmental geology maps of Glasgow (parts of National Grid sheets NS55, 57 and 65). Technical Report, British Geological Survey, Edinburgh.

Gostelow, T. P. & Browne, M. A. E. 1981. Engineering geology of the Upper Forth estuary. Technical Report, British Geological Survey, Edinburgh.

Gostelow, T. P. & Browne, M. A. E. 1986. Engineering geology of the Upper Forth estuary. Report of the British Geological Survey, 16, 8.

Gostelow, T. P. & Tindale, K. 1980. Engineering geology investigations into the siting of heavy industry on the east coast of Scotland: the north side of the Cromarty Firth. Technical Report No. EG S80/5. British Geological Survey, Edinburgh.

Henry, J. J. 1987. The geological input into land use planning in Lothian Region, Scotland. In: Culshaw, M. G., Bell, F. G., Cripps, J. C. & O'Hara, M. (eds.), Planning and Engineering Geology. Engineering Geology Special Publication No 4, Geological Society, London. 583-587.

Lee, E. M., Doornkamp, J. C., Griffiths, J. S. & Traghiem, D. G. 1988. Environmental geology mapping for land use planning purposes in the Torbay area. Proceedings of the Ussher Society, 7, 1, 18-25.

Nickless, E. F. P. 1982. Environmental geology of the Glenrothes district, Fife Region. Description of 1:25 000 sheet NO 20. Report of the British Geological Survey, 82, 15.

Simpson, B., Blower, T., Craig, R. N. & Wilkinson, W. B. 1989. The engineering implications of rising groundwater levels in the deep aquifer beneath London. Special Publication 69. Construction Industry Research and Information Association, London.

Williamson, I. T. & Giles, J. R. A. 1984. Geological notes and local details for 1:10 000 sheets: Sheet SE 32 NW (Rothwell). Technical Report, British Geological Survey, Keyworth.

Worth, D. H. 1987. Planning for engineering geologists. In: Culshaw, M. G., Bell, F. G., Cripps, J. C. & O'Hara, M. (eds.), Planning and Engineering Geology. Engineering Geology Special Publication No 4, Geological Society, London. 39-46.

Environmental mapping applied to the planning of urban and natural park areas in the north coast of Spain

Cartes géoscientifiques pour la planification de parcs naturels et zones urbaines sur la côte nord de l'Espagne

E. Francés, J.R. Díaz de Terán & A. Cendrero
DCITTYM (Ciencias de la Tierra), Univ. de Cantabria, Spain

D. Gómez Orea & T. Villarino
Departimento Proyectos, E.T.S.I. Agrónomos, Univ. Politécnica, Madrid, Spain

J. Leonardo & L. Saiz
Taller de Arquitectura y Urbanismo, Santander, Spain

ABSTRACT: Two examples of geoscientific maps of the environment and their application to the establishment of planning regulations are presented. The methodology used is based on the mapping of integrated homogeneous units, defined on the basis of their geoenvironmental features. In a natural park area a hierarchical subdivision was made and the units thus obtained were transformed into "diagnosis categories", according to those qualities most significant from the point of view of the preservation of the natural environment. These categories were then used to establish guidelines for planning and management. In an urban area, morphodynamic units were mapped and described in terms of variables and qualities significant for planning. Units were then translated into the "planning categories" contemplated within land-use norms, using in each case the most limiting factor. Planning categories were used to define the regulations affecting land-use, within the master plan for the town.

RÉSUMÉ: On présente deux exemples de cartes de l'environnement, appliquées a l'élaboration de plans pour une zone urbaine et un parc naturel dans une zone côtière. La méthodologie utilisée se base sur la définition d'unités homogènes integrées. Dans le parc naturel on a fait une division echelonnée du territoire, á trois niveaux. Les unités homogènes ont été transformées en "categories de diagnostique" selon leur qualités plus significatives du point de vue de la conservation. Ces categories sont la base pour l'établissement des normes pour réguler l'amenagément du parc. Dans la zone urbaine on a établi une division en unités morphodynamiques, lesquelles ont été aprés transformées en "categories d'amenagément", sur la base des facteurs les plus limitants de leur utilisation et d'accord avec les normes habituelles du pays. Le type d'occupation du terrain dans chaque catégorie d'aménagément fût alors établi dans le plan urbain.

INTRODUCTION

Geoscientific maps of the natural environmental potential (Lüttig, 1978) or environmental geological maps (Fisher et al., 1972) constitute an instrument that can be used as a basis for very diverse types of plans regarding land-use.

In this paper, two case studies of the application of such maps to planning in the coastal zone of northern Spain, using a methodology based on the definition and diagnosis of integrated synthetic units (Brown et al., 1971; Cendrero, 1975; Cendrero and Díaz de Terán, 1987) are presented.

PLANNING AIMS

The aims of the plans in the two study areas (Fig. 1) were very different. The

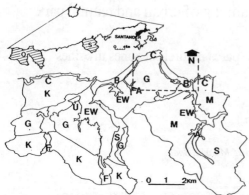

Fig. 1. Location map (S, Suances; O, Oyambre) and map of morphodynamic systems of Oyambre. The area within the dashed line is represented in Fig. 2.

municipality of Suances is a resort area which constitutes a suburb of the industrial town of Torrelavega, with 60,000 inhabitants, located 6 km away. This results in a rapid urban growth for residential and tourism activities, originating conflicts between urban uses and conservation of the natural environment, as well as increasing natural hazards. In this case the aims of the master plan were to establish a zonning of the municipal area that would regulate urban growth protecting natural values, maintaining agricultural productivity and avoiding or reducing natural hazards.

The area of Oyambre encompasses several municipalities and includes cliffs, beaches, dunes, estuaries, wetlands, karstic massifs and some natural forests. It also contains several natural sites of scientific and educational interest (Quaternary marine terraces with well preserved sedimentary sequences, good peat sequences, botanical endemisms, nesting areas of different birds species, etc.) This area has been recently (26, Oct., 1988) declared a Natural Park by the Regional Legislature of Cantabria. The aims of the study carried out here were to determine the levels of protection for the different zones within the park and surrounding areas and to define the basis for a management plan.

METHODOLOGY AND PLANNING RESULTS

The methodological approach used in both cases was based on the identification, definition and mapping of integrated units (equivalent to the land systems method of Christian and Stewart 1968; resource capability units of Brown, et al., 1971; units of the natural environmental potential, Lüttig, 1972; morphodynamic units of Cendrero and Trilla, 1983) which were then transformed into "diagnosis or planning categories".

In the area of Oyambre mapping was carried out at the 1:10,000 scale and a three-level hierarchical subdivision was made. First, "morphodynamic systems" were defined on the basis of lithology, surficial deposits and morphology. The systems identified (Fig. 1) were: coastal cliffs, beaches, dunes, raised marine terraces, karstic massifs, gentle reliefs over sandstones-siltstones-claystones, moderate reliefs over sandstones-siltstones-claystones, strong reliefs over sandstones-siltstones-claystones, estuaries and associated wetlands, alluvial plains, fluvial valleys with well-defined adjacent slopes. These systems were divided into "morphodynamic units", using active processes, slope and orientation as criteria. Finally, units were divided into "morphodynamic elements" on the basis of vegetation and features of scientific or educational vale (Fig. 2).

A diagnosis of morphodynamic systems, units and elements was made in order to define levels of protection and to establish regulations and management plans in different parts of the study area. This was done by determining in each case the most relevant features from the point of view of conservation and management. Through the evaluation of integrated units, on the basis of such features, a series of "diagnosis categories" were obtained (Gómez Orea, 1978). Diagnosis categories correspond to either "systems" (S), "units" (U) or "elements" (E). The following are the main categories which were established:

A-coastal cliffs (S) and adjacent raised

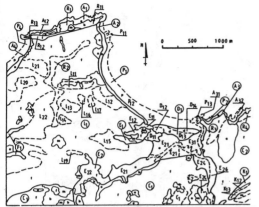

Fig. 2. Map of morphodynamic units and elements of part of Oyambre. Capitals indicate systems A) cliffs, R) raised marine terraces, P) beaches, D) dunes, F) fluvial valley, E) estuaries and wetlands, L) gentle reliefs over sandstone–siltstone–claystone, C) moderate reliefs over sandstone-siltstone-claystone, K) karstic massifs. Numbers indicate units within a system. Small letters correspond to elements.

marine terraces (U).

B- Estuaries and coastal wetlands (S)

B_1- well preserved, with natural flora and fauna (E).

B_2-degraded, with allochtonous flora (E).

C-Dunes and beaches (S)

C_1-dunes with well preserved morphology (U).

C_2-degraded dunes (U).

C_3-beaches (U).

D-Karstic massifs (S).

D_1-with well preserved green oaks (E).

D_2-with degraded green oak bushes (E).

E-Gentle to moderate reliefs over sandstones, siltstones and claystones (S).

E_1-areas affected by aggregate extraction and quarrying; active/abandoned (U).

E_2-Eucalyptus (E).

E_3-heath and pasturelands (E).

E_4-grasslands (E).

E_5-natural decidious forests (F).

F-Strong reliefs over sandstones, siltstones and claystones (S).

F_1-areas subject to slope instability (U).

F_2-natural decidious forests (E).

F_3-American oak (F).

F_4-Eucalyptus (E).

F_5-Pinus (E).

F_6-grasslands (E).

Once morphodynamic systems, units and elements were transformed into diagnosis categories, they were used to define the basis for the legal norms that should regulate the use of the area under consideration, and also for establishing a series of actions that should be part of the management plan for the zone. These norms and actions were adapted, in each case, to the environmental characteristics of the diagnosis units described.

Examples of such norms and guidelines for management are:

Diagnosis category A: coastal cliffs and adjacent raised marine terraces.

General criteria: the aim in this case should be to preserve the existing ecological, geomorphological and landscape values, so that minimum human intervention should take place.

Activities to promote: recreation without permanent structures, educational.

Acceptable activities: line fishing, existing grass and fodder cultivation, shepherding, under control to ensure the preservation of ecological values.

Forbidden activities: all activities implying excavation of any sort, installation of permanent or semi-permanent structures, hunting.

Management actions: no specific actions required.

Diagonosis Category B_2: degraded estuaries and wetlands with allochtonous flora.

General criteria: the aim is to restore the original condition of the estuary eliminating disturbing influences and recuperating the natural biological productivity.

Activities to promote: ecosystem and landscape regeneration, controlled shellfish harvesting, educational.

Acceptable activities: controlled fishing, hunting and recreation.

Forbidden activities: all activities implying disturbance of the surficial sedimentary layer and vegetation, installation of permanent or semi-permanent structures, or excessive exploitation of living resources.

Management actions: control of fishing and shellfish harvesting by the park authorities; elimination of existing <u>Eucalyptus globulus</u> and <u>Bacharis hallimifolium</u> and substitution by autochtonous marshland vegetation; removal of existing artificial fillings; removal of the existing walls which restrict circulation in the estuary.

<u>Diagnosis category E_1</u>: areas affected by aggregate extraction and quarrying.

General criteria: the aims are to maintain the resource under production, where compatible with the preservation of ecological conditions, and to eliminate or reduce the impact of abandoned exploitations.

Activities to promote: controlled aggregate extraction, restoration of abandoned exploitations.

Acceptable activities: reforestation, re-implantation of agriculture, constructions of low visual impact.

Forbidden activities: large scale construction.

Management actions: relandscaping and revegetation of abandoned excavations and spoil heaps.

Thus, the mapping units initially defined on the basis of descriptive features were evaluated and transformed into diagnosis categories and, finally, into prescriptive, regulatory units for the establishment of management guidelines (Fig. 3).

In the case of Suances, homogeneous units (equivalent to the morphodynamic units described above) were mapped at the 1:5,000 scale. These homogenous units were defined on the basis of bedrock, slope, landform, active processes and vegetation. Each morphodynamic unit was described in terms of the following parameters: location, slope, orientation, bedrock, regolith type

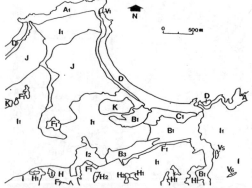

Fig. 3. Plan of management actions for part of Oyambre. A) management actions for cliffs and adjacent raised marine terraces, B_1) control of fishing and shellfish harvesting in estuaries, B_3) elimination of barriers to tidal circulation and regeneration of vegetation, C_1) cleaning and re-vegetation of dunes, D) control of recreation in beaches, F_1) conservation and improvement of autochtonous vegetation, H_2) reforestation with oak, I_1) improvement of grass and foder cultivations, I_2) drainage and improvement of grass production, J) limitation for structures of high visibility, due to the protection of visual landscape, K) re-location of lampsites, V) protection of sites of scientific interest.

and thickness, landform, soil type and soil capability, vegetation, ease of excavation, bearing capacity, slope stability, hazards (flooding, water stagnation, slope instability, collapse, rockfall, aquifer vulnerability, coastal erosion).

The units thus defined and described were then classified according to the categories contemplated within the usual urban and rural land-use regulations in Spain. This was done identifying in each unit the most limiting factor from the point of view of land-use.The categories established were: 1) areas where high-density building is allowed; 2) areas where low-density building is allowed but requires a special detailed plan; 3) areas where building is not

98

allowed; 4) areas of ecological and/or landscape protection; 5) areas with building restrictions due to natural hazards; 6) areas reserved for agriculture; 7) areas with building restrictions for the protection of aquifers; 8) Areas with building restrictions due to high visual incidence and potential slope instability, but where building could be allowed after the preparation of a detailed project.

Figure 4 (Cendrero, 1989) shows part of the area in which the types of homogeneous units mapped are represented, as well as their transformation into planning categories. As the master plan for the area must establish clearly what kind of regulations affect each property, the natural boundaries between homogeneous units are often not suitable as a basis for such regulations, because those limits are in many cases transitional or they correspond to the interpretation of features, such as bedrock or geomorphology, not directly visible at the surface, especially by non-specialists. Therefore, the boundaries of the units initially defined were modified, adapting them to existing roads, tracks, footpaths and property limits, so that they would constitute clearly defined limits of easy application within the master plan. When possible, the criterion followed in the transformation of boundaries was to extend the limits of units requiring protection or certain restrictions.

The main points of the regulations established in the final master plan for the different planning categories were:

1) areas where high-density building is allowed; 2) areas where low-density building is allowed but requires a special detailed plan; 3) areas were building is not allowed. These three categories have regulations derived mainly from considerations not related to the natural environment.

4) areas of ecological or landscape protection (includes coastal cliffs, beaches and wetlands): no construction or activity is allowed, without a prior environmental impact assessment proving that no undesirable effects will be produced. In any

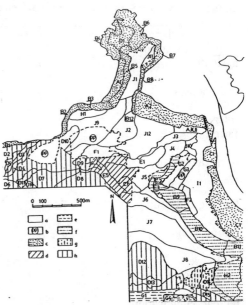

Fig. 4. Homogeneous units and planning categories in Suances. Units: A, beaches and dunes; B, cliffs; C, wetlands; D, karstic zones; E, units with unstable slopes; F, flood-prone units; G, units with good quality soils; H, units with vegetation or landscape of interest for conservation; I. artificial fill; J, units totally or partially built-up. Planning categories: a) areas of high-density building; b) areas of low-density building; c) areas of ecological and/or landscape protection; d) restrictions due to natural hazards; e) areas reserved for agriculture; f) areas of potential instability and high visual incidence; g) no building allowed; h) aquifer protection.

case, no alteration whatsoever will be allowed in the wetlands and any intervention in other areas must include the regeneration of natural vegetation.

5) areas with building restrictions due to natural hazards (they include flooding, water stagnation, different types of slope instability, cavity collapse, coastal erosion): parts of these areas are reserved for public parks and recreation. In the rest, building will only be permitted if the project includes adequate safeguards against existing hazards.

Table 1   Activities allowed in areas affected by different types of restrictions.

Activities / Planning Categories	ecological	landscape	agriculture	aquifer	hazards	other(human)
agriculture	-	=	+	=	=	+
greenhouses	-	-	+	=	=	+
intensive dairy farming	-	-	+	=	=	+
open range dairy farming	=	=	+	=	=	+
hunting and fishing	=	-	=	=	=	+
reforestation	-	-	-	-	=	+
landscape regeneration	+	+	=	+	+	=
recreation-education	-	-	=	=	=	=
camping	-	-	-	-	-	=
solid waste disposal	-	-	-	-	=	=
isolated houses	-	-	=	=	=	=
mining and quarrying	-	-	-	-	-	-
industry	-	-	=	=	-	=

+ Recommended uses; = Permitted, with some restrictions; - Not allowed

6) areas reserved for agriculture (they correspond to high productivity soils): no constructions are allowed, except those directly related with agricultural activities.

7) areas with restrictions due to the protection of groundwater (they are the recharge areas of karstic aquifers): constructions will only be allowed if a totally impervious sewage system is included. Waste disposal of any kind and use of agrochemicals are not allowed.

8) areas with building restrictions due to slope instability and high visual incidence (corresponding to former, presently inactive cliffs): building can be allowed but a detailed project must be presented including safeguards against the potential hazard and providing measures to eliminate visual impacts.

Table 1 summarizes the kinds of activities which, according to the plan, are allowed in areas with different types of restrictions.

So, in this case the procedure followed started with the identification and mapping of homogeneous units, followed by their description in terms of variables and qualities significant for planning (Cendrero and Trilla, 1983, Lüttig, 1987). Those qualities were used for the evaluation of units and for their transformation into planning categories. Finally, the boundaries of the units were adapted to clearly visible limits of easy incorporation into the master plan, and the planning regulations were established.

CONCLUSIONS

The procedure described enables the integration of different environmental features for the preparation of geoscientific maps, representing homogeneous units, and the transformation of these units into different categories which can be directly incorporated into various types of plans.

The method proposed allows the detailed definition and evaluation of units from a geoscientific and environmental point of view and provides the means to translate this information into maps which can be used by planners to establish land-use guidelines and regulations.

REFERENCES

Brown, L.F., Fisher, W.L., Erxleben, A.W. and McGowen, J.M. (1971) Resource capability units; their utility in land and water use management, with examples from the Texas Coastal Zone. Geol. Circ. 71-1, Bur. Econ. Geol., Univ. of Texas at Austin.

Cendrero, A. (1975) Environmental geology of the Santander Bay Area, northern Spain. Env. Geol., 1, 97-114.

Cendrero, A. (1989) Mapping and evaluation of coastal areas for planning. Ocean and Shoreline Mgt., 12(5-6), 427-462.

Cendrero, A. and Díaz de Terán, J.R. (1987) The environmental map system of the University of Cantabria, Spain. In: P. Arndt and G. Lüttig, eds., Mineral resources extraction, environmental protection and land-use planning in the industrial and developing countries. E. Schweizerbart Verlag, Stuttgart, 149-181.

Cendrero, A. and Trilla, J. (1983) La geología ambiental en la evaluación del territorio para usos agrícolas. II Reunión Nac. Geol. Ambiental y Ord. Territorio, Vol. Ponencias, Lérida, 11-57.

Christian, C.S. and Stewart, G.A. (1968) Methodology of integrated surveys. In: Aerial surveys and integrated studies, UNESCO, Nat. Resources Research, Paris.

Fisher, W.L., McGowen, J.M., Brown, L.F. and Groat, C.G. (1972) Environmental geologic atlas of the Texas Coastal Zone: Galveston-Houston Area. Bur. Econ. Geol., Univ. of Texas at Austin.

Gómez Orea, D. (1978) El medio físico y la planificación. Vols. I-II, Cuadernos del CIFCA, Madrid.

Lüttig, G. (1972) Naturraumliches Potential, I,II und III. Niedersachsen Industrieland mit Zukunft: 9-10.Nds. Min. Wirtschaft u. off Arbeiten, Hannover.

Lüttig, G. (1978) Geoscientific maps of the environment as an essential tool in planning. Geol. en Mijnbouw, 57(4), 527-532.

Lüttig, G. (1987) Large scale maps for detailed environmental planning. In: F.Ch. Wolff, ed., Geology for environmental planning. Special Publ., 2, Geol. Survey of Norway, Trondheim, 71-76.

Enseignements tirés de la comparaison des données des cartes géotechniques et d'essais postérieurs à l'exécution des cartes
Comparison of geotechnical maps with more recent data

E. H. G. Goelen – *Commission de Cartographie Géotechnique*

J. P. Dam – *Centre de Cartographie Géotechnique interuniversitaire de Bruxelles, Belgium*

S. Ghiste – *Centre de Cartographie Géotechnique de l'Université Catholique de Louvain, Belgium*

Cl. Polo Chiapolini – *Centre de Cartographie Géotechnique de l'Université de Liège, Belgium*

I. Bolle – *Centrum voor Grondmechanische Kartering van de Rijksuniversiteit Gent, Belgium*

RESUME : Les auteurs ont choisi quelques exemples typiques pour illustrer dans quelle mesure ou avec quelle précision les cartes géotechniques fournissent les données recherchées quant aux caractéristiques du sous-sol. Sous forme d'une courte synthèse ils présentent les résultats d'une comparaison portant sur un grand nombre d'informations réparties sur une aire de 120 km^2.

ABSTRACT : The authors chosed some typical examples to illustrate in what measure and with what precision the geotechnical maps give the sought data of the sub-soil characteristics and of the stratigraphy. In the form of a short synthesis they present the results of a comparison made for a great number of informations spread on an area of 120 km^2.

INTRODUCTION : Lorsqu'on envisage de comparer les données tirées de cartes géotechniques et les résultats des reconnaissances postérieures à l'élaboration de la carte on pourrait songer d'emblée à un traitement statistique. La plupart du temps et sauf cas particulier, on ne dispose cependant pas d'un nombre suffisant d'informations pour valider cette approche de la question.

Toutefois, la comparaison effectuée pour quelques sites où bon nombre d'essais ont été exécutés après l'établissement de la carte permet une certaine évaluation de la qualité de cette dernière.

Si dans la plaine ou dans les formations subhorizontales il ne faut guère s'attendre à des écarts importants entre les zones cartographiées et des renseignements postérieurs nouveaux, il est des cas où d'emblée et dès l'élaboration on sait que la carte ne peut donner que le contexte géologique général sans rendre compte de tous les détails lithostratigraphiques. C'est le cas notamment des amoncellements de colluvium en pied de versants et des substratums plus ou moins altérés.

Mais, même dans des structures géologiquement simples il n'est pas non plus toujours possible de disposer des données nécessaires pour rendre compte en détail de toutes les particularités. C'est le cas notamment des criques, des dépôts alluvionnaires, d'anciens bras de cours d'eau remblayés, de modifications résultant de l'intervention de l'homme.

La carte a donc avant tout un but prévisionnel. Elle a pour objet de faire connaître la lithologie et la stratigraphie, mais aussi d'attirer l'attention sur l'existence de caractères spécifiques - colluvium, substrats altérés... ou d'anomalies - criques, anciens bras de cours d'eau qu'il n'est pas possible de reproduire dans tous leurs détails dans une carte géotechnique.

C'est donc dans ces cas d'espèce que les auteurs se sont attachés à montrer comment se compare la prévision - la carte géotechnique - et la réalité - reconnue en un ou plusieurs points ponctuels mais toujours postérieurs à la carte.

Cartographie dans la région de Bruxelles.

Dans la région de Bruxelles deux cas spécifiques ont été considérés (fig. 1). Le premier est situé dans la vallée de la Senne (coupe A-B, fig. 2). La succession des formations d'origine alluvionnaire est subhorizontale et présente une continuité assez homogène. Le second est situé avenue Louise dans un secteur où un ancien vallon,

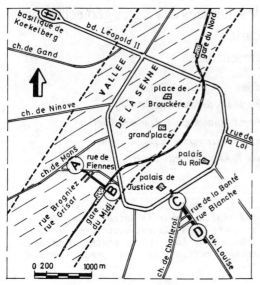

fig. 1 : Bruxelles : Plan de situation
des coupes.

tête d'un affluent secondaire de la Senne,
existait et fut comblé avant l'élaboration
de la carte (coupe C-D, fig. 3).

Dans les deux cas les données venues à
être disponibles postérieurement à
l'élaboration de la carte (fig. 2a et 3a)
proviennent des études effectuées pour des
travaux d'infrastructure importants, en
l'occurence les études préliminaires à la
réalisation du métro. Celles-ci nécessitent
un grand nombre de forages et d'essais de
pénétration de sorte que la coupe qui en
résulte présente une bonne précision.

Il est évident que les tracés des
différentes planches qui composent l'atlas
de la carte géotechnique résultent par
contre de l'interprétation de forages et
d'essais plus ou moins nombreux et
inégalement répartis et qu'ils présentent
dès lors par la force des choses une
fiabilité moindre.

Dans le cas de la première coupe (fig.
2b) considérée, le nombre de données
initiales, sans être très élevé (6 forages
situés à moins de 50 m du tracé et plus ou
moins bien répartis) peut être considéré
comme suffisant compte tenu du type de zone
investiguée et de l'échelle même de la
représentation (1/5000).

Dans le cas de la seconde coupe (fig. 3b)
on ne disposait lors de l'établissement de
la carte que d'un seul forage exécuté dans
la zone. Cette seconde coupe est située
dans un secteur où un ancien vallon, tête
d'un affluent secondaire de la Senne,
existait et fut comblé avant l'élaboration
de la carte. Dans le secteur considéré les

interprétations étaient d'autant plus
délicates que seule la topographie ancienne
(1860) était disponible pour apprécier la
répartition des remblais et des limons de
couverture. Compte tenue de l'échelle de la
carte et de son aspect de synthèse
générale, ces données de base peuvent être
considérées comme satisfaisantes pour les
besoins du tracé de la carte.

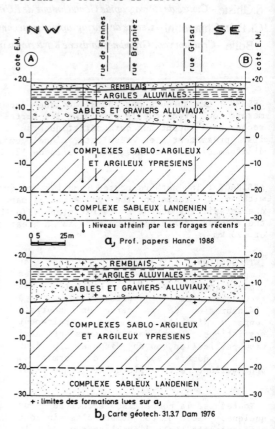

fig. 2 : Coupe dans la vallée de la Senne.

Dans les comparaisons effectuées,
deux types de formations doivent être
distingués : les formations dites "de
couverture" d'âge quaternaire, composées
soit de remblais et d'alluvions, soit de
remblais et de limons et les formations du
substratum d'âge tertiaire dont la
stratification est subhorizontale.

Comme on peut le voir, si les coupes A-B
(fig. 2a et 2b) présentent quelques
différences quant aux épaisseurs des
formations, celles-ci se situent néanmoins
autour de 2 m (équidistance des courbes de
la carte géotechnique).

Par contre, les formations de couverture
des coupes C-D (fig. 3a et 3b), présentent

des différences plus importantes. Elles résultent non seulement d'un manque de données initiales mais aussi de la présence d'un ancien vallon dont la position précise du thalweg et les pentes des versants rendent inévitablement les interprétations moins précises. L'effet de pente est évidemment prépondérant du fait qu'un léger écart horizontal peut influencer de façon importante l'appréciation de l'épaisseur ou de la profondeur.

Quelles que soient les différences qui peuvent apparaître entre les deux phases de réalisation des coupes, on doit noter, comme prévu dans la planche de zonage de l'atlas des cartes géotechniques (non figurée ici) que la succession des formations prévue a été rencontrée au cours des reconnaissances géotechniques actuelles.

On note encore que dans la coupe C-D (fig. 3), le niveau de base des sables bruxelliens est quasi identique dans les deux cas. Pour la coupe A-B (fig. 2) cette comparaison n'est pas possible car dans les forages récents la base des complexes sablo-argileux et argileux yprésiens n'a pas été atteinte.

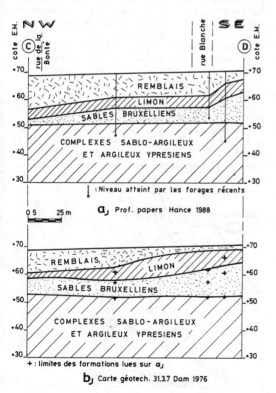

fig. 3 : Coupe avenue Louise.

Cartographie dans la région de Liège.

Dans l'agglomération liégeoise, le colluvium, les limons de pente et les cônes de déjection apparaissent en bordure de la plaine alluviale où ils peuvent atteindre des épaisseurs supérieures à 10 mètres (zones d'accumulation au pied des versants). C'est le cas notamment du cône de déjection de la Légia qui s'étend sous le Palais des Princes Evêques et sous la place Saint Lambert.

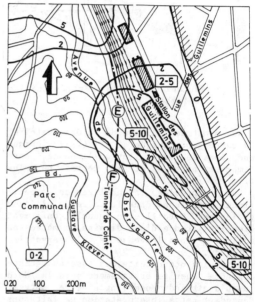

fig. 4 : Liège : Plan de situation de la coupe.

Les limons de pente recouvrent indifféremment le substratum primaire plus ou moins altéré ou les alluvions de la Meuse. Ils contiennent une nappe aquifère de versant à écoulement lent, ce qui leur confère des caractéristiques dangereuses.

Un exemple caractéristique de comparaison entre les prévisions fournies par la cartographie géotechnique et les conditions réelles rencontrées à la faveur des travaux se situe à l'entrée septentrionale (Guillemins) du tunnel de Cointe (liaison autoroutière E40-E25) (coupe E-F, fig. 4).

105

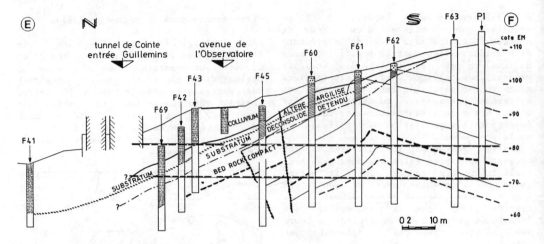

fig. 5 : Tunnel de Cointe - Entrée Guillemins. Carte géotechnique 42.6.1 -
    Calembert et al. 1977.

Au droit de l'avenue de l'Observatoire,
la carte géotechnique 42.6.1 (1977) indique
une épaisseur de colluvium/limon de pente
comprise entre 5 et 10 mètres ou plus
(fig. 5).

Les nombreuses reconnaissances spécifiques réalisées dans le cadre de l'étude
géologique détaillée du tunnel confirment
bien l'existence d'une dizaine de mètres de
colluvium sous l'avenue de l'Observatoire
(fig. 5).

La puissance déduite des essais récents
augmente ensuite brutalement vers le Nord
jusqu'à atteindre une vingtaine de mètres.
Cependant cette interprétation ne fait pas
de distinction entre le colluvium et le
sommet très argilisé et remanié du bed-rock
houiller sous jacent, assimilable dans ce
cas à une formation meuble.

A l'entrée du tunnel, la puissance du
bed-rock altéré se situe entre 0.5 et 3.5 m
tandis que le rocher déconsolidé détendu a
une épaisseur de 6 à 7 m.

Cet exemple nous semble bien illustrer le
rôle de la cartographie géotechnique qui
présente un caractère prévisionnel certain,
mais qui ne peut se substituer à une
prospection détaillée dans le cadre de
projets ou de travaux localisés.

Cartographie dans la région d'Anvers.

Dans l'agglomération anversoise deux coupes
ont été choisies en guise de comparaison
(coupes G-H et I-J, fig. 6). La coupe G-H
est située dans l'axe de la
Blancefloerlaan, recoupe l'Escaut et
s'étend jusqu'au Schoenmarkt. La coupe (I-
J) suit une partie de la Frankrijklei. Dans

les deux cas beaucoup d'observations ont
été faites après l'établissement de la
carte géotechnique 15.3.6 qui date de 1978.
Ces données proviennent d'un grand nombre
de forages et d'essais de pénétration,
exécutés dans le cadre d'études pour la
réalisation de travaux d'infrastructure.

Le long de la coupe G-H le nombre
d'observations a presque triplé tandis que
le long de la coupe I-J 24 essais ont
permis de contrôler le tracé établi sur
base de 4 observations.

La comparaison entre les tracés fait
apparaître le caractère très variable des
dépôts quaternaires et matériaux
anthropogènes et la continuité des
sédiments tertiaires d'origine marine. En
effet les incisions et la lithologie des
dépôts quaternaires varient très rapidement
et sont de ce fait très difficilement

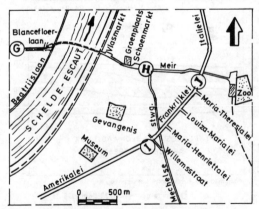

fig. 6 : Anvers : Plan de situation des
    coupes.

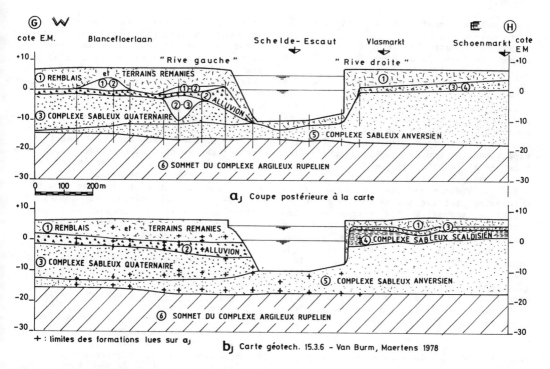

fig. 7 : Coupe Blancefloerlaan - Schoenmarkt.

prévisibles. Même avec des essais de pénétration il n'est pas aisé de faire la distinction entre dépôts quaternaires et terrains remaniés ou entre dépôts holocènes et pléistocènes.

Dans la coupe G-H (fig. 7a et 7b) le sommet de l'argile de Boom (Rupélien) est situé pratiquement au niveau prévu dans la carte. Il en est de même de la base des dépôts quaternaires.

Il est à noter que cette base présente une différence de niveau de plus de dix mètres entre la rive droite et la rive gauche, ce qui était déjà mentionné sur la carte. Les écarts principaux entre les profils des figures 7a et 7b se remarquent dans le Quaternaire où, dans la coupe postérieure à la carte on distingue des différences lithologiques non mentionnées sur la carte.

Il est possible que dans deux zones sur la rive gauche, indiquées par ① - ②, les restes d'anciennes fortifications aient été rencontrés. L'incision profonde (② - ③) remplie d'alluvions aurait pu être causée par un tourbillon à l'occasion d'une rupture de digue. Sur la rive droite la différence entre les dépôts quaternaires et les terrains remaniés est moins précise sur la carte. Il est évident que dans cette zone, urbanisée depuis des siècles, il est

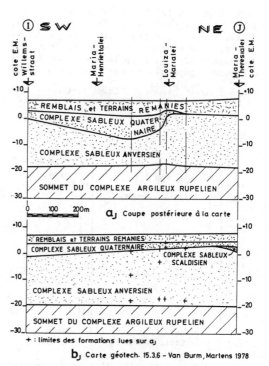

fig. 8 : Coupe Frankrijklei.

pratiquement impossible de détailler la carte davantage.

Dans le profil I-J (fig. 8a et 8b) le sommet de l'argile de Boom (Rupélien) est de nouveau au niveau prévu par la carte. La base du Quaternaire par contre s'écarte assez fortement du niveau déduit de la carte. Toutefois on rencontre la même succession, notamment les terrains remaniés, le complexe sableux quaternaire, le complexe sableux anversien et l'argile rupélienne. Le complexe sableux scaldisien, apparaît à l'extrême droite dans la carte, mais n'est pas détecté par après.

Il est à noter que dans les deux cas la succession des formations indiquée par la carte de zonage a été retrouvée. Les différences entre la carte et la réalité se situent surtout dans les terrains remaniés et quaternaires aussi bien en ce qui concerne leur extension que leur composition. Il en résulte que la carte géotechnique est un excellent document pour concevoir et élaborer le programme d'observations de terrain sur base d'une connaissance globale du sous-sol.

Cartographie dans la région de Mons.

Les cartes géotechniques de la région de Mons couvrent 120 km^2 et sont établies à partir de l'étude de 4.750 dossiers au total. Depuis leur réalisation de nombreux dossiers sont venus s'ajouter au fichier. Contrairement aux exemples précédents, la présente comparaison est établie sur un

ensemble d'essais appartenant à 145 nouveaux dossiers. La réalité des faits est comparée aux données figurant sur les cartes en vue d'une étude statistique de la fiabilité de ces documents. Parmi les critères retenus nous présentons ici ceux se rapportant à la lithostratigraphie et aux courbes isohypses des différentes formations rencontrées.

La méthode de travail a été la suivante. Chaque dossier est étudié à partir des cartes par un observateur indépendant des chantiers traités, et il établit une coupe type du site, renseignant la nature géologique, la superposition des formations et les cotes de niveau du sommet des formations (valeurs interpolées à partir des courbes isohypses). Cette coupe type est ensuite comparée à la coupe réelle transmise par le responsable du chantier. Les concordances ou les différences entre ces deux coupes sont alors analysées (fig. 9).

L'étude lithostratigraphique porte sur 319 observations dont 302 (94,7 %) se révèlent en accord avec les cartes, tant au point de vue de la nature des formations que de leur disposition. Il reste 17 observations (5,3 %) se décomposant ainsi :
- 5 observations (1.6 %) ne peuvent être prises en compte vu l'imprécision des données fournies (modifications topographiques importantes, localisation douteuse des essais, etc.) ;
- 5 observations (1,6 %) apportent une précision dans la composition des alluvions, la situation réelle étant cependant plus favorable que celle figurant sur les cartes ;
- 6 observations (1,8 %) signalent la présence de zones limoneuses dans des terrains quaternaires renseignés sur les cartes comme à majorité sableuse ;
- 1 observation (0,3 %) marque une extension des alluvions débordant légèrement de la zone cartographiée pour ces terrains.

Par conséquent seules ces 7 dernières observations peuvent présenter un désagrément pour le constructeur en cas d'un usage exclusif des cartes, ce qui est bien sûr à déconseiller.

Les courbes isohypses du sommet des formations sont cartographiées à l'équidistance de 5 m, parfois 10 m pour les formations profondes. Le critère de fiabilité retenu a été de ± 2.50 m, ce qui correspond au tracé des courbes de niveau à l'échelle retenue de 1/10.000. Sur 391 observations : 289 observations ont une précision dont l'erreur est inférieure à ± 2.50 m (soit 73,9 %) 370 ont une erreur inférieure à ± 5 m (94,6 %). Il

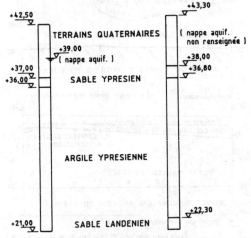

Coupe à partir de la carte

Coupe réelle
(postérieure à la carte)

+42,50

+43,30
(nappe aquif. non renseignée)

TERRAINS QUATERNAIRES

+39.00
(nappe aquif.)

+38,00
+36,80

+37,00
+36,00

SABLE YPRESIEN

ARGILE YPRESIENNE

+22,30

+21,00

SABLE LANDENIEN

fig. 9 : Mons : Exemple d'un point spécifique de la carte géotechnique 45.7.1.à4 - Ghiste 1980.

reste ainsi 21 cas (5,4 %) que l'on peut résumer ainsi :
- 7 cas concernent des conditions nouvelles depuis la réalisation des cartes : par exemple : déblais ou remblais importants, affaissements ou glissements de terrain, etc. ;
- 3 cas relèvent d'imprécisions dans les données fournies par le sondeur, par exemple : craie rencontrée entre 30 à 40 m ou socle primaire rencontré sans indiquer le degré d'altération ;
- 6 cas se rapportent à des poches de dissolution dans les craies, ce risque étant par ailleurs renseigné dans le texte explicatif accompagnant les cartes :
- 5 cas sont à ranger dans les divers : zones très dérangées en carrières, en terrils, ou zones à faible densité de renseignements.

Si nous éliminons les cas indépendants de la cartographie proprement dite, nous constatons que la marge d'erreur est de nouveau inférieure à 3 %.

Rappelons le principe de base de la cartographie géotechnique qui veut que l'utilisation de ces cartes ne dispense nullement l'utilisateur de se livrer à des études de laboratoire et de chantier au droit du site étudié.

Au vu des résultats analysés (1.268 observations sur 145 sites) et dont quelques uns sont présentés ici, il a été déterminé que statistiquement parlant, les cartes géotechniques de Mons ont une probabilité de fiabilité d'au moins 95 %, ce qui est remarquable compte tenu de la densité des informations de base.

Conclusions.

Les comparaisons effectuées entre d'une part des cartes géotechniques existantes et d'autre part des résultats d'essais postérieurs à l'élaboration de la carte ont montré que la "précision" que l'on peut attribuer à une carte n'est pas uniforme sur l'étendue des cartes mais est différente selon le contexte géologique dans lequel on se trouve.

En plaine et dans les cas de couches subhorizontales - et comme il fallait s'y attendre - une bonne précision est obtenue, nonobstant une entredistance éventuellement importante entre les points de référence ayant servi à l'établissement de la carte.

Par contre dans certaines situations - pied de colline, transition versant plaine avec présence de colluvium - des écarts peuvent être observés allant croissant avec le caractère plus complexe du cas. Mais même en ce cas les écarts observés ne relèvent pas de différences fondamentales mais bien d'une part de problèmes géométriques - échelle des cartes - précision de report, incidence même des épaisseurs des tracés - et d'autre part de la géologie elle-même - colluvium par exemple ne se distinguant guère de la roche d'origine très altérée et argilisée, - L'utilisateur doit tenir compte de l'un et de l'autre.

Les cartes géotechniques sont et restent le résultat d'interpolations. Dans le cas de la cartographie générale d'une région celle-ci reste basée pour une grosse part sur les résultats d'essais disponibles. Multiplier les essais supplémentaires pour préciser systématiquement les données sur la superficie de la carte est prohibitif. Cette multiplication n'est d'ailleurs en général pas nécessaire comme le montrent les exemples traités.

Les cartes géotechniques sont un merveilleux outil pour connaître, dès avant le début d'une étude, la constitution du sol, qui permet de guider le choix du site d'une construction et d'élaborer en connaissance de cause le programme de détail des investigations à exécuter. C'est à cet égard certes un outil de précision.

REFERENCES

CALEMBERT L. et al., 1977, Carte geotechnique de Liège 42.6.1 - Institut Géotechnique de l'Etat.

DAM J.P., 1976, Carte géotechnique de Bruxelles 31.3.7. Institut Géotechnique de l'Etat.

DEROO L., 1986, Etude statistique sur la fiabilité des cartes géotechniques de la région de Mons. Mém. fin d'études - Graduat en secrétariat technique - ISICH - Mons.

GHISTE S. et al., 1970 à 1980, Cartes géotechniques de la région de Mons - 45.3.5 à 8 - 45.4.5 à 8 - 45.7.1 à 4. Institut Géotechnique de l'Etat.

HANCE L., 1988, Géologie du Métro de Bruxelles. Professional Paper 1988/3 n° 233. Service Géologique de Belgique.

VAN BURM Ph., MAERTENS J., 1978. Carte géotechnique Antwerpen centrum 15.3.6. Rijksinstituut voor Grondmechanica.

An investigation into the origin and engineering geology of periglacial slope deposits in Wenlock Shale, Ironbridge, UK using interactive digital terrain modelling

Une étude sur l'origine et la géologie de l'ingénieur des dépôts de pentes périglaciaires dans le Wenlock Shale (Silurien), Ironbridge, UK, utilisant la modélisation de terrain par interaction digitale

T.P.Gostelow
Engineering Geology Research Group, British Geological Survey, Keyworth, Nottingham, UK

ABSTRACT: Landslip geology is represented usually as a series of two-dimensional cross sections in most publications. However, this can be misleading in complex ground conditions where the three-dimensional nature of geological boundaries may influence engineering design. Interactive digital terrain modelling can be used in these circumstances. The technique provides the geologist with isometric plots and the means of overlaying different combinations of boundary conditions in landslipped slopes. This helps to establish a) critical areas of instability b) overall three dimensional relationships of the subsurface geology and c) models for engineering design. An example of how the technique was used in a study of a periglacially disturbed valley slope in Wenlock Shale near Ironbridge UK, is given here. The slope forms the foundation to a large (300 m long, 25 m high) earthfill road embankment and the paper describes how the terrain modelling assisted in geological interpretation and led to an explanation for the origin of the deposits in terms of frost mounding.

RESUME: Dans la plupart des publications, la géologie des glissements de terrain est généralement représentéee par une série de coupes transversales à deux dimensions. Cependant, ceci n'est plus vrai pour des conditions de sol complexes où les limites géologiques naturelles à trois dimensions viennent influencer les plans d'ingénieur. En de telles circonstances, on peut utiliser la modélisation de terrain par interaction digitale. Cette technique permet des tracés isométriques aimsi que de superposer différentes limites de situations dans les pentes dues aux glissements de terrain. Ceci permet d'établir a) des zones critique d'instabilité b) les relations tri-dimensionelles de la géologie de sub-surface c) des modèles pour les plans d'ingénieur. L'exemple donné ici montre comment cette technique a été utilisée dans une étude sur l'inclinaison d'une vallée après perturbation périglaciaire dans le Wenlock Shale (Silurien) près de Ironbridge UK. L'inclinaison forme le soubassement d'un grand terrassement de remblais d'une route (300 m de long, 25 m de haut) et l'on peut voir comment la modélisation de terrain a facilité l'interprétation géologique et a permis d'expliquer l'origine des dépôts par des tertres de gelée.

1 INTRODUCTION

Two-dimensional cross sections and contoured plans are used frequently to portray engineering geological boundaries. However, they can obscure the true three-dimensional nature of the geology, particularly where boundaries lack lateral continuity, or are closely spaced. Typical situations, where this problem might arise, include rock mass assessment, estuarine or marine sediments, glacial materials, slope deposits, and landslipped ground. There are also circumstances, for example during construction, where a boundary might need to be updated, or interfaced with time-dependent data such as piezometric elevations, strain vectors, or, in the case of landsliding, movements along shear surfaces. This can be a repetitive, and time consuming process, but the development of interactive, computer based techniques which are able to combine rapidly data sets for three-dimensional analysis have provided a solution to the problem.

This paper describes an example of how computer modelling contributed to a practical study of landslips and periglacial valley slope deposits, derived from Wenlock Shale, near Ironbridge, UK. Here, a combination of Intergraph and Interactive Surface Modelling (ISM) software packages were used to investigate the geological conditions associated with landslip reactivation below a road embankment.

2 SITE LOCATION AND GENERAL ENGINEERING GEOLOGY

Figure 1 shows the Ironbridge area and the relation of the valley slope (known as Holbrook Coppice) with the River Severn. Gostelow et al. (1990) suggested that the valley sides and adjoining Ironbridge gorge (Fig. 1) were formed by river downcutting towards the end of the last, Devensian, glaciation.

The slope rises from an elevation of 75 m OD to 125 m OD and consists of Silurian Wenlock Shale (a clayey siltstone) which dips to the south east at between 5° and 14°. The crest is capped by a 20 m scarp of unconformable, faulted, Carboniferous (Lydebrook) sandstone and a thin till layer.

The shale is blue-green to grey in its unweathered state and is a closely jointed, moderately strong to strong rock with an unconfined compressive strength in the range 15 to 30 MPa. However, within 2 to 3 m of the ground surface oxidation and physical weathering have produced a firm to stiff grey-brown clayey silt of intermediate plasticity (PI 20-25%).

Frequent, thin (up to 400 mm) layers of pale grey soft, unlithified silty-clays are an important engineering feature of the formation. These follow the bedding and contain chlorites, kaolinite and variable percentages of potassium-bentonite (a mixed layer mica-montmorillonite) which has been derived from the in situ alteration of volcanic ash bands. X-ray diffraction has shown the percentages of potassium bentonite range from 15% to 40%.

Residual shear strength measurements, using a ring shear apparatus suggest the shale and clay seams have upper and lower bound residual friction angles, 'r, of c. 20° and c. 10° respectively (assuming a linear failure envelope). The low values were obtained from the bentonitic clay seams and show the formation is susceptible to landsliding, although a wide range of particle size and clay content percentages are also present (Gostelow et al. 1990).

3 SLOPE DEVELOPMENT AND MASS MOVEMENT

The Devensian Late-glacial climate encouraged erosion and weathering of the shales in the valley and they now have the typical hummocky appearance of landslipped ground. Holbrook Coppice is strongly

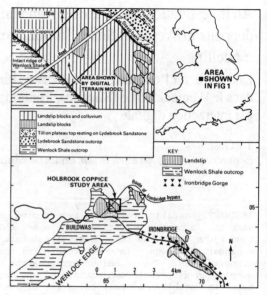

Figure 1   Location of study area

asymmetric, with average side slopes of between 7° and 9° to the south (in the direction of dip) and 20° to the north. The latter forms part of an intact ridge which separates Holbrook Coppice from the landslipped slopes next to the main River Severn floodplain (Fig. 1).

Boreholes and trial pits revealed three main groups of materials involved in down slope mass movement:

1. Blocks of Wenlock Shale which have apparently moved on bedding surfaces presumably involving the soft bentonitic clay layers.

2. Periglacial soils of solifluction origin, referred to here as head, consisting of grey clayey silts, silty clays and pebbly deposits, and reworked bentonitic clay seams.

3. Post glacial soils, referred to here as colluvium, including brown clayey silty sands with cobbles of Carboniferous sandstone and reworked bentonitic clay seams.

It became apparent during the borehole investigation that the three-dimensional relationship between these groups was complex. The head was discontinuous and layers of bentonite and shear planes were recorded within and between each group. In addition, thin layers of inorganic, laminated, water-lain deposits were found. Figure 2 is a summary of the different materials recorded in the boreholes, their sequence, and significant engineering boundaries.

112

BOREHOLE SEQUENCE

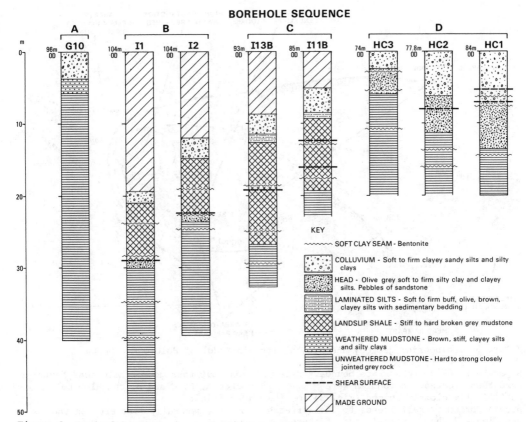

Figure 2 Typical borehole logs of slope deposits

This heterogeneity was also reflected by the hydrogeological conditions. For example, although groundwater levels were generally 2 to 3 m below ground surface, artesian pressures with piezometric head levels 2 to 3 m above ground level had developed in a small area, (Fig. 3). Groundwater discharge, or upward flow, was therefore associated with the landslipped deposits.

4 EMBANKMENT CONSTRUCTION AND SLOPE DEFORMATION

The earthfill embankment was designed as part of a road by-pass to Ironbridge and crosses the valley from the intact ridge of Wenlock Shale to the Carboniferous sandstone at the Holbrook Coppice slope crest (Fig. 1). It is approximately 300 m long and 25 m high with side slopes of 2:1 upslope and 3:1 downslope. Figure 3 shows the embankment orientation (the base outline) in relation to the valley side, suggesting that if landslide reactivation occurred, movements would tend to be across rather than downslope.

The foundation was treated with counterfort drains prior to construction and instrumented with inclinometers and pneumatic piezometers. Nevertheless, a small area developed large porewater pressures (Fig. 3). This increase was sufficient to trigger local inclinometer movements below the embankment towards the south east (Fig. 3). However, the depths at which deformation was occurring suggested movements were not taking place on a single, pre-existing shear surface which could be related easily to the known distribution of slope deposits. Interactive surface modelling techniques were therefore used to help in the interpretation of the deformation and its relation with a) the surrounding materials, b) the ground surface topography, and c) the groundwater conditions.

5 SURFACE MODELLING AND ENGINEERING GEOLOGICAL BOUNDARIES

The ground and embankment surface contours (0.5 m vertical interval) were digitised, on an Intergraph cartographic workstation.

113

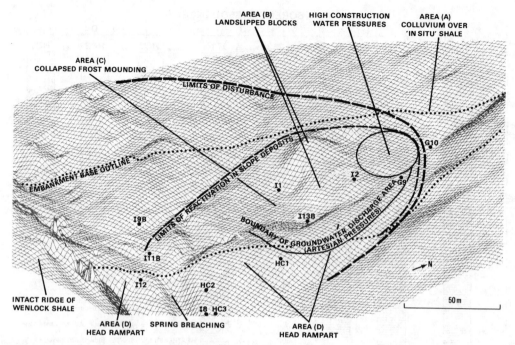

AREA (B)
LANDSLIPPED BLOCKS

HIGH CONSTRUCTION
WATER PRESSURES

AREA (A)
COLLUVIUM OVER
'IN SITU' SHALE

AREA (C)
COLLAPSED FROST MOUNDING

LIMITS OF DISTURBANCE

G10

EMBANKMENT BASE OUTLINE

LIMITS OF REACTIVATION IN SLOPE DEPOSITS

I2

I1

I9B

I13B

BOUNDARY OF GROUNDWATER DISCHARGE AREA G9
(ARTESIAN PRESSURES)

I11B

HC1

N

I12

HC2

INTACT RIDGE OF
WENLOCK SHALE

50 m

I8 HC3

AREA (D)
HEAD RAMPART

SPRING BREACHING

AREA (D)
HEAD RAMPART

Figure 3   Engineering geology and Digital Terrain Model of Holbrook Coppice

Separate files of subsurface boundary data were then created. Boreholes were sufficiently closely spaced (Fig. 3) to provide broad overall trends to the surfaces and Figure 2 shows typical logs of those examined in detail. The boundaries selected for modelling were:

1. Intact Wenlock Shale surface - unslipped ground
2. Base of Colluvium
3. Base of solifluction (head) deposits
4. The first appearance of an identified shear surface
5. The second appearance of an identified shear surface
6. The first appearance of a bentonite seam
7. The second appearance of a bentonite seam
8. The level of deformation recorded in the inclinometers
9. Piezometric levels recorded during construction

6 DIGITAL TERRAIN MODEL OF THE VALLEY SIDE

Figure 3 shows the digital terrain model (DTM) of the valley side with the embankment outline, borehole and inclinometer positions using ISM (Dynamic Graphics Ltd, 1986). The slope crest is to the right of the figure and intact ridge to the left. The 'hummocky' nature of the ground is clearly

seen, but four geomorphological features marked A, B, C and D, can also be identified:

A - A generally flat area on the upper slopes
B - Three subdued peaks to the south and west of inclinometer I2
C - A comparatively flat area between inclinometers I9B, I11B and I13B
D - A north-south trending lobe or mounded feature which starts at inclinometer I8 and extends north and south towards inclinometer I12 and borehole G9.

These features were not easily distinguishable in the field or on the conventional contoured map. However, the selected borehole logs in each of these geomorphological areas, shown on Fig 2, illustrate that the geological sequences also differ.

7 RELATION OF THE DTM TO OTHER GEOLOGICAL SURFACES

The area where inclinometers recorded shear movements extends from I8 to I2 and I1 (Fig. 3). Inclinometers I4, I11, I12 did not show significant deformation, suggesting the topographic features (B) were directly involved with landslip reactivation. Figure 4 shows the surface of the in situ shale within this area (with borehole positions and a 5 x vertical exaggeration)

114

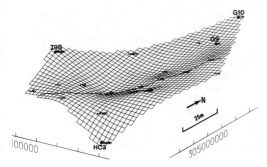

Figure 4   ISM - 'In situ' Wenlock Shale
Surface

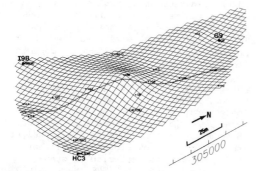

Figure 5   ISM - Surface through points on
base of Colluvium

confirming that beneath the embankment the
(B) area is underlain by a hollow filled
with slope deposits.   The hollow appears to
rise to a 'lip' beneath the topographic
mound (D) then falls away into the valley.
Figure 5 is a similar plot showing the base
of the colluvium which rises as a mounded
feature over areas (B) and (C).   Figure 6 is
a section through these two surfaces in the
direction of slope movement.   They suggest,
together with the borehole evidence, that
the hollow is filled with up to 15 m of
landslipped blocks of rock, with possibly a
thin layer of head (0.8 m) below area (B).
Figure 7 is a best fit surface through the

Figure 6   Cross section through Wenlock
Shale Surface (lower line) and Colluvium
surface

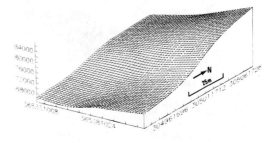

Figure 7   ISM - Surface through points on
base of 'head'

base of the head which shows it is
comparatively planar.   Therefore, although
the head may be discontinuous beneath areas
(B) (C) and (D) (Fig. 2) it suggests the
start of head formation and deposition
occurred on a uniform slope surface prior to
the landsliding in (B).
   The shape of these various surfaces can be
used in combination with topographic
features A to D and the borehole information
to develop a three dimensional picture of
the ground conditions.   The important
engineering geological features can be
summarised as:
   1. A 'hollow' within the Wenlock Shale
infilled with landslipped/broken rock,
overlain and underlain by very thin, or no
'head'.
   2. A topographic mound beyond the hollow
which consists of a pebbly 'head' deposit
and colluvium.
   3. A comparatively flat area behind the
mound underlain by colluvium, water-lain
ponded sediments, and up to 15 m of
disturbed/landslipped rock but with no
'head'.
   4. Three topographic mounds beyond the
flat area consisting of up to 8 m of
landslipped rock and c. 3 m of colluvium.
   5. A continuous layer of colluvium up to
5 m thick lying parallel to the slope
surface, but generally thickening towards
the valley.

8 RELATIONSHIP OF THE DEFORMATION PLANE TO
   THE SLOPE DEPOSITS

Figure 8 is a best fit, contoured surface
through the movement elevations identified
by the inclinometers.   It clearly shows
there are two separate planes.   The upper or
first plane approximately follows the
bedding (and dip) at about 10 m depth,
underlies the landslipped blocks in (B) but
passes through them in (C), while the second
plane is at a higher elevation (in relation
to ground level) below (D) within the head

115

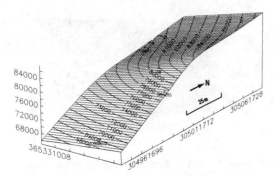

84000
80000
76000
72000
68000

Figure 8 ISM - Contoured surface through
shear - surface

layers (landslip is absent). The junction
corresponds with the edge of the embankment
and the rise or 'lip' in the in situ Wenlock
Shale surface shown in Figures 4 and 6.
Evidence discussed in more detail in
Gostelow et al. (1990) suggests that a 10 m
long rising shear surface (possibly a first
time failure) in a passive zone connects the
two planes. This zone is not shown by Fig.
8, but corresponds in position to the 'lip'
(Fig. 7) and appears to have been strongly
influenced by the shape of the in situ
Wenlock Shale surface, ie. the landslip
blocks are restrained by a passive
resistance from in situ material downslope
of area (C). However, the rockslides in (B)
were probably left in a 'loose' condition
and movements to the southeast were
sufficient to induce reactivation of shear
surfaces at higher levels in the adjacent
head deposits.

The deformation planes do not follow the
first or second appearance of a bentonite
seam or earlier shears in the boreholes and
hence their position could not be predicted
prior to construction. The best fit
surfaces through the bentonitic layers are
also complex, (eg. Fig. 9) suggesting that
most are now discontinuous. Reactivation

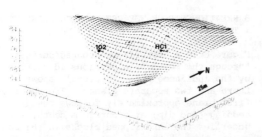

Figure 9 ISM - Surface through points on
base of 1st bentonite seam

was associated with a clay seam, below area
B, which may have been continuous but this
layer was not found at the same relative
position in the boreholes. The shear
surface was associated with the area with
high construction water pressures (Fig. 3)
and occurred at the boundary between the
hard, landslipped rock and soft clay, rather
than within a seam.

9 ORIGINS OF SLOPE DEPOSITS

9.1 General

The association of the locally derived
pebbly head deposits, landslipped shale,
water-lain sediments, high construction pore
water pressures with an area of groundwater
discharge may be more than coincidence.
Toth (1971) drew attention to the latter as
an independent geological agent, capable of
producing a wide range of features. He
pointed out that springs in a cold,
periglacial climate can give rise to frost
mounds, frost blisters and pingos.
Contemporary examples, found in northern
Canada are commonly circular to oval areas
25-40 m across, reach up to 5 m in height
and are found below slopes with either
spring discharges, artesian or standing
water. (Van Everdingen 1978, Mackay 1977 and
Pollard and French 1984). Flemal (1976) in
a worldwide review also suggested that they
are most prevalent in areas where the
substrate is wet, for example at places of
emerging groundwater. Therefore, plains,
valley bottoms, and lower valley sides are
the most common topographic environments for
the formation of frost mounds.

9.2 Mechanism of head formation and frost mounds

Pollard and French (1984) measured water
pressures in excess of hydrostatic within
active seasonal frost mounds in Canada.
This pressure was sufficient to cause doming
of soil layers, and they found that a cavity
of water developed below the frozen soil
layer. During the summer months the frozen
layer melted and the sediment collapsed,
moving down slope because of low effective
strengths. An accumulation of sediment on
the downslope sides of the mound was thus
built up.

9.3 Relict frost mounds in the UK

The distribution and form of relict frost
mounds or pingos in the UK were been

116

reviewed by Bryant and Carpenter (1987).
Most examples have been found towards the
base of slopes or on flood plains and have a
central depression surrounded by a ramparted
accumulation of solifluscted slope deposits.
These head ramparts range from 1 m to 7 m in
height and vary from between 40 m and 160 m
in length. They are the most easily
recognisable feature of relict frost mounds
and Watson (1977) suggested that underlying
slope angles tend to control their shape.
Circular types are more common on flat
ground while linear forms develop on slopes.
Flemal (1976) noted that on 'steep ground' a
ridge and furrow morphology within 'head'
also has been found.

9.4 Evidence for past frost mounding below Holbrook Coppice

The computer modelling and subsurface
investigation has identified a linear
mounded structure (D) consisting of pebbly
head, which curves into the valley bottom
(Fig. 3). The original height was probably
about 5 m (without a colluvium cover) and
length up to 125 m. Careful examination of
boreholes within this feature suggested
there were distinct layers of head which
could be distinguished by slight differences
in fabric, clay seams, or shear surfaces.
At least four such layers were identified.
The mound is thus best described as a
rampart, that has been subsequently
breached by a spring (Fig. 3). According to
Flemal (1976) this is a characteristic
feature of many ramparted structures.

A plausible explanation for the
geomorphological features at Holbrook
Coppice is that frost mounding initiated
landslipping from the upper parts of the
slope. Local steepening and the generation
of high water pressures would be encouraged
as the mound thawed and collapsed into
itself. Thus, the three subdued peaks (B)
can be interpreted as the most recent
landslipped blocks, while the flat area (C)
(which includes the laminated ponded
sediments) behind the rampart may be the
remains of earlier slides and a collapsed
mound. The break of slope midway on the
valley side (Fig. 3) probably marks the
limit of disturbed ground. The area
affected by mounding thus lies in a shallow
bowl-shaped depression of about 125 m x
150 m, which extends to the valley bottom
and includes a number of slipped Wenlock
Shale blocks, fronted by a linear rampart of
head. Figure 10 (a to d) illustrates the
sequence of slope development.

The processes controlling the slope form
prior to frost mounding are unknown and a

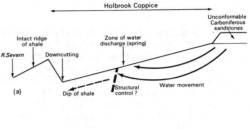

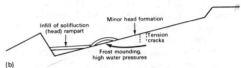

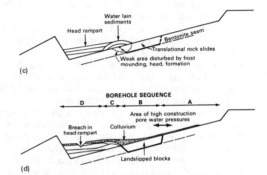

Figure 10 Inferred sequence of slope
development

uniform slope angle is assumed for the start
of mass movement (Fig. 10a). The thin layer
of disturbed head recorded below area (B)
and earlier rock slides are difficult to
explain with this geometry, but the initial
slope topography was probably steeper and
more irregular than shown, particularly
behind the disturbed mound, in area (A)
where the head is absent. This would have
allowed a greater downslope movement of
shale blocks and overiding of the head layer
in (B).

The area of current groundwater discharge
and high construction pore water pressures
lies upslope of the landslipped blocks at
the edge of area (B). The underlying
materials were not sampled in detail, and
hence the reasons for the poor drainage and
high pressures were not fully established
(Gostelow et al. 1990). It is likely that a
change of permeability caused by a slip
plane, a bentonitic layer, or faulting
'sealed' a layer of landslip from
surrounding materials. Whether this was due
to the Late Glacial processes or the
original rock structure is unknown.

10 CONCLUSIONS

The interactive computer modelling proved to be a useful technique, firstly for encouraging a selection of significant boundaries for presentation and analysis and secondly, for creating a three dimensional site model of complex geology. The geological surfaces were easily modified during the model development to ensure compatibility, and they assisted with overall interpretation of the site. Borehole data were progressively stored for printout and reference.

Vertical exageration of the DTM emphasised geomorphological features which were not easily seen in the field or on conventional contoured maps. This display led directly to an explanation of the slope deposits encountered in the boreholes.

The evidence reviewed here suggests that an area of groundwater discharge during the Late glacial led to 'open system' frost mounding, head (solifluction) formation and rock sliding. The surface expression of these features was obscured subsequently by post glacial colluvial formation.

An embankment constructed across the slope reactivated shear surfaces below the most recent of the rock slides and in the head rampart. Surface modelling of the inclinometer data in relation to the geological boundaries within the deposits indicated that movement was initiated within a bentonite seam beneath the rock slides, but was constrained by the edge of a frost mounded hollow within in situ Wenlock Shale. A passive zone formed at this position, but a slip plane at a higher elevation was reactivated within the rampart of soliflucted head beyond the rock slides.

This case record emphasises the complexity of slope deposits which can form under periglacial conditions. It also shows that slope processes and deposits may lack lateral continuity. Potential failure mechanisms are difficult to predict, identify or model in these circumstances. Therefore, engineering design of earthworks should include careful monitoring, which takes into account the three dimensional relationship of the underlying materials.

ACKNOWLEDGEMENTS. The author would like to acknowledge helpful discussions with Dr Brian Kelk, Tony Clifton, Kevin Becken and Keith Adlam of BGS, who assisted with overall interpretation and the interactive computer modelling. Published by permission of the Director of the British Geological Survey (NERC).

REFERENCES

Bryant, R H and Carpenter, C I (1987). Ramparted ground ice depressions in Britain and Ireland in periglacial processes and landforms in Britain and Ireland, edited by J Boardman. Cambridge Univerisity Press. Cambridge.

Flemal, R C (1976). Pingos and Pingo scars: their characteristics, distribution and utility in reconstructing former permafrost environments. Quaternary Research, 6, 37-53.

Gostelow, T P, Hamblin, R J O, Harris, D I and Hight, D (1990). The influence of Late and Post glacial slope development on the engineering geology of Wenlock Shale, near Ironbridge, Salop. Proc. 25th Annual Conf. Eng. Group of Geol. Soc. Edinburgh. (In press)

Mackay, J R (1977). Pulsating Pingos, Tuktoyaktuk Peninsula NWT. Can. J. Earth Sci. 14, 209-222.

Pollard, N H and French H M (1984). The groundwater hydraulics of seasonal frost mounds, North Fork Pass, Yukon Territory. Can. J. Earth Sci. 21, 1073-1081.

Toth, J (1971). Groundwater discharge: A common generator of diverse geologic and morphologic phenomena. Bull. Int. Sci. Hyd. 16, 1.3 7-24.

Van Everdingen, R O (1978). Frost mounds of Bear Rock near Fort Norman, N.W. Territories 1975-1976. Can. J. Earth Sci. 15, 263-276.

Watson, E. (1977). The periglacial environment of Great Britain during the Devensian. Phil. Trans. Roy. Soc. Lond. B, 280 183-197.

The engineering geology of the Samanalawewa dam site

La géologie de l'ingénieur du site du barrage de Samanalawewa

H.L.Gunaratna
Central Engineering Consultancy Bureau, Sri Lanka seconded to Sir Alexander Gibb & Partners Ltd

ABSTRACT: The engineering geology of the Samanalawewa dam site is presented. Pre-construction investigation results are discussed and their interpreted results are compared with the geological conditions encountered during construction. Particular reference is made to slope stability, foundation levels and curtain grouting below the main dam. The comparison shows a good correlation between data interpreted from site investigation results and final construction records.

RESUME: La géologie de l'ingénieur du site du barrage Samanalawewa est présentée. Les résultats d'investigation de la préconstruction sont discutés et comparés aux conditions géologiques recontrées pendant la construction. Le rapport se concentre sur les questions de stabilité des pentes, des niveaux de fondations du barrage et des écrans d'injections sous le barrage principal. La comparaison montre une bonne corrélation entre less résultats d'interprétation des investigations préliminaires du site et les relevés definitifs, faits durant la construction.

1.0 PROJECT DESCRIPTION

The Samanalawewa dam is part of the Samanalawewa Hydro-Electric Project and is located on the Walawe Ganga, which flows south from the Central Highlands some 160km south-east of Colombo.

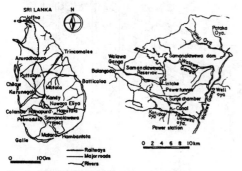

Fig. 1 LOCALITY PLAN

The project, which is currently under construction, consists of a 100m high rockfill embankment dam with a central clay core, a 5.2km long, concrete lined 4.5m diameter power tunnel, a surface steel penstock and a surface powerstation, with two 60MW units.

Due to the limited space available in this paper the Engineering Geology of the Dam site alone will be discussed.

1.1 History of the Project

A number of feasibility studies have been carried out since 1958 by the Irrigation Department and Central Engineering Consultancy Bureau of Sri Lanka in association with a number of foreign consultants as shown in Table 1.

Table 1

Date	Consultant	Refraction Seismics carried out	Total Length of Cored Holes(m)
1958-1960	Engineering Consultants Inc.	No	1710
1972-1973	Snowy Mountain Eng. Corp.	No	277
1975-1978	Technopromexport	Yes	3212
1986-1987	Nippon Koei Co. Ltd.	Yes	2464

In addition a number of review reports have been prepared since 1982.

Construction of the project started in 1986, with the award of the Tunnel Contract when funding from Sri Lanka, Japan and U.K. became available, and is due for completion in 1991.

2.0 SUMMARY OF INVESTIGATIONS AND ANTICIPATED CONDITIONS

Investigations for the Samanalawewa Project have occurred at intervals over the thirty year period 1958-1987 under the guidance of different consultants. This has resulted in a large amount of data being available but of varying quality and format. It was therefore necessary at the pre-tender stage to combine all the available site investigation data and produce a reasonable geological framework and geomechanical classification from which project design could proceed. The following section describes the geological framework and geomechanical classification that were used during design.

2.1 Geological framework

2.1.1 Regional geology

The Samanalawewa Project is located in the Highland Series of the Sri Lankan Precambrian Complex. These are a group of high grade metamorphic rocks, which in the Project Area, have been ascribed to the Kaltot Formation and occur as an open synform, plunging towards the NW.

2.1.2 Stratigraphy

From the investigation drilling and geological mapping carried out at the dam site a broad stratigraphy was recognised before construction begun and this is presented in Table 2.

2.1.3 Geological structure

The dam site is located on the north western edge of the Balangoda synform, where the foliation on the left bank dips generally 25°-35° towards the river, while on the right bank it dips 30°-60° into the hillside. These foliation angles result in an asymmetric valley with the left bank sloping at 30°-35° and the right bank sloping at 35°-45°. In addition to the fold trench a number of fault systems were identified. These are orientated NW-SE, WNW-ESE and NNE-SSW. A summary geological section along the dam axis is presented as Figure 2.

Table 2 - SAMANALAWEWA DAM SITE STRATIGRAPHY

Unit	Description	Occurrence	Thickness
Upper Granulite (GG)	Garnetiferous granulitic GNEISS in places interfoliated with amphibole and mica rich bands. Some calcareous components.	Outcrops on the upper slopes of the right bank.	>100m
Calcareous Band (CAL)	Mainly CHARNOCKITES and crystalline LIMESTONE interbanded with thin Granulitic GNEISS bands.	Exposed on the lower slopes of the right bank, in the dam foundations and at the toe of the left bank.	30m-50m
Lower Granulite (GR)	Massive to faintly gneissic garnetiferous GRANULITE. This unit is unique and was therefore chosen as a marker band.	In the valley bottom and the dam foundations.	10m-25m
Lower Charnockite (CHA)	Mainly charnockitic GNEISS interbanded with crystalline LIMESTONE and thin granulitic GNEISS bands.	Exposed in the dam foundation on the left bank and river bed. Also in the quarry site.	>100m

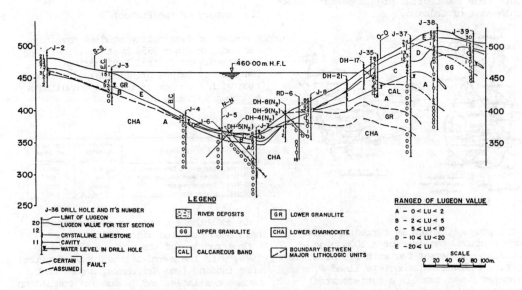

Fig. 2. GEOLOGICAL CROSS SECTION ALONG DAM AXIS SHOWING PREDICTED GEOLOGICAL CONDITIONS

TABLE 3.

INITIAL ROCK MASS CLASSIFICATION SYSTEM
(Derived from Site Investigation Data)

GEOLOGICAL CONSIDERATIONS				ROCK GRADE		GEOTECHNICAL	CLASSIFICATION	
Solid Core Recovery %	R.Q.D %	Seismic Velocity m/sec	Weathering Grade. BS 5930:1981	Original Design Grade	Strength and Deformability of Rock Mass and Discontinuities	Slope Design	Suitability of Foundations	Permeability and Groutability
90-100	75-100	>3500	Fresh to faintly weathered	A	Where discontinuity orientation is: i) Favourable: use Hoek-Brown Failure Criterion for this rock type and quality. ii) Unfavourable: analyse as rock failure on disconti-nuity planes, using $\phi=43$.	Where discontinuity orientation is: i) Favourable: 4V:1H with 2.5m wide berms at 8-10m vertical intervals. ii) Unfavourable: As for i) above but slopes at 2V:1H reinforced with 2.5m long dowels on a 2m x 2m grid.	Suitable for all foundations envisaged.	Permeability generally less than 1 Lu. Some scattered zones with 10-100Lu associated with weathered LIMESTONE and open joints. Grouting should achieve <2Lu
50-90	25-75	2000 to 3500	Slightly to moderately weathered	B	Where discontinuity orientation is: i) Favourable: use Hoek-Brown Failure Criterion for this rock type and quality. ii) Unfavourable: analyse as rock failure on disconti-nuity planes, using $\phi=38$.	As for Grade A	Suitable founda-tion for rockfill shoulders and minor structures.	Permeability generally <10Lu with scattered zones of 10-100Lu associated with weathered LIMESTONE and open joints. With the exception of weathered LIMESTONES, cement grouting should achieve 5Lu. Weathered LIMESTONES are thought to be internally erodible and will need jetting and grouting.
<50	<25	<2000	Moderately to completely weathered	C	Where discontinuity orientation is: i) Favourable: use Hoek-Brown Failure Criterion for this rock type and quality. ii) Unfavourable: analyse as rock failure on disconti-nuity planes, using $\phi=32$.	0-3m depth: slopes of 1V:2H, no support, but interception drain if long slope above. Below 3m: where discontinuity orientation is: i) Favourable: 2V:1H with 2.5m wide berms at 8-10m vertical intervals. ii) Unfavourable: As for i) above but reinforced with 2.5m long dowels on a 2m x 2m grid.	Not suitable.	Frequent zones with permeability > 100Lu. Cement grouting may achieve 10Lu. Jetting and grouting required as for Grade B.

NOTES: 1. Favourable discontinuity orientation = Failure of intact rock is required for movement to occur.
2. Unfavourable discontinuity orientation = Failure along naturally occurring discontinuities is feasible.
3. Lu = Lugeon Value.
4. Hoek-Brown Failure Criterion = Empirical Failure Criterion for Jointed Rock Masses by Hoek and Brown. (See references)

2.2 Geomechanical classification

By asessing all the available site investi-gation and laboratory test data the rock massel classification presented in Table 3 was derived for design purposes.

3.0 ENGINEERING GEOLOGY

The geological conditions at the dam site have had a significant effect in three main areas of Design as anticipated:
- Slope stability.
- Foundation quality and **levels**.
- Cut-off requirements and grouting.

During construction Table 3 has been refined and modified to suit the actual conditions encountered and the rock mass classification at present in use is presented as Table 4.

3.1 Slope stability

Initial slope designs were based on the parameters given in Table 3. During construction design parameters were modified to suit the site conditions encountered and this led to the redesign of some slopes as shown on Table 4 and described below.

(a) Left bank excavations

During excavation it was found that the foliation on the left bank dipped 15°-45° towards the river with the steeper foliation occurring above El. 430m (approx.).

A number of slope failures occurred during construction and these can in general be ascribed to

1. Steeper foliation than anticipated occurring above El. 430m.
2. The presence of continuous micaceous bands, the result of complete weather-ing of the limestones.
3. A very rapid change from completely or highly weathered rock to fresh rock.

121

TABLE 4. MODIFIED ROCK MASS CLASSIFICATION SYSTEM
(Modified during Construction)

GEOLOGICAL CONSIDERATIONS				ROCK GRADE		GEOTECHNICAL CHARACTERISTICS			
Solid Core Recovery %	R.Q.D %	Seismic Velocity m/sec	Weathering Grade. BS 5930:1981	Original Design Grade	Modified Grade	Strength of Deformability of Rock Mass and Discontinuities	General Slope Design	Suitability for Foundations	Permeability and Groutability
100	90-100	>5000	Fresh	A	A	Where discontinuity orientation is: i) Favourable: Analyse using Hoek-Brown Failure Criterion for this rock type and quality. ii) Unfavourable: Analyse as rock failure on discontinuity planes using φ = 40-50 and C = 20-50 kPa	Where discontinuity orientation is: i) Favourable: 2V:1H with 4m wide berms at 8-10m vertical intervals. ii) Unfavourable: As for i) above but reinforced with dowels at 4m centres. Length and diameter to suit slope height and geometry.	Suitable for all foundations envisaged.	Fissure flow is the main mechanism. Permeability generally <1 Lu. Highly fractured zones may have permeabilities upto 10 Lu. Cement grouting should achieve <3 Lu.
90-100	75-90	>5000	Faintly weathered	A	B				
75-90	50-75	3500 to 5000	Slightly weathered	B	CH	Where discontinuity orientation is: i) Favourable: Analyse using Hoek-Brown Failure Criterion for this rock type and quality.	As for A and B, except further protection of 100mm thick, mess reinforced shotcrete, with drainage, applied to slopes containing unfavourable discontinuities.	Suitable foundation for minor structures and core zone.	Fissure flow. Permeability 1-10 Lu. Cement grouting should achieve <3 Lu.
50-75	25-50	2000 to 3500	Moderately weathered	B/C	CM	ii) Unfavourable: Analyse as rock failure on discontinuity planes using φ = 30-40 and C = 2-20 kPa	Where discontinuity orientation is: i) Favourable: As for A and B but protected with 100mm thick, mesh reinforced shotcrete and drainage. ii) Unfavourable: 1V:1.5H with 4m berms at 8m vertical intervals. Protected with shotcrete and mesh as in i) above.	Suitable foundation for rockfill shoulders.	Fissure flow. Permeability 10-100 Lu. Increasing permeability reflects joint spacing and openness. Cement grouting should achieve <3 Lu.
25-50	0-25	1000 to 2000	Highly weathered	C	CL	Where discontinuity orientation is: i) Favourable: Analyse using Hoek-Brown Failure Criterion for this rock type and quality.	As for Grade CM, but turf used for protection instead of shotcrete.	Not suitable.	Combination of porous and fissure flow. Reflects wide range of permeability 1-100 Lu. Cement grouting may achieve 5-10 Lu. Jetting and grouting required.
0-25	0	<1000	Completely weathered	C	D	ii) Unfavourable: Analyse as rock failure on discontinuity planes using φ = 20-30 and C = 0 kPa			Predominantly porous flow. Permeability 10-100 Lu. Cement grouting may achieve 5-10 Lu. Jetting and grouting required.

Notes: 1. Favourable discontinuity orientation = Failure of intact rock is required.
2. Unfavourable discontinuity orientation = Failure along discontinuities is feasible.
3. Lu = Lugeon Value.
4. Hoek-Brown Failure Criterion = Empirical Failure Criterion for Jointed Rock Masses by Hoek and Brown. (See references)

As a result of this the upper slopes were redesigned to 1V:1.5H with 4m wide berms at 8m intervals. In highly and completely weathered ground these slopes were protected by turfing. In moderately weathered ground shotcrete, mesh and drainage were used for protection.

(b) Right bank excavations

Excavation of the very large slope between El. 463m and El. 533m showed conditions that were generally as anticipated with a favourable orientation of foliation. A small slip caused by sliding on a random joint occurred at El. 487m, as a result of this and due to the scale of this excavation immediately above the dam crest, it was decided to turf the upper slopes and protect the lower slopes from erosion by applying mesh reinforced shotcrete, with abundant drainage holes.

3.2 Dam foundations

At the design stage it was anticipated that slightly weathered rock would be generally suitable for the impervious core foundation. These conditions were anticipated as follows:

Location	Depth below ground level
Left Bank	15m to 17m
River Section	1m to 2m
Right Bank	10m to 20m

Excavation of the dam foundation has encountered conditions almost as anticipated.

3.3 Leakage cut-off requirements

3.3.1 Ground water levels

During the investigations ground water levels rather than piezometric pressures were monitored in 38 boreholes during 1976-1977 and in 31 holes during 1986-1987. These showed two distinct ground water regimes at the dam site.

(a) The left bank where water levels remained practically constant in the boreholes over the periods of monitoring. These groundwater elevations produce a smooth phreatic surface, similar in shape to the left bank topography, but varying in depth from river level at the bottom of the valley, to some 25m below ground level at El. 460m. There is no indication that there is a perched water table on the left bank.

(b) The right bank where water levels in the boreholes below surface El. 400m remained practically constant during the period of monitoring. However, above 400m, recorded water levels dropped with time. The indication were that underdrainage was occurring below the right abutment.

3.3.2 Permeability

Permeability testing was carried out to some extent during all the investigations. The most reliable data is available from the Japanese work in 1986-1987, and these results are presented in Figure 3.

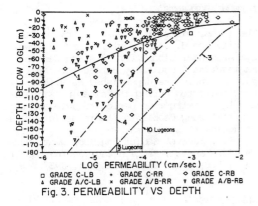

Fig. 3. PERMEABILITY VS DEPTH

KEY TO LINES
Line
1. Variation of permeability with depth-
   LEFT BANK and RIVER
2. Variation of permeability with depth-
   RIGHT BANK, Grade A and B rock.

3. Variation of permeability with depth-
   RIGHT BANK, Grade C rock.
4. Expected lowest permeability achievable in Grade A or B rock, with cement based grout. (3 Lugeons)
5. Expected lowest permeability achievable in Grade C rock, with cement based grout. (10 Lugeons)

The data has been plotted according to the location, left bank (LB), right bank (RB) or river (RR) and also according to the weathering grades (A, B and C). A number of lines have been superimposed on this plot. The intersection of lines 1 and 4 gives the depth at which, for initial grouting purposes, it is unlikely that further grouting will significantly reduce the permeability of the surrounding rock mass. At Samanalawewa, this is a depth of some 60m below original ground level, where grade A/B rock is within 25m of the ground surface. This applies to the left bank and river sections of the dam site. On the right bank, deep weathering, associated with faulting and karstic LIMESTONE features has penetrated to depths in excess of 150m below ground level. For this reason the special remedial measures of jet grouting and grouting adits in the right abutment were included in the design (see Figure 4).

It was also realised at the design stage that the assessment of watertightness at a dam site with some variably altered and solution affected limestones must continue throughout construction and even then may only be fully evaluated on first impounding. For this reason provision was maintained for the Engineer to modify the depth and layout of the grouting measures during construction as more information becomes available.

3.3.3 Cut-off requirements

From the information obtained during site investigations the grout curtain was designed on a single row basis. However due to the known different conditions on the left and right banks, different techniques were to be used for each side.

On the left bank conventional drilling and grouting, using the split spacing method was to be employed. The grout curtain is to be taken down to low permeability rock (Lugeon Value <3), which was thought to occur at depths of 40-60m. Primary holes, at 4m centres, with compulsory secondaries will be drilled and grouted in 5m stages.

On the right bank, where calcareous rocks affected by solution occur in the dam foundations more extensive measures were planned. Four grouting adits (as shown in Figure 4) have been driven into the right abutment in order to reach the calcareous areas where special treatment is necessary. In these

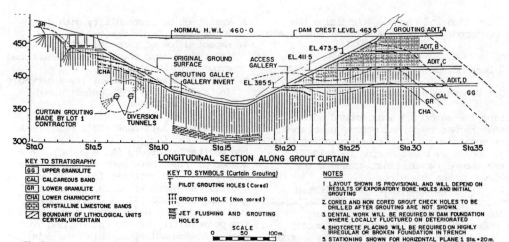

Fig. 4 SECTION ALONG GROUT CURTAIN (DAM AXIS) SHOWING GEOLOGICAL
CONDITIONS ENCOUNTERED DURING EXCAVATION

areas the single row grout curtain contains
additional investigation holes for locating
cavities. In the areas where no cavities
are encountered, conventional grouting, as
on the left bank, is anticipated. In the
areas where cavities are encountered, closely
spaced holes are to be drilled to investigate
the cavities before jet flushing and grouting,
to create either a diaphragm wall, or completely grout filled cavities. On completion
of jet flushing and grouting conventional
drilling and grouting will be required to
finish the grouting programme.

Excavation of the grouting gallery and
adits encountered conditions very much as
expected between STA 0 to STA 30 (See Figure
4). However beyond STA 30 the poor rock
conditions anticipated have continued for a
significantly greater distance than expected,
and at present further investigations are
underway to delimit this zone of poor ground
and hence the extent of the right bank cut-
off.

3.3.4 Curtain grouting

To date only 1600m of primary and secondary
curtain grouting in a 40m section of the
river bed has been completed. These have
shown very low permeabilities as expected.

In the adits where the calcareous units
have been encountered, infill material has
varied between very loose micaceous material,
which can easily be treated by jet flushing,
to loose/medium dense sandy clayey silt,
which may need other forms of treatment. To
date no treatment of the cavities encountered
has been carried out, this is programmed to
start in the next few months and will require
the careful use of the check drilling and
water pressure testing to ensure its
adequacy as a leakage cut-off.

4.0 CONCLUSIONS

This discussion shows that for detail design
consistent investigation data of high quality
is required to produce a satisfactory rock
mass assessment that will require little
modification during construction. At Samanalawewa, because of its long history of
investigations, this did not occur and therefore a number of changes have been required,
particularly in the area of slope design.
For foundation levels and the design of the
grout curtain, below the main dam, the site
investigation data has produced consistent
and accurate results, which more than
justifies the work carried out.

ACKNOWLEDGEMENTS

The author wishes to thank the Ceylon Electricity Board, Joint Venture Samanalawewa,
Sir ALexander Gibb & Partners and Central
Engineering Consultancy Bureau for their
permission to use unpublished data that has
been collected during the site investigations
and the construction period to date and also
the following people for their assistance in
producing this paper :
Mr P.O. Squire, Senior Eng.Geologist, GIBB
Mr S.Takahashi, Section Eng.,JVS/Nippon Koei
Mr V.F.Pereira, Eng. Geologist, JVS/CECB

REFERENCES

Back P.A.A and Westwell J.R.,(1988). The
design of Sri Lanka's Samanalawewa
Project. International Water Power & Dam
Construction. June, 46-50.

Balfour Beatty Ltd, GEC Energy System Ltd.,
Sir ALexander Gibb & Partners and EPD
Consultants Ltd.(1984). Technical report -
Unpublished property of CEB.

BS 5930:1981 Code of Practice.for Site
Investigations. British Standards Institute.

Electrowatt Engineering Services Ltd.(1984-
1985). Review report - Unpublished property
of CEB.

Engineering Consultants Inc., U.S.A.(1966).
Technical report - Unpublished property
of CEB.

Hoek, E and Bray, J.W.(1981). Rock Slope
Engineering. Third Edition, IMM, London.

Hoek, E and Brown, E.T.(1980). Empirical
Strength Criterion for Rock Masses. Journal
of The Geotechnical Engineering Division.
ASCE 106 GT9, 1013-1035.

Nippon Koei Ltd., Electrowatt Engineering
Services Ltd., Central Engineering Consultancy Bureau (1987). Additional Geotechnical Investigation for Samanalawewa Dam -
Unpublished property of CEB.

Nippon Koei Ltd.(1982). Reconnaissance
report - Unpublished property of CEB.

Snowy Mountain Engineering Corporation/
Mahaweli Development Board (1973). Technical report- Unpublished property of CEB.

Technopromexport,Moscow/Central Engineering
Consultancy Bureau, Sri Lanka (1978).
Detailed Project report - Unpublished
property of CEB.

Engineering geological mapping and related geotechnical problems in the wider industrial area of Thessaloniki, Greece

Cartographie géotechnique et autres problèmes géotechniques de la région industrielle des alentours de la ville de Thessalonique, Grèce

I. Hadzinakos, D. Rozos & E. Apostolidis
Institute of Geology and Mineral Exploration, Greece

ABSTRACT: The Engineering Geological Map of the wider industrial area of the town of Thessaloniki is presented and the geotechnical characteristics of the formations structuring the area are evaluated. The evaluation shows that the geotechnical problems affecting the area are mainly connected with the presence of a certain Quaternary horizon with unfavourable geomechanical behaviour which when connected with the extensive water extraction leads to the manifestation of severe subsidence phenomena.

RESUME: Se presentent ici la carte géologique technique de la région industrielle des alentours de la ville de Thessalonique et l'évaluation des characteristiques géotechniques des formation participantes à la structure de la même région. L'évaluation faite montre que les problemes géotechniques interessant la région sont lié surtout à la présence d' une certain couche Quaternaire a comportement géomechanique favotisant la manifestation de phenomenes de subsidence, à la suite d' une extraction continu d'eau.

1 INTRODUCTION

The area under study is located to the west of the town of Thessaloniki, the second largest town in Greece, which shows a continually increasing industrial development. It is an almost flat area with smooth morphology and low mean elevation (4-5m) mainly consisting by marine-lacustrine deposits. Three main rivers with an intense deposition rate, are or were converging in the area (Aliakmonas, Axios and Gallikos), causing significant changes to the morphology. Thus present morphology has only recently been developed (before the 2nd World War) with both the opening of Loudias river for the drainage of Giannitsa lake and the deviation of Axios river (Valalas 1988). The recent action of the above rivers in combination with other factors also caused significant and visible fluctuations in the seashore line during the last 40 years (Fig. 1).

Because of the above unfavourable conditions the area has only recently been developed as an industrial zone for the necessities of the town of Thessaloniki.

Therefore, the exact knowledge of the prevailing engineering geological conditions constitutes one of the main factors for the required fast and economic development of the latter.

2 GEOLOGICAL SETTING

The broader area lies in the geotectonic zone of Axios and more specifically in the Paionia sub-zone which is characterized by the presence of Mesozoic semi metamorphic to metamorphic rocks such as phyllites, gneisses, crystalline limestones and conglomerates as well as ophiolitic bodies of gabbric composition (IGME 1978).

The area under study constitutes a part of an old and extended basin, in which the rocky basement is covered by Neogene deposits and Quaternary formations, which acquire considerable thickness towards the coastal area (Fig.1).

The Neogene deposits outcrop in the north and north-eastern parts of the examined area and consist of two main horizons. Quaternary formations cover the Neogene deposits throughout the rest of the area acquiring a significant thickness towards the coast which exceeds that of 200m. The upper layers of these, mainly lagoon or deltaic loose formations, can be distinguished into three horizons, which show changes in both the horizontal and

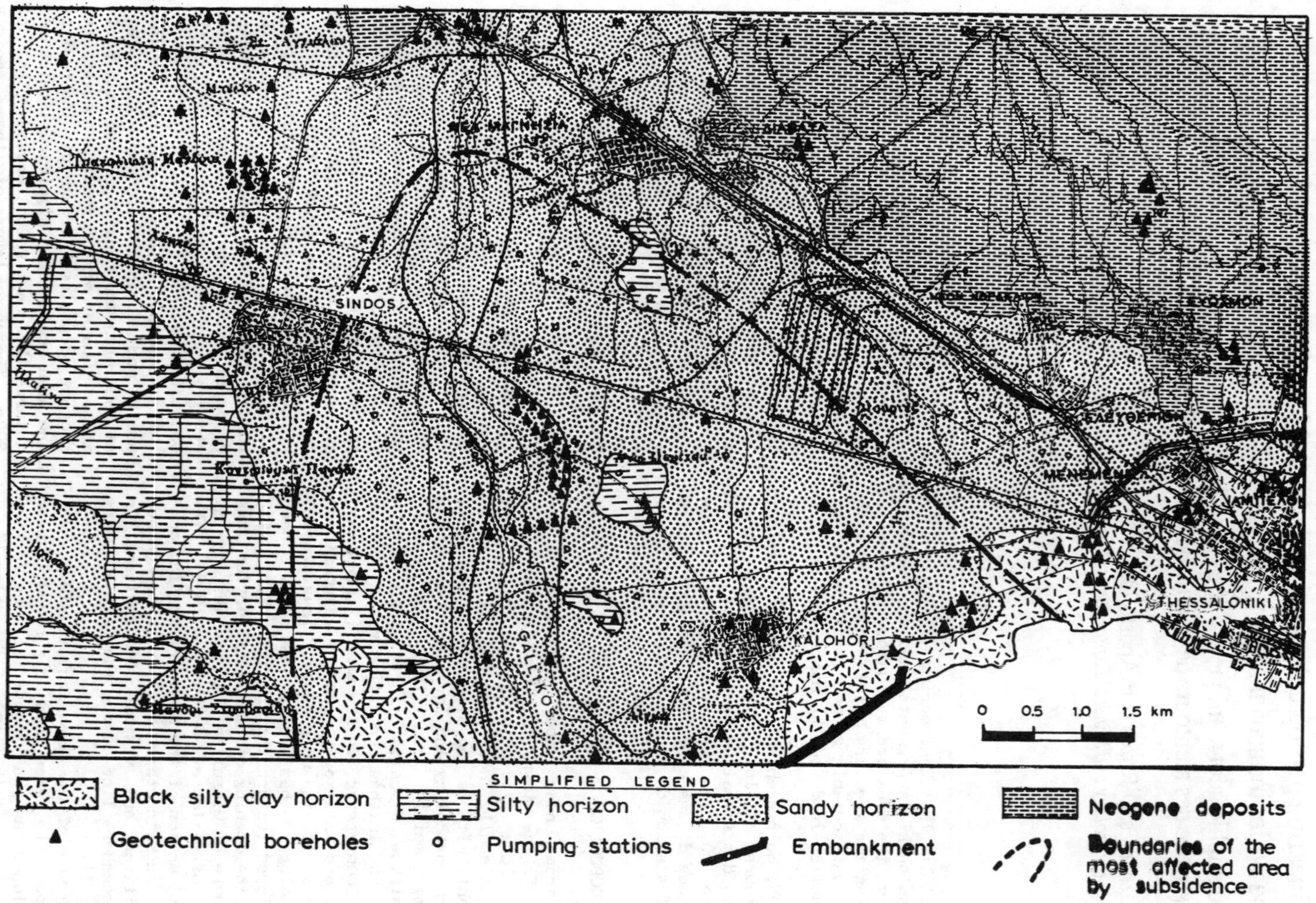
SINDOS
KALOHORI
GALLIKOS
THESSALONIKI
0 0.5 1.0 1.5 km
SIMPLIFIED LEGEND
Black silty clay horizon
Silty horizon
Sandy horizon
Neogene deposits
Geotechnical boreholes
Pumping stations
Embankment
Boundaries of the most affected area by subsidence

Figure 1. Engineering geological map of the wider industrial area of Thessaloniki vertical directions and different compositions which are attributed to the conditions, prevailing in such a changeable sedimentation environment. One of the above horizons consists mainly of sand, the other mainly of silt and the third of silty clay with thin sandy intercalations. The latter is also characterized by the presence of shell fragments mainly occurring in the sandy intercalations.

3 ENGINEERING GEOLOGICAL MAPPING

As far as the Greek territory is concerned, there is a general lack of engineering geological maps referring to wider areas and supplying useful geotechnical information (Koukis 1980). Therefore, the preparation of such a map for the industrial area of Thessaloniki will provide an extremely useful tool for land use planning purposes.

The engineering geological distinction and unification of the geological formations was made on a 1:10,000 topographical map, in accordance with the up to date international practice. Necessary adjustments for the peculiarities of the Greek territory (Koukis 1988) and especially of the area under study (geomorphological characteristics, subsidence phenomena, etc.) were also taken into account.

According to the international views and recommendations (Anon 1972, UNESCO-IAEG 1976, Dearman and Matoula 1976, Report of Commission of Eng. Geol. Mapping of IAEG 1979), the above map is characterized as a multi purpose, synoptic and large scale engineering geological map, and is shown with a simplified legend in Fig.1. In the above map the surface development of the four lithological types is given, while their analytical geotechnical description is presented below.

The data used in the description of the lithological types were obtained from a geotechnical borehole data base developed in IGME which includes boreholes executed by a number of investigators from both the public and private sectors.

3.1 Quaternary formations

They include three of the four horizons distinguished in the engineering geological map of the area, namely sandy, silty and black silty clay horizons. Grain size distribution ranges for these are shown in Fig.2, while their physical characteristics and mechanical properties values are given in Table 1.

3.1.1 Sandy horizon

The horizon acquires the largest surface development, consists mainly of fine to medium grained sand with a varying percentage of silt, clay not exceeding 20%, and an abundance of mica at places. Atter-

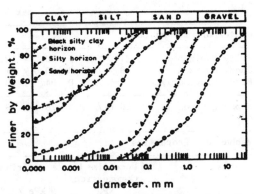

Figure 2. Particle-size limits for the three horizons of the Quaternary deposits

berg limits, and especially liquid limit, show a rather large variation, but in most cases the horizon shows no plasticity. Moisture content also varies significantly, between 3.1% and 39.1%. A barchart of the formation's US classification is shown in Fig.3.

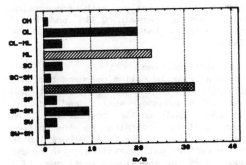

Figure 3. AUSCS barchart for the sandy horizon

Compressive strength values show a soft to stiff soil formation (B.S. 1975, Lambe and Whitman 1979, B.S. 1981) with cohesion ranging between 1 and 50 kPa and angle of friction values, with a very small variation, between 28° and 30°. Finally, the consolidation test produced

Table 1. Minimum and maximum values for the physical characteristics and mechanical properties of the four lithological types

Physical Characteristics

	Atterberg Limits WL / WP min	Atterberg Limits WL / WP max	Water Content w % min	Water Content w % max	Dry and Wet Density kN/m³ min	Dry and Wet Density kN/m³ max
Quaternary formations						
Silty Horizon	21.0 / 17.5	49.0 / 32.2	23.2	53.0	12.3 / 16.5	15.2 / 19.9
Sandy Horizon	8.7 / 11.8	39.0 / 21.4	3.1	39.1	11.8 / 14.8	18.4 / 21.7
Black Clay Horizon	17.8 / 10.0	94.8 / 38.0	4.0	69.4	8.7 / 14.8	18.2 / 23.0
Neogene Deposits	20.0 / 8.0	75.1 / 26.6	8.0	44.6	6.34 / 10.6	21.1 / 13.8

Mechanical Properties

	Unconfined Compression kPa min	Unconfined Compression kPa max	Triaxial Test kPa / ° c_u / φ_u min	Triaxial Test kPa / ° c_u / φ_u max	Consolidation Test C_c / e_o min	Consolidation Test C_c / e_o max
Quaternary formations						
Silty Horizon	17.0	25.0	5.0 / 7	80.0 / 28	0.10 / 0.677	0.47 / 1.332
Sandy Horizon	14.0	117.0	1.0 / 28	50.0 / 30	0.12 / 0.622	0.29 / 0.990
Black Clay Horizon	4.0	120.0	3.0 / 1	170.0 / 37	0.08 / 0.413	1.01 / 2.047
Neogene Deposits	44.0	1340.0	43.0 / 8	170.0 / 58	0.03 / 0.460	0.38 / 0.800

values such as those expected from this kind of soil formation, with compression index varying between 0.116 and 0.293 and void ratio between 0.622 and 0.990 (Table 1).

3.1.2 Silty horizon
It is a brown to yellow coloured horizon, with silt as the major constituent and mica in abundance, while sand and clay also participate in the composition. The classification according to the Unified System is shown in Fig.4. Atterberg limits show a rather small variation, with ranges equal to 28 for the liquid limit and 14.7 for the plastic limit. Moisture content also shows a range of 29.8, which in comparison with the other horizons, can be considered small. The above should be attributed to the rather large proportion of mica particles participating in the silt and clay fractions of the horizon.

Unconfined compression values characterize a soft to very stiff soil formation with cohesion ranging between 5 and 80 KPa and angle of friction between 7° and 28°

(Table 1). Consolidation test results show a soil formation with compression index lower than 0.470 and void's ratio ranging between 0.677 and 1.332.

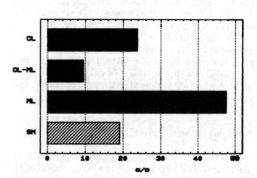

Figure 4. AUSCS barchart for the silty horizon

3.1.3 Black silty clay horizon
It consists of a black to black-grey silty

clay with an abundance of mica. Tests carried out also show the presence of organics at places. The horizon includes scattered sandy intercalations with a thickness not exceeding 1.00m. It is a loose horizon, mainly composed of silt and clay, with a restricted surface development in the examined area. However, it acquires an extended depth development since it is found at a small depth under the above mentioned horizons, with a thickness of up to 30.00m. Atterberg limits present a rather large range of values with a plasticity ranging from low to high. A barchart of the US classification is given in Fig.5. Moisture content also varies from low (sandy intercalations) to high (for the silty clay).

Taking into consideration the results of the tests for the determination of the mechanical characteristics of the horizon, uniaxial compressive strength shows that the black silty clay horizon can be characterized as a very soft to stiff soil formation.

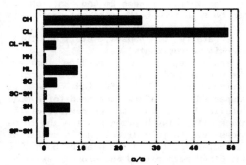

Figure 5. AUSCS barchart for the black silty clay horizon

Cohesion and angle of friction values vary significantly, depending on whether the test was carried out on the sandy intercalations (c_u=2.5-90 kPa, φ_u=5-37°) or the silty clay part of the horizon (c_u=10-170 kPa, φ_u=1-20°). Finally, consolidation tests carried out in samples from this horizon, revealed high values for the compression index (C_c=0.305-1.014). Only when the test was executed on samples from the thin sandy intercalations of this horizon lower C_c values (0.075-0.300) were obtained. These high values of the thick silty clay part indicate that special problems, connected with this part, are to be expected.

3.2 Neogene deposits

The Neogene deposits are found in the north and north-western parts of the examined area, consisting mainly of sandy silts to silty sands, with a varying small proportion of clay and scattered grits which are locally classified into layers of small thickness. Coarse-grained horizons of loose to weakly cemented conglomerate composed mainly of limestone grits and cobbles cover the above formations at places. The latter are composed by grits and cobbles, mainly of limestone origin, with calcitic to calcitic-marly cementing material. The above described neogene horizons, acquire a thickness which varies between 1.5m and 10m, and cover clayey-marly deeper sediments of the Neogene, of green-grey to brown-grey colour.

Therefore, Neogene deposits as a whole, show a widely variable composition, with clay not exceeding 25%, while the rest of the constituents vary between the grain size distribution curves shown in Fig.6. A barchart of the US classification is also given in Fig.7. Atterberg limits, as expected, show rather large variations, the ranges being 55.1 and 18.6 for the liquid and plastic limit respectively. Water content varies between 8.0% and 44.6% (Table 1).

Tests carried out for the determination of the mechanical properties (Table 1) in-

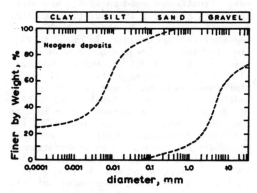

Figure 6. Particle-size limits for the Neogene deposits.

dicate a soft to hard soil or weak rock formation with cohesion ranging between 43 kPa and 170 kPa and angle of friction between 8° and 58°.

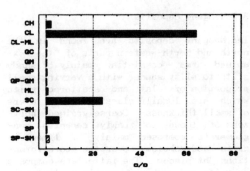

Figure 7. AUSCS barchart for the Neogene deposits

Compression index values are considered generally low ($C_c \leq 0.38$), with void ratio values ranging between 0.460 and 0.800.

4 GEOTECHNICAL PROBLEMS IN THE AREA

From the above description of the lithological types participating in the engineering geological map, is made clear that no geotechnical problems are to be expected from the presence of the sandy and silty horizons of the Quaternary formations and the presence of the Neogene deposits, which in general present satisfactory geomechanical behaviour.

However, the area is affected by the manifestation of extensive subsidence phenomena which cause serious obstacles to its industrial development. The phenomena were firstly observed in 1955, and since then were gradually increasing, resulting to sea intrusion up to the last line of Kalohori village houses. In order to face the problem, a sea embankment was constructed by the local technical authorities in the beginning of the 70's. Since then the embankment keeps increasing in height. Today the sea level is above that of the ground and a pumping system is used for the drainage of the area. The Engineering Geology Department of IGME started a programme for the examination of the phenomena in 1985 (Andronopoulos et al 199C), while other references in the frame of the need for the improvement of the area west of Thessaloniki town also exist (Andronopoulos 1979, Technical Problems in the Development of the area west of Thessaloniki 1988).

From the engineering geological mapping and the above mentioned investigation programme of IGME, certain characteristics were found to be associated with the subsidence phenomena:

1. The affected area is mainly focused in the Gallikos river and Kalohori village wider area, and extends up to the village of Sindos (Fig. 1).

2. In all the affected area the black silty clay horizon is found in a small depth from the surface, covered by either the sandy or silty horizons, and for a considerable thickness.

3. Extensive underground water pumping was carried out for both irrigation of the area and water supplying for the town of Thessaloniki and the nearby villages, in the 50's, 60's and 70's causing considerable lowering of the underground water level although the area was continuously supplied with water from the converging rives. Nowadays, water extraction is restricted to irrigation purposes and water supplying for only a small number of villages, which had as a result a certain recovery of the underground water table and a considerable reduction in the subsidence rate.

4. No differential settlements are observed in the area. Instead, the ground is uniformly subsiding and the greatest compaction varies between 2m and 3m, as it can be seen from both technical works in the area (protrusion of well casings, outside stairs of houses, etc.) and topographic measurements.

From the above, a connection between the presence of the thick black silty clay horizon and the affected area is suspected. Therefore, the need for a more detailed look into the properties of the black silty clay horizon was thought to be necessary. To this direction, a number of boreholes drilled in the area by both IGME and other authorities was closely examined up to the depth of 50m, and the following underground structure was deduced: A silty or sandy horizon with satisfactory mechanical behaviour extends up to a maximum depth of 5m, followed by the highly compressible silty clay with sand intercalations, up to the depth of 30-35m from the surface. A layer of sand with grits and good geomechanical characteristics, which acquires a maximum thickness of 14m is then observed, followed by a relatively thin layer of the black silty clay in the depth of 40-45m. Below that depth, the horizons consist of yellow to brown coloured sands to sandy and clayey silts with grits and gravels at places with a very satisfactory geomechanical behaviour. With regard to the hydrogeological conditions the above formations develop a rich aquifer, with a today's underground water level at the depth of 3-4m. During the period of heavy pumping in the area this water table had been gradually reduced to a maximum depth of 40m (according to information by the local authorities). It should be mentioned that before the above period, part of the area was flooded. This reduction of the underground water level resulted to a draining of the formations above the depth of 40m, causing them to become unsaturated or partly saturated.

Therefore, it can be concluded that most of the 3m of compaction has occurred in the top 45-50m below land surface, chiefly in two very highly compressible silty clay beds of 25-30 and 5-7m thick respectively. The dominant fraction in this horizon is clay, and is characterized by a high to very high plasticity. However, its activity value (0.74) lies between that of illite and kaolinite, which characterises a non-active material (Fig.8).

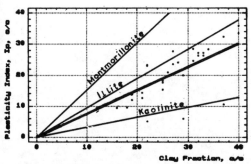

Figure 8. Relation between plasticity index and clay fraction for the black silty clay horizon (Activity 0.74).

So the high compressibility of this material should be mainly attributed to the presence of mica particles in abundance. These particles in such unconsolidated formations which were deposited in a salty and kept remaining in an aqueous environment show a scattered arrangement thus resulting in a floculent structure of the formation (e₀ up to 2.047).

Bringing the sediments in a dry or wet condition instead of a fully saturated one, during the period of heavy water extraction, caused a reduction in the volume of the above loose formations, that accompanies the process of compaction, by which particles become more closely packed and the amount of pore space is reduced (Allen 1984, Hunt 1984).

This gradual lowering of the under-ground water table was followed by a corespondent compaction of the black silty clay horizon which was expressed as a uniform subsidence in the area. Today, with the continuous decrease of water extraction rates, which started in 1980, and the subsequent ground water level rebound, it seems that the subsidence phenomena have either stabilised or eliminated, although this can only be certified by further geodetic measurements in the area.

5 CONCLUSIONS

From the engineering geological mapping in the wider industrial area of Thessaloniki, the following general remarks are drawn:

1. Four lithological types are distinguished which belong to both Quaternary formations and Neogene deposits and were drawn in the multi purpose, synoptic and large scaled map of the area.

2. Three of the above types show a satisfactory geomechanical behaviour while the fourth one is faced with serious engineering geological problems. It is a black to dark grey coloured silty clay formation which dominates the area and can be found in a small depth from the surface. High compressibility, mainly attributed to the presence of mica particles in abundance, and both low uniaxial and shear strength parameters characterize this horizon.

3. Serious subsidence phenomena affecting the area, are connected with the presence of this horizon in combination with the intense underground water extraction from the rich aquifer developed in the area. Therefore, the industrial development of the whole area requires the construction of serious technical works for the protection by sea intrusion. Such measures are concentrated to the improvement of the existing embankment and the quicker and more effective drainage of the lower parts of the area.

4. The presence of this horizon should be seriously taken into account for every construction of heavy structures, which are usual in industrialised areas. Such structures will certainly cause further uniform settlements, in their foundation area, in an already problematic engineering geological environment.

REFERENCES

Allen, A.S. (1984). Types of land subsidence. Guidebook to studies of land subsidence due to ground water withdrawal. p.133-142. Poland, J.F. (ed). UNESCO.

Andronopoulos, V. (1979). Geological and geotechnical study in the Kalohorion (Salonica) area. Engineering Geology Investigations No 10, IGME, Athens.

Andronopoulos, V., Rozos, D., Hadzinakos I. (1990). Geotechnical study of the subsidence phenomena in Kalohori area. Unpublished Report, IGME.

Anon. (1972). The preparation of maps and plans in terms of Eng. Geology. Q. Jl. Eng. Geol., Vol.5, pp 293-381.

B.S.1377. (1975). Methods of test for soils for civil engineering purposes. 143 p. British Standards Institution. England.

B.S.5930. (1981). Code of practice for Site investigation, 147 p. British standards Institution. England.

Dearman, W.R.& Matoula, M. (1976). Environmental aspects of Eng. Geological Mapping. Bulletin of I.A.E.G., Vol.14, pp 141-146.

Hunt, R.E. (1984). Geotechnical engineering investigation manual. 951pp. McGraw-Hill. USA.

IGME. (1978). Geological map of Greece. Thessaloniki sheet. Scale 1:50,000.

Koukis, G. (1980). Geological - Geotechnical maps and their use in technical works. Mining and Metallurgical Annals, τευχ.44, pp.29-40. Athens.

Koukis, G. (1988). Slope deformation phenomena related to the engineering geological conditions in Greece. Proc. of the 5th Int. symposium on Landslides, Vol. 2, pp 1187-1192, Laussanne.Balkema Publ. Rotterdam.

Matoula, M., Hrasna, H., Vleko, J. (1986). Regional Engineering geological maps for land use planning documents. Proceedings of 5th International I.A.E.G. Congress, pp 1821-1827. Buenos Aires. Balkema, Rotterdam.

Report of the Commission of engineering geology mapping of the I.A.E.G. (1979). Classification of rocks and soils for Engineering Geological Mapping. Part I: Rock and Soil Materials. Bulletin of I.A.E.G., Vol.19, pp 364-371.

Technical Problems in the Development of the area west of Thessaloniki. (1988). One day Symposium. Technical Chamber of Greece. Thessaloniki

UNESCO - IAEG. (1976). Engineering geological maps. A guide to their preparation, 79 p. The Unesco Press, Paris.

Valalas, D. (1988). An introduction to the symposium for the development of the area west of Thessaloniki. in Technical problems in the development of the area west of Thessaloniki.

Special zoning for pumped-storage plant on Plešivská planina karst plateau
Le zonage spécial pour la centrale à accumulation sur le plateau de Plešivec

R. Holzer & J. Vlčko
Comenius University, Bratislava, Czechoslovakia

ABSTRACT: The aim of the investigation for the optimum site location of individual pumped-storage plant units, comprising aerial photography, field mapping and testing, was the analysis of the regional and local karst-tectonic deformation of the rock mass which led to the preparation of special engineering geological zoning map

RESUME: Le but de l'investigation pour le site optimal de l'emplacement des centrales à accumulation, y compris les photographies aériennes, le levé de carte et l'essai sur terrain, était l'analyse de la deformation de la tectonique karstique régionale et locale du massif rocheux, conduisant à la preparation de la carte zonée spéciale géologique de l'ingénieur.

1 INTRODUCTION

The pumped-storage plants(PSP) are a very complex system of structures. The functional and technological nature of the scheme requires specific engineering geological conditions. Besides suitable topography, the siting of PSP units is closely connected with the "lithologically and structurally homogeneous and geodynamically stabilized block of the bedrock" (Matula et al. 1985).

The neotectonically uplifted karst plateaus of the Slovenský kras Mountains, situated in the southeastern part of Slovakia(Fig.1), exhibit favourable morphological conditions for the selection of several sites for the construction of the pumped-storage plant, although the karstic terrain is generally considered as an engineering hazard. The enginee-ring geological investigation of the Plešivská planina Plateau near Kunova Teplica, aimed at optimum location of a PSP, has been focused in the preliminary stage of investigation (without expensive exploratory works) on the following items:
1. Desk study and interpretation of aerial photographs
2. Field mapping and the analysis of regional and local karst-tectonic and other geodynamic phenomena
3. Field and laboratory testing
4. Preparation of a special zoning map for the optimum site location of individual PSP units.

2 LOCATION AND DESCRIPTION

The proposed Kunova Teplica pumped-storage plant consists of an impervious circular-shaped rockfill dam of the upper reservoir with a capacity of 2.6 mil. m^3, located on the Plešivská planina Plateau (836 m a.s.l.), two underground pressure tunnels, an underground power station, two 1 100 m long tailrace tunnels and the lower reservoir, situated in the Štítnik River valley, with a 12 m high earthfill dam (263 m a.s.l.) and a capacity of 2.6 mil. m^3.

Figure 1
Location map

3 SITE GEOLOGY

The geological tectonic evolution of the Plešivská planina Plateau is connected with the origin of other karst plateaus of the Slovenský kras Mts. Prealpine regional faults of N-S to NNW-SSE orientation (the Štítnik fault system), Alpine folds and overthrusts of E-W orientation are of great importance. In the late Pliocene (Rhodanian tectonic phase) uplifting and subsequent tilting to the SW took place along the Rožňava overthrust fault line (E-W to NE-SW) resulting in disintegration of the karstic paleo-relief into a mosaic of different disjunctive structures mostly of WNW-ESE orientation (the Gombasek fault system).

The Plešivská planina Plateau is built of two tectonic units: the Meliata Group and the Silica nappe. The first unit consists of slightly metamorphic, mostly pelitic rocks, outcropping in the NW part of the plateau, the latter is formed by up to 1 000 m thick strata of lower and middle Triassic carbonates underlain by a lower Triassic flyschoid complex with marly limestones, shales and sandstones (Fig.2).

All PSP units will be founded in a relatively monotonous complex of massive carbonates, which may constitute a risk in terms of tectonic karstic and tectonic joint nets and underground cavities along the faults.

The superficial sediments deposited in the Štítnik and Slaná River valleys are composed of Tertiary lacustrine gravels, sands and clays and of Quaternary alluvial sands and gravels. Residual soils of terra rossa character are exposed only on the plateau. Its slopes are covered with more or less cemented blocks of carbonates of colluvial character, exhibiting some degree of instability.

The karst landscape is characterized by a dense network of dolines and irregular sinkholes, ranging from 50 to 300 m in diameter and reaching up to 50 m in depth.

4 TECTONIC STRUCTURE AND GEODYNAMIC PHENOMENA

The interpretation of aerial photographs and field mapping has shown a distinct regular pattern of the

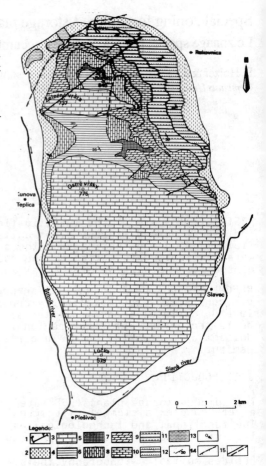

Figure 2
Geological map(after Mello,1988)
Legend:1-cross section,2-slope sediments; Middle Triassic:3-Wetterstein limestones,4-Reifling limestones,5-Schreyeralm limestones,6-Steinalm limestones,7-Guttenstein dolomites,8-Guttenstein limestones; Lower Triassic:9-marly limestones, 10-sandstones and shales,11-Meliata Group,12-dip direction/dip,13-karst springs,14-overthrust fault line, 15-faults

arrangement and forms of karst phenomena on the surface as well as in the underground.This has been caused not only by differential dissolution of carbonate rocks but also by the tectonic dissection of the rock mass. It has been assumed that

136

the most expressive tectonic lines
are controlled by the oldest overthrust lines of individual structural blocks orientated WNW to ESE and
E to W. The orientations inside
structural blocks are less expressive and the N to S orientation with
deviation to the W or E prevail.

In the basic tectonic plan of the
Plešivská planina Plateau three sets
of structural lines (exhibiting horizontal and vertical displacement)
were observed:

1. The Štítnik fault zone, striking NNW to SSE (dip direction/dip
250^0 to $260^0/60^0$) and N to S (80^0
to $90^0/85^0$ to 90^0)

2. The Gombasek fault zone, striking WNW to ESE (10^0 to $30^0/80^0$ to
85^0)

3. The Slaná fault zone, striking
NE to SW (300^0 to $330^0/80^0$ to 90^0)
and NNE to SSW (110^0 to $120^0/70^0$ to
85^0).

It is assumed that the Slaná fault
zone developed under the influence
of an important tension deformation
(no hiatus in the strata occurred),
resulting from differential vertical
movements and contemporary tilting
of the study area.

Associated karst phenomena and
their patterns in the form of "sinkhole rows" and "doline strings" are
structurally controlled and they
parallel the identified fault zones
and other important local joints.
The maximum density of karst geomorphic features, including cavities, is
the intersections of all structural
elements (Fig.3).

The "favourable" geological structure (the deformation of underlying
pelitic strata subjected to uneven
loading by carbonates), the neotectonic activity up to the late Quaternary (the release of horizontal
stresses) as well as the surface
and subsurface karst, produce a specific stress distribution in the
rock mass.

Due to the base-level cave passages a deep-seated deformation at
the western part of plateau margin
occured. An associated feature of
this phenomenon is a "sinkhole row",
reaching 100 m in width, located up
to 500 m from the plateau edges.

Other factors of instability, lining the study area, are the ope-
ning-up of near vertical joints,
weathering and slope deformations.
The deformations, predominantly in

the form of rockslips, toppling
and rock falls, are progressive e-
specially along the plateau edges
and scarps.

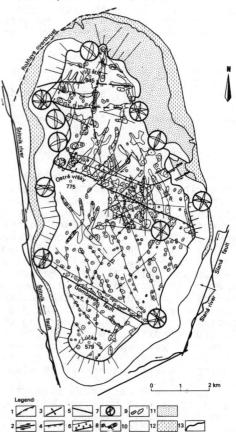

Figure 3
Tectonic map
Legend:1-overthrust zone,2-faults
with horizontal displacement,3-
faults with vertical displacement,
4-Alpine overthrust,5-local joints,
6-ruptured zones,7-stereograms,8-
karst dolines,9-sinkholes,10-car-
bonate rocks, 11-non-carbonate
rocks,12-Quaternary deposits,13-
plateau margin

5 HYDROGEOLOGICAL CONDITIONS

At this stage of investigation it
was impossible to locate the subsurface karst which is the controlling factor of hydrogeological conditions. The only thing predictable about the hydrogeological regime
is that it is unpredictable (Cul-

shaw et al.1987,says it about caves). This was also proved by tracer studies when the expected ground water flow toward SW was not confirmed.The underground flow rate from karst springs lining the plateau in soluble limestones is 8 $l.sec^{-1}.km^{-2}$(Orvan 1983).

The surface runoff practically does not exist on the plateau because most of the precipitation water percolates rapidly along the tectonic-karstic joints or sinkholes into the subsurface karst system.

6 ENGINEERING GEOLOGICAL ZONING

To present comprehensive detailed information on engineering geological conditions of a proposed PSP construction site,three groups of relevant geofactors have been selected then subdivided into several groups (Table 1 to 3) and finally divided into three categories of suitability. On the basis of these data,a special zoning map for optimum site selection of individual PSP units has been prepared(Fig.4). The geofactors controlling engineering geological conditions on the Plešivská planina Plateau are: the rock mass tectonic-karstic inhomogeneity, lithology, karst intensity and slope instability.

From the engineering geological point of view it has been proved that the identified fault zones and other important joints sets combined with a dense network of sinkholes produce great rock mass inhomogeneity and contribute to the foundation weakness (sinkhole cavity collapse, potential settlement, leakage paths etc.). This has been considered as the prime criterion for the delineation of zoning units on the special zoning map (the assigned symbols I to III).

The evaluation of the inhomogeneity of the rock mass (Table 1) is based on the adapted classification by Golodkovskaya, Matula and Shaumyan (1982).

Lithology reflects the relative proportions of natural solution features associated with main lithological types of the study area. Special solution rate characteristics have not been studied, and the chemical composition, namely the CaO content, has been the main criterion for the subdivision of the lithological types into three groups. The groups have been assigned symbols A,B,C,D, E (Table 2).

The karst intensity rating(Table 3) represents the ratio of karst area to the total area in a given region and is expressed by the amount of karst phenomena per unit area in %. The symbols assigned are 1, 2, 3.

Slope instability phenomena have not been ranked because of lack of exact data, nevertheless their presence is significantly marked in the special zoning map (Fig. 4).

According to final suitability ratings the zones assigned I A 1 and I B 1 represent suitable areas; III E 3 non-suitable areas. All other combinations of geofactors are assigned conditionally suitable areas.

7 CONCLUSIONS

The initial stage of the investigation of the designed Kunova Teplica pumped-storage plant has shown that due to the presence of the Alpine overthrust line in the E-W orientation south from the Vlčí štít Mt. (845 m a.s.l.) and the accompanied

Table 1. Rock mass tectonic - karstic inhomogeneity.

I - 1st order	II - 2nd order	III - 3rd order
Lithological-structural rock mass forming geological bodies homogeneous in lithological types; low presence of local dislocations or joint sets	Lithological-structural rock mass forming geological bodies homogeneous in lithological types; medium concentration of fault or joint zones with karst and mass movement phenomena	Rock mass with tectonic dislocations forming geological bodies in intense fault and crushed zones with a net of karst geomorphological features

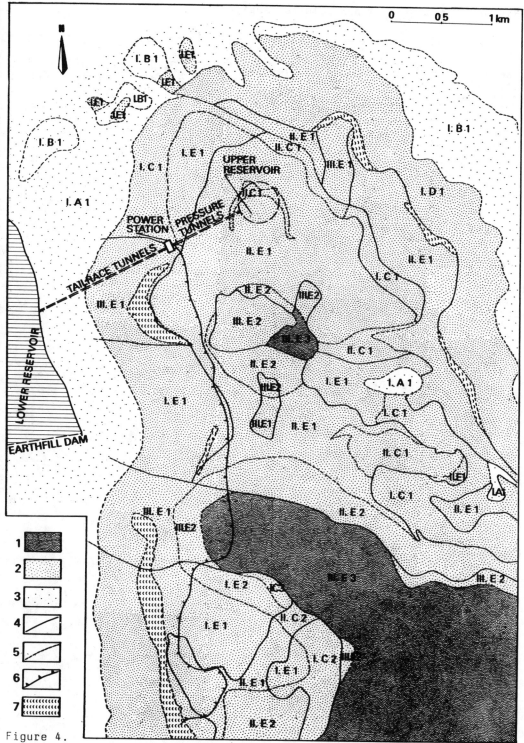

Figure 4.
Zoning map. Legend:1-non-suitable areas; 2-conditionally suitable areas; 4-zones;5-subzones;6-deep-seated deformation;7-rockslip,toppling,rockfal?

Table 2. Lithological types solution feature

A,B 1% CaO	C,D 30 to 50% CaO	E 50% CaO
Unconsolitated Quater- nary and Tertiary de- posits, pelitic rocks of the Meliata Group	Dolomites,limestones of Schreyeralm and Reifling type, flyschoid rock of the Silica nappe	Limestones of Wetter- stein, Guttenstein and Steinalm type

Table 3. Karst intensity

Symbol	1	2	3
Karst inten- sity rating(%)	10	10-20	20

crushed zone as well as the presence
of a deep-seated deformation, unexpected stresses and groundwater inflow may occur during the construction of underground power station
from zoning unit III.E1 300m north-
wards to a lower hazard zone, zoning
unit I.C1 (Fig.4).

The presented special zoning map
(as a semiquantitative, predictive
model) reflects the spatial distribution of geofactors limiting the
engineering geological conditions
for the proposed type of construction. This kind of maps have become
increasingly popular in recent years
because they provide full information
to engineering geologists as well as
non-geologists. Moreover, they represent economically efficient tools
for the early stages of the comprehensive decision-making process.

REFERENCES

Culshaw, M.G.,A.C.Waldham 1987. Natural and artificial cavities as
ground engineering hazards.Quater-
ly J. of Eng.Geology 20:139-150.
Golodkovskaya,G.A.,M.Matula,L.V.Shau-
mian 1982. Engineering-geological
classification of rock masses.
Proc. 4th Congress IAEG.Vol.II,
p.25-32,New Delhi.
Matula, M.,R.Ondrášik, R.Holzer,
A.Hyánková 1985. Regional evaluation of rock mass conditions for
pumped-storage plants. Bull. IAEG
31,Paris,p.89-94.
Mello, J. 1988. Geology of Plešivská
planina Plateau. Ochrana prírody
6, p.39 (in Slovak).

Nešvara, J. 1974. Deformation of
karst massifs in the West-Carpa-
thians. Hydrogeology and Engineering Geology. J. of Geol. Sciences,Praha, p. 177-189 (in Czech).
Orvan, J. 1983. Hydrogeological
conditions of Plešivská planina
Plateau. Research report,Liptovský
Mikuláš (in Slovak).

Application of engineering geological investigation and mapping techniques for stability assessment of an urban slope in weathered rocks in Hong Kong

L'application des techniques de recherches et de cartographie géotechniques pour l'évaluation de la stabilité d'un versant urbain de rochers dégradés à Hong Kong

T.Y. Irfan
Geotechnical Control Office, Hong Kong

A. Cipullo
Formerly of Geotechnical Control Office, Hong Kong

ABSTRACT : Site investigations for slope stability assessments in weathered rocks in urban areas such as Hong Kong require careful planning, execution and interpretation because of the complex and inhomogeneous nature of the weathering profiles and the constraints imposed by the urban development. A number of engineering geological investigation and mapping techniques specially adopted for weathered rocks were used to develop an engineering geological model of a slope adjacent to a heavily trafficked urban street in Hong Kong.

RESUME : L'etude du terrain pour l'évaluation de la stabilité des versants en rochers dégradés en milieu urbain tel que Hong Kong demande une planification, une réalisation et une interpretation soigneuses à cause de la nature complexe et inhomogène des profils de dégradation et des contraintes imposées par l'aménagement urbain. Un nombre de techniques de recherches et de cartographie géologiques d'ingénieur specialement adaptées pour les rochers dégradés ont été employées pour développer un modèle géologique d'ingénieur d'un versant voisin d'une rue urbaine ayant une circulation dense.

1 INTRODUCTION

Hong Kong has intensely developed hilly terrain with numerous high and steep cut slopes and retaining walls behind high rise buildings and adjoining heavily used urban roads and highways. In the early years of rapid post-war development in Hong Kong, little importance was attached to slope height or ground conditions. Slopes in the saprolitic and residual soil cover were constructed using a common rule of thumb angle of 73° prior to 1950 and 10 on 6 (60°) in the 1950's. From the mid-1960's onward 0.9 to 1.5 m wide berms were provided at about 7.5 m intervals, resulting in gentler overall slope angles of 50° to 55°. Slopes in the partially weathered and fresh rocks (i.e. rock slopes) were constructed as steep as possible, generally at or over 73°.

Following disastrous landslides in 1972 and 1976 during intense rainfall which resulted in heavy human losses, the Geotechnical Control Office (GCO) was established, and legislative and administrative controls were imposed. Slope design now generally conforms to the guidance given in the Geotechnical Manual for Slopes, first published in 1979. The 'Manual', which is now in its second edition (GCO 1984) describes various methods of

slope design and stability assessment, and specifies minimum factors of safety for new cut slopes and retaining walls in relation to the likely consequential loss of life or economic loss.

Hong Kong has a humid sub-tropical climate with an average annual rainfall of 2250 mm, more than 80% of which falls during the period May to September, with intensities of 50 mm per hour and 200 mm in 24 hours being common. Many landslips occur in cut slopes formed prior to 1977 and also in natural slopes during periods of exceptionally heavy rainfall.

Since 1977, the Geotechnical Control Office has been carrying out a programme of stability studies of all existing fill slopes, cut slopes and retaining walls to identify those requiring stabilization measures to bring them to current standards of safety. The slopes and walls are selected and studied in four stages on a priority basis determined by relative risk and consequence of failure (Brand 1988): preliminary stability assessment (Stage 1 Study), detailed site investigation and engineering assessment of stability (Stage 2 Study), design of preventive works (Stage 3 Study), and construction of preventive works.

Site investigation in Hong Kong prior to

the establishment of the GCO was not of a very high standard. Since the publication of the Geotechnical Manual for Slopes which provides guidance on the investigation, design, construction and maintenance of slopes and site formation works, there has been significant improvements in the site investigation practice. The current site investigation practice including field and laboratory testing closely follows British Standards BS 5930 (BSI 1981). A geoguide : A Guide to Site Investigation (GCO 1987), based on BS 5930 but with amendments adopted for local conditions presents a recommended standard of good practice for site investigation in Hong Kong.

2 SITE INVESTIGATION TECHNIQUES FOR SLOPE STABILITY STUDIES

A wide range of surface and subsurface investigation techniques are available (Clayton et al 1982, IAEG 1981, BSI 1981). The choice of the most suitable techniques depends upon the character and variability of geological conditions and groundwater, and the technical requirements as well as the site physical characteristics, availability of equipment and personnel, the cost of the methods and the amount of existing information.

Site investigation methods can be grouped under the following headings :
  (a) Desk study of existing records
  (b) Aerial photographic interpretation and remote sensing images
  (c) Geological mapping
  (d) Engineering geological mapping
  (e) Geophysical investigation methods
  (f) Boreholes and excavations
  (g) Laboratory testing
  (h) Field testing and instrumentation
(a) to (f) are required to determine the nature of materials of which ground is composed and their distribution within the volume of ground that will be influenced by, or influence, the engineering work.

The geological data of engineering significance, including those geological features of rocks and soils which are closely related to engineering properties such as strength, deformability and permeability, can be represented on engineering geological maps. The type and scale of mapping depends on engineering requirement, the complexity of the geology, and the staff and time available. The basis and techniques of engineering geological mapping and the types of maps which can be prepared are described in the authorative document published by the UNESCO (1976).

Engineering geological mapping is an integral part of a site investigation,

usually carried out prior to the start of subsurface investigation to permit efficient planning and the anticipation of engineering problems. Engineering geological mapping is also carried out during construction to record the actual ground conditions exposed, which may be of use later if design modifications are made to the engineering works, or if there are claims based on actual ground conditions, or in the design of remedial works following construction.

2.1 Slope stability assessments in weathered rocks

In slope stability assessments, it is important to develop a reasonably accurate three dimensional site engineering geological and groundwater model and to determine the engineering properties which are critical to stability, as well as the potential mode and mechanism of failure. Adequate engineering geological information facilitates the choice of a suitable method of analysis.

Site investigation and slope stability assessments in areas underlain by tropically weathered rocks can be complicated due to extreme variability and inhomogeneity. A whole range of materials is often present in deeply weathered rocks. Adequate means of describing and categorising the elements of the weathering profile at a site are therefore very important.

In areas of complex geology, it is seldom economical to develop sufficient data to permit a complete stability analysis. In such cases design is carried out to the best of data available economically using experience with similar materials. The design must be accompanied by careful observation during construction and redesign must be carried out if the geological conditions revealed are significantly different than those assumed on design.

2.2 Classification of the weathering profile

Numerous classifications have been used to describe and characterize the weathering profile and to grade the weathered rock for various engineering uses. In tropical countries, the emphasis has been towards the classification of the upper soil zones, i.e. residual soil and saprolite as it is these materials that are most commonly encountered in engineering works. In subtropical countries, with distinct wet and dry seasons (e.g. Hong Kong), the soil zones are generally thinner and the fresher rock zones are more often encountered in excavations.

In such regions, the behaviour of the weathered and jointed rock mass, as well as that of the soil mass needs to be known and attention has been given to the description of the full weathering profile.

A sixfold mass weathering grade classification scheme (Dearman 1976) has found favour with various international engineering bodies (e.g. IAEG, ISRM) and has been recommended for various engineering purposes (BSI 1981). A comparison of this scheme with other common zonal classification schemes is shown in Table 1. This paper follows the scheme which was successfully applied to various rock types and engineering situations in Hong Kong and U.K. by the first author (e.g. Irfan and Powell 1985). The term saprolite (Sowers 1963) has also been used as a general term to describe the soil zones still retaining evidence of the original rock texture and structure.

3 THE SLOPE AND THE GEOTECHNICAL STUDIES

The subject slope is located along an urban street which carries heavy traffic all day long in the residential Mid-levels district of Hong Kong island (Figure 1). No pedestrian pavement exists at the slope location. The slope was formed in the early 1900's to provide access between the idyllic Mid-levels and the business centre, Central.

The slope is 70 m long and 25 m high. The often abrupt change of slope gradient and the generally uneven irregular topography reflect the past instability which has resulted in the present day alternating spurs and depressions. The gradient hence varies across the slope from a steep 60° to 70°, mainly along the toe and the spur locations, to 25° to 50° at other places. A dense vegetation of trees and shrubs covers most of the slope. Two multistorey residential buildings and the western end of a private hospital are located along the crest (Figure 1). The caisson foundations of one of the piers for a flyover is situated at the toe of the northwestern end of the slope.

Preliminary investigation of the slope was carried out in 1985. In view of high risk to both life and property in the event of a failure and a calculated factor of safety lower than the recommended value of 1.2 for existing slopes, the slope was subjected to a more detailed geotechnical study to establish whether the slope was liable to become unstable during a heavy rainstorm.

An engineering geological investigation by surface and subsurface techniques was undertaken to establish an accurate model of the subsurface geology and groundwater for a detailed stability analysis. A laboratory

Table 1. Comparison of engineering classification of weathering profiles and typical slope failure modes.

IDEALIZED PROFILE	SOWERS, 1963 IGNEOUS & METAMORPHIC	DEERE & PATTON 1971 ALL ROCKS		DEARMAN 1976 * ALL ROCKS	FAILURE CHARACTERISTICS MODE	FAILURE CHARACTERISTICS TYPE
	SOIL	I RESIDUAL SOIL	IA A-HORIZONS	VI SOIL OR TRUE RESIDUAL SOIL	Soil Fabric Controlled	Erosion Circular Non-Circular
			IB B-HORIZONS			
	SAPROLITE		IC C-HORIZONS (SAPROLITE)	V COMPLETELY WEATHERED		
	PARTIALLY WEATHERED ROCK	II WEATHERED ROCK	IIA TRANSITION FROM SAPROLITE TO WEATHERED ROCK	IV HIGHLY WEATHERED	Relict Discontinuity Controlled	Planar Wedge Toppling Boulder falls
			IIB PARTLY WEATHERED ROCK	III MODERATELY WEATHERED		Rockfalls Planar Wedge Toppling and Combinations
	SOLID ROCK	III UNWEATHERED ROCK		II SLIGHTLY WEATHERED	Discontinuity Controlled	Circular (Closely Jointed)
				I FRESH ROCK		

* Rock Mass Weathering Grading Scheme. Grades do not necessarily occur in the sequence indicated. All other schemes are zonal (layer) schemes.

and field testing programme was devised to determine the properties, mainly of shear strength, of materials composing the slope.

4 ENGINEERING GEOLOGICAL INVESTIGATION

4.1 Existing design/construction records and previous investigations

Several government departments in Hong Kong possess information, in the form of report files, on design and construction of projects that were carried out under their control. The Geotechnical Control Office holds catalogues of all known retaining walls, cut slopes and some natural slopes, totalling over 8,000 in number. Since its inception in 1977, the GCO has checked the geotechnical designs for all civil engineering projects and building development regarding slopes, site formation works and deep excavations. In addition, the Geotechnical Information Unit (GIU), which is operated by the GCO, holds reports of site investigations and results from laboratory testing of soils and rocks, basic details of all landslips attended by the GCO engineers and a large amount of other information of direct relevance to site investigation.

The earliest record of the site extracted from the above sources is that of 1958 when the slope crest was investigated in connection with the foundation design for the development of two residential buildings. Only dynamic probing was carried out and these indicated the presence of a very dense "hardpan" soil horizon and reportedly "rock" at depths of between approximately 2 m and 7 m. In 1976/77, thirty three drillholes were sunk for the ground investigation of the road junction at the slope toe, eight of which were of direct relevance. A preliminary stability analysis of the slope assuming it to be formed entirely in residual soil was also carried out. Further information was obtained from five drillholes sunk as part of the investigation for a proposed private redevelopment at the slope crest and from the site investigation carried out in 1987 by the GCO on the slopes to the east of the subject slope (Figure 1).

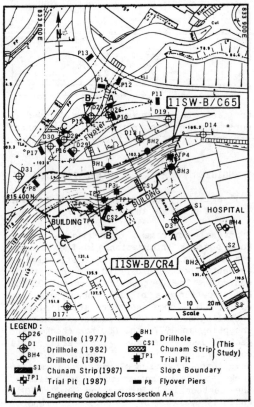

Figure 1. Site investigation plan.

4.2 Aerial photographic interpretation

Aerial and terrestrial photography and imagery from other remote sensing techniques are used in engineering geology for direct mapping as well as for obtaining supplementary information on the geology, geomorphology, hydrology and vegetation cover of a particular site. By studying sequences of aerial photographs taken at different times valuable insight may be gained into the distribution and thickness of natural and fill materials as well as instability history of a particular site.

There is an excellent aerial photographic coverage of Hong Kong in black and white photographs dating back to 1924 with annual photographic flights undertaken since 1950. It is a standard practice in the GCO to carry out an examination of all relevant aerial photographs to establish the site history of a slope in terms pre-slope formation topography, and subsequent modifications, surface water regimes, previous instability, distribution of rock-soil types and stages of urban development.

The detailed examination of aerial photographs dating back to 1924 revealed that the slope was originally a side valley to a main valley running steeply down from the Peak area. By 1960, the low-rise building at the crest area was replaced by high rise-rise residential buildings with considerable removal of mainly the colluvial material from the crest area and some filling near the crest edge. Detailed observation of past instability was not possible due to thick vegetation growth from 1950's onwards. However, high reflectance areas due to the presence of chunam cover (a cement-lime stabilised soil traditionally used in Hong Kong to protect slopes from infiltration and erosion) were seen in the photographs of 1963 and 1972 as evidence of minor landslips.

4.3 Engineering geological mapping and ground investigation

The most common direct ground investigation techniques in Hong Kong are trial pits, slope strippings and drilling. Indirect geophysical methods, either by themselves or in combination with other techniques found limited use in slope stability assessments in Hong Kong because of extreme variability of ground conditions due to weathering state of the rock mass.

The intensity and location of conventional subsurface investigation techniques were severely limited for the subject slope due to the irregular and steep nature of the terrain, the need to minimise road traffic disturbance and the limited access allowed by the owners of the residential buildings at the crest. This situation meant that emphasis had to be placed on engineering geological mapping of the slope and the adjacent areas.

4.4 Ground investigation

Trial pits and chunam strips. Hand excavated shallow trial pits are commonly used in Hong Kong to investigate the ground conditions in the upper soil zones of the weathered rocks and colluvium. They permit a full assessment of the material and mass properties of the soil to be made including measurement of relict joint orientations in saprolite and thickness of fill or colluvium deposits which often blanket slopes in urban areas. In saprolites in Hong Kong most slope failures are confined to a few metres of the ground and many are controlled directly or indirectly by relict joints (Irfan and Woods 1988). The trial pits are also useful for taking bulk, tube and undisturbed block samples and for carrying out insitu density tests.

Many soil cut slopes in Hong Kong are protected by chunam. One of the common investigation methods is to remove strips of chunam or other protective material and log the exposure.

In the investigation of the subject slope six trial pits were excavated to a maximum depth of 3.4 m in the gentler upper slope area. The maximum thickness of fill of loose to medium density was found to be about 2.5 m in the central portion with only a patchy distribution at other places. Colluvium of very bouldery composition was present only on the upper southwestern portion. The presence of boulders in this area was probably responsible for the shallow depth to "bedrock" in some of the probes sunk in connection with the crest development.

Two chunam strips located in previous landslip areas on the slope were mapped in detail. Some of the joints in the weathered granite were open indicating onset of surficial instability in the landslip areas. This conclusion was also confirmed from the surface exposures of completely and highly weathered granite in other steep middle portions of the slope.

Probing. The Standard Penetration Test (SPT) is routinely used in Hong Kong for assessment of subsurface soil conditions. The test is carried out to high blow counts, sometimes in excess of 200 although this practice is now discouraged in the recently published Geoguide (GCO 1987) limiting SPT to blow counts of maximum 100.

In this investigation, in addition to SPT carried out at 2 m intervals in drillholes, a GCO probe was used to investigate the soil characteristics near the surface. The GCO probe (GCO 1987) is generally used as a subsurface investigation tool to depths of 15 m or so. Probe results are found to be useful in assessing the depth and compaction of fill as well as for making comparative semi-quantitative assessments of ground characteristics and in supplementing the information obtained from drillholes. The depth of penetration on the slope was limited to a maximum of 1.5 m in most cases because of presence of large corestones in the weathered soil mantle and boulders in the colluvium.

Drillholes. The most common method of subsurface investigation used in Hong Kong is rotary drilling. The hand and mechanical auger boring, light cable percussion boring and wash boring methods found little use in slope investigations in Hong Kong due to :

(a) the very coarse nature of the colluvial deposits blanketing most slopes, and common occurrence of corestones in the weathered rock,

(b) the sampling purposes for testing, and

(c) the need to core into and prove rock.

Although water is the most common flushing medium, air foam flushing has been used when investigating potentially unstable slopes or when drilling on failed slopes.

For the subject slope, drilling of two of the three boreholes, which were limited by the site conditions, was carried out overnight from the road at the toe and the third on the slope crest. A number of drillhole records from past investigations in the area were available, but none were on the slope itself.

Sampling. Sampling of ground is carried out in drillholes, trial pits and other excavations in order to supplement data obtained from surface mapping and to obtain disturbed and undisturbed materials for laboratory testing. For slopes in weathered rocks, particularly where the slope behaviour is to be dominated by structure, it is important that sampling of the discontinuities be also carried out. In Hong Kong, large diameter Mazier triple tube retractor barrels (100 mm) are commonly used in conjunction with air-foam as a flushing medium to sample the soil zones (Phillipson and Chipp 1982).

The identification of individual relict discontinuities in core samples is of limited value for design of new slopes in saprolites since such important factors as persistence, waviness and orientation cannot be adequately determined. Irfan and Woods (1988) review the methods available for inspection and sampling of saprolites and recommend that trial pits or trenches, preferably excavated at different orientations to avoid directional bias in discontinuity measurements, are very useful.

For the subject slope, Mazier samples followed by SPT with liner samples were taken in the completely and highly weathered zones for the identification of soil conditions and laboratory testing. A number of block samples were also taken from the trial pits for direct shear testing.

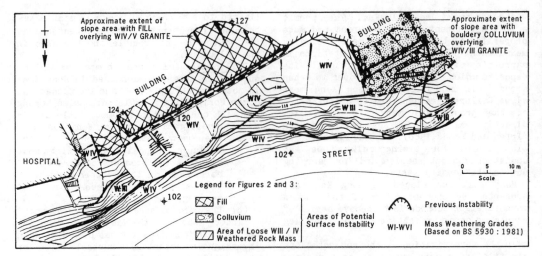

Figure 2. Mass weathering grade distribution map.

4.5 Engineering geological mapping

In Hong Kong, general purpose engineering geological maps have been produced at 1:20 000 scale for the whole Territory, but also at large scale of 1:2 500 for special areas of engineering significance as part of the Geotechnical Area Studies Programme by the GCO (Burnett and Styles 1982, Brand 1988). These maps are based entirely on aerial photograph interpretation, field reconnaissance and existing sources of information. Although useful for general development planning and broad engineering feasibility evaluations these maps are of little use in slope stability evaluations, since they are mainly based on broad generalizations of landform and geology, which require site specific information at scales of 1:1 000 or larger. Although not routinely carried out, detailed special engineering geological maps have also been produced.

Another way of representing engineering geological information for stability investigations of existing cut slopes and newly formed slopes, particularly useful for remedial works design, is a series of overlays on good quality colour or black and white photographs of the faces. Areas and types of instability, rock mass characteristics, discontinuity measurements, seepage conditions, etc. can be shown on transparent overlays (Powell and Irfan 1986). A further overlay showing remedial works can also be prepared.

An engineering geological mapping of the subject slope was undertaken to supplement the data obtained by the limited subsurface investigation. Access bamboo scaffolding was erected to cover the entire slope face,

shrubs and superficial vegetation cover were removed and mapping was carried out to assess the following :

(a) identification and distribution of surface materials and their weathering state,
(b) properties of discontinuities,
(c) type and extent of past and current instability, and possible potential modes of instability,
(d) groundwater seepage, and
(e) other features critical to stability.

A series of engineering geological maps prepared as overlays to a base topographical plan at 1:100 scale included :

(a) A Site Investigation Plan (Figure 1).
(b) A Surface Features and Materials Distribution Map (not shown).
(c) Mass Weathering Grade Distribution Map (Figure 2).
(d) Discontinuity Zoning Map (not shown).
(e) Instability Map, showing areas of past and current instability, type of instability, etc. (Figure 3).

4.6 Laboratory and field testing

Numerous types of field and laboratory testing techniques available for soils and rocks and their applicability to various ground and engineering conditions are briefly summarised in IAEG (1981). The technical papers published in Brand and Phillipson (1985) review the international practice in sampling and testing of saprolitic and residual soils in countries such as Hong Kong, United Kingdom, Australia, Brazil and South Africa.

Selection of one or many tests from a very extensive list depends upon the data

146

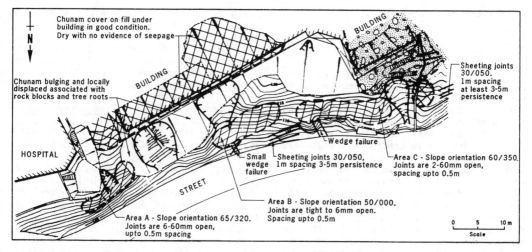

Figure 3. Instability map showing areas and type of past and current instability.

required for each project, ground conditions, type of testing equipment available, national practice and economic considerations. Field tests are generally desirable when it is considered that the mass characteristics of the ground would differ appreciably from the material characteristics determined by laboratory testing.

In Hong Kong, laboratory testing is preferred to insitu testing for the determination of strength properties of rocks and soils, although an increasing number of field tests are being carried out, particularly in connection with the determination of bearing capacity of for foundations.

The commonly carried out laboratory tests in soils are single and multistage triaxial testing on saturated specimens and the related classification tests. Direct shear tests using relatively large size 100 mm square samples are becoming more commonly used for strength assessment of coarse grained granitic soils. Direct shear technique is also used to determine strength of rock discontinuities.

The soils formed from weathering of rocks in tropical climates do not exhibit properties normally associated with sedimentary soils. The material behaviour of these soils appears to be dominated by the microfabric of the soil which is related to the original rock fabric and weathering history (Vaughan 1988, Massey et al 1989).

In rock, point load index and uniaxial compressive strength tests according to both ISRM and ASTM recommended methods are carried out for the determination of intact strength of rock and Schmidt hammer tests are used for both rapid quantitative

assessment of rock hardness and classification purposes (Irfan and Powell 1985).

Amongst the common field tests used in Hong Kong are the constant head and falling head permeability tests which are carried out in drillholes in addition to SPT probes. The Triefus borehole impression packer device is used in drillholes to determine the subsurface discontinuity orientations in fresh and weathered rocks. The application of impression packer test to determine the discontinuity characteristics of saprolitic soil zones has not been very successful (Irfan and Woods 1988). The pressuremeter has been infrequently used because of interpretation difficulties in saprolitic soils.

A programme of triaxial testing was carried out on the Mazier samples recovered from the drillholes in the slope to determine the shear strength parameters to be used in the detailed stability analysis by traditional methods (e.g. Janbu 1973), commonly applied in Hong Kong. No testing was done on colluvial or fill material because of their limited occurrence on the slope and because of their very bouldery nature. Single stage and multistage consolidated undrained triaxial compression tests with pore pressure measurements of saturated specimens were carried out in line with the common procedures for testing such soils in Hong Kong. Recently determined test results from the adjacent slopes were used in the stability analysis to supplement those determined on the materials from the slope. A discussion of the testing programme and the results of the laboratory tests and their applicability to the mass strength of the slope are outside the scope of this paper.

4.7 Groundwater monitoring

Since groundwater pressures are the dominant factor controlling stability of slopes by giving rise to hydraulic forces as well as causing reduction in shear strength, the determination of groundwater levels and pore pressures are very important in slope investigations. Pore pressures in weathered rocks are typically complex and show seasonal variations, superimposed on which are much more rapid changes taking place during or after periods of heavy rainfall. Geological features such as clay-bearing discontinuities, relatively impermeable dykes, natural pipes and zones of differing permeabilities within weathering profiles, together with the variation in permeability with degree of saturation give rise to pore pressure distributions which are very difficult to predict and to measure in the field. The state of knowledge in the prediction of pore pressures in saprolitic and residual soil slopes has been assessed as "very poor" by Brand (1985).

The monitoring of pore pressures, and their response to rainfall, is carried out routinely by using piezometers in Hong Kong. The Casagrande type standpipe piezometers are frequently used in drillholes (GCO 1987). Closely spaced Halcrow buckets are commonly installed in standpipe piezometers to measure the peak groundwater response during rainfall. Electronic pressure transducers and automatic bubbling recorders are also used to record transient water levels.

Open tube standpipe piezometers together with Halcrow buckets were installed in each of the three drillholes and monitored over one wet season to obtain information on the groundwater levels within the subject slope.

5. RESULTS : AN ENGINEERING GEOLOGICAL MODEL OF THE SLOPE

An engineering geological model of the slope encompassing the mass and material characteristics of rocks, including the discontinuity and weathering states, groundwater conditions and existing and potential modes of instability, is presented below :

Geology. The slope is dominantly composed of two types of granite weathered to various degrees; the lower part being composed of a medium-grained granite (grain size of 2 to 5 mm) and the upper parts and the western corner a finer grained granite (grain size of 1 to 3 mm). Localized variations in

grain size to the main granite types are common. Loose fill is confined to the slope crest area and has an uneven distribution. A relatively thick bouldery colluvium (up to 6 m or more), which originally blanketed most of the slope area, was mostly removed by the building developments except at the western end (Figure 2).

Weathering state of the rock mass. Surface mapping indicated that the majority of slope face is composed of highly weathered granite (WIV) with a moderately weathered granite (WIII) occurring in the mid-slope region. Completely weathered granite (WV), with a possible thin or missing overlying residual soil, is present locally below the colluvium layer. The distribution of various grades of rock within the slope is complex as revealed by a few boreholes. A simplified model is presented along a cross-section through the eastern part of the slope in Figure 4. The slightly weathered to fresh granite is well below the existing ground level. It is considered that, in addition to other factors (e.g. topography and climate), the variation in grain size of the granite had a significant effect in the complex distribution of the mass weathering grades seen at this site. Thin kaolin veins of hydrothermal origin were also encountered in the joints with some increased alteration effects adjacent to the veins. The effect of hydrothermal alteration does not appear to be as intense as observed at some other sites in Hong Kong.

Discontinuities. Five joint sets, three subvertical sets with average dips of about 75° and two subhorizontal sets with average dips of about 30°, were identified from the discontinuity measurements carried out on the slope. The joints were particularly prominent along mid-slope region with the subvertical sets being typically closely to medium spaced (less than 0.5 m) and the gentler subhorizontal sheeting joints widely spaced. Most of the subvertical joints were open near the surface, particularly along the mid-slope and upper slope regions including the existing landslip scars.

Groundwater. No evidence of active or previous seepage signs was observed during mapping of the slope in September and October. Groundwater monitoring records derived from piezometers and buckets installed for the study and from the previous investigations indicated that the base groundwater table was at a significant depth below the surface (Figure 4). Persistent perched water tables were not evidenced. The colluvium layer, which usually develops a perched water table, underlying the paved area is mostly isolated as a result of platform formation thus reducing the possibility of groundwater migration from the upper areas. Infiltration from the upslope platform would in any case be minimal as it is mostly paved with concrete.

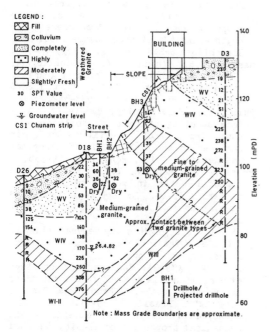

Figure 4. Engineering geological cross-section A-A.

Instability. Areas of previous instability were clearly identifiable on the steep slope face (Figure 3). The loose rock mass with open joints in the landslip scars in the Grade WIII and WIV weathered granite areas indicated that the failures were at least partly discontinuity controlled. In the flanking areas adjacent to the slip scars, bulging of chunam and opening up of joints, partly associated with the wedging action of tree roots and small rock falls, pointed to a continuous deterioration of the slope face and subsequent potential failures. The previously failed area (Figure 3) which involved both the weathered granite and the overlying colluvium was also considered to be joint controlled. The failure which was probably initiated in the highly weathered granite resulted in undermining and eventually collapse of the overlying colluvium.

A simple kinematic analysis carried out (not given) using the stereonet indicated no possibility of a large scale planar failure, but small scale wedge failures along the intersection of steep joints of two of the subvertical sets. Toppling failure of

numerous loose rock blocks bounded by subvertical joints is likely to be the main type of potential failure mode as was also observed on the slope.

6 SLOPE STABILITY ANALYSIS AND RECOMMENDATIONS

The engineering geological study concluded that conventional stability analyses normally adopted for slopes formed in saprolitic and residual soils, assuming failure through weathered material of uniform shear strength parameters (Brand 1985) are not entirely appropriate in this case. This is because most of the slope is formed in highly to moderately weathered and jointed rock mass, except for the upper portion in colluvium and fill. A more realistic and meaningful analysis should consider the shear strength of the weathered rock mass by incorporating the structural discontinuities.

Based on the results of the engineering geological investigation and the detailed stability analysis (not reported here), it was concluded that a major deep seated instability is remote, however the slope shows signs of localized joint controlled surficial instability and if left uncontrolled may result in failures.

7 CONCLUSIONS

Slope stability assessments in weathered rocks, particularly in urban areas, require careful planning, execution and interpretation of site investigation because of complexity and inhomogeneous nature of weathering profiles and the restrictions imposed by the urban development. It is important to develop a reasonably accurate three dimensional site engineering geological model and determine the engineering properties including groundwater which are critical to stability as well as the potential mode and mechanism of failure. This can be done by surface and subsurface investigation methods including mapping and laboratory and field testing specially adopted for weathered rock conditions.

A detailed engineering geological investigation was undertaken as a part of the Geotechnical Control Office's on-going programme of geotechnical studies on a slope in weathered granite adjacent to a heavily trafficked urban street. The investigation concluded that the slope shows signs of past and current dominantly joint controlled surficial instability and if left unattended could result in failures which could pose a direct risk to life. It was also concluded that the detailed stability analysis should take into account the jointed nature of the weathered rocks and the saprolite composing the slope and that the conventional analysis normally adopted for slopes in such soils are not entirely appropriate in this case in view of the observed joint controlled instability.

ACKNOWLEDGEMENTS

This paper is published with the permission of the Director of Civil Engineering Services, Hong Kong Government. The specification of the ground investigation was prepared by W.K. Lai.

REFERENCES

Brand, E.W. (1985). Predicting the failure of residual soil slopes. (Theme Lecture). Proc. 11th Int. Conf. ISSMFE, San Francisco, 5, in press.

Brand, E.W. (1988). Landslide risk assessment in Hong Kong. Proc. 5th Int. Symp. on Landslides, Lausanne, 2, 1059-1074.

Brand, E.W. & Phillipson, H.B. (eds) (1985). Sampling and Testing of Residual Soils : A Review of International Practice. Scorpion Press, Hong Kong, 194 pp.

BSI (1981). Code of Practice for Site Investigations (BS 5930 : 1981). British Standards Institution, London, 147 pp.

Burnett, A.D. & Styles, K.A. (1982). An approach to urban engineering geological mapping as used in Hong Kong. Proc. 4th Int. Cong. IAEG, New Delhi, 1, 167-176.

Clayton, C.R.I., Simons, N.E. & Mathews, M.C. (1982). Site Investigation. A Handbook for Engineers. Grenada, London, 424 pp.

Dearman, W.R. (1976). Weathering classification in the characterisation of rock : a revision. Bull. IAEG, No. 13, 123-127.

Deere, D.U. & Patton,. F.D. (1971). Slope stability in residual soils. Proc. 4th Pan-Am. Conf. SMFE, Puerto Rico, 4, 87-170.

GCO (1984). Geotechnical Manual for Slopes (2nd ed.). Geotechnical Control Office, Hong Kong, 295 pp.

GCO (1987). Geoguide 2 : Guide to Site Investigation. Geotechnical Control Office, Hong Kong, 362 pp.

IAEG (1981). Report of the commission on site investigations. Bull. IAEG, No. 24, 185-226.

Irfan, T.Y. & Powell, G.E. (1985). Engineering geological investigations for pile foundations on a deeply weathered granitic rock in Hong Kong. Bull. IAEG, No. 32, 67-80.

Irfan, T.Y. & Woods, N.W. (1988). The influence of relict discontinuities on slope stability in saprolitic soils. Proc. 2nd Int. Conf. on Geomechanics in Tropical Soils, Singapore, 1, 267-276.

Janbu, N. (1973). Slope stability computations. In Embankment Dam Engineering (Casagrande Volume), ed. R.C. Hirschfield & S.J. Poulos, Wiley, 47-107.

Massey, J.B., Irfan, T.Y. & Cipullo, A. (1989). The characterization of granitic saprolitic soils. Proc. 12th ISSMFE Conf., Rio de Janeiro, 6, 533-542.

Phillipson, H.B. & Chipp, P. (1982). Air-foam sampling of residual soils in Hong Kong Proc. ASCE Specialty Conf. on Engineering and Construction in Tropical and Residual Soils, Honolulu, 339-356.

Powell, G.E. & Irfan, T.Y. (1986). Slope remedial works in weathered rocks for differing risks. In Rock Engineering and Excavation in an Urban Environment, IMM, London, 347-355.

UNESCO (1976). Engineering Geological Maps. A Guide to Their Preparation. The Unesco Press, Paris, 79 pp.

Sowers, G.F. (1963). Engineering properties of residual soils derived from igneous and metamorphic rocks. Proc. 2nd Pan-Am. Cong. SMFE. Brasil, 1, 39-61.

Vaughan, P.R. (1988). Characterising the mechanical properties of in-situ residual soil. Proc. 2nd Int. Conf. on Geomechanics in Tropical Soils, Singapore, 2, in press.

6th International IAEG Congress / 6ème Congrès International de AIGI, © 1990 Balkema, Rotterdam. ISBN 90 6191 130 3

Etude des sols naturels pour inventariser les possibilités d'implantation de centrales hydroélectriques dans la région Amazonienne

Study of natural soils of borrow areas in the inventory and feasibility stages of hydroelectric power plants in the Amazon region

Alarico A.C.Jácomo
Geotechnics Division, Hydroelectric Schemes and Studies Department, Monasa Consultoria e Projetos Ltda, Brazil

Antonio Carlos Oliva Ribeiro
Geotechnics Division, Hydroelectric Schemes and Studies Department, Eletronorte - Centrais Elétricas do Norte do Brasil S.A., Brazil

Resumé: Cet article concerne une recherche relative a une etude du methodologie e les investigacions sur naturale terrains dans le inventarie et viabilité du Usine Hidreletric dans le region Amazonas. Dans le inventarie les etude du investigacion sous jour caracterize tout les materiale prés du barrage. Dans le viabilité, les etudes sons 3,0 km du rayon du barrage dans le foret vierge.

Abstract: This paper discusses the methodology and kind of investigations required to study natural soils in the inventory and feasibility stages of power plants in the Amazon Region. During the inventory phase, the investigations are directed towards characterizing all materials around the site of the earth dam. During the feasility phase, the study is more towards borrow areas in a 3,0 km radius around the earth dam axis in the Amazon rainforest.

INTRODUCTION

The ELETRONORTE's perfor-mance area is around 4,994,000 km² covering the states of Acre, Ama-zon, Rondonia, Mato Grosso, Pará, Maranhão, Tocantins and the terri-tories of Roraima and Amapá and corresponds to 58.7% of all Brazilian territory. The Equator passes through the area and the boundaries are the latitudes 5°6' to the North and 18° to the South (fig.1).

Around 80% of the vegetation in the ELETRONORTE area (fig.2) is forest. Scrubland occurs on the forest boundaries in the eastern, southern and southeastern parts and in some isolated areas to the North and Centre. As a result of the advancing, accelerating set-tlement in the Mato Grosso and Rondonia states (new farming fron-tiers), both the scrubland and forest are being drastically re-duced.

The climate in the region is well-defined, divided according to the Koppen classification into su-per-humid, humid and dry areas. In the regions the annual precipita-tion is 2,500 mm, 2,000 mm and 1,500 mm respectively. In general, the rainy seasons are well marked, varying from 6 to 10 months (fig.3).

The geological history of the Amazon region can be understood in two stages. The longest is related to the evolution of the earth's crust, from the Archean to the end of the Mesozoic eras when the for-mation of the large geological de-mains occurred. The shortest cor-responds to the end of the past, cold dry periods accompanying the Pleistocene glaciations followed by the recent hot, wet period, have influences on the vegetation and origin of the soils.

GEOLOGICAL ASPECTS

Geologically, the Amazon can be subdivided into two broad units - the crystalline basement and sedimentaty basins. The crystal-lyne basement consists of the Craton or Amazonian Shield in the Center, the Guyana Shield to the North and part of the Central Brazilian Shield in the Southeast. The sedimentary basins are the Amazon in the Central West, the Parnaíba in the East and Paraná in the South (fig.4 and 5). Each of these units present different geological and geotechnical characteristics (age, lithology and structures).

The geological history of the region is very old, beginning in the Archean era with the formation of the greenstone belts in the region of the Carajás and Navio mountain ranges, which probably occurred around 2.6-3 billion years ago. Later, during the trans-Amazon development around 2 billion years ago, the region was deeply affected by granitization and migmatation phenomena. This event is quite noticeable in the eastern and notheastern part of the region.

In the Middle Proterozoic era around 1.7 billion years ago, a tectonic reactivation began to affect the Central part of the Amazon, causing extensive granitization, acid to intermediary volcanism and sedimentation. This occurrence, by the name of Pará, was in a belt to the NNW over 600 km in width. Then, continuing in a western direction, the region was affected by another two reactivation occurrences, the Madeira around 1.4 billion years ago and the Rondonia 1 billion years ago. 500 million years ago saw the start of the process, in an approximately E-W direction, that gave rise to the Amazon sedimentary basin, which is the largest intercratonic basin in South America, covering an area of approximately 2 million km², and shows the most varied development. This basin is elongated in shape and runs parallel to the Amazon valleys. Cross structural heights enable the basin to be divided into five sub-basins called from West to East: Acre, Upper Amazon, Middle Amazon, Lower Amazon and the Amazon estuary. The sedimentary sequences that fill the Amazon basin rarely outcrop, except for the units older than the post-Cretaceous period. The Paleozoic units outcrop only in the North and South flanks of the Middle and Lower Amazon sub-basins, between the Negro and Xingu rivers, exposing sandstones, shales and local limestones in a long narrow band.

The Paleozoic Parnaíba basin, mainly in the states of Maranhão and Tocantins, consists of a transgressive sequence, beginning with littoral and metalittoral belts, moving into calm environments of tidal plains. The tectonism of the basin lowered blocks to further the medo-Devonian sea penetration in linking up the southern part of the Parnaíba basin to the North-Northeast part of the Paraná basin, and also in opening up the Northwest part to connect with the Amazon Basin (Carozzi et al, 1975). The lithological components are sand-stones, shales, arkoses and basalts.

The Paraná Basin occurs in the states of Mato Grosso and Rondonia, represented by its north-northeastern boundary. It consists of Paleozoic sediments which, in most of the area, are re-covered by Mesozoic rocks. The Mesozoic deposits cross over the Paleozoic on the northeastern border. Both area made up of sandstone, siltites and shales.

GEOTECHNICAL ASPECTS

From the geotechnical point of view, given the various existing lithologies, the region can first be separated into two extensive domains. One is of rocks belonging to the Paleozoic sedimentary basins with their Meso and Cenozoic cover, and the other of Archean and Proterozoic rocks. The geomorphological occurrences, followed by the stages of intensive physical-chemical weathering during the Cenozoic period, especially in the Quaternary stage, created different soil profiles in both domains.

In the Archean and Proterozoic rock region, which is mostly crystalline rocks, with various degrees of metamorphism, there developed thick, structured, sandy-clayey soils, generally 15 to 40 meters in depth, with strong laterization. This profile starts as a mature residual soil, gradually becoming structured residual soil, saprolite and altered rock until reaching unaltered rock. Boulder concentrations occur locally.

The soils from the sedimentary rocks in the Paleozoic basins are thin, generally clayey or sandy, not very structured, with low laterization. The change from residual soil to fresh rock is sudden, occurring at a maximum of 5 meters and generally less than one meter.

The larger drainage area presents different features in both regions. In the riverbed over crystalline rocks, rocky outcrops are found, and on the river banks the rocky top dips to lower elevations, giving an inverse relief to the rock substratum. Such a fact was noticed at the Tucuruí and Samuel hydroelectric power plants and at the Juruá dam on the Kararaô plant.

The sedimentary rocks present rocky outcrops both in the riverbed and on the banks, generally in the form of escarpments, as occurs at the Cachoeira Porteira, Manso, Barra do Peixe and Kararaô power plants.

The different levels reached by the sea during the Cenozoic period caused situations at times of intense erosion and others of large lake-river deposits.

During the erosion stage, the valleys were deepened and the hills smoothed down; consequently, in some of the drainage basins, the river channels became blocked up and sometimes overflowed. In this way there emerged the broad plains of the Amazon river channel and its main tributant the large lowland areas such as the Tocantins, Xingu, Tapajós and Madeira; and the plateau belts that, in general, are great watersheds.

Deposits occurred locally on the floodplains of drainage basins, forming extensive alluvial terraces (e.g. Tucuruí and Samuel), or consist of paleovalleys (e.g Moju river at Tucuruí). These terraces generally consist of sequences of sand, gravel and clay levels.

Due to the various kinds of environment in the Amazon scenary during the Cenozoic period, the separate and joint work of vegetation, organisms and micro-organisms, as well as chemical processes, caused soil laterization that is not always very easy to explain. This combination of phenomena was the reason for lateritic soils with high permeability and concentration of laterized nodules at Samuel, Tucuruí and Balbina plants.

METHODOLOGY FOR GEOLOGICAL-GEOTECHNICAL INVESTIGATIONS

The dam studies and designs are prepared in four stages — inventory, feasibility, basic and final design, in accordance with ELETRONORTE (1978) and ELETROBRÁS (1983 and 1984).

Inventory — At this stage, the aim is to determine the best fall division of the basin from the multiple usage viewpoint. Physical limitations such as towns, roads, mineral deposits, national parks and Amerindian reserves are analyzed.

Feasibility — This is the in-depth study of the hydroelectric scheme, with emphasis on the alternatives for the best position of the dam axis and the definition of the project's general lay-out, with cost-benefit index optimization.

Basic design — This covers the detailed study of the project, descriptive records, technical specifications, quantities, maps, sections of structures and fixed equipment, time schedules and budgets — for the purpose of bids and respective adjudications.

Final design — This is the basic design in detail, including all necessary details to carry out civil works, erection of fixed equipment, inspection, operational testing and commissioning.

The hydroelectric projects in the Amazon region in general have the following characteristics — wide rivers, smooth topography on the abutments, high rainfall, far from towns and land access, thick vegetation, in addition to the unhealthy and dangerous conditions.

During the inventory phase, the studies for natural building materials takes into account the general geotechnical aspects mentioned above and uses the following methodology for each hydroelectric project site:

153

- photographic interpretation of radar, satellite images and air photographs to mark the homogeneous areas;
- geological-geotechnical reconnaissance of the ground surface using augur drilling and geophysics for sampling and surface mapping, on both banks;
- taking 4 or 5 sampling points, in the shape of a cross, for each potential area, tauch-visual classification of the soils sampled per point and a selection of the sampples the laboratory characterization testing;
- researched volumes of the soil material area cubed per area and bank of occurrence, and must satisfy the criteria of having at least double the volume investigated on each bank;
- the minimum investigation radius to start the investigation is fixed at around 5 km, covering all types of material per bank;
- if necessary, use inspection wells and/or mechanical drilling for a better evaluation of the materials there;
- the laboratory characterization tests are grain size, Atterberg limits and density;
- due to the adverse conditions of the site-vegetation and rainfall index, it must be born in mind that the investigations are to be minimized and the study period for each site limited.

During the feasibility phase, the studies for the natural building materials - soil, sand and rock - must be conclusive regarding the aspects of location, volume and geotechnical characteristics. The soil investigations must be intensive, all existing materials considered, and a laboratory set up on the site of the studies. The borrow area investigations must be considered from the transport distance viewpoint, with intervals from 0.5 km to an average of 3.0 km on each bank. In principle, the investigation mesh is 400 m in the more homogeneous materials and can reach 200 m in heterogenous material. The investigation's process - augur drilling, wells and possibility geophysics and percussion drilling - must be performed so that the volumes and characteristics of the material be compatible with and

reliable for the work study. Research into natural building materials is closely linked to the balance of materials, interruption of the building sequence, availability of obrigatory excavation material, local rainfall, site lay-out and final cost of the stuides and construction. The laboratory characterization tests must be done in groups of samples with similar tauch-visual characteristics to minimize and rationalize the number of tests. Once the types of materials on the worksite have been defined, and the lay-out, building sequence and local rainfall known (analysis of the workable days in one year with compacted earthfill), a detailed study is made of the definition and choice of dam section. Once this stage is complete, then special test must be performed and sent to laboratories outside the worksite.

CONCLUSIONS

In ELETRONORTE's area of activity there are two important aspects to be considered: the rainfall index, vegetation and unreliable support services, and the compartmentation of two large geological-geotechnical regions, of which the former is more relevant.

Due to the high rainfall index and its duration, it is always necessary to investigate and find sandler, more drainable soils with more suitable compaction characteristics. Very often this is not possible and, therefore, an analysis of alternative dam sections considering any kind of material has to be made.

The long rainy seasons mean that fieldwork has to be carried out during the dry season and requires a large number of personnel, and this, together with unreliable logistic support - absence of roads, distance from towns and villages - leads, in general, to a false synchronism between the field and office teams and a longer period to accomplish the work.

ACKNOWLEDGEMENTS

We give our thanks to
ELETRONORTE - Centrais Elétricas
do Norte do Brasil S.A. for their
permission to publish the concepts
used in their studies, also Eliza-
beth H. Cardoso for drawing the
figures, and Maria Terezinha Gomes
Sampaio for her attention and
type.

BIBLIOGRAPHY

Amaral, Gilberto, 1978. Geology of
the Pre-Cambrian period in the
Amazon Region. Thesis of "Livre
Docência", 20-25.
Anais, 1980. 1st Symposium on the
geological-geotechnical characteristics of the Amazon Region.
Carozzi, A.V. et al., 1975. Envi-
ronmental analysis and tectonic
evolution of sedimentary series
of the Silurian-Eocarboniferous
Section in the Maranhão Basin.
Petrobrás CEMPES - SEPEX - 7th
publication.
Centrais Elétricas Brasileiras
S.A. - ELETROBRÁS 1984. Handbook
of the hydroelectric inventory
of the hydrographic basin.
Centrais Elétricas Brasileiras
S.A. - ELETROBRÁS 1983. Instruc-
tions Manual for hydroelectric
scheme feasibility studies.
Centrais Elétricas do Norte do
Brasil S.A. - ELETRONORTE 1978.
Handbook for studies of hydro-
electric power plants.
Petri-Setembrino Fulfaro, Vicente
Jose, 1983. Editora USP 1-577.
Pettena, José Luiz, 1981. Hydro-
logical diversity of the Brazil-
ian Amazon. Symposium on the hy-
droelectric development of the
Amazon Region.
Projeto Radam Brasil, 1982. Bole-
tim Técnico, Projeto Radam
Brasil - Série Vegetação, 32-41.

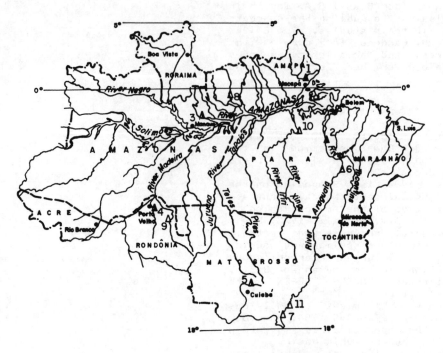

- State Capital
- ▲ Hidroeletric Power in Operation
- ▲ Hidroeletric Power under Construction
- △ Hidroeletric Power under Basic Project or Feasibility Stage

1 — UHE Coaracy Nunes
2 — UHE Raul Garcia Lhano
3 — UHE Balbina
4 — UHE Samuel
5 — UHE Manso
6 — UHE Santa Isabel
7 — UHE Couto de Magalhães
8 — UHE Cachoeira Porteira
9 — UHE Jiparaná
10 — UHE Kararaó
11 — UHE Barra do Peixe

Fig. 1 — Eletronorte's Performance Area.
Eletrobras' Hidroeletrical Development
Plan to 2010

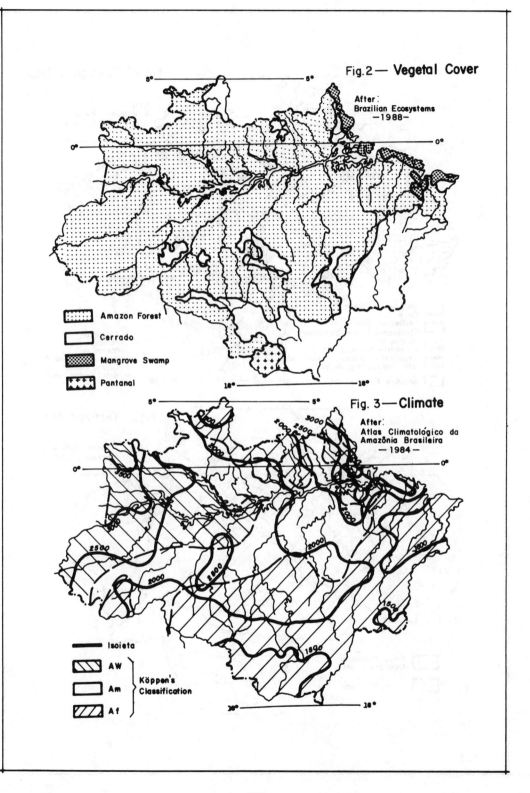

Fig.2 — Vegetal Cover

After :
Brazilian Ecosystems
—1988—

Amazon Forest

Cerrado

Mangrove Swamp

Pantanal

Fig. 3 — Climate

After :
Atlas Climatológico da
Amazônia Brasileira
— 1984 —

Isoieta

AW

Am   } Köppen's
      Classification

Af

157

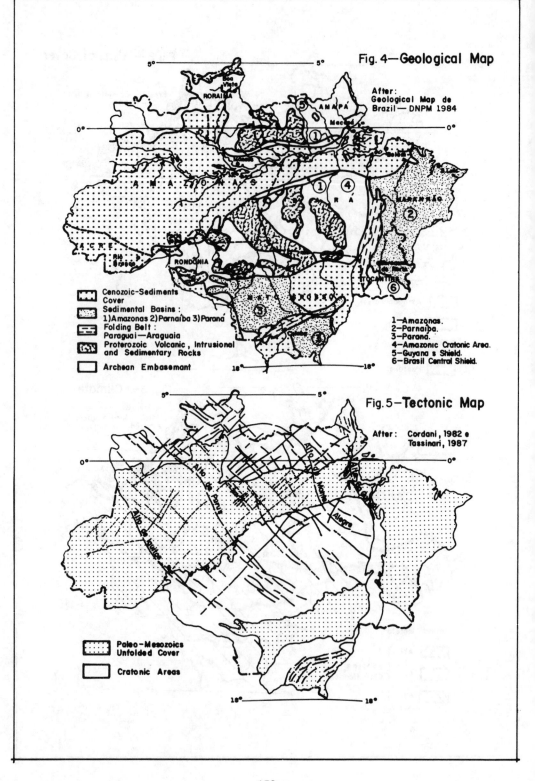

Fig. 4—Geological Map

After:
Geological Map de
Brazil—DNPM 1984

Cenozoic-Sediments
Cover
Sedimental Basins :
1) Amazonas 2) Parnaíba 3) Paraná
Folding Belt :
Paraguai—Araguaia
Proterozoic Volcanic, Intrusional
and Sedimentary Rocks
Archean Embasement

1—Amazonas.
2—Parnaíba.
3—Paraná.
4—Amazonic Cratonic Area.
5—Guyana s Shield.
6—Brasil Central Shield.

Fig. 5—Tectonic Map

After: Cordani, 1982 e
Tassinari, 1987

Paleo-Mesozoics
Unfolded Cover
Cratonic Areas

The engineering geological mapping 1:25 000 of Nordrhein-Westfalen, FRG
A regional engineering geological map

Cartographie géotechnique à l'èchelle 1:25 000 de Nordrhein-Westfalen, RFA
Une carte géotechnique régionale

J. Kalterherberg
Geologisches Landesamt Nordrhein-Westfalen, Krefeld, FR Germany

ABSTRACT: Since 1980 the Geologisches Landesamt Nordrhein-Westfalen (Geologic Survey of North-Rhine-Westphalia) has been publishing engineering geological maps at a scale of 1 : 25 000 (IK 25) for the urban areas on the Rhine and Ruhr. These maps are designed to provide comprehensive initial information concerning the underground for planning purposes of all kinds at or near the surface of the earth. For this purpose the engineering geological characteristics of the soils occurring are described. The distribution and thickness of soil units are shown for an average area of approximately 130 km per map.

Several maps and cross-sections, soil-mechanical data, results of soil-mechanical tests, information on underground water levels and their presentation in hydrographic curves, contour maps and isolines, a documentation map with locations and depths of 2 000 to 10 000 drillings and exposures provide a three-dimensional strata record.

The maps and diagrams are being published in two sheets. Short explanatory notes describe the geological, hydrogeological and soil-mechanical conditions. Combined with the relevant German standards (DIN) they give practical advice for planning. The maps are intended to support expert planning and to further the integration of special investigations into a larger frame of reference in order to elucidate regional connections and differences.

RESUME: Depuis 1980, le Geologisches Landesamt Nordrhein-Westfalen publie des cartes géotechniques à l'échelle 1/25 000 des agglomérations urbaines et industrielles du Rhin et de la Ruhr. Ces cartes veulent fournir une information aussi étendue que possible sur le sous-sol pour toutes sortes de planifications sur et près de la surface de la terre.

A cet effet, les types de sol dont on présente l'extension et l'épaisseur, sont caractérisés d'une façon géotechnique. La description des couches à trois dimensions comprend plusieurs (4 à 6) cartes à différentes échelles, 3 à 7 coupes verticales jusqu'à une profondeur de 30 m, les résultats des essais de mécanique des sols, des constantes du sol, des cartes et diagrammes des niveaux de la nappe aquifère ainsi qu'une carte qui montre la situation et la profondeur d'environ 2 000 jusqu'à 10 000 forages et sondages sur une surface de 130 m² par carte.

Les cartes et les diagrammes sont publiés sur deux feuilles en couleurs. Des commentaires brefs décrivent les conditions géologiques, hydrogéologiques et géotechniques. De plus, le texte comprend une coupe schématique pour expliquer le principe de la description. En relation avec les standards allemands (DIN), les commentaires donnent des recommandations et des informations pratiques pour des planifications.

Les cartes géotechniques aident à faciliter les planifications du constructeur et à classifier les résultats de ses essais spéciaux des sols, et elles rendent clair les connexions et les différences régionales.

The quaternary loose rocks in the Nieder-rhein area are characterised by thick terrace gravels with silt cover. In the Ruhr area river regimes with lower water velocities and sediment transports predominate. Here one also has to regard the Quaternary sediments of the inland ice sheet and the preloading of older Pleistocene strata by several hundreds of metres of inland ice during the Saale glaciation.

In the Ruhr area pre-Quaternary solid rocks of Upper Carboniferous and Upper Cretaceous age approach the surface; in the West, near the Rhine valley, there are Tertiary loose rocks, which occur in the Rhine valley only in depths of 10 to 45 m.

In the south and south-west of the urban Rhine area near Bonn there are Tertiary strata near the surface above the variscan folded Devonian solid rocks. They show a greater variety due to intercalations of basalts and trachyte tuffs.

Human influence on the subsoil is obvious in the old cities on the Rhine and particularly in the Ruhr area, where a classic industrial landscape was formed.

The underground mining of coal has had particular influence on the overlying strata up to the surface. Land subsidence up to 20 m, e.g. in the city of Dortmund, is common. These effects are not shown in our maps, but the mining companies and mining inspectorate are quoted as competent for relevant information. However, there are indirect consequences for the design of the maps, as in some cases it is difficult to obtain actual m.s.l. heights for the construction of cross sections.

There are other big problems concerning the presentation of fills: from archive material only site and dimension of fills can be determined, but rarely is their composition and thickness known. In particular cases archive drilling results yield the local thickness of fills, but no information about their contents. In the Ruhr area large slag heaps are obvious, where the waste rock from coal mining (and processing) is dumped.

Under certain conditions this waste can be used as earthwork material.

Distribution and thickness of fills gives information about important stages in the historical evolution of the old cities on the Rhine (e.g. Düsseldorf, Cologne, Bonn). The fills are shown - as far as they are known - without claim to completeness.

For mapping purposes the strata of the top 20 to 30 m can be summarised by three engineering geological units:

Unit 1 contains beds with medium to low load-bearing capacity, which are usually not suited for earthworks except under particular conditions. This unit includes unpreloaded Holocene to Upper Pleistocene silts (flood plain loam, high flood loam, loess loam, organic/humic silts), peat, partly loose, uniform sands (aeolian sands, flood plain sands) and fills. On the main map (1 : 25 000) unit 1 is shown in thickness intervals of 1 m. Frequently old alluvial river meanders of the Rhine appear this way. Thick fills near the Rhine often show site and shape of former harbours of Roman or medieval origin (e.g. Duisburg, Düsseldorf, Cologne).

Unit 2 consists of fluvial sand and gravel (terraces), fluvioglacial sand and gravel, till or residual loam of Upper Carboniferous or Upper Cretaceous rocks with very high to medium load bearing capacity, which are moderately or well suited for earthworks. Their total thickness and the nature and distribution of the underlying unit 3 are shown in the Quaternary thickness map at a scale of 1 : 50 000 or 1 : 100 000.

Unit 3 follows underneath the Quaternary strata and consists of compact sands and well consolidated clays of Tertiary age or of diagenetically lithified beds of the Upper Cretaceous, Upper Carboniferous or Devonian. The Upper Cretaceous and Paleozoic strata are mainly to be classified as solid rocks. In many regions they occur near the surface of the earth.

Unit 2 can approach the surface next to

Unit 1, or it can be entirely missing, so that Unit 1 lies on top of Unit 3. The different units are explained in the following schematic cross section (Fig. 1) for the area of Duisburg.

measures. From 1989 onwards an additional map of the fills will be published at a scale of 1 : 50 000 (maps of Herne, Essen, Mülheim/Ruhr) with a higher differentiation (of the fills). The fills will

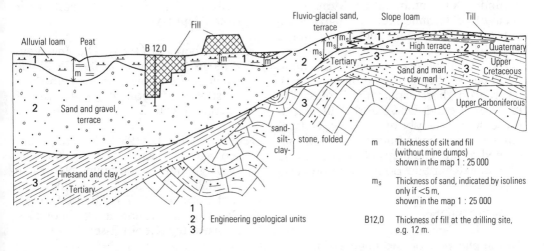

Fig. 1: Explanation of engineering geological units for mapping purposes (schematic cross section)

The hydrogeologic part comprises an underground water contour map and a map of depths to the water table for a certain short time (of a ground-water high-level) or for a mean underground water level. The maps are based on data of the Landesamt für Wasser und Abfall Nordrhein-Westfalen (Hydrologic and Waste Disposal Survey of Northrhine-Westphalia) and the hydrologic map 1 : 10 000 of the Rheinisch-Westfälischer Steinkohlenbezirk (Rhenish-Westphalian coal district) edited by the Westfälische Berggewerkschaftskasse in Bochum. In addition several hydrograph curves inform about the long-term behavior of the groundwater table for as long a period as possible, usually for two to four decades.

In some cities of the Ruhrgebiet mining areas and sites with fills of more than 1 m thickness cover 25 to 30 % of the surface. They often cause problems for the foundation of buildings. They regularly demand special investigation and

be subdivided according to the following aspects:

- fill (general, unknown content)
- rubble, excavated soil
- slag heap
- domestic waste (solid)
- industrial waste (solid)
- mud settling pond, settling basin, sewage mud, other muds
- coal mining area, factory site
- brick works, quarries, excavation sites.

This information may contribute to the problem of "Altlasten" (previously contaminated land). The compilation is based on old topographic maps (from approximately 1890 on), on data from the Staatliches Amt für Wasser- und Abfallwirtschaft (State Office for Water and Waste Control), the Kommunalverband Ruhrgebiet (Municipal Union of the Ruhr Area), the coal mining companies and if possible from the municipal administration affected.

The map of fills - combined with the map of depths to the water table and the main map 1 : 25 000 - may serve for the evaluation of areas with possible underground water contamination. Moreover it is easy to determine areas with geological barriers for the construction of waste deposits.

The engineering geological maps 1 : 25 000 are being printed on two sheets, partly with short explanatory notes. The maps of Krefeld, Düsseldorf, Düsseldorf-Kaiserswerth, Duisburg, Cologne, Gelsenkirchen and Dortmund have been published so far. The maps of Herne, Essen, Mülheim/Ruhr and Bonn are in preparation.

CONCLUSIONS

The application of the engineering geological maps 1 : 25 000 - particularly because of their clarity - appears to be efficient for questions of superior municipal and regional planning, e.g. for:

- site evaluation of development areas, residential and industrial sites
- evaluation of suitable areas for dumps
- planning of traffic routes, public supply systems (water, gas, electricity)
- excavation of loose and solid rocks
- evaluation of underground water levels and thickness calculation of the uppermost aquifer.

Apart from these primary planning purposes the map can be consulted for special building projects. Combined with valid German standards (DIN) first reliable assumptions can be stated in a preliminary draft for the foundation of buildings (DIN 1054). With these data the planning expert is able to develop a foundation concept, which has to be extended and supported by specific local investigations.

REFERENCES

DIN 1054 (1976). Baugrund, Zulässige Belastung des Baugrundes. Subsoil; permissible loading of subsoil - mit Beibl., 30 S., 15 Abb., 10 Tab.; Berlin, Köln (Beuth-Verl.)

Kalterherberg, J. (1985). Die Ingenieurgeologische Karte 1 : 25 000 (IK 25) des Geologischen Landesamtes Nordrhein-Westfalen - Geol. Jb., C 41, 21 - 68, 22 Abb., 1 Tab., 2 Taf., Hannover

Kalterherberg, J. (1985). Die Ingenieurgeologische Karte 1 : 25 000 des Geologischen Landesamtes Nordrhein-Westfalen - Ber. 5. Nat. Tag. Ing.-Geol. Kiel 1985: 207 - 217, 6 Abb., 1 Tab. (Dt. Ges. Erd- u. Grundb. u. Dt. Geol. Ges., Fachsek. Ing.-Geol.) Essen

Kalterherberg, J. & Lüthen, M. & Schmidt, K.-D. (1989). Die Ingenieurgeologische Kartierung 1 : 25 000 des Geologischen Landesamtes Nordrhein-Westfalen im Ruhrgebiet - Mitt. Geol. Ges. Essen, 11, 38 - 41, 1 Abb., Essen

6th International IAEG Congress / 6ème Congrès International de AIGI, © 1990 Balkema, Rotterdam. ISBN 90 6191 130 3

The engineering geological map of Mainz/Rhine, FRG
La carte géotechnique de Mayence/Rhin, RFA

E. Krauter & J. Feuerbach
Geological Survey of Rhineland-Palatinate, Mainz, FR Germany

M. Witzel
Department of Urban Development of Mainz, FR Germany

ABSTRACT: The Engineering Geological Map of Mainz -a pilot project- is going to be published in August 1990. The map is beeing produced in cooperation with the Geological Survey of Rhineland-Palatinate, the Johannes Gutenberg-University Mainz and the Department of Urban Development of Mainz.
All engineering geological and hydrological data and maps are computer stored. Soil mechanical units are represented in a three-section model. In addition to these maps, there is a slope stability map and an erosion risk map. Applied engineering-geophysical methods, foundations of old buildings, the chemistry of ground water and problems of ground-water increase will be discussed as well. Man-made ground and the fortifications built under the ground during the last century caused unusual problems.

RESUME: La Carte de Géologie Appliquée de Mayence -un projet premier- est publiée en août 1990. Elle est developpée en coopération du Service Géologique de la Rhénanie-Palatinat, du Johannes Gutenberg-Université à Mayence et du Service d'Urbanisme de Mayence.
Toutes les indications de la géologie appliquée-hydrogéologique sont enregistrées digitalement. Des unités de mécanique de sol sont présentés dans un modèle à trois section. En complément à cettes cartes il y a une carte de stabilité d'inclination et une carte de danger d'erosion. De plus des méthodes de géologie appliquée-géophysiques, des fondations de vieux bâtiments, le chimisme de l'eau souterraine et des problèmes resultés par la montée de l'eau souterraine sont discutés. Des problèmes spéciales pour le sol de fondation sont causés par des remplais et des fortifications souterraines, construits dans le dernier siècle.

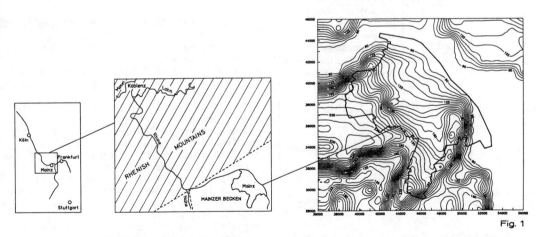

Fig. 1

1 INTRODUCTION

The cooperation of the Regional Geological Survey of Rhineland-Palatinate in Mainz, the Johannes Gutenberg-University of Mainz, the Authority for Environmental Protection and Urban Development and several engineer's offices made the completion of the engineering geological map of the municipal area of Mainz possible within 3 years. All data is available to planning offices and authorities. The maps have been stored digitally, can be updated at all times and can be combined with other thematic maps. New engineering geophysical methods are tested and implemented successfully for building ground investigations.

2 A GEOLOGICAL AND GEOMORPHOLOGICAL SUMMARY

2.1 Geology

The municipal area of Mainz is situated in the northeastern corner of the "Mainzer Becken" (Fig. 1). The development of this basin is closely correlated to the formation of the Upper-Rhine Graben. Both were formed when the European crust collided with the Austro-alpine-adriatic crust. The earliest detectable sedimentation found today, took place at the beginning of the Tertiary (Fig. 2).

Quarternary	Holocene	Mio.	Auffüllungen
		0.01	Auensedimente
	Pleistocene		Flugsand
			Löß
		2.0	Haupt/Mittel/Niederterrasse
Tertiary	Pliocene	5.1	arvernensis-Schotter
	Miocene		Dorn-Dürkheim-Schichten
			Dinotheriensand
			Hydrobienschichten
			Corbicula-Schichten
		24.6	Obere Cerithienschichten
	Oligocene		Mittlere Cerithienschichten
			Untere Cerithienschichten
			Süßwasserschichten
			Cyrenenmergel
			Schleichsand
			Rupelton
		38.0	Mittlere Pechelbronn-Schichten
	Eocene	54.9	Eozäner Basiston

after ROTHAUSEN & SONNE (1984)

Stratigraphic Sequence of the city area of Mainz Fig. 2

Transgressions from the south and north deposited marine, limnic and brackish sediments; these still exist today as an interbedding of limestone banks, and clayish to calcareous marl with sandy zones Fluvial sediments from the original Rhine have been deposited there since the beginning of the Upper Miocene. Tectonic movements, fluvial sedimentation and erosion during the Pleistocene formed terraces in the municipal areas which can be subdivided into a lower terrace (up to 95 m above sea level), a middle terrace (up to 150 m above sea level) and an upper terrace (up to 230 m above sea level). Rock dust was blown off alluvial covers and deposited on leeward slopes (east) during the various Pleistocene glacial periods. This loess covers a large part of the municipal area of Mainz. Sand was blown out of the lower terrace and was deposited as flying sand in the northwestern part of the municipality. Holocene valley plain sediments cover a large part of the Rhine valley overflow area.

2.2 Morphology

The municipal area of Mainz can be divided into 3 morphological units, which are separated by steep inclines:
- Rhine valley/lower terrace from 80 m to 95 m above sea level
- middle terrace from 110 m to 150 m above sea level
- upper terrace from 170 m to 230 m above sea level.

There are many erosional channels which drain the municipal area and flow into the receiving stream.

3 HYDROGEOLOGY

3 ground-water storeys can be differentiated in the municipal area of Mainz:
- the impermeable Eocene to Upperoligocene marl Tertiary
- the Oligocene to Miocene calcous Tertiary, a zone with interbedded impermeable beds and aquifers
- the Uppermiocene to Quaternary pore water aquifers.

The Uppermiocene to Quaternary pore water aquifers are of great importance for the building ground in Mainz because some of the superficial ground-water horizons reach the terrain surface (e.g. Rheinaue-downtown area). A large number of the wells within the municipal area have been closed down. As a result, most of the drinking water is supplied by water plants outside the city borders. For this reason and higher levels of precipitation between 1980 and 1987,

164

the ground-water level has risen by 6 m. This in turn caused an inflow of ground-water into basements in parts of the down-town area. Water analyses from the Regional Geological Survey show that the superficial water has a concrete aggressive character which, in the long run, endangers the foundations of the buildings.

4 ENGINEERING GEOPHYSICAL INVESTIGATIONS

The engineering geophysical investigations include:
- refraction seismic
- geoelectric
- georadar
- CO2-soilgasmeasurements
- radon-soilgasmeasurements.

The following will be clarified with these engineering geophysical measuring methods:

1. The depth of the Quaternary/Tertiary border.
Low seismic wave velocities (<1000 m/s) mark the Quaternary sediments (loess, flying sand, terrace-sediments), while the Tertiary limestones and marls have much higher seismic wave velocities (1500 m/s). The electric conductivity of these two units are also very different. Clayish marls, for example, are very good electric conductors (15 ohm meters), but dry loess has very high electrical resistance (50 ohm meters).

2. The depth of the ground water level
The aquifers show typical seismic wave velocities. The electric conductivity increases dramatically and allows the usage of geoelectric and electromagnetic measuring methods.

3. Localisation of slip bodies
The border of each sliding body, the depth of the moving zone and as well as the maximal depth of the sliding surface were determined with the help of refraction seismic in the southern part of the municipality. Soilgasmeasurements (radon, CO2) were implemented to ascertain zones with high gas concentrations. These zones correspond to an increased number of migration-paths and indicate movements within one landslide. Airphoto-linears were verified at the same time. They mark the borders of each slip body.

5 AIRPHOTOANALYSIS

Airphotos on a scale of 1:13 000 were interpreted stereoscopically. Linear analysis was implemented to localize the zones of weakness which mark the zones with more migration-paths. Infrared photos on a scale

of 1:13 000 were also interpreted for the airphotoanalysis of the southern part of the municipal area. Many different building ground investigations, traffic-planning and waste-deposit mapping questions have been clarified with the help of these airphotos.

6 DRILLINGS / SOUNDINGS / EXCAVATIONS

6000 drillings and soundings were looked in order to find useful data. This data belongs to various authorities, institutes, and offices in the surrounding area of Mainz and we were kindly given access. Some 3100 soundings/drillings and their soil physical values were stored in a databank. Another 1500 drillings were implemented to obtain a more dense network of drillings. Statistically speaking, there is a filed boring profile every 147 m in the 100 sq.km large municipal area.

7 SOIL PHYSICAL VALUES

The informative material was used to ascertain and statistically evaluate the soil physical values and were itemized for each building ground unit (Fig. 6).

8 DATA BANK

A DBASE-data bank was set up for the drillings/soundings/excavations, in which the following data is stored:
- file number
- right values
- high value
- sounding/drilling/excavation
- location
- bore-location above sea level
- horizons 1-8 (description of the layers according to DIN 4023 and 18 196 (norm of german industry)).
The data was formated to be partly or completely transferable to other programs, e.g. for the construction of isoline maps (Fig. 1) and bore location maps (Fig. 3).

9 CONSTRUCTION OF MAPS

The scale selected for the bore location maps and the engineering geological maps is 1:5000 (41 ground maps) and 1:20 000 for the general map. All maps have been digitized and stored with a special CAD-program. They can therefore be updated at all times, combined with other thematic maps or scaled differently.

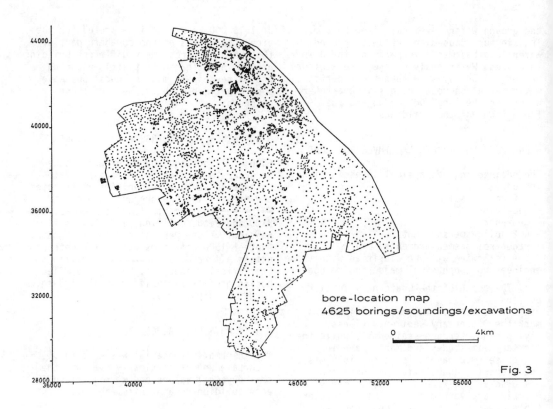

bore-location map
4625 borings/soundings/excavations

0   4km

Fig. 3

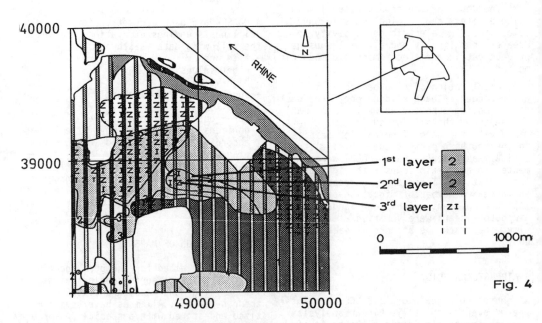

RHINE

1st layer
2nd layer
3rd layer

0   1000m

Fig. 4

Fig. 4. Engineering Geological Map 1:5000

10 REPRESENTATION

A maximum of 3 layers from 1.0 to 5.0 m below the terrain surface are represented in the municipal area of Mainz because there are only 3 layers. The first layer is represented as a 1.5 cm wide vertical beam, the second is 0.5 cm wide, and the third is represented with hatching according to DIN 4022. The thickness of the first layer is normally indicated with isolines. When there is a third layer, the thickness of the second layer is indicated on the thin vertical beam as well. The minimum thickness of the third layer can be ascertained when the thickness of the first two layers are added (5 m minus thickness no. 1 and no. 2). A layer has only been taken in consideration when it has a minimum thickness of 0.5 m. The following building ground units were included:
- made ground / accretion
- meadow loam / meadow clay
- flying sand
- loess loam
- loess
- lower terrace (sand, gravel)
- middle and upper terrace (sand, gravel)
- Miocene sands
- marl and limestone.

The typical grain-size distribution of these layers is shown in Fig. 5. The different depths of the ground-water table are depicted on a special contour map of the water table on a scale of 1:20 000.

		A	AT	FLS	LÖL	LÖ	NT	HT	TRT
plasticity	Ip	15.6	13.6		13.2	12.4			33.9
unitweight in moist state	kN/m³	18.2	18.8	18.3	18.6	18.2	19.3	19.9	18.6
dry density	kN/m³	16.1	14.9	14.7	15.7	16.6	17.0	17.8	15.2
liquid limit	%	36.7	41.5			34.2	27.8		56.9
plastic limit	%	18.3	20.2		18.4	18.2			25.1
void ratio	e	0.71	1.09	1.05	0.67	0.59	0.45	0.41	0.30
porosity	Vol%	39.4	39.6	47.2	39.9	39.0	32.5	30.6	45.1
angel of internal friction	(°)	27.1	22.2	30.5	25.7	28.9	33.3	31.5	21.0
coefficient of compressibility	MN/m²	12.4	8.50	39.5	12.8	12.3	59.6	43.4	15.1
cohesion	kN/m²	6.25	22.5		21.1	17.3			24.9
consistency	Ic	0.57	0.62		0.79	0.93			0.74
water content	%	22.3	27.1	10.2	20.4	15.9	33.7	11.1	28.8

A   - made ground
AT - meadow loam
FLS - flying sand
LÖL - loess loam

LÖ - loess
NT - lower terrace } sand, gravel
HT - upper terrace }
TRT - marl

Fig. 6

spersed with the Tertiary schluffs and clays. These sandy layers form potential sliding zones because they carry water well. These landslides have been mapped engineering geologically and geophysically and depicted on the maps.

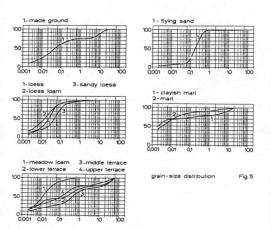

grain-size distribution   Fig.5

11 LANDSLIDES

There were several landslides in 1981/82 in the southern part of the municipality (viniculture area). Fine sandy layers, which vary in their thickness, are inter-

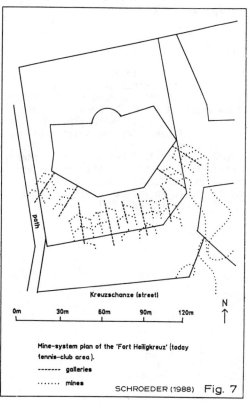

Kreuzschanze (street)

0m   30m   60m   90m   120m   N↑

Mine-system plan of the 'Fort Heiligkreuz' (today tennis-club area).

------- galleries
········ mines

SCHROEDER (1988) Fig. 7

12 HISTORICAL BUILDING GROUND

Artificial underground passages and cavities (e.g. in Fig. 7: "Fort Heiligkreuz") have been built as protection against enemy attacks since the 17th century in the municipality of Mainz. The position of the damaged underground constructions has been depicted on a general map for planned and existing buildings. The foundations of historical buildings (e.g.: the pile foundations of the Mainz cathedral in the Aquifer of the Rheinaue) have also been described. An almost complete exchange of the natural soil to a depth of 10 m below the terrain surface has taken place over the centuries. As a result, recommendations for building-foundations have been given.

13 CONCLUSIONS

Slope-incline maps, slope-stability maps and erosion risk maps have been constructed with digital height models, the existing engineering geological maps and soil-physical values since July 1989. The maps and the file for drillings are constantly being updated in order to construct a more extensive soil information system for the municipality of Mainz.

REFERENCES

Kreuder, S. 1988 . Ingenieurgeophysikalische Untersuchungen im Stadtgebiet von Mainz. Diplomarbeit (unpublished), Mainz.
Rothausen, K. & Sonne, V. 1984. Mainzer Becken. Samml. geol. Führer 79. 203 p., 21 fig., 47 tab.. Berlin, Stuttgart: Borntraeger.
Schroeder, U. 1988. Ingenieurgeologischer Aufbau im südwestlichen Stadtgebiet von Mainz und der Einfluß unterirdischer Festungsanlagen auf den Baugrund. Diplomarbeit (unpublished), Mainz.

Engineering geological mapping of natural slopes
Classification géotechnique de versants naturels

A. Lakov & K. Anguelov
Higher Institute of Mining and Geology, Sofia, Bulgaria

ABSTRACT: One of the main problems for the design of buildings in slope regions is the accessment of their influence on the slope stability. It is suggested that the problem can be solved with back recalculations taking in account the natural stability and accepted minimum allowable value of the safety factor. Basing on the received results maps for the additional external loading can be compiled for the region which may be used for architectural planning and design.

RESUME: Un problème important pedant le processus d'élaboration des projets des consrtuctions sur les terrain inclinés est l'influence sur leurs stabilité. On propose une solution en utilisant des calcus inverses de la stabilité ajaint en vue le coéficient minimal de securité accepté. Sur la base des résultats obtenus on propose une carte de capacité portanté supplémentaire des versants qui sera utilisée pour le placement des constructions sur le terrain.

Many towns in Bulgaria are situated in hilly regions, comparetively narrow river valleys and intermountain lowlands. Their relief conditions are quite unfavourable for their expansion over new areas. In that cases most suitable for future building are the surrounding natural slopes.
One of the major problems in sich cases is the determination not only of the natural stability of the slopes but also the possibilities for additional loading with building and constructions. The engineering geological conditions are of great importance for correct location and design of the structures. It must be mentioned that in many cases the architecture plans for the future development of towns are made in advance or independently from the results of the necessary engineering geological investigations. That may result in projects which may be quite unsuitable for the region concerning the preservation of the slope stability. The most unfavourable situation will be if such supposition cames true during building activities.
In the present paper is described an idea for compiling engineering geological maps of slopes as a base of architectural planning according the ability of the massif for additional loading. The problem is solved over two main suggestions:

1. There are no absolute stable slopes, which means that human activity can cause the failure of each naturally stable slope.
2. The safety factor F is not the best parameter for evaluation the possibility for additional loading of a slope. For example two slopes with equal values for F but different length of the sliding surfaces and equal other conditions obviously will bear different additional loads on the surface.
It means that for each natural slope could be determined preliminary combination of additional loads presented as forces acting vertically over the terrain surface which the slope can bear without loosing its stability on account of the difference between the shear resistance and the sliding forces acting on the slid plane. One of the methods for solving the problem are back recalculations based on the natural stability and providing its preservation at certain minimum allowable levels. Similar approach has been already accepted for determination of the additional loading of of open-mine levels (Fissenko 1965, Manual for ... 1981) and suggested from Bromhead (1986).
The present method is developed mainly for natural slopes with expected sliding planes with arbitrary shape. For the stability

calculations is used a simplified method of the horizontal projections of the forces (Anguelov 1988). The factor of safety is derived as a ratio between the total sums of the horizontal projections of the resistance forces (ΣR_i) and the sliding forces (ΣT_i):

$$F = \frac{\Sigma T_i}{\Sigma R_i} = \frac{\Sigma \{(G_i \cos^2 \alpha_i - u_{wi} l_i) tg\psi' + C' l_i\}}{\Sigma G_i \sin\alpha_i \cos\alpha_i + \Sigma A_i} \quad (1)$$

where G_i is the weight of the slice, α_i is the inclination of the slip surface at base of the slice, u_{wi} is the pore water pressure along the slip surface, b_i is the width of the slice, ψ' and C' are the effective shear parameters of the soil, A_i is the horizontal resultant of other sliding forces-seismic forces, seapage forces and others-see Fig. 1.

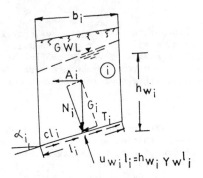

Fig.1 Forces on a typical slice.

Each additional force applied over the surface P which changes the safety factor to a certain minimum value F_{min} will modify the equilibrium equation (1) as:

$$F_{min} = \frac{\Sigma R_i + P\cos^2\alpha_p tg\psi'}{\Sigma T_i + P\sin\alpha_p \cos\alpha_p} \quad (2)$$

where α_p is the angle of the sliding surface under the force P.
Solving together (1) and (2) in respect to P gives:

$$P = \frac{(F - F_{min})\Sigma T_i}{\cos\alpha_p (F_{min} \sin\alpha_p - \cos\alpha_p tg\psi')} \quad (3)$$

From (3) is obvious that P>0 in two cases:
1. If $F > F_{min}$ (stable slope) and $tg\alpha_{no} > tg\psi'/F_{min}$. This means that there is practical restriction to the maximum value of α_p and P can be applied over the most steep part of the sliding surface.

2. The other case is when $tg\alpha_p \leq tg\psi'/F_{min}$ or the additional load P acts over sectors of the sliding surface with very low or even negative inclination. As $P \geq 0$ it is necessary $F \leq F_{min}$. This means that in such cases the friction resistance created by the additional force overcomes the sliding resultant and in fact increases the stability of the slope. This is often used approach for slope stabilization with overloading at suitable place (mainly in the passive blocks with $\alpha_p \leq 0$).

In the future matter will be considered only the first case as more unfavourable. From (3) is clear that the excess of shear strenght over a sliding surface corresponding to F_{min} can be expressed by a single value of P. It can be accepted that for each point of that surface corresponds part of this force depending on its coordinates and the shear strenght $dP = f(x, y, \tau_f)$. If through one point in the massif pass more than one potential slip planes different values for dP should be expected. This generally is due to the different stress conditions in the point created dy the different interaction between the slices in the various cases. It is not necessary the lowest value of dP to be associated with the sliding surface with F_{min}.
The toal stability of natural slopes is normally determined by a single sliding surface passing through weak layer, bedding plane, joint set, contact plane with bedding rocks (Fig.2a).
The additional force for point A dP_A will depend on the stress conditions created over the examined sliding surface and by the volume of the involved masses (for example planes $a_1 \div i-1$, $a_2 \div i-1$ etc.).

In order to determine the lowest value of dP_A must be examined practically endless number of slidimg planes. In order to decrease the amount of calculations it is accepted to determine the additional external loads ΔP_i corresponding to the sectors of the sliding planes under each of the slices to each of the slope can be devided for stability analisys. In this case the determination of ΔP_i is for all reasonable combinations between the slices. As there are no active and passive blocks occuring in the massif they are drawn for each examined plane as circular sectors tangent to the main plane and occuring at the terrain at angles $45^0 \pm \psi/2$ (Fig.2c). As we are discussing the first case from the comments to formula (3) the force P is

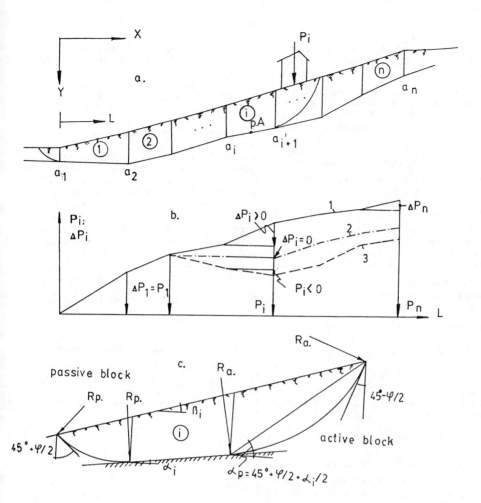

Fig.2 - a. potential sliding plane along a typical natural slope, b. types of distribution of the additional external load ΔP_i, c. drawing of active and passive blocks.

applied over the active block and α_p is equal to the mean inclination which is approximately $45^0 + \emptyset/2 + \alpha_i/2$. If sliding is expected over weak layer or set of joints with much lower shear strenght than the average in the massif then it must be accepted $\alpha_p = \alpha_i$ for the slice. As general criterion must be realized the condition $tg\alpha_p > tg\emptyset'/F_{min}$ when additional load is applied always in zone where creates higher sliding than resisting forces.

The calculations begin from slice 1 (Fig.2a) for which using (3) is calculated a value $P_1 = \Delta P_1$ according the selected F_{min}. Then the calculations are carried out for sliding body consisting of both slices 1 and 2 and a value for P_2 is received. The additional external force corresponding to the second slice then is $\Delta P_2 = P_2 - P_1$. In such a way are received number of values ΔP_i^1 for i = 1 to n (the index marks the number of the starting slice). Then the same procedure is made again but beginning from slice 2. Another group of values ΔP_i^2 is received from i = 2 to n. The routines are stopped when reaching the top slice n. The results may be arranged in a triangular matrix shown in the example at table 1. As a design value for ΔP_i is taken the lowest one from the obtained for a slice. That value is not necessarily corresponding to

171

the lowest factor of safety. The additional bearing capacity generally for the slope is characterized by the increasing of the additional load $P_d = \Sigma \Delta P_{di}$ from the toe to the top (index d means design values) - Fig.2b. They show the allowable sum of the external loads for the section if they are evenly distributed on the terrain. In that respect must be given some explainations:

i - if some slid planes from the slopes are with $F<F_{min}$ then their region must be excluded as unsuitable for building.

ii - for some slices may occur that $\Delta P_i \le 0$ with $F \ge F_{min}$ for all surfaces passing through them (curves 2 and 3 in Fig.2b). In such cases these zones are not recommended for major constructions and the total value for P_{di} corresponding to that distance from the toe must be accepted inspite the higher values for the previous slices.

iii - in the calculations for ΔP_i for each slice over one slip plane the action of the previous loads ΔP_j (j=i-1, i-2, etc.) on sliding sectors with inclinations $\alpha_j < \alpha_{pj}$ is not cosidered. In such cases if $tg\alpha_j < tg\psi'/F_{min}$ they will have stabilizing effect on the slope. However more accurate calculations will lead to unsteadiness of the interpretation of the results because the values of ΔP_{di} are maximum allowable but not necessary to be realized.

iv - the values of P_{di} are corresponding to more or less evenly distributed on the terrain loads. The maximum allowable single force (i.e. the maximum weight of a single building) must be determined by the soil bearing capacity determined by the method of soil mechanics according to the project of a typical building.

Significant problem is the interpretation of the results from a number of sections regarding the whole slope area. The stability solutions are assumed to be made in two-dimentional stress conditions over sections with unit weight. The interpolation between two neighbouring sections is most accurate when solving a three-dimentional problem. Unfortunately this is quite complicated problem using analytical solutions. Therefore it is accepted such interpolation to be based on general geological priciples. For initial zero line for P_d interpolation is taken the geomorpho-lgical base of the slope where usually occur natural sliding surfaces on the terrain.

After generalization of the additional load capacity for a slope region a specialized engineering geological map can be drawn which takes in account all geological factors influencing stability. It may be used compiling architectural plans (location of constructions, hight of buildings etc.) and for design. In these cases the responsibility for the overall stability of the slopes will be taken both from designers and engineering geologists.

An exapmle for such mapping of a slope area is shown at Fig.3 and 4. The region is situated southern from the town of Pernik (South-West Bulgaria) and is build up by delluvial yellow-brown clays with thickness up to 12 meters. Their parameters are $\gamma=18$ kN/m³, $\psi'=10^0$ and C'=46 kPa. They are reclining over paleogenic marls and sandstones. The general sliding surface is drawn through the contact between clays and bedrock. The stability calculations are made for sections passing through the highest inclination of the slid plane. The natural stability factors vary from 1.79 to 4.60 (see Table 1) and the accepted value for F_{min} is 1.20.

Table 1. Additional external loads $\Delta P_i{}^j$.

Starting slice, j	$\Delta P_i{}^j$, kN/m³ and safety factor (F) for slice number "i"			
	1	2	3	4
1	2217*	3523*	3465*	4412
	(1.79)	(2.32)	(2.62)	(2.59)
2	–	4818	3557	4408*
	–	(3.04)	(3.38)	(2.98)
3	–	–	5993	4387
	–	–	(4.27)	(3.59)
4	–	–	–	7358
	–	–	–	(3.01)
ΔP_{di}, kN/m'	2217	3523	3465	4408
P_{di}, kN/m'	2217	5740	9205	13613

* Minimum design values for ΔP_i.

The zero line for P_d is drawn through the toe of the sections where the terrain changes its slope to almost horizontal.

According the map the total sum of the weights of the buildings for unit width over the most loaded section should not exceed the value of P_{di} corresponding to the distance from the beginning or for the whole slope not more than 12 000 up to 13 613 kN/m'.

Such maps have been compiled for other regions near the towns Pernik and Blagoevgrad after which new districts have been projected.

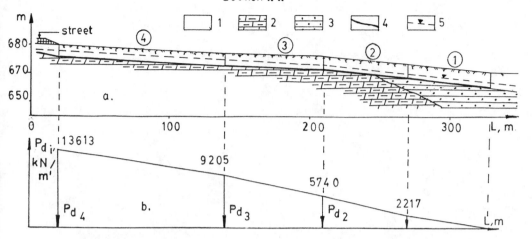

Section II-II

Fig.3 - a. Geological section II-II from the investigated slope region, 1-delluvial yellow-brown clays, 2-paleogenic marls, 3-paleogenic sandstones, 4-general sliding surface, 5-free water level; b. relationship between ΔP_{di}, P_{di} and distance from the slope toe L.

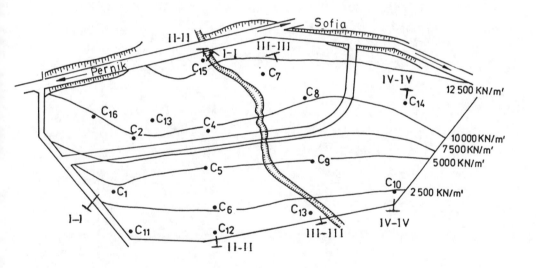

Fig.4 Map with lines of equal possible additional loading with external loads.

At the same time according us the proposed method at the present stage can be developed in the following directions:

1. Receiving more accurate criteria for intersections interpolation based on three-dimentional models.

2. Appling more accurate and complicated analytical solutions .

3. Using reasonable strenght values and F_{min} corresponding to the types of structure.

REFERENCES

Anguelov, K.A. 1988. Handbook of engineering geology, p.47-49. Sofia (in Bulgarian).
Bromhead, E.N. The stability of slopes, p. 332-334. Surrey Press, Glasgow. 1986.
Manual for stability calculations for open-pit mines. 1981. Minproject. Sofia.
Fissenko, G.L.1965. Stability of open-pit cuts and wastes. Nedra.Moscow (in Russian).

The principles and practice on drawing the maps of urban environmental engineering geology

Théories et pratiques dans l'établissement de cartes géotechniques de l'environnement urbain

Li Shenglin & Shi Bin
Department of Earth Sciences, Nanjing University, People's Republic of China

Ren Runhu
Hydrology and Engineering Geology Team, Henan Province, People's Republic of China

ABSTRACT: The relationship between urban geologic environment and urban development and the principles of drawing environmental engineering geological maps are discussed in this paper. The urban environmental engineering geological maps should include basic element maps, man—made changes maps and rational utilization maps. The series of maps of Pingdingshan city is taken as an example in thes paper, which include 6 basic element maps, 1 man—made changes map, 5 rational utilization maps.

1 THE RELATIONSHIP BETWEEN URBAN GEOLOGIC ENVIRONMENT AND URBAN DEVELOPMENT

The urban geologic environment where a city is built is an important part of the whole natural environment and the substantial basis of urban development, Table 1 gives the list of the basic elements of the urban geologic environment. The urban geologic environment implies the superficial part of the lithosphere of the urban area, including rock, soil, water(surface and underground water), geologic structure, geomorphic features and geologic processes and phenomena. The upper limit of the urban geologic environment is the sur- face of the earth, and the lower limit is the depth of influence of human activities. In this sense, with the development of science and technology and the increasing forcefulness of human activity, the range of the urban geological environment will become deeper and deeper. The effect of human activities upon the geologic environment is already no less than that of natural geologic processes, so that we can say that the effect of human activities is a kind of special geologic process. It includes two aspects: in some cases the effect of human activities is creative and favourable but in other cases, it is destructive and disturbs the natural balance of the geologic environment resulting in geologic disaster.

Table 1 Basic elements of the urban geological environment

Material constituent	Geological structure	Dynamic process
rock, soil, surfacewater, groundwater	geologic tectonics and neotectonics, geologic structure, surface shape, geomorphic form	inner—dynamic geological processes, geo—physical and geo—chemical processes, physo—chemical and biochemical processes in rock soil and water

A city is a densely populated area and the center of human activities. A vatiety of geologic disasters and a great deal of economic loss have come to our cities due to the unfavourable effect of human activities upon the geologic environment such as land subsidence caused by unplanned extraction of groundwater, landslides due to unreasonable excavation of the slope, pollution of surface and groundwater by sewage and garbage and suface collapse by mining etc.

One of the strategical measures for developing a city is how to coordinate the relations between human engineering and economic activities and geological environment. It is unwise to boundlessly enlarge urban area. In the future, the way to develop a city should be toward underground and high altitude. Certain measures are necessary to rational use and protection of geologic environment, such as to investigate and grasp correctly the characteristics and regularity of the urban geologic environment; to predict the changing trend of urban geologic environment under the effect of human activities; to establish the data—storage system and to draw urban environment engineering geological maps. The maps should show the features of the geologic environmental elements, man—made changes, and geological data and information and should provide data for the urban planners. Maps are efficient method to coordinate the relation between human activities and geologic environment.

2 PRINCIPLES OF DRAWING ENVIRONMENTAL ENGINEERING GEOLOGICAL MAPS

Maps utilized should include the following:

1. Basic element maps of geologic environment. These maps cover the environmental geologic features and regularity of the geologic elements. Maps such as geologic, geomorphic, rock masses, soil masses and water system maps are main maps of the basic elements. The number of basic element maps depends on the complexity of the urban geologic environment. For example, an independent geomorphic map is not necessary on a coastal plain because the areas on the plain are the same geomorphic unit. If there are soft soil, expansive soil or other soils with special property, the number of maps needs to be increased.

2. Man—made changes maps of the geologic environment. These maps must show the various changes of the geologic environment under the effect of human activities, including the scale, characteristics and changing trends of the various environmental engineering geological problems. Man—made changes maps are those maps which should reflect the relation between geologic environment and human activities such as land subsidence and surface collapse maps etc.

3. Rational utilization maps of geologic environment. These are maps which should show how to make rational use of the geologic environment and point out the strategical goals which people should be aiming toward making rational use of and protecting geologic environment.

In drawing maps, the data and information must be correct. If the information is not very clear in some areas, maps are better left blank in these areas. They should be amended after the detailed works of these areas are done. We should fully use colour, coloured lines and symbols to show important parts. In order to cover much information, Figures and Tables should be used in drawing the maps.

3 PRACTICE ON DRAWING THE SERIES OF MAPS OF PINGDINGSHAN CITY

The city of Pingdinshan in Henan Province is situated on the joint area of Funiu Mountain and Huanghua Plain and is a developing energy—producing city. There are 14 coal mines which produce 17 million tons of raw coal per year and 1.3 million kilowatts of electricity. Pingdinshan has the proper condition for drawing standard maps of urban environmental engineering geology. The series of maps of Pingdingshan cityinclude the following:

3.1 The basic element maps

Basic element maps of geological environment of Pingdingshan city, which are:

    1. Geomorphological map
    2. Geological map
    3. Map of rock masses
    4. Map of soil masses
    5. Map of water system
    6. Map of distribution of expansive soil

The geomorphological and geological maps have covered the urban geologic history, geologic tectonic and information about regional stability.

176

The map of distribution of expansive soil shows not only the distribution but also the depth of the top surface of expansive soils, which is very important to engineering construction. The structure, stablity problems and engineering characteristics of rock and soil masses are the main content of the rock and soil masses maps. The map of water system, including surface water and groundwater, mainly shows the water quantity and quality of Baiguishan reservoir, Sa river and Zhan river and a falling funnel of groundwater.

3.2 The man—made changes map

The map shows the areas of gangue, surface collapse and fly ash which are caused and produced by mining and the power plant. The map indicates serious, harmful and positive developing trends of the main environmental geologic problems.

3.3 The rational utilization maps

Maps of rational utilization and protection of geological environmentin Pingdingshan city include the following:

1. Map of allowable bearing capacity of solis deep from 0 to 2 meters
2. Map of allowable bearing capacity of solis deep from 2 to 5 meters
3. Map of allowable bearing capacity of solis deep from 5 to 10 meters
4. Planning map of Pingdingshan city
5. Map of natural building material distribution

The first, second, and third map shows the distribution and foundation bearing strength of the shallow soil masses (above 10m) in three intervals: 0–2m, 2–5m and 5–10m thick. The fourth map is planning map, which include 3 regions and 9 subregions.

The region of the coal mines (I). It is divided into two subregions: hilly subregion and surface collapse subregion. It is suggested that the collapse subregion should be refilled by gangue and be converted into a fishfarm and building areas.

The region of urban engineering (II). On the basis of the characteristics of foundational soils, it is divided into 3 subregions: the subregion of expansive soils and serious pollution, the subregion of general civil engineering and the subregion for tall buildings.

The region of water resources protection and utilization (III). The region is divided into four subregions according to the types of water resources. The subregion of Baiguishan reservoir must be protected because it is the main water supply area. The hilly subregion along the reservoir should be planned into a forest area and scenic spot to provent pollution of the reservoir. The subregion below the reservoir should not be planned into the industrial area, but into the cultural and scientific area. The subregion of groundwater protection, which mainly is the flat area of an acient riverbed, is the main area of groundwater resource of Pingdinshan city. The area should not be used as building area but as a vegetable farm and be protected and used in a planned way.

The fifth map is the map of natural building materials. There are plenty of natural sandy material along Sa river. The map shows also the distribution and storage of clays which can be used to make brick and cement.

The scale of all the maps is 1: 5000

REFERENCES

Li Shenglin: Rational utilization of the urban geological environment. The Role of Geology in Urban Development Geological Society of HongKong Bulletin No. 3, October 1987.

Li Shenglin: Problem on rational utilization of geological environment. Journal of Nanjing University 1987 No.1

Hydrology and engineering section of Nanjing University "Engineering geology" Geology Publishers 1982, pp 326–332

MAPS OF ELEMENTS OF GEOLOGICAL ENVIRONMENT IN PINGDINGSHAN CITY HENAN PROVINCE (No.2)

Geological Map

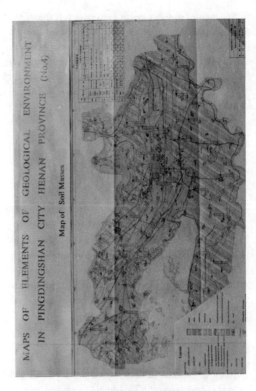

MAPS OF ELEMENTS OF GEOLOGICAL ENVIRONMENT IN PINGDINGSHAN CITY HENAN PROVINCE (No.4)

Map of Soil Masses

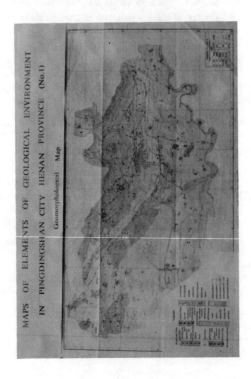

MAPS OF ELEMENTS OF GEOLOGICAL ENVIRONMENT IN PINGDINGSHAN CITY HENAN PROVINCE (No.1)

Geomorphological Map

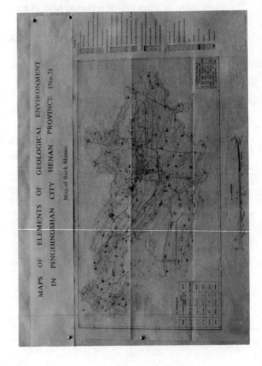

MAPS OF ELEMENTS OF GEOLOGICAL ENVIRONMENT IN PINGDINGSHAN CITY HENAN PROVINCE (No.3)

Map of Rock Masses

178

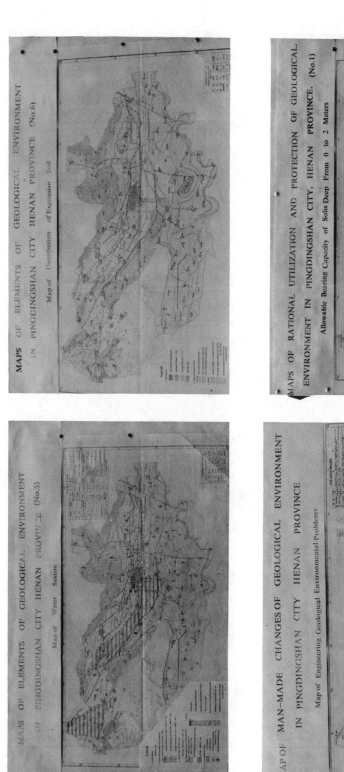

MAPS OF ELEMENTS OF GEOLOGICAL ENVIRONMENT IN PINGDINGSHAN CITY HENAN PROVINCE (No.6)

Map of Distribution of Expansive Soil

MAPS OF ELEMENTS OF GEOLOGICAL ENVIRONMENT IN PINGDINGSHAN CITY HENAN PROVINCE (No.5)

Map of Water System

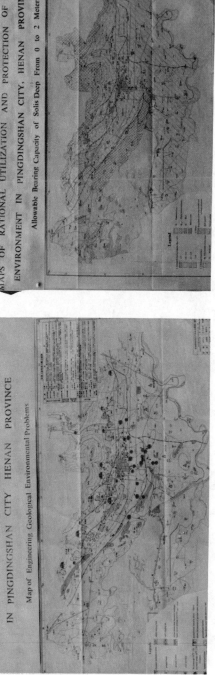

MAPS OF RATIONAL UTILIZATION AND PROTECTION OF GEOLOGICAL ENVIRONMENT IN PINGDINGSHAN CITY, HENAN PROVINCE. (No.1)

Allowable Bearing Capacity of Soils Deep From 0 to 2 Meters

MAP OF MAN-MADE CHANGES OF GEOLOGICAL ENVIRONMENT IN PINGDINGSHAN CITY HENAN PROVINCE

Map of Engineering Geological Environmental Problems

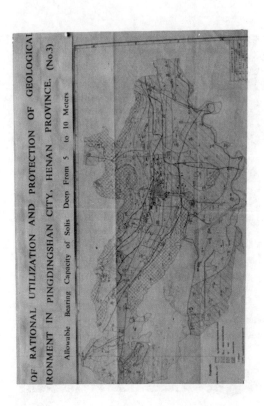

MAPS OF RATIONAL UTILIZATION AND PROTECTION OF GEOLOGICAL ENVIRONMENT IN PINGDINGSHAN CITY, HENAN PROVINCE. (No.3)

Allowable Bearing Capacity of Solis Deep From 5 to 10 Meters

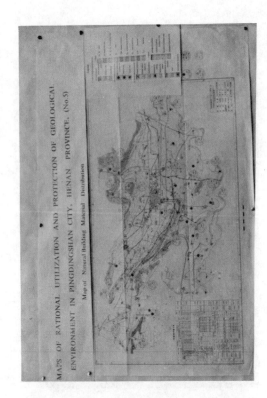

MAPS OF RATIONAL UTILIZATION AND PROTECTION OF GEOLOGICAL ENVIRONMENT IN PINGDINGSHAN CITY, HENAN PROVINCE. (No.5)

Map of Natural Building Material Distribution

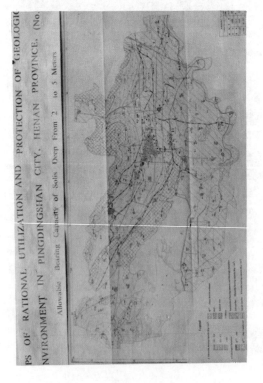

MAPS OF RATIONAL UTILIZATION AND PROTECTION OF GEOLOGICAL ENVIRONMENT IN PINGDINGSHAN CITY, HENAN PROVINCE. (No.)

Allowable Bearing Capacity of Solis Deep From 2 to 5 Meters

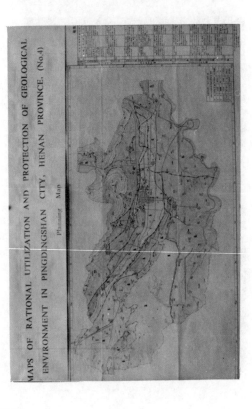

MAPS OF RATIONAL UTILIZATION AND PROTECTION OF GEOLOGICAL ENVIRONMENT IN PINGDINGSHAN CITY, HENAN PROVINCE. (No.4)

Planning Map

180

Destruction and protection of geological environment in urban areas
Discussion on Belgrade example

Destruction et protection de l'environnement géologique dans des zones urbaines
Discussion sur Belgrade

Petar Lokin
Faculty of Mining and Geology, Belgrade, Yugoslavia

Milica Bajić-Brković
Faculty of Architecture, Belgrade, Yugoslavia

SUMMARY: The geological environment as a part of the human environment is mostly endangered in the urban areas. This paper, having Belgrade as an example, deals with various aspects of degradation of the geological environment and the necessity of its rational use and protection in function of the development planning and designing construction projects.

RESUME: Le milieu géologique, en tant que partie intrégrante de l'environement naturel, est le plus menacé dans les zones urbaines. L'étude, avec l'exemple de Belgrade, examine les différentes formes de dégradation du milieu géologique, ainsi que les nécessités d'utilisation et de protection rationnelle dans le cadre de la planification du développement de l'espace urbain.

1. INTRODUCTION

Geological environment as an important part of the human environment has particularly been exposed to degradation in urban areas. There it has due to the conditions of utilization been exposed to the most intensive and different harmful effects. Among them both natural processes and various technogenous activities emerge. These areas suffer from specific interaction between nature and human being and therefore many of the processes develop and appear in a different shape than in a free area.

As the space and ecological capacities of the geological environment, particularly in urban conditions, are nevertheless limited, it even more acquires treatment of an important natural resource that should be utilized rationally, with due care, but with necessary measures of precautions. Accordingly, the relation between the rational utilization of geological environment and regional and urban planning is stressed in this paper.

Belgrade has been taken as an example to illustrate our considerations, fairly from the reason that we have excess to the information on many problems the city is faced with, but also from the certain specificalities it is characterized by. Just to mention, primarily, quite an extensive development it has undergone during the last few decades, particularly the development of underground construction as well as specific geological conditions evidenced by the fact that the majority of the town is located on the terrain, the surface part of which in fact is loess, highly sensitive to water and containing neogene sediments with numerous "fossile" landslides beneath.

2. BASIC DATA ON THE DEVELOPMENT OF BELGRADE

Belgrade with its 16 municipalities makes regional community and represents the largest settled agglomeration with the highest population density in Serbia. Besides, Belgrade is economic, traffic, as well as social and cultural events center in Serbia.

The wider area – Belgrade region stretches over 3 244 km^2 and has

approximately 1,5 milion of inhabitants. The city of Belgrade by itself
covers 20% of the whole territory, of
population, and 72% of the total number
of job positions.

Quite an intensive construction
distinguishes the whole post-War
period, particularly two last decades.
Construction of residential housing as
well as supportive public utility
projects, infrastructure facilities,
industrial plants, sports recreation
grounds and others had a dominant part
in the overall construction. Due to
that fact the town rapidly extended its
limits and thus the city area has
doubled each 30 years in an average.

Development in the near future
according to the present expectations
will be characterized by an increased
volume of necessary actions in the
existing city facilities (reconstruction, revitalization, change in land
use, etc), construction of traffic
corridors and infrastructure network
(pipeline and similar) as well as new
construction in the industrial areas.

Having in mind the efficient use of
geological environment it is important
to point out that so far within the
inner city area more than 40 km of
railway, road and hydrotechnical
tunnels has been constructed as well as
a great number of underground warehouses, shelters, and garages; that a
hundred of kilometers of shallow dug,
public utility installations have been
made and also a great number of high-
rise projects with one or more underground floors. There is an intention in
Belgrade to utilize even more the
underground space of the city for its
further development. The construction
of more road and hydrotechnical tunnels, extended underground network of
heating and gas pipelines is in the
course or is at the very beginning.
Also construction of metro is planned.

3. DEGRADATION OF GEOLOGICAL ENVIRONMENT
   IN THE PRESENT DEVELOPMENT OF THE TOWN

Out of the quite numerous and various
aspects of geological environment degradation within the Belgrade area, only a
small number has been demonstrated and
analysed in this paper but the examples
chosen were those which due to the town

characteristics and geological conditions
have the greatest influence to the
present and will probably be of great
importance for the future development of
the town.

Terrain instability occurences

Terrain instability occurences - land
slides within the Belgrade have been very
frequent. Predominantly, two main types
of slides appear:

1) Landslides in eluvial-deluvial
layers of mesosoic sediments, generally
less in depth and thickness are mostly
located outside the proper city area.

2) Landslides in neogene sediments
stretch mainly along the weathering zone
of those sediments (Sarmatian-Panon).
Numerous investigations show that during
the loess precipitation ($Q_1^?$, Q_2) almost
all the slopes made of neogene sediments
were caught by sliding processes. Loess
layer, thickness of up to 20 m had a
double influence on the sliding process.
It insulated neogene clay-marl sediments
from further action of exogene agens and
represented the natural contra for on the
instable terrain. It lead to an almost
complete stabilization of the slopes and
prevented further development of the
process. Previous slides presently
situated beneath the looss are called
"fossil". The errosion of rivers and
springs where a part of loess layer has
been destroyed made many of those slides
active again. It is our estimation that
within the wider city area such slides
active or temporary inactive exceed today
several hundredths. Some of them being of
extremeely large dimensions. Such is, for
example, the slide "Duboko" with a
surface of about 25 ha, depth up or 26 m
and volume of around 5 millions of m^3.
Head scarp of this slide is high about 30
m. Building of a highway over this slide
increases the average price by 10 times.

Many of the slides have been reactivated by inadequate human activities -
ground works during foundation works,
construction of infrastructure projects,
portal parts of tunnels etc. Thus, the
known landslides have been formed during
the construction of the highway through
Belgrade, on railways in the vicinity of
the city, slides on portal parts of
basic-road and water supply tunnel,
slides found during excavation carried

out for "Crvena Zvezda" stadion as well as during excavations of foundation pits for many facilities. Such numerous, unfortunately bad experiences, led to a smaller number of such kind of "surprises" today as the technical solutions for construction works have mostly been adjusted to such geological conditions.

Changes in natural regime and chemical properties of underground waters

In the old "core" of the town the public utility facilities being made 50-70 years ago, became worn out and tehrefore a large quantity of water running through water-supply and sewer network gradually disappears and leaks into the underground layer. As per certain estimates the figure is even more than 40% of water running through the water supply network. This fact mostly changes the natural regime of underground waters and creates a new unfavourable natural regime as it stimulates formation of slides and has a destructive effect on loess sediments. This creates the degradation of loess structure, considerable decrease in bearing capacity and settlement of the existing facilities. In the recent years a great number of the old facilities even ater 60-70 years of normal usage underwent considerable damages in this way. Among them are many facilities of public use. It sounds like a joke, but it is true that in this group the buildings of Serbian Academy of Sciences and Arts, University Library, Faculty of Technical Sciences, Faculty of Mining and Geology, Federal Geological Institute and many othes may be classified.

This problem now-a-days proves to be very present not only for rebuilding of the damaged facilities but for the protection of other ones being still in good order. At the same time, while planning and carrying out the revitalization of the old city "core" being in full swing, the degree of disturbances of geological conditions should be determined and possible new disturbance be estimated.

Besides, such changes in underground regime in many cases conditioned dampness and even penetration of water in the underground floors of certain buildings. Changes in chemical properties of underground waters also occured and they, as a rule, became aggressive to concrete and other building-construction materials. This mostly appears to be the aggressiveness of sulphite-type.

The question of polution of ground and underground waters within the wider city area, particularly in the zones of water-supply wells in Belgrade is also very present. Polution of underground waters is mostly direct and usually is made through poluted river waters which became collectors of almost all the quantities of waste and other poluted waters (sewerage for faeces, rain and mixture of both). So, for example, only within the proper area of the main Belgrade well based on exploitation of underground waters is from the alluvion of the Sava river, eleven large flows of waste waters directly into the Sava river have been registered near to water-catchment facilities. Besides, there is a series of internal sewer ourflows belonging to either smaller or larger facilities. The existing degree and type of polution of the Sava water (oil, phenol, etc.) condition the temporary disconnection of wells from the water supply system and endanger successful operation of the plant for river water purification (processing).

Direct polution of underground waters is also caused by various city dumps waste disposal and on the occasion of different accidents. Thus, one of the former city dumps was located on the river island on the Danube river. Tehere were several accidents that conditioned polution of the underground waters, and the most dangerous one was the outflow of xylene from vaggon-tank at marshalling yard "Makiš" near to the very city well. Even after five years traces of xylene in the underground layers ate still present and therefore a regular control of concentration and widespreading of contaminated zone is carried out.

Waste Disposal

For dumping of approximately 1000 tons of city waste per day Belgrade disposes of only one and inadequately equipped city waste disposal.

Area for collecting waste is 76 500 ha approximately. For the purpose of the long-lasting solution, Belgrade chose waste disposal site as the most simple and cheep solution. All other methods of upgraded technology are considerably expensive.

Thorough investigations have been made in sellecting the new waste disposal sistes within which, based on the specific model, a complex evaluation of each alternative location has been carried out. Out of the total number of 46 evaluation criteria, the very half refers to the properties of geological environment.

A serious problem represents elimination of dangerous matters - various chemicals, radioactive waste and others that cannot be damped along with the city waste. Since recently this problem has been delt with by many experts and geologists as well.

Earth Works

Similar to any other urban area daily a large number of earth works - excavation and filling have been carried out in the Belgrade area. The largest extent of such work in the overall postwar period was the filling of the wide surface on the left bank of the Sava river in front of the mouth into the Danube by pumped sand from the Sava and the Danube beds. The filled sand stretches over the surface of aounr 20 km^2 the thickness of which is up to 5 m. This was done for the purpose of improvement of urbanization conditions of so far porous and morass soil. This example clearly indicates that technogenous activity reasonably directed does not necessarily lead to the degradation of the geological environment but to the contrary, as evidenced in this very case, may contribute to its improvement.

The largest excavations within the wider Belgrade area are carried out at coal and building construction materials (quarries and clay pits for brick industry). The fact that in the very vicinity of the town there is one of the largest Yugoslav coal mines with accompanying energy capacities, for sure burdens additionally the ecological conditions of this area.

However, excavations, fillings and transportation of earth materials performed in the daily building construction activities at excavation of foundation pits, construction of roads, terrain levelling and others may not be neglected. The best illustration on the quantity of materials in question is the example of "Bežanijska Kosa" housing project where reclamation of 2.000 million of m^3 of land has been projected. Dependent on the urban sollutions that covers 100-200 and in particular cases even up to 500 m^3 of land per one flat. In case of inadequate solutions of distribution of excavated soil, the situation occurs that throughout the town thousands of m^3 of excavated soil circulates daily thus causing the traffic difficulties ruining the pavement, roads, dirtying the town and etc.

Consequences caused by building and exploitation of underground facilities

As already indicated, a great number of underground projects both shallow installations and deep tunnels have been constructed. Shallow installations - underground floors and city installations generally influece the decrease of the total surface capacities of that part of the geological environment. In cases of inadequate works or usage there appears the stability of soil and surrounding facilities particularly due to the specific properties of loess forming the surface parts of the terrain in the major part of the town territory.

At building the deep underground facilities - tunnels for different purposes, shelters, warehouses and others more serious problems appear. As the majority of slopes is instable or labile at excavation of the portal parts activating of the slides occurs. More or less this problem was very present regarding some tunnels that are being built (basic tunnel, water supply tunnel at Franse D'Epere square, Kneževac railway tunnel and others). Therefore today in similar conditions of building the portal parts it is projected that it be carried out by using reinforced - concrete diafragms considerably influencing the price of building but diminishes the risk of demaging the surrounding facilities.

The second problem that is frequently faced with is the surface subsidence above underground facilities. Its speed and quantity have, besides the size of profile and facility depth, strictly been dependent to the geological structure of the terrain. The most unfavourable conditions appear in the weathering zone of the neogene sediments. Even usage of full-face machines failed to secure the degree of subsidence that could be neglected.

Besides all the above stated series of other changes in the geological environment appear (change in level and underground waters regime, change in dampness and temperature, dynamic influence on the soil and others) but those when Belgrade is concerned are of a minor importance.

In this paper we have stated and briefed those occurances that in our opinion have the greates influence of the geological environment degradation. The total number of the same is for sure considerably higher but their effects are frequently interdependent, intertwine and doubled.

4. NECESSITY FOR MORE COMPLEX TREATMENT OF THE GEOLOGICAL ENVIRONMENT AND ITS PRESERVATION IN THE FURTHER DEVELOPMENT OF BELGRADE

It is evident that there is a necessity for more elaborate and thorough treatment of the geological environment in regard to furhter urban development. This fact under the existing planning practice is not taken into consideration to the favourable extent, but the rising degradation of the environment and more vigorous effects of the existing development that are today emerging, remind on the importance of this relation and oblige to undertake necessary changes. It is our opinion that taking the proposals explained herein into consideration would make this condition partly satisfied.

The starting point of our argument is that the geological environment is a resource, a limited property that must be preserved, protected and rationally utilized. While studying the matter of urban development and different construction works adopting of these postulates would lead to changes as regard to:
. information data for development planning and construction,
. criteria for development/project evaluation,
. criteria for decision making, and
. planning and designing method itself.

The following items are to be considered to be of the highest importance:

(1) Determining the geological environment capacity

Urban development compatible with the environment in a broader sence and geological environment in a strict sence, is based upon its real potentials (possibilities), and in close connection with local geological characteristics and conditions. As for the protection and preservation, there are quite specific requirements that have to be included into the information system as the basis for future development and construction. In case of Belgrade, these would mean:
(a) Designation of the areas that are to be preserved and protected,
(b) Evidencing the areas of limited capacity and intensity of use,
(c) Evidencing the ares that require specific treatment with noting the usage conditions as well, (d) Evidencing the areas where new developmnet may take place, with stated preconditions that are to be undertaken and specific requirements that are to be fulfilled. That would mean that a new zoning is needed, both as a part of a data bank as well as in the form of a geological environment capacity zoning map. This would be followed by determined criteria and standards that are to be taken into account. These all together would lead to a more responsive planning and construction.

(2) Environmental impact assessment study (EIAS) as the integral part of the plan

The current practice in Yugoslavia is not familiar with this requirement. Rough estimations about impacts that are usually made so far prove not to be sufficient. Thus, the new thought in the profession is pushing this question further. The more precise analyses and elaboration of more complete studies is supported at least by following reasoning:
(a) The development plans and projects have usually multiple and complex im-

pacts. Certain groups may have synergetic effects: being of minor importance taken one by one, taken together may produce major impact.

(b) Certain impacts are growing by time. If the expected behaviour is not systematically investigated and stated in advance, the insight in possible consequences and related requirements (further resource engagement, investements etc.) is not available. And this is very important when decisions about further urban development are to be made.

(c) Some of the impacts are likely to require detailed consideration. The environment impact study is made to help to determine these questions.

(d) Better decisions can be made if the various parties involved have a clear understanding of the likely impacts of development or project.

(e) The urban development or project cnostruction have the real cost. This cost is combined of both the direct and indirect spending. The environmental impact assessment helps to determine the relevant components that should be taken into account.

(3) Monitoring the cumulative effects

Many of the effects are expected and as such are to be included in the elaboration of the EIA studies. Based on results, the decision on development and construction is passed as on the simultaneous activities with the aim to minimaze the side effects. Taking into consideration the degree of reliability of the predictions we count on today, it can be confirmed that as for the questions of different construction works, a large part of consequences and future states may be anticipated and included into the project ex ante.

Quite an opposite situation is, however, in case of long range development plans. Here, the projections cover only certain items, and usually in a general manner. The long range plan is not a detailed plan, and geological environment is correspondingly treated. As for the global plan, a five year screening is made, on which ocassion in accordance with changes occured in the meantime the corrections of the plan are adopted, namely main marks for the detailed and structure plans. Such a possibility, however, does not cover

the adequate input as for the geological environment, to the extent and quality which geologist are able to offer. It is our strong belief that the professional practice and quality of decisions will be improved if periodical screening concerning geological environment would be made as well as regular monitoring of changes.

(4) Determining the development cost that includes protection of the geological environment as well

Urban development and different construction works have a real price. In the existing urban planning practice it is respected in a quite simple way: for smaller units and some aspects of the development, estimation of investments are the integral part of the plan. As for a town or a region such mechanisms have not been more thoroughly developed.

The price of preservation and protection of the geological environment is complitely missing here. The absence of the obligation to EIA study makes it quite understandable. The latest efforts in the profession, however, insist just on this segment of the plan. Requirement for more rational and efficient planning and rational and efficient use of space give full support to this opinion. A good planning is one that maximizes positive and minimizes negative effects, and at the same time minimizing the overall development costs.

As for the environment protection, determination of the developing price will be closely connected to the elaboration of the study of effects on the environment. The matters of concern are: (I) degradation costs, (II) preservation and compensation costs, and (III) costs that will be emerged by the lapse of time – exploitation costs. It is quite clear that here we have to take into account the preciseness of the input data and choice of effects that are included but this does not diminish the importance of decision making. The greatest harm is when they are completely omitted. Given even generally they complete the "image" of the problems we face and extent of engagement of those to which they relate.

5. CONCLUSION

The constant tendency towards the degradation of the geological environment in the Belgrade urban area is unacceptable as it could lead to the deterioration of the living conditions and considerable material damages.

The more complete treatment of the geological environment in the practice of the town development planning and construction as elaborated in the previous chapter, puts certain specific tasks in front of the geologists.

1) Eliminating of practice to make so-called "geological base" for plans and instead making the full integration on geological investigations in the field of planning. This for sure means including the experts from the science of geology into the planning teams.

2) Carrying outh the registration and the analysis of all the aspects and cases of the existing degradation of the geological environment as well as the possible consequences and rehabilitation conditions.

3) Arranging the complex monitory of the geological environment for the purpose of observing all the processes and occurences both natural and technogenous that may lead to its degradation.

4) Protecting of the geological surrounding from further degradation, recultivation of the destroyed parts and possibly the improving should be projected through the plans of the surface development of the town and relevant projects and carried out in all the parts of the activity - construction, transportation, exploitation of raw materials and energy resources and others. This represents the tast of state namely city and municipal bodies to control carrying out of such a policy and although those bodies had been constituted a long time ago, in our opinion they are not yet quite efficient.

5) In Yugoslavia and in Belgrade as well it is indispensable to update the relevant legal and technical regulations, training of the relevant stuff and educating the citizens to create quite a different attitude towards the living environment as a whole in order to protect and improve it more successfully.

Surface geotechnical mapping of São Carlos, Brazil

Carte geotechnique de surface de São Carlos, Brésil

Reinaldo Lorandi, Adail Ricardo Leister Gonçalves & Maria Lúcia Calijuri
Federal University of São Carlos, Brazil

ABSTRACT: Surface geotechnical map of São Carlos (São Paulo, Brazil) urban and suburban area was elaborated with basis on the interaction among pedological associations and their mechanical properties. From these parameters it was possible to identify four homogenous pedogeotechnical units: Soak soils area; Litholic soils area; Sand soils area and Lateritic soils area. The comparative study of an application of two mechanical technics on soils of this geographic area has shown that these technics have limited applications on lateritic or non-lateritic soils and that it necessary to carry out complementary studies involving geomorphological, pedological and genetic criteria.

RESUMÉ: Le établissment des unités geotechniques homogène of fait par la interation entre las associations pedologique et de las caractéristiques mecanique des sols de chaque échantillon superficial. Pour cette méthode on été possible la identification de quatre unités regional pedogeotechnique homogène: des sols hidromorphique; des lithosols; des sols sableuse et des sols latéritique. De le étude comparative on peut déduire qu'il y a une restriction de la application de ces deux technique dans cette region geographique et de la nécessité de development d'étude sur la seletion des critères geomorphologique, pedologiques et the génèse des sols pour leur application.

1 INTRODUCTION

This paper which presentes the surface geotechnical map of São Carlos area was elaborated with basis on the interaction among pedological associations and their mechanical properties.

The purpose of the investigations was to lay down a simple geotechnical classification of tropical soils in São Carlos area and to illustrate it in a simple form on a surface geotechnical map for the use of Civil Engineers, especially for general planning.

The area of interest is located in the central region of the Brazilian State of São Paulo. It approximated limits are $22^0 00'$ of south latitude and $47^0 55'$ and $47^0 55'$ of west longitude.

2 THE PEDOLOGICAL UNITS - THE PEDOLOGICAL ASSOCIATION

The interpretation of the soil classification and accompanying map of São Carlos municipality, situated in the Brazilian State of São Paulo, was made in 1985 (Lorandi, 1985) to obtain data for urban and suburban development planning.

The soil survey refers to an area of $1,20725.10^8 m^2$ situated in the São Carlos Plateau. The soil units (see Fig. 1) were identified by the study of nine soil profiles, which were morphologically described and sampled for physical, chemical and mineralogical analysis.

Six cartographic units were identified: five of them refer to a single taxonomic unit and one is an association. The simple units are: Typic Haplorthox, Distropeptic Quartzipsamment, Paleudultic Haplorthox, Distropeptic Haplacrox and Histic Psammaquent. The association is formed by a Lithic Hapludent and Oxic Distroptet.

Based on physical, chemical and mineralogical data of urban and suburban soils of São Carlos city area, Lorandi & França (1986/87) were able to identify four pedological associations and two geopedological environments (colluvial and alluvial processes) as a function of their geological origins. Regarding to their evolution conditions, the soils are quite advanced, specially the lat

Figure 1. Pedological map

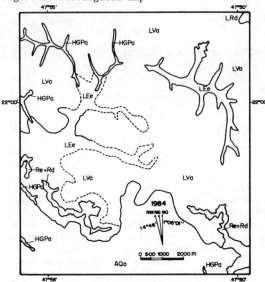

LEGEND

LVa – Red Yellow Latosol
LEa – Dark Red Latosol
LRd – Dusky Red Latosol
Re+Rd – Litholic soils
AQa – Sandy soils
HGPa – Hidromorphic soils

osols, which can be seen on the observed wheathering levels and on the clay mineralogy represented by the kaolinite group.

The above-mentioned four pedological associations are:

a. Association 1: Litholic soils (Lithic Hapludent and Oxic Distroptet)

b. Association 2: Soal soils (Histic Psammaquent)

c. Association 3: Red yellow Latosol and Sandy soils(Typic Haplorthox and Distropeptic Quartzpsamment)

d. Association 4: Dusky Red Latosol and Dark-Red Latosol (Distropeptic Haplacrox and Paleudultic Haplorthox).

3 GEOTECHNICAL CLASSIFICATION OF SOILS

The soils classification presented in this paper consists of determining for each pedological units, one or two of the following mechanical properties:

. Classification for tropical soils(MCT)- this mechanical test was based on a new classification, developed by Nogami & Villibor(1980),for tropical soils which uses the results of a simplified test process. This process involves compacting small size specimens(0,05m diameter) made according to the principle developed by Parsons(1976)for the determination of MCV (Moisture Condition Value).

. Unified Classification System - U.S.C.

4 THE PEDOGEOTECHNICAL UNITS - THE SURFACE GEOTECHNICAL MAP

From the description of pedological units and the determination of their geotechnical properties, it has been deduced that one can divide the investigated area into four pedogeotechnical units. Each unit corresponds to one or more pedological units and has more or less equal geotechnical properties.

The description of the pedogeotechnical units is summarized in tables 1 and 2 and in the legend of Fig.2, an estract of which is attached; they are numbered from 1 to 4.

The studies performed showed two different situations when applied to Casagrande and Nogami & Villibor (1980) metodologies:

a. The geotechnical and pedological unities are equivalent-this corresponds to the geotechnical unities number 1 and 2 respectively, represented by the pedological unities Histic Psammaquent and Distropeptic Quartzpisamment. Thus, for the geotechnical unity 1, soaked soils predominate all over the year, being texturally contituded by a true sandy clay loam matrix and also mineralogically contituded by clay minerals from the group 2:1 (expansives) and presenting medial plasticity.

On the other hand, the Distropetic Quartzpsamment are constituted by low-evoluted soils whose textural and mineralogic characterisitics keep tight relations with the subjacent material of the profile, whitout important variation as far as the studied depth.

b. One geotechnical unity corresponds

Table 1. Textural and mechanical characteristics of the pedogeotechnical unities 1,2,3

Pedogeotechnical unit	Pedological unit	Horizon	Depth (m)	sand	silt	clay	unified
1	Histic Psmmaquent	A_1	(0-0,07)	68,0	6,0	26,0	SC
		A_3	(0,07-0,35)	62,0	7,0	31,0	SC
		B_1	(0,35-0,45)	68,0	4,0	28,0	SC
2	Distropeptic Quartzpsamment	B_{1g}	(0,45+)	63,0	5,0	32,0	SC
		A_{11}	(0-0,35)	89,0	0,2	10,8	SP
		A_{12}	(0,35-0,70)	87,9	0,2	11,9	SP
		C_1	(0,70-1,25)	86,8	0,3	12,9	SP
	Distropeptic Quatzpsamment	C_2	(1,25-1,85)	86,2	0,2	13,6	SP
		A_1	(0-0,65)	92,7	0,2	7,1	SP
		C_1	(0,65-1,40)	93,3	0,2	6,5	SP
		C_2	(1,40-1,80+)	93,2	0,3	6,5	SP
3	Oxic Distrop tet	A_1	(0-0,40)	47,4	47,7	4,9	SP
	Lithic Haplu-dent	A_1	(0-0,30)	32,0	32,9	35,1	CL

Table 2. Textural and mechanical caracteristics of the pedogeotechnical unity 4

Pedogeotechnical units	Pedological units	Horizon	Depth (m)	sand	silt	clay	Unified	W %	ρ Kg/m³	MCT
4	Typic Haplorthox	A_1	(0-0,42)	65,0	3,0	32,0	SC	17,00	1780	LG'
		B_{21}	(0,42-1,15)	65,5	6,5	28,0	SC	17,40	1765	LG'
		B_{22}	(1,15-200+)	57,6	3,8	38,6	SC	18,25	1735	LG'
	Paleudultic Haplorthox	A	(0-0,40)	73,5	12,7	13,8	SC	11,90	1973	NA'
		A_3/B_1	(0,40-0,75)	77,3	8,4	14,3	SM	14,25	1880	LG'
		B_{21}	(0,75-1,20)	74,2	·7,6	18,2	SC	18,00	1812	LG'
		B_{22}	(1,20-1,70)	67,4	7,0	25,6	SM	18,00	1812	LG'
		B_{23}	(1,70-1,90+)	65,5	7,2	27,3	SC	17,75	1800	LG'
	Distropeptic Haplacrox	A_1	(0-0,35)	19,3	25,2	55,5	CL	20,50	1700	LG'
		A_3	(0,35-0,70)	21,2	24,5	54,3	CL	20,50	1704	LG'
		B_{21}	(0,70-1,40)	20,6	26,9	52,5	CL	21,54	1740	LG'
		B_{22}	(1,40-2,10)	20,2	26,5	53,3	CL	20,75	1730	LG'

Figure 2. Surface geotechnical map

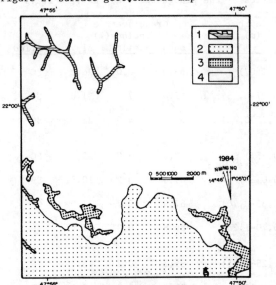

to two or more pedological unities-this situation refers to the Lithosoils and Latosoils which correspond respectively to geotechnical 3 and 4.

Although it had been possible to separate the litholic soils(Rd and Re) by morphological, chemical, physical and mineralogical properties, this was not possible in geographical terms, because of the intercalation among the sandstones layers from Botucatu Formation and the basical intrusions from Serra Geral Formation. Therefore, they constitute a geopedologic association (Lorandi & França, 1986/87) characterized by litholic soils associated to basalt levelling (Oxic Distroptet) and sandstones (Lithic Hapludent) with textural aspect from clay loam to sandy loam from medial plasticity to non-plasticity and from expansive to non-expansive respectively.

The mapping units classified as Latosoils, is despite of belonging to two distinct geopedological associations (Lorandi & França, 1986/87), were enframed into the same geotechnical unity. When the results from the mechanical properties are compared(table 2) we can observe that although we would expect from the Unified Classification of Casagrande having pedological unities belonging to several geotechnical ones,this never happens when they are classified by the methodology proposed by Nogami&Villibor(1980) . Except for the horizon Ap from the Dark Red Latosol (Paleudultic Haplorthox) classified as sandy non-lateritic the other horizons have the same mechanical properties , being classified as clayed lateritic.

The pedogeotechnical units are plotted on the São Carlos City urban and suburban map, which forms-with the pedological map as a background - the surface geotechnical map.

5 CONCLUSIONS

a. The comparative study of an application of two mechanical technics on soils of this geographic area has shown that these technics have limited applications on lateritic or non-lateritic soils and that it is necessary to carry out complementary studies involving geomorphological, pedological and genetic criteria.

b. It has been shown,for São Carlos urban and suburban area, that different pedological associations can be grouped by a combination of pedological and mecanical characteristics in a unique pedogeotechnical units.

c. It has been shown, for São Carlos urban and suburban area, that different pedological associations can be grouped by a combination of pedological and mechanical characteristics in a unique pedogeotechnical units.

d. The map, as presented, is designed for urban and suburban general planning. Detailed planning requires more information but the existence of this map should simplify others works.

ACKNOWLEDGEMENTS

The authors are indebted to FAPESP (Funda-
ção de Amparo à Pesquisa do Estado de São
Paulo),that supplied the financial support
to carry out part of this work.

REFERENCES

Lorandi, R. (1985). Caracterização dos so-
    los das áreas urbana e suburbana de São
    Carlos(SP) e suas aplicações. (Soil char
    acterization of São Carlos urban and sub
    urban Areas). 181p. PhD thesis. Universi
    dade de São Paulo, Brazil.
Lorandi, R. & França, G.V.(1986/87). Inter
    pretação geopedológica das áreas urbana e
    suburbana de São Carlos,SP. I. Ambientes
    Geopedológicos. (Geopedological Interpre
    tations of São Carlos urban and Suburban
    Areas. Geopedological Environments). Geo
    ciências 5/6:131-150.
Nogami, J.S. & Villibor, D.F. (1981). Uma
    nova classificação de solos para finali-
    dades rodoviárias. (A New Soil Classifi-
    cation for Highway Purposes). Anais Sim-
    pósio Brasileiro de Solos Tropicais em En
    genharia, 30-41. Rio de Janeiro, Brazil.
Parsons, A.W. (1976).The Rapid Measurement
    of the Moisture Condition of Earth Mov-
    ing Material, L.R. 750, Transport and
    Road Research Laboratory, Crowthorne, U.
    K.

Application d'une cartographie géoenvironnementale au littoral du Roussillon

Application of a geoenvironmental cartography to the Roussillon's littoral

M. Masson & P. Buquet
Centre d'Etudes Techniques de l'Equipement Méditerranée Aix-en-Provence, France

C. Allet
Direction Départementale de l'Equipement Perpignan, France

RESUME : Sur le littoral du Roussillonnais, l'urbanisation intensive doublée d'aménagements lourds crée des problèmes croissants de cadre de vie, de pollutions, de risques d'inondation et de disparition des zones naturelles. Afin de repenser l'aménagement à un niveau supra-communal et en prévision du long terme, une approche à base géo-environnementale, expérimentée partiellement par ailleurs, a été programmée.
En phase analytique, une cartographie géomorphologique appliquée a permis d'identifier et de délimiter des unités homogènes dont les caractéristiques ont été analysées en termes d'aptitudes et de risques, par rapport à l'eau (inondations - perméabilité - nappes), à la géotechnique (compressibilité - ressources en matériaux), à l'agriculture et aux écosystèmes. Puis ont été évaluées les corrélations entre ces caractéristiques et les modes d'utilisation des sols. Des inadéquations ont été répertoriées, en termes de planification de l'espace et de techniques d'aménagement.
Des propositions argumentées économiquement ont été formulées pour la planification et pour l'application de techniques alternatives adaptées aux caractéristiques des unités géomorphologiques (pour la lutte contre les inondations, l'épuration des eaux, la réalimentation des nappes).
La préservation et la mise en valeur de l'Environnement et des paysages constituent également des retombées positives de l'étude, et permettent d'envisager un essor équilibré du tourisme.

ABSTRACT : On the Roussillon's littoral, the intensive urbanization with leavy developments creases increasing problems concerning environment pollutions, flood risks, disappearance of natural areas. In order to reconsider the development on a supra communal level and in long term anticipation, a geo-environmental approach wich was partly tested, has been planned.
In analytical phase, an applied geomorphological cartography enabled us to identify and define homogeneous units whose characteristics were analysed in terms of aptitudes and risks in relation to water (floods, permeability, sheets), geotechnic (compressibility, materials resources), agriculture and ecosystems.
Then, the corelation existing between these caracteristics and the methods concerning soils utilization were estimated. Inequations were listed in terms of space planning and technics of development.
Proposals, argued from an economic point of view, were formulated for the planning and the application of alernative technics adapted to the characteristics of geomorphological units (fight against flood, water purification, sheet supply). The preservation, the environment and the landscape development also constitue positive repercussions of the study and emable us to consider a balanced rise of tourism.

1 INTRODUCTION

La plaine littorale du Roussillon présente des caractéristiques communes avec l'ensemble du secteur côtier du Languedoc Roussillon :
- milieu littoral et lagunaire sensible, en relation de proximité avec un arrière pays montagneux,
- forte pression humaine, agricole et surtout urbaine, liée au développement touristique.
- contraintes naturelles importantes, liées en particulier aux risques d'inondation induits par le climat méditerranéen.

L'expérience que nous relatons provient d'une étude réalisée dans le cadre d'une démarche de planification, en l'occurence celle de la révision des Plans d'Occupation des Sols (POS), pour une

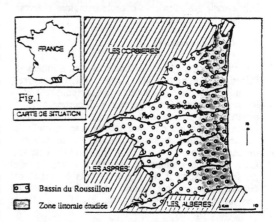

Fig.1

CARTE DE SITUATION

[⊟ ⊟] Bassin du Roussillon
[▦] Zone littorale étudiée

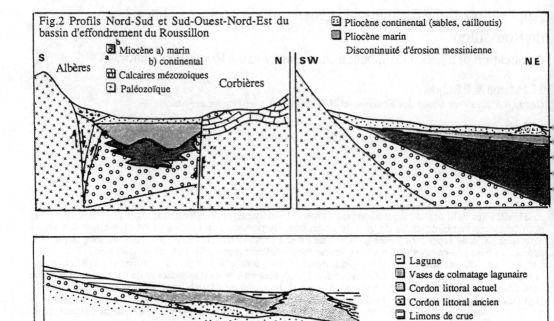

Fig.2 Profils Nord-Sud et Sud-Ouest-Nord-Est du bassin d'effondrement du Roussillon

Miocène a) marin
b) continental
Calcaires mézozoiques
Paléozoïque

Pliocène continental (sables, cailloutis)
Pliocène marin
Discontinuité d'érosion messinienne

S   Albères   N   SW   NE
Corbières

Fig.3 Profil du complexe littoral

Lagune
Vases de colmatage lagunaire
Cordon littoral actuel
Cordon littoral ancien
Limons de crue
Alluvions sablo-graveleuses

meilleure prise en compte de l'Environnement. Elle concerne une bande côtière de 35 kilomètres sur une largeur d'une dizaine de kilomètres, correspondant à 25 communes. (Fig.1).

Compte tenu de la modicité des moyens consacrés à cette étude, la méthodologie adoptée a fait une large part à l'approche géomorphologique et à la télédétection, avec un objectif affirmé de déboucher sur des applications concrètes au niveau de la planification, des techniques d'aménagement et des options de mise en valeur.

2 METHODE D'ETUDE

Elle associe des approches complémentaires, couvrant un champ pluridisciplinaire :

2.1. La documentation

La synthèse documentaire concerne les données de l'Environnement au sens large (milieu naturel, pollutions, paysage, protections réglementaires) et les données de l'Aménagement : documents de planification, réalisations et projets d'aménagement. En ce qui concerne le milieu naturel, une priorité a été donnée au milieu physique, considéré comme facteur de causalité par rapport à l'occupation de l'espace. C'est ainsi que la connaissance des caractéristiques et du fonctionnement de ses différents biotopes est considérée comme clé d'interprétation de la répartition des écosystèmes naturels ou artificiels. Cette

connaissance a été acquise en grande partie par synthèse documentaire, qui fournit en particulier les principales informations relatives à l'évolution des biotopes, sous l'angle :

a) de leur genèse : historique de la formation du complexe littoral, liée à la période d'érosion - sédimentation Mio-Pliocène, à la subsidence du bassin du Roussillon (Fig.2), à la formation d'un cordon littoral et des lagunes, puis au colmatage partiel de ces lagunes au cours de l'Holocène (Fig.3).

b) de leurs caractéristiques et de leur dynamique actuelles : érosion des versants, colmatage des lagunes ; inondabilité par les crues, et aménagements hydrauliques destinés à maîtriser celles-ci ; compressibilité des sols de colmatage lagunaire, etc...

2.2. La synthèse cartographique

Cette approche documentaire permet de mettre au point un modèle de terrain schématisé par des profils géomorphologiques. La généralisation de ce modèle a été rendue possible par la réalisation d'une cartographie géomorphologique appliquée au 1/25.000ème. Celle-ci, établie par photointerprétation et vérifications de terrain, a permis une délimitation précise des unités géomorphologiques différenciables par observations de surface. Les autres unités ont pu être analysées à partir des coupes de sondages anciens et des résultats de diverses études particulières.

196

A partir de cette cartographie, a été établie une correspondance entre les unités géomorphologiques et les caractéristiques utiles à l'aménagement, à deux niveaux successifs :

a) *Analytique* : chaque unité peut être caractérisée en termes topographique et de succession lithologique type, susceptible de fluctuer dans une fourchette connue. Lui sont associées également des caractéristiques hydrogéologiques, en termes de perméabilité et de présence, vérifiée ou potentielle, de nappe.

b) *Appliquée* : des données précédentes peuvent être déduites :

- l'existence de risques naturels tels que les inondations en lits majeurs, circonscrites à une unité géomorphologique bien différenciée. Ces risques ont fait l'objet d'études spécifiques prises en compte par la synthèse documentaire, et sont donc connus au plan quantitatif probabiliste. Les risques de tassement liés à la compressibilité des sols sont également répertoriés et reliés à l'unité correspondant au colmatage des lagunes,

- la présence de ressources naturelles exploitables : eau des nappes et matériaux d'origine alluvionnaire. Les ressources en eau font déjà l'objet d'une exploitation intensive dans le secteur d'étude, alors que l'extraction des alluvions, très florissante dans un passé récent, est aujourd'hui arrêtée,

- les aptitudes diverses à l'aménagement, en particulier à :
. la mise en valeur agricole, liée à la nature des sols, à la topographie, aux conditions hydrologiques et aux possibilités d'amélioration,
. l'urbanisation : en zones non inondables, peu compressibles, dans lesquelles l'assainissement pluvial peut s'effectuer par gravité,
. l'assainissement : les sols sableux ou sablo-graveleux des collines Pliocènes et des terrasses alluviales permettent d'envisager le développement de l'assainissement autonome, solution à prendre en compte pour les extensions urbaines futures,
. les loisirs et le tourisme : outre la frange littorale, les lagunes et leurs abords, les lits mineur et moyen des cours d'eau présentent un intérêt certain pour la population résidente ou saisonnière,
. le maintien en "zone verte" : on constate que les unités géomorphologiques les moins favorables à l'aménagement sont aussi les plus riches au plan écologique. Ce constat rapide a été vérifié et précisé par une synthèse écologique cartographique réalisée par l'Association Charles Flahault, de Perpignan, qui a permis de délimiter, à l'intérieur de certaines unités, les zones où subsistent des associations végétales de grand intérêt, dont certaines, exceptionnelles dans la région, méritent une protection totale.

A partir de ces informations a pu être réalisée une carte de synthèse du milieu physique, dont la Fig.4 représente un extrait, et un tableau de correspondance, présenté également partiellement.

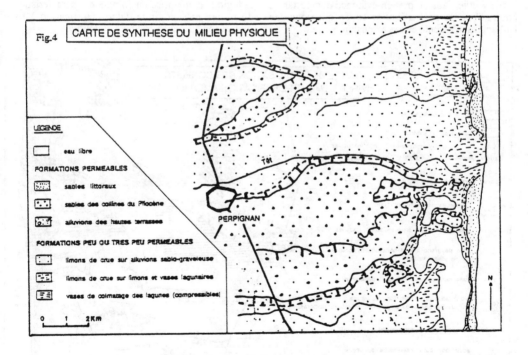

Fig.4 CARTE DE SYNTHESE DU MILIEU PHYSIQUE

UNITES	LITHOLOGIE	TOPOGRAPHIE	PERMEABILITE EN SURFACE	NAPPE	RESSOURCES		RISQUES		APTITUDES			
					EAU	MATERIAUX	INONDA-TIONS	COMPRESSI-BILITE	MISE EN VALEUR AGRICOLE	URBANISA-TION	ASSAINIS-SEMENT TERTIAIRE	LOISIRS TOURISME
2 - COLLINES												
Pliocène	Sables	Pentes moyennes à faibles	2 à 3	Profondes	2 à 3	1	0	0	1 à 2	2 à 3	2 à 3	0 à 1
Terrasses	Graviers											0
3 - BASSES TERRASSES	Alluvions grossières	Faiblement incliné	3	Semi profondes	2	2	0	0	1	2 à 3	2	0
4 - PLAINES ALLUVIALES MODERNES	Limons de crue sur graviers Graviers	Faiblement incliné	1	Superfi. à semi-prof.	1 à 2	0	3	0	3	1	1	0 à 1
Lit majeur												
Lit moyen		Faiblement incliné	2 à 3	Superfi.	1 à 2	2	3	0	1	0	0	2
Lit mineur		Faiblement incliné	/	Superfi.	2	2	3	0	0	0	0	2

2.3. Le constat de l'état actuel de l'aménagement

L'état actuel de l'aménagement est le résultat d'évolutions anciennes, liées à la mise en valeur agricole de bonnes terres, et du développement urbain, initialement concentré, puis étendu en tâche d'huile, et linéairement sur la bordure littorale.

Cette évolution se traduit en termes :

a) *Spatial* : l'occupation des sols actuelle et les orientations de la planification retenues dans les Plans d'Occupation des Sols ont fait l'objet d'une analyse spécifique. La comparaison de ces éléments avec les résultats de la synthèse cartographique présentée par la Fig.4 et le Tableau met en évidence d'importantes contradictions, en particulier en ce qui concerne l'urbanisation récente et son extension future :

- en zones inondables,
- en zones compressibles,
- en zones naturelles de grand intérêt écologique.

Par contre, les unités géomorphologiques les plus favorables techniquement à l'urbanisation sont de moins en moins concernées par les extensions urbaines.

Cette évolution s'explique évidemment par l'attractivité de la frange littorale. Elle s'effectue sans tenir compte de ses conséquences négatives, censées être résolues, du moins, pour les inondations, par les programmes d'aménagement hydraulique.
La carte de la Fig.5 montre l'importance des zones concernées par ces "conflits" ou contradictions entre la logique d'une prise en compte des caractéristiques du milieu et la logique d'une planification à forte motivation balnéaire, foncière et politique.

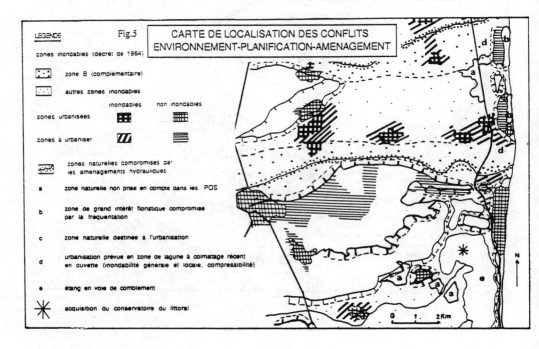

LEGENDE — Fig.5 — CARTE DE LOCALISATION DES CONFLITS ENVIRONNEMENT-PLANIFICATION-AMENAGEMENT

zones inondables (décret de 1964)

zone B (complémentaire)

autres zones inondables

     inondables   non inondables

zones urbanisées

zones à urbaniser

zones naturelles compromises par les aménagements hydrauliques

a   zone naturelle non prise en compte dans les POS

b   zone de grand intérêt floristique compromise par la fréquentation

c   zone naturelle destinée à l'urbanisation

d   urbanisation prévue en zone de lagune à colmatage récent en cuvette (inondabilité générale et locale, compressibilité)

e   étang en voie de comblement

✳   acquisition du conservatoire du littoral

0   1   2 Km

b) *De méthodes d'aménagement* : la multiplicité des contradictions et des conflits relevés au niveau spatial résulte en grande partie de l'approche trop monofonctionnelle des méthodes présidant à la conception des aménagements. Les limites de ce type d'approche apparaissent particulièrement en ce qui concerne les aménagements hydrauliques. On constate ainsi :

1. La persistance du risque d'inondation, malgré l'importance des moyens techniques et financiers mis en oeuvre. On constate ainsi, comme ailleurs en régions méditerranéennes, que l'aménagement hydraulique consiste avant tout à évacuer les flux exceptionnels le plus vite possible vers l'aval au moyen du recalibrage des lits mineurs. Efficace ponctuellement, pour protéger une zone urbanisée, cet objectif a été peu à peu étendu à une grande partie du réseau hydrographique. Il s'ensuit l'accélération des écoulements, la perte de l'effet d'écrêtage dû à l'expansion des eaux sur les champs d'inondation initiaux, la diminution du temps de concentration des eaux sur tout le bassin versant, et l'accroissement de la probabilité de superposition des crues unitaires. Compte tenu de l'intensité des pluies (600 à 1.200 mm suivant l'altitude, mais jusqu'à 400 mm en 24 heures), de l'imperméabilité et de la faible couverture végétale des versants, la maîtrise des inondations des plaines aval ne peut plus être assurée, malgré l'importance des travaux déjà réalisés. Ainsi, le barrage de Vinça n'est-il efficace que pour des crues de fréquence trentennale au maximum. On constate de plus l'état de vétusté des anciennes digues, faisant craindre la pire pour des ruptures possibles, alors que les cours d'eau s'écoulent en toit dans la plaine du Roussillon.

La contradiction relevée quand aux urbanisations prévues en zones inondables n'est donc pas justifiée par un programme d'aménagements hydrauliques dont l'achèvement nécessiterait la mobilisation de moyens financiers très importants.

La poursuite des orientations actuelles de la planification et des aménagements hydrauliques conduit donc à une impasse technique et financière.

2. Des conséquences graves en matière d'environnement, en particulier :

- le tarissement des nappes : le recalibrage des lits mineurs entraîne le drainage et, à terme, le tarissement des nappes phréatiques (Fig.6). De plus, le recalibrage risque de provoquer le court-circuitage des eaux qui jusqu'ici, provenant des versants imperméables, s'étalaient sur les sables Pliocènes, impluvium des nappes profondes exploitées dans les plaines du Roussillon (Fig.7). Il peut donc s'ensuivre un déficit d'alimentation de ces nappes dans l'avenir,

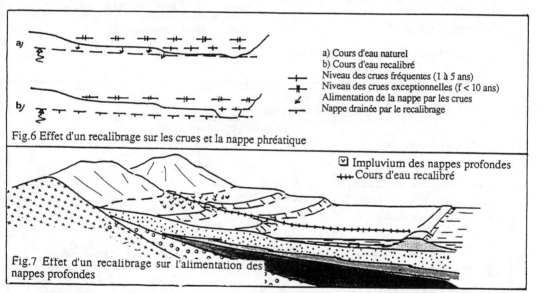

a) Cours d'eau naturel
b) Cours d'eau recalibré
Niveau des crues fréquentes (1 à 5 ans)
Niveau des crues exceptionnelles (f < 10 ans)
Alimentation de la nappe par les crues
Nappe drainée par le recalibrage

Fig.6 Effet d'un recalibrage sur les crues et la nappe phréatique

☑ Impluvium des nappes profondes
+++ Cours d'eau recalibré

Fig.7 Effet d'un recalibrage sur l'alimentation des nappes profondes

- l'accélération du colmatage des lagunes littorales par les matières en suspension qui ne peuvent plus se déposer sur les champs d'inondation des cours d'eau en période de crue,

- la perte du pouvoir auto-épurateur des cours d'eau, par baisse de leur débit hors période de crue, du fait de l'arrêt de la réalimentation par les nappes phréatiques. Cette cause peut expliquer en partie les pollutions enregistrées sur certaines plages du littoral en été,

- la disparition du patrimoine naturel, déjà fortement amputé par les urbanisations : destruction des biotopes aquatiques et terrestres liés aux cours d'eau lors des opérations de recalibrage ; disparition des écosystèmes lagunaires après colmatage ; destruction des associations végétales endémiques des dunes littorales par piétinement et par la moto dite verte, etc...

3. De mise en valeur : la disparition de zones naturelles s'accompagne d'une banalisation des paysages et d'une perte de zones de loisirs et de tourisme complémentaires de la seule façade littorale en partie saturée en période estivale. Cette évolution favorise à son tour la concentration de la fréquentation touristique sur la frange littorale, jusqu'à une surdensification et une saturation, facteurs de dégradation de la qualité de la vie. A terme, ce processus risque donc de conduire à une impasse, avec perte progressive de l'attractivité touristique de l'ensemble du secteur. Les facteurs de développement économique trop sectorisés en oeuvre jusqu'ici risquent donc d'entraîner une récession. Les effets de celle-ci seront aggravés par un accroissement des coûts d'équipements nécessaires pour s'affranchir des contraintes (risques d'inondation, de tassement, tarissement des nappes, pollutions) résultant d'une insuffisante prise en compte des caractéristiques du milieu naturel. Le traitement curatif de l'ensemble de ces contraintes risque donc de ne plus être suffisamment efficace pour contrebalancer les altérations qualitatives à la qualité de la vie.

Il apparaît donc urgent d'envisager des solutions mieux adaptées aux contraintes inhérentes au milieu naturel.

3. PROPOSITIONS

Les propositions élaborées dans le cadre de l'étude concernent les trois aspects de l'aménagement présentés ci-avant :

3.1. Planification

- renforcement de la politique de protection d'espaces naturels sensibles, en application de la récente Loi Littoral. Il s'agit en particulier des ZNIEFF (Zones Naturelles d'Intérêt Floristique et Faunistique), les zones à protéger et à gérer devant le plus possible présenter une continuité entre elles,

- renforcement des prescriptions concernant les zones soumises au risque d'inondation, dans le cadre par exemple de la mise en oeuvre de Plans d'Exposition aux Risques (PER),

- incitation à un développement des zones à urbaniser en retrait du littoral, sur les collines Pliocènes, dépourvues de risques naturels ou géotechniques,

- affirmation de la vocation de certaines zones à servir exclusivement soit à l'agriculture, soit de zones tampon et de liaison : par exemple pour l'écrêtage des crues et la sédimentation des matières en suspension, et pour les loisirs en bordure des cours d'eau.

La carte d'orientation (Fig. 8) permet de proposer une représentation spatiale à l'échelle (initiale) du 1/100.000ème, de la planification qui serait souhaitable, compte tenu des connaissances acquises concernant en particulier le milieu physique. Ce modèle idéal, inapplicable bien sûr dans son intégralité, doit servir de base aux négociations qui vont être entreprises avec les élus.

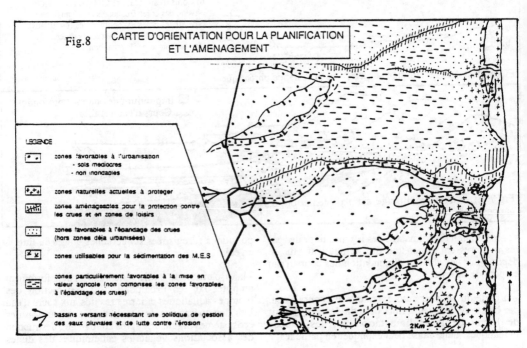

Fig.8   CARTE D'ORIENTATION POUR LA PLANIFICATION ET L'AMENAGEMENT

3.2. Méthodes et techniques d'aménagement

Compte tenu de l'analyse présentée ci-avant quant aux conséquences inquiétantes des méthodes et techniques mises en oeuvre pour satisfaire aux objectifs actuels de planification et d'aménagement, il est clair que la réorientation de la planification doit s'accompagner de la proposition de solutions alternatives et complémentaires à celles citées jusqu'ici.

Ces solutions ont en règle générale déjà été expérimentées dans d'autres régions ; leur efficacité est globalement reconnue, mais, pour diverses raisons, elles n'ont encore fait l'objet d'aucune application systématique à l'échelle d'un espace comparable aux plaines du Roussillon. Elles impliquent le recours à une démarche basée moins sur les grands travaux que sur des actions multiples, réparties sur tout le territoire, et conçues précisément à partir des connaissances acquises quant aux comportements et aux processus d'évolution des unités naturelles.

Dans le cas des plaines du Roussillon, les principales orientations suivantes ont été proposées :

a) *Aménagement et gestion des eaux pluviales* ; la protection des plaines du Roussillon passe par une gestion rationnelle des eaux pluviales de l'ensemble des bassins versants. Les principaux moyens à mettre en oeuvre concernent :

- la revégétalisation intensive des nombreux versants actuellement soumis à l'érosion,

- la réactivation des champs d'inondation dans la partie moyenne des grands cours d'eau et sur leurs affluents. Bénéfiques sur le lit moyen, par irrigation de la forêt riveraine (ripisilve), les inondations exceptionnelles sont supportables en lit majeur cultivé, à condition de maîtriser les courants, causes d'érosions et de destructions,

- la lutte contre l'érosion des sols dans les secteurs de viticulture,

- la réactivation de l'impluvium des nappes profondes,

- la réhabilitation des berges, particulièrement sur les cours aval, avec reconstitution du milieu naturel et, dans certains cas, aménagement en zones de loisirs et parcs suburbains,

- la mise en place, à l'aval, de protections rustiques contre les courants : levées de terre et haies freinant l'écoulement sans le bloquer,

- dans les petits bassins très urbanisés des Albères, le recours à des techniques alternatives au ruissellement urbain (bassins d'orage, structures réservoir, etc...).

b) *Épuration des eaux* : La plaine du Roussillon comprend encore de vastes espaces non bâtis permettant une utilisation extensive des caractéristiques des sols. Afin de résoudre les épineux problèmes cités ci-avant posés par les effluents des stations d'épuration, deux solutions, dont la maîtrise technique est actuellement assurée, peuvent être mises en oeuvre :

- les formations sableuses ou sablo-graveleuses du Pliocène et des terrasses alluviales sont propices à l'assainissement autonome. Cet argument conforte le raisonnement tendant à orienter l'urbanisation future vers les zones de collines,

- les étendues plates et en cuvettes correspondant à l'unité géomorphologique des anciennes lagunes aujourd'hui colmatées par des vases, mais non encore recouvertes de limons de crues, sont suffisamment vastes pour accueillir des bassins de lagunage permettant l'épuration tertiaire des effluents de stations d'épuration.

Les technologies correspondantes ont été expérimentées avec succès dans des conditions similaires, dans le Languedoc. Ces lagunes peuvent sans difficulté cohabiter avec des espaces de nature protégés pour le développement de la flore et de la faune. Ce type de solution peut donc résoudre une grande partie des problèmes posés par l'assainissement des zones déjà urbanisées en front de mer.

3.3. Mise en valeur

Afin de faire cesser le gâchis d'espaces naturels actuellement constaté, il est nécessaire de partir des spécificités du milieu naturel pour proposer une nouvelle politique de mise en valeur susceptible à la fois de favoriser le développement touristique et de protéger les zones les plus sensibles. Ainsi, l'aménagement des cours d'eau permettrait-il de créer de nouvelles zones de loisirs et de nature, en améliorant le cadre de vie hors frange littorale. Sur celle-ci, la valorisation informative des zones à protéger permettrait de compenser l'application de règles plus strictes de protection des zones et des sites les plus remarquables.

4. CONCLUSION

L'analyse globale de la structure et du fonctionnement du milieu naturel des plaines littorales du Roussillon a fourni une nouvelle approche des problèmes d'aménagement qui se posent à ce secteur. En particulier, la méthode d'étude utilisée pour l'analyse du milieu physique permet de mieux comprendre les interrelations existant entre divers facteurs de l'environnement et de l'aménagement, de mettre en évidence les dysfonctionnements et les conflits, et d'appréhender le coût global des aménagements mal adaptés à leur milieu d'accueil.

Dans un deuxième temps, cette approche géoenvironnementale permet de proposer des solutions nouvelles, plus souples, permettant de répondre simultanément à diverses préoccupations.

L'approche géomorphologique et géotechnique, combinée avec des techniques de génie de l'environnement, permet de réaliser cette démarche sans moyens d'étude trop importants. Elle permet aussi aux aménageurs de disposer d'une vision globale, concrète et très explicite, des problématiques auxquelles ils sont confrontés.

Conçue en prolongement de la cartographie géotechnique, cette approche devrait permettre une meilleure prise en compte du milieu naturel dans les politiques de planification et de mise en valeur, dont la relance s'avère aujourd'hui nécessaire.

Bibliographie sommaire

G. Clauzon
Le détritisme néogène du bassin du Roussillon
(Pyrénées Orientales, France)
Géologie Alpine, Mém h.s. n°13, 1987

C. Duboul - Razavet, R. Martin,
La sédimentation holocène de trois étangs du littoral du Languedoc Roussillon
Bull. Soc. Languedocienne de Géographie 1981

M. Masson
Cartographie géotechnique de l'agglomération rouennaise
1er congrés international de Géologie de l'Ingénieur 1970

Carte géotechnique de la région d'Alger – Cas d'étude: La zone de l'Aéroport Houari Boumédiène et de l'Université de Bab Ezzouar

Geotechnic map of Algiers region – Case study: The Houari Boumédiène airport area and the University of Bab Ezzouar area

O. Mimouni & I. Saidoun
USTHB/IST, Algiers, Algeria

ABSTRACT: The extensive urbanization of Algiers city, the growth and extension of industries in the suburbs and the lack of what was considered as good constructions soils lead urbanists to take a closer loock to areas of known problems (swamps, settlements, weak geotechnical characteristics...). A regional geological and hydro-geological study of Algiers region will be given with a focus on the geotechnical characteristics. Our work will consist of the following: a collection of all borings data, an identification of the different soils, a classification of these soils. Geo-technical laboratory tets of thes soils, determining all geotechnical characteristics (densities, water content, saturation ratio, plasticity index, granulometry, conso-lidation pressure...). Geotechnical map of four area will drawing based upon corre-lation between borings.

RESUME : l'urbanisation intensive de la ville d'Alger, la croissance et l'expansion des industries et l'abscence de ce qui était considéré comme bon sol de fondation, a conduit les urbanistes à s'interesser aux zones à problémes (marécages, tasse-ment, faibles caractéristiques géotechniques etc...). Une étude géologique et hydro-logique de la région d'Alger sera entreprise en soulignant l'aspect géotechnique. Notre travail consistera en une collecte des données de forages, une identification des différents sols et leur classification. Des tests en laboratoire de ces sols don-nant leur caractéristiques géotechniques (densités, teneur en eau, degré de satura-tion, index de plasticité, granulométrie, pression de consolidation etc...).
Des corrélations faites à partir des différents sondages permettront de tracer les   r cartes géotéchniques de 4 zones.

1 INTRODUCTION

Le but de l'étude consiste en une iden-tification de l'ensemble des sols de la région d'Alger en vue de leur utilisation comme assise de fondation.
Sur le plan géotechnique régional, les principaux sols rencontrés et étudiés sont
- le socle métamorphique
- le miocéne et les marnes plaisanciennes
- la molasse astienne à dominance sa-bleuse et gréso-carbonatée
- les argiles vaseuses et tourbeuses pro-venant d'anciens marécages
- les marnes et cailloutis du comblement de la Mitidja
- les grès dunaires du Quaternaire.

2 CADRE GEOLOGIQUE ET GEOTECHNIQUE

Le secteur étudié, à savoir la région de Bab Ezzouar (Alger) sera divisé en quatre zones A, B, C et D (Fig.1).
La synthése géologique établie pour cha-cune de ces zones a été réalisée sur la base de sondages carottés, de pénétro-métres et de pressiométres.
A partir des documents du laboratoire Na tional de l'Habitat et de la Construction (LNHC), du Laboratoire des Travaux Publics du Centre (LTPC) et du contro-le technique des Travaux Publics (CTTP). Les informations géotechniques sur la zo-ne d'étude ont été recueillies, classées et mises sous forme de deux tableaux (Fig.2) avec pour l'un des paramétres d'identification et pour l'autre les pa-ramétres mécaniques.
L'ensemble des faciés rencontrés dans cha que zone a été regroupé dans des unités distinctes pour lesquelles nous avons dressé un tableau des résultats d'essai de laboratoire. Ils seront par ailleurs traités de maniére statistique.

Les paramétres d'identifications classés
dans le tableau ci-dessous: (Fig.2)

Paramétres d'identification	Unité	Symbole
Poids volumique du poids sec	T/m3	d
Teneur en eau naturelle	%	W
Degré de saturation	%	Sr
Passant au tamis de 2mm	%	/ 2mm
Passant au tamis de 80microns	%	/ 80 u
Passant au tamis de 20 microns	%	/ 20 u
Passant au tamis de 3 microns	%	/ 3 u
Limite de liquidité	%	Wl
Limite de plasticité	%	Ip
Indice de consolidation	%	Ic

Les paramétres mécaniques sont classés
dans le tableau ci-dessous:

Paramétres mécaniques	Unité	Symbole	Essai
Pression de consolidation	bars	Pc	Oedométre
Coefficient de compressibilité	%	Cc	Oedométre
Coefficient de gonflement	%	Cg	Oedométre
Cohésion (consolidé drainée)	bars	CD	Triaxial
Cohésion nonconsolidée non drainée	bars	UU	Triaxial
Cohésion (consolidé non drainée	bars	CU	Triaxial
Angle de frottement interne	degré	ϴ	Triaxial

3 CARTES D'APTITUDE DES SOLS AUX FONDATIONS

Elles seront établies en tenant compte des coupes de sondages carottés et des résultats des études géotechniques.
Du fait que notre secteur a une topographie relativement plate, les facteurs morphologiques ne seront pas pris en compte Il en sera de même pour l'aspect hydrogéologique pour lequel le nombre réduit de forage n'aura pas permis l'approfondissement de l'étude.

L'information à traiter se présentant sous forme de description lithologique de coupes de sondages, de résultats d'essai in situ et en laboratoire, est dense et variée. Néanmoins, elle présente des insuffisances et incompatibilités avec la réalité du fait:
- d'une bréve description lithologique des coupes de sondage
- d'une imprécision dans le positionnement des sondages (pas de coordonnées Lambert).
L'étude des différentes coupes de sondages de la zone A a permis de distinguer les horizons suivants:
- les remblais et terres végétales, d'épaisseur n'excédant pas les 3 métres,
- les argiles bariolées ou versicolores

avec ou sans concrétions calcaires d'âge Rharbien ou Soltanien,
- les argiles sableuses et sables argileux d'âge Tensiftien,
- les sables, graviers, galets et conglomérats d'âge Tensiftien,
- les marnes à proportion non négligeable d'argiles d'âge villafranchien,
- les grès à ciment calcaire d'âge Tensiftien à Astien.

4 DISTRIBUTION GEOGRAPHIQUE DES FACIES ET ASPECT SEDIMENTOLOGIQUE

La distribution des faciés dans le sous--sol de la zone A montre que :

- les marnes se localisent à l'Ouest et à l'Est de la zone (épaisseur faible),
- les vases et tourbes se présentent en dépôts lenticulaires à différentes profondeurs, à travers toute la zone sauf au Nord,
- les faciés sableux et graveleux se présentent en dépôts lenticulaires sauf à l'Ouest,
- les faciés gréseux se rencontrent dans les parties Nord et Sud,
- les autres faciés s'observent en général à travers toute la zone à des profondeurs variables.

Vu l'aspect sédimentologique des dépôts de forme lenticulaire d'extension verticale et horizontale, leur nature dépend essentiellement de la localisation géographique des affluents (paléochenaux) et des fluctuations du niveau de base des eaux. L'ensemble de ces facteurs a favorisé un agencement particulier qui peut être assimilé a une superposition de lentilles de nature et de dimension diverses.

5 SYNTHESE GEOTECHNIQUE

Les différents faciés ont été, pour des raisons d'homogénéité de la lithologie regroupés en quatre unités:
- UNITE I : vases tourbes et argiles vaseuses,
- UNITE II: argiles sableuses et sables argileux,
- UNITE III:sables, graviers, galets, grès et conglomérats,
- UNITE IV: argiles et marnes bariolées.

Dans chacune de ces unités, les caractéristiques géotechniques seront énumérées, excepté pour l'unité IV. Elles sont présentes dans toute la zone A, où nous ferons une différenciation verticale par tranches de 5 métres.
Les caractéristiques géotechniques de chaque unité sont reportées dans le tableau suivant:

tableau 1. caractéristiques géotechniques de chaque unité.

	U I	U II	U III
d	0.85 t/m3	1.83	1.83
W	66%	16	14
Sr	97%	93.5	84
granulo 80 u	$>$ 50%	$>$ 50	$<$ 50
Ip	36%	23%	
Ic	0.8	1.15	
cohésion C	0.4 bars	0.75	0.4
angle ɸ	9°	15	19
Pc	1 bar		
Rp	5 bars	55	100
Cc	0.654%	10.3	
C		2.2%	

avec: U= UNITE
Rp= Resistance en pointe
C = Coefficient de gonflement
W = Teneur en eau

tableau 2. synthése de l'UNITE IV.

	0 à 5m	5 à 10m	10 à 15m	15m
d	1.77t/m3	1.76	1.75	1.67
W	19%	19.5	20	20
Sr	97%	97	95	90
granu Lométric	$>$ 50 %	$>$ 50	$>$ 60	$>$ 50
Ip	28%	26	25	23
Ic	1.16	1.18	1.23	0.87
C	1.5 bars	1.3	1.3	1.25
ɸ	11°		9	15
Pc	2 bars	2	2.6	3
Cc		11.5%	10.9	14.3
Cg	4.4%	5.4	4.2	5.8
Rp	40 bars	50		

On remarque que les caractéristiques des différentes tranches de terrain de l'unité IV n'ont pas évolué avec la profondeur.

Les horizons rencontrés dans le sous-sol des zones B, C et D sont à peu près semblables à ceux de la zone A. Les seules différences du point de vue géologiques sont:
- au niveau de la zone B, les sondages carottés ont recoupés de nouveaux horizons représentés par des limons argileux des argiles et sables limoneux et des limons sableux
- la distribution des terrains dans le sous-sol est différente d'une zone à une autre
- dans la zone C, les vases et les tourbes sont inexistantes ainsi que les marnes au niveau des zones B, C et D

Nous procéderons de maniére identique pour l'étude des différentes zones B, C et D et des caractéristiques géotechnique de leurs différents terrains.
Il en résulte les cartes géotechniques des 4 zones (Fig.3, 4, 5 et 6).
L'élaboration de la carte géotechnique de la région de l'Aeroport Houari Boumédiéne a été basée sur le même principe.

6 CONCLUSION

Ce travail de cartographie géotechnique dans la région d'Alger est une premiére étude ponctuelle faisant partie d'une étude régional initiée par d'autres géotechniciens. Le résultat obtenu nous permet d'avoir une idée sur l'emplacement d'une bonne assise de fondation, et constitue déjà un premier document géotechnique d'une zone en pleine expansion urbaine.

REFERENCES BIBLIOGRAPHIQUES

Glangeaud, L. 1932. Etude géologique de la région littorale de la province d'Alger.
DEMRH .1973. Notice explicative de la carte hydrogéologique de la région d'Alger.
Bennie and Partners. 1983. Géologie et hydrogéologie de la Mitidja.
LTPC, LNHC, CTTP. Ensemble des études de sols entreprises dans le secteur de Bab Ezzouar par ces différents laboratoires.
Sanejouand, R. 1972. La cartographie géotechnique en France. Publication du LCPC.
Liquéfaction des sables. Bulletin du LTPC

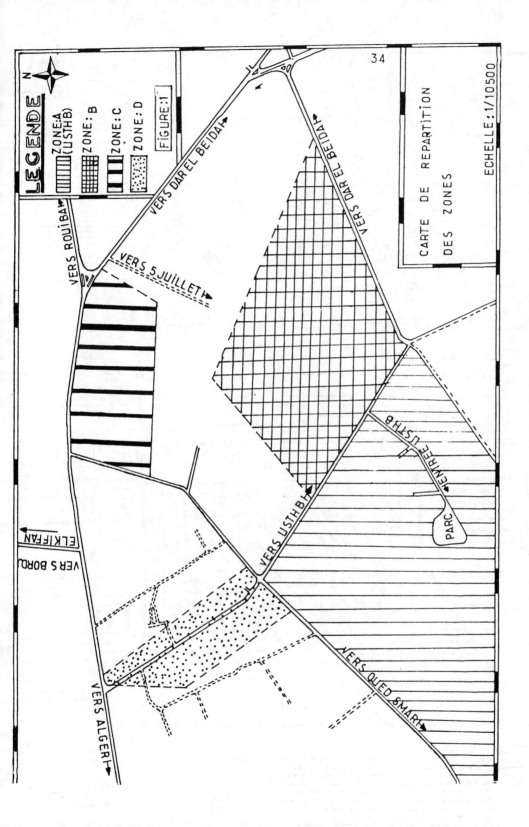

CARTE DE REPARTITION DES ZONES

ECHELLE : 1/10500

LEGENDE

ZONE A (USTHB)
ZONE : B
ZONE : C
ZONE : D

FIGURE : 1

VERS DAR EL BEIDA
VERS DAR EL BEIDA
VERS ROUIBA
VERS 5 JUILLET
VERS BORDJ ELKIFFAN
VERS ALGER
VERS USTHB
VERS OUED SMARI
USTHB
PARC
ENTREE USTHB

34

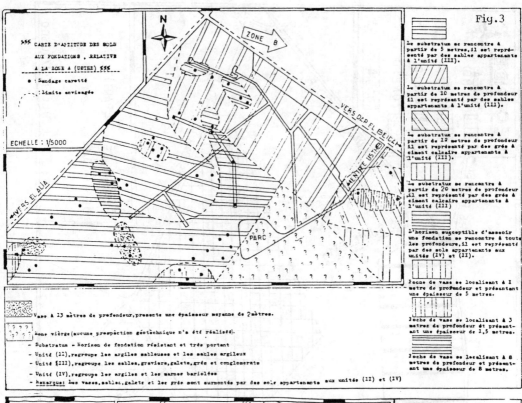

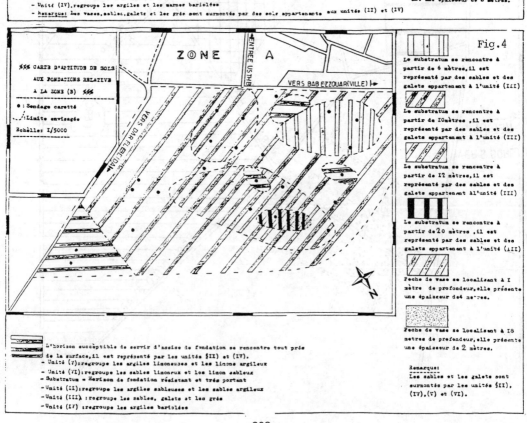

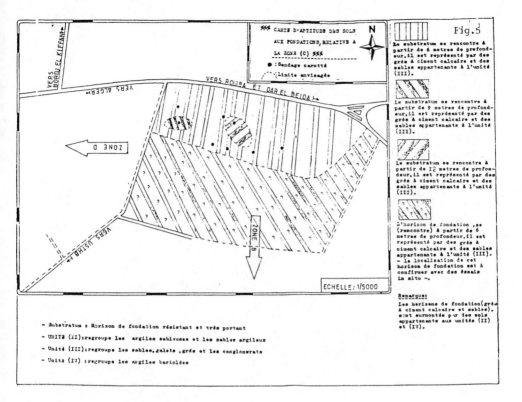

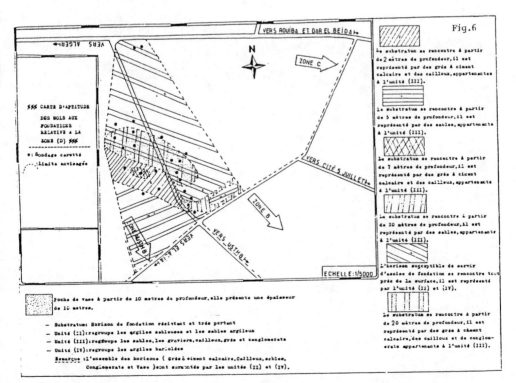

Alger: Cartographie et problèmes géotechniques
Algiers: Geotechnical mapping and problems

K.Ourabia & K.Benallal
L.T.P.C., Hussein-Dey, Alger, Algeria

ABSTRACT : In 1987 , Algerian government had accepted our project of a geotechnical mapping in the Algiers'region which concerns an area about 500 km² . In this paper , we present : - Algiers'region through its geological , hydrogeological and geotechnical points of view ;
- The steps of the geotechnical mapping ;
- The geotechnical problems of the principal formations and their relations with town planning .

RESUME : En 1987 , les autorités algériennes acceptèrent le projet de la carte géotechnique qui concerne une superficie de 500 km² . Dans cette note , nous présentons :
- La région d'Alger sur les plans géologique , hydrogéologique et géotechnique ;
- Les étapes d'élaboration de cette carte géotechnique ;
- Les problèmes géotechniques des principales assises en liaison avec les aménagements réalisés et projetés .

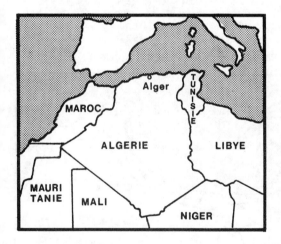

Alger , capitale de l'Algérie , est une ville méditerranéenne localisée à mi-distance entre la Tunisie et le Maroc. Depuis le début du siècle , Alger est confrontée à de graves difficultés d'urbanisation pour faire face à un exode rural massif et à une démographie très élevée . Cette situation est aggravée par une configuration topographique contraignante et par des conditions géotechniques qui n'ont pas toujours été prises en considération lors des phases successives de planification urbaine .

Actuellement , il semble que les matériaux et techniques sophistiqués acheminent les décideurs à improviser une planification , le plus souvent anarchique et inadaptée, aboutissant à de véritables erreurs urbanistiques et économiques (BENALLAL & OURABIA , 1988 c.) .

Pour tenter de solutionner , en partie (et de manière fiable et économique) ces difficultés , le Ministère des Travaux-Publics et le Haut Commissariat à la Recherche acceptèrent en 1987 notre projet de cartographie géotechnique de la région d'Alger ; projet dont nous esquissons les grandes lignes .

1 PRESENTATION DE LA REGION D'ALGER

1.1 Les grands ensembles géographiques

* Le massif de la Bouzaréah (ou massif d'Alger). Il s'étend sur une longueur de 20 km selon une direction Est-Ouest et sa largeur maximale est de 6 km . Boisé dans son ensemble , il est le "poumon" d'Alger et un lieu de loisirs (Parc de la forêt de Baïnem) . Son sommet culmine à 407 m d'al-

titude et sa topographie est très acciden-
tée . Les piémonts Nord et Est de ce massif
forment la côte déchiquetée entre le port
d'Alger et Aïn-Benian .
   * Le Sahel d'Alger . Il englobe les petits reliefs qui s'étendent entre le massif
de la Bouzaréah (au Nord) et les rives gauche de l'Oued El Harrach (au Sud et au Sud-
Est) et droite de l'Oued Mazafran (au Sud-
Ouest) . On y distingue :
   - Les collines à dominante marneuse de
Chéragas / Douéra / Dely Ibrahim ;
   - Le plateau mollassique d'Hydra bordé par
les grandes falaises du Hamma , d'El-Biar ,
d'Hydra et de Birmandreis ;
   - Le piémont Sud du Sahel dont les formations argilo-caillouteuses offrent des pentes assez fortes prouvant une faible éroda-
bilité .
   * La plaine littorale orientale . Elle
s'étend entre la mer (au Nord-Est), la cor-
niche mollassique du Hamma (au Sud-Ouest),
la Place du 1er Mai (au Nord-Ouest) et la
rive gauche de l'Oued El Harrach (au Sud-
Est) . Son altitude varie entre 2 et 15 m.
   * Le cordon littoral dunaire . Il se dé-
veloppe entre la rive droite de l'Oued El
Harrach et Bordj El Kiffan (où il sépare
le rivage marin des zones basses septen-
trionales de la Mitidja) et entre Aïn benian et Zéralda . Il s'agit d'une barrière
sablo-gréseuse généralement parallèle au
rivage et structurée en plateaux étagés
("marches d'escaliers") .

   ‡ La plaine de la Mitidja . Jusqu'au début du siècle , la zone septentrionale de
cette plaine était occupée par de grands
marécages . La récupération des terres
avait nécessité de grands travaux d'assainissement . La topographie plane de la Mitidja a très tôt inspiré les aménageurs .

1.2 Géologie (cf carte)

   La région d'Alger est constituée d'un
socle métamorphique (le massif de la Bouzaréah ou massif d'Alger) entouré par des
séries sédimentaires discordantes (d'âges
miocène , pliocène et quaternaire) for-
mant le Sahel et limitées en leur partie
méridionale par le bassin mio-plio-quater-
naire de la Mitidja .
   * Les terrains métamorphiques . Ils forment le massif d'Alger et ses prolongements
vers l'Ouest (presqu'île de Sidi-Fredj) et
vers l'Est (Bordj El Bahri) . " Ce sont des
formations de nature pélitique et carbonatée qui ont subi un métamorphisme monopha-
sé épi à mésozonal . Cette série a été re-
coupée par des manifestations magmatiques
(filons doléritiques , niveaux tuffacés
acides , intrusion granitique de Si Zouak).
L'édifice outre un plissement synmétamor-
phique porte les traces d'un écaillage intense s'accompagnant de rétromorphose ,
mylonitisation et chevauchements qui décou-
pent les séries en 7 unités tectoniques .

CARTE DES ENSEMBLES GEOGRAPHIQUES DE LA REGION D'ALGER

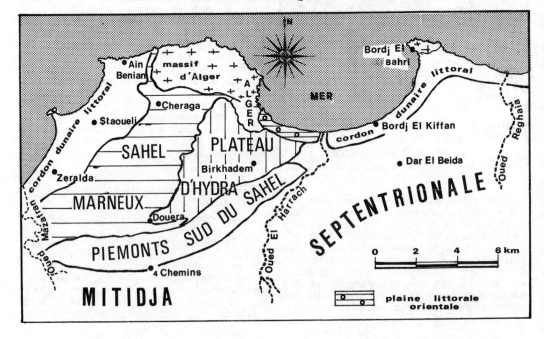

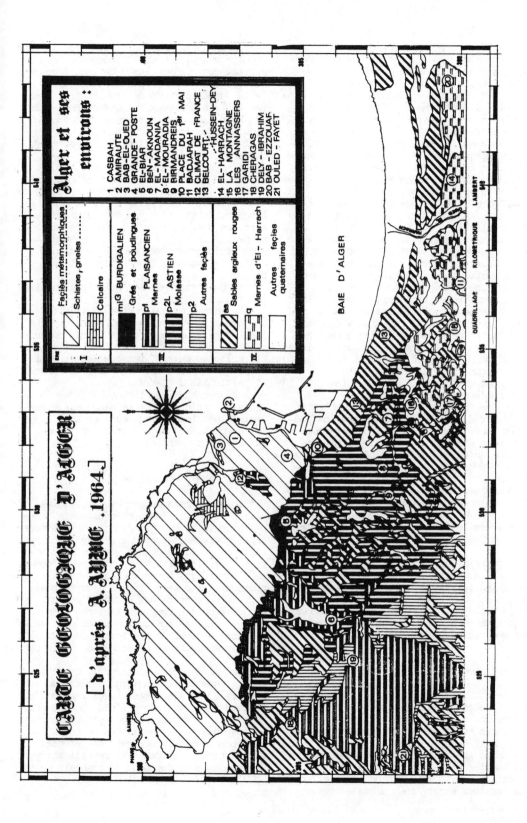

CARTE GÉOLOGIQUE D'ALGER

[D'après A. AYME .1964]

Alger et ses environs :

1 CASBAH
2 AMIRAUTÉ
3 BAB-EL-OUED
4 GRANDE - POSTE
5 EL-BIAR
6 BEN - AKNOUN
7 EL - MADANIA
8 EL-MOURADIA
9 BIRMANDREIS
10 PLACE DU 1er MAI
11 BADJARAH
12 CLIMAT DE FRANCE
13 BELCOURT - HUSSEIN-DEY
14 EL - HARRACH
15 LA MONTAGNE
16 LES ANNASSERS
17 GARIDI
18 CHERAGAS
19 DELY - IBRAHIM
20 BAB - EZZOUAR
21 OULED - FAYET

I — Facies métamorphiques
Schistes , gneiss ....
Calcaire

II — m1G BURDIGALIEN
Grés et poudingues
p1 PLAISANCIEN
Marnes
p2L ASTIEN
Molasse
p2 Autres facies

IV — Sables argileux rouges
Marnes d'El - Harrach
Autres facies quaternaires

BAIE D'ALGER

213

Il est possible que la série ait subi un métamorphisme panafricain et une rétromorphose alpine (SAADALLAH , 1981) " .

* La région d'Alger est marquée par les grandes lacunes du Secondaire et de la base du Tertiaire (Eocène et Oligocène) .

* Les terrains tertiaires .
- Le Burdigalien (Miocène) borde localement le massif métamorphique . Trois termes pétrographiques ont été définis , de bas en haut : le conglomérat de base , les grès grossiers et les marnes argileuses bleutées à Clypeaster , Amphiope et Schizaster (FLANDRIN , 1952) . Des formations miocènes ont été signalées dans plusieurs quartiers de la capitale : au Climat de France , au Chemin Sidi Brahim , au Chemin Laperlier et à Beni-Messous (AYME & MOUSSU , 1963) .
- Le Pliocène débute par une épaisse assise de marnes bleues du Plaisancien surmontée par une passée argilo-glauconieuse très fossilifère recouverte par la mollasse astienne sablo-gréso-carbonatée. L'épaisseur des marnes dépasse 200 m dans le Sahel et celle de la mollasse peut atteindre 100 m . Le sommet des marnes bleues est d'âge pliocène inférieur et la mollasse est pliocène moyen (YASSINI , 1973) .

* Les terrains quaternaires . Les plus particuliers sont :
- Les marnes et cailloutis du comblement de la Mitidja qui affleurent dans les piémonts Sud du Sahel et qui sont largement représentés dans le bassin de la Mitidja ;
- Les faciès sablo-gréseux de Birkhadem , d'Oued Ouchaiah et les grès dunaires littoraux ;
- Les sables argileux rouges du Villafranchien largement répandus au-dessus des formations mollassiques et dunaires ;
- Les argiles vaso-tourbeuses de la bordure Nord de la Mitidja .

Du point de vue structural , un axe anticlinal a été mis en évidence dans le Sahel (AYME , 1964) et un axe synclinal orienté Sud-Ouest/Nord-Est marque le bassin de la Mitidja . De petits indices néotectoniques ont été signalés dans les termes métamorphiques (SAADALLAH , 1981) .

1.3 Hydrogéologie

* Le complexe métamorphique . Il existe une présence d'eau dans le socle métamorphique qui se manifeste sous forme de résurgences (sources) et de nappe exploitée (par puits dans la Casbah) . Cette eau circule dans les fissures et diaclases et peut s'accumuler dans les zones d'altération superficielle (plus ou moins épaisses) et les écailles karstifiées de calcaire .

* La mollasse astienne . Elle est le plus important aquifère de l'Algérois et son mur est marneux plaisancien . Cet aquifère est subdivisé en deux nappes distinctes :
- La nappe profonde de la Mitidja ; captée par de très nombreux forages d'AEP , elle est bien connue et a même fait l'objet d'un modèle mathématique , de simulations d'exploitation et de construction de stations de captages ;
- La nappe du plateau mollassique ; c'est une nappe libre dont les eaux ont alimenté Alger à partir des sources captées du Hamma , du Telemly , de Birtraria et de Ben-Aknoun . Cette nappe possède de grandes surfaces d'affleurement et une grande perméabilité . Dans le Sahel , le contact Astien/Plaisancien est jalonné de sources : Douéra , Crescia , Oued Romane , Telemly et Hamma .

* Le Quaternaire . Ses principaux aquifères sont :
- Les marnes et cailloutis de la Mitidja où de nombreux horizons aquifères ont été décelés par forages - cette formation surmonte l'assise astienne et leurs relations hydrauliques ne sont pas connues et sont compliquées par l'hétérogénéité des marnes et cailloutis ;
- Les grès dunaires littoraux . De dimensions restreintes , cette nappe est surexploitée par les agriculteurs et très sujette aux problèmes de pollution et surtout aux invasions marines .

1.4 Contexte géotechnique

* Le socle métamorphique . Ses faciès offrent des qualités géotechniques : bonne portance (élevée), grande stabilité même en zones pentues (sous réserve de petites précautions de soutènement) , faible sensibilité à l'érosion et à l'eau et surtout non-agressivité vis à vis du béton . Dans la pratique , les calcaires sont exploités pour la protection des ouvrages portuaires et des remblais routiers (Autoroute de l' Est) , pour la fabrication du ciment (Raïs Hamidou) et pour la confection de granulats (carrières Jaubert, de Baïnem et d'Aïn Benian) .

* Les marnes plaisanciennes . Cette formation essentiellement argilo-marneuse est sujette à une altération (argilisation) qui la rend très sensible à l'eau (gonflement, baisse des caractéristiques géo-techniques et instabilité) . Ces marnes possèdent un caractère évolutif qui rend dangereuses la réalisation de talus et la réutilisation des produits de décapage en remblais . Cette formation est identifiée par :

Paramètre	Valeur minimale	Valeur maximale	Nombre de valeurs
ρ_s	2,65	2,66	2
$<80\mu$	93	99	199
$<3\mu$	40	52	188
Sr	96	98	301
W_L	52	57	209
I_P	26	34	209
I_C	1	1,4	197
Lr	13	16	9
$CaCO_3$	20	24	14
$CaSO_4, 2H_2O$	0,4	0,5	13

I_P: Indice de plasticité ;
I_C: Indice de consistance ;
Lr : Limite de retrait (%);
$CaCO_3$: Teneur en carbonate de calcium (%);
$CaSO_4, 2H_2O$: Teneur en sulfate de calcium
(%) .
Ces valeurs caractérisent un sol fin ,
proche de la saturation et de forte plasti-
cité .

* La mollasse astienne . Sableuse à sablo-
gréseuse carbonatée , cette assise très
épaisse et homogène présente les qualités
geotechniques suivantes : portance élevée,
insensibilité à l'eau (massif drainant) et
terrassements aisés en toute période de
l'année . Les terrains mollassiques de l'
Algérois sont utilisés comme pierres de
taille (cas des passées grésifiées) dans
les vieilles batisses ou bien comme matériaux de construction routière et d'aménagement d'aires de jeux (cas des horizons
peu consolidés ou sableux) .

* Les marnes et cailloutis du comblement
de la Mitidja . Ils sont bien appréciés des
géotechniciens pour leur bonne portance ,
leur chimisme (non-agressif) et la stabilité de leurs talus . Seuls les faciès les
plus argileux sont sensibles à l'eau .
Ces qualités ont motivé les opérateurs à y
implanter les grandes zones urbaines de
Beaulieu , de Badjareh et d'Aïn-Nadja .

* Les argiles vaseuses et tourbeuses .
Ces dépôts caractérisent les anciens marécages du Nord de la Mitidja . Baignant dans
une nappe aquifère , peu consolidés , de
faible perméabilité et sujets à des tassements excessifs et au fluage , ces terrains
sont marqués par des hétérogénéités de faciès et de structure géologique . De plus,
les eaux de la nappe sont subaffleurantes
et possède un chimisme agressif dû à la
pollution industrielle (rejets) et agricole
(pesticides, engrais et herbicides) .

Paramètre	valeur moyenne	Paramètre	valeur moyenne
ρ_s	2,57	I_P	27,
ρ_d	1,40	Pc	1,4
W	37,6	Cc	30,5
Sr	96	Cg	10
$<80\mu$	96	e_o	0,9
$<3\mu$	60	Matière organique	6 %
W_L	55		

* Les grès et sables dunaires quaternaires . Leurs propriétés géotechniques s'ap-
parentent à celles de la mollasse astienne:
portance élevée et bonne stabilité des talus . Les blocs de grès étaient utilisés
dans la construction ancienne d'habitations
et de murs de cloture . Actuellement , le
sable est exploité à Staouéli , Zéralda et
Beni-Mered pour la confection de béton hydraulique , pour les remblais routiers et
de l'aérodrome ainsi que pour la construction des rampes d'accès des ouvrages d'art
(autoroute de l'Est et Rocade Sud) .

* Les sables argileux rouges . Affleurant
largement dans l'Algérois, ils sont protégés de l'urbanisation pour des raisons agricoles (sol de culture) . Ils servent parfois de niveau de fondation : cas des villas de Mohamedia et d'Hydra . Cette formation peu épaisse (dépassant rarement 10m)
est très homogène sur les plan géologique
et géotechnique et possède les caractéristiques moyennes suivantes :

Paramètre	valeur moyenne	Paramètre	valeur moyenne
ρ_d	1,79	I_P	18
W	14,7	I_C	1,1
Sr	78	Pc	2,4
$<80\mu$	47	Cc	11,9
$<3\mu$	23	Cg	2
W_L	36		

Il s'agit d'un sol grenu (classification
LCPC), relativement surconsolidé , moyennement compressible et non gonflant . Ses
bonnes caractéristiques lui confèrent une
bonne stabilité et une bonne portance .
Les valeurs moyennes sus-citées concernent
la tranche de profondeur comprise entre 0
et 5 mètres .

1.5 Sismicité

Dans les règles parasismiques algériennes, l'Algérie est subdivisée en 4 zones d'activité sismique (O,1,2 et 3 de la plus faible à la plus forte) et Alger est classée en zone 2 (C.T.C., 1984) . Dans ce document , est donnée la liste des principaux séismes algériens parmi lesquels nous avons sélectionné ceux des :

Date	Epicentre	Intensité Magnitude	Observations
03-02-1716	Alger	I=X	Alger détruite
02-03-1825	Blida	I=X	7000 morts
02-01-1867	Mouzaia	IX⟨I⟨X	100 morts
06-01-1888	Mouzaia	I=VIII	
05-11-1924	Alger	M=5	
09-09-1954	El Asnam	I=X , M=6,7	1243 morts
10-09-1954	El Asnam	I=X , M=6,2	
04-02-1955	El Asnam	I=VIII	
05-06-1955	Beni Rached	I=VIII , M=5,7	
07-12-1959	Bou Medfa	I=VIII , M=5,5	
11-03-1973	Tenès	M=5,7	
10-10-1980	Beni Rached	X⟨I⟨XI , M=7,5	
	" " "	M=6,5	
08-11-1980	" " "	M=5,6	

    I : Intensité Mercali (chiffre romain)
    M : Magnitude (Richter) (chiffre arabe)

Une autre liste de séismes est donnée dans les règles parasismiques algériennes 1988 (C.G.S., 1988) :

Date	Epicentre	Magnitude
30-06-1981	Ain Benian/Chéraga/ Staouéli	M=4,5
23-05-1982	Bordj Menaïel	M=4,5
07-12-1983	Alger/Staouéli	M=4,5
15-06-1984	Gouraia/Tenès	M=4,3
Janv.-1986	Blida/ El Affroun/	M=4,6
19-12-1986	Blida/Alger	M=4,4

Alger est située dans une région sismique active et ressent surtout les manifesta-i tions provenant de Blida/Mouzaia/El Affroun (extrémité occidentale du bassin néogène de la Mitidja) , d'Ech Chlef (bassin néogène du Chélif) et de Tenès .
Tout récemment, plusieurs séismes ont été ressentis et enregistrés à Alger :

Date	Epicentre	Magnitude
29-10-1989	Nador (Tipasa)	M=5,6
04-02-1990	Ain Defla	M=4,3
05-02-1990	Ech Chlef	M=4,2
09-02-1990	Tipasa/Blida	M=4,9
	Oued Djer	10 répliques 3⟨M⟨3,9
17-02-1990	Oued Djer	M=3,9
18-02-1990	" "	M=3,3

Le séisme du 29-10-1989 a causé de grands dégâts dans la ville d'Alger ; de nombreuses constructions anciennes (datant de plus d'un siècle) furent endommagées .

CARTE DES PRINCIPAUX EPICENTRES

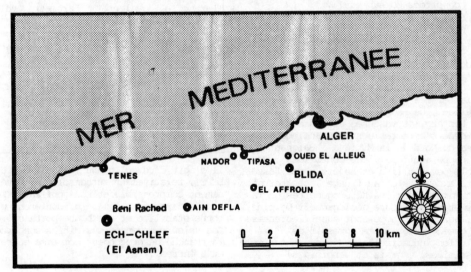

2 ETAPES D'ELABORATION DE LA CARTE GEO-TECHNIQUE D'ALGER

La phase bibliographique comprend la recherche, le tri des documents topographiques, géologiques, hydrogéologiques et géotechniques et comporte aussi la mise sur pied du fichier informatisé des données (BENALLAL & OURABIA, 1988 b).
Une banque des sondages (élaborée en 1988 sur un micro-ordinateur compatible IBM) permet le stockage et la recherche des informations selon un ou plusieurs critères tels que : le numéro du dossier géotechnique, les coordonnées kilométriques Lambert, la nature lithologique et l'âge des terrains. A ce jour, plus de 800 sondages carottés ont été récupérés dans les archives du L.T.P.C. ; sans compter les nombreux forages hydrauliques. La sauvegarde et la préservation de ces renseignements trouvent leur grand intérêt du fait qu'ils appartiennent à des zones urbanisées ou en cours de construction, donc dans des zones d'accés de plus en plus difficile. Cette banque des sondages sera complétée par des fichiers de caractéristiques géotechniques. Les traitements statistiques de ces dernières aideront à mieux identifier les sols et à quantifier les corrélations entre paramètres pour chaque formation.
La phase complémentaire comprendra les travaux de terrain à entreprendre dans les «secteurs vierges», les travaux de laboratoire (essais spéciaux) et la mise en forme du rapport de synthèse et des documents graphiques (BENALLAL & OURABIA, 1988 b). Des cartographies géologique et géotechnique sont prévues au 1/25 000. Les zones problématiques ou de contact entre différentes formations seront levées au 1/10 000 et même au 1/5 000. Ces cartographies actualiseront les données classiques grâce aux renseignements apportés par les centaines de sondages et les affleurements nouveaux conséquents aux terrassements routiers.
Une attention toute particulière sera accordée à la cartographie des mouvements de terrains, à leur description, à la prescription de remèdes pour éviter une aggravation des phénomènes et de recommandations strictes pour contrer l'apparition (ou la création) de nouvelles instabilités.
Une classification des unités géotechniques (selon leurs qualités de portance et stabilité) ainsi qu'une cartographie spatiale de ces unités aideront les aménageurs à :
- Opter et concevoir les meilleurs projets en liaison avec les spécificités naturelles des sites ;
- Contourner les problèmes propres à certaines unités.

3 PROBLEMES GEOTECHNIQUES

3.1 Le socle métamorphique

Dans le cas des gneiss et des micaschistes, l'altération donne naissance à des sols argileux peu portants et très sensibles à l'eau : cas des "gneiss pourris" du Telemly (BENALLAL & OURABIA, 1988 a).
La construction de l'Hotel Aurassi a tout d'abord nécessité une réduction du nombre initial d'étages et une injection massive des terrains de fondations.
Mitoyen à cet Hotel, le Théatre de Verdure est surplombé par de grands talus rocheux verticaux dont la stabilité a posé de sérieux problèmes à cause des plans de schistosité et de diaclases, des éboulis et des parties altérées (argilisées) du socle.
Une partie du tracé du Métro d'Alger est localisée dans les faciès métamorphiques qui présenteront des difficultés de :
- Creusement en liaison avec la fissuration qui provoquera des éboulements et nécessitera un soutènement d'autant plus conséquent que la zone traversée (par la galerie) est urbanisée et ancienne -Cette zone urbaine est donc très sensible aux ébranlements induits par les explosifs (de creusement) et aux ruptures des parois du tunnel- ;
- Venues d'eaux qui circulent dans les fissures et diaclases ;
- Activation, par les séïsmes, des nombreuses failles qui parcourent le massif; des indices ont montré que certaines failles avaient rejoué durant le Quaternaire récent (SAADALLAH, 1981).

3.2 Les marnes plaisanciennes

Leurs grands affleurements dans le Sud-Ouest d'Alger ont suscité la convoitise des aménageurs urbains. C'est ainsi qu'est née l'idée de développer Alger vers cette zone (plan CNERU) et que l'on a assisté à l'extension des petits villages de Chéraga, de Dely-Ibrahim et de Douéra. De nombreux quartiers "résidentiels" ont ainsi été implantés sur les talus marneux. Si l'urbanisation de ces collines semble aisée dans une première approche, il n'en réside pas moins que des problèmes entraveront cet aménagement. Pour exemple : le cas du quartier du Telemly qui possède un panorama exceptionnel (dominant la baie) mais qui est classé "inconstructible" à cause des glissements de terrains argilo-marneux.
Les principales difficultés géotechniques de ces marnes sont :

- Les phénomènes de retrait/gonflement qui induisent des fissurations et des altérations à la faveur d'un climat à saisons très contrastées (saison sèche prolongée et saison humide courte à pluies violentes et de grande intensité) ;
- Les glissements de terrains qui naissent dans la zone saturée et altérée des marnes (l'équilibre fragile des talus acquis au cours des temps géologiques est très sensible aux terrassements ; les talus de la Rocade Sud sont marqués de nombreuses loupes de glissement) .

Certains affleurements marneux sont couronnés par des corniches mollassiques (Telemly , Ben Aknoun et Climat de France) . Ces corniches agissent en "réservoirs d'eau" dont les résurgences favorisent la saturation des marnes et la naissance d'instabilités à la faveur d'une topographie inégale .

Afin d'éviter la zone de saturation , les ouvrages d'art de la Rocade Sud (Zéralda/ Ben Aknoun) et de l'Autoroute de l'Est sont fondés sur pieux ancrés profondément dans la marne saine .

3.3 La mollasse astienne

Reconnue comme un bon terrain de fondation et un bon matériau de construction routière , la mollasse est sujette à un problème de rupture de pans de falaise : cas du Hamma et de Birmandreis . Au Telemly , à Ben Aknoun et au Climat de France, ces ruptures ont été favorisées par les glissements des argiles marneuses qui se trouvent à l'aplomb des corniches .

Lorsque le sommet des corniches est boisé, le développement du système radical des arbres favorise l'élargissement des fissures de la mollasse et participe donc à la rupture de cette assise .

Il est urgent et impératif d'étudier et de réaliser le confortement des corniches qui dominent les zones habitées par des contreforts (en béton armé) entrecroisés et ancrés (par tirants) dans la masse rocheuse . Les corniches du Hamma , de Birmandreis et du Telemly ont la priorité dans ce plan d'urgence .

3.4 Le Quaternaire de la Mitidja

La plaine de la Mitidja a toujours suscité la convoitise des aménageurs par sa topographie et sa grande superficie . Outre la valeur agricole de ses sols, la Mitidja cache des difficultés géotechniques liées à :
- La complexité de la structure géologique des faciès alluvionnaires ;
- La présence d'une nappe aquifère dans laquelle baignent des terrains vaso-tourbeux de portance faible et de compressibilité forte ;
- De nombreux forages d'exploitation des eaux souterraines qui, en provoquant des rabattements , peuvent favoriser la naissance de tassements (suite à une dissipation de la pression interstitielle) dans les faciès vaso-tourbeux .

Ces spécificités ont obligé à opter (dès 1950) pour une conception des pistes, de l' aérodrome de Dar El Beida , en béton précontraint . Lors de l'extension des pistes (en 1988/1989) , les paléochenaux tourbeux des Oueds Saïd et Boutrik génèrent sérieusement ce projet . Des expérimentations de drainage , des mesures de tassements et la mise en place d'un remblai de préchargement ont été menées dans ce site .

La nouvelle aérogare internationale d'Alger possède une structure lourde inadaptée au site et nécessitera 1800 pieux pour ses fondations . Les sous-sols prévus seront touchés par des venues d'eau car la nappe est subaffleurante dans ce site . D'ailleurs, les sous-sols de l'Université de Bab Ezzouar sont régulièrement inondés à cause des remontées de la nappe et des stagnations des eaux météoriques .

La majorité des ouvrages d'art (viaducs, ponts) localisés dans la zone septentrionale de la Mitidja sont fondés sur pieux frottants .

3.5 Les sables dunaires quaternaires

Leur épaisseur peut atteindre 25 à 30m dans la zone du Caroubier (plaine littorale orientale) . Peu consolidés, ces sables possèdent une granulométrie redressée (80% des grains ont des diamètres compris entre 80μ et 0,2mm et 10% entre 0,2 et 2mm) et baignent dans une nappe côtière . Toute sollicitation sismique pourrait y provoquer un phénomène de liquéfaction avec toutes ses conséquences pour les ouvrages fondés sur ces sables .

Pour parer à ce risque, un traitement dynamique par vibrations a été mené en 1987 sur le site de la future gare routière du Caroubier .

Pour les ouvrages d'art de l'Autoroute de l'Est, les opérateurs ont préféré des fondations profondes ancrées dans les marnes sous ces sables .

La nappe littorale de ces sables est , en général , polluée par les eaux marines ce qui oblige à prévoir des ciments spéciaux pour les fondations profondes .

4 CONCLUSION

Les problèmes géotechniques variés de cette région située dans un environnement sismo-tectonique actif , la restructuration planifiée des vieux quartiers , la démographie galopante qui induira un accroissement des besoins immobiliers et de la construction et la politique de préservation des terres agricoles montrent l'urgence de la carte géotechnique d'Alger .

L'absence de celle-ci peut expliquer les erreurs suivantes :

- L'inadaptation de certains projets aux conditions géotechniques des sites (exemple des structures lourdes de l'aérogare internationale et de l'Université de Bab Ezzouar ; un parallèle peut être fait entre les bâtiments R+4 de Badjarah -fondés dans des assises marno-caillouteuses de bonne portance- et les batiments R+10 des cités Soummam et 5 Juillet -fondés dans des sols argilo-vaso-tourbeux-) ;

- Les mauvais choix de sites tels que :
   * Les affectations des terrains métamorphiques au Parc des loisirs de Baïnem et des assises mollassiques au Parc zoologique ;
   * L'attribution des sols marneux et gonflants pour des quartiers résidentiels (petites villas) de Chéraga , Dely Ibrahim et Douéra ;
   * La construction dans des terrains affectés par les glissements du Telemly et de Ben Aknoun ;
   * Les zones d'habitations urbaines nouvelles (ZHUN) prévues dans les vergers de Birkhadem .

Ces inadéquations sont compliquées par l'indisponibilité de matériel adéquat de réalisation . Tel est le cas des fondations profondes qui sont plus fonction du matériel de forage (le diamètre 1,2m est le plus couramment usité) plutôt que d'un dimensionnement classique .

Traduites en termes économiques (surtout en période de crise avouée), ces erreurs et inadéquations justifieraient amplement la nécessité et la faisabilité des cartes géotechniques pour toutes les grandes villes d'Algérie .

REFERENCES

AYME A. 1964. Carte géologique au 1/50 000, feuille n°21 : ALGER . Publ. SCGA , Alger.

AYME A., AYME J.M., MOUSSU H. & ROYER L. 1961 . Carte géologique au 1/50 000 , feuille n°20 : CHERAGA . Publ. SCGA, Alger.

BENALLAL K. & OURABIA K. 1988 a . Géologie et géotechnique : aspects généraux de la région d'Alger . In : Monographie géologique et géotechnique de la région d'Alger . 25-43 , O.P.U., Alger .

BENALLAL K. & OURABIA K. 1988 b . Projet de la carte géotechnique d'Alger . In : Monographie géologique et géotechnique de la région d'Alger, 81-84 , O.P.U., Alger.

BENALLAL K. & OURABIA K. 1988 c. Alger : géologie et planification urbaine dans le cadre historique . In : Symp. intern. AIGI/IAEG , Athènes Sept. 1988 , Thème V, vol.3, 1419-1426 , A.A. BALKEMA publ., Rotterdam .

C.G.S. 1988 . Règles parasismiques algériennes RPA 88 . Publ. C.G.S., Alger .

C.T.C. 1984 . Règles parasismiques 1981 (version 1983) . ENAL , Alger .

DERVIEUX F. 1948 . Etude géotechniques des glissements de terrains des côteaux d'El Biar . Terres et Eaux, n°1, 48-57, Alger.

FLANDRIN J. 1952. Les chaînes atlasiques et la bordure Nord du Sahara . Publ. XIXème C.G.I., Alger, Monogr. région., 1ère sér. , n°14 .

GLANGEAUD L. et al 1952 . Histoire géologique de la province d'Alger . Publ. XIXème C.G.I., Alger , Monogr. région., 1ère sér., n°25 .

OURABIA K. & BENALLAL K. 1989. La carte géotechnique d'Alger : état d'avancement des travaux . In : 7ème Séminaire Nation. Sc. Terre , 124 , Alger .

ROZET M. 1830. Description géologique des environs d'Alger . Voyage dans la Régence d'Alger . Journ. Géol., 3 : 360 .

SAADALLAH A. 1981. Le massif cristallophyllien d'El Djezaïr (Algérie) : Evolution d'un charriage à vergence Nord dans les Internides des Maghrebides . Thèse Doct. 3ème cycle, USTHB , Alger .

YASSINI I. 1973. Nouvelles données stratigraphiques et microfaunistiques sur la limite Pliocène inférieur-Pliocène moyen (Plaisancien-Astien) dans la région d'Alger . Rev. Micropal., 4 , 229-248 , Paris .

6th International IAEG Congress / 6ème Congrès International de AIGI, © 1990 Balkema, Rotterdam. ISBN 90 6191 130 3

Collapsible soil hazard mapping along the Wasatch Range, Utah, USA

Cartographie des sols à danger d'affaissement le long des Montagnes Wasatch dans l'état d'Utah aux USA

Russell L. Owens
US Bureau of Reclamation, Provo, Utah, USA

Kyle M. Rollins
Civil Engineering Department, Brigham Young University, Provo, Utah, USA

ABSTRACT: Along the southern Wasatch Range, Utah, collapsible soils are associated with alluvial fan and colluvium deposits. By mapping these deposits, running laboratory tests on deposit samples, and correlating the deposits to areas known to contain collapsible soils, a collapsible soil hazard map can be developed. On the hazard map each deposit is numerically ranked based on the likelihood of the presence of collapsible soils in the deposit. The ranking depends on the properties of the deposit, the characteristics of the drainage basin of each deposit, and on correlations with similar deposits known to contain collapsible soil.

RESUME: Le long de la chaîne des Montagnes Wasatch, en Utah, des sols susceptibles à l'affaissement sont associés à des plaines alluviales et à des gisements de sédiments. Il est possible de créer une carte des sols susceptibles de s'affaisser, en marquant ces gisements sur une carte, en faisant des analyses en laboratoire de prises issues de ces gisements et en mettant ces gisements en corrélation avec les régions dans lesquelles la présence de sols susceptibles de s'affaisser a été établie. Sur la carte des sites dangereux chaque gisement est numéroté et classé d'après un système évaluant la possibilité de présence de sols d'affaissement dans ce gisement. Le classment dépend des propriétés du gisement, des caractéristiques du bassin d'écoulement (source des sédiments) et de corrélations avec des gisements similaires dont on a établi le contenu en sols d'affaissements.

1.0 INTRODUCTION

Collapsible soils are relatively dry, low density soils, which undergo a decrease in volume when they become wet for the first time since deposition. The volume decrease depends upon the soil structure, the previous stress history of the soil, and the thickness of the soil layer involved.

Collapsible soils are found throughout the world, particularly in semi-arid and arid environments. Associations with loess deposits, alluvial fan deposits comprised largely of mudflows and debris flows, or with unconsolidated, colluvium deposits are common. In Utah, alluvial fan or colluvium deposits produce collapsible soils.

1.1 Purpose and scope of study

Soil collapse has caused significant damage to numerous structures along the southern Wasatch Range. Human activities that introduce water into a relatively dry environment induce problems with collapsible soil. These activities include irrigation, construction of canals, or disposal of waste water. Although soil collapse is not life threatening, it can cause severe damage. Canals, dams, pipelines, roads, buildings, or agricultural fields are most susceptible to damage, resulting in substantial remedial costs (Prokopovich, 1984).

The purpose of this study is to produce a collapsible soil hazard map for the southern Wasatch Range. This map delineates areas, particularly alluvial fans, and ranks the areas according to their potential for containing collapsible soil. It also delineates and ranks additional areas, such as colluvium and weathered Lake Bonneville deposits, which are known to contain collapsible soils.

Collapsible soil hazard maps are beneficial for three main reasons. First, they heighten awareness to the fact that geologic environments with the potential for soil collapse do exist along the southern Wasatch Range. Second, they help in land-use planning by delineating areas likely to contain collapsible soils. In this way if extreme problems with collapsible soils are anticipated, avoidance of the area may be desirable or alternative uses of the land considered. Third, anyone desiring to build a structure can use the maps to see if collapsible soils are likely to be present in the area. If a high collapse potential is indicated, then a site-specific investigation should be undertaken to determine the extent of the problem.

2.0 DESCRIPTION OF COLLAPSIBLE SOILS

Collapsible soils are unsaturated soils which undergo a radical rearrangement of particles upon wetting, resulting in a significant decrease in volume. For alluvial soils in Utah, this volume decrease usually occurs without any additional load being applied to the soil. In order for significant volume decrease (collapse) to occur, the soil must meet the following criteria:

1. it must possess a potentially unstable, unsaturated, structure with a large void ratio. The void ratio may be the result of clay bonds which connect the bulky grains, bubble cavities formed by air entrapped during deposition, interlaminar openings between beds, unfilled desiccation cracks, or voids left by the decomposition of vegetation (Bull, 1964).

2. it must have some type of force, bond, or cementing agent which stabilizes the soil structure. Capillary tension, silt or clay bonds, or soil cement (usually calcium carbonate) may provide the force, bond, or cementing agent. However, the force, bond, or cementing agent must be susceptible to removal or reduction by the addition of water, allowing the soil structure to collapse.

3. it must be subjected to a high enough stress to develop an unstable condition, and exist in a climate or state which prevents spontaneous collapse.

Soil collapse occurs when the magnitude of the shear stresses between the bulky grains exceeds the shear strength of the bonding or cementing agents. As moisture is allowed access to the bonding or cementing agents, they tend to soften, weaken or dissolve. Eventually they reach a state where they can no longer resist the existing compressive stress. Thus the bulky grains are able to slide (shear) on one another, moving into the void spaces. Complete saturation of the soil is not necessary to trigger soil collapse, especially in cases where external loads are applied (Hunt, 1984).

3.0 GEOLOGIC ENVIRONMENT OF COLLAPSIBLE SOILS

Collapsible soils are found in a variety of geological environments. Loess, colluvium, mudflows, alluvial fans, residual soil, and man-made fills have all yielded collapsible soils. Along the southern Wasatch Range, alluvial fan and colluvium deposits produce collapsible soils.

The fact that collapsible soils are associated with certain geologic environments does not mean that the existence of a particular environment insures the presence of collapsible soil. However, once the presence of collapsible soils is confirmed, regional correlations can be made to similar environments, and the presence of a particular geologic environment may alert developers or planners to the necessity of more detailed investigations.

3.1 Alluvial fans

Alluvial fans are formed where streams emerge from steep mountain ranges and deposit their sediment load at the mouth of stream channels. Sediment may be deposited as streambed deposits confined to drainage channels, sheet flow deposits resulting from braided streams, or mudflow deposits (includes debris flows). The proportion of each deposit in the fan varies according to the amount, frequency, and intensity of runoff. Debris and mudflows are prevalent in semi-arid and arid regions, and are most common in fans with collapsible soil.

Mudflows are poorly sorted and have a complete gradation of material ranging from clay to large boulders. Large mudflows deposits have been essentially "dumped" on the fan and is material that has simply come to rest. The internal water has drained away or evaporated, and the deposit may be thicker than the depth of subsequent wetting. In this case, the lower part of the deposit is never subjected to wet-dry cycles after deposition.

Once deposited, the material is not reworked and the clay-size particles are not winnowed out. The clay has a high dry strength, and if enough clay is present, it may act as a binder within the deposit. The dry clay helps the deposit withstand subsequent overburden pressures. When water does percolate through the deposit, the clay adsorbs the water. As a result, the clay loses its binding strength. This allows the overburden load to compact the soil structure and produce collapse. For this reason, thick mudflows are the most severely collapsing part of alluvial fan deposits (Bull 1964, Friedman and Sanders 1978).

3.1.1 Drainage basin characteristics

Alluvial fans along the southern Wasatch Front, which exhibit a high potential for containing collapsible soil, originate from drainage basins composed of the Manning Canyon or Arapien Shale. Because the Wasatch Front is a semi-arid region, mechanical weathering of the drainage basins prevails over chemical weathering; therefore, the lithology of the drainage basin determines the lithology of the associated alluvial fan (Blissenbach, 1954). Consequently, shale-rich alluvial fans arise from drainage basins composed of shale-rich formations, such as the Manning Canyon or Arapien Shale. These formations provide sediment with enough clay binder to give the alluvial fan a high dry strength, and the shale-rich formations also provide the bulky grains necessary for a high void ratio. As a result of the high void ratio, the alluvial fan is susceptible to collapse.

Drainage basins composed of Manning Canyon or Arapien Shale produce considerably more sediment than drainage basins of comparable size composed of harder, more resistant rocks. Subsequently, the sediment is deposited in the alluvial fan resulting in a larger fan and in an increased potential for the formation of collapsible soil. In fact, the likelihood of the fan to contain collapsible soils increases as the ratio of the size of the drainage basin to the size of the alluvial fan decreases.

3.1.2 Role of regional tectonics

The size of an alluvial fan is dependent on the tectonic history of the region. Prominent relief is essential for fan formation. Tectonic movement along fault zones provides a continual source of sediment, and increases the capability of sediment transport by runoff (Friedman and Sanders, 1978). This is particularly true for the southern Wasatch Range where normal faulting along the Wasatch Fault zone has resulted in displacements ranging from 4000 to 7000 feet. Because of the significant relief, virtually every canyon that drains the Wasatch Range has an associated alluvial fan.

3.1.3 Climate

Semi-arid and arid climates with a mean annual precipitation of 9-12 inches produce the highest sediment yields. In this climate there is little vegetation to inhibit erosion, but sufficient runoff from intense storms to transport large amounts of sediment (Beckwith and Hansen, 1989). Because of the large sediment load available, mudflows are common in such climates. The average annual precipitation amounts for the thirty year period between 1951 and 1980 for cities along the southern Wasatch Range varied from a high of 18.02 inches to a low of 10.66 inches. Consequently, the area has a high sediment yield resulting in a proportionally high occurrence of mudflows.

3.2 Residual soils and colluvium

Along the southern Wasatch Range collapsible soils are associated with the Silt, Sand, and Clay Members of the Lake Bonneville Group. These members are composed of easily weathered silt, sand, or clay originally laid down as terrace deposits with horizontal surfaces against the steep mountain range. Soluble and fine grained material is leached out by water from the range resulting in a high void ratio and residual deposits susceptible to collapse. Continual erosion of the terraces has also produced thick colluvium deposits at the base of the slopes. These deposits are unconsolidated and composed of bulky grains with sufficient clay binder to produce an open, potentially collapsible structure.

4.0 DEVELOPMENT OF A COLLAPSIBLE SOIL HAZARD MAP

The thrust of this study was to map geologic deposits, particularly alluvial fans along the southern Wasatch Range, and rank their potential of containing collapsible soil. Initial maps were developed from recent aerial photo mapping supplemented by previous geologic maps. The aerial photo mapping was performed on 1:20,000 and 1:40,000 inch vertical aerial photos dated 1984 and 1980 respectively. The area mapped on the aerial photos was transferred to overlays on 1:24,000 inch orthophoto maps. The orthophoto overlays were then used in transferring the data to 7 1/2 minute topographic quadrangle maps. The alluvial fans mapped were Late Pleistocene or Holocene in age.

Once the alluvial fans were mapped, the lithologies of the drainage basins associated with the fans were determined. Geologic quadrangle maps, special-use maps, university theses and dissertations, and professional papers were used to determine the drainage basin lithologies.

A data search of geotechnical firms and state agencies was then undertaken to determine where collapsible soils had previously been identified along the southern Wasatch Range. Correlations were made between the areas known to contain collapsible soils and the alluvial fans previously mapped. Correlations included soil series, soil properties, drainage basin characteristics, presence of high water

table or previous flooding, surface gradients, and fan composition. Fans with no previous geotechnical investigations, which had characteristics similar to fans known to contain collapsible soils, were targeted for further investigations. In addition, a variety of fan sizes and drainage basin lithologies were investigated to determine whether collapsible soils were present. The initial maps were revised to reflect the results of field investigations and testing. Laboratory testing consisted of consolidation tests, gradation analyses, Atterberg limits, and in-place density and moisture determinations.

The colluvium deposits derived from, or the residual weathering of, the Silt, Sand, and Clay Members of the Lake Bonneville Group were also mapped. The characteristics of these deposits were evaluated and correlations were made to similar deposits along the southern Wasatch Range. These deposits were then investigated to see whether collapsible soils were, in fact, present. County soil maps were used to locate areas of similar soil types.

Because of the size of the study area, approximately 65 miles in length and 5 miles in width, the 7 1/2 minute quadrangle maps at a scale of 1:24,000 inches were reduced to a scale of 1:48,000 inches. At this scale, three maps were necessary to cover the southern Wasatch Range (see Figure 1 for a typical section of one of the collapsible soil hazard maps).

5.0 RANKING SYSTEM FOR THE STUDY AREA

Once the areas containing collapsible soils were delineated, a ranking system was devised based on the preliminary data search, field investigations, and subsequent laboratory testing. Areas were ranked according to their potential to contain collapsible soil (see Figure 1). This ranking was not intended to provide data as to the amount of collapse but only to alert the user of areas where collapsible soils were likely to be found.

The ranking provides a numerical designation as follows:

1. - indicates areas of very low
   collapse potential
2. - indicates areas of low
   collapse potential
3. - indicates areas of moderate
   collapse potential
4. - indicates areas of high
   collapse potential
5. - indicates areas of very
   high collapse potential

The collapse potential designations are based on the following parameters:

very low (1)
-areas with high water table -coarse grained deposits not susceptible to collapse -areas of very low gradients (0-5%) and prior flooding -bedrock formations other than the Manning Canyon and Arapien Shale

low (2)
-predominantly very coarse grained fans -fans with perennial stream drainage, prior flooding -low gradients (5-10%) -areas previously irrigated

moderate (3)
-fans with mixed deposits -intermittent or ephemeral stream drainage -moderate gradients (5-15%) -low water table (>10 feet) -correlated with similar areas of known collapse

high (4)
-predominantly fine grained -colluvium/alluvium from Manning Canyon, Arapien Shale, or Members of Lake Bonneville Group -ephemeral stream drainage -low water table (10-15 ft) -high gradients (10-25%) -known areas of collapse

very high (5)
-predominantly fine grained fans derived from Manning Canyon or Arapien Shale -ephemeral stream drainage -low water table (>15 ft) -known problem areas

6.0 CONCLUSIONS

Collapsible soils along the southern Wasatch Range are associated with alluvial fan and colluvium deposits. The area has prominent relief with high mountain fronts which provide a continual source of sediment and the area also has a climate conducive to high sediment yields. Consequently, nearly every drainage along the range has an associated alluvial fan.

Alluvial fans with shale-dominated drainage basins have the highest potential to produce collapsible soils. In addition to alluvial fans, colluvium derived from the Silt, Sand, and Clay Members or the Lake Bonneville Group, as well as residual weathering of these members, produce collapsible soil.

A collapsible soil hazard map was developed which delineates areas likely to contain collapsible soils. A ranking system was devised and each area was ranked according to its potential of containing collapsible soils.

7.0 REFERENCES

Beckwith, G.B., and Hansen, L.A., Identification and Characterization of the Collapsing Alluvial Soils of the Western United States, in Foundation Engineering: Current Principles and Practice, ed. Fred H. Kulwahy (ASCE,1989), pp. 143-157.

Blissenbach, E., 1954, Geology of Alluvial Fans in Semiarid Regions, Bulletin of the Geological Society of America, Vol. 65, pp. 175-190.

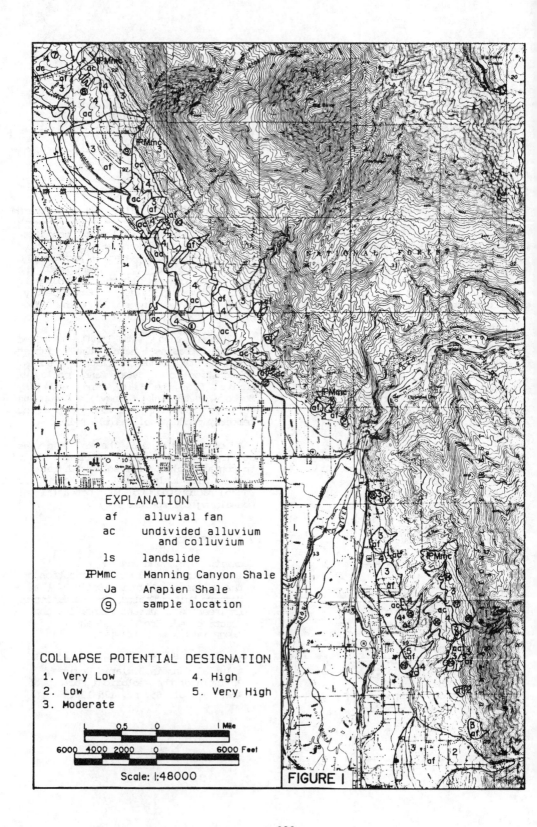

EXPLANATION

af   alluvial fan

ac   undivided alluvium
         and colluvium

ls   landslide

ₗPMmc   Manning Canyon Shale

Ja   Arapien Shale

⑨   sample location

COLLAPSE POTENTIAL DESIGNATION

1. Very Low   4. High
2. Low   5. Very High
3. Moderate

Scale: 1:48000

FIGURE I

Bull, W. B., 1964, Alluvial Fans and Near-Surface Subsidence in Western Fresno County, California, United States Geological Survey Professional Paper 437-A: U.S. Government Printing Office, Washington, DC, 70 p.

Friedman, Gerald M., and Sanders, John E., Principles of Sedimentology (John Wiley and Sons, 1978), pp. 202- 207.

Hunt, Roy E., Geotechnical Engineering Investigations Manual (Mcgraw-Hill Inc., 1984), pp. 813-836.

Prokopovich, Nikola P., 1984, Validity of Density-Liquid Limit Predictions Hydrocompaction: Bulletin of Association of Engineering Geologists, Vol. XXI, No.2, pp. 191-205.

Results of the pilot study of Ingeokaart Amsterdam
Les résultats de l'étude épreuve d'Ingeokaart Amsterdam

M.A.Philippart
Grondmechanica Amsterdam, Netherlands

ABSTRACT: Ingeokaart Amsterdam is a project of six Dutch institutes, operating in the field of engineering geology. The projects goal is to produce engineering geological maps of the city of Amsterdam for use in management and planning.

A pilot study has been carried out for two districts both of which haven an area of 4 km². One of the districts is situated at the centre of the town where relatively much information on subsurface conditions is available while the other is situated at the outskirts of Amsterdam where relatively little information is available.

The data consist of Dutch cone penetration tests, borings, laboratory results and groundwater level meauserments, all of which have been derived mainly from the records filed by Grondmechanica Amsterdam, a division of Amsterdam City Authority.

The data have been manually processed with great care to produce a wide variety of engineering geological maps on a scale of 1:10.000.

RESUME: Ingeokaart Amsterdam est un projet des six institutions néerlandaises qui s'occupent aux travaux géotechniques. L'objectif de cette projet est la production des cartes de la géologie de l'ingenieur de la ville d'Amsterdam à l'usage de la planification et d'aménagement urbain.

L'étude épreuve est realisé pour deux regions qui ont chacune une surface de 4 km² environ. Une region est située aux centre de la ville où se trouvent relativement beaucoup d'information de la constitution de la sol tandis que l'autre region est située au quartier extérieur où se trouvent peu d'information.

Les données se consistent en les essaies de sondages, les forages, les résutats de les essaies de laboratoire et les mesures des nappes aquifère.
Tous sont derivés des archives de la Grondmechanica Amsterdam, un departement de la municipal d'Amsterdam.

Les données sont soigneusement convertis, à la main, en une grande diversité des cartes de la géologie de l'ingenieur à l'un echelle de 1:10.000.

1. INTRODUCTION

Ingeokaart Amsterdam is a project of six Dutch institutes, initiated to produce a set of engineering geological maps of the city of Amsterdam. The six institutes, all operating in the field of engineering geology, are:
- Rijks Geologische Dienst (RGD)
- Grondmechanica Amsterdam (GRM)
- Grondmechanica Delft (GD)
- Technische Universiteit Delft (TU)
- Vrije Universiteit Amsterdam (VU)
- Institute for Aerospace Surveys and Earth Sciences (ITC)

The production of engineering geological maps starts with the inventory of available data (borings, dutch cone penetration tests (CPT) and groundwater level measurements) which are then displayed on location maps. The data are interpreted to a 3-D model from which some geological features are distilled into maps of main geological or geotechnical levels and profiles. Input of geotechnical and or hydrological parameters to the different layers lead to the production of thematical maps.

To estimate the amount of work, time and money needed, a pilot study has been carried out.

The pilot study dealt with two districts, each having an area of 4 km². The districts are selected on their geographic position and data density (see figure 1).

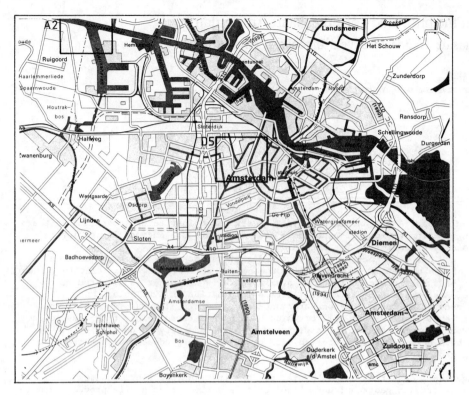

figure 1 - district D5 and A2

For each district a set of engineering
geological maps on a scale of 1:10.000
together with a description have been
produced.

District D5 is situated in the centre of
the city of Amsterdam. Its history dates
back to 1613 when urbanisation started in
the eastern part of the district. The city
council had issued housing for the rich
along the Herengracht, Keizersgracht en
Prinsengracht while the area between the
Prinsen- and Lijnbaansgracht, known as the
Jordaan, was destinated to be an industrial
area.

It was not until 1880 that the
urbanisation of the western part of the
district started. Urbanisation of district
D5 was completed in 1930 with the land fill
of the former Middelveldsche polder.

District A2 has been part of the IJ-
estuary until about 1850 when the IJ was
empoldered and thus the Groote-IJ polder
was created. The Noordzeekanaal was dug in
the years 1872 and 1876. Because the ports
had to expand the polder was heightened to
a level of NAP + 1,0 m and the Amerika
harbour was dug in 1960-65.

2. GENERAL MAPS

The oldest data on subsurface conditions
filed at the archives of the city of
Amsterdam is a boring from about the year
1605. Since then a large amount of data
(borings, CPT's) have been carried out and
filed.

For the pilot study of district D5, 250
borings to an average depth of about
NAP - 13 m, 700 cone penetration tests to
an average depth of NAP - 22 m and the
measurements of 250 wells mainly monitoring
the phreatic water level have been used.
The soil data amount to a density of about
170 points/km^2.

For district A2, 30 borings to an average
depth of about NAP - 17 m and 35 cone
penetration tests to an average depth of
about NAP - 24 m have been used. The soil
data amount to a density of about 22
points/km^2.

The locations of the data are displayed
on location maps (Ingeokaart Amsterdam,
appendix D5-1 and appendix A2-1).

While the data of district A2 are equally
distributed, the data of district D5 occur
partially in clusters.

Interpretation of the data resulted in

230

geological single value maps and cross sections. The geological and geotechnical features do not exceed a depth of NAP - 30 m because information beneath this level is sparse.

For both districts the depth of the Holocene-Pleistocene boundary is displayed in <u>geological map A</u> (Appendix D5-2 and A2-2). <u>Geological map B</u> deals with a geotechnical feature namely the top of the second sand layer (Appendix D5-3 and A2-3). Appendix D5-3 is shown in figure 2.

clay and sand composition (Calais deposits) and a thin layer of peat (Lower peat) which marks the boundary to the pleistocene deposits.

The first pleistocene deposit encountered is the First Sand Layer which is made up of fine eoalian sands deposited during the last glacial of the Weichselian.

A somewhat warmer period during this glacial resulted in the deposition of very fine sand, loam and clays. These sediments make up the intermediate layer found at a

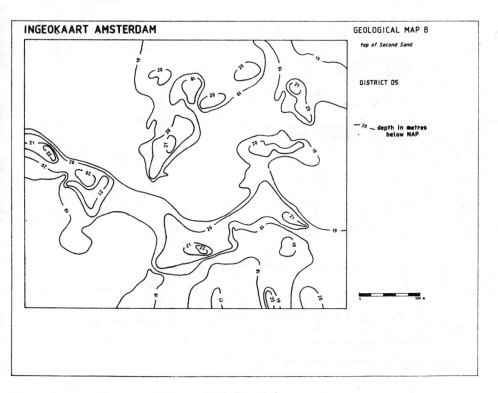

figure 2 - Ingeokaart Amsterdam; appendix D5-3

The <u>cross-sections</u> have been taken more or less along an east-west line (Appendix D5-4 and A2-4) while for district D5 also a cross-section along a north-south line (Appendix D5-5) has been produced. Appendix D5-4 is shown in figure 3.

To understand the value of the maps a global description concerning the subsoil of Amsterdam is given.

Starting at the top a man-made layer of variable composition to a depth of about NAP - 3 m is encountered. Beneath a depth of about NAP - 3 m to a depth of about NAP - 13 m holocene deposits are found. The holocene deposits consist of a layer of peat (Holland peat), a layer of varying

depth of about NAP - 15 m to a depth of about NAP - 19 m.

Beneath the intermediate layer a second sand layer is encountered. The top of this layer consists of fine sands deposited during Weichselian while to a depth of NAP - 30 m the second Sand Layer consists of medium coarse sand of marine origin deposited during Eemian.

There is only little information available on subsoil conditions beneath a depth of NAP - 25 m, but of geotechnical importance are the marine and glacial clay found to a depth of about NAP - 60 m and beneath this level a third sand layer.

The above description aplies to district

D5 except for the fossil tidal channels and gullies which are incised into the First and partially into the Second Sand Layer to a maximum depth of about NAP - 22 m. The channel deposits consist of thin clay and sand layers.

For district A2 the general description of the subsoil is also valid, although the Lower peat is scarcely encountered.

For the first aquifer there are not enough monitoring wells availabe and with respect to the second aquifer no monitoring wells are available at all hence no hydrological maps regarding these aquifers have been produced.

In district A2 no monitoring wells are placed either so no hydrological maps of this district are available.

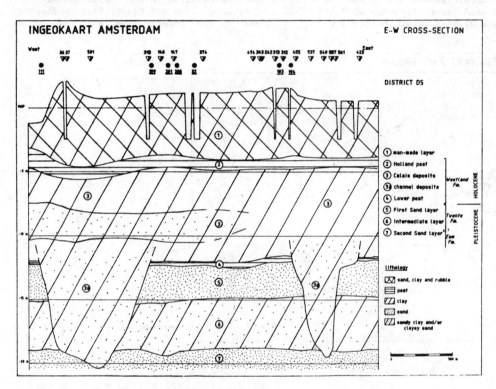

figure 3 - Ingeokaart Amsterdam; appendix D5-4

3. HYDROLOGICAL MAPS

In the subsoil of Amsterdam a phreatic water level and two aquifers are distinguished.

The phreatic water level is in the man-made layer provided the layer contains permeable fill. The first aquifer is composed of the First and Second Sand Layer while the second aquifer is composed of the Third Sand Layer.

District D5 comprises about 250 wells monitoring the phreatic water level. Based on the average results of the year 1986 a hydrological map regarding the phreatic water level is produced (appendix D5-6). The average water level of the district is about NAP - 0,4 m which is the level of the canals maintained by the City Authority.

4. THEMATICAL MAPS

Based on the geological information and relevant geotechnical parameters the next thematical maps have been produced.

Pile foundation map A (appendix D5-7 and A2-5) is a map based on the use of concrete piles with a dimension of 0,32 m x 0,32 m and a safe bearing capacity of 300 kN. In Amsterdam this agrees to foundation requirements for houses and small buildings.

The calculated foundation depth in district D5 is at a level of about NAP - 14 m in the First Sand Layer except for the locations where the channel deposits are encountered. At these locations the calculated foundation depth varies between NAP - 14 m and NAP - 22 m.

The calculated foundation depth in district A2 is in the First Sand Layer at a level of about NAP - 14 m.

Pile foundation map B (appendix D5-8 and A2-6) is based on the use of concrete piles with a dimension of 0,38 m x 0,38 m for a safe bearing capacity of 1000 kN. In Amsterdam this agrees to foundation requirements for buildings.

The calculated foundation depth in district D5 varies between about NAP - 18 m and about NAP - 22 m (see figure 4) that is in the Second Sand Layer (see figure 3). The deeper foundation levels coincide with the holocene tidal channel deposits.

The calculated foundation depth in district A2 varies between NAP - 17 m and NAP - 22 m.

Pile foundation depths are calculated using the method Koppejan whereas only positive friction has been taken into account and no negative friction.

period of 10.000 days.

The calculated settlements for district D5 vary between 0,15 m and 0,55 m. The highest settlements occur at the locations of the fossil holocene channel deposits.

Settlements for district A2 vary between 0,15 m and 0,35 m.

5. CONCLUSIONS

The results of the pilot study are of high quality due to the large amount of data available and due to the large amount of time and energy spent by all participants of the Ingeokaart project.

However the production by hand of the engineering geological maps on this scale is too much time and money consuming. Therefore the policy has changed and research has started on the use of data bases.

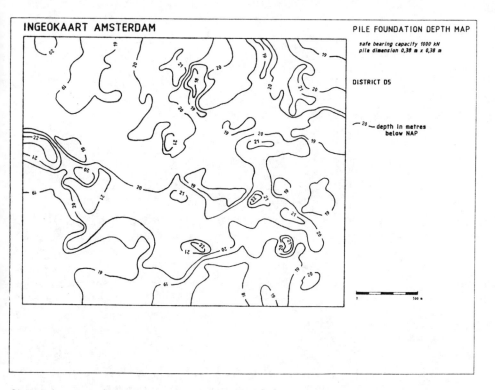

figure 4 - Ingeokaart Amsterdam; appendix D5-8

Settlement prediction map (appendix D5-9 and A2-7) is a map based on a equally distributed static load of 25 kN/m2 whichs conforms to a land fill of 1,5 metres of sand. The settlements are calculated with the combined method Terzaghi-Buisman for a

Grondmechanica Amsterdam has initiated the Ingeobase project and uses in its research the hand-made maps produced of District D5 as a comparison and control of the computer generated maps. The results of the project are presented by ir. J Herbschleb.

233

REFERENCES

Grondmechanica Amsterdam (1976) 25 jaar
 Grondmechanica Amsterdam - Werk in
 uitvoering 10 & 11, Dienst Openbare Werken,
 Amsterdam: 371-404.
Herbschleb, J. (1990) Ingeo-base, an
 engineering geological database -
 Proceedings of the sixth international
 congress of IAEG: 7 pp.
Ingeogroep (unpublished) Ingeokaart
 Amsterdam, resultaten van de pilotstudie:
 24 pp.
Keverling-Buisman, A.S. (1944)
 Grondmechanica - Waltman, Delft.
Terzaghi, K. & Peck, R.B. (1967) Soil
 mechanics in engineering practice - Wiley,
 New York: 729 pp.

Large-scale engineering geological mapping in the Spanish Pyrenees
Cartographie géotechnique à grande échelle dans les Pyrénées espagnoles

Niek Rengers & Robert Soeters
ITC, Enschede, Netherlands

Paul A.L.M.van Riet & Edwin Vlasblom
Dow Benelux, Terneuzen, Netherlands

ABSTRACT: A compilation at a scale of 1:12,500 is presented of engineering geological maps prepared during a number of years by students from ITC and Delft University of Technology in an area of 65 km^2 in the central Spanish Pyrenees. The methodology is based on mapping of terrain units in large-scale aerial photography and subsequent geotechnical description of the units in the terrain with field classification techniques.

RESUME: Il est présenté à l'échelle de 1:12.500 une compilation des cartes de géologie de l'ingenieur, préparées au cours de plusieurs années par les étudiants de l'ITC et de l'Université Technique de Delft. La région couvre 65 km^2 et est située dans le centre des Pyrénées espagnols. La méthodologie repose sur le lever, à partir de photographies aériennes à grande échelle, d'éléments de terrain suivi d'une description géotechnique "in situ" de ces unités, en faisant appel à des techniques de classification sur le terrain.

1 INTRODUCTION

1.1 Background of the mapping project

In the framework of the educational programmes of the International Institute for Aerospace Surveys and Earth Sciences (ITC) in Enschede and Delft and of the Section Engineering Geology of the Faculty of Mining and Petroleum Engineering of Delft University of Technology, field mapping exercises have been carried out with Dutch and foreign students for a large number of years.

The main objective of the field mapping exercises was the practical application of aerial photo interpretation techniques and training of the methodology for engineering geological description and classification of soils and rocks in the field.

The results of these mapping exercises have been recorded in a large number of reports and MSc theses with accompanying maps. In the framework of a TU Delft research study, these reports, theses and maps were compiled and presented in one large map and report (Van Riet & Vlasblom 1988) and will be published with a colour printed map at a scale of 1:12,500 in one of the next ITC Journals (Rengers, Soeters, Van Riet & Vlasblom 1990).

1.2 General description of the area

The mapped area is situated in the northern part of the province of Lerida in the Spanish Pyrenees. The main localities in the study area are Pont de Suert and Vilaller.

Concerning the geology of the area, intensive use was made of the excellent mapping done by geologists of the University of Leiden (the Netherlands), and in particular of the publication of Meij (1968).

The area covers a large variety of lithological formations as it includes the southern part of the axial zone of the Pyrenees, the Nogueras zone and the northernmost part of mesozoic Pyrenean nappes.

The axial zone consists in the study area mainly of a low-grade metamorphic sequence of carboniferous shales/slates, siltstones and calcareous sandstones, which are overlying Devonian calc schists and nodular limestones outcropping in the extreme northeast of the area. The sequence is unconformably overlain by Permo-Triassic redbeds of the Bunter formation.

The Nogueras zone consists in the first place of highly disturbed gypsiferous marls of the upper Triassic Keuper, which acted as a sliding horizon for the Pyrenean nappes. Large slabs of dark grey micritic limestones (Muschelkalk) and masses of basic ophitic intrusive rocks are distributed irregularly throughout the gypsiferous marls. Overturned tectonic blocks composed of Palaeozoic calc schists and slates and post-Hercynian volcanics and sedimentary rocks are overlying the gypsiferous marls and are capping most of the higher parts in the Nogueras zone.

The Mesozoic Pyrenean nappes consist mainly of Jurassic and Cretaceous limestones alternated with marly formations.

The lithological photo interpretation of the area is difficult due to the complexity of the geological conditions. The tectonically strongly disturbed structure makes that in many cases lithological contacts are faults. Added to the geological complexity also the changeable geomorphological conditions make the photo interpretation difficult.

The present relief forms of the Pyrenees are in the first place the result of the glacial erosion during the Pleistocene, when large glaciers scoured deep valleys. In the area the valleys of the Ribagorzana and the Tor rivers were occupied by glaciers. The landscape on the interfluves was modelled by periglacial mass movements and fluvio-glacial processes into a fairly undulated rounded relief. The warmer and perhaps somewhat wetter climates during the interglacials resulted in a rather intensive weathering on the rolling relief of the interfluves.

After the regression of the glaciers, the main rivers re-incised their valleys and the tributaries started a fast headward erosion on the valley slopes.

The present landscape consists for this reason of three morphologically different elements: an old erosional surface, a rejuvenated surface of fast headward erosion with mass movement, and valley floors with alluvial deposits (figure 1).

2 ITC METHODOLOGY OF ENGINEERING GEOLOGICAL MAPPING

2.1 Introduction

The engineering geological mapping in the Pont de Suert area was carried out with the main objective of

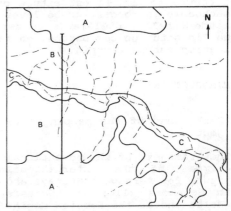

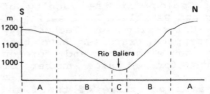

Figure 1. Schematic map and cross section of the Western part of the study area at a scale of approximately 1:20,000.
A: old erosional surface
B: rejuvenated surface with strong erosion and mass movement
C: valley floor with accumulation of alluvial and colluvial deposits.

making an inventory and classification of the various soils and rocks outcropping in the area with a description of their geotechnical characteristics.

The mapping was carried out as a reconnaissance mapping at a scale of 1:5,000 or 1:10,000 with the help of an existing geological map at a scale of 1:50,000 available.

The resulting compilation is an analytical general purpose engineering geological map not specifically prepared for one application. It can be used as a base for further special-purpose maps such as maps for urban and infrastructural planning, recreation, waste disposal or for the selection of borrow areas for construction materials. The various map units each have their specific characteristics for different applications. In the reports special emphasis has been given to the application as construction material.

The ITC methodology of production of engineering geological maps consists of the following main steps:

A <u>Interpretation of aerial photography</u>. A suitable scale of aerial photography is usually double as large as the final required map scale. For a discussion about the scale of aerial photography and the details which can be mapped with it, reference is made to Sissakian, Soeters and Rengers (1983). One of the conclusions of these authors is that the photo scale can be enlarged up to a factor of 6x with an important improvement of the resolution and of the mapping facility.

B <u>Field check and collection of field data</u> on the outcropping soils and rocks in order to describe and classify these for engineering geological purposes.

C <u>Reinterpretation</u> of the aerial photographs, using the acquired knowledge about the field circumstances.

D <u>Transfer of the photo overlay data</u> to a topographical base map, using a simple mirror stereoscope or an analytical plotter.

2.2 Photo interpretation of rock units

If there is no detailed geological map available at the required map scale and if detailed geological mapping of the area is not possible, then aerial photography can be used to delineate parts of the terrain where the engineering characteristics of the soil and rock are so uniform that they can be mapped as homogeneous zones.

Homogeneous rock zones (units with uniform 'rock mass strength') are delineated on the aerial photo on the basis of uniform weathering and erosion characteristics, which are reflected in uniform morphological aspects, such as slope and stream gradients and drainage patterns. Tone, texture and vegetation are other photocharacteristics which may be used.

The recognition of homogeneous rock mass strength zones may be obscured due to differences in geomorphological regime (mature, erosional, and depositional) in different parts of the area (compare figure 1), and particularly due to the human influence on landform (terracing), landuse and vegetation.

Boundaries between homogeneous rock mass strength zones often coincide with boundaries between lithological units, but in many cases rock mass strength differences within one lithological unit and the influence of tectonics makes subdivision necessary of one lithologic unit into a number of different rock mass strength units.

2.3 Photo interpretation of soil units

Homogeneous soil zones are outlined on aerial photographs mainly on the basis of the genesis of the soil (such as alluvial, colluvial, glacial, etc.), which can usually be revealed by the landform and by the position of the soil in the landscape.

The distinction between soil and rock is based on a simple and practical criterion: materials which can be excavated directly are considered as soil, while they are considered to be rock when ripping or blasting is necessary before excavation. In the mapping area this means that all quaternary geological materials are mapped as soil, although colluvial scree may be lo-

cally cemented (so as to be transformed into a rock) due to the evaporation of carbonate containing
groundwater when reaching the terrain surface by seepage.

All geological materials of pre-
quaternary age are mapped as rock
unless they are weathered so intensively that they have been transformed into a residual soil.

```
┌─────────────────────────────────────────────────────────────────────────┐
│                                                                           │
│   ITC - Engineering Geology        ┌──────────────────────────┐          │
│                                     │ ROCK MAP UNIT DESCRIPTION │          │
│   Name of observer:                 └──────────────────────────┘          │
│                                                                           │
│   Date:                                                                   │
│   Location:                          Coordinates:......../.........       │
│   Rock map unit name:                                                     │
│                                                                           │
│   Dominant lithology:                (.... %)                             │
│                                                                           │
│   Sublithology 1:                    (.... %)                             │
│                                                                           │
│   Sublithology 2:                    (.... %)                             │
│                                                                           │
│   Weathering characteristics:                                             │
│                                                                           │
│                                                                           │
│   Dominant systematic discontinuity plane: absent/faintly developed/present│
│   orientation:......./........                                            │
│                                                                           │
│   Discontinuity pattern: regular/slightly irregular/irregular/chaotic     │
│   (incl. bedding)                                                         │
└─────────────────────────────────────────────────────────────────────────┘
```

	type of disc.	orientation	spacing	persistence	roughness	
			cm	cm	meter scale[x]	mm scale[xx]
set 1		/				
set 2		/				
set 3		/				
set 4		/				

unit block shape   massive/blocky/tabular/columnar/irregular/crushed

unit block size   .... x .... x .... cm^3 to .... x .... x .... cm^3

Rock material strength

	Dominant lithology	Sublithology 1	Sublithology 2
Mean Schmidt Hammer strength value	.... MPa	.... MPa	.... MPa
Mean point load strength value	.... MPa	.... MPa	.... MPa

x) large roughness (meter scale): stepped, undulating or planar
xx) small roughness (mm scale)   : rough, smooth or slickensided

Figure 2. Standard form used for the field description of rock mass
strength units.

Soil is only outlined as a separate unit if its vertical thickness
exceeds 1 m. If the soil is less
thick the underlying rock is mapped
as being outcropping.

2.4 Photo interpretation of geody-
namic phenomena

The geodynamic phenomena which were
mapped in the area are mass movement and erosion. Due to the large
scale of the aerial photography
(enlargements up to 1:5000) mass
movement phenomena of even a limited size of less than 20 x 20 m
could be detected due to their morphological expression and outlined
on the photo interpretation overlay. Erosion phenomena of even
smaller sizes (down to 5 m) could
be detected on the aerial photography due to the strong spectral
contrast between bare eroded ground
with very high reflectance of the
visible light and the surrounding
vegetated ground with much lower
reflection characteristics. The
classification and mapping methodology for mass movement and erosion
as used in this mapping is based on
Nemčok, Pašek and Rybář (1972) and
on Carrara and Merenda (1974).

3 ENGINEERING GEOLOGICAL MAP OF THE
PONT DE SUERT AREA

3.1 Rock mass strength units

The units which were outlined during the photo interpretation stage
as having uniform rock mass
strength are described and tested
in order to determine their geotechnical characteristics. The figure 2 shows the forms which were
used in the field to ensure that
the most essential engineering
geological aspects are collected.
A complicating factor for the
field description and testing of
outcrops of rock mass strength
units which were outlined and classified in aerial photography is
that natural outcrops usually show
strong surface weathering. During
the execution of engineering works
this surface weathering layer is
usually removed, so it is not representative for the engineering
behaviour of the rock mass strength

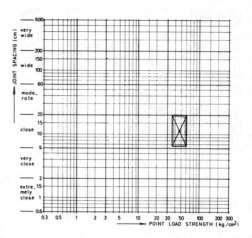

Figure 3. Rock mass strength diagram used to characterize rock
units on the basis of material
strength and discontinuity spacing.

unit. For this reason extreme care
should be given to the selection of
representative locations for the
description and sampling for testing of the rock mass strength
units.
Erosional environments within the
landscape will show more fresh rock
outcrops than such parts of the
area which are parts of an older
landscape (see figure 1).
The discontinuity spacing value
and the rock material strength
value are used for the characterization of rock mass strength in
the field. The two aspects can be
used to represent the unit in a
rock mass strength diagram (see
figure 3), which is also contained
in the legend of the map. The position of a rock mass strength unit
in the diagram will give an indication of its suitability for engineering applications (Fookes,
Dearman and Franklin 1971). Rengers
and Soeters (1980) describe the
details of this procedure.
Figure 4 shows a part of the
compiled map. Four rock mass
strength units are distinguished:
with very high, high, medium and
low rock mass strength. An additional lithological index number is
attached to indicate which lithological unit is forming the rock
mass strength unit.
The table of figure 5 gives a
summary of the photo recognition

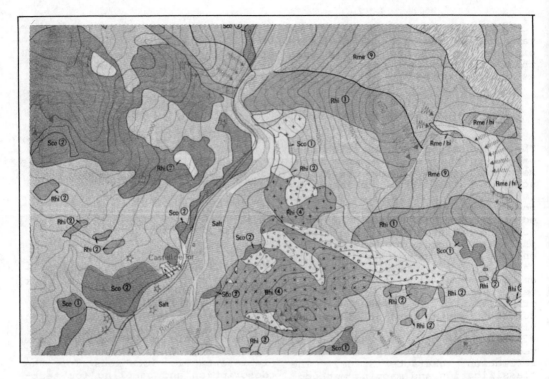

Figure 4. Part of the compiled map at a scale of approximately 1:15,000. Copies of the colour printed maps can be obtained from the authors (see section 3.4 of this paper).

characteristics, the resistance against erosion and mass movement processes as well as an indication of the geotechnical characteristics of the four rock mass strength units. The table shows that rock mass strength units do not always coincide with lithological units, although some correlation is apparent.

3.2 Soil units

The engineering geological soil units which are outlined during the photo interpretation stage on the basis of their position in the landscape and their micromorphological characteristics, are described during fieldwork in terms of their geotechnical characteristics and are sampled for a more detailed determination of the grainsize distribution and plasticity in the laboratory. With the obtained data the Unified Soil Classification

System can be used for the classification of the soils (ASTM, 1988).

Figure 6 shows the form which was used in the field for description and classification.

Figure 4 shows a photocopy of part of the map in which most of the soil units are exposed. In the legend to the map an indication is given of the USCS soil classes as determined from samples taken from these units.

Seven different soil types were distinguished during the mapping: three alluvial soil types, two colluvial soil types and one of glacial origin. The residual soil type is the product of in situ weathering.

In the table of figure 7 the photo recognition and morphological characteristics, and the geotechnical description of the different soil types are given as well as an indication of the suitability of the soil types for a number of applications as construction material.

240

	Very high rock mass strength	High rock mass strength	Medium rock mass strength	Low rock mass strength
Photo interpretation characteristics	very pronounced higher parts of the area, steep-slopes without erosion	higher parts of the area	undulating slopes	gentle sloping to almost flat areas
Lithological units	sandstones & conglomerates, limestones & dolomites ophitic intrusions	limestones, calcareous & quartzitic slates, sand & silt stones	mudstones, shales, silt stones, marly limestones	gypsiferous marls, shales
Characteristic rock material strength (point load strength)	4 - 15 MPa	2.5 - 10 MPa	2 - 7 MPa	3 MPa
Joint spacing characteristics	40 - 100 cm	20 - 60 cm	10 - 30 cm	25 cm
Susceptibility for mass movement & erosion	very high resistance, only bedding plane slides & toppling; no erosion	high resistance, only bedding plane slides & rock fall; rill & sheet erosion	susceptible to mass movement when under-cut; rill & gully erosion	unstable when exposed in steep slopes, very active gully & badland erosion
Use as construction material	limestones & dolomites & in some cases ophitic suitable to be crushed for high quality aggregate	to be crushed for low quality aggregate	no use	no use

Figure 5. Summary of photo recognition and geotechnical characteristics of the rock mass strength units in the mapping area.

3.3 Geodynamic phenomena in the mapping area

The geodynamic phenomena which were mapped in the area are mass movement and erosion. The mass movement subdivision is based on previous work by Nemčok, Pašek and Rybář (1972) and by Carrara and Merenda (1974).

Abundant mass movement phenomena occur in the mapped area. Usually a combination of more competent material overlying less competent material, which weathers or erodes, or a combination of colluvial materials of higher permeability overlying clayey bedrock in combination with a strong relief and active erosion cause abundant mass movement.

Erosion in the mapping area is restricted to soil units and units of low rock mass strength which are usually related to the lithology of slates and shales specially in those areas where headward erosion is cutting into the old landscape. Some erosion will occur in the drainage gullies of the whole area, due to the strong relief, but erosion is more strong where these gullies are located in units of lower rock mass strengths and in the soil units.

3.4 Map production

The transfer of photo overlay information to a base map is a labour intensive work for which usually the interpretative transfer is used: the interpreter views the stereo model under the stereoscope and transfers the information from the stereo model to the topographic map by comparing the three dimensional form of the stereo model with the contour lines, drainage channels, and the crest lines in the map.

For the map of the Pont de Suert - Vilaller area the transfer of photo overlay data to the base map was done in a very accurate new approach with help of the Zeiss Planicomp analytical photogrammetrical equipment. The stereo photography is introduced into the Planicomp and after relative and absolute orientation of the photography the use of a floating mark allows the interpreter to determine the xyz-terrain-coordinates of point locations on the aerial photographs. With the Planicomp all unit boundaries and other mapping features were digitized. With the CAD program/system Intergraph the digitized information was plotted at the scale of the final map and

SOIL MAP UNIT DESCRIPTION

Name of observer:

Date:
Location: Coordinates:
Soil unit in map: Parent material:

Field observations

layering   bedding: clear/poor/absent
                          laterally: regular/irregular
cementation   : absent/little/medium/high/very high
moisture condition   : dry/moist/wet/saturated
estimated permeability   : very high/high/medium/low/very low
surface infiltration   : very high/high/medium/low/very low
form of coarse particles (>5mm): angular/subangular/subrounded/rounded
hardness of coarse particles   : very hard/hard/medium/weak/very weak
present slope angle   : .... maximum (critical) slope angle:....°

Field estimation of grainsize distribution

weight percentage boulders (> 20 cm)   = ... % A (from scanline)
weight percentage cobbles   (> 7.5 cm) = ... % B (from scanline)
weight percentage gravel   (> 4.75 mm) = ... %
weight percentage sand   (> 0.075 mm) = ... %
weight percentage fines   (< 0.075 mm) = ... %

Laboratory data on fraction < 7.5 cm determined in field lab from samples

weight percentage gravel = ... % C $C' = \frac{100-(A+B)}{100} \times C = ...$ %

weight percentage sand   = ... % D $D' = \frac{100-(A+B)}{100} \times D = ...$ %

weight percentage fines = ... % E $E' = \frac{100-(A+B)}{100} \times E = ...$ %

Note: A + B + C' + D' + E' must be 100 %

if fines percentage higher than 5 % of C + D + E:

dry strength = none/slight/medium/high/very high
toughness   = none/medium/slight/high
dilatancy   = none/very slow/slow/quick

Conclusion on name in USCS: _____

Sand equivalent test: sand reading (s) =   Sandequivalent
                      clay reading (c) = $\frac{s}{c}$ x 100 =
                                                 =========

Figure 6. Standard form used during fieldwork to describe homogeneous soil units.

superimposed on the contour line map of the area which had already been prepared by the ITC Photogrammetry department using the same aerial photography.

The final coloured map has been drafted by the ITC Cartographic department. A printed copy of that map can not be included with this publication in the congress proceedings, but will be published in one of the next numbers of the ITC Journal. Those who are interested can receive a copy of the map with a more detailed report on the area by writing to N. Rengers or R. Soeters, c/o ITC, P.O. Box 6, 7500 AA Enschede, the Netherlands.

	Riverbed (Sal)	Alluvial fan (Salt)	Terrace (Salt)	Slope deposit (Scol)	Footslope dep. (Sco2)	Glacial (S gl)	Residual (S res)
Location in the landscape	along main river	where tributary enters valley	elevated over valley floor	in middle part of long slopes	along foot of slope	isolated remnants	flat parts interfluves
Weight % 7.5 cm	30-50 %	5-20 %	20-40 %	0-15 %	< 5 %	5-20 %	0
Weight % gravel	20-40 %	30-60 %	20-40 %	20-60 %	5-40 %	10-40%	0-40 %
Weight % sand	10-30 %	20-50 %	10-50 %	15-40 %	20-60 %	30-60 %	10-30 %
Weight % fines	< 5 %	5-25 %	0-20 %	10-30 %	15-40 %	10-40 %	30-60 %
USCS soil class	GP, SP	GM, GW, GP, SP, SM, SW	GM, GP, GW, SM, SP, SW	GW, GC, GM, SC, SM	SC, SM, GC, GM	SW, SC, GM, GW, GC	SC, SM, CL, CH
Suitability for fill	good	very good	reasonable	reasonable	poor	poor	poor
Suitability for aggregate	very good after sieving	too many fines	good after sieving	poor	not suitable	not suitable	not suitable

Figure 7. Summary of photo recognition and geotechnical characteristics of the soil units in the mapping area.

REFERENCES

ASTM, 1989. Annual Book of ASTM Standards. American Society for Testing and Materials.

Carrara, A. & L. Merenda 1974. Métodologia per un censimento degli eventi franosi in Calabria. Geologia Applicata e Idrogeologia Vol. IX, pp. 237-255.

Fookes, P.G., W.R., Dearman & J.A. Franklin 1971. Some engineering aspects of rock weathering with field examples from Dartmoor and elsewhere. Qu. Jl. Engng. Geol. 4, pp. 139-185.

Meij, P.H.W. 1968. Geology of the Upper Ribagorzana and Tor Valleys, central Pyrenees, Spain. Leidse Geologische Mededelingen Vol. 41, pp. 229-292.

Nemčok, A., J. Pašek & J. Rybář 1972. Classification of landslides and other mass movements. Rock Mechanics 4, pp. 71-78.

Rengers, N. & R. Soeters 1980. Regional Geological Mapping from Aerial Photographs. Bull. IAEG No. 21, pp. 103-111.

Rengers, N., R. Soeters, P.A.L.M. van Riet & E. Vlasblom 1990. A large scale engineering geological map of the Pont de Suert - Vilaller Area, Central Spanish Pyrenees. ITC Journal 1990-..... (in print).

Sissakian, V., R. Soeters & N. Rengers 1983. Engineering geological mapping from aerial photographs: the influence of photo scale on map quality and the use of stereo-orthophotographs. ITC Journal 1983-2, pp. 109-118.

Riet, P.A.L.M. van & E. Vlasblom 1988. Engineering geological map Vilaller - Pont de Suert and Castarnés - Llesp, Spain. Memoir 55 of the center of Engineering Geology in the Netherlands. Delft

Analytical photogrammetry for measurement and plotting of steep rock slopes

Utilisation de la photogrammétrie analytique pour la mesure et la représentation de versants rocheux escarpés

Niek Rengers & Michael Weir
ITC, Enschede, Netherlands

Jan Willem Rösingh
Ballast Nedam Engineering BV, Amstelveen, Netherlands

ABSTRACT: The development of compact, fast and inexpensive personal computers and connected plotting devices has brought the use of analytical photogrammetric systems within the reach of non-photogrammetrists. This paper describes the measurement of point co-ordinates and the preparation of distortion-free large-scale topographical base maps with horizontal or vertical reference plane of steep rock slopes.

RESUME: Le développement de calculateurs personnels compacts, rapides, peu onéreux et reliés à des périphériques de restitution à mis l'usage des systèmes photogrammétriques analytiques à la portée de personnes non-spécialisées. Cette communication décrit la mesure des coordonnées des points et la préparation de cartes topographiques de base exemptes de distorsion pour représenter, sur des plans de reference horizontal ou vertical, des versants rocheux escarpés.

1 INTRODUCTION - THE PRINCIPLE OF PHOTOGRAMMETRIC MEASUREMENT

Photogrammetry is the science of obtaining accurate measurements from photographs. Although primarily applied for map production from aerial photographs, photogrammetry can also be carried out using "terrestrial" photography. Applications of terrestrial photogrammetry can be found in architecture, medicine and various branches of engineering. Engineering geology employs both aerial and terrestrial photogrammetry.

To understand the principles of photogrammetric measurement, it is necessary to examine the way in which the terrain or some other object is imaged on a photograph. Each object point Pi can be considered as the origin of a light ray which travels in a straight line through the lens of the camera to form an image on the film at the point pi (figure 1). We are therefore dealing with two co-ordinate systems, one in the photograph and the other in the object space.

Photogrammetry is concerned with

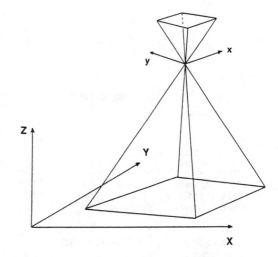

Figure 1. Basic relationship between object and photo co-ordinate systems.

establishing a relationship between these two co-ordinate systems so that object co-ordinates X,Y,Z can be determined from measured photo co-ordinates x,y. In order to de-

termine the transformation parameters it is necessary to know the
co-ordinates of at least three discrete points in both systems together with the focal length of the
camera. These data enable the camera orientation and the scale factor (depending on the distance to
the object) to be determined and
used to obtain the co-ordinates of
all other points measured on the
photograph. This simple relationship holds, however, only as long
as Z is constant, or, in other
words, the object (or terrain) is a
flat surface. Since this is rarely
the case, it is necessary to obtain
additional data in order to determine Z. Normally these additional
data are obtained by double (stereoscopic) measurement of the photo
co-ordinates on two overlapping
photographs (figure 2).

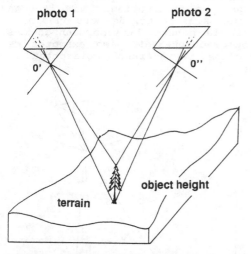

Figure 2. Determination of the
third dimension (height) Z from
double (stereoscopic) measurement
on overlapping photographs.

2 ANALYTICAL PHOTOGRAMMETRIC IN-
STRUMENTS

Traditionally, photogrammetry has
employed complex optical-mechanical
instruments to establish the relationship between photographs and
object in an analogue way.
In recent years, the availability
of inexpensive but powerful computer systems equipped with high-resolution graphics monitors and
plotting devices have made analytical approaches a practical alternative to conventional stereo plotting on analogue instruments. Most
analytical plotters are high-preci-
sion instruments which are employed
in topographic and cadastral mapping where high accuracy is required. For many other applications,
however, such equipment is still
prohibitively expensive. It is only
recently that manufacturers have
begun to produce relatively low-
cost instruments with limited precision for use in fields such as
forestry (Weir & Bartelink 1990)
and geology. Examples include:
- the APY system designed for
thematic mapping and map revision
from aerial photographs (Yzerman
1984);
- the Carto Instruments AP190,
also designed for mapping from aerial photography, and which has been
particularly employed by forestry
agencies (Warner 1989);
- the Topcon PA-1000, originally
developed at ITC as a training instrument (Goudswaard 1988);
- the Adam Technology MPS-2, designed for plotting from small-for-
mat aerial and terrestrial photography (Weir 1988) (fig. 4).
Apart from minor differences, all
analytical photogrammetric instruments are designed along similar
principles (figure 3). The two photographs forming a stereo pair are
held on carriers and observed
through a fixed optical system as a
single three-dimensional "model".
Two small measuring marks are incorporated into the optical system.
The two photographs can be shifted
in the x and y directions by means
of servo-motors driven by hand-
wheels, a joystick or a simple
free-hand motion. A differential x
shift of the two photographs is observed as a change in elevation
(Z). In this way, the measuring
marks can be kept fused together as
a single "floating" mark for measuring in three dimensions. The
shifts of the two photographs in
the x and y directions are encoded
and transmitted to the host computer as two sets of co-ordinate pairs
(x',y' and x",y").
Before the actual measurement
process, it is necessary to carry
out an orientation procedure, including observing the co-ordinates
of at least three known control

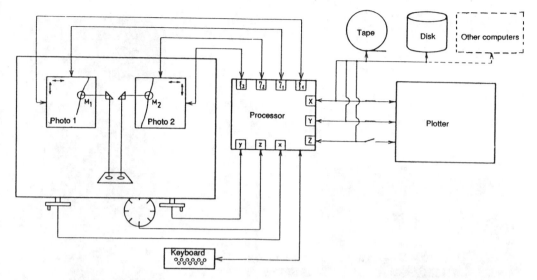

Figure 3. Principles of the design of analytical photogammetric instruments (adapted from Krans 1982, Photogrammetrie, Ümmler, Bonn).

points. These data are used to derive parameters for the transformation from photo to object co-ordinate system during the subsequent measurement phase. All measured photo co-ordinates are transformed in real time to object co-ordinates, displayed graphically on a monitor and stored in a file for subsequent plotting.

Figure 4. The MPS-2 Micro Photogrammetric System with a connected personal computer and plotter.

3 AERIAL AND TERRESTRIAL PHOTOGRAPHY OF STEEP SLOPES

A large range of cameras can be used for photogrammetric purposes. A main distinction is between metric cameras and non-metric cameras. Metric cameras have no or very little optical and film (or glass-plate) distortion, a fixed focal length ("principal distance") and well-defined corner marks ("fiducial marks"), which are used to define the photo centre.

Non-metric cameras have a variable focal length (for "focusing") and have inaccuracies due to lens distortion and lack of film flatness. Negative sides and corners are identical in every negative, but not well defined in terms of photo co-ordinates.

Analytical photogrammetric instruments allow the use of non-metric cameras (with little loss of accuracy if good-quality lenses are used). Focal length and lens distortion as well as corner positions of

Figure 5. Metric (distortion-free) cameras. a) Zeiss aerial camera RMK with 230x230 mm negative size; b) Wild P32 terrestrial camera connected with a theodolite, negative size 60x90 mm.

the negative can be calibrated in the analytical instrument for use during the numerical calculations.

Figure 5 shows two metric cameras: a Zeiss RMK for aerial photographs with a 230x230 mm negative size (cost US$ 150,000) and a Wild P32 photo-theodolite camera for terrestrial photography with a negative size of 60x90 mm (cost US$ 10,000).

Figure 6 shows two non-metric professional cameras with 60x60 mm negative size.

If the slope to be photographed is not obscured by vegetation at its foot and suitable camera locations can be found for a complete stereo coverage (preferably with more or less parallel camera axes for good stereoscopic vision), then terrestrial photography is to be preferred over, and is much less expensive than, aerial photography. A visit to the site is anyhow necessary to survey the terrain coordinates of a number of control points. A photo-theodolite camera or a professional (60 mm negative)

camera can be used to make the photographs.

If the slope is not well visible because of vegetation or if no suitable camera locations are available for full stereoscopic vision, photographs can be taken from an aircraft. Normal survey aircraft have a large-format camera mounted vertically in the floor. This means that steep slopes can only be photographed obliquely in a strongly tilted aircraft position while flying in a curve (see fig. 7). This makes it very difficult to obtain full stereo coverage.

For this reason it is attractive to use a professional (60 mm negative) camera for oblique photography through the window or door of the aircraft or by mounting the camera in a tilted position underneath the aircraft. With a parallel oriented video camera the field of vision of the photo camera can be determined by the camera operator inside the aircraft (figs. 8 and 9). The camera must be equipped with remote control and motorized

Figure 6. Non-metric professional cameras. a) Hasselblad 60x60 mm negative; b) Bronica 60x60 mm negative.

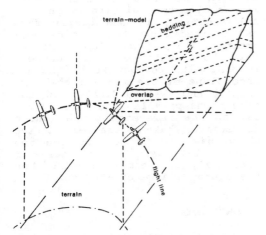

Figure 7. Oblique photography with an inclined standard aircraft for aerial survey.

expressed as the percentage of the distance from the measured point in the terrain to the camera at the moment of exposure.

Tests have shown that the use of an MPS-2 instrument for terrestrial stereo photography which is taken on normal roll film in a Wild P32 camera can lead to a relative accuracy of an average 0.2% if the control points are well visible in the photographs and their position in the local co-ordinate system has been accurately determined with usual surveying accuracy.

film transport. The aircraft used can also be a light or ultralight aeroplane, but experience has shown that the mounting of the camera should be such that vibrations of the aircraft transferred to the camera are reduced to a minimum (Rösingh 1987) in order to obtain good-quality, sharp negatives.

4 ACCURACY OF POINT MEASUREMENTS

A number of factors influence the final accuracy of the calculated position of points in the terrain. The relative accuracy of the X, Y and Z vertical co-ordinates can be

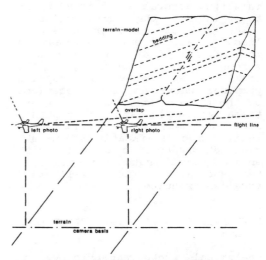

Figure 8. Oblique photography with a tilted camera mounted on the exterior of a light aircraft.

Figure 9. Non-metric photo cameras (Hasselblad and Canon) with parallel video camera mounted on the exterior of a light aircraft (Dimona 1136).

This means that in a terrestrial survey of a slope with a height of 40 m, where a distance between camera baseline and rock wall of 50-100 m is used, relative accuracies in X, Y and Z direction will occur of 15-25 cm. For aerial survey of steep slopes with oblique photography, distances of 200 to 300 m are usually necessary, which leads to relative accuracies of 60-90 cm in te position of the terrain points.

The use of non-metric cameras leads to an important increase in the inaccuracy of the point measurements, so the accuracy of the final plan will reduce. For mapping purposes where absolute locations of the map elements are not of extreme importance and main emphasis lies on photo interpretation and rock mass classification. Non-metric photography in combination with an analytical photogrammetric system, such as the MPS-2, will yield excellent results.

5 EXAMPLE OF A PLAN OF A STEEP ROCK FACE

Analytical photogrammetric systems cannot be used only to calculate point co-ordinates, but when compiled with a plotter they can also be used to produce plans of a rock slope where a vertical or horizontal datum plane can be chosen. The handling of a micro-photogrammetric system such as the MPS-2 does not require detailed photogrammetric background knowledge and can be learnt by an engineering geologist within a few days time.

Figure 10 is a terrestrial photograph of one of the walls of a quarry in Upper Devonian Sandstones near the river Bocq in the central part of Belgium. This photo was taken with a Wild P32 photo theodolite and has a scale (enlarged) of approximately 1:250.

Figure 11 shows a plan with vertical reference plane of the quarry wall of figure 10 which was prepared with the MPS-2 (Rösingh 1987). In the plot the relative heights of all locations in the quarry face can be read along the vertical co-ordinate axis of the plan. Horizontal photo axis and the contour reading with respect to the reference plane of the map are the terrain X and Y co-ordinates in the directions parallel and perpendicular to the reference plane (which is usually taken parallel to the photo baseline connecting the two camera stations). In the case of figure 11, the camera baseline has the contour value of +100 m.

6 CONCLUSION

For a relatively limited cost, a complete system for photogrammetric surveying of steep rock slopes can be acquired. The basic elements are a non-metric professional camera and a micro-analytical photogrammetric system to which a standard personal computer and plotter must be connected. Such a system is a powerful tool for the preparation of plans of steep rock slopes.

An important aspect for the choice of camera type for the photogrammetric survey is the scale and point accuracy which is required in the resulting plan. For reconnaissance surveys the use of non-metric cameras can yield good results.

Figure 10. Terrestrial photograph made with the Wild P32 terrestrial camera of a rock quarry in Belgium.

REFERENCES

Berghuis, R. 1987. The possibilities of the construction of an engineering geological map from oblique aerial photographs with the stereocord apparatus.

Goudswaard, F. 1988. A new concept for quantitative interpretation of stereo photographs by non-photogrammetric specialists. Int. Archives of Photogrammetry and Remote Sensing Vol. 27, B. 8 II, pp. 9-14.

Rengers, N. 1967. Terrestrial photogrammetry: a valuable tool for engineering geological purposes. Rock Mechanics and Engineering Geology Vol. V/2-3, pp.150-154 (1967).

Rösingh, J.W. 1987. Use of photogrammetry in engineering geological mapping of steep slopes. Final thesis at Delft University of Technology, November 1987.

Warner, W.S. 1989. A complete small format aerial photography system for GIS data entry. ITC Journal 1989-2, pp. 121-129.

Weir, M.J.C. 1988. Evaluation of the MPS-2 analytical plotter. Internal report of the International Institute for Aerospace Survey and Earth Sciences (ITC), 1988.

Weir, M.J.C. & H. Bartelink 1990. Computergesteunde en analytische fotogrammetrie - toepassingen in de bosbouw. Nederlands Bosbouw Tijdschrift Vol 62(1), pp. 12-18.

Yzerman, H. 1984. The APY system for analytical photogrammetry. Photogrammetric Record XI(64), pp. 407-413.

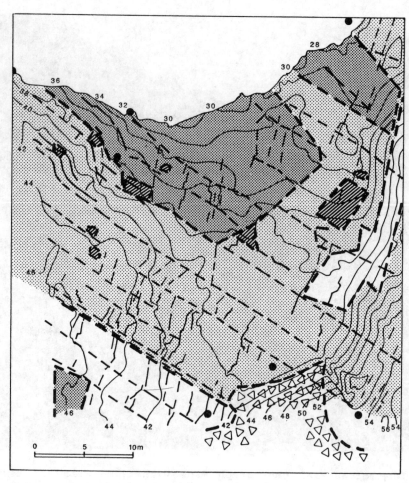

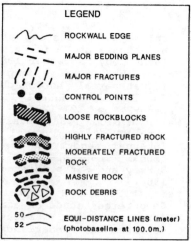

LEGEND

～～～ ROCKWALL EDGE

－ － － MAJOR BEDDING PLANES

⎛⌇⎞⌇⎞ MAJOR FRACTURES

● ● CONTROL POINTS

▨▨ LOOSE ROCKBLOCKS

HIGHLY FRACTURED ROCK

MODERATELY FRACTURED
ROCK

MASSIVE ROCK

ROCK DEBRIS

50 ⎯⎯
52 ⎯⎯ EQUI-DISTANCE LINES (meter)
(photobaseline at 100.0m.)

Figure 11. Plan with vertical reference plane of the quarry in figure 10.

252

Evaluation of suitability of the North Armenia territories for major construction works

L'estimation de l'aptitude des terrains de l'Arménie du nord pour les grands travaux de construction

G.G.Strizhelchik
Ukrvostok Güntiz, Kharkov, Ukraine, USSR

ABSTRACT: The evaluation and division into seismic districts of the North Armenia territories to determine their suitability for major construction works enables to take optimum design decisions, increase the effectiveness of expenditures on restoration works and make the efficient use of land resources

RESUME: L'estimation de l'aptitude du terrain du nord de l'Arménie dans le but des grands travaux de construction a permit d'accepter des pro - jets optimum d'augmenter le rendement d'investissements de capitaux pour les travaux de remise en état et les travaux neufs, ainsi que l' utilisation rationnelle de ressources terrestres

1 DISCUSSION AND OBSERVATIONS

The tragic aftermath of the earthquake in Speatak, which shocked the entire world by its huge num - ber of victims and staggering amount of destruction, has made obvious a demand both to increase earthquake resistance of structures and revi- se the existing traditions in eva- luating suitability of formely- and newly-developed lands for ma - jor construction works. At the sa- me time the situation which has taken form has caused an urgent need to take prompt and reliable enough measures.

The earthquake of 1967 (M=5), 1975 (M=3.8), 1978 (M=4), which had anticipated the earthquake in Speatak in 1988 (M=7), indicated the increased activity of the Ba- zhum-Sevang system of deep fractu- res. The intensity of the last earthquake measured according to the scale adopted in the USSR was: force 9-10 - in Speatak, force 8- 9 - in Leninakhan, force 7-8 - in Kirovokhan and Stepanovan, force 5-6 - in Yerevan. The historical and geological background shows that these territories had repea- tedly suffered seismic catastro - phic effects with some of them be- ing even more disastrous than tho- se caused by the Speatak earthqu- ake.

On specifying the information of the extremely high initial seismic activity (force 9) throughtout the entire North Armenia territory,the research workers and experts coor- dinated their efforts in the ende- avour to reveal and calculate the limit values with respect to those local factors which increase seis- mic and accompanying effects on buildings and other structures. This investigation was of decisive importance in view of the fact that the rated seismicity of con - siderable areas of the existing settlements and zones preselected for new construction projects ex- ceeded the values (force 9) allow- ed by the building code.

The above-mentioned negative factors are as follows:
-netting of active deep fractures and close-to-surface breakages with the registered rock displace- ments and dislocations, increased vibration intensity and expected intensification of exogenous pro- cesses;

- steepness of slopes (15°) with the vibration intensity on their ledges being also increased and gravitation processes inevitably intensified;
- presence of such hazardous phy - sical-geological processes and fe- nomena as landslides, creeps, ava- lanches, earth flows, suffosion, karst;
- close-to-surface subsoil water occurence or possible formation of technogenous water-bearing levels;
- subsidence, salinity, thixotropy or suffosion instability of high- porous soils in active zones.

The study of these factors was mainly conducted by traditional methods of division into microsei- smic districts and engineering ge- ology research for construction purposes. The additional surface levelling, radioactive emanation and biological detection servays were conducted only in the epicen- tre (Leninakhan-Speatak-Kirovokhan) in order to estimate the rate of activity of numerous damages caus- ed by breakages.

The practical effect of each fac- tor was specified by superpositi - oning the data concerning macrosco- pic effects of heavy earthquakes on the large-scale maps of division into microseismic districts.

The matrix form of information generalization was used in the fac- tor analysis which enabled to con- siderably simplify processing and estimating the significance of each factor.

There are three classes of areas determined on the maps in accord - ance with technical and economic possibilities for preventing addi- tional negative effects of each factor on buildings and structures. These classes are: "suitable", "relatively suitable" and "unsuit- able" areas for major construction works.

The suitable areas were related to those with rated seismicity of not greater than force 9. These areas may be characterized by the following general features:
- the areas located beyond the bounds of influence of damages caused by active breakages;
- steepness of slopes of less than 15°;
- depth of subsoil water occurence

greater than 6 m and no conditions for increase of the level or for - mation of new water-bearing levels;
- absence of conditions for develo- ping dangerous physical-geological processes on the surface and in the mass of soil.

Foundation bases for the build- ings and structures in such areas are rocks, fragmental rocks, firm sandy and dust-clay soils with their seismic properties relating to the first and second categories.

To the relatively suitable areas belong those with rated seismicity of, at least, force 9, i.e. the areas where the factors causing the increase of seismic effects are present or may come into rea - lity. Within the bounds of these areas the specific subareas were selected in accordance with the presence of individual or combined factors. This enabled the sharply oriented construction work recom - mendations worked out on the basis of standard prediction, i.e. alter- native description of the ways to achieve the preset targets, to be included in the typological scheme. The subareas were classified in accordance with the complexity of the required measures. It comes na- tural that the least complicated subareas were found to be those with gentle slopings and volcanic plateaux, where due to proper engi- neering ground preparation (arran- gment) of surface run-off, clean- ing of natural drains, provision of retaining and protective walls) the seismic and accompanying ef - fects were reduced. The more com- plicated subareas were found to be those having in the active zone a thickness of high-porous delluvial loesses possessing properties of subsidence. The thickness of such soils in certain zones of the Shi- rak hollow and delluvial accumula- tions along the Bazhum mountain ridge exceeds 15 m. The subareas where these high-porous soils were flooded with the danger of sliding displacements of huge masses of soil were associated with special difficulties.

The complex of the engineering requirements for such subareas in- cluded draining operations, cutt- ing or substitution of soft ground, arrangement of packed mixed-soil cushions etc.

The minimum thickness of cushi - ons was to meet the requirement for the residual thickness of soft grounds (third category with res - pect for seismic properties) not to exceed 5 m in the thickness of 10 m. If the thickness of soft grounds was large, the cushions were to overpass the limits of the building outline by 3-5 m with compaction of soil by rolling in the clerances around the building.

The envisaged actions ensured reduction of the intensity of seis- mic effects on buildings and struc- tures down to force 9, which, in its turn, enabled to accomplish mass construction works.

The areas found unsuitable for major construction works were re- lated to those with the influence of active tectonic fractures and where due to technical and econo- mic reasons, the reduction of seis- mic effects down to force 9 could not be practically achieved.

It is quite natural, that the use of these areas was excluded from the plans of new building de- velopment, as well as, from the plans of reconstruction works wi- thin the bounds of the existing populated areas. In special cases certain areas were allowed to be used for the construction of tempe- rary light-weight installations (playgrounds without stands, unco- vered parking areas, stores etc.).

2 CONCLUSION

The principles stated above enabl- ed to maximally take into account the complex combination of seismic and engineering-geological factors to reduce the risk of living in one of the regions of our planet considered to be dangerous with respect to seismic conditions. The evaluation and division into seis- mic districts of the North Armenia territories to determine their sui- tability for major construction works enabled to take optimum de- sign decisions, increase the effec- tiveness of expenditures on resto- ration works and make the effici - ent use of land resources.

6th International IAEG Congress / 6ème Congrès International de AIGI, © 1990 Balkema, Rotterdam. ISBN 90 6191 130 3

Engineering geological mapping for urban planning in developing countries
Cartes géotechniques pour l'urbanisme dans des pays en voie de développement

A.van Schalkwyk
Department of Geology, University of Pretoria, South Africa

G.V. Price
Jeffares & Green Inc., Pietermaritzburg, South Africa

ABSTRACT: Rapid urbanization due to population growth in developing countries requires infrastructure and housing development over large areas. Problems experienced as a result of unfavourable geological conditions have shown that engineering geological mapping and site classification is an essential prerequisite for the planning of such development.

Due to a shortage of time and manpower, conventional detailed engineering geological mapping has to be replaced by a rapid method of regional assessment based on aerial photographic land facet analysis, detailed investigation of representative mapping units and extrapolation to similar areas.

Three geological influencing factors, namely foundation conditions, drainage conditions and slope stability conditions are rated individually in terms of their effect on residential development, and the total rating is used to classify a site as good, fair or poor. Such a sub-division, provided as an overlay for the engineering geological map, can readily be understood and used by planners, local authorities, homeowners and builders.

RÉSUMÉ: L'urbanisation rapide due à l'accroissement démographique dans les pays en voie de développement nécessite le développement de l'infrastructure et des logements sur de grandes étendues. Les problèmes rencontrés dûs aux conditions géologiques défavorables, ont montré qu'il était absolument nécessaire de préparer des cartes de géologie de l'ingénieur et de classement du site dans la planification d'un tel développement.

Par suite du manque de temps et de main d'oeuvre, on a dû remplacer la cartographie de géologie de l'ingénieur détaillée par une méthode rapide d'évaluation régionale reposant sur des analyses à facettes de terrain faites à partir de photos aériennes, des études détaillées de cartes représentatives type et d'extrapolation à des régions semblables.

Trois facteurs d'influence géologique, c'est-à-dire les conditions des fondations, les conditions de drainage et les conditions de stabilité des pentes sont évaluées individuellement selon leurs effets sur le développement résidentiel, et l'on utilise l'évaluation totale pour classer le site comme bon, assez bon ou mauvais.

Une telle subdivision, fournie au dessus de la carte de géologie de l'ingénieur est facilement comprise et utilisée par les planificateurs, les autorités locales, les propriétaires et les entrepreneurs.

1 INTRODUCTION

Rapid population growth in developing countries results in the urgent need for increased housing and associated services such as water, sanitation, electricity and transportation. In accordance with traditional lifestyle in southern Africa, single storey, detached units on plots varying in size between 300 square metres in urban areas and 10 000 square metres in rural settlements are preferred.

This type of development enables people to construct their own homes and to add extra rooms or outbuildings when required. In certain areas it also allows self-sufficiency in terms of basic foodstuffs such as meat, vegetables and maize. Disadvantages are the additional cost of the spread-out infrastructure and the large areas of land required for township development.

Unfavourable geological conditions such as slope instability, poor drainage, sinkholes and doline formation and volumetrically unstable foundation soils, can cause extensive damage, especially to low cost housing where funds for individual site investigations and special foundation design are generally not available.

In an attempt to overcome these problems, a simple site classification system, based on regional scale engineering geological mapping was developed. The purpose of this classification is as follows:

(1) to aid the Planner in selecting areas for different usage according to geological conditions,

(2) to aid Local Authorities in prescribing precautionary measures for the reduction of risk of damage to structures, and

(3) to provide the prospective owner or builder with information regarding the general site conditions and the need for additional investigations or precautionary measures.

2 ENGINEERING GEOLOGICAL MAPPING

Engineering geological mapping is required to present geological data for regional planning, proper land use, choice of investment siting, zoning of problem areas and utilization of natural resources (Dearman & Fookes, 1974 and Melnikov, 1979). In regions where rapid development of new growth areas is occurring, regional engineering geological mapping is an important element of the planning and development process.

The term "regional" generally refers to areas larger than 100 square kilometres at a scale of between 1:10 000 and 1:50 000 (Rockaway, 1976). The choice of scale depends on the size of area involved, the complexity of engineering geological conditions, the accuracy required and the scale of available base maps. In South Africa, the most detailed topo-cadastral and geological maps that are generally available, are on a scale of 1:50 000 and, therefore, that scale was considered most appropriate.

The recommended mapping procedure is based on the well-known land facet system described by Brink, Partridge and Matthews (1970) and is presented as a flow diagram on Figure 1 and further illustrated by means of an area mapped to the east of Pretoria.

The mapping procedure is normally carried out in well-defined stages as summarised in the following paragraphs.

2.1 Desk Study

During the desk study, all available data on the area is collected and a brief aerial photographic interpretation is conducted in order to control the data and

to familiarize the area.

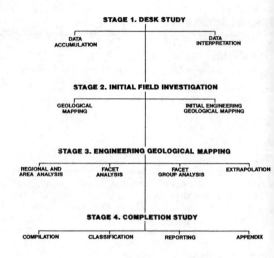

Figure 1. A recommended procedure for engineering geological mapping on a regional scale (modified after Price, 1981).

2.2 Initial field investigation

The initial field investigation is aimed primarily to check the accuracy of geological maps, to improve them where necessary and to obtain preliminary information on engineering geological conditions, potential problem areas and availability of construction materials.

2.3 Engineering geological mapping

For the actual engineering geological mapping, the region is sub-divided into various type areas and for each type area, a representative aerial photograph is selected for land facet analysis.

A type area is selected so that its surface features such as landform and landuse, its climate and its bedrock geology are unique.

One aerial photograph, or part thereof, representing each type area is sub-divided into its land facets according to the method of Brink et al (1970). A land facet is defined as part of the landscape, usually with superform consisting of a particular rock or surficial deposit and with soil and water regime that is either uniform over the whole facet, or if not, varies in a simple and consistent way. Typical examples are ridge crests, free faces, talus slopes, pediments, gullies, convex and concave slopes, terraces, alluvial fans, point bars, etc.

Facets that are too small to be represented on a regional map, may be grouped together as facet groups. Such a facet group or mapping unit should be similar in regional landform, bedrock geology, soil profile and engineering behaviour. An example is solid rock outcrops, comprising the land facets free face, cliff crest and ridge plateau. A portion of the representative aerial photograph with the boundaries of facet groups shown, for an area to the east Pretoria is presented on Figure 2.

The various land facets and facet groups represent different mapping units, and once they have been identified and delineated, each mapping unit is allocated a code consisting of letters and/or numbers. Each mapping unit is then investigated by means of field mapping, test pitting, sampling and laboratory testing. A brief description of each mapping unit is compiled.

Engineering geological data from each mapping unit on the representative aerial photograph is now extrapolated to similar

facets or facet groups on other photographs of the same type area. This procedure is repeated for all the type areas until the entire area has been mapped.

Advantages of analysis through type area, facet and facet group is that it is comprehensive; similar units can be recognised and boundaries extrapolated by analogy; it is inexpensive and can be done in a short period of time; it can be presented through a standardized and easily understood format and, most of all, it provides a starting point from which the inexperienced person can work.

2.4 Completion study

During the completion study, mapping unit boundaries on the aerial photographs are transferred by hand to orthophoto's which are free from distortion and from there it can be photographically reduced to fit on 1:50 000 topo-cadastral base maps (Figure 3). Each mapping unit is allocated a map symbol and is described on the legend which accompanies the map (Figure 4).

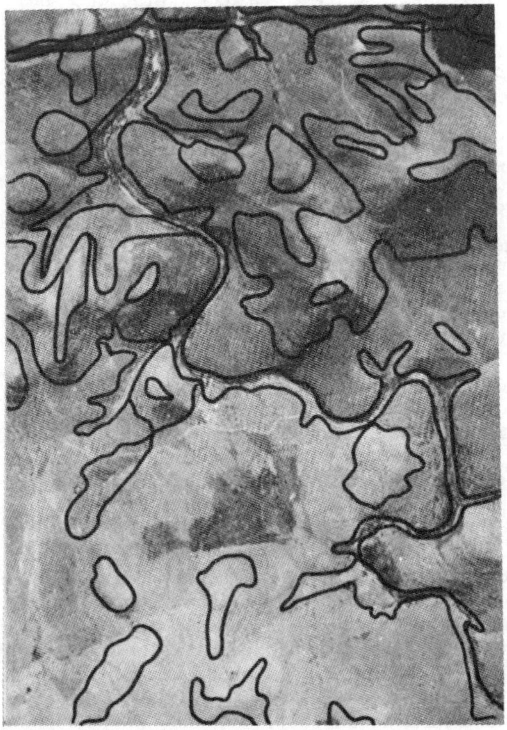

3 SITE CLASSIFICATION

In order to be of use to the planner, en-

Figure 2. Land facets and facet groups delineated on a portion of the representative aerial photograph (after Price, 1981).

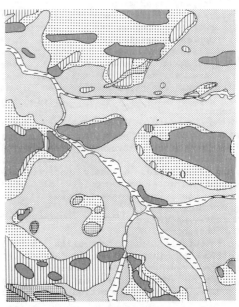

Figure 3. Regional engineering geological map of an area to the east of Pretoria (after Price, 1981).

GROUP OR SUBGROUP	FORMATION	MAPPING UNIT	MAP SYMBOL	FACET GROUP DESCRIPTION AND UNIFIED CLASSIFICATION
● RECENT DEPOSITS				
RECENT UNCONSOLIDATED DEPOSITS		A1	~ ~ ~ ~	SILTY SAND, SANDY CLAY, GRAVEL, CLAY–VARIABLE
		S1		COLLUVIAL SILTY SAND (SM)/VARIETY OF ROCK TYPES
		S2		RED COLLUVIAL SILTY CLAY (SM)/COLLUVIAL SILTY SAND (BUFF COL.)/ROCK
● TRANSVAAL SEQUENCE				
TRANSVAAL	PRETORIA RAYTON	R1		CONTINUOUS OUTCROP –QUARTZ – OFTEN HYBRISED
		R2		SCATTERED OUTCROP WITH INTERSTITIAL SILTY SAND
		R3		COLLUVIAL RED-BROWN SAND/QUARTZITE
● IGNEOUS INTRUSIVES				
MARICO DIABASE SUITE		D2	VVVVVVVVV VVVVVVVVV VVVVVVVVV	SCATTERED OUTCROP WITH INTERSTITIAL RED SANDY CLAY
		D3		RED RESIDUAL SANDY CLAY (CL)/GREEN-GREY RESIDUAL DIABASE (CL)/DIABASE

Figure 4. Legend for regional engineering geological map (modified after Price, 1981).

gineer, local authority, architect, home owner and builder, the engineering geological map must be interpreted and translated in terms of a particular application or type of development. Each application is influenced by a large number of factors, only few of which are directly related to the geology.

For housing development, the most important geological influencing factors are:
(1) foundation conditions,
(2) slope stability conditions and
(3) drainage conditions
(Transvaal Provincial Administration, 1989)

Other factors such as the availability of construction materials, possible sterilisation of strategic mineral deposits, availability of waste disposal and burial sites etc., are all related to the geological environment, but these aspects need to be dealt with as part of the regional planning process.

In order to keep the site evaluation system as simple as possible, each of the three geological influencing factors is classified in terms of three conditions, namely favourable, slightly unfavourable and unfavourable. Rating points allocated to these conditions are as follows:
Favourable (1)
Slightly Unfavourable (2)
Unfavourable (5)

A favourable condition implies that a specific geological influencing factor will not affect residential development, provided that the standard building regulations prescribed by Local Authorities are adhered to.

A slightly unfavourable condition indicates a geological problem which will cause damage to residential buildings unless certain standard precautionary measures are taken. Examples of such standard measures are the removal of a surficial layer of expansive clay, pre-compaction of collapsible sand, surface drainage, etc. The additional cost of such measures per stand, is generally less than 10 per cent of the cost of the building.

An unfavourable condition is indicative of severe geological problems which cannot be solved by means of standard inexpensive precautionary measures. For each stand, a special geotechnical investigation will be required and the cost of construction will generally rule out any residential development. Development of high-rise buildings and industrial structures may be feasible, or otherwise, such areas must be used as parklands or recreational areas.

Guidelines for identification of the different conditions for each of the geological influencing factors, are presented in Tables A1 - A3 in the Appendix.

In order to classify foundation conditions, the term volumetric stability is used to describe the behaviour of swelling or shrinking clays, collapsible soils and compressible soils. The various terms are defined in Table A4 in the Appendix.

The three influencing factors are individually rated for each mapping unit and the total rating for each unit is obtained by adding the points for each factor. The site is then classified as Good, Fair or Poor, according to Table 1.

This classification implies that for a site to be good, all three geological factors must be favourable. One or more slightly favourable conditions place the site in

260

Table 1. Site classification in terms of total rating.

SITE CLASS	TOTAL RATING
Good	3
Fair	4 – 6
Poor	7 – 15

the fair class, while one or more unfavourable conditions classify the site as poor.

An example of the rating and classification of an area to the east of Pretoria is presented in Table 2.

5 SITE SELECTION AND PRECAUTIONARY MEASURES

A special map or overlay for the engineering geological map is now produced for use by the Planner. On this map, distinction is made between good site class areas (blank), fair site class areas (lightly dotted) and poor site class areas (heavily dotted). As an example, such a map for the area east of Pretoria is shown on Figure 5.

Good site class areas can be zoned for any type of residential development, fair site class areas are generally for higher income groups and special precautionary measures are compulsary, while poor site class areas are generally not suitable for residential development.

The precautionary measures to be prescribed for fair site class areas depend on the type and degree of geological problem that is involved. Typical precautionary measures for the different influencing factors are given in Table 3.

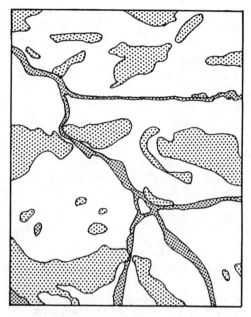

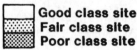
Good class site
Fair class site
Poor class site

Figure 5. Site classification overlay for regional engineering geological map.

CONCLUSIONS

Planning for the selection of large areas of land suitable for rapid expansion of low-cost housing projects in urban and rural areas of developing countries, requires engineering geological input in

Table 2. Rating and classification of mapping units in an area to the east of Pretoria.

MAPPING UNIT	FOUNDATION	RATING DRAINAGE	SLOPE	TOTAL	SITE CLASS	POTENTIAL PROBLEM
A1	2	5	1	8	poor	expansive
S1	1	1	1	3	good	
S2	1	1	1	3	good	
R1	2	1	2	5	fair	slope
R2	2	1	1	4	fair	excavation
D2	2	2	1	5	fair	drainage
D3	5	2	1	8	poor	expansive

Table 3. Typical precautionary measures for residential development on sites with slightly unfavourable geological conditions.

INFLUENCING FACTOR	TYPICAL PRECAUTIONARY MEASURES
Foundations	Prevent ingress of surface water into subsoil Pre-treatment of volumetrically unstable soil Removal of volumetrically unstable soil Use of split construction methods Light reinforcement in foundations and walls Blasting required for foundations and services
Drainage	Local landscaping to improve surface drainage Shallow sub-surface drainage
Slope instability	Avoid deep cuttings and trenches Provide surface drainage Install shallow sub-surface drainage Construct low retaining walls

order to avoid unnecessary damage to buildings as a result of unfavourable geological conditions.

Regional engineering geological maps on a scale of 1:50 000 can be produced quickly and inexpensively by means of aerial photographic land facet analysis, detailed investigation of representative mapping units and extrapolation to similar areas. Mapping units can be identified and described in terms of their engineering geological properties.

The major geological factors which influence residential development are
(1) foundation conditions,
(2) drainage conditions and
(3) slope stability conditions.

By rating each of these conditions as favourable, slightly unfavourable or unfavourable, attention can be drawn to those geological factors which may affect the proposed development.

Based on the above ratings, sites can be classified as good, fair or poor. Good class sites can be used for all types of residential development, fair class sites will require certain standard precautionary measures in order to minimize damage to buildings and poor class sites can generally not be utilized for residential development.

ACKNOWLEDGEMENTS

The contributions of Dr. J.P. Venter and Mr. D.J. de Villiers as co-authors of the guidelines for the Transvaal Provincial Administration (1989) are acknowledged with thanks.

REFERENCES

Brink, A.B.A., T.C. Partridge & G.B. Matthews 1970. Airphoto interpretation in terrain evaluation, Photo Interpretation, 5:15-23.
Dearman, W.R. & P.G. Fookes 1974. Engineering geological mapping for civil engineering practice in the UK. Quat. Jnl. Engng. Geol. 7:223-256.
Melnikov, E.S. 1979. The main principles for the national engineering geological survey in the USSR. Bull. Int. Ass. Engng. Geol. 19:93-96.
Price, G.V. 1981. Methods of engineering geological mapping and their application on a regional scale in South Africa. Unpublished MSc dissertation, University of Pretoria, 121 p.
Transvaal Provincial Administration 1989. Township Development Guidelines for Geological Reports. Unpublished document distributed to members of the SA Institute of Engineering Geologists. 10 p.

APPENDIX: Tables A1 - A4

Table A1. Guidelines for the identification of different foundation conditions.

CONDITION	RATING	DESCRIPTION
Favourable	1	No risk for sinkhole or doline formation >500mm of volumetric stable* topsoil
Slightly unfavourable	2	Low risk for sinkhole or doline formation <1 500mm unfavourable layer of volumetrically very unstable topsoil >1 500mm layer of volumetrically moderately unstable topsoil Scattered or continuous rock outcrop
Unfavourable	5	Medium to high risk for sinkhole or doline formation >1 500mm layer of volumetrically very unstable topsoil

*   See Table A4

Table A2. Guidelines for the identification of different drainage conditions.

CONDITION	RATING	DESCRIPTION
Favourable	1	Good surface drainage - no ponding Deep groundwater table Highly to moderately permeable topsoil and bedrock
Slightly unfavourable	2	Satisfactory surface drainage - occasional surface ponding Seasonal groundwater level fluctuations Poor draining topsoil on permeable bedrock
Unfavourable	5	Poor surface drainage - standing water Permanent shallow groundwater table - marshy area Located in valley, below 1:50 year flood line Poor draining topsoil on impermeable bedrock

Table A3. Guidelines for the identification of slope stability condition

CONDITION	RATING	DESCRIPTION
Favourable	1	Low surface gradient (<10 degrees) Deep groundwater level Good sub-surface drainage Dense granular topsoil Sound bedrock with favourable bedding dip
Slightly unfavourable	2	Moderate surface gradient (10 - 20 degrees) Fluctuating groundwater level Reasonably good subsurface drainage Unstable topsoil <500mm thick Evidence of soil creep Sound bedrock with favourable dip
Unfavourable	5	Steep surface gradient (>20 degrees) Shallow groundwater table Poor sub-surface drainage Unstable topsoil >500mm thick Evidence of hummocky ground or slip scars Discontinuous bedrock with unfavourable dip

Table A4. Definition of volumetrically unstable soils.

VOLUMETRIC STABILITY	SHEAR STRENGTH (Cu:kPa)	TOTAL MOVEMENT (mm)
Stable	>200	<6
Moderately unstable	50 - 200	6 - 50
Very unstable	<50	>50

6th International IAEG Congress / 6ème Congrès International de AIGI, © 1990 Balkema, Rotterdam. ISBN 90 6191 130 3

Mountain hazard analysis using a PC-based GIS

Analyse des risques naturels en terrain montagneux, un domaine d'application pour SIG sur PC

C.J.van Westen
International Institute for Aerospace Survey and Earth Sciences (ITC), Enschede, Netherlands

J.B.Alzate Bonilla
IGAC, Bogotá, Colombia

ABSTRACT: The use of Geographical Information Systems (GIS) opens new possibilities for mountain hazard analysis, but also confronts us with very specific requirements where the input of data is concerned. A description is given of the structure of input data in a GIS, and the various techniques for small, medium and large-scale analysis, with some examples from the Alps and the Andes.

RESUME: L'utilisation des Systèmes d'Information Géographique (SIG) ouvre des nouvelles perspectives quant à l'analyse des risques naturelles en terrain montagneux. Mais elle nous confronte également aux exigences spécifiques à l'élaboration des bases de données. Sont décrites dans cet article, la structure de ces bases de données pour un SIG, et les différentes techniques d'analyse, que ce soit à petite, moyenne ou grande échelle. Certains cas provenants des Alpes et des Andes sont reprès comme exemples.

INTRODUCTION

Geographical Information Systems are increasingly recognized as a powerful tool in earth sciences. They allow the user to store, retrieve and analyze large quantities of data needed in a relatively short time, and are very helpful for planning. The UNESCO has declared the use of GIS in environmentally sound management of natural resources one of their major goals for this decade, that has been designated as the "International Decade for Natural Disaster Reduction (IDNDR)" by the UN. Recently UNESCO and ITC have started a project that aims at assisting the national earth science organizations in the Andean countries through the development of mountain hazard mapping methodologies and techniques using GIS, and the transfer of this knowledge to the organizations responsible for hazard mapping.

Although the use of computers for modelling and statistical analysis of landslides is very advanced

(Bonnard 1988), the use of GIS is still rather scarce in this field. The first coarse raster based GIS systems were used for landslide hazard susceptibility mapping in the late seventies (Newman et al 1978). A much more elaborate method, using field checklists and multivariate analysis was presented by Carrara et al (1978). More recently examples of the use of GIS for mountain hazard mapping can be found in Stakenborg (1986), Wagner et al (1988), Kienholz et al (1988) and Wadge (1988).

1 PC-BASED GIS: ILWIS

The Integrated Land and Watershed Information System (ILWIS) was recently developed at the International Institute for Aerospace Survey and Earth Sciences (ITC), Enschede, the Netherlands. It combines conventional GIS procedures with image processing capabilities and a relational database. All operations can be performed on a 80286-based computer (IBM-AT), with a coprocessor, a digitizer, a Matrox graphics card, high-resolution monitor and output devices (colour printer, plotters). The system was developed for a PC, taking into consideration the situation of most developing countries, where mainframes or minicomputers are still very rare (Valenzuela 1988). Modelling can be done in the separate modules, in the relational database and in the GIS software kernel (described by Gorte et al 1988).

2 THE TEST AREA

The test area for the UNESCO-ITC project on mountain hazard analysis using GIS is located in the catchment of the Chinchina river, in the highest part of the Central Cordillera, Colombia (see fig. 1).

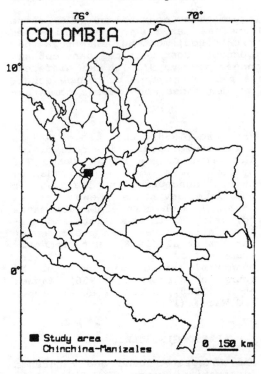

Figure 1. Location of the study area for the UNESCO-ITC project. The sample areas for each scale are indicated.

The area is subjected to a variety of hazards:
- seismic hazard, due to the presence of an active fault system;
- volcanic hazards, caused by the active Ruiz volcano; and
- erosion and mass movement hazards, caused by the combination of deep weathering, strong uplift and incision, high rainfall intensities and active human interference.

3 INPUT DATA

The use of a GIS such as ILWIS requires a different methodology of data capture and display than the conventional mapping methods as used in geomorphology (see Demek & Embleton 1978). Unlike the conventional maps, that are mostly symbol maps with all information in one layer, the ideal maps in a GIS are areal type maps, with different information stored in different layers. The four basic types of maps within a GIS are point maps (for which the table is more important than the map), line maps, polygon maps (units) or pixel maps (different values per pixel).

In table 1 a list is given of the way the input maps for mountain hazard analysis were stored in ILWIS. The list is an extensive one, and all types of information will only be available in an ideal case. However, as can be seen from the right column of the table, the information levels available determine the type of hazard analysis that could be used, going from a qualitative one where little information is available to complex statistical models. In table 1 also the type of map is given, the way it is obtained, for which scale it can be made (regional, medium or large) and the type of attribute table linked to the map. The last columns of table 1 display which maps are used for which type of analysis.

For the information layers 4, 13, 14, 15, 16, 17, 18, 25 and 31 a series of parameters will have to be sampled in the field in a number of observation points using checklists or field computers.

Table 2 gives the list of parameters to be sampled. For levels 13, 14 and 31 laboratory testing is needed. The information layers 19, 25, 26, 27, 28, 29, 30 and 31 require a series of continuous mea-

Table 1. The structuring of input data for mountain hazard analysis in ILWIS.

Nr	Map data-base	Type	Scale R M L*			Made from	Tables	Analysis QL HD HI SS SP SC RF SS CS DM									
	Geomorphological																
1.	Main Geomorphological units	Units	+	+	-	API+fieldcheck	DC	+	-	-	-	-	+	+	-	+	-
2.	Geomorphological subunits	Units	+	+	+	API+fieldcheck	DC	+	-	-	-	-	+	+	-	+	+
3.	Morphography/morphometry	Line	-	+	+	API+fieldcheck	D	-	-	-	-	-	+	+	-	+	+
4.	Denudational features (1986)	Point	-	+	+	API+fieldwork	FD	-	+	+	+	-	+	+	-	+	+
5.	Denudational features (1969)	Point	-	+	+	API	D	-	+	+	+	-	+	-	-	+	+
6.	Denudational features (1945)	Point	-	+	+	API	D	-	+	+	+	-	+	-	-	+	+
	Topographical																
7.	Digital Elevation Model	Pixel	+	+	+	Contour interpol.	-	-	-	-	-	-	+	+	-	+	+
8.	Slope map (degrees or %)	Pixel	-	+	+	From 7	-	-	-	-	+	+	+	+	-	+	+
9.	Slope direction map	Pixel	-	+	+	From 7	-	-	-	-	+	+	+	+	-	+	+
10.	Hillshading map	Pixel	+	+	+	From 7	-	for display only									
11.	Anaglyph map	Pixel	+	+	+	From 7	-	for display only									
12.	Block diagrams	Pixel	+	+	+	From 7 + SPOT	-	for display only									
	Geotechnical																
13.	Engineering soil type map	Unit	-	+	+	API+fieldwork	RFD+RLD	+	-	-	+	+	+	+	+	+	+
14.	Engineering rock type map	Unit	+	+	+	Geology+fieldwork	RFD+RLD	+	-	-	+	+	+	+	+	+	+
15.	Isopach map of pyroclastics	Pixel	-	+	+	Field data+model	RFD	-	-	-	-	-	-	+	+	+	+
16.	Isopach map of weathering	Pixel	-	+	+	Field data+model	RFD	-	-	-	-	+	+	+	+	+	+
	Structural geology																
17.	Faults and lineaments	Line	+	+	<	API+SPOT+fieldwork	FD	+	-	-	+	+	+	-	+	+	+
18.	Structural zones	Unit	-	+	+	API+fieldwork	RFD	+	-	-	-	+	+	-	+	+	+
19.	Seismic events	Points	+	+	<	Existing data	RSD	-	-	-	+	-	+	-	+	+	+
	Landuse/vegetation/infrastructure																
20.	Present land use	Unit	+	+	+	API	D	+	-	-	-	+	+	-	+	+	+
21.	Landuse 20 years ago	Unit	-	+	+	API	D	+	-	-	-	+	+	-	+	+	+
22.	Landuse 40 years ago	Unit	-	+	+	API	D	+	-	-	-	+	+	-	+	+	+
	Hydrological																
23.	Drainage (different orders)	Line	+	+	+	API+topomap	D	+	-	-	-	+	+	-	+	+	+
24.	Catchment (different orders)	Unit	+	+	+	API+topomap	D	+	-	-	-	+	+	-	+	+	+
25.	Watertable	Pixel	-	+	+	Model	RFD	-	-	-	-	+	+	-	+	+	+
26.	Isohyets	Pixel	+	<	<	Model	RFD	-	-	-	-	+	+	-	+	+	+
27.	Isotherms	Pixel	+	<	<	Model	RSD	-	-	-	-	+	+	-	+	+	+
28.	Evapotranspiration	Pixel	+	<	<	Model	-	-	-	-	-	+	+	-	+	+	+
29.	Specific discharge	Pixel	+	+	+	Model	-	-	-	-	-	+	+	-	+	+	+
30.	Climatic zones	Pixel	+	-	-	Model	RSD	-	-	-	-	+	+	-	+	+	+
	Volcanological																
31.	Sampling points	Point	-	+	+	Existing data	SD+FD+LD	-	-	-	-	-	+	+	-	+	+

Explanation:
* Scale of analysis
  R= Regional (1:100.000)   M= Medium scale (1:25.000)
  L= Large scale (1:5.000)
  += Made on this scale, -= Not needed on this scale,
  <= Data from the smaller scale is used.
** Table types:
  D= Descriptive   FPD= Field point-data
  LPD= Laboratory point-data   SPD= Station point-data
  RFD= Reworked field data   RLD= Reworked laboratory data
  RSD= Reworked station data   DC= Data from map crossing

*** Type of analysis
  QL= Qualitative analysis
  HI= Hazard isopleth
  SP= Probability
  RF= Rockfall calculation
  CS= Complex statistics
  HD= Hazard distribution
  SS= Simple hazard susceptibility
  SC= Complex susceptibility
  SS= Slope stability models
  DM= Complex deterministic models.

surements over a long period, using more or less expensive equipment. The gathering of these types of data will be the most expensive and time consuming in a hazard analysis project.

4 QUALITATIVE ANALYSIS

Basically this type of analysis is intended as a desk study on a regional scale when only the general geomorphological information (nos. 1, 2) and a general rock type map (no. 14), without any point observation data, or a general geological map are available. In most cases a topographical map at a 1:100.000 scale will be available too so that a coarse DEM can be made, however, without the necessary accuracy to create a slope map (8). From the topographical map, together with airphoto or SPOT interpretation layers 17, 20, 23 and 24 can also be made.

The method is such that for each Terrain Mapping Unit of level 1 a series of attributes from the other available maps are sampled in a semi-automatic way with GIS, such as lithological types, land use, altitude range, internal relief, drainage density, slope steepness distribution, slope length, slope and crest forms and occurrence of various mass movement and erosion processes (described in Meijerink 1988).

5 QUANTITATIVE ANALYSIS

When the analysis is done on a medium or large scale, more data can be gathered in the field, and other type of hazard analysis techniques with GIS come within reach (see table 1).

Table 2. List of parameters gathered in the field and laboratory for different input maps

Map	Parameters
4	Recent denudational features. Type, area affected, age, development, activity, dimensions, source, transport or accumulation, material, slope angle, slope angle before slope direction, slope form, contact, damage, possible causes, risk, vegetation, land use, drainage, water table
13	Engineering soil type map Code, thickness, % in profile, texture, sorting, layering, cementation, moisture content, permeability, consistency, vane shear strength, weight%of boulders, cobbles, gravels, sand andfines, dry strength, toughness, dilatancy, Atterberg limits, density For a limited number of samples: cohesion, angle of internal friction, clay types, pF, permeability
14	Engineering rock type map Code, thickness in profile, % of profile, weathering class, weathering susceptibility, layer thickness, spacing, characteristic block size, material strength, Schmidt hammer rebound value, tilting angle, point load strength
15/16	Isopach map of pyroclastic deposits and weathering soils Depth, slope angle, signs of erosion
17	Faults and lineaments Width, orientation, displacement
18	Structural zones Strike and dip, bedding, foliation, discontinuity
24	Depth of freatic water level Date, depth

5.1 Hazard susceptibility maps

Most of the map based quantitative analysis techniques are based on map crossings. In this way the relations between the various types of mass movements and a whole series of other parameters can be displayed, either as a relation between two, three or more variables. Therefore it is of crucial importance that the mass movement phenomena are mapped accurately. In figure 2 the relation between mass movements and slope angle for an earlier test area in Austria is displayed as an example.

For the hazard susceptibility analysis a varying number of maps can be used, depending on those that have a clear relation with mass movements. Therefore it is advised to evaluate their importance first by crossing them individually with the maps displaying denudational features.

In a simple hazard susceptibility analysis only five maps are taken into account: 8, 13, 14, 20 and 4. The procedure is shown in figure 3. The four input maps (rock type,

soil type, slope class and vegetation/land use) are crossed with each other resulting in a combination map and a cross-table. This combination map is then crossed with the maps for the various denudational features, and for each of these the percentage for the combination of the other four maps is stored (%1, %2, etc.). For each type of process the combination map is then renumbered with the respective values of these columns, resulting in a susceptibility map with a scale of 1 to 100. This is reclassified to fewer classes and the actual occurrences of mass movements are displayed in this map with the highest susceptibility class.

If more maps are to be used, then this method will not be suitable due to the large number of possible combinations. In that case it is advised to design decision rules with IF statements, using the frequency distributions of the map crossings by pair.

The method can de improved by taking the time factor into account. By overlaying the maps 4, 5

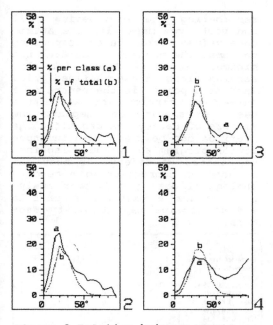

and 6, maps of active phenomena within a 20 years' period can be obtained.

5.2 Deterministic models

When sufficient field data is a-vailable on geotechnical proper-ties, deterministic models can be used in connection with the infor-mation levels stored in the GIS (see table 1). Here, two examples are given.

Together with the University of Amsterdam a neighbourhood analysis program was developed for ILWIS that can be used for rockfall haz-ard mapping. Based on a DEM, the program calculates the rockfall ve-locity for each pixel, in relation to the velocity of the neighbouring pixel with the greatest height dif-ference. The calculation starts from pixels indicated as rockfall source areas, obtained from the geomorphological map (2) and/or a slope map. Figure 4 shows a result of the calculation, also based in this case on the decelerating ef-fect of forest. With some modifica-

Figure 2.Relation between mass movement types and slope angle
A: % per class B: % of total
1= surficial sliding
2= flow type mass movement
3= deep-seated sliding
4= rockfall accumulation

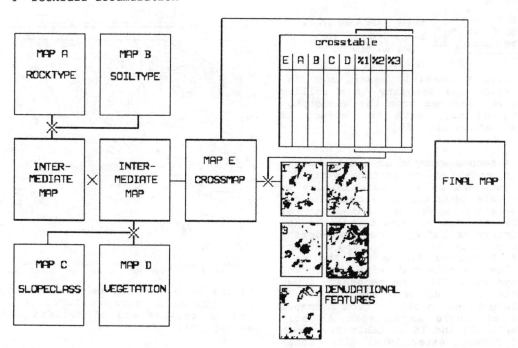

Figure 3. Procedure for hazard susceptibility analysis using 5 input maps

tions the neighbourhood program can also be used for other hazard analyses, such as avalanche run-out distance, or areas affected by a lahar flow of given volume.

LEGEND

▦ 0 - 30 m/s	
▨ 30 - 60 m/s	
▰ > 60 m/s	

N △

Mapsize: 400 × 500 pixels
Contour interval: 100 m.
pixelsize: 20 m.

Figure 4. Rockfall hazard map displaying the velocity of a rolling block. Derived from the geomorphological map, with the effect of vegetation included.

A second example of a deterministic model in connection with GIS input data is a slope stability analysis program, developed in cooperation with the University of Utrecht. The program calculates the Factors of Safety along user-defined profile lines with the methods of Bishop or Fellenius. The profiles are selected with the cursor keys on a DEM. The necessary input data are then read simultaneously along those profiles from a number of attribute maps, such as soil depth (15 and 16 in table 2), depth of freatic water level (25). Cohesion, angle of internal friction and bulk density maps are made by

renumbering the engineering soil and rock type maps with the attribute values from the corresponding columns. The program calculates the minimum Factor of Safety for slip circles along a user-defined grid. Each grid point is the centre of a number of circles, of which the user can specify the depths. The Factors of Safety values along the profiles can be used in an interpolation program to produce isolines of Factors of Safety, or they can be used in connection with geomorphological data for further analysis. Figure 5 displays a profile and an isoline map of Safety Factors.

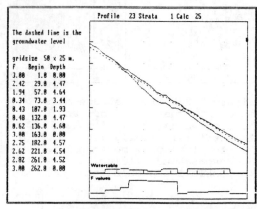

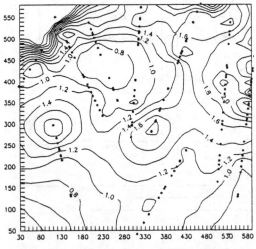

Figure 5. A profile as it is read from the input maps in ILWIS (a), and an isoline map of Factors of Safety (b).

CONCLUSIONS

The use of GIS can be an important tool in disaster management. However, the gathering of data and the way the various input maps are stored are different from most of the conventional techniques. The structure of the databases (maps and tables) can be made in such a way that a variety of hazard analysis techniques can be applied in a relatively short time.

As the UNESCO-ITC project still is in the phase of data gathering, not all of the analysis methods could be tested yet. In the near future emphasis will be put on hazard index maps, multivariate analysis and probability maps based on models, including the effects of seismic events and extreme rainfall.

REFERENCES

Bonnard, C. (Ed.) 1988. Proc. 5th ISL, Lausanne, Volume 1, 809 p. Balkema, Rotterdam.

Carrara, A., E. Catalano, M. Sorriso-Valvo, C. Realli & I. Osso 1978. Digital terrain analysis for land evaluation. Geologica Applicata e Idrogeologia, Vol. XXIII, 69-127.

Demek, J. and C. Embleton (Eds) 1978. Guide to medium-scale geomorphological mapping. IGU Commission on Geomorphological Survey and Mapping, Brno, 348 p.

Gorte, B., R. Liem and J. Wind 1988. The ILWIS software kernel. ILWIS, Integrated Land and Watershed Information System. ITC Publ. No. 7, ITC, Enschede, the Netherlands, pp.15-22.

Kienholz, H., P. Manni & M. Klay 1988. Rigi Nordlehne. Assessment of natural hazards and priorities of forest management. Proc. INTERPRAEVENT 1988 Graz, Bd. 1, pp. 161-174.

Meijerink A.M.J. 1988. Data acquisition and data capture through terrain mapping units. ITC Journal 1988-1, pp.23-44.

Newman, E.B., A.R. Paradis & E.E. Brabb 1978. Feasibility and Cost of Using a Computer to Prepare Landslide Susceptibility Maps of the San Francisco Bay Region, California. Geological Survey Bulletin 1443, 24 p.

Stakenborg, J.H.T. 1986. Digitizing alpine geomorphology. ITC Journal 1986-4, pp.299-306.

Valenzuela, C.R. 1988. ILWIS overview. ITC Journal 1988-1, pp.4-14.

Wadge, G. 1988. The potential of GIS modelling of gravity flows and slope instabilities. Int. J. GIS, Vol. 2, No. 2, pp.143-152.

Wagner, A., R. Olivier, & E. Leite 1988. Rock and Debris Slide Risk Maps Applied to Low-Volume Roads in Nepal. Proc. 4th Int. Conf. on Low-Volume Roads. Transportation Research Record 1106, pp.255-267.

Geotechnical mapping: A basic document to urban planning
La cartographie géotechnique: Un document fondamental pour la planification urbaine

L.V.Zuquette
University of São Paulo, Campus of Ribeirão Preto, Brazil

N.Gandolfi
University of São Paulo, Campus of São Carlos, Brazil

ABSTRACT: Planning is the best alternative for a better use of environmental resources attending to the populational and industrial expansion as well as recreational, educational and social needs, viewing and economic and efficient supply of these requirements. The geotechnical mapping optimizes all this process and provides the information for the rational ocupation of land.

RESUMÉ: La planification est la meilleure alternative pour l'emploi adéquat des recours ambiants dans le but de répondre (de façon efficiente et économique) à l'expansion populationnelle et industrialle mais aussi aux nécessités de récréation, éducationnelles et sociales. La cartographie géotechnique optmise ces processus et fournit les informations pour l'occupation rationnelle du sol.

1 INTRODUCTION

The occupation of the Earth surface is increasing fast and seeks mainly social and material needs of mankind (food, habitation, manufactures, health, education, etc.).This occupation may occur in both organized and disorganized manner: the first makes the occupation respecting the environment and there fore not changing its conditions, and the second makes the occupation without respectiving the environmental limits (advantage or disadvantage), causing damage directly to environmental conditions. For an organized occupation there must be a process of global planning, involving general aspects (regional planning) and local aspects (urban planning). It is important to emphasize that in such process, the word environmental is ment in various aspects: physical environment,biological environment, social-economic conditions and the relationship between them (Zuquette, 1987).
It may be understood that planning is the choice between best alternatives existent, for environmental utilization, viewing the stimatives of populational and industrial increase, as well as recreational,educational and social needs in general.
Such planning has to: assist a positive economy, with the least cost/benefit rate and a long term growing production; respect technical potentialities; foresee the variations that may occur and consider the populational and social changes, as well as an efficient supply of various needs of the community (energy, food, land).
All forms of occupation are based either upon, inside or in the interface with the environment; therefore, the managers/or planners when planning, authorizing, implementing and fiscalizing the occupation,must know the variations of environmental components, as well as the demands of each form of occupation related to environment,to other forms of occupation, to atmosphere conditions and the methods to implement them (constructives, etc.)

2 URBAN PLANNING

The urban planning applies to both, the portion already fixed and the

urban center expansion limit.
The urban center development always occurs with the implementation of several related forms of occupation and, therefore with different demands. In many countries, the federal, state and municipal governments have the responsibilities of the occupation pre-requirements. They utilize as basic elements for their decision, graphic documents (maps, charts, plants) and reports showing environmental variations (recomendations and limitations), known as geotechnical mapping.
The geotechnical mapping may present different informations depending on the scale it presents. In its utilization as urban planning, where adequate scales are between 1:25.000 and 1:10.000, the aim is to plarticipate in the forms of occupation guidance process, inventory, analyse, selection and the best indication for local investigations. In surveys with scales larger than 1:10.000, the documents present local investigations characteristics, and do not make part of the geotechnical mapping process.
Amongst the forms of occupation of urban center development the following may be considered:
- residential areas, popular habitation;
- access roads (streets, avenues, etc.);
- potencial aquifer exploitation;
- foundations;
- industrial areas;
- potencial areas building materials exploitation;
- recreational areas;
- waste disposal areas (industrial and domestic);
- building lots;
- reservoirs, dams and embankments;
- sewerage pipe nets;
- natural drainage channels;
- public areas (gardens, parks).
The geotechnical mapping may be used as a scope to furnish informations for a set of forms of occupations or an specific one.

3 STUDY OF PHYSICAL ENVIRONMENT

The geotechnical mapping must show the attributes variations of the following components of the physical environment:
- Rock substratum - formed by hard and soft rock;

- Unconsolidated material - formed by materials derived from different rock materials alterations (soils) and sediments (transported materials);
- Water - superficial and/or deep;
- Relief - forms, declivity and dynamic.
The graphic documents which an produced should always show the environmental conditions, approaching homogeneous space units according to the level of attributes considered.
As a subproduct of the geotechnical mapping, a data bank may be organized which will mantain a central record of information and will enable a continuous renew of graphics doccuments through an increase of information because of its dynamic aspect.
According to these considerations, the process of geotechnical mapping is considered of fundamental importance to urban planning optimization, (Varnes, 1973; I.A.E.G.,1976).

4 SOME FORMS OF OCCUPATIONS

In urban centers the forms of occupation are normally implemented at the same time and in the same area (Table 1), with interferences amongst them. As a consequence of this condition, it is important to know the demands of each occupation in relation to its attributes (Table 2). Basic demands related to the attributes of some forms of occupation are presented as follows:

a. Residential area, popular habitation:

Declivity - The definition of classes of lot slope should condition the distribution of cities blocks, the size of each lot, the adequate position of buildings (viewing the reduction of embankments and cuts)or define particular constructive conditions in areas with more than 30% declivity. The declivity map should be presented in classes, according to the scales: 1) scale 1:25.000 classes of: 0 to 2%; 2 to 5%; 5 to 10%; 10 to 15% and > 15%; 2)scale 1:10.000 classes of: 0 to 1%; 1 to 2%; 2 to 5%; 5 to 8%; 8 to 10%; 10 to 15%; > 15%.

Table 1 - Forms of occupation X needs and/or interferences

	1	2	3	4	5	6	7	8	9	10	11	12	13	14	15
1. Residential areas	X	X			X		X	X	X	X	X	X		X	
2. Access ways	X		X			X		X	X	X		X		X	X
3. Dams/earthfills													X	X	X
4. Aquifer potential	X					X		X	X	X		X			X
5. Foundations	X						X	X				X	X	X	
6. Industrial areas		X	X		X		X	X	X	X	X	X	X	X	X
7. Building materials	X											X			
8. Floods						X			X	X	X	X	X		X
9. Recreational areas		X	X		X	X		X	X	X	X	X	X	X	X
10. Risks areas	X					X		X	X	X	X	X	X		
11. Waste disposal	X					X		X	X	X	X	X	X	X	X
12. Building lots	X	X	X		X	X	X	X	X	X	X	X	X	X	X
13. Reservoirs			X				X		X					X	
14. Underground constructions	X		X			X		X	X	X	X	X	X		X
15. Channelizations				X		X		X	X	X	X	X		X	

275

Table 2 - Attributes X Forms of occupation (0 - basic; X - secondary)

Forms of occupation \ Attributes	Type of material	Profile variation	Water level depth	Rock substratum depth	Declivity	C.E.C.	Drainage	Altitude	Salinity/corrosiveness	Spatial units disposition	Expansibility	Collapsibility/compressibility	Landforms extension	Bearing capacity	Urban centers distance
Residential area	X	X	O	O	O		O	X			X	X	O		
Access roads	O	X	O	O	O		O				O	O			
Dams/earthfills	O		O				O				O	O			
Aquifer potential	O		O						O	O					
Foundations	O	O	O	O	X				X	O	O	O		O	
Industrial areas	X	X	O	O	O		O				O	O	O		
Building materials	O		X	O						O					O
Floods			O	X	O		O								
Recreational areas			O		O		O	O							
Risk areas	O		O		O		O				O	O			
Waste disposal	O	O	O	O	O	O	O			O	O	O	O		
Building lots	X	X	O	O	O		O	X			O	O	O		
Reservoirs								O	O	O					
Underground constructions				O		O					O	O			
Channelizations	O	O	O	O	O			X			O	O			

276

Water level depth - When the water level is deep less than 2m during great part of the year special care is demanded, viewing the probability of temporary flood occurence which may cause damage to underground works, cuts, etc.

Rock substratum depth - Depths less than 2m, will condition the execution of all cuts and embankments before the construction of any building because the utilization of explosives may cause damages, and also will condition the type of foundation, forms of investigation and unables the development of areas that utilize septic tanks and pits.

Drainage - This attribute represents the facility with which water flows in surface and subsurface; therefore, areas with bad drainage conditions should be avoided, specially if high costs for drainage systems are involved.

Extension of landforms - It constitutes a homogeneous unit of the physical environment and it is appropriate to this occupation; if the extension is small (< 5 ha) it is not reconmended when surrounded by unappropriate units; however, when it is surrounded by appropriate units, but with differences in some attributes, it might be recommended after a global analyse of the cautions required.

Type of material/Profile variation- The occurence of organic clay material is a strong limitimg factor for this occupation; other materials may be acceptable if the technological resources for the occupation do not require high costs.

Altitude variation - It constitutes a problem when related to unappropriate declivities.

Expansibility/Collapsibility/Compressibility - They should represent a limiting factor for low costs habitational designs because technological resources as special foundations may make them impracticable; in the other hand, such situation may cause an alteration in the inicial design (example:option for vertical buildings).

b. Building materials exploitation

Type of material - The basic attribute is that it has to be studied considering all mineralogical, chemical and physical aspects that may make its utilization impracticable as sand, clay, pebbles or crushed rock.

Rock substratum depth - The thickness of the overlayed unconsolidated material when significant(> 20m) may make impracticable a rock layer to coarse aggregate.

Spatial arrangement - The spatial relations of an unit may make them technically and economically impracticable when the conditions for exploitation have high costs.

Water level depth - When the water depth is < 5m or is above the level of the superior layer it will be necessary to drain the area and sometimes forbid the exploitation; however, it is not fundamentaly a limiting attribute.

Distance related to the urban center - Distance is one of the most important economic factor for building material exploitation; its location far from urban centers is appropriate when related either for environmental protection or urban center expansion.

c. Buildings lots

Declivity - The building lots implantation usually occurs in large environments (about 50 ha) where different declivity classes are included. During the design approval it is necessary to know the limiting extension of declivity so that the government may require preventive and protective measures in the occupation process, specially with respect to lot size and services.

The declivity classes for geotechnical mapping are:

0 to 8% - the lots size and conditions for implantation are considered common.

8 to 20% - there should be a standardization of the lots, not less than 600m², for the various subclasses.

20 to 30% - if there is no other option, it is recomended the division of the lots in large extensions (> 1000m²) and heterogeneously distributed.

> 30% - sepecial conditions.

Observations:

1) In areas with declivity superior

to 15%, positions, depths and forms of construction must attend legisla tion;
2) In high declivities (> 20%) the cuts and embankements (heights, retaining walls, etc.) must be specified when the building lot design is authorized;
3) Avoid groups of lots (blocks) with different declivity classes.

Water level depth - Areas with a water level depth less than 2 m should be destinated to natural parks, avoiding constructive proces ses. If the depth is in the order of 5m, the occupation will be normal, avoiding though building lots where the waste disposal will requi re septic tanks and pits. Special cares must be taken when the place is characterized as a superficial aquifer or recharge area.
Rock substratum depth - When the depth is less than 2m, excavations, cuts, embankments should always be demanded as well as any form of environmental alteration before any building in the lot. If the rock substratum is until 5m and the declivity of the lot is about 10% spe cial cares are necessary at the pla ces where connections of underground pipes occur.
Drainage - Areas with declivities close to 0% and with the rock substratum near the surface (< 2m), drainage conditions present difficulty to lots implantation, and spe cific constructions are necessary for an appropriate water flow.
Extension of ladforms - Areas where the landforms present small extention (< 5 ha) problems with the dis tribution of blocks are frequent, since different landforms define drainage channels or different posi tions in slopes. It is important to define the lots conditions for each slope as well as fix the specific protetion area boundaries since the landform is very common.
Expansibility/Collapsibility - When possible lots with such forms of behavior must be avoided, mainly if the building lots present low cost characteristics. In such lots it is quite normal that buildings present cleavages no matter if they are low or high costs building lots.

REFERENCES

Coulon, F.K. 1976. Mapa geotécnico de Morretes e Montenegro (RS).Por to Alegre, Dissertação de Mestrado. Instituto de Geociências/UFRGS, pág. irreg.
Haberlehner, H. 1966. Princípios de mapeamento geotécnico. In Congresso Brasileiro de Geologia, 20, Rio de Janeiro, SBG, p.37-39 (Publica ção SBG, 1).
International Association of Engine ering Geology. 1976. Engineering geological maps. Paris, The UNESCO Press, 79p. (Earth sciences, 15).
Varnes, D.J. 1973. The logic of engineering geological and related paper. U.S. Geological Survey.
Zuquette, L.V. 1987. Análise críti ca da cartografia geotécnica e pro posta metodológica para as condições brasileiras. São Carlos, 3v. (Doutorado) - Escola de Engenharia de São Carlos-USP.

1.2 Sampling and testing in boreholes
   Echantillonnage et essais dans les trous de sonde

The geotechnical investigation of geologically aged, uncemented sands by block sampling

L'investigation géotechnique, par le moyen d'échantillonnage en bloc, des sables sans ciment géologiquement mûris

M. E. Barton
Department of Civil Engineering, University of Southampton, Southampton, UK

S. N. Palmer
Petrofina (UK) Ltd, Epsom, UK

ABSTRACT: This paper is concerned with uncemented sands where cohesion arises from the development of an interlocked fabric due to pressure solution. A measurable cohesion can develop in sands even as young as the Pleistocene and becomes substantial in older sands where the resulting material is described as a "locked sand". Thus natural sands even with only a slight degree of geological ageing should not be considered as cohesionless but as cohesive soils. The cohesion allows geologically aged sands to be sampled as intact, undisturbed block samples. The procedure of block sampling followed by laboratory testing, which is the classical, Terzaghi-inspired Site Investigation procedure for clay soils, can thus also be used for geologically aged sands. The paper describes the block sampling technique used by the authors. The advantages of testing undisturbed material are illustrated by comparison of the microfabric and properties of the undisturbed and recompacted materials.

RESUME: Dans cet article il s'agit des sables sans ciment où cohésion se doit au développement de textures entrelacées à la suite de la dissolution sous pression. Une cohésion mensurable peut se développer aux sables aussi récents que la Pléistocène, et devenir plus considérables à l'égard de sables plus mûrs où la matière produite s'appelle "un sable entrelacé". Aussi devrait-on décrire comme "sols cohérents", pas comme "sans cohésion", les sables naturels qui ont âgé très peu géologiquement. La cohésion permet aux sables géologiquement mûris d'être échantillonnés en blocs intacts. La méthode d'échantillonnage en bloc, avec examination au laboratoire, est la méthode classique inspirée par Terzaghi pour l'investigation sur place des sols argileux. On peut ainsi se servir de ce procès au cas des sols géologiquement mûris. L'article décrit la méthode d'echantillonnage en bloc dont se sont servis les auteurs. Si l'on compare la micro-texture et les propriétés des matériaux intacts avec celles des matériaux remoulus, on percoit les avantages d'utiliser les matériaux non-remués.

INTRODUCTION: COHESION IN SANDS

The diagenetic processes experienced by natural sands lead to profound changes in the original microfabric. From their originally cohesionless state, sands acquire cohesion. This may be the result of either the bonding of grains (due to intergranular welding and/or the introduction of a cement) or the development of an interlocked fabric (due to pressure solution of detrital grains and/or the crystallisation of grain overgrowths). Thus the cohesion is either cement generated (as in cemented sands) or interlock generated (as in locked sands). This paper is concerned with uncemented sands which have acquired interlock cohesion.

The development of interlock cohesion is associated with geological ageing. Initially the cohesion will be relatively small as in "weakly locked sands" (Barton et al, 1986a) but becomes much more pronounced in sands which have experienced more severe diagenesis as in "locked sands" (Dusseault and Morgenstern, 1979). A slight interlock cohesion has been found even in sands of Quaternary age and in general it is reasonable to expect a significant degree of diagenetic alteration in all natural sands other than those of Recent or, at most, late Pleistocene ages.

The interlock cohesion is highly sensitive to moisture. It is important to note, however, that interlock cohesion is a different phenomenon from the suction (frequently called "apparent") cohesion generated by moisture tension in partly saturated sands. The magnitude of the interlock cohesion in locked sands can be several orders of magnitude greater than the suction cohesion of partly saturated, recent sands. Thus geologically aged, diagenetically altered sands can be regarded as soils possessing both frictional and cohesive strength whereas recent sands (like sand fills) are essentially cohesionless soils.

UNDISTURBED SAMPLING IN SANDS

Truly cohesionless sands defy attempts at undisturbed sampling although many techniques have been tried (Marcuson and Franklin 1979; Mori 1986; Mori and Koreeda 1979 and Seed et al 1982) including improved tube samplers (Hanzawa and Matsuda 1977; Ishihara and Silver 1977 and Seko and Tobe 1977) and new methods involving freezing (Singh et al 1982 and Yoshimi et al 1977). Tube sampling suffers the inherent weakness of densifying loose sands but loosening dense sands with a change-over at about 40 to 50% relative density (Horn 1979). Freezing is more promising for clean sands but suffers the disadvantage of being expensive.

Attempts have been made to undertake undisturbed sampling by relying on the suction cohesion of partly saturated sands (Terzaghi and Peck 1967; Marcuson 1978; Horn 1979). However, both of the latter authors found disturbance in the form of a change of density between their block samples and the in-situ sand as measured by in-situ testing.

Undisturbed sampling using intact block samples is a practical proposition providing the sand possesses a measure of true cohesion. Thus it has long been used to investigate the characteristics of the "collapsing sands" encountered in arid regions (Jennings and Knight 1957; Dudley 1970). It has been successfully used in clayey sands (Tohno 1977) although in this latter case the samples were found to deteriorate with storage because of the swelling characteristics of the clayey matrix.

Undisturbed sampling can be carried out in weakly cemented sands (Salamone et al 1978; Saxena and Lastrico 1978).

In the case of Salamone et al (1978) this was carried out in quite fragile sands (relative density given as 50%) and the procedure used was to cut out a cylindrical block around which was lowered a protective tube, filling the intervening space with wax to prevent jostling during transport. Rotary coring can be used in well cemented sands but is apt to destroy the fabric of the more weakly cemented layers (Salamone et al 1978; Clough et al 1981).

BLOCK SAMPLING PROCEDURE FOR LOCKED SANDS

The interlock cohesion of locked sands is sufficient to permit the direct excavation and removal of intact blocks. The process is facilitated with the use of a collapsible-sided box (Fig. 1) whose construction allows for the easy positioning of the brittle sample and its subsequent transport to the laboratory (Barton et al 1986b). With weakly locked sands, great care is required to avoid any strains which could destroy the very brittle fabric. To avoid frequent and frustrating failure it is necessary to follow a careful sequence of operations such as those depicted in Fig. 2.

Sources of difficulty are jointing, coarse grained layers and dampness in the sand. The cohesive strength of locked sands permits joints to remain open although its trace may be obscured by loose material. The problem is

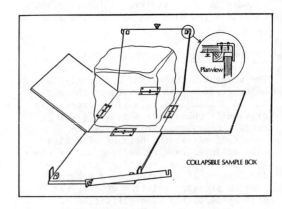

Fig.1 Collapsible sampling box used for transport of the brittle intact sand samples.

analogous to that encountered in clay soils where samples can be either intact or fissured. In the case of geologically aged sands only intact samples can be obtained and tested.

The influence of grain size is important in that where the same amount of pressure solution has occurred, the degree of interlocking and hence cohesion will be less in a coarse sand than a fine sand. The result can be intuitively appreciated from a consideration of the actual volumetric loss from grains of different sizes as has been discussed by Houseknecht (1987 and 1988). Experience in the U.K. confirms that interlock cohesion in coarse sands is far less well developed than in fine and fine-medium sands. The latter, however, tend to be the more dominant sizes in the geological record.

The strength of natural sands is markedly moisture sensitive and the best results are obtained in dry exposures. Nevertheless, the field state is unlikely to be absolutely dry and hence on returning to the laboratory the samples should be placed in a cool oven and left to dry completely. Rapid changes in heating and cooling are undesirable as the thermal stresses may cause cracking of the blocks. When completely dry the blocks can be cut and specimens for laboratory testing prepared by hand trimming. Fig. 3 shows samples prepared for direct shear box and triaxial tests.

LABORATORY INVESTIGATIONS

The undisturbed samples can be used for a wide variety of laboratory investigations which would be either impossible or pointless with only disturbed samples. Lack of space restricts discussion here to only four such investigations.

(a) In-situ density and porosity

Since the block samples are undisturbed, measurement of their density and porosity will give the in-situ values for the intact material and with more accuracy than can be obtained from routine in-situ tests. The density is obtained either by measuring the volume of a prismatically shaped block or by impregnating with an epoxy resin and using water displacement. It has been shown (Dusseault and Morgenstern 1979; Barton and Palmer 1989) that the in-situ dry density of locked sands and most

Fig.2 Recommended procedure for block sampling of sands. The sides and top are cut using the chisel end of a geological hammer and deepened with a narrow bladed trowel (a). The most critical stage is when the base is progressively undercut while the rear is also cut using a thin bladed tool such as a spatula (b). With care the sample is made to break away from the remaining area and transferred to the waiting collapsed box. The box is then closed (c) and the gap between sample and box sides infilled with loose sand or, more preferably, with polyurethane foam.

weakly locked sands is greater than can
be reproduced in the laboratory (i.e.
relative density is greater than unity
in most cases). A major implication of
this is that charts which correlate
engineering properties against relative
density are inapplicable to geologically
aged sands if the charts end at 100%
relative density. This currently
applies to most of the correlation
charts in the geotechnical literature
although the authors have recently
provided an extension beyond 100%
relative density for the standard
penetration test (Barton et al 1989).

(b) Microfabric

Thin sections can be prepared by
impregnating the intact blocks with
epoxy resin, using a dye to enhance the
colour contrast between grains and pores
(Palmer and Barton 1986). Fig. 4 shows
microphotographs of thin sections of the
weakly locked Barton Sand. The
microfabric of the undisturbed material
(Fig. 4B) is compared with the fabric
produced by recompacting, either by
pouring (Fig. 4A) or by tamping (Fig.
4C). The poured specimen (dry density
95% in-situ) has less straight and
concavo-convex inter-particle contacts
and more tangential contacts than the
undisturbed material. By contrast, the
tamped sample (dry density 110% in-situ)
has achieved a high state of packing at
the expense of considerable fracturing
and breakdown of the grains.

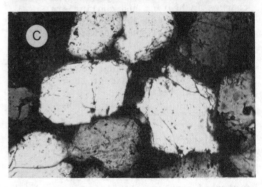

Fig.4 Microphotographs of Barton Sand
(Eocene) under crossed nicols.
(4A) Recompacted by pouring, n = 38.9%.
(4B) Intact sample, n = 35.6%.
(4C) Recompacted by tamping, n = 29.1%.
Magnification in 4A and 4B = X105, in
4C = X112.

(c) Uniaxial compressive strength

The strength of the intact samples can
be demonstrated by the uniaxial
compressive strength of standard 38mm
diameter x 76mm long cylindrical
samples. The tests are readily
performed on the oven dry samples. Two
important aspects of the results are as
follows.

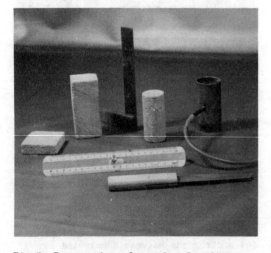

Fig.3 Preparation of samples for direct
shear box and triaxial testing.

(i) The strength tends to increase with decrease in porosity as shown in Fig. 5. This trend is as expected since both a porosity reduction and an increase in the interlock must occur with an increase in the amount of pressure solution. However, as noted previously, the grain size can be expected to influence the results. Fig. 5 is a trend noted in matrix-free fine and fine-medium sands in the U.K.

(ii) The strength is anisotropic as expected from previous work with recompacted sands (Arthur and Menzies 1972; Arthur and Phillips 1975; Oda 1981; Oda and Koishikawa, 1977). The largest differences between perpendicular and parallel samples (i.e. the orientation of the bedding with respect to the cylindrical long axis) has been noted in sands of lowest porosity (in accordance with the trend noted by Arthur and Phillips, 1975). Thus a pale grey locked sand from the Grantham Formation (Jurassic) gives a ratio of 19 for the perpendicular to parallel uniaxial strengths.

The measurement of the uniaxial compressive strength of saturated samples presents difficulties owing to the readiness of the material to slake and also to its free-draining characteristics. Nevertheless it has been observed that moisture drastically reduces the interlock cohesion. The reason why this is so is unknown and further research is required.

The compressive strength developed by geologically aged, uncemented sands has been largely ignored in geotechnical literature. Thus although they first introduced the term "locked sands", Dusseault and Morgenstern (1979) did not provide an indication of the unconfined compressive strength of their intact block samples and carried out only confined tests on saturated material. Nevertheless, the uniaxial strength of locked sands can be very considerable as indicated by the Cretaceous and Jurassic sands in Fig. 5.

(iv) Shear strength

Samples suitable for direct shear box testing can be readily carved from the intact blocks and tested with the bedding either perpendicular or parallel to the plane of shearing. Drained tests on 60 x 60mm samples with normal stresses up to 900kPa have been carried out. Tests with zero normal load were used to determine the cohesive intercept for the dry samples.

In the case of the weakly locked Barton Sand, the shear strength versus normal stress shows only a slight curvature and the envelope can be assumed linear (Barton et al 1986a). However for dense granular materials a degree of curvature is to be expected and for more accurate analysis it is preferable to interpret the envelope in terms of a power law in the form:-

$$\tau = c' + A \, \sigma_n'^b$$

where τ = shear strength, c' = effective cohesion, σ_n' = effective normal stress and where A and b are power law constants. For the dry samples of Barton Sand the best fit envelopes give:-

Perpendicular : $\tau = 10 + 1.0 \, \sigma_n'^{0.971}$

Parallel : $\tau = 10 + 1.06 \, \sigma_n'^{0.983}$ with c' and σ_n' given in kPa.

Tests on dry samples of Barton Sand re-compacted by pouring to different porosities have been carried out by

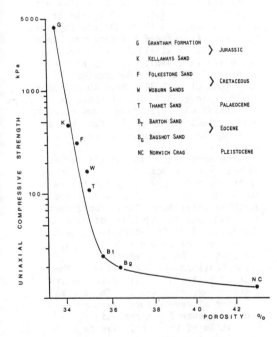

Fig.5 Uniaxial compressive strength tests on 38mm dia x 78mm cylindrical samples (perpendicular orientation) of U.K. fine and fine medium, weakly locked and locked sands versus porosity. Log-scales used for both axes.

Maddison (1979). The dense samples, as expected, gave curved shear strength envelopes. To compare the re-compacted samples with the intact samples, the results have been plotted in Fig. 6 as secant ϕ' values for a normal stress of 200kPa. Sands re-compacted by pouring also possess an inherent anisotropy but only the parallel samples (i.e. with the plane of deposition parallel to the shear box plane) have been tested.

The in-situ porosity cannot be reproduced without considerable grain fracturing and breakdown and hence such results would be unreliable for comparative purposes. In Fig. 6 the line joining the re-compacted sample results has been extrapolated and it can be seen that the result for the parallel orientated intact sample fits this extrapolation very neatly. The perpendicular sample result falls well below the extrapolation. The perpendicular intact sample (i.e. bedding normal to the shear box plane) is weaker than the parallel intact sample (by about $3\frac{1}{2}^\circ$ in the secant ϕ' values). No perpendicular re-compacted samples have been tested but it appears likely that these could be similarly weaker than the parallel samples.

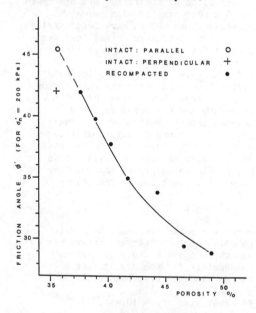

Fig.6 Direct shear test results on intact and recompacted samples of Barton Sand (Eocene). Values of Friction Angle ϕ' (derived as secant ϕ' for a normal stress of 200kPa) are plotted against porosity.

Differences between perpendicular and parallel poured samples in direct shear of about 5° have been reported (Phillips and May, 1967) although for the remoulded Leighton Buzzard Sand used in these tests the perpendicular samples were the strongest.

APPLICATIONS

The ability of geologically aged sands to be sampled as undisturbed blocks adds considerable scope to Site Investigation practice. Where the sand bed is exposed on site, sampling can be easily undertaken as has already been carried out for a large quarry development in Surrey (C. P. Thomas, private communication). The sand bed here is the Folkestone Sand (Cretaceous) from which block samples were obtained and tested in direct shear to obtain parameters for the engineering design of long-term slopes.

If the sand bed is at depth and also below the water table, the problem can be resolved either by locating a nearby exposed outcrop (ensuring that significant lithological change has not occurred) or by pitting and/or shaft sinking. Such deliberate exposure has been recommended for cemented sands by Salamone et al (1978) and Dobereiner and Oliveira (1986). The exposure can be sampled after site dewatering (Salamone et al 1978; Horn 1979). Horn (1979) also records a case where an excavation was deliberately dewatered to permit block sampling of a sand bed as part of the Site Investigation procedure.

Block sampling of the weakly locked Barton Sand has been carried out by the author to determine the permeability of the intact material on behalf of the Southern Water Authority. The latter are examining the aquifer potential of the Barton Sand formation to augment the water supplies on the Isle of Wight. The cost of the laboratory investigations are a fraction of the cost of full-scale field permeability tests using trial wells. The block sampling procedure thus makes good economic sense for a preliminary evaluation of the likely yields from this sand formation.

Block sampling of geologically aged sands must also be considered as a highly recommended procedure for the evaluation of the dynamic properties of these materials. Horn (1979) and Walberg (1978), among others, have pointed to the significant differences between undisturbed and re-compacted sands obtained during dynamic tests. Clearly, it would be most unreliable to assume the dynamic properties of these materials from tests carried out on the re-compacted material.

Although undisturbed sampling is being recommended here as being of especial value for geologically aged sands, this is meant not as an alternative, but as an addition, to in-situ tests. The latter are very important for the evaluation, for instance, of mass properties and can avoid sampling difficulties. Nevertheless in-situ tests need to be interpreted with full knowledge of the characteristics of the material (see Barton et al, 1989 in respect of penetration tests). Undisturbed block sampling is the way that this full knowledge can be provided.

CONCLUSIONS

1. Geologically aged sands even if uncemented possess cohesion. This interlock cohesion permits undisturbed block sampling and hence makes it possible to carry out laboratory testing of intact samples.

2. The in-situ fabric cannot be reproduced by re-compaction and laboratory tests on intact samples are essential for evaluating their geotechnical characteristics. Of particular significance is the anisotropic fabric which crucially influences their engineering properties and behaviour.

3. The geotechnical investigation of sites containing geologically aged sands can be substantially augmented by either trial pitting, shaft sinking (with dewatering as necessary) or by sampling at nearby exposed outcrops.

ACKNOWLEDGEMENTS

The authors wish to record their thanks for discussion with Mr. R. M. Harkness (who first suggested the phrase "interlock cohesion") and for field and laboratory assistance to Messrs. A. Brookes, T. Pickett and Y. L. Wong. Thanks are also recorded to Dr. C. P. Thomas of Geotechnical Engineering Ltd. for information and to Mrs. M. Binnie for typing.

REFERENCES

Arthur, J. R. F. and Menzies, B. K. 1972. Inherent anisotropy in a sand. Geotechnique 22, 115-128.

Arthur, J. R. F. and Menzies, B. K. 1975. Homogeneous and layered sand in triaxial compression. Geotechnique 25, 799-815.

Barton, M. E., Palmer, S. N. and Wong, Y. L. 1986a. A geotechnical investigation of two Hampshire Tertiary Sand Beds: are they locked sands? Quart. Journ. of Engineering Geology, 19, 399-412.

Barton, M. E., Brookes, A., Palmer S. N. and Wong, Y. L. 1986b. A collapsible sampling box for the collection and transport of intact block samples of friable uncemented sands. Journ. of Sedimentary Petrology, 56, 540-541.

Barton, M. E. and Palmer, S. N. 1989. The relative density of geologically aged, British fine and fine-medium sands. Quart. Journ. of Engineering Geology, 22, 49-59.

Barton, M. E. Cooper, M. R. and Palmer, S. N. 1989. Diagenetic alteration and micro-structural characteristics of sands: neglected factors in the interpretation of penetration tests. In Penetration Testing in the U.K. Thomas Telford, London, 57-60.

Clough, G. W., Sitar, N., Bachus, R. C. and Rad, N. S. 1981. Cemented sands under static loading. Journ. Geotech, Eng. Division: Proc. ASCE 107, 799-817.

Dobereiner, L. and Oliveira, R. 1986. Site investigations on weak sandstones. Proc. 5th Int. I.A.E.G. Congress, Buenos Aires, 1, 411-421.

Dudley, J. H. 1970. Review of collapsing soils. Journ. Soil Mechs. and Found. Division, Proc. ASCE, 96, 925-947.

Dusseault, M. B. and Morgenstern, N. R. 1979. Locked Sands. Quart. Journ. Eng. Geology. 12, 117-131.

Hanzawa, H. and Matsuda, F. 1977. Density of alluvial sand deposits obtained from sand sampling. Proc. 9th Int. Conf. Soil Mechs. & Found. Eng. Tokyo. Speciality Session No. 2: Soil Sampling, 7-14.

Horn, H. M. 1979. North American experience in sampling and laboratory dynamic testing. Geotechnical Testing Journ. 2, 84-97.

Houseknecht, D. W. 1987. Assessing the relative importance of compaction processes and cementation to reduction of porosity in sandstones. Am. Assoc. of Petroleum Geologists 71, 633-642.

Houseknecht, D. W. 1988. Intergranular pressure solution in four quartzose sands. Journ. of Sedimentary Petrology, 58, 228-246.

Ishihara, K. and Silver, M. L. 1977. Large diameter sand sampling to provide specimens for liquefaction testing. Proc. 9th Int. Conf. Soil Mechs. & Found. Eng., Tokyo. Speciality Session No. 2: Soil Sampling, 1-6.

Jennings, J. E. and Knight, K. 1957. The additional settlement of foundations due to a collapse of structure of sandy subsoils on wetting. Proc. 4th Int. Conf. Soil Mechs. and Found. Eng., London 1, 316-318.

Maddison, S. P. 1979. The characteristics of a fine sand. Unpublished M.Phil. Thesis, University of Southampton, U.K.

Marcuson, W. F. III. 1978. Determination of in-situ density of sands. Dynamic Geotechnical Testing, ASTM STP 654, American Soc. for Testing Materials, 318-340.

Marcuson, W. F. III and Franklin, A. G. 1979. State of the art of undisturbed sampling of cohesionless soils. Proc. Int. Symp. of Soil Sampling, Singapore (State of the Art on Current Practice of Soil Sampling). 57-71.

Mori, H. 1986. Recent development in sampling of granular soils in Japan. 4th Int. Geotech. Seminar (Field Instrumentation and In-situ Measurements), Nanyang Technological Institute, Singapore, 13-20.

Mori, H. and Koreeda, K. 1979. State of the art report on the current practice of sand sampling. Proc. Int. Symp. of Soil Sampling, Singapore (State of the Art on Current Practice of Soil Sampling). 73-93.

Oda, M. 1981. Anisotropic strength of cohesionless sand. Journ. Geotech. Eng. Division Proc. ASCE 107, 1219-1231.

Oda, M. and Koishikawa. 1977. Anisotropic fabric of sands. Proc. 9th Int. Conf. Soil Mechs. Found. Eng. Tokyo, 1, 235-238.

Palmer, S. N. and Barton, M. E. 1986. Avoiding microfabric disruption during the impregnation of friable, uncemented sands with dyed epoxy. Journ. of Sedimentary Petrology 56, 556-557.

Phillips, A. B. and May, Ph.D. 1967. A form of anisotropy in granular media. Special Task Report, Dept. of Civil & Munic. Engng., Univ. Coll., London.

Salamone, L. A., Singh, H., Miller, V. G. and Fischer, J. A. 1978. Improved sampling methods in variably cemented sands. Am. Soc. Civ. Eng. Convention and Exposition, Chicago, U.S.A. Session No. 79, Preprint Vol., 320-349.

Saxena, S. K. and Lastrico, R. M. 1978. Static properties of a lightly cemented sand. Journ. Geotech. Eng. Div., Proc. ASCE 104, 1449-1464.

Seed, H. B., Singh, S., Chan, C. K. and Vilela, T. F. 1982. Considerations in undisturbed sampling of sands. Journ. Geotech. Eng. Div., Proc. ASCE 108, 265-283.

Seko, T. and Tobe, K. 1977. An experimental investigation of sand sampling. Proc. 9th Int. Conf. Soil Mechs. & Found. Eng. Tokyo. Speciality Session No. 2: Soil Sampling, 37-42.

Singh, S., Seed, H. B. and Chan, C. K. 1982. Undisturbed sampling of saturated sands by freezing. Journ. Geotech. Eng. Div. Proc. ASCE 108, 247-264.

Terzaghi, K. and Peck, R. B. 1968. Soil Mechanics in Engineering Practice, 2nd Edition. Wiley, New York.

Tohno, I. 1977. Methods to evaluate quality of undisturbed samples of sands. Proc. 9th Int. Conf. Soil Mechs. & Found. Eng., Tokyo. Speciality Session No. 2: Soil Sampling. 29-35.

Walberg, F. C. 1978. Freezing and cyclic triaxial behaviour of sands. Journ. Geotech. Eng. Division, Proc. ASCE, 104, 667-671.

Yoshimi, Y., Hatanaka, M. and Oh-Oka, H. 1977. A simple method for undisturbed sand sampling by freezing. Proc. 9th Int. Conf. Soil Mechs. & Found. Eng., Tokyo. Speciality Session No. 2: Soil Sampling. 23-28.

6th International IAEG Congress / 6ème Congrès International de AIGI, © 1990 Balkema, Rotterdam. ISBN 90 6191 130 3

Considerations on subsoil exploration by means of instantaneous drilling parameters recording

Considérations sur la reconnaissance du sous-sol par enregistrement instantané des paramètres de forage

Giulio C. Borgia, Giovanni Brighenti & Ezio Mesini
Istituto di Scienze Minerarie, Università di Bologna, Italy

ABSTRACT: The method of subsoil exploration based on the interpretation of drilling parameters recording is here presented. After examining the possibility of interpreting such information following an approach of current use in petroleum industry, the differences in the interpretative techniques in the geognostic–geotechnical field are discussed. These differences are due to both the diverse conditions and different equipments used in these two distinct fields. The possibilities of the method are examined through the discussion of some diagraphies carried out in Italy.

REŚUMÉ: Cet exposé vise à examiner la méthode d'exploration du sous–sol basée sur l'interpretation des enregistrements des paramètres de forage. Après avoir examiné la possibilité d'interpréter les données à l'aide des methodes déjà employées dans le secteur pétrolier, les différences dans les techniques de interpretation des diagraphies dans les secteurs géotechnique et géognostique ont été éxaminées. Des conclusions sont tirées quant à l'emploi de ces moyen sur la base de l'examen de quelques diagraphies enregistrées en Italie.

1 INTRODUCTION

The subsurface exploration by means of continuous sampling and subsequent laboratory testing imply a very high cost; on the other hand the technical and economical benefits given by a good knowledge of the subsoil during planning and work execution stages are known. Moreover, a good knowledge of the subsoil is required because of the wide diffusion of the geological and geotechnical thematic maps. The development of those techniques that increase the low–cost information on the subsoil is therefore necessary. Here we will examine the parameter recordings during destructive drilling; it is a technique which may give useful physical and stratigraphycal information. This technique has given good results (Bru et Al. 1983, Desloovere 1980, Deveaux et Al. 1983, Lutz 1983, Richez 1981) and is spreading in Italy, too. The optimization of its employment may be pursued by means of an accurate evaluation of the method to detect both the most representative parameters and the best techniques to interpret the data in view of what is done in the field of oil drilling.

2 GENERAL CONSIDERATIONS

The drilling performance depends not only on the characteristics of the underground but also on the type of drilling rig equipment, on the bit type and its wear, on drilling fluid, on the degree of experience of rig personnel and on drilling parameters (weight on the bit, rotary speed, etc.). These factors being equal, the drilling performance depends on the characteristics of the soil. Under such conditions, for example, the penetration rate is affected by the formation drillability (related to compressive strength, rheological properties, etc.).

The rate of penetration is the most studied parameter; there are also other parameters that may give useful information on lithology, such as the bit torque, the power needed to drill the unit volume of rock, the pressure drop in the mud circuit, the "reflected percussion" in rotopercussion, etc.

Provided one single parameter is not enough to univocally recognize a soil (e.g., the drillability may greatly vary on the same lithotype and different lithotypes may have, converserly, the same drillability) several parameters are usually recorded and compared (Desloovere 1980, Lutz 1981, Pfister 1980).

In the oil field the study of the influence of drilling parameters on drilling performance has been in progress for more than 25 years (Burgoyne 1986) and it is employed both to minimize drilling cost and to detect formations characterized by abnormal pressures (usually underconsolidated and therefore with less mechanical strength than similar formations with normal pressure at the same burial depth). This latter case study, whose aim is similar to the one of soil mechanics field tests, has been based on the analysis of drilling performance data. The study is based on prediction from a mathematical model of drilling process. Obviously, the complexity of the mechanisms of rock cut and cutting removal has hindered the complete modeling of the drilling process (Warren 1987). Yet, as the qualitative knowledge of fundamental principles is fairly developed and based on a remarkable amount of experimental data, it is possible to yield schematic models to be validated by means of experimental correlations. Such models do not completely describe the drilling process, but allow to quantify the relative effects of changing various parameters that may be altered during drilling.

The model proposed by Burgoyne and Young (1974) is perhaps the most general among those proposed in literature, since it examines the effects of depth, formation characteristics, mechanical parameters (weight on bit and rotary speed) and mud properties.

The approch usually followed is to assume that the effects of the different parameters on penetration rate are independent from one to the other, and that the composite effect can be computed by:

$$R = f_1 f_2 \ldots\ldots f_n. \qquad (1)$$

In eq.(1) R is the rate of penetration and f_j

are functional relations between R and various variables. In particular eq.(1) can be expressed by:

$$R = exp(a_1 + \sum_{j=2}^{8} a_j x_j), \qquad (2)$$

where x_j are functions of the considered parameters (expressed in the adimensional form obtained by dividing their real value by the mean value), a_1 is a constant which mainly takes into account both the formation strength and the effect of all those factors that are not included in the model explicitly (i.e., bit type, mud solid content, etc.). Constants a_2 through a_8 consider formation depth, formation compaction, pressure differential across the hole bottom, bit diameter and weight on the bit, rotary speed, bit wear, and bit hydraulics, respectively. According to Graham and Muench (1959) and in view of the possibility to linearize eq.(2) by writing it in a logarithmic form, a_i ($i = 1, 8$) constants are determined through a multiple regression analysis on the basis of the measurements carried out during the drilling of short intervals, where the soil characteristics (and therefore a_1) may be considered as uniform. For many cases in eq.(2) only the effects of the weight on bit and the rotary speed (that are taken into account by terms $a_5 x_5$ and $a_6 x_6$ respectively) are made explicit while all the other effects are included in a_1 constant. In these particular cases eq.(2) becomes:

$$R = exp(a_1 + \sum_{j=5}^{6} a_j x_j), \qquad (3)$$

with

$$x_5 = ln W_d, \qquad x_6 = ln N_d, \qquad (4)$$

where W_d and N_d represent the weight on bit and the rotary speed respectively, expressed in an adimensional form. Eq.(3) taking into account eq.(4) becomes:

$$R = exp(a_1) W_d^{a_5} N_d^{a_6}, \qquad (5)$$

from which, constant being known, the normalized rotary speed is obtained:

$$\bar{R} = \frac{R}{W_d^{a_5} N_d^{a_6}}, \qquad (6)$$

which is independent from W and N, and from the drillability index

$$K' = a_1 = ln\bar{R}. \qquad (7)$$

Other drillability indices have been similarly obtained starting from different drilling models. Among these we remember the "d exponent" obtained by Jorden and Shirley's (1966) on the basis of Bingham model (Bingham 1964-65).

3 EQUIPMENT

In geognostic and geotechnical surveys and in drillings for civil works, oleodynamic system rigs are usually employed; by means of shift, speed, pressure and acceleration transducers the following parameters can be continuously recorded:
- the top drive position (from which the depth and the rate of penetration are obtained);
- the pressure in the oleodynamic circuit that gives the weight on bit (hence the strength on the drilling battery);
- the pressure in the oleodynamic circuit which rotates the battery (hence the torque applied to the drill pipes);
- the rotary speed;
- the mud pump pressure (and therefore the pressure drop of the mud circuit);
- the part of energy reflected through the battery in case of rotopercussion drilling.

Not all these parameters are always measured. Since such measurements are taken on surface, some of them differ from the ones that could be obtained if the actions applied at the bottomhole could be directly detected. For example, the weight on bit depends, not only on the weight applied to the battery (measured on surface), but also on the battery weight, on the friction at the hole walls, and on the battery and oleodynamic circuit inertia. Similar considerations can be made for other parameters, too.
Recording and measurement devices are customized to be assembled on the normal marketed rigs. A characteristic of this equipment concerns its sturdiness and facility of handling when considering often hard working conditions.
In modern equipment the recording is digitalized and stored on magnetic cassettes. All quantities are usually recorded either every minute (for low rate of penetration), or every centimeter of penetration.

4   MEASUREMENT INTERPRETATION

The recording of the drillability index is enough to detect the consolidation state of deep layers in the oil field; in the geological–geotechnical field it is necessary to know more indices for a better subsoil characterization. In this respect one has to consider that:

1 – The normalized rate of penetration depends, not only on the formation strength, but also on the bit wear and drilling fluid, and also on the employed equipment; an example concerns the presence of cavities when drilling: in this case the soil strength is null. However, the bit penetrates according to the cavity size and to the inertia of the battery and of the hydraulic circuit (Cailleux & Toulemont 1983). Under these conditions it is difficult to realize the difference between an empty cavity and a cavity filled with low–strength materials. Should the presence of a cavity be assumed, it might be convenient to carry out a systematic calibration of logs through cavity simulation by partially lifting the battery and letting it fall down again.

2 – The bit torque mainly depends on the weight on the bit and on the soil rheological behaviour. Usually, it is high in plastic rocks (clay, marl, etc.) where the bit teeth tend to penetrate into the rock and essentially exert a cutting action, whereas it is low in elastic rocks where the tooth mainly exerts a compression action. Furthermore, since the measured torque is applied at the wellhead, a part of the torque may be due to mechanical sticking of the strings or to shale sloughing.

3 – Mud chemical–physical characteristics and mud flowrate being equal, the pump pressure depends on the circuit pressure drop: distributed pressure drops in the circular bore of the drillstring and in the annulus space between the drillstring and the hole (which gradually increase with depth), and concentrated pressure drop across bit (that remarkably increase when low permeability and plastic behaviour formations are drilled).

4 – In rotopercussion, a part of the power given

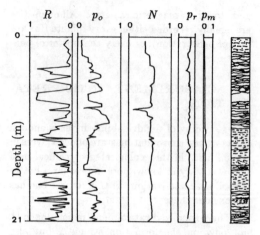

Figure 1. Case A. Normalized drilling parameters vs. depth, and relative stratigraphy.

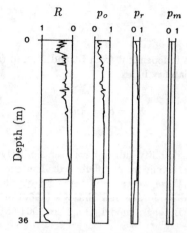

Figure 3. Case C. Normalized drilling parameters vs. depth, and relative stratigraphy.

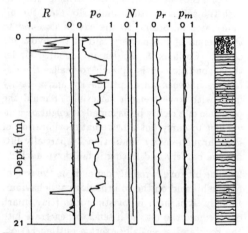

Figure 2. Case B. Normalized drilling parameters vs. depth, and relative stratigraphy.

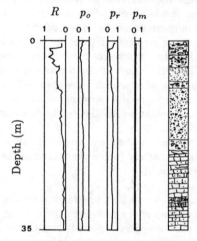

Figure 4. Case D. Normalized drilling parameters vs. depth, and relative stratigraphy.

by the hammer to the rock breaks the rock itself, a part spreads in the rock and a part is reflected onto the surface through the battery and is measured. Obviously, the value of this last part of power depends on the elastic behaviour of the formation.

5 RESULTS AND CONCLUSIONS

Several surveys have been carried out in Italy using the method of the instantaneous drilling parameters recording. Figures 1 through 4 show the recordings carried out during four drillings. In the first two cases (A, B), the following quantities have been recorded: rate of penetration (R), hydraulic pressure in the bit–load circuit (p_o), rotary speed (N), hydraulic pressure giving the torque to the battery (p_r) and pump pressure (p_m). In the other cases (C and D) the rotary speed has not been measured.

The data reported in Figs. 1–4 have been corrected for the noise (which is due to the recording technique). Moreover, they have been normalized dividing them by an assumed maximum value. Linear correlation matrices among measured parameters have been calculated and the results can be summarized as follows:

– in almost all cases there are weak correlations among measured parameters and this leads to the multicollinearity problem (Al–Bebiri et Al. 1988). So, it is not fully respected the assumption that in the proposed drilling model there are no correlations among various parameters. This makes it difficult to detect the relative influence of every parameter and it limits the validity of the model itself;

– the correlation coefficients among different quantities vary case by case: this depends on the variable influence of the factors. Only in case A, for example, where there are no large bit drop pressures, the mud pressure is strictly correlated to depth;

– in all cases there is a negative correlation coefficient between rate of penetration and bit load. This apparently anomalous relationship occurs because during the drilling the driller tends to keep the rate of penetration within a reasonable range by decreasing the load on bit when the rate of penetration increases.

The examination of the results went on through the calculation of three important indices: the drillability index (K'), the hardness index (D) and the cohesion index (H).

As far as the first index is concerned, it is worth noting that if the battery weight is neglected with respect to the weight on the bit and frictions, W_d value is equal to the adimensional value of the pressure in the load–bit circuit (p_{od}). Therefore, the normalized rate of penetration is given by:

$$\bar{R} = \frac{R}{p_{od}^{a_5} N_d^{a_6}}. \tag{8}$$

As far as a_5 and a_6 constants are concerned, it was impossible to get acceptable values by means of the multiple linear regression, at least for the cases under examination; however, giving them values within the usual range ($a_5 = 1 \div 1.25$, $a_6 = 0.4 \div 1$) (Smalling & Myers 1988) the trend of K' as a function of depth (z) did not vary. Calculations have been made by assuming $a_5 = a_6 = 1$.

According to what proposed by Hamelin et Al. (1983), the following expression, proportional to the power given to the bit torque to drill the hole unit length, has been chosen as a hardness index:

$$D = \frac{p_{rd} N_d}{R_d} \tag{9}$$

whereas for the cohesion index it has been chosen:

$$H = \frac{p_{rd}}{p_{od}} \tag{10}$$

(proportional to the battery–torque pressure for an unit pressure in the bit–load circuit).

The comparison among the values of K', D, H and the mud pump pressure p_m are reported in figs. 5 through 8. Their examination and comparison with the stratigraphic profile, if any, show that, while the mechanical and rheological

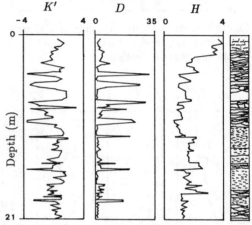

Figure 5. Case A. Drillability index (K'), hardness index (D) and cohesion index (H) vs. depth, and relative stratigraphy.

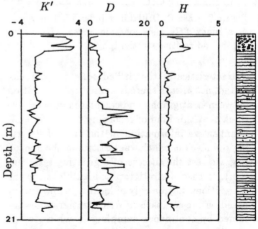

Figure 6. Case B. Drillability index (K'), hardness index (D) and cohesion index (H) vs. depth, and relative stratigraphy.

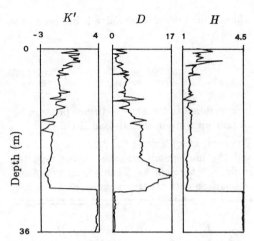

Figure 7. Case C. Drillability index (K'), hardness index (D) and cohesion index (H) vs. depth, and relative stratigraphy.

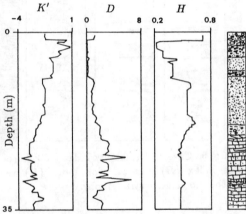

Figure 8. Case D. Drillability index (K'), hardness index (D) and cohesion index (H) vs. depth, and relative stratigraphy.

characteristics of the drilled soils are well detected, not always there is a univocal correlation between stratigraphy and calculated indices. We think that this kind of recording may give semi-quantitative information in the case of detecting the mechanical characteristics of the subsoil (e.g, to test the success of a grouting intervention). In case of stratigraphic identifications the correlations may be only of local type and always based on the comparison with stratigraphies obtained by continuous sampling in which, when possible, low–cost services are required (cuttings examination, electrical, radioactive or sonic logs, etc.). In all cases the mud flowrate must be al-

ways recorded in addition to the previously mentioned six parameters. We believe it is useful to extend such recordings; in this connection Public Authorities should examine the possibility to extend them systematically by requiring them for all drillings (including those concerning the water wells) in order to get information on the subsoil characteristics all over the territory. They would be of the utmost importance especially in the case of regional studies when the use of the surface and underground planning are concerned.

ACKNOWLEDGEMENTS

The authors acknowledge ISMES S.p.A. (Bergamo, Italy) that released part of the field data. This reasearch was supported by Italian Ministero della Pubblica Istruzione and by Italian Consiglio Nazionale delle Ricerche.

REFERENCES

Al–Betary E.A., M.M. Moussa & S. Al–Atabi 1988. Multiple regression approach to optimize drilling operations in the Arabian Gulf area. Soc. Petr. Eng. Drilling Eng., 83–88.

Bingham M.G. 1964–65. A new approach to interpreting rock drillability. Oil & Gas J., Nov. 2 – April 5.

Burgoyne Jr. A.T. & F.S. Jr. Young 1974. A multiple regression approach to optimal drilling and abnormal pressure detection. Soc. Petr. Eng. J., 371–384.

Burgoyne Jr. A.T. et Al. 1986. Applied drilling engineering. Soc. Petr. Eng., Richardson, 502 pp.

Bru J. et Al. 1983. Les diagraphies et les essais de mecanique des sols en place. Bull. IAEG, 26–27:25–32.

Cailleux J.B. & Toulemont 1983. La reconnaissance des cavites souterraines par méthodes diagraphiques. Bull. IAEG, 26–27:33–42.

Desloovere P. 1980. Reconnaissance du sous-sol par enregistrement instantane de divers paramètres au cours de la perforation d'un forage destructif. Bull. IAEG, 21:43–48.

Deveaux et Al. 1983. Diagraphies instantanées en recherche d'eau. Bull. IAEG, 26–27:59–63.

Graham J.W. & N.L. Muench 1959. Analytical determination of optimum bit weight and rotary speed combinations. Soc. Petr. Eng., 34-th Annual Fall Meeting, Dallas.

Hamelin J.P. et Al. 1983. Enregistrement des paramètres de forage: nouveaux developments. Bull. IAEG, 26–27:84–88.

Jorden J.R. & O.J. Shirley 1966. Application of drilling performance data to overpressure detections. J. Petr. Tech., 18:1387–1394.

Lutz J. 1981. Enregistrement des paramètres de forage. Travaux, 84–89.

Smalling D.A. & R.L. II Myers 1988. A drilling model for young offshore Louisiana and Texas trends. Soc. Petr. Eng. Drilling Eng., 141–152.

Pfister P. 1980. Interpretation des enregistrements de paramètres de forage. Bull. IAEG, 21:49–52.

Richez P. 1981. L'enregistrement des paramètres de forage. Travaux, 77–83.

Warren T.M. 1987. Penetration–rate performance of roller–cone bits. Soc. Petr. Eng. Drilling Eng., 9–18.

Determining the spatial variation of CPT properties

Détermination de la variation latérale des propriétés CPT

A.J.Coerts
University of Utrecht, Netherlands

ABSTRACT: A method is presented to detect geological features of recent deposits by determining the spatial variation of properties that are deduced from Cone Penetration Test (CPT) curves. Multivariate statistical and geostatistical techniques play an important role in the analysis.

The analysis has been done for a small area west of Leiden, the Netherlands. The spatial variation of CPT deduced properties considering a segment of the CPT curve that represents a certain layer of holocene deposits appears to be a reflection of the geology of this layer.

RESUME: Cet article présente une méthode à découvrir des charactéristiques géologiques des sédiments récents par la détermination de la variation latérale des propriétés qui sont dérivées des graphiques pénétrométriques. Surtout des techniques statistiques et géostatistiques font partie de l'analyse.

L'analyse a été executé pour une petite territoire à l'ouest de Leiden, Hollande. La variation latérale des propriétés pénétrométriques se trouve être une équivalent des charactéristiques géologiques de la strate considerée.

1 INTRODUCTION

The Cone Penetration Test (CPT) is often used as a geotechnical method for determining strength characteristics of soil. It is a useful tool for (engineering) geological purposes.

A method is presented to determine the spatial variation of CPT deduced properties in an objective way. Whether this method can be useful to predict the geological characteristics of recent deposits has been investigated in a case study. The spatial variation of CPT characteristics will be compared with the geological features of a layer of holocene deposits in an area west of Leiden in the western part of the Netherlands.

2 METHOD

The CPT-curve is divided in a number of more or less homogeneous segments. This can be done by the method described by Gill (1970). It may be necessary to stabilize the variance by a log-transformation before the segmentation. The segments of interest are described by quantiative properties.

To get a first impression of any spatial differentiation within the considered layer of deposits a reconnaissance study is carried out. First, a factor analysis is executed to get information about the correlation structure within the set of CPT deduced properties (Jöreskog et al., 1976). Moreover, a number of factors is extracted. The scores of the significant factors might be used in further analyses.

The second step in the reconnaissance study is a k-means cluster analysis. The principles of this type of cluster analysis are described in Anderberg (1973). The choice of the properties to be involved in the cluster analysis depends on the correlation structure within the original set of properties. A map on which the resulting clusters are indicated gives rough information about the spatial structures that may exist within the considered layer.

Each of the relevant properties is then subjected to a geostatistical analysis to determine its spatial variation. The geostatistical analysis comprises a semivariance analysis and an optimal inter-

polation, or kriging, procedure. In the
semivariance analysis the spatial correlation structure of a property is determined. This information is used in the
kriging procedure. The outputfile of the
kriging procedure contains for each grid
cell the estimated value Z as well as its
estimation error.

The theory of geostatistics is discussed
in more detail in Journel and Huijbrechts
(1978), Burgess and Webster (1980),
Burrough (1986), Davis (1986) and Oliver
and Webster (1986).

Contour maps that display the kriged
estimates of the selected CPT deduced properties may show typical patterns that
reflect the geological characteristics of
the represented layer.

3 CASE STUDY

3.1 Geological setting

The study area is situated in the coastal
zone of the Netherlands, west of Leiden as
is indicated in figure 1. It is part of a
region where the holocene deposits belong
to the Westland Formation.

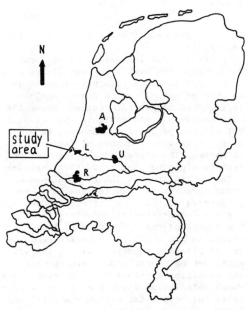

Figure 1. Location of the study area

At about 6000 BP the region became part of
the Rhine-Meuse delta. Its geological development has been strongly influenced by

the (relative) sea level rise that has
taken place from the beginning of the Holocene.

Of interest in this case study are the
lagoonal and tidal flat deposits of Calais
and sands and clays deposited in and from
tidal channels which belong to either the
Calais or the Dunkirk deposits. Moreover,
dunesands, perimarine deposits from the
river Old Rhine (Tiel-0) and peat may be
found in the considered layer (De Mulder
and Pruissers,1984; Zagwijn,1986; Pruissers and De Gans,1988).

3.2 Data collection

The study area is indicated on the map of
figure 3. Its dimensions are about 275 m
in E-W direction and 250 m in N-S direction. In this area 16 hand borings have
been done, most of which reach to a depth
of 8 m below surface level. Surface level
is at mean sea level (MSL). In a more or
less regular grid with spacings of about
30 m in both directions mechanical Cone
Penetration Tests had been done by Delft
Soil Mechanics. The locations of the
borings and the CPT's are indicated in
figure 3. 59 CPT's have been selected for
further study. All CPT curves showed a
characteristic segment between 3 and 7 m
below MSL. This segment, an example of
which is shown in figure 2, has been quantitatively described by the five properties that are listed in table 1.

Table 1. Properties describing a segment
of a CPT curve

Property	S.I Unit	Symbol
Depth of top surface	m	d
Total thickness	m	t
Mean cone resistance	MNm^{-2}	r_m
Linear trend	MNm^{-3}	f
Curvature of CPT curve	MNm^{-4}	c

The meaning of the properties f and c is
illustrated by the equations (1) and (2)
respectively.

$$l(z) = fz + e \quad (1)$$

In (1): z = depth; l(z) = linear trend
function; e = constant; f = regression
coefficient of linear trend;

$$q(z) = cz^2 + bz + a \quad (2)$$

In (2): q(z) = quadratic trend function;
a,b = constants; c = regression coefficient of quadratic trend, or curvature.

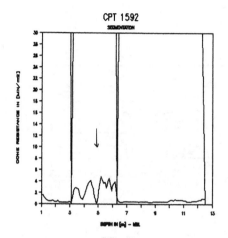

Figure 2. Part of a CPT curve representing
the Holocene sequence. Arrow indicates the
segment that has been described with quantitative properties.

3.3 Reconnaissance study

The five properties mentioned in table 1
were used in a factor analysis, that was
caried out using the CSS package. Two factors, F1 and F2, were extracted. They explain a greater part of the total variance
than any of the original properties. Table
2 shows that F1 is a measure of the
geometry of the unit and F2 reflects the
strength properties of the unit. The spatial variation of the scores of F2 was
determined by the geostatistical analysis,
which is discussed later in this paper.

Table 2. Loadings of the first two factors

	F1	F2
d	−0.844	0.176
t	0.805	0.303
r_m	0.042	0.693
f	0.137	−0.338
c	0.138	−0.571

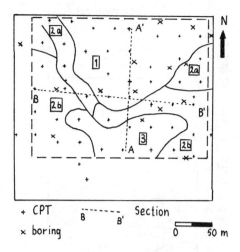

+ CPT   ------- Section
           B   B'
× boring   0   50 m

Figure 3. Division of the study area in
zones. Division based on cluster analysis.

In a k-means cluster analysis the standardized values of the variables t, r_m, f and c
were involved. The variable d was excluded
because of its significant correlation
with t, which may affect the cluster analysis in a way that makes the resulting
classification less reliable.
  The analysis resulted in six clusters.
This number had been arbitrarily chosen
before the analysis. Clusters 2 and 5 contain only one case. The results of the
analysis of variance showed that for all
concerned properties the classification is
significant.
  Figure 3 shows a map of the study area
on which is indicated the cluster each CPT
belongs to. The map shows a typical pattern with the CPT's belonging to cluster 1
mainly situated in the northern part, in
zone 1, and the ones belonging to cluster
3 in the southern part of the area, in
zone 3. These zones are separated by zone
2, which contains both CPT's of cluster 4,
in zone 2a, and CPT's of cluster 6, in
zone 2b.

3.4 Geostatistical analysis

Semivariograms were computed for the variables t and F2. The lag increment h was
15 m. The maximum lag distance was 180 m.
To check for anisotropy directional semivariograms were computed for the four sectors indicated in figure 5.

299

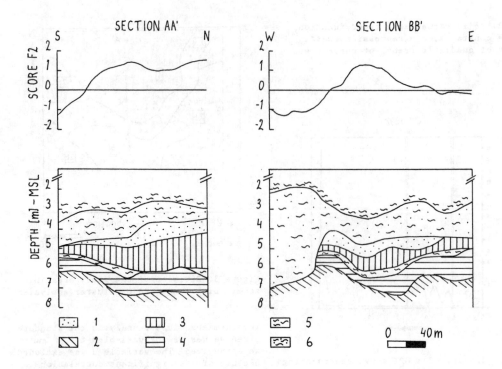

Figure 4. Stratigraphic sections and corresponding diagrams of F2 scores of recent deposits in the study area. The sections are based on borings, CPT's and information from earlier investigations by the Geologic Survey of the Netherlands. 1=dunesand; 2=lagoonal deposits,clay; 3=tidal deposits,sand; 4=tidal flat deposits,clayey sand; 5=perimarine deposits,clay; 6=perimarine deposits,clayey sand. Sections indicated in figure 3.

Figure 5. The four sectors and their standard numbering.

The results of the semivariance analyses are listed in table 3. F2 shows anisotropy. The azimuth of a_{max} is NNE. The variable t even showed a clear trend in NNE direction. The trend was approximated by a linear equation:

$$M(x,y) = 0.00053x + 0.0046y + 0.3604 \quad (3)$$

In (3) x and y are the distance in m in EW and NS direction with respect to the lower left corner of the (1km*1km)-grid cell in which the study area is situated. The semivariance analysis was applied to the residuals $\{Z(x,y) - M(x,y)\}$.

Table 3 Resulting semivariogram parameters

Var	Model	c_0	c	a_{max}[m]	a_{min}[m]	$\theta^1[°]$
t^2	expon[3]	0.17	0.27	100	100	0.0
$F2^4$	spher[5]	0.42	1.37	122	82	67.5

1. azimuth of a_{max};
2. residuals from first order trend surface;
3. exponential semivariogram model;
4. scores of factor 2.
5. spherical semivariogram model;

The resulting semivariogram parameters
have been used in the kriging procedure
that was carried out using the package
GeoEAS. The mean values of the properties
within (10m*10m)-blocks and their standard
deviations have been estimated by ordinary
block kriging. These blocks correspond
with the grid cells in a (26*20)-grid.

In the case of t the kriging estimates
of the residuals were re-added to the values of this property as was estimated by
the trend surface analysis. By means of a
contouring program contour maps of t and
F2 were drawn. These are presented in
figure 6. One example of a map of estimation errors is shown in figure 7.

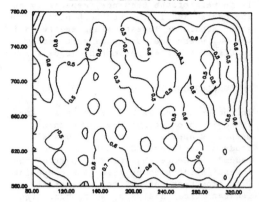

ESTIMATION ERRORS SCORES F2

Figure 7. Contour map of the estimation
errors of the scores of F2.

4 DISCUSSION

In figure 6a the isopach map of the concerned unit is presented. The map shows a
typical pattern of concentric isopachs
except in the SW and SE parts of the area.

The scores of F2 are dependent on r_m and
c, but t has some influence as well.
Figure 6b shows resemblance to the map
based on the cluster analysis (figure 3).
The 0.2 contour on the map coincides with
the boundaries of zone 1 and the -0.4 contour -which is not drawn- coincides with
the boundary of zone 3. Within zone 1 the
more or less radial increase of the scores
of F2 is in agreement with a gradual increase of thickness of the dunesands and
the sandy tidal deposits of Calais as can
be seen in figure 4. Within zone 3 dunesands are almost totally absent, but instead tidal and perimarine deposits with a
high clay content and hence a low cone
resistance are abundant.

Between the -0.4 and the 0.2 contour
there is a zone that is less homogeneous.
Most CPT's in this zone belong to clusters
4 and 6. CPT's from cluster 4 reflect the
presence of rather clayey perimarine deposits lying on dunesands, while at the locations of CPT's from cluster 6 the perimarine deposits are more sandy and the
dunesands are absent.

Figure 7 shows a map of the estimation
errors of F2. It is obvious that the estimation error increases with increasing
distance from the CPT locations. Hence the
largest errors are at the edges of the
map.

Referring to figure 4 it is obvious that
the F2-scores are largest at the locations

THICKNESS OF CONCERNED CPT-UNIT

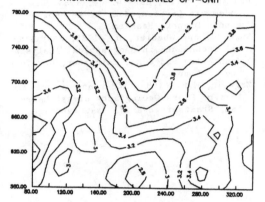

F2 SCORES OF CONCERNED UNIT

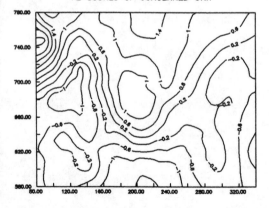

Figure 6a. Isopach map of the concerned
unit; b. Contour map of the scores of F2
within the concerned unit.

where the dunesands form a relatively large part of the sequence, especially at the center and in the northern part of the study area. The lowest values are found in the SW, where the concerned unit mainly consists of clayey perimarine deposits.

5. CONCLUSIONS

In the presented method CPT data are analysed in an objective way. When this method is applied an extensive visual study of individual CPT curves becomes redundant. Hence the analytic part of the investigation becomes less time-consuming.

An important part of the analysis is a geostatistical analysis, which generates useful information about the spatial variation of CPT deduced properties. When the geostatistical analysis is applied in combination with a cluster and a factor analysis typical spatial patterns that possibly reflect geological features are easier to discover and boundaries are determined more exactly. The method helps the geologist to generate hypotheses with respect to the geological structure of the considered layer within the area.

The way the CPT-curves have been objectively described is not yet optimal. This is the main reason why the analysis and interpretation of zone 2 as indicated on map c in figure 4 was problematic. Hence the method to describe a CPT-curve or part of it must be improved.

REFERENCES

Anderberg, M.R., 1973, Cluster Analysis for Applications, Academic Press, New York.

Burgess, T.M. and Webster, R., 1980, Optimal Interpolation and Isarithmic Mapping of Soil Properties, Part II Block Kriging. Jour. Soil Sci. 31, 333-341.

Burrough, P.A., 1986, Principles of Geographical Information Systems for Land Resources Assessment, Monographs on soils and resources survey no. 12, Clarendon Press, Oxford.

Davis, J.C., 1986, Statistics and Data Analysis in Geology, 2nd edition, Wiley, New York.

Gill, D., 1970, Application of a Statistical Zonation Method to Reservoir Evaluation and Digitized-Log Analysis. Am. Assoc. Petr.Geol. Bull. V. 54, no. 5, 719-729.

Jöreskog, K.G., Klovan, J.E., R.A. Reyment, 1976, Geological Factor Analysis, Methods in Geomathematics no. 1, Elsevier Scientific Publishing Company.

Journel, A.G. and Huijbrechts, A., 1978, Mining Geostatistics, Academic Press, New York.

Matheron, G., 1971, The theory of regionalized variables. Les Cahiers du Centre de Morphologie Mathematique de Fontainebleu, 5, 1-210, Ecole Nationale Superieure des Mines de Paris.

Mulder, E.F.J. de, Pruissers, A.P., Zwaan, H., 1984, Kwartairgeologie van 's-Gravenhage, in: De Mulder (ed.) Mededelingen Rijks Geologische Dienst 37-1, 13-43.

Oliver, M.A. and Webster, R., 1986, Combining Nested and Linear Sampling for Determining the Scale and Form of Spatial Variation of Regionalized Variables. Geographical Analysis 18, 227-242.

Pruissers, A.P., Gans W. de, 1988, De bodem van Leidschendam. Rijks Geologische Dienst, Haarlem.

Shaw, G. and Wheeler, D., 1985, Statistical Techniques in Geographical Analysis. John Wiley and Sons, Chicester, New York etc. 364 pp.

Zagwijn, W.H., 1986, Nederland in het Holoceen. Rijks Geologische Dienst Haarlem, Staatsuitgeverij, 's-Gravenhage.

Piezocone and other measurements in an overconsolidated glacial clay
Piezocone et autres mesures dans une argile glaciaire préconsolidée

G.Greeuw
Delft Geotechnics, Delft, Netherlands

F.Schokking
Geological Survey of the Netherlands, Haarlem, Netherlands

ABSTRACT: Piezocone tests provide more information on of the lithology of the soil and on geotechnical parameters than conventional cone tests. Piezocone tests, with a filter just behind the cone, were performed in a deposit consisting of pleistocene glacial tills, sands and clays in the north of The Netherlands. The Sounding reached a depth of 95 m. The geological sequence tested has been subjected to sub-glacial deformation during the Saalien glaciation.
Near the sounding location, a 120 m deep borehole was drilled and samples from this hole were subjected to various geotechnical tests.
In some parts of the overconsolidated clay, negative excess pore pressures were registered. This effect was attributed mainly to fissuring of the clay layer. Similar effects were observed in an earlier piezocone test in a comparable geological sequence. In addition, piezocone data were interpreted with regard to soil identification and shear strength.

RESUME: Des tests piézocones donnent plus d´information sur la lithologie du sol et sur les paramètres géotechniques que les tests pénétrométriques statiques habituels.
Les tests piézocones, avec un filtre juste derrière le cône , ont été éffectués dans des couches contenant des moraines de fond, des sables, des argiles, dans le nord des Pays-Bas. Le test pénétrométrique a atteint une profondeur de 95 m. La séquence géologique testée a été soumise à une déformation sous-glaciaire durant la glaciation Saalienne. Près de l´emplacement du test pénétrométrique un sondage de 120 m à été éffectué et des échantillons de ce sondage ont été soumis à divers tests géotechniques. Dans certaines parties des argiles préconsolidés nous avons enregistré un excès négatif pression d´eau interstitielle.
Cet effet est largement dû à la fissuration des couches d´argile. En outre, les données des piézocones ont été analysées en vue des caractéristiques du sol et de la résistance au cisaillement.

INTRODUCTION

A piezocone test was performed by Delft Geotechnics to a depth of 95 m near Noordbergum in the northern Netherlands (Fig. 1) as part of a joint research project of the Geological Survey of the Netherlands and the University of Edinburgh, in which the geotechnical properties of glacially overconsolidated clays and sub-glacial sediment deformation was studied. A borehole was drilled at the same location to a depth of 115 m. A selection of undisturbed samples from this borehole was visually inspected and subjected to geotechnical laboratory tests. On two intervals: from 28 to 38 m depth in an upper clay layer and from 48 to 71 m in a lower clay sequence negative excess pore pressures were observed. A study was

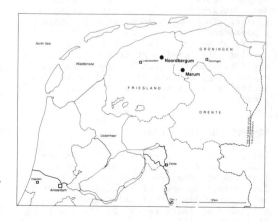

Fig.1. Location map.

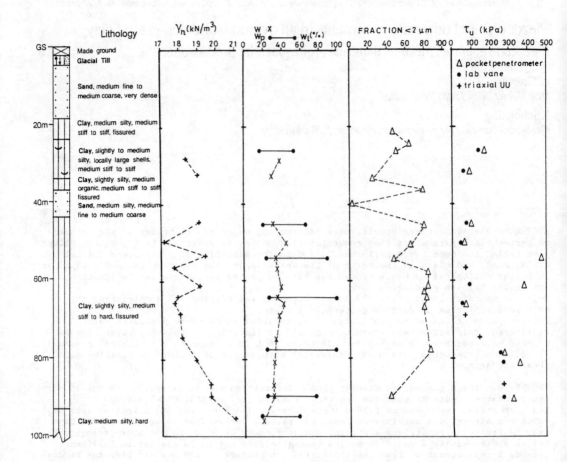

Fig.2. Borehole lithology and geotechnical data.

performed in these intevals to compare structural and geotechnical properties with the piezocone results.

Strucural features such as fissures seem to play a significant role in pore pressure distribution during piezocone testing. A similar study which combined a borehole and a piezocone test, has been carried out at before at Marum. This yielded similar results. The results of this study are applicable in the field of geotechnical engineering, but are also of interest for geological and sub-glacial deformation studies.

BOREHOLE MEASUREMENTS

Undisturbed and orientated tube samples with a diameter of 70 mm were collected from the Noordbergum borehole at 3 m intervals. A number of samples were cut in half lengthwise for visual inspection and geotechnical index tests. On the remaining samples triaxial strength and consolidation tests were performed.

Geological section

The geological section that was encountered in the Noordbergum borehole consists of Pleistocene sediments (Fig. 2): a glacial till, an upper sand layer, an upper clay layer, a coarse sand layer and finally a thick clay deposit at the bottom. The sequence is overlain by 1.7 m earth that forms the dike on which the borehole and the piezocone test are located.

The top 10 m of the upper clay layer is partly of fluviatile and partly of estuarine origin (Urk Formation, Late Pleistocene). Its composition ranges from a clay with high clay content at the top becoming more sandy toward the bottom. In this part of the section horizons containing large shells are encountered.

At aproximately 34 m depth the sandy clay layer changes abrupty to a clay deposit of lacustro-glacial origin, which has been deposited at the end of the Elsterian glaciation (Peelo Formation, Late Pleistocene). This clay, best known by its local name "Pot Clay", has equivalents the

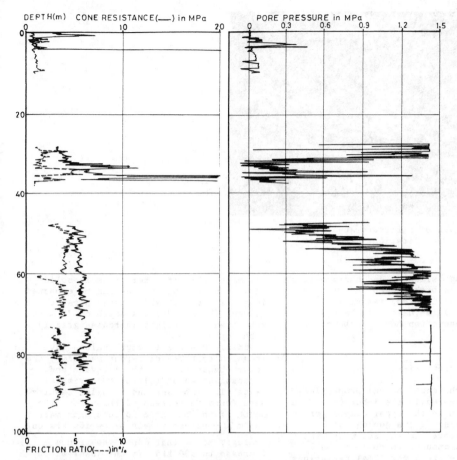

Fig.5. Piezocone test results. Left diagram: cone resistance and
friction ratio. Right diagram: pore pressure.

North Sea area (Swarte Bank Clay) and in
northern Germany (Lauenburger Ton). The top
5 m of this clay is rich in organic matter,
it is underlain by a 7 m-thick sand layer.
The lower thick Pot Clay layer is fairly
homogeneous throughout its 67 m, thickness
but silty and fine sandy clay layers occur
locally. Below a depth of 95 m the sand
content of the clay increases.

In a geological section near Marum, where
the top of the the Pot Clay occurs near the
surface, we encountered shearing, folding,
hydrodynamic consolidation and fissuring of
the clay (Schokking, 1990a,b). Similar
features were observed in the sequence at
Noordbergum. These phenomena result from
loading by Saalian ice sheets, causing
consdidation and deformation of the
underlying soils.

Structural features

Both in the upper clay layer and in the

lower Pot Clay, fissures were found. In the
cut samples, these fissures are visible as
liniations. In general the fissures are
continuous over the width of the sample.
The orientation of these fissures shows a
definite pattern. In the upper clay layer
the fissure planes dip in a southerly
direction. In the lower Pot Clay fissures
are northerly oriented. A conjugate set of
fissures was also found. These fissures are
not continuous and abut at the continuous
ones. The spacing of the fissures ranges
from 5 to 30 mm. The deeper fissures are
further apart and continuous and pronounced. In the triaxial tests on the samples
in which fissures occur the clay failed
along the continuous fissures. These
fissures have been formed as a result of
the thrusting force of the land ice during
the Saalian glaciation. The SEM image of a
fissure in a sample from 77.7 m depth shows
the displacement that has occurred in the
clay structure, which curves on both sides
of the fissure plane (Fig.3). The fissures

305

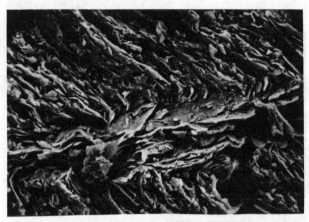

Fig.3. SEM image of a fissure at a depth of
77.7m. Magnification: 3200 times.

have a shining, slickensided surface in a
sample which is broken open. Apart from
the orientated fissures, randomly
orientated fissures have also been
observed. These often have a sub-vertical
direction.

Geotechnical properties

Fig.2 shows the geotechnical properties,
that were measured in the laboratory. The
fraction <2 μm in the upper clay layer
varies according to the geological
description above. In the Pot Clay, the
clay fraction amounts to 80 % on the
average. There are a few local deviations,
that reflect siltly layers. The plasticity
values, when plotted in a Cassagrande
chart, indicate that the Pot Clay is an
inorganic clay of high plasticity. The
plasticity index vs. % <2 μm ratio
indicates that the activity ranges from
medium to high (from 1.0 to 1.25). This
property, which is related to the ability
of the clay particles to attract water, is
probably a result of the relatively high
montmorillonite content, which has been
established for Pot Clay samples from other
locations (Breeuwsma,1984).

The saturated unit weight of the Pot Clay
varies considerably, between 17.2 and 19.4
kN/m . The natural water content is
relatively stable over this interval and
increases from 30 to 40 % with depth. These
features are attributed to the occurence of
the fissures described above. After the
sample tube is opened the fissures, which
have no tensile strength, allow the clay to
expand, thereby releasing the stress. Since
the mass of the clay remains the same, this
increase in volume results in a lower unit
weight in the zones where fissures occur.

The natural water content, which equals

the mass ratio of water and soil solids,
shows a stable picture as can be expected
in a clay body in which water pressures are
in equilibrium. Below a depth of 75 m the
saturated unit weight increases greatly
with depth.

Undrained shear strength values have been
measured by three different methods : the
pocket-penetrometer, the lab-vane and
unconsolidated undrained (UU) triaxial
tests. The lab-vane and triaxial results
were found to correspond throughout the
depth of the borehole up to approximately
75, m τ_u — values remain below 100 kPa on
avarage. Below that depth the values
increase to 300 kPa. In the interval in
which the greatest number of fissures was
observed, down to approximately 65 m, the
pocket-penetrometer values are considerably
higher than the lab-vane and triaxial-test
values (Fig.2). This difference might be
expained by the assumption that fissures do
not influence pocket-penetrometer
measurements. In the lab-vane and triaxial
tests the fissures cause shear planes at
low strains. The same type of phenomenon
was observed in the Marum borehole
(Schokking, 1990a), where differences were
found between the values from the fall-cone
tests and from lab-vane and triaxial tests.

Two consolidation tests, performed on
samples from depths of 53.9 and 68.7 m,
yielded an average overconsolidation ratio
of 2.5. In two of the UU triaxial tests,
performed on samples from 56.5 and 68.5 m
depth, a pore pressure decrease was
observed during deviatoric loading.
However, a sample from 74.5 m showed an
increase in the pore pressure. This last
sample exhibited the highest shear
strength, but this was still far below the
value corresponding to the cone resistance
(about 300 kPa). This suggests that all
the samples failed along fissure planes,

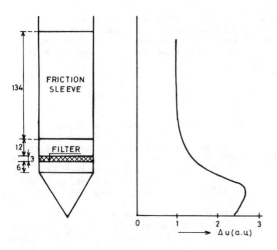

Fig.4. Delft Geotechnics piezocone with idealized pore pressure profile.

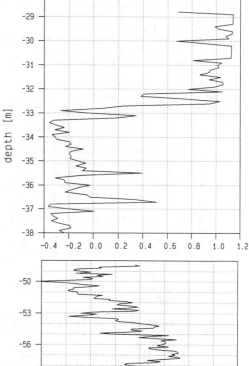

Fig.6. Excess pore pressure profiles.
a): 28 to 38 m depth.
b): 48 to 71 m depth.

but that the upper part of the Pot clay exhibits a more rapid dissipation pore pressure and more dilation. This is probably related to the degree of fissuring (see below).

PIEZOCONE TEST RESULTS

At a few metres from the borehole, a 95-m deep cone test was performed. A Delft Geotechnics standard 10 cm2, 600 apex cone was used with the filter element 6 mm above the edge of the cone and a filter height of 3 mm (Fig.4).

In Figure 5 the cone resistance , pore pressure and friction ratio (ratio of friction shear stress on sleeve and cone resistance) have been plotted versus depth. The missing parts of the curves correspond to layers that were too stiff to penetrate. To perforate these layers, drilling and cone testing were alternated (Schokking et al., 1990).

Generally speaking, a high cone resistance q_c corresponds to a sand layer, in this case these sand layers are very dense. High pore pressures (symbol:u) indicate impermeable clay layers, a high friction ratio FR corresponds to clay or peat. It should be noted that as a result of the dimensions of the filter (3 mm high) and the friction sleeve (134 mm high), the pore pressure is much more sensitive to rapid variations in lithology.

In this section, two depth ranges will be discussed, viz. from 28 m to 38 m (range A) and from 48 m to 71 m (range B), since in these ranges interesting low pore pressures were registered. It should be noted that below 71 m the pore pressure exceeded the maximum limit of the transducer, in this

case 1.44 MPa.

For range A, the excess pore pressure Δu has been plotted in Figure 6a. The excess pore pressure is equal to the difference between the measured pore pressure and the hydrostatic pressure. At a depth of 32.5 m a sharp decrease occurs to negative values. A negative excess pore pressure may indicate a dense silty or sandy layer or a highly overconsolidated clay layer (Houlsby, 1988; Zhang, 1990).

From cone resistance, friction ratio and two borehole samples the following is inferred. From 32.5 to 34.0 m the clay is sandy and shells of cm size are present. From 34.0 to about 36.0 m the clay fraction is much higher (around 78%). In a sample from this layer many fissures were observed. From 36 to 38 m the sand fraction increases again. Normally, the pore pressure would rise in the layer with high clay content. Therefore, we tentatively conclude that a major reason of the negative excess pore pressure is the extensive fissuring in this lacrustine clay. This is discussed below.

Range B (48-71m) exhibits low excess pore pressures in the upper part around 50 m depth (Fig.6b). In this case there is no indication of a significant variation in the grain size (Fig.2). Again we conclude that the pore pressures measured reflect fissuring. The effect of fissures might be twofold:
- the permeability around the cone could increase;
- the fissures could act as shear planes or split planes, which can result in the creation of voids, which in turn may have a dilating effect.

In both cases the pore pressure would be lower than normal.

The number of fissures decreases with depth, while, at the same time, the average excess pore pressure increases. A similar effect was observed in the piezocone test near Marum (Schokking, 1990a), where negative pore pressures were observed in fissured Pot Clay over a range of more than 15 m.

The cone resistance and the friction ratio were found to be more or less independent of the number of fissures. The value of q_c is about 6 Mpa, which

corresponds roughly to an undrained shear strength of 300 kPa. As mentioned above, the laboratory tests, except for the pocket-penenetrometer, show lower shear strength values because of fissuring and because of the release of the internal stress.

We note that the apparent increase in the friction ratio after predrilling is pro-bably an artefact, caused by the lubri-cating effect of the drilling fluid.

CONCLUSIONS

-Piezocone tests provide valuable geological information on the litho-logical section and geotechnical parameters. The pore-pressure signal is especially sensitive to rapid variations in lithology.
-The correlation between low pore pressures and degree of fissuring led to the con-clusion that fissures in clay may cause a significant reduction in excess pore pressure.

-The shear strength calculated from the cone resistance in the fissured Pot Clay is higher than the shear strength resulting from lab-vane and triaxial tests.

REFERENCES

Houlsby G.T., (1988). Penetration Testing in the UK. Proceedings Geotechnology Conference, Birmingham, 6-9 July 1988.

Zhang C.,Greeuw G.,Jekel J.,Rosenbrand W., (1990). A new classification chart for soft soils using the piezocone test. Engineering Geology, Vol. 29, no.2, July 1990.

Breeuwsma A., Balkema W., Zwijnen R., (1984). Kleimineralogisch onderzoek van enkele kwartiare en tertiare afzettingen. Stichting voor Bodemkartering Rapport nr.198 (in Dutch).

Schokking F., (1990a). A sub-glacial sediment deformation model from geotechnical and structural properties of an overconsolidated lacustro-glacial clay. Geol. & Mijnbouw, Vol.69, No.3.

Schokking F., (1990b). On estimating the thickness of the Saalian ice sheet from a vertical profile of preconsolidation loads of a lacrusto-glacial clay. Geol. & Mijnbouw, Vol.69,No.3.

Schokking F., Van de Graaf H.C., Hoogendoorn R., (1990). A new method for deep static cone penetration testing in overconsolidated clays. This conference, Theme 1.

Automated stratigraphic classification of CPT data

Classification automatique et stratigraphique de données pénétrométriques

G. P. Huijzer

Free University, Amsterdam, Netherlands

ABSTRACT: This paper describes a method for automated processing of cone penetration test (CPT) data. It can be used in subsurface modelling as it classifies data according to their CPT characteristics and stratigraphic position. The first stage of the method segments single CPT data sequences. Subsequently, corresponding segments in distant sequences will be correlated. Both procedures yield acceptable results but a final algorithm has yet to be devised to complete the method.

RESUME: L'article décrit une méthode de traitement automatique de données de sondage au pénétrometre. On peut l'employer à la production des modèles géotechniques parce que la méthode classifie les données selon leurs caractéristiques pénétrometriques et selon leurs positions stratigraphiques. La première phase se compose de la division de toutes les sondages. Ensuite, la corrélation entre ces segments est determinée. Ces opérations produisent des résultats acceptables mais un algorithme final, qui n'est pas developpé encore, faut compléter la méthode.

1. INTRODUCTION

Lithologic subsurface models, for example geotechnical cross sections, are normally generated manually. During the production of the model a large number of cone penetration test (CPT) data will be converted into soil types. Conversion is frequently based on few, arbitrary comparisons of CPT and core data. Moreover, intuition and knowledge of regional geology may play a substantial role. Consequently, the production of a lithologic subsurface model is a time-consuming and subjevctive process and the final model will not be reproducible. Automatic CPT dataprocessing and application of statistical methods may reduce these drawbacks of manual processing and modelling.

In this paper, a method for automated dataprocessing, which may be useful in subsurface modelling, will be introduced. Section 2 discusses some requirements that should be satisfied by a such a method. In Section 3, a prototype method will be described and some results will be presented. Section 4 evaluates the method with respect to the requirements stated in section 2 whereas section 5 presents the conclusions.

2. REQUIREMENTS FOR AUTOMATED DATA-PROCESSING

Subsurface modelling can be seen as a way of classification: a subsurface model comprises several units each representing a part of the original data. These units should be internally homogeneous or, to put it otherwise, the original data in a specific unit should show strong similarity. Currently, some methods for automated CPT dataprocessing classify data according to their CPT characteristics. Due to their automated nature, these will reduce the time needed for model production. Nevertheless, subjectivity may still remain. Tsuchiya et al. (1988) for example, manually classify a subset of their CPT data into four groups. Then, they allocate the remainder of the data to these groups in an automated way.

Their method has three drawbacks. First, the initial classification was done manually and hence will be subjective. Secondly, the number of groups is determined at the beginning and is fixed during the analysis. The introduction of more data will never result in a new class though this might be necessary. Thirdly, class boundaries taken from the soil classification system and converted into

CPT characteristics serve as class boundaries for the CPT data. This operation assumes that a classification system derived from the soil classification system will also suit the CPT data. Ghinelli and Vannucchi (1988) apply this idea in an even more strict way. After having concluded that classification is necessary, they impose a classification system without any regard for the data.

Consequently, an automated classification method should not refer to a pre-existent classification system but should provide a classification based on both the specific category of data and all available data.

A method characterized by these properties will place observations that are similar in their CPT characteristics, in a specific unit. Units in a subsurface model however, will also be internally homogeneous with respect to the location of the original data. Or, to be more precise, the original observations posess an equivalent stratigraphic position. Thus, a method which will have any utility in subsurface modelling should classify not only according to CPT characteristics but also according to stratigraphic level. To be classified as a single unit, observations in distant data sequences should correlate in a stratigraphic sense i.e. correspond in character and stratigraphic position (Hedberg, 1976).

Summarizing this section, less subjective and more rapid subsurface modelling should be done by an automated method that classifies all available data according to their CPT characteristics and stratigraphic position without making reference to a pre-existent classification.Throughout the paper, this type of classification is called penetrostratigraphic classification, by analogy with e.g. magneto-, chemo- and seismostratigraphy.

3. NUMERICAL STRATIGRAPHIC CLASSIFICATION

3.1 Introduction

Automatic classification and subsurface modelling of data sequences is a familiar problem in earth sciences. It has been dealt with frequently but it has not been solved yet. This research focusses on two main aspects:

1 segmentation of a single data sequence and
2 matching and correlation of two data sequences.

Segmentation splits a data sequence into several internally homogeneous layers or

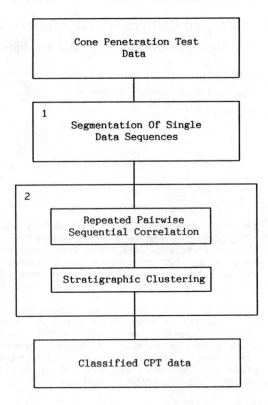

Figure 1. Outline of the proposed method for automated penetrostratigraphic classification.

segments. Both matching and correlation, according to the definition of Shaw (1982), determine the correspondence between units in distant data sequences. Correlated units however, may be interpreted as stratigraphic units whereas matched units may be closely similar without being a stratigraphic unit. Schwarzacher (1985) and Hoyle (1986) review research on these subjects; Tipper (1988) gives an comprehensive bibliography.

The proposed classification method first segments every single data sequence. Subsequently, a two step procedure will correlate the segments of distant data sequence to form penetrostratigraphic units. The main elements of the method are shown in figure 1. The following sections will give a general description of the method.

3.2 Segmentation

The segmentation procedure is largely

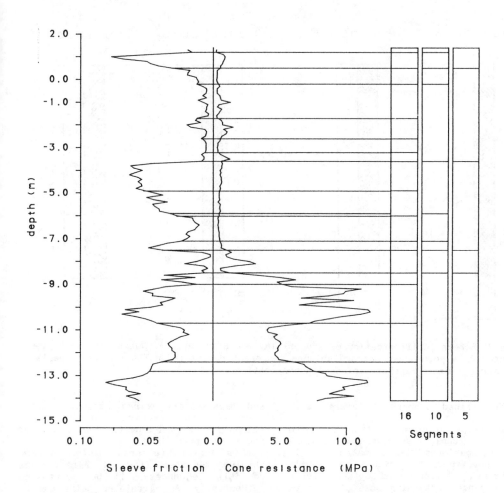

Figure 2.   A data sequence in which segmentation is based on two variables. Optimal solutions for segmentation into 5, 10 and 16 layers are shown.

based on the work of Hawkins and Merriam (1973). As in all automatic classification methods, two aspects play an important role: the similarity measure, which indicates the similarity between observations, and the algorithm which optimizes the similarity measure.

The proposed method applies a minimum variance criterion as similarity measure. Although details go beyond the scope of the paper, it should be mentioned that the criterion is calculated for every variable that serves as input. Normally, these will be cone resistance and sleeve friction or only one of them. The use of either friction ratio or piëzocone data should be avoided because the former is a dependant variable and the latter usually possesses a trend. The specific minimum variance criterion may only be applied if several

statistical assumptions are satisfied. In most cases, some of these assumptions will not be correct and the final segmentation will be useless. To improve the result several data transformations, which will not be described here, has been applied.

The algorithm described by Hawkins and Merriam (1973) is very efficient and identifies the global optimum i.e. the best solution will be taken out of all posible solutions. This efficiency is due to the fact that the algorithm is sequentially constrained. This means that groups are made to consist of subsequent observations only. Thus, groups cannot be split by other groups and hence groups can be called layers or segments. The algorithm offers an array of solutions ranging from the optimal segmentation into two layers to segmentation into a pre-

Sequence A   Sequence B   Sequence C   Sequence D

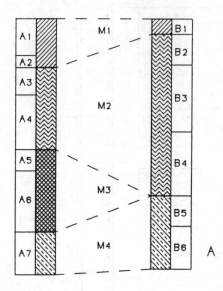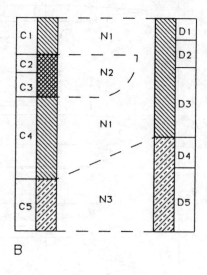

Figure 3. Schematic illustration of several layer configurations. Outer columns represent segmented data sequences. Hatched columns represent correlated units and are labelled in the central column. Further explanation see text.

viously defined number of layers (e.g. 16).

A prototype program, written in Turbo pascal and running on a MS-DOS personal computer, performs the computations. It takes approximately a minute to provide the optimal solution up to 16 layers for a data sequence containing some 150 observations. Figure 2 shows that all segments identified by the method are, at least visually, relevant. Some segment boundaries, appearing in the 10 segment solution, do not appear in the 16 segment solution because every solution is optimal. Subsequent optimal solutions are not hierarchically nested although disappearing boundaries may reappear at a higher level of solution.

3.3. Correlation

Introduction

The method of Hawkins (1984) for correlation of several data sequences is closely related to the segmentation method. Unfortunately, for this method the computational load increases seriously with the number and length of data sequences. In order to reduce the computational load, Hawkins applies a heuristic approach which is mathematically complex. For this reason

and because his method has limitations from a stratigraphic point of view, attempts have been made to develop another heuristic approach.

This heuristic correlation method involves two steps (fig. 1). First, pairs of data sequences will be correlated sequentially. As correlated pairs can be seen as new data sequences, the same method can be applied to subsequently correlate these new data sequences. Repetition of the procedure will finally result in a single 'supersequence'. During repeated pairwise correlation some data have to be discarded. These data should be incorporated into the supersequence in the second step. In the next subsections the pairwise sequential correlation method will be discussed as well as some aspects of the multiple sequence correlation procedure which is still under investigation.

Pairwise sequential correlation

The pairwise sequential correlation procedure strongly resembles the segmentation method: it segments two data sequences at the same time. As these segments will generally consist of observations in both sequences, the observations can be said to correlate. The algorithm is se-

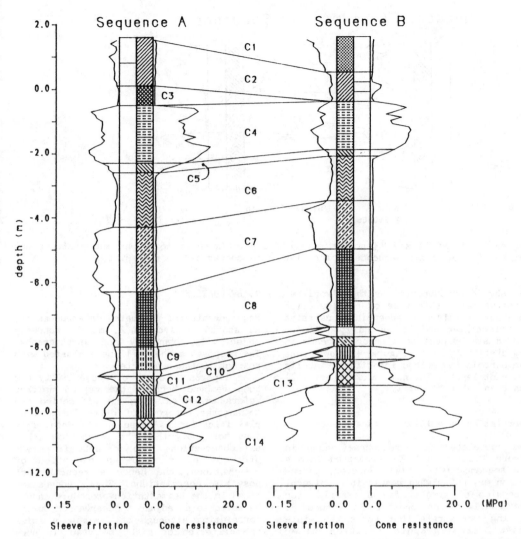

Figure 4. Two pairwise sequential correlated data sequences showing the optimal solution for 14 layers. Column layout is comparable to fig. 3.

quentially constrained and optimizes the same minimum variance criterion. Again, for a range of solutions the best set of correlations will be found. Because the length of the data sequences strongly influences the computational load and memory requirements, segmented data sequences serve as
input rather than the original sequences of observations. According to the sequential constraint, several segments in a single data sequence can only be taken together when they are contiguous.

A practical implication is that the algorithm allows several segments in the first data sequence to correlate with several segments in the second data sequence (A3,A4 v. B2,B3,B4; figure 3.a). If one of these sets contains zero segments, the specific layer will disappear laterally thus representing a discontinuity (M3; figure 3.a). This feature is important because discontinuities occur in many actual geologic sequences.

The algorithm does not cover the type of configuration shown in figure 3.b. In contrast to figure 3.a, a stratum (N2) does not disappear between two other strata but within a single stratum (N1). From a geologic point of view this type of configuration is important, it represents intercalations (e.g. tongues and lenses),

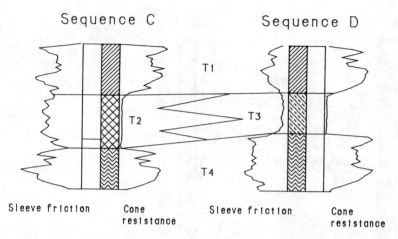

Sequence C   Sequence D

Sleeve friction   Cone resistance   Sleeve friction   Cone resistance

Figure 5. Units T2 and T3 posess equivalent stratigraphic positions and obstruct conversion into a single sequence. Column layout is comparable to fig. 3.

and should be supported by the algorithm. Unfortunately, this type of configuration does not satisfy the sequential constraint stated earlier and it can only be implemented at the cost of a major increase in computational load. Figure 4 shows that the correlation method identifies visually relevant layers. The number of layers (14) has been set arbitrarily.

Correlation of multiple data sequences

Two correlated data sequences shown in figure 4 can readily be converted into a new sequence (C1 - C14). However, correlation may not always result in a univocal succesion of strata. As an example, for the configuration present in figure 5, showing two strata that posess an equivalent stratigraphic position after correlation, no single sequence can be established. This type of sequence as well as sequences containing intercalations, obstruct the repeated pairwise correlation as it was proposed in the introduction of this section.

A solution to this problem could be to discard one of the equivalent strata and the intercalated layers for the time being. Thus, two 'true' sequences will result and repeated pairwise matching will be possible. Once the final supersequence has been obtained, the temporarily discarded strata have to be related to this supersequence. This process, provisionally called stratigraphic clustering (fig. 1), will be very complicated and has not been devised yet. Future research will hopefully result in the completion of this final step.

4. EVALUATION

Requirements for automated dataprocessing, as stated in section 2, mainly concerns objectivity, rapidity and stratigraphy. The proposed method will be evaluated with respect to these aspects.

An important source of subjectivity has been removed because the method neither refers to an existing classification nor to another category of data. Futhermore, classification is based on all data. This does not imply that the method has taken all observations into account. It merely proceeds by a stepwise reduction of observations, and hence a reduction of possible correlations. Thus, boundaries fixed in the segmentation procedure cannot be moved to an adjacent observation during correlation although this might optimize the criterion or might be visually more relevant. Nevertheless, it is felt that most important comparisons have been made. The method will at least yield reproducible results.

Another aspect of objectivity concerns the criterion which strongly influences the final shape of the units. In fact, the choice of the minimum variance criterion is mainly due to practical and computational considerations. The minimum variance criterion posess two properties which should be kept in mind. First, the within-group meanvalue for every variable is an important factor. Two different units with equal meanvalues but with different standard deviations, which may be due to larger fluctuations in one of the units, will be correlated. Secondly, units that posess a trend may be cut into several layers (e.g. C5; fig. 4) during segmentation. Some may feel that these effects are not desirable, but they are inherent to the criterion.

The method is computationally demanding and it may take quite some time to reach the final classification. However, if further research shows that all computations can be performed using a personal computer, the costs will be very small compared to manual classification and subsurface modelling. Incorporation of heuristic search methods may reduce computer time in future.

In general, the method takes stratigraphy into account and the final units will be penetrostratigraphic units. It covers most of the geologically relevant configurations. Nevertheless, its utility in subsurface modelling has to be proved yet as the final correlation algorithm must be extremely flexible to cover actual stratigraphic configurations which can be very complex. Faults will not be considered because they are merely an interpretation of a specific configuration of layers.

5. CONCLUSION

A method for automated penetrostratigraphic classification, which satisfies most of the requirements stated, is proposed. The segmentation and pairwise correlation procedures provide acceptable results. Development of a final step ihould complete the classification method which enables the engineering geologist to process CPT data with reduced subjectivity and to use the classified data in subsurface modelling.

6. REFERENCES

Ghinelli, A. & G. Vannucchi (1988).Statistical analysis of cone penetration testing: An application. In De Ruiter, J. (ed.), Penetration Testing 1988. Proc. ISOPT-1, 757-769. Rotterdam, Balkema.

Hawkins, D.M. (1984). A Method for stratigraphic correlation of several boreholes. Mathematical Geology 16, 393-406.

Hawkins, D.M. & D.F. Merriam (1973). Optimal zonation of digitized sequential data. Mathematical Geology 5, 389-395. Hedberg, H.D. (ed., 1976). International stratigraphic guide, 200 pp. New York, John Wiley and Sons.

Hoyle, I.B. (1986). Computer techniques for the zoning and correlation of well logs. Geophysical Prospecting 34, 648-664.

Schwarzacher, W.S. (1985). Lithostratigraphic correlation and sedimentation models. In Gradstein, F.M., F.P. Agterberg, J.C. Brower & W.S. Schwarzacher (eds.), Quantitative stratigraphy, 361-386. Dordrecht, Reidel.

Shaw, B.R. (1982). A short note on the correlation of geologic sequences. In Cubitt, J.M. & R.A. Reyment (eds.), Quantitative stratigraphic correlation, 7-11. Chichester, John Wiley and Sons.

Tipper, J.C. (1988). Techniques for quantitative stratigraphic correlation: a review and annotated bibliography. Geological Magazine 125, 475-494.

Tsuchiya, H., T. Muromachi, Y. Sakai & K. Iwasaki (1988). A soil classification method using all three components of CPTU data. In De Ruiter, J. (ed.), Penetration Testing 1988. Proc. ISOPT-1, 757-769. Rotterdam, Balkema.

Geostatistical characterisation of the SPT

Caractérisation géostatistique du SPT

R.A.N. Mackean
Engineering Geology Ltd, Godalming, UK

M.S. Rosenbaum
Imperial College of Science, Technology and Medicine, London, UK

ABSTRACT: The Standard Penetration Test (SPT) is a traditional method in the UK for determining the relative density of loose to dense granular soils and gives a measure of pile bearing capacity. The lack of sensitivity of this test to small-scale variations in relative density across a site can lead to inaccuracy and a broad spread of test results. Geostatistical analysis has been successfully applied to the SPT results obtained from Tertiary sediments at Aldershot which are typical of ground conditions found in many parts of southern England. This analysis has permitted a quantitative measurement of variability and an assessment of the correlation likely to be present between neighbouring sites.

RESUME: Le Standard Penetration Test (SPT) est une methode traditionelle au Royaume Uni pour déterminer le densité relative des sols granulaires meubles à denses. Cette methode donne une mesure de la capacité portante de pieu. Le manque de sensitivité de ce test aux faibles variations de densité relative à travers un site conduit à des imprécisions et à une grande dispersion des résultats. L'analyse géostatistique a été appliquié avec succès aux resultats SPT dans les sédiments Tertiaire d'Aldershot qui sont typiques des conditions des sols trouvés en de nombreux régions d'Angleterre du sud. Cette analyse a permis une mesure quantitative de la variabilité et une appréciation de la corrélation entre les régions avoisinantes a été faire.

1 INTRODUCTION

The Standard Penetration Test (SPT) is a traditional method in the UK for determining pile bearing capacity on loose to dense granular soils (Terzaghi & Peck 1948). The test provides a measure of the relative density of the soil by driving a cone-tipped probe into the ground at the base of a borehole. The test is not very sensitive to small-scale variations in relative density and this can lead to inaccuracy when the bearing capacity is estimated (Skempton 1986). There has been a considerable effort aimed at reducing the broad spread of test results by encouraging standardised procedures for rationalising the soil strengths derived from such tests (Nixon 1982).

There is now a large database of SPT results upon which empirical methods for determining foundation settlement can be based, following the practical guidelines initially set out by Terzaghi and Peck (1948).

The basic source of such geotechnical parameters is the borehole log, yet the retrieval of descriptive information can be rather difficult unless such logs are produced to a standard format (Raper & Wainwright 1987). Retrieval is assisted considerably if the data has been systematically recorded and microcomputers can be particularly effective for such a task (Norbury et al 1984).

To evaluate the microcomputer-based approach, ground investigation data for the Aldershot area has been chosen for examination. This is illustrated here using the SPT results obtained from the Tertiary sediments which have permitted a quantitative determination of the variability likely to occur within a selected area together with an assessment of the correlation likely to be present on a local scale between neighbouring sites.

2 THE SITE

The geology of the Aldershot area comprises Tertiary sediments, principally of Eocene age (locally subdivided into the Bagshot, Bracklesham and Barton Beds), which are locally overlain by fluvial Pleistocene Terrace Gravels (Dines et al 1929). The Bagshot Beds consist largely of current-bedded sands comprised of well rounded quartz grains; thin seams of reworked kaolin, occasional concretions, and lignite are also reported as being present. The Bracklesham Beds occur across much of the area and are rather more heterogeneous since they also contain clays, glauconitic sands and pebble beds. These occur as lenticular horizons and make lithological identification from boreholes difficult to establish. The Barton Beds consist of fine grained sands which have more frequent clayey horizons towards the top. The overall depositional environment is of marine near-shore conditions, perhaps at times deltaic whilst at others estuarine; towards the end came a transgression over the adjacent landmass.

Aldershot lies just north of the Hogs Back monocline at the southern edge of the London Basin. Here a fairly well marked anticline trends east-west through Burley Hill bringing the Bagshot Beds to the surface in the south and also in a small inlier by the reservoir at Burley Bottom. A gentle anticlinal fold, at right angles to the general trend of folding, runs along the Blackwater Valley (Clarke et al 1979).

The stratigraphic aspect of these strata significant to their engineering geological behaviour is the occurrence of small scale channels in which fine grained over-bank deposits (notably compressible organic-rich muds) inter-digitate with more clayey marine deposits.

3 SPT RESULTS

The results for the Standard Penetration Test have been readily managed using a relational database (dBASE IV) which can access multiple files simultaneously. The borehole records were obtained from both the British Geological Survey at Keyworth and from the Property Services Agency in Croydon and their distribution is shown in Figure 1.

The clayey deposits within the Tertiary sediments are unlikely to have yielded reliable SPT results. It was ini-tially envisaged that this problem could be averted with the use of either U100 blows (suitable corrected using a conversion factor) or a semi-quantitative assessment based upon the sample description logs. However, choice of an acceptable conversion factor proved to be an insur-mountable problem. Neither could the sample description logs give reliable strengths since their relative inaccuracy distorted the outcome of the directly determined SPT results. This provides an illustration of the need to appre-ciate the geotechnical variability of the material in order to recognise the limitations of any model based upon it.

The strength profiles were extracted from the database using the query facility for each of the geological hori-zons. This was undertaken at one metre intervals from ground level down to 10 metres depth in a similar manner to that employed for the probabilistic studies concerning offshore investigations carried out by Wu et al. (1986). The Barton Beds showed the most uniform increase in strength with depth (Figure 2) as had been expected since these sediments are relatively uniform. Some exceptional-ly high values were nevertheless present and a few of these could possibly be ascribed to errors of recording.

The majority of the SPT values in the Bracklesham Beds showed a steadily increasing strength with depth, but there was some scatter of points away from the main trend, notably with values greater than 80; these were associated with gravel horizons. The maximum reasona-ble value for the SPT is governed by the capacity of the drop hammer available to drive the tool into the ground. Straightforward shear failure of the soil beneath the SPT cone is assumed to be the mode of penetration rather than other failure mechanisms or distortion of the apparatus (Kovacs et al 1981). Little significance could therefore be placed 'N' values greater than 80 since these would repre-sent encounters with obstructions or significant energy losses with apparatus distortion. A few points were also found to lie below the main trend and appear to occur primarily within clay horizons. Such low values are likely to have been caused by softening of the material associat-ed with swelling accompanied by the ingress of water.

In order to investigate the spatial distribution of relative

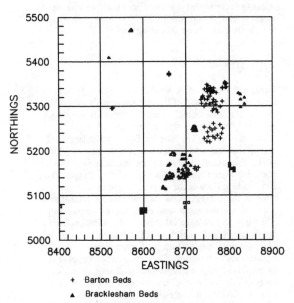

Figure 1. Locations of boreholes in the Aldershot study area and the main Tertiary stratigraphic units encoun-tered.

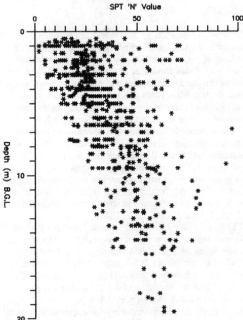

Figure 2. Depth profile of SPT "N" values within the Barton Beds.

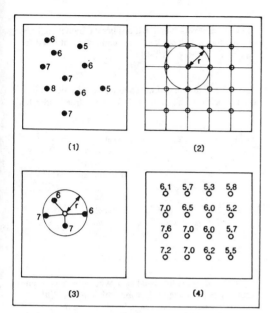

(1)   (2)

(3)   (4)

Figure 3. The four stages for calculating the values at the grid nodes from irregularly spaced control points, based on Davis (1986).

density, data description was conducted using the public domain Geo–EAS software on a PC (Englund & Sparks 1988, Rosenbaum & Stevens 1990). The data files were obtained from dBASE as free format ASCII files which could be read directly by Geo–EAS. Generally visual presentation of irregularly spaced data points is achieved using contours. The interpolation and extrapolation of such data requires calculation of a regular grid (Figure 3). The node values of the grid define the polynomial model which describes the surface, and this can be contoured at chosen intervals. Although hand–drawn contours are still in vogue in some quarters, mathematical modelling produces rather more consistent results despite their relative insensitivity to geologically "correct" interpretations (Dahlberg 1975).

The real problem with computer–based contouring is how to best assign values to the nodes (Green & Sibson 1978). A variety of algorithms are available but the suitability of each varies according to the given circumstances so care must be taken over the choice (Dagbert 1981). The main difference between algorithms arises from the method of starting the interpolation process. Each estimated value will depend in some way on its nearest neighbours; the assumption is that this dependency diminishes with increasing distance. Depending on the software used, some further control over the gridding algorithm may be offered, commonly:

1. The search distance, within which the data points will be used to calculate values at each node, can be limited.

2. Anisotropic dependency of data upon distance may be accounted for by giving two values for the radius so that an ellipse may be defined.

3. The maximum and minimum number of data points may be chosen to consider a reasonable number of points for the calculation of each node value.

4. A smoothing algorithm can be applied to improve the presentation (but not the accuracy) of the contours.

Whichever interpolation method is chosen, initial exploratory data analysis should first be carried out and this can be achieved using colour coded dot maps. These can quickly pick out trends within the data and identify possible rogue points. The distribution of SPT results at Aldershot indicate a steady increase in value with increasing depth. The broad distribution of values (Figure 1) shows that the data can be considered as regionalised random variables so that stochastic modelling using geostatistics may be applied to characterise their spatial distribution.

4 GEOSTATISTICAL ANALYSIS

For interpolation to be successful it must be assumed that each estimated value will depend in some way upon its neighbours and that this dependency is likely to diminish with increasing distance. This assumption is implicit for contoured plots and can be extended by trend surface methods to develop a three dimensional polynomial defined by the set of control points, yet such a deterministic modelling approach is really only applicable to data sets for which the variables may be described as variates (Davis 1986).

When the data is suspected to consist of random variables, the degree of homogeneity and then the spatial dependency should be determined by the construction of semivariograms (Cressie 1988). A stochastic model can then be selected and cross–validated with the control points by point kriging (Rosenbaum 1987). If the residuals are low then the model can be considered as a good estimator. The advantage of this approach is that it gives a measure of the degree of accuracy with which the estimations are made for a chosen degree of statistical confidence.

The characteristics of geostatistical interpolation must be considered in order to appreciate the difficulties associated with developing a reliable modelling technique:

1. The value at any point comprises both a fixed component of the trend and a random variation about that trend.

2. The relationship between the values at any two points depends upon their distance of separation and possibly also on the direction. If the weighting of samples can be controlled not by distance but by similarity then some of the restrictions imposed on such models may be eased.

3. The model assumes continuity of geotechnical units and cannot be applied across boundaries except under special circumstances.

Generally a regionalised random variable will have a lower value of semivariance at small lag distances than at larger ones, and this will increase with increasing distance to a more or less constant value called the sill. Commonly the sill will have a value numerically equal to the overall population variance. The case when the distance between two variables is zero would be expected to give a difference between the two values of zero unless

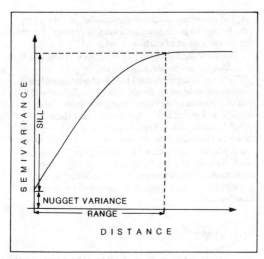

Figure 4. A semivariogram with a spherical model.

In general it was found that a spherical model could be created for each depth below ground level that fitted the data well, as shown in Table 1.

TABLE 1. Summary of the spherical stochastic models used to describe the Tertiary SPT results from Aldershot.

Depth m	Nugget	Sill	Range km	Mean of Actual - Predicted SPT values
1.0	.20	.320	0.5	-0.014
3.0	.10	.280	1.1	0.003
5.0	.05	.170	1.2	-0.008
6.0	.08	.184	1.0	0.011
9.0	.70	.001	1.2	0.031

there is an inherent measurement or sampling error or a natural micro–variability; this is revealed by extrapolation of the semivariogram to intersect the y–axis so giving the nugget variance as shown in Figure 4.

The fitting of a model curve to the semivariogram is not always a straightforward task with small data sets since spatially dependent variability may be difficult to distinguish from random disturbance. For example it was found that for the SPT values at one metres depth that when a large search distance was applied there appeared to be no spatial data dependence. However, once this search distance had been reduced to about 500 m the spatial component became obvious.

The next stage of a geostatistical study uses the semivariogram model to produce estimates based on kriging (Clark 1979). This approach assumes that the data may be treated as a regionalised random variable and to have no regular trend; if a trend is present then this needs first to be removed, for example by fitting a trend surface and then basing the model on the residuals to that surface. Kriging estimates the value at a specific point by applying weighting coefficients to the neighbours such that the estimation variance is minimised.

Once the best model has been identified, block kriging can be carried out to model the distribution of the parameter. This is generally achieved by subdividing the study area into packets and estimating the value of the parameter within each. This produces a grid of estimates for the site together with the associated standard errors of estimation.

The advantages of stochastic modelling based upon kriging include:

1. Smoothing based upon actual data variability (Dubrule 1983, Watson 1984).

2. Declustering of concentrations of observation points since the kriging weight assigned to a sample is lowered according to the degree that its information becomes duplicated by nearby, but highly correlated, samples.

3. Incorporating the effects of anisotropy.

4. Enhanced understanding of spatial variability.

Since the range is the distance at which the maximum semivariance is reached, beyond which there is little spatial dependence. The range is influenced by the sedimentary environment, distribution of sample points, and the method of obtaining the test results.

The sedimentary environment in which the material was deposited provides the fundamental control on the lithological distribution of the material. Nearshore and deltaic facies tend to be more variable than offshore deposits on

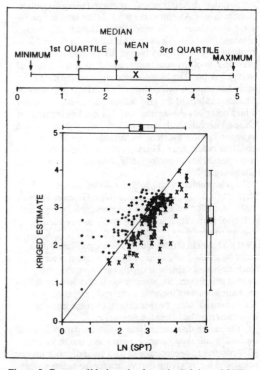

Figure 5. Cross–validation plot for point kriging of SPT "N" values at 1 m depth.

a relatively small scale due to their lenticular channels and over–bank splays. Offshore the control of the environment is exerted by the depth of water and the influence of waves and tends to favour sheet–like deposition, hence more extensive ranges in excess of one kilometre can be readily achieved. This is borne out by the 1.0 to 1.2 km range of the models summarised in Table 1. The shorter range for the shallowest (1 m) depth was necessary owing to the presence of made ground and occasional confused identification within the borehole logs between Tertiary and River Terrace Gravels.

The concentration of testing points towards the centre of the Aldershot area would be expected to affect the range of influence. Beyond a certain distance, dependent upon the borehole spacing, the lags will contain very few points and significant sampling errors can then arise giving calculations of semivariance based upon occasionally spurious results. This effect can be reduced by increasing the minimum number of points acceptable for determining the semivariance at each given lag, or by increasing the search distance for selection of points to be used for each calculation. However, such measures tend to smooth the data and although they are satisfactory for establishing average ("normal") behaviour of the variable, they lead to less sensitivity when considering the extreme values.

The average size of each site used for the Aldershot study was in the order of one square kilometre. From Table 1 it can be seen that the mean of the differences is very close to zero, suggesting a high degree of accuracy using the stochastic modelling. The cross–validation scatter plots (Figure 5) reveal that the degree of over-estimation equals the degree of under–estimation. Further increase in the degree of accuracy cannot be achieved since this scatter is inherent within the natural variability of the SPT data, as also shown by the moderately high nugget variance.

The search areas used for the block kriging were initially set at one kilometre so that the modelling could be limited to each of the specific sites within the study area. These estimations appeared to give reasonable results and so they were then extended to beyond the original site boundaries. The results gave realistic estimations (Figure 7) and proved to be far more stable than could have been obtained using conventional contouring techniques or applying deterministic algorithms. As expected there was an associated increase in the standard error of estimation which was in proportion to the distance from the data points (Figure 8).

5 CONCLUSIONS

The stochastic modelling was based upon a study of the spatial distribution of semivariance for the SPT results. At the shallowest depths it was necessary to use a range of just 500 m to provide an acceptable fit owing to the presence of made ground and the influence of River Terrace Gravels occasionally being misidentified as Tertiary sediments. At greater depths the range could be extended to around one kilometre.

Large discrepancies in SPT value were generally found to be due to the presence of localised cemented horizons,

but occasionally high readings had been caused by the SPT tool jamming in the side of the borehole or by energy absorption in the drill rods.

To prevent unreasonable extrapolations, four limitations were imposed on the range of the models:

1. Limiting the pair comparison distance and using a comparable value when kriging.
2. Limiting the sill value so that it did not exceed the overall variance.
3. Limiting the krige search distance to within the range where the semivariance was still rising.
4. Respecting the geological boundaries beyond which continuity of geotechnical units could not be expected.

The models were tested by cross–validation using point kriging so as to predict the SPT at each of the sample points in turn and comparing the outcomes with the actual values. In general the models were found to be

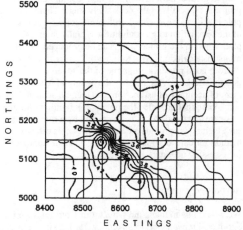

Figure 7. Contoured map for the estimated means for SPT "N" values at a depth of 9 m across the Aldershot study area based on block kriging.

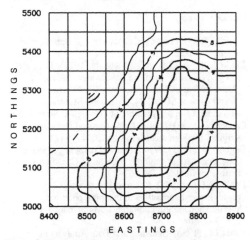

Figure 8. Contoured map for the standard errors of estimation for SPT "N" values at a depth of 9 m across the Aldershot study area based on block kriging.

good estimators and showed that:–

1. There is a general increase in the value of SPT "N" value with depth.

2. There is a significant spatial dependence of the data across the Aldershot area, responsive to the known structural deformation and exhumation history.

3. The errors of estimation are lowest in the areas where there is the highest concentration of data and increase steadily towards the periphery where the data is scanty.

The gridded results can then be presented either as contoured plots or as surface projections in which case the standard errors of estimation may be superimposed upon the estimated values to give a visual representation of their associated reliability.

The setting up of a database with subsequent geostatistical analysis is considered by some to be a lengthy and hence costly business. However such methods have been used to great benefit in medium sized commercial projects, especially at the desk study stage. There is a potential danger associated with some software because of the temptation to go straight for visually impressive plots wherein the gridded data can be produced by deterministic calculations. Whilst the written report will benefit from good presentation, none of the advantages of geostatistical analysis, notably the quantitative assessment of accuracy of estimation, can then be achieved. It is considered that provided the operator is familiar with the software and the data has been entered by a typist then such techniques provide a commercially cost effective method for data analysis and provide a usable approach for assessing new data as it is acquired.

Acknowledgements

We would like to thank the Director of Civil Engineering Services, Property Services Agency, for his kind permission to use the data for the analysis presented in this paper and for permission to publish.

REFERENCES

Clark,I. (1979). Practical Geostatistics. 129pp. Applied Science, London.

Clarke,M.R., A.J.Dixon & M.Kubala (1979). The sand and gravel resources of the Blackwater Valley (Aldershot) area. Description of 1:25,000 sheets SU 85, 86 and parts of SU 84, 94, 95 and 96. Mineral Assessment Report, Institute of Geological Sciences, 39.

Cressie,N. (1988). Spatial Prediction and Ordinary Kriging. Mathematical Geology, 20, 405–421.

Dagbert,M. (1981). Simulation and mapping of space-dependent data in geology. Bulletin of Canadian Petroleum Geology, 29, 267–276.

Dahlberg,E.C. (1975). Relative effectiveness of geologists and computers in mapping potential hydrocarbon exploration targets. Mathematical Geology, 7, 373–394.

Davis,J.C. (1986). Statistics and Data Analysis in Geology. 646pp. John Wiley, 2nd Edition.

Dines,H.G. & F.H.Edmunds (1929). The Geology of the country around Aldershot and Guildford. Explanation of sheet 285. Memoirs of the Geological Survey of England and Wales.

Dubrule,O. (1983). Two Methods with Different Objectives: Splines and Kriging. Mathematical Geology, 15, 245–257.

Englund,E. & A.Sparks (1988). Geo–EAS User's Guide. Environmental Monitoring Systems Laboratory, P.O.Box 93478, Las Vegas, Nevada 89193–3478, USA.

Green,P.J. & R.Sibson (1978). Computing Dirichlet tessellations in the plane. Computer Journal, 21, 168–173.

Kovacs,W.D., L.A.Salomone & F.Y.Yokel (1981). Energy measurements in the Standard Penetration Test. U.S. National Bureau of Standards, Building Science Series, 135.

Nixon,I.K. (1982). Standard Penetration Test, state–of–the–art report, and recommended standard for the SPT by the Subcommittee on Standardisation of Penetration Testing in Europe. Proceedings of the 2nd European Symposium on Penetration Testing, Amsterdam, 1, 3–24.

Norbury,D.R., G.H.Child & T.W.Spink (1984). A critical review of section 8 of British Standard 5930 – soil and rock description. Site investigation practice: assessing British Standard 5930. Geological Society Engineering Group, 20th Regional Meeting, Guildford.

Raper,J.F. & D.E.Wainwright (1987). The use of the geotechnical database 'Geoshare' for site investigation data management. Quarterly Journal of Engineering Geology, London, 20, 221–230.

Rosenbaum,M.S. (1987). The use of stochastic models in the assessment of a geological database. Quarterly Journal of Engineering Geology, London, 20, 31–40.

Rosenbaum,M.S. & R.L.Stevens (1990). The application of stochastic modelling to the evaluation of ground conditions in Late Quaternary sediments in Göteborg, Sweden. Proceedings of the 25th Annual Conference of the Engineering Group of the Geological Society (Forster,A., M.G.Culshaw, J.A.Little, J.C.Cripps, C.Moon & S.Penn (eds.)), Edinburgh 1989, 81–90.

Skempton,A.W. (1986). Standard Penetration Test procedures and the effects in sands of overburden pressure, relative density, particle size, aging and overconsolidation. Géotechnique, 36, 425–447.

Terzaghi,K. & R.B.Peck (1948). Soil Mechanics in Engineering Practice. New York, John Wiley, 1st Edition.

Van de Veen,C. & L.Boersma (1957). The bearing capacity of a pile predetermined by a cone penetration test. Proceedings of the 4th International Society of Soil Mechanics and Foundation Engineering, London, 2, 72–75.

Watson,G.S. (1984). Smoothing and Interpolation by Kriging and with Splines. Mathematical Geology, 16, 601–615.

Watson,G.S. (1971). Trend–Surface Analysis. Mathematical Geology, 3, 215–226.

Wu,T.H., J.C.Potter & O.Kjekstad (1986). Probabilistic analysis of offshore site exploration. Journal of Geotechnical Engineering (ASCE), 112, 981–1000.

6th International IAEG Congress / 6ème Congrès International de AIGI, © 1990 Balkema, Rotterdam. ISBN 90 6191 130 3

Standard penetration tests on eolian sands of Somalia
Essais standard de pénétration sur les sables éoliens de la Somalie

O. Mohamed Ahmed, O. Shire Yusuf & A. Dahir Hassan
Somali National University, Mogadishu, Somalia

R. Mortari
Rome University 'La Sapienza', Italy

ABSTRACT: The most widespread soils outcropping in the Benadir region of Somalia are sands of eolian origin. At least three kinds of sands were differentiated, and were called "red", "yellow" and "white" depending on their colour. On each of these sands two standard penetration tests and density and moisture content determinations were carried out to depths of a few meters. So penetration resistence was related to dry density. Although under static loads these sands have a satisfactory behaviour, under the blows of the SPTs they behave as loose or very loose soils. Because of capillarity, dry densities near surface are often greater than in depth. Both yellow and red sands tend to give constant typical values of penetration resistance at different depths; such behaviour is probably in connection with granulometric characteristics and geological history of the two sands.

RESUME: Dans la région du Bénadir, en Somalie, la plus part des affleurements est constituée de sables éoliens. On a distingué au moins trois types de sable, qu'on a appelé "rouges", "jaunes" et "blancs" selon leur couleurs. Sur chacun de ces sables on a exécuté deux essais standard de pénétration et on a déterminé, jusqu' à 2 ou 3 m de profondeur, la teneur en eau et la densité. La resistence aux SPT a été mise donc en relation avec la densité sèche. Ces sables ont un comportement géotechnique satisfaisant sous des charge statiques, alors que dans les SPT ils se montrent de laches à très laches. A cause de la dessiccation près de la surface topographique, les densités sont souvant plus grandes qu'un peu plus en profondeur. Aussi bien les sables jaunes que les sables rouges ont la particularité de présenter valeurs de resistence qui sont parfaitement constantes pendant quelques mètres de profondeur. Peut etre ce fait est en relation avec les caractéristiques de granulométrie et l'histoire géologique de ces matériaux.

1 INTRODUCTION

Standard penetration tests are a very widely used tool to get information about sand density in an indirect way, this parameter being generally difficult to assess in this type of sediment. The results of these tests are referred to relative density values of selected sands, which are generally alluvial in nature, seeing that this type of deposit is frequently encountered in geotechnical investigations.

Eolian sands of the nearabouts of Mogadishu, in the region of Benadir, are mostly good foundation soils and for this reason they are not much investigated from a geotechnical point of view. Results of a few SPTs on these soils are not in agreement with whatever one would expect

either from actual observations or from literature data. Therefore we think that our results, even if not numerous, can contribute to a better geotechnical knowledge of this type of sediments.

2 GEOLOGICAL DATA

A relief, about 200 m high and about 20 km wide, extends for more than 800 km near the eastern coast of Somalia. In the first meters of this structure we always find eolian sands of more or less reddish colour; therefore we will call this relief the "red dune". Almost everywhere in the red dune, the first layer of 2 to 4 m is formed bypaler sands, ranging in colour from yellowish to orange, more simply cal-

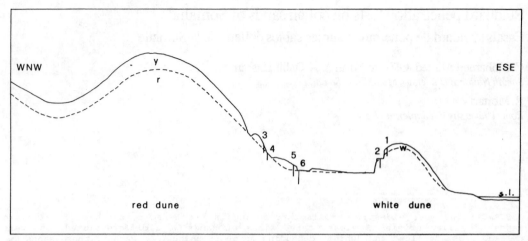

Figure 1. Schematic relationship between the studied sands: y = yellow sands, r = red sands, w = white sands. Numbers and vertical stretches refer to standard penetration tests. Test 3 was carried out on the foot of a mobil dune, tests 2 and 6 respectively in a quarry and road excavation.

led here "yellow sands". These are underlain by intensely reddish sands, called here "red sands". The separation surface between the two is very sharp.

Between the red dune and the sea, we frequently find a smaller dune, only a few dozen meters high and some hundreds of meters wide, composed of white eolian sands, which we will call the "white dune". Another separation surface can be found inside the white dune and is recognized because there is some thin layer of red sand along it or because the "white sand" is gently cemented just underneath it. The two dunes, the white and part of the red, with their separation surfaces, are shown schematically in Fig.1.

The sand of the red dune is almost completely formed of quartz coming from the crystalline outcrop of the Bur region, in the interior. Instead, the sand of the white dune is carbonatic in nature and comes from the ocean, from the dismantling of organic marine structures.

Besides the colour the granulometric composition too is different between the three kinds of sands: the white sands are medium to medium-fine without silt, the yellow ones are medium-fine with 1 to 3 % passing 0.074 mm and the red ones are medium-fine too but with 3 to 10 % passing 0.074 mm. Red sands are more cemented because of a longer weathering that produced clayey minerals.

Both dunes must have formed during glaciations, when the sea was lower (Sauro 1980) and climate was drier than nowadays (Pokras and Mix 1985). Very likely red sands are of the middle Pleistocene while

yellow and white ones are of the upper Pleistocene.

3 STANDARD PENETRATION TESTS - RESULTS

Two standard penetration tests for each kind of sand were carried out using a standard split-barrel sampler in holes obtained each time by means of a manual auger of 15 cm in diameter. After every single stage of a test a pothole was dug deeper and deeper in order to execute a density determination in place alongside the same vertical stretch of 30 cm where the blows of the test were considered. For this determination the same base plate of the sand-cone method was used in order to dig a hole with a regular mouth, then this hole was lined with a very thin plastic sheet and filled with water to the mouth. A possible error of 1 % was estimated to account for hole volumes, that were of at least 3000 cm^3.

The positions of the six SPTs are shown in Fig. 1, where we can note that tests with odd numbers crossed the whole thickness of the more recent sands and also part of the sands underlying the separation surface. Instead, SPTs with even numbers started from this surface and crossed the older white sands in one case (test 2) and the red sands in the other two cases (tests 4 and 6). There is therefore a certain degree of stratigraphic overlapping between even and odd tests, involving part of the older sands. In the case of test 3, we poit out that the crossed yellow sands have been completely mobilized by wind because they are in effect found in an area of currently mobile dunes.

The results of the standard penetration tests are plotted in Fig. 2. The vertical dashes indicate the 30 cm intervals into which the number of blows necessary for that deepening were counted; these numbers are shown in abscisses. The ordinates show depths, referred to the top of the more recent sands. Tests 2, 4 and 6, wich only deal with the oldest sands, are shown in the same diagrams as the odd tests they accompany. The representation of their results begins at the depth at which, in the odd tests, the separating surface was encountered.

Densities were determined to depth of 1.5-2 m and, only for the SPT 6, to over 3 m. Water contents varied from 2 to 8 %. Dry densities were calculated and then correlated with the penetration resistances of their respective locations in the diagram of Fig. 3. Here the curves connect consecutive results of a same test and the beginning of each curve is marked by a dashed line. Dry density values of sands tested in SPTs 2 to 6 vary in the same range as found for eolian Kalahari sands by Schwartz and Yates (1980) at the same depths.

The low values of dry density are not so low if we consider that the minimum values obtained in laboratory by re-sedimentation were also less than unity, while maximum values, attained using the compaction procedure according to ASTM 1557 method, were of about 1.35 to 1.5 g/cm^3. Relative density was close to 100 % for SPT 1, always greater than 100 % for the other tests. It is our opinion that other procedures, which Tavenas and La Rochelle (1972) report, could give more moderate results.

4 DISCUSSION

Granulometric and other technical characteristics of the sands confirm the first distinction made on the basis of their colour. It was also confirmed the distinction between sand of red and white dunes: at equal conditions the former ones have less penetration resistance, even if they have equal or greater dry densities.

Among the white sands a further distinction may be done between those over and below the separation surface as the former ones show dry density values of

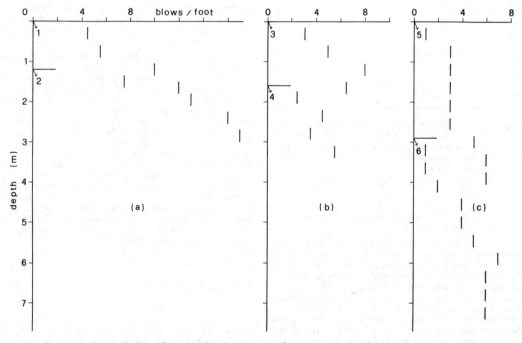

Figure 2. Results of the six standard penetration tests carried out on the white sands (a), the yellow and red sands (b) (c). Tests are represented two by two in order to show overlapping occurring near the same surface separating higher and lower sands of Fig. 1. Arrows start from the beginning position of each test.

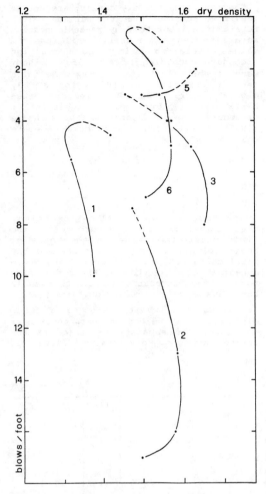

Figure 3. Blows per foot versus dry density for five of the six SPTs. Values of the blows generally increase with depth: the lines, starting dashed, connect consecutive data.

1.3- 1.4 g/cm³ against values of about 1.5 of the older sands.

Both dry density and penetration resistance of all these sands show very low values, and according to the classification of Terzaghi and Peck (1948) or Mitchell and Katti (1981) these values would range in the classes of very loose and loose sands. But this situation is normal for eolian sands which are less dense than alluvial sands at a parity of conditions: even the minimum values that we can obtain in laboratory for the studied sands are of about 1.0 to 1.1 g/cm³ against values of 1.2-1.6 g/cm³ for the alluvial ones (Gibbs and Holtz 1957;

Marcuson and Bieganousky 1977). And it is common to find for eolian sands relative density values even very higher than 100 % (Mortari 1973).

If under static loads these eolian sands have to be considered dense enough, their behaviour in the standard penetration tests shows that they are sensitive to dynamic stresses.

The various tests carried out pointed out also that technical characteristics of these sands do not always become better with depth, as one would expect. In the SPTs 1, 5 and 6 density is greater near to the topographic surface than just down, while for the tests 2 and 6 the same thing happens more in depth. Likely these facts depend on capillary phenomena.

Another observation deals with the number of blows per 30 cm found in the depth interval in which the stratigraphic overlapping of the tests occurs. The most evident instance of this is found in tests 5 and 6: in test 5 the red sand underlying the yellow sand shows values of 5-6 blows/foot while in test 6 the sand occupying the same stratigraphic position - but with its top exposed on the surface - shows values of 1-2 blows/foot. The thickness of red sands in which this happens is greater than 2 m, while in the white sands a similar thing happens at about 50 cm. This difference, due to the nearness to the topographic surface, is more important in the red sands, because of the presence of a not negligible fine fraction which created a swelling process.

A third observation may be made on the results of the SPTs 4, 5 and 6 in Fig. 2: both the yellow and the red sands tend to take on constant values of blows/foot over a wide depth interval, at values of 3 and 6 blows/foot respectively. This is far from normal in the sands, which tend to show a constant increase with depth, as was indeed found in tests 1, 2 and 3. This fact may be due to the presence of a certain clayey-silty fraction within the yellow and red sands and to their geological history: a similar behaviour is found in Pleistocene and Pliocene clays, which often show great thicknesses with the same value of preconsolidation pressure (Mortari 1977). A similar behaviour was found in sands of Lagos by Balagun (1980).

5 CONCLUSIONS

The tests carried out showed that eolian sands of different origin tend to behave analogousely but they show differences in the relations between penetration resistance and relative density. This is mainly motivated by their full geological history.

REFERENCES

Balagun, L.A. (1983). Foundation design for some tall buildings in Lagos area. Proc. 7th Reg. Conf. for Africa on SMFE, Accra, 2, 627-647.

Gibbs, H.J. & Holtz, W.G. (1957). Research on determining the density of sands by spoon penetration testing. Proc. 4th Int. conf. SMFE, London, 1, 35-39.

Marcuson, W.F. & Bieganousky, W.A. (1977). Laboratory standard penetration test on fine sand. ASCE, JGED, 103, GT6.

Mitchell, J.K. & Katti, R.K. (1981). Soil improvement. State of the art report (preliminary). Proc. 10th Int. Conf. SMFE, Stockholm, General reports, p. 264.

Mortari, R. (1973). Effetto nel tempo della capillaritö sulla densitö di alcune sabbie eoliche. Proc. 9th Conv. Geot. AGI, Milano, 19 pp.

Mortari, R. (1977). Elmenti per una nuova interpretazione della preconsolidazione delle argille. Geologia applicata e Idrogeologia, Bari, 12, 189-200.

Pokras, E.M. & Mix, A.C. (1985). Eolian evidence for spatial variability of late Quaternary climates in tropical Africa. Quaternary Res., 24, 137-149.

Sauro, U. (1980). Appunti sulla morfologia costiera della Somalia (zona di Mogadiscio). L'Universo, 60, 617-646.

Schwartz, K. & Yates J.R.C. (1980). Engineering properties of aeolian Kalahari sands. Proc. 7th Reg. Conf. for Africa on SMFE, Accra, 1, 67-74.

Tavenas, F. & La Rochelle, P. (1972). Accuracy of relative density measurements. GÉotechnique, 22, 549-562.

Terzaghi, K. & Peck, R.B. (1948). Soil mechanics in engineering practice. John Wiley & S., New York, 327 pp.

6th International IAEG Congress / 6ème Congrès International de AIGI, © 1990 Balkema, Rotterdam. ISBN 90 6191 130 3

A new method for deep static cone penetration testing
Une nouvelle méthode de pénétrométrie statique profonde

F. Schokking & R. Hoogendoorn
Geological Survey of the Netherlands, Haarlem, Netherlands

H.C. van de Graaf
Mos Soil Mechanics, Rhoon, Netherlands, (Previously: Delft Geotechnics, Delft, Netherlands)

ABSTRACT: Static cone penetration testing in overconsolidated soils has a limited penetration depth. In overconsolidated clays friction along the rods and sometimes the effect of additional friction resulting from swelling of clays are the principal reasons for not reaching greater depths. Known methods for overcoming rod friction use large diameter shell and auger drilling alternating with cone penetration testing. This requires wing casing along the connecting rods to reduce friction between the rods and the casing and buckling of the rods. This paper describes a new method in which rotary straight flush drilling with standard DCDMA BQ-size drill rods is alternated with cone penetration testing. A successful cone penetration test was carried out in the north of The Netherlands in glacially overconsolidated clays using a 200 kN penetration machine mounted on a ballast truck and a hyrdraulic drilling rig. Penetration stages from 10 to 12 m could be achieved in clays, with cone resistances ranging from 5 to 7 MPa and local friction ranging from 0.1 to 0.2 MPa at depths ranging from 45 to 100 m. An average rate of approximately 1.3 m/h could be reached during the entire test.

RESUME: La pénétrométrie statique dans des argiles préconsolidés a des limites en ce qui concerne la profondeur de pénétration. Les raisons pour lesquelles des grandes profondeurs ne sont pas atteintes sont le frottement le long des tiges et parfois l'effet de frottement additionel résultant du gonflement des argiles. Les méthodes connues pour vaincre le frottement sur les tigesutilisent un forage tube à la soupape en grand diamètre alterné de pénétrométrie statique. Ceci requiert un tube à ailes lelong des tiges reliés pour reduire le frottement entre les tiges et le tubage et le flambage des tiges. L'article décrit une nouvelle méthode dans lequel le forage a rotation a circulation directe avec une foreuse standard DCDMA - BQ et la pénétrométrie sont alterné. Un essai de pénétration fructueux fut exécuté dans le nord des Pays-Bas dans des argiles préconsolidés glacialement, usant une machine de pénétration 200 kN montée sur un camion lesté et une sondeuse hydraulique. Des passes de pénétration de 10 à 12 m ont été effectuées dans les argiles avec une résistance à la pointe variant de 5 à 7 MPa et un frottement latéral variant de 0.1 à 0.2 MPa a des profondeurs variant de 45 à 100 m. Un avancement moyen de 1.3 m/h a été atteint durant l'essai.

INTRODUCTION

A cone penetration test (CPT) was planned in near Noordbergum in the north of The Netherlands as part of a research project of the Geological Survey of The Netherlands (GSN) and the University of Edinburgh studying the geotechnical behaviour of subglacially deformed sediments.

The principal aim of this test was to obtain data on the glacially overconsolidated Pleistocene clays ('Pot' Clay) that occur from approximately 45 to 115 m below the surface.

On the basis of experience with techniques for reducing friction (Van de Graaf, 1988) and earlier cone penetration testing in the Pot Clay near Marum (Schokking, 1990) it was expected that penetration in clay would not exeed 15 m. A method of cone penetration testing alternating with destructive drilling, would therefore be necessary to test the complete 70 m of Pot Clay, but also to be able to pass through the heavily overconsolidated sands above the clay.

The conventional procedure (Fig. 1) was to first perform a CPT stage (1) until refusal and pull out of the rods. Subsequently (2) a conventional casing is brought down to the penetrated depth by shell and auger drilling and a wing-casing is run down inside it (3) for horizontal support of the CPT rods. After that a second CPT stage is performed (4) and the procedure can be repeated down to the target depth.

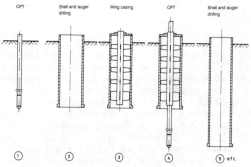

Fig.1 Sequence of alternating cone penetration testing and shell and auger drilling using wing-casing for support

A much faster known procedure is the downhole penetrometer method (Zuidberg & Windle, 1979, Fig. 2). The penetrometer consists of a small diameter jacking

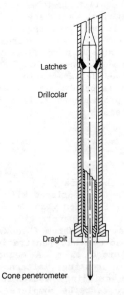

Fig.2 Downhole penetrometer method (after Zuidberg & Windle, 1979)

system, lowered down inside a drillstring, like a wireline tool. It latches automatically onto the bottom section of the pipe. This system is used widely for offshore site investigation. However, the limited stroke of the jacking system results in a CPT diagram that shows small discontinuities every 3 m, which is a disadvantage for the interpreting the results. Moreover, the system is expensive in onshore investigations, since it requires heavy operating equipment.

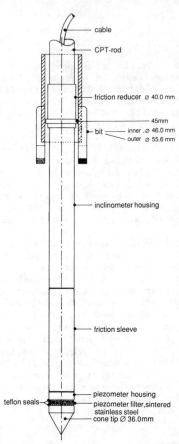

Fig.3 Detail of cone, friction sleeve, inclinometer housing and friction reducer through open bit on BQ-drill rod

These considerations lead to the development of an alternative method, using a standard DCDMA BQ-size drill rod to support the penetration rods.
It was assumed, that alternating cone penetration testing and straight flush drilling using a polymer based mud as a friction reducing agent, intervals of between 10 and 20 m could be achieved.

A CPT in which this method was tried out was performed by Delft Geotechnics in cooperation with the Geological Survey of The Netherlands who carried out the drilling.

The Department of Exploration Geophysics of the Institute of Earth Sciences of the Universty of Utrecht, performed a seismic VSP test to investigate the anisotropies in the soils (Douma et al., 1990) at three levels during the penetration test. The test was done by keeping the cone, equiped with a geophone for three-dimensional recording, in one position during for a few hours, while a shear wave signal was generated at various locations at the surface with a sledgehammer device. For the stages in which no VSP measurements were performed, a cone was used equipped with a biaxial inclinometer instead of a geophone.

The new method, described in this paper, enables deep cone penetration tests, that find applications in the fields of geotechnical engineering, engineering geology and glaciology.

EQUIPMENT

For the drilling a Unimog UB 1700 truck mounted hydraulic rig, suitable for top drive rotary drilling was used. Standard BQ wireline rods (ID 46 mm) with a fully open bit (Fig. 3) were used for straight-flush drilling and support of the penetration rods (OD 36 mm). A hydraulically driven piston pump (0.001 m^3/s at 30 bar) was used for mud circulation. The drilling mud consisted of a polymer/fresh water-mix using approximately 0.3% of the liquid polymer Cebo-Sol problems with clay such as swelling, sloughing, bit balling and flow plugging were effectively prevented. Moreover, it was assumed that the mud would reduce friction between connecting CPT and drill rods. Special attention was given to preventing penetration of the drill string during cone testing caused by internal friction by using a rod clamp (Figs. 4b and 5).

For the CPT a modern truck mounted 200 kN CPT rig was used. The dead weight

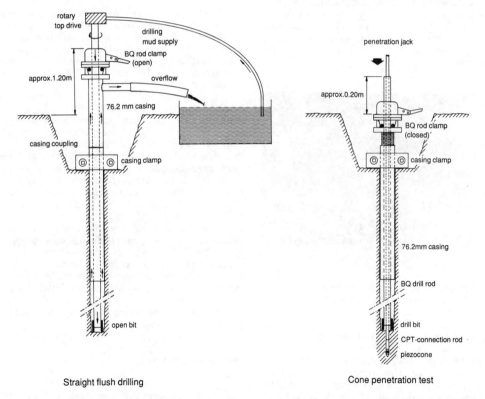

Straight flush drilling

Cone penetration test

Fig.4 (a) Set-up for straight-flush drilling with mud-circulation system

(b) Set-up during CPT; BQ-size drill rod fixed by clamp at top of casing

of the truck was 230 kN. The piezocone designed by Delft Geotechnics was equiped with a friction sleeve and an inclinometer (Fig. 3). The CPT data were recorded at 20 mm. depth intervals on magnetic tape.

A shape and size of the friction reducer at 200 mm above the friction sleeve was selected on the basis of the results of a comparative study of various friction reducers (Van de Graaf, 1988).

METHOD

As first stage, a CPT is performed until refusal. Then, a 90mm-diameter casing, is run through the top soil layer down to a depth of approximately 2 m. The function of the casing is to permit the drilling mud and cuttings to flow to above the groundsurface and into the mud tank (Fig. 4a). Straight flush drilling is performed down to refusal depth (170 kN) of the previous CPT stage. After moving away the truck-mounted drilling rig, the CPT truck is positioned over the hole and a second CPT stage is carried out until refusal. Then, the CPT string is retrieved, the CPT truck is moved away and straight flush drilling continues down to the depth of CPT refusal.

Fig.5 Casing clamp and drill rod
clamp in position for CPT

The maximum thrust applied during the test is 170 kN to prevent problems in extracting the rods. Experience has shown that friction on the rods may be higher during extraction then during penetration as a result of swelling of the Pot Clay.

It was found that placing the trucks with their fronts facing gave optimum manoeuverability for positioning over the borehole (Fig.6)

Fig.6 CPT-truck in postion for testing and drilling truck, fronts facing

CONE PENETRATION TEST

Geology and geotechnical profile

The test was carried out near the location of a borehole which had been drilled previously for the same research project. This borehole was sampled with undisturbed tube samples at 3 m intervals on which geotechnical laboratory tests were performed.

The geological profile (Fig. 7) contains boulder clay, sands and clays of Pleistocene age that are overconsolidated as a result of an ice load during the Saalian glaciation (Schokking, 1990).

The upper clay layer, in which the CPT was performed consists of fluviatile clay with a 5m thick lacustro-glacial clay layer at the base. The consistencies range in this upper layer from medium stiff to stiff (S_u = 50 - 240 kPa).

The lower part of the sequence consists entirely of the 70m thick lacustroglacial Pot Clay. The consistency of the clay ranges from medium stiff to stiff (S_u = 50 to 100 kPa)in the upper 5m and from very stiff to hard (S_u = 300 to 500 kPa) in the remaining part.

The layers in between these clays consist of very dense sands.

Test performance

The test was carried out during July 1989 in excellent weather conditions.

Table I shows the schedule of 7 stages of alternating drilling and cone penetration testing.

Table I

Stage	Depth in m	Penetration length in m	Lithology	Drilling time in h	CPT time in h
1	0 - 10.0	10.0	boulder clay and sand		0.5
	0 - 10.0	10.0		4	
2	10.0 - 10.7	0.7	sand		0.1
	10.7 - 29.0	18.3	sand	8	
3	29.0 - 38.6	9.6	clay		3.75
	29.0 - 48.3	19.3	clay and sand	8	
4	48.3 - 60.7	12.4	clay		4.5
	48.3 - 61.0	12.7	clay	4.5	
5	61.0 - 71.6	10.6	clay		5
	61.0 - 74.0	13.0	clay	4.25	
6	74.0 - 84.5	10.5	clay		4.5
	74.0 - 86.4	12.5	clay	4.5	
7	86.4 - 95.2	8.8	clay		4.5

The test was performed during the day. The average drilling rate was 2.7 m/h and the average cone penetration rate, including lowering and extracting rods, was 2.2 m/h.

This means a total rate of 1.3 m/h for the combined operation.

Testing in the sand layers was stopped after the connecting CPT rods got jammed in the drill bit, as a result of an inflow of sand. Some problems were encountered during retraction of the CPT rods, probably due to the fact that the tolerance between the friction reducer (Fig. 3) and the inside of the drill bit does not permit any sand to enter the system. As further testing in the sand layers would have been very time consuming and the information obtained would be of minor importance only in the research project, no cone penetration testing was done in the sand layers.

Friction between the connecting CPT rods and the inside of the drill rods increases with depth, which can be deduced from the decrease in the penetration length in clay with increasing depth (Graph in Fig. 8). Extrapolation of this graph indicates, that the total friction inside the drill rods will theoretically enable penetration

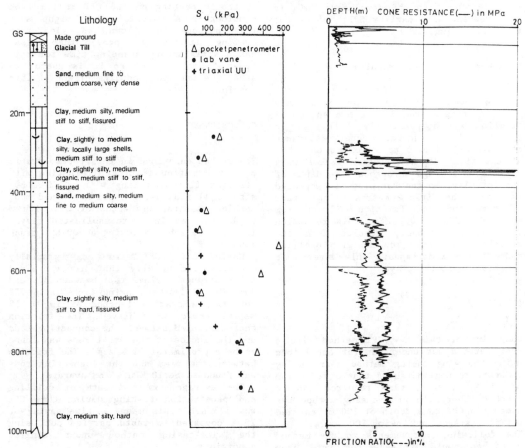

Fig.7 Geological sequence, undrained shear strength (S_u) and CPT results

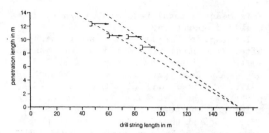

Fig.8 Graph showing penetration-length versus length of drill-string

up to a string length of 160 m, assuming a thrust of 10 kN for the cone resistance and local friction and a maximum total thrust of 170 kN. The friction inside the drill string therefore amounts to approximately 1 kN/m'.

Back calculation of the friction between connecting CPT rods and clay gives an average value of 10 kN/m'. Since the Pot clay has an average S_u-value of 400 kPa the friction between the clay and the rods of $S_u = 40/m'$ must be concluded. This is in agreement with earlier CPTs in the Pot Clay (Schokking, 1990).

Cone Penetration Test results

In the upper clay layer, cone resistance (Fig. 7) ranged from 3 to 6 MPa and local friction was approximately 0.05 MPa. In the Pot Clay sequence, cone resistance rose steadily from 5 to a maximum of 7 MPa at the final depth. Friction values were 0.2 MPa on average throughout the depth of the Pot Clay. In the Pot Clay, it appeared that over the first 2 metres at the start of stage, the local friction built up from about 0.1 MPa to 0.2 MPa. This reduction in friction probably results from the friction shaft being coated with drilling mud. This mud dissapears only after some penetration in the clay.

DISCUSSION

Comparing of this new CPT method with a test carried out under similar conditions using the wing casing method near Fawley in southern England shows that a reduction of 25 % in total time required can be reached. There the wing casing method was used in a CPT to a depth of 100 m, carried out in 7 stages and taking 100 hours. The following improvements to the method described could increase the rate even more.

During drilling with the open bit the drilling rate has been kept low , to prevent the problem of 'coring' of the clay, causing clay to get plugged inside the bit. This can be overcome by using a wireline tool, that closes the bit during drilling and can be removed to allow the cone to pass through the bit for testing. This will increase the drilling rate considerably.

The clamping action during lowering and extracting of the CPT rods was done in an inprovised way, using a device that had to be tightened by hand for each rod. A hydraulic clamping system on the CPT rig will decrease the time required for lowering and extracting the rods.

A factor limiting the maximum penetration depth is the friction between the connecting CPT rods and the inside of the drill rods, which appeared to be approximately 1 kN/m' at 170 kN.

To further decrease this friction, which occurred despite the friction- reducing effect of the drilling mud, the use of a synthetic coating on the CPT rods may be considered. A material that might suit this purpose is Rilsan, a nylon type coating. It has been proven that Rilsan has a satisfactory standing time for this type of application, so it is useful to study the friction reducing effect of this coating.

CONCUSIONS

The method described in this paper: static cone penetration testing alternating with straight flush drilling with DCDMA BQ-size drill rods using a polymer based mud, enables penetration depths of over 100 m to be reached in overconsolidated clays. This can be achieved using a 200 kN, truck mounted CPT rig.

During a CPT in a glacially overconsolidated clay that has undrained shear strength values (S_u) between 300 and 500 kPa at depths ranging from 45 to 115 metres, penetration lengths of 10 to 12 metres could be achieved. The friction that developed between the connecting rods and the inside of the drill rods was found to be approximately 1 kPa/m'. The friction between the clay and the connecting rods was found to be 10 KPa/m' on average.

The average combined rate of drilling and penetration testing during this CPT was 1.3 m/h. This means a time saving of 25 % compared to tests carried out using the wing-casing method under similar conditions.

The drilling rate may be increased by using a closed wireline bit, and the penetration rate can be improved by mechanical rod handling, resulting in further time saving.

A nylon coating on the connecting CPT rods will reduce the friction with the inside of the drill rods and consequently increase the final penetration depth that can be reached by this method.

REFERENCES

Douma, J., K. Helbig, F. Schokking & J. Tempels, 1990, Shear-wave splitting in shallow clays observed in a multi-offset and walk-around VSP. Prep. for publ. in First Break.

Schokking, F., 1990, On estimating the thickness of the Saalien ice sheet from a vertical profile of preconsolidation loads of a lacustro-glacial clay. Geol. en Mijnb. Vol.69, No.3.

Van de Graaf, H.C. & P. Schenk 1988. The performance of deep CPTs, International Symposium On Penetration testing, Orlando.

Zuidberg, H.M. & D. Windle 1979. High-capacity sampling using a drill-string anchor, International Conference on Offshore Site Investigation, London.

1.3 Laboratory and in-situ testing and instrumentation
    Expérimentation et instrumentation en laboratoire et sur le terrain

Prediction of swelling pressure of soil from suction measurements

Prédiction de la pression du gonflement des sols par la mesure de la suction

Namir K.S.AL-Saoudi & Fahmi S.Jabbar

Building & Construction Department, University of Technology, Baghdad, Iraq

ABSTRACT: The object of this paper is to correlate the swelling pressure of the soil with its moisture suction. Samples of highly expansive soil were compacted at different initial moisture contents and dry densities The swelling pressure and the corresponding suction of these samples were measured using the suction plate apparatus. The suction plate used was similar in principle to Alpan's apparatus (1957) with modifications to account for the swelling pressure at different suction levels. The soil suction was also measured using the filter paper technique. Special tests were carried out to correlate the swelling pressure of the soil with its final degree of saturation.
An empirical equation was developed relating the swelling pressure of the soil to the degree of the soil suction (Uu)(defined as the ratio of change in soil suction to the initial value), in terms of the initial dry density.

RESUME:La but de cet article est la determination d'one relation entre la pression du gonflement du sol et as humidite de la succion. Des echant-tillons du sol de tres haut gonflement ont ete compactes en quelque valeurs du tenear en eau et de la densite sec. la pression du gonflement et la succion des echantillons ont ete mesurees par l'appariel du plateau de la succion. La plateau de la succion utilise a la meme principe de l'appariel d'Alpan avec quelque modifications pour compter la pression du gonflement aux nireaux differents de la succion, la succion du sol a ete aussi mesuree par la technique du tittre de papier.
Des testements speciaux nous ont amenes a relier la pression du gonfle-ment du sol avec sa degre de saturation finale.Une formule a ete developp-pee pour relier la pression du gonflement du sol a sa degre de la succion (Uu) par la terme de la densite sec initiale (a difinir comme un rapport du changement de la succion du sol a la valeur initiale).

1 INTRODUCTION

Several methods and techniques are available for predicting the swel-ling characteristics of the soils. Some of these methods are based on direct measurements of swelling pressure while others are based on indirect methods Chen (1975).
For the present work a method bas-ed on soil suction measurements is proposed to predict the swelling pressure. The soil moisture suc-tion was measured by two methods,
the suction plate method and the filter paper method.
Samples of expansive soil , brought from the north part of Iraq were prepared and compacted at initial dry densities ranging from 12-17 kN /m³ at moisture contents ranging from the hygroscopic moisture con-tent up to a moisture content near saturation. Several relationships were found correlating the soil moisture suction with the swelling

pressure depending on the value of
the initial dry density.

2 EXPERIMENTAL WORK

2.1 Materials

The soil used in this work was
brought from the north part of
Iraq with the following properties.
L.L = 65%
P.L = 25%
P.I = 40
S.L = 17.8%
S.I = 47.2
Percentage < 2 micron   = 32%
Activity   = 1.25
Specific Gravity   = 2.78
Optimum moisture content = 20%
Maximum dry density   = 16kN/m³
(Standard Proctor)

2.2 The suction plate device

This device was designed and manufactured to measure the soil suction,in principle it is similar to
Alpan's suction plate (1957) with
modifications to account for the
swelling pressure at different
suction levels.A schematic diagram
for the device is shown in figure
1. The principles of this device
is based on the equalization of
the moisture suction of the soil
sample with the water tension imposed inside the water chamber and
controlled by the manometer level.

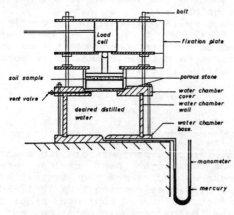

FIG.(1) DEVICE OF SOIL MOISTURE SUCTION AND SWELLING
PRESSURE MEASUREMENTS.

2.3 Test procedure for suction
measurements

Soil samples were prepared at
moisture contents ranging from the
hygroscopic moisture content to a
moisture content near saturation.
Each soil sample was compacted
statically inside a steel ring
(diameter 75 mm, height 20 mm), to
the required initial dry density.
the sample moisture suction was
then measured by either the suction plate method or the filter
paper method.

2.3.1 The suction plate method

The soil sample was placed inside
the suction plate device and the
load cell was then fitted and fixed in position properly and rapidly. The sample started sucking
water from the chamber. The manometer level was lowered slowly and
contineously so that no swelling
pressure was recorded in the load
cell.The procedure continued until
equilibrium was achieved.

2.3.2 The filter paper method

The filter paper method was used
to measure soil suction larger
than 98 kPa. The filter paper used
was type (Whatman cad No 1001 105),
it was calibrated using the conventional consolidation equipment and
according to the procedure proposed by Chandler and Gutierrez(1986).
The relationship between the filter paper moisture content and the
soil suction is shown in figure 2.
It is represented by the following
equation.

$$\text{Log } \tau° = 3.982 - 0.022 \ w \quad \text{for}$$
$$\tau° > 80 \text{ kPa} \quad ...(1)$$

where $\tau°$ = Total Suction,
      w = filter paper moisture
          content.

2.4 Test procedure for swelling
pressure measurements

The compacted soil sample was
placed inside the suction plate
device and water was then added to
the sample.The gradual development
of the swelling pressure was recorded by the load cell at different
time intervals until equilibrium
state was achieved.

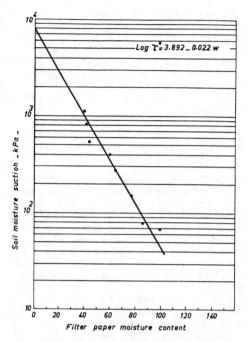

FIG·(2) RELATIONSHIP BETWEEN SOIL MOISTURE SUCTION
AND FILTER PAPER MOISTURE CONTENT.

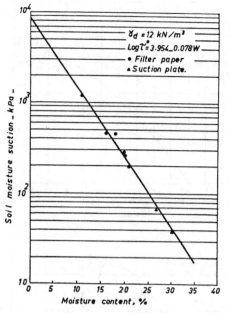

FIG.(3) RELATIONSHIP BETWEEN SOIL MOISTURE SUCTION
AND MOISTURE CONTENT.

3. PRESENTATION AND DISCUSSION OF RESULTS

The following relationships have been investigated.

3.1 Soil suction-moisture content relationship

Snethen(1980) found that the relationship between the soil suction and the moisture content can be represented by the equation

$$\text{Log } \bar{\tau}^\circ = \bar{b} - \bar{a}w \qquad \ldots(2)$$

where

τ° = Total suction
w = moisture content
$\bar{a}$ & $\bar{b}$ = specific soil constants

Equation (2) has been arrived at experimentally for the tested soil using two techniques for measuring the soil suction. A typical relationship for $\tau d = 12$ kN/m³ is shown in figure 3. Similar results were obtained for other values of dry densities.

3.2 Swelling pressure-moisture content relationship

The swelling pressure is directly proportional to the amount of water entering the soil. Typical relationships between the swelling pressure and the final moisture contents for soil samples with initial dry densities of 12 and 16 kN/m³ are shown in figure 4. Samples were prepared at initial moisture contents of 8,12, and 20% and were allowed to suck water for time intervales corresponding to 25%, 50%, 75% and 100% of the final swelling pressure. It was observed that at constant value of dry density, the initial moisture content has significant effect on the final value of the swelling pressure. Also at higher values of initial dry densities $\tau d = 16$ kN/ m³ the swelling pressure continued to develope until a high degree of saturation was reached (Sr ≈ 95 %). At lower values of dry densities $\tau d = 12$ kN/m³ a shorter period of time was required to achieve the final swelling pressure and the degree of saturation was approximately 85%. This different behaviour can be explained in terms of the

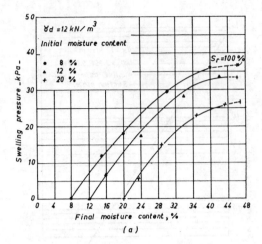

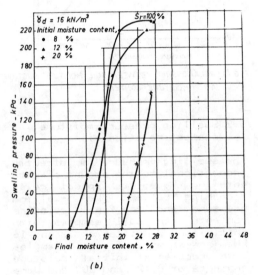

FIG.(4)RELATIONSHIP BETWEEN FINAL MOISTURE CONTENT
AND SWELLING PRESSURE.

diffuse double layer.The repulsive
forces between the clay particles
will continue to develop due to
the close distances between the
particles at higher values of dry
density, while at lower dry densities the repulsive forces can be
equalized with the residual cohesion and external stresses at a
specific degree of saturation.
Yong and Warkentin (1966).
Typical relationships between the
initial moisture content and the
final swelling pressure is shown
in figure 5. The swelling pressure

decreases with increasing initial
moisture content and also the rate
of decrease of the swelling pressure with respect to the change in
initial moisture content was found
to be low at lower initial moisture contents and increases at
higher values of initial moisture
contents.

3.3 Swelling pressure-soil suction relationship

According to the previous discussion relating the swelling pressure with the initial and final
moisture contents, a relationship
between the swelling pressure and
the decrease in soil suction was
obtained as shown in figure 6.This
figure shows typical results for
samples compacted at intial dry
density of 12 kN/m³ with intial
moisture contents of8%,12%,and 20%.
Kassiff and Ben Shalom(1971) found
that such relationship can be
divided into three parts. At high
suction values there is a constant
gradual increase in swelling pressure per unit decrease in soil
suction,subsequently the relationship gradually changes hyperbolic-
ally followed by a rapid increase
in the swelling pressure with
decreasing soil suction until the
maximum swelling pressure is attained.
The swelling pressure is related
to the degree of soil suction.This
term has been used to take into
account the combined effect of the
initial soil suction and the rate
of decrease of soil suction. The
degree of soil suction (Uu) is
define d as:

$$Uu = \frac{U_i - U_f}{U_i} \qquad ...(3)$$

Uu= degree of soil suction (Simi-
lare to the degree of consolidation)

U_i = Initial soil suction

U_f = Final soil suction

342

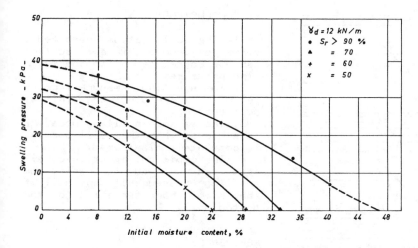

FIG.(5) RELATIONSHIP BETWEEN SWELLING PRESSURE, INITIAL MOISTURE CONTET AND FINAL
DEGREE OF SATURATION.

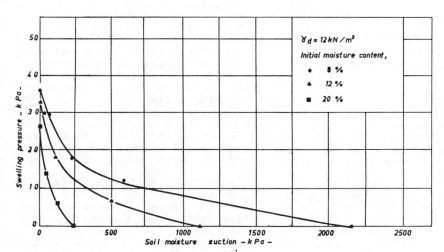

FIG.(6) RELATIONSHIP BETWEEN SWELLING PRESSURE AND DECREASING IN SOIL MOISTURE SUCTION.

Figure 7. shows the swelling pressure plotted against the degree of soil suction on a semilogarithmic scale for the four initial dry densities. The relationship can be expressed in the form.

$$Log\ P_s = I\ U_u + L \qquad ...(4)$$

where

P_s = Swelling pressure

U_u = Degree of soil suction

I & L = Constants

The coefficient $I = 1.35$ for the range of densities used while the coefficient L depends on the initial dry density according to the following equation

$$L = 0.235\ \tau_d - 2.65 \qquad ...(5)$$

The general equation relating the swelling pressure to the degree of soil suction can be written as:

$$Log\ P_s = 1.35 U_u + 0.235\tau_d - 2.65 \qquad ...(6)$$

343

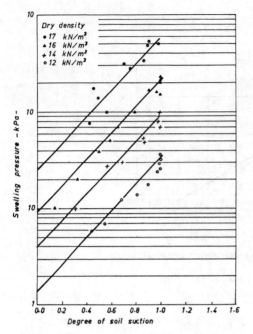

FIG. 7 EXPERIMENTAL RELATIONSHIP BETWEEN DEGREE OF SOIL SUCTION AND SWELLING PRESSURE.

Comparison between the measured swelling pressure with the calculated swelling pressure using equation 6 is shown in figure 8, the coefficient of correlation r =0.86

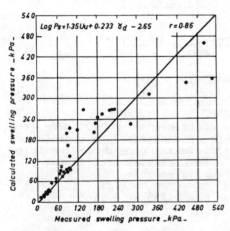

FIG.(8) MEASURED (VS) CALCULATED VALUES OF SWELLING PRESSURE.

4- CONCLUSIONS

1. The soil suction presented in terms of degree of soil suction can be correlated with the swelling pressure.

2. The swelling pressure increases with increasing the initial soil suction.

3. The maximum swelling pressure is achieved when the soil suction is completely dissipated.

4. The empirical equation developed can be used to predict the swelling pressure of any soil of the same characteristics of the tested soil.

REFERENCES

Alpan , I 1957 "An apparatus for measuring the swelling pressure in expansive soil"
Proceedings of the 4th I.C.S.MF.E. Vol.1 pp 3-5 London.

Chandler R.J and Gutierrez C.I. 1986 "The filter paper of suction measurements". Geotechnique Vol.36 No.2 pp 265-268.

Chen F.H. 1975 "Foundations on Expansive soils".
Elsevier, Amsterdam.

Kassiff, G. and Shalom, A.B.,1971 "Experimental Relationship between swelling pressure and suction". Geotechnique Vol.21, No.3.

Snethen D.R.1980 "Characterization of Expansive soils using soil suction data". Proceedings of the 4th I.C. on Expansive soils pp. 54-75.

Yong, R.N.and Warkentin, B.P.1966, "Introduction to soil behaviour macmillan, New York.

6th International IAEG Congress / 6ème Congrès International de AIGI, © 1990 Balkema, Rotterdam. ISBN 90 6191 130 3

Developments in measuring the deformability of sandstones

Développements dans la mesure des déformabilités des grès

L. Dobereiner
Department of Civil Engineering, PUC-RJ, Catholic University of Rio de Janeiro, Brazil
(Visiting Researcher at the Laboratoire Central des Ponts et Chaussées, France)

A. L. L. Nunes
Department of Civil Engineering, PUC-RJ, Catholic University of Rio de Janeiro, Brazil

C. G. Dyke
Department of Min. Res. Engineering, Imperial College of Science, Technology and Medicine, UK

ABSTRACT: Deformability measurements of sandstones are, in general, made as routine tests following several methods; the most common being the ISRM suggested method. Based on the experience gained from testing several different types of sandstones, ranging from weak to strong in strength, it was observed that a few aspects of the current testing methods need to be reviewed. For instance, by considering volumetric strain variations, in order to better understand the stress-strain behaviour, it is apparent that the weaker rocks display a behaviour markedly different from the stronger rocks. In the weaker rocks, the elastic range is limited to very low stress levels. The sensitivity of the strength and volumetric strain, in relation to the sample moisture content is also evaluated, and shown to be an important variable in rock deformability behaviour. Mineralogy and textural parameters, including grain contacts and cementation, are considered in terms of strength and deformability of the different sandstone types. The development of a new high sensitivity strain transducer, used in several of the tests performed, is also presented.

RESUME: En général, la déformabilité des grès est mesurée par des essais de routine suivant différentes méthodes, souvent selon la méthode suggérée par la SIMR. L'analyse d'un grand nombre d'essais sur différents types de grès de faible à haute résistance montre que certains aspects de la méthodologie des essais courants doivent être révisés. Par exemple en considérant les variations des déformations volumiques en vue de mieux comprendre le comportement contrainte-déformation, il apparaît que les roches de faible résistance ont un comportement bien différent des roches de haute résistance. Dans le cas des roches de faible résistance le domaine élastique est limité aux bas niveaux de contraintes. La dépendance de la résistance et de la déformation volumique avec l'humidité des échantillons est aussi évaluée, c'est une variable importante dans le comportement de déformabilité de la roche. Les paramètres de minéralogie et texture, incluant les contacts des grains et les caractéristiques de matériaux de cimentation, sont considérés en terme de résistance et déformabilité de différent types de grès. Le développement d'un nouveau capteur de déformation de haute sensibilité, utilisé dans quelques essais, est aussi présenté.

1 INTRODUCTION

The uniaxial compression strength test is the most widely used rock mechanics test in engineering practice. It has the advantage of being simple in principle and requiring relatively unsophisticated equipment. Therefore, in most cases, the deformability and elasticity modulus of rocks are obtained by uniaxial compression, this being the technique recommended by the International Society of Rock Mechanics (ISRM, 1978).

Standards for sample preparation and testing of rock core samples in uniaxial compression have been proposed by several institutions including the ASTM (American Society of Testing Materials), SAA (Standards Association of Australia), ABNT (Associaçao Brasileira de Normas Técnicas). These are in most cases similar to the ISRM recommendations. Testing methods should be standardized as much as possible, so as to allow the comparison of test results from different rocks from several countries around the world.

The ISRM recommendations for uniaxial compression testing were prepared in 1978, and as highlighted by several authors including; Pells & Ferry, (1983), Dobereiner (1984) and Nunes (1989), have been developed primarily for hard rock. Further developments are needed to standardise the testing of weaker rocks.

Test results on a wide range of sandstones, including weak to hard rocks are presented in this paper, with discussion of the problems and advantages of the different testing procedures that were used. The influence of the intrinsic rock properties, such as mineralogy, texture and sample moisture content, on the deformability of sandstones are considered, together with testing procedures, including sample preparation, rate of loading and strain measurement devices. Suggestions in relation to the deformability test methods are presented as the conclusions.

2 TYPICAL STRESS-STRAIN BEHAVIOUR

The stress-strain behaviour of sandstone is frequently regarded as being predominantly elastic. However, test results have shown that this is not always the case. In fact, most stress-strain curves obtained from sandstones, ranging in saturated uniaxial compression strength from around 70 to 0.5 MPa, show the typical behaviour displayed in Figure 1, with only a limited range of elastic behaviour. Three main phases of deformation are observed:

a- Initial closure of microfissures - this stage is observed at low stress levels, with significant strains in the more fissured samples,

b- Elastic behaviour - this region is not always well defined; in many cases the volumetric strain curves do not present a linear segment. At higher stresses the sandstones show a "hardening" behaviour which is related to the closure of the texture, through the breakdown of a few grain contacts and a subsequent increase in grain contact area.

c- Fissure propagation - Initially, new microfissures propagate at low velocities and are relatively stable. With an increase in stress, the velocity of fissure propagation increases, until the macroscopic failure of the sample.

As observed by Dobereiner (1984), Dyke (1984) and Nunes (1989) the onset of dilatancy in the weaker sandstones, and even some of the saturated strong sandstones, occurs at low stress levels, frequently below 30% of peak strength. Therefore the determination of the elasticity modulus at 50% of peak strength, as recommended by the ISRM, could lead to the inclusion of non elastic strains that occur after the onset of dilatancy, within the measurement of the elastic modulus (Dobereiner & Oliveira, 1986).

In practice the systematic evaluation of volumetric strain curves during deformability determinations, would greatly help in promoting a better understanding of the material behaviour.

3 INFLUENCE OF SAMPLE MOISTURE CONTENT

The typical stress-strain behaviour observed for sandstones, is dependant on sample moisture content. It is well known that the sample saturation has a great effect on the strength of rocks, as described by several authors listed in Table 1. The rate at which strength is reduced with increase in moisture content is very variable, however there is a trend that indicates that the weaker materials are more sensitive to sample moisture content. As shown in Figure 2 (Dyke, 1984 and Dobereiner & Dyke,1986), the largest variations in rock strength occur at low moisture contents, in most cases below 1% (for sample characteristics see Table 2). In practice this range of critical sample moisture contents represents the variation within the air dry range, due to variations in environment humidity.

As described by Dobereiner & De Freitas (1986), the reduction in strength due to an increase in sample moisture content must involve the usual variations in cohesion and friction of granular materials. They also mentioned that the presence of water increases the velocity of crack propagation in silicates, by replacing strong silica oxygen bonds with much weaker hydrogen bonds within the silicate lattice (Atkinson, 1984). When this phenomenon occurs at the tip of a microcrack propagating under tension, the stress required for failure at the tip of the crack is reduced

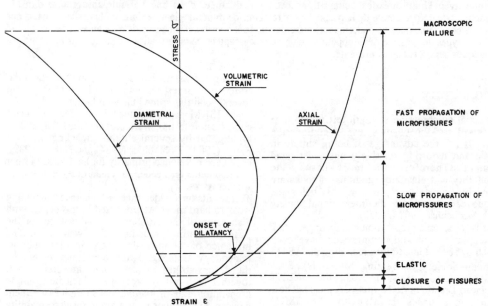

Figure 1. Typical stress-strain behaviour of sandstones.

Table 1. Influence of sample moisture content on the uniaxial compression strength of some sedimentary rocks (Nunes, 1989)

Reference	Rock	Sat. strength/ dry strength
Colback&Wiid, 1965	sandstones and shales	50%
Wiid, 1970	sandstones	60%
Kitaowa et al.,1977	sandstones	64%
Bell, 1978	Fell sandstones	68%
Hassani et al., 1979	sandstones	70%
Ferreira et al., 1981	Bauru sandstone	50%
Priest&Selvakumar,1982	Bunter sandstone	65%
Koshima et al., 1983	Bauru sandstone	50%
Pells & Ferry, 1983	sandstone	41%
Dobereiner, 1984	Waterstones	83%
Dyke, 1984	Waterstones	49%
	Bunter sandstone	71%
	Penrith sandstone	66%
Gunsallus & Kulhawy, 1984	coarse crystaline sandstone	98%
	coarse clastic sandstone	81%
	fine grained sandstone	49%
Denis et al., 1986	Limestones	48%
Howarth, 1987	sandstones	67%

by weakening the strength of the crystal lattice ahead of the crack tip. This phenomenon is termed "stress corrosion" and is responsible for the increase in velocity of crack propagation. It explains the large strength reduction at moisture contents at 0 to 1%, since a change in environment from one almost free of hydrogen ions, to one in which free hydrogen is present at the crack tip is sufficient for stress corosion to occur.

The stress levels of the onset of dilatancy at the different moisture contents are also displayed in Figure 2, with the propagation of microfissures starting at lower stress levels at higher moisture contents.

The volumetric strain versus stress curves, at different sample moisture contents, for the sandstones characterized in Table 2, are presented in Figure 3. The onset of dilatancy occurs at different stress levels, depending on sample moisture content. The range of stress over which the sample volume decreases becomes smaller as the moisture content increases. This means that the range of elastic behaviour is larger within dry samples, since the propagation of microfissures starts at higher stress levels.

Figure 4 shows the variation of the elasticity modulus with sample moisture content for the same samples of sandstones. It is observed that the the tangent modulus, measured at 50% peak strength, decreases following a similar trend as the peak strength and the onset of dilatancy. However the tangent modulus determined below the onset of dilatancy does not show any significant change in relation to sample moisture content. The small variations are interpreted as errors related to measurement accuracy and variability between samples. These results clearly show that the elasticity modulus should be determined at stress levels below the onset of dilatancy, and not at a fixed percentage of the peak strength as recommended by the ISRM.

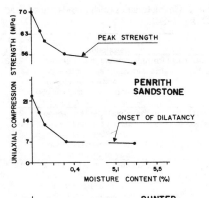

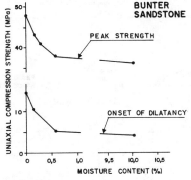

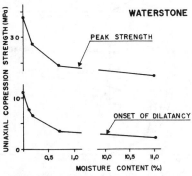

Figure 2. Variation of uniaxial compression strength and stress level at onset of dilatancy with sample moisture content - for sample characteristics see Table 2 (Dyke, 1984).

4 INFLUENCE OF CEMENTATION AND GRAIN CONTACTS

The effects of mineralogy and texture on the deformability of sandstones are difficult to define exactly. Rock, being a composite material consisting of more than one mineral, should have an effective deformability dependant upon the contribution of these individual minerals, their relative proportion, shape and orientation. The variations in mineralogy of the sandstones referred to in this paper are

Table 2. Characteristics of the sandstones tested to evaluate the influence of moisture content of Figures 2, 3 and 4 (Dyke, 1984).

Material	Age	Petrographic description	Porosity (%)	Sat. compressional wave vel. (m/s)	Dry density (Kg/m3)	Sat. uniaxial comp. strength (MPa)	Triaxial compression C(MPa)	Φ(°)
Waterstones (U.K.)	Permo-triassic	medium to fine grained quartz sandstone with secondary overgrowth of quartz, occasional clay or iron hydrox. cement	22.3	2060	2020	24	5.8	45
Bunter sandstone (U.K.)	Triassic	Medium to fine grained quartz sandstone with secondary overgrowth of quartz and occasional clay or iron hydrox cem.	21.1	3210	2070	38	-	-
Penrith sandstone (U.K.)	Permian	medium grained quartz sandstone with secondary quartz overgrowth and occasional iron hydrox. cement	12.3	4240	2300	54	38.5	45

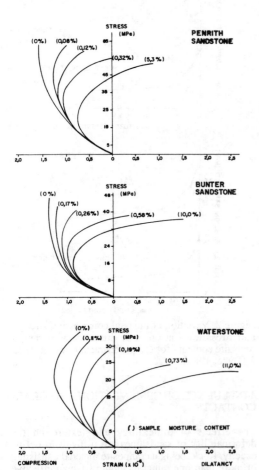

Figure 3. Variation of volumetric strain with sample moisture content for different sandstone types - for sample characteristics see Table 2 (Dyke, 1984).

relatively small, since most are quartz sandstones with different types and quantities of cementing materials, such as carbonates and clays.

The amount of cementing material is important with respect to strength and deformability, for some types of cement. If the cementing material is silica, as strong as the quartz grains forming the rock, the cement should increase considerably the strength and stiffness of the sandstone. However, where weak cement is present, for example clay or carbonate, it is not the dominating factor for influencing strength and deformability.

Dobereiner (1984) studied sands and sandstones (see Table 3 for their characteristics), with clay matrix in the range of 3 to 35% of total solids, and showed that the amount of cement did not relate with either strength or deformability. More recently, Nunes (1989) has also performed a large number of strength and deformability tests on sandstones from deep boreholes drilled within several coastal sedimentary basins in Brazil. The amount of carbonate and/or clay cement/matrix within these materials ranged from 0 to 60%, and could not be directly correlated with their geotechnical properties. As observed in Figure 5, the stress-strain curves obtained from sandstones possessing a wide range of carbonate contents (see Table 4 for characteristics), show a great scatter in terms of their strength and deformability behaviour. The convex shape of the stress-strain curves is probably due to the "hardening" of the material caused by the closure of the texture, which is related to the formation of microfissures within the cementing material or at the grain contacts, at high stress levels below sudden macroscopic failure. The closure of the texture will therefore increase grain contacts and as described by Dobereiner (1984), there is a direct relationship between grain contact area, and the strength and deformability of weak sandstones. Figure 6 shows the relationship obtained between the deformability modulus, measured at stress levels below the onset of dilatancy, with the percentage of grain to grain contact for the weak clay cemented sandstones of Table 3.

348

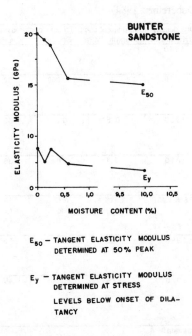

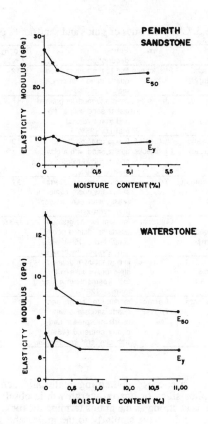

Figure 4. Relations of elasticity modulus with sample moisture content - for sample characteristics see Table 2 (Dyke, 1984).

E_{50} – TANGENT ELASTICITY MODULUS DETERMINED AT 50% PEAK

E_y – TANGENT ELASTICITY MODULUS DETERMINED AT STRESS

LEVELS BELOW ONSET OF DILATANCY

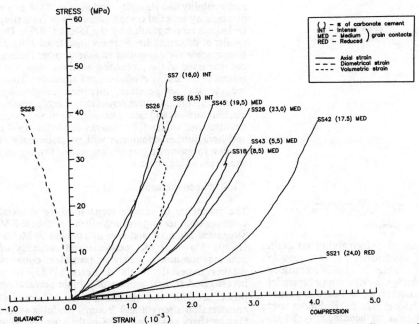

Figure 5. Stress-strain curves of carbonate cemented sandstones from coastal sedimentary basins in Brazil - for sample characteristics see Table 4 (Nunes, 1989).

Table 3. Characteristics of sands and sandstones of Figure 6 and 7 (Dobereiner, 1984).

Material	Age	Petrographic description	Porosity (%)	Sat. compressional wave vel. (m/s)	Dry density (Kg/m3)	Sat. uniaxial comp. strength (MPa)	Triaxial compression C(MPa)	ϕ(°)
Coina sand (Portugal)	Pleistocene	fine to medium grained quartz sand with clay matrix (23%)	32	–	1960	–	0.04*	39.9*
Castanheira sand (Portugal)	Jurassic	coarse to medium grained quartz sand with a clay and iron hydroxide matrix (35%)	35	400	1970	–	0.0003	32
Ferrel sand (Portugal)	Jurassic/ Cretaceous	fine to coarse grained quartz sand with a clay and iron hydroxide matrix (30%)	33	1760	1930	0.5*	0.2	36
Kidderminster sandstone (U.K.)	Triassic	medium grained quartz sandstone poorly cemented (<8%) with clay and/or iron hydroxide	31	1000	1810	2.1	2.6	36
Bauru sandstone (Brazil)	Cretaceous	medium to fine grained quartz sandstone poorly cemented (<15%) with clay and iron hydroxides	27-35	2100-3100	1750-1980	0.4-5.5	0.13 to 1.0	30 to 34.5
Lahti sandstone (Turkey)	Pliocene	fine to medium grained lithic graywacke with rock fragments in a clay matrix (25%)	28	2030	1990	3.6	1.1*	54.5*
Waterstones (U.K.)	Permo-Triassic	medium to fine grained quartz sandstone with quartz overgrowth and poorly cementd (<5%) with clay and iron hydrox.	25	2060	1970	23	5.8	45

* samples not fully saturated

If the presence of cementing material, such as amorphos silica or quartz overgrowth is observed, which is as strong as the grains forming the rock, the cement will behave similarly to the grains and will increase the area of grain contact.

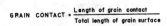

$$\text{GRAIN CONTACT} = \frac{\text{Length of grain contact}}{\text{Total length of grain surface}}$$

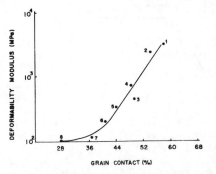

Figure 6. Dependence of deformability on grain contact in weak sandstones - for sample characteristics see Table 3. 1- Waterstones (perpendicular to bedding); 2- Waterstones (parallel to bedding); 3- Lahti sandstone; 4- Bauru sandstone B3; 5- Bauru sandstone B2; 6- Kidderminster sandstone (perpendicular to bedding); 7- Bauru sandstone B1; 8- Kidderminster sandstone (parallel to bedding), (Dobereiner, 1984).

5 TESTING PROCEDURE

The study of the behaviour of rocks over the pre-failure stress range is as important as the study of their behaviour at failure. In most cases, deformability and elasticity moduli of sandstones are obtained by uniaxial compression; this is the testing technique recommended by the ISRM (1978). The results of deformability tests obtained from uniaxial compression are dependant on several other factors, not discussed in the preceeding sections, which are related to the testing conditions and procedure. These include the temperature, direction of loading, specimen volume, specimen-platen contact, testing machine stifness and rate of loading. All these aspects are discussed in detail by Hawkes & Mellor (1970), and here further comments will be made only in relation to specimen preparation, rate of loading and strain measuring devices.

5.1 Specimen Preparation

The procedure of sample preparation for uniaxial compression testing is described in the ISRM Suggested Methods and also by Hawkes & Mellor (1970). However the recommended accuracy of sample preparation is not always possible to obtain in weaker materials. Pells & Ferry (1983) have investigated several aspects of sample preparation procedure for the determination of uniaxial compression strength and Youngs modulus in the Hawksesbury sandstone (UCS in the range of 10 to 65 MPa) from the Sydney Basin. The main parameters investigated were:

Table 4. Characteristics of sandstones from coastal basins from Brazil of Figure5 (Nunes, 1989).

Sample	Petrographic description	Porosity (%)	Dry density (Kg/m3)	Sat. uniaxial comp. strength (MPa)	Grain diam. Av. (mm)	% Carbonatic cement
SS6	fine grained subarkose poorly cemented with carbonates	15.4	2260	41.2	0.15	6.5
SS7	medium to coarse grained lithic graywacke cemented with carbonates	6.0	2530	47.0	0.6	16.0
SS18	fine grained subarkose poorly cemented with carbonates	13.6	2320	31.2	0.15	8.5
SS21	medium grained feldspatic graywacke cemented with carbonates	7.0	2330	7.8	0.25	24.0
SS26	medium grained feldspatic graywacke cemented with carbonates	18.5	2190	40.4	0.35	23.0
SS42	medium grained subarkose cemented with carbonates	9.2	2490	37.8	0.20	17.5
SS43	medium grained subarkose poorly cemented with carbonates	12.7	2360	32.4	0.30	5.5
SS45	coarse grained sublitharenite cemented with carbonates	8.4	2480	41.3	0.50	19.5

- effects of non-parallel ends
- effect of L/D < or > 2.0
- effects of end flatness.

The results obtained permitted the following conclusions to be drawn: for determination of Youngs modulus, the use of only two axial gauges on the specimens did not make any significant difference when the specimens had only a L/D ratio of 2.0 or were cut with non parallel (but flat) ends. For uniaxial compression tests, it was shown that ends non-parallel by up to 2° did not influence greatly the results. However strengths obtained from the group of samples with non flat or rounded ends, were significantly reduced.

The sample preparation for the weaker materials referred to earlier in this paper, was carried out by the "hand" use of fine sandpaper on a flat surface to flatten the ends. These delicate samples could not be ground to the required standards with a conventional thin section surface grinder, due to their sensitivity to water and consequent risk of desaggregation. The results obtained from these tests were reproducable and confirmed the conclusions outlined by Pells & Ferry (1983).

The standard of sample preparation that is obtainable is directly related to the quality of the rock. For very weak and friable rocks, it is very difficult to obtain high standards of sample preparation, however, it has been observed that on these materials there is no need to follow strictly the ISRM recommendations.

5.2 Rate of Loading

The rate of loading suggested by the ISRM (1978) is either, in the range of 0.5 to 1.0 MPa/sec., or alternatively at a constant loading rate such that macroscopic failure will occur in 5 to 10 minutes. However, researchers tend to use their own loading rate, which in general depends on the available equipment. Table 5 reveals a predominance of tests performed with constant strain rates of the order of 10 microstrain/sec.

With increasing loading rate, strength tends to increase. Sangha & Dhir (1972) on testing the Laurencekirk sandstone (UCS= 70MPa) at various strain rates, ranging from a time to failure of 3 seconds to 1 month, obtained a strength variation of around 20%. More recently, Singh et al (1989) presented results in which the strength of sandstone increased from 74MPa at 10E-4 strain/sec. to 112MPa at 10E3 strain/sec. This is a change of 50% over 10E7 orders of magnitude of strain rate. The Young's modulus determined from these same samples, showed an increase of 30 to 40% over the same strain rate ranges.

The importance of the rate of loading appears to increase as the strength of the rock decreases. The deformation of weaker rocks is more highly influenced by time dependant deformation. For instance, the Waterstones (UCS=23MPa) were tested

351

at three different strain rates, resulting in a strength that was reduced by approximately 10% for a strain rate variation of 25 times (1.6E-2 to 4.1E-1 strain/sec.). Figure 7 shows the uniaxial compression stress-strain curves for saturated Waterstones (for material characteristics see Table 3), at the different strain rates. Dilatancy starts at lower stress levels at lower rates of strain, and the deformability modulus, as well as the ultimate strain increase slightly.

These results indicate that the rate of loading can affect the strength and deformability of sandstones. However, at the narrow range of strain rates recommended by ISRM, only small differences are observed, and thus for the routine testing of sandstones, this range of loading rates can be considered to be adequate.

5.3 Strain Measurement

The ISRM (1978) suggested method for the determination of the deformability of rock materials in uniaxial compression mentions the use of various strain measuring devices, and recommends that they should be robust, stable and with a strain sensitivity of the order of 10E-6.

The most common device for measuring strain in rocks under applied load is electrical strain gauges. However, the use of these can sometimes present a few inconveniences:
- they require careful preparation of the rock surface to be glued, which can be a difficult task in weak or friable rocks,
- a high sample moisture content can cause difficulties in gluing the strain gauges,
- the limited area of measurement can induce erroneous readings if microfractures develop away from the measurement area,
- the impossibility of measuring post-peak deformations,
- the cost of the gauges and the long sample preparation time.

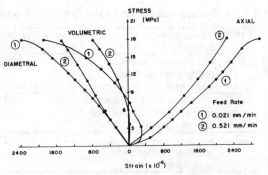

Figure 7. Stress-strain curves at different loading rates for saturated Waterstones - for material characteristics see Table 3 (Dobereiner, 1984).

Several strain measuring devices have in the past been developed with the aim of producing a reliable and accurate means of measuring sample deformations (Priest & Selvakumar,1982; Height, 1983; Attinger & Koppel, 1983; Jardine et al, 1985). The development of an additional strain measuring device, of the clip-gauge type, by Nunes (1989) represents one more effort in this direction.

The clip-gauge strain measuring device is a simple and practical tool and is constructed by bonding strain gauges to the upper and lower sides of a solid piece of channel-shaped high strength steel, as shown in Figure 8. This device is mounted directly on the core sample, which deforms and deflects the arms of the device causing a variation in strain gauge resistivity. The displacements and geometrical proportions of the arms are dependant on the magnitude of the sample deformations.

The clip-gauge device has been shown to possess the required sensitivity and accuracy that is necessary for rock testing, and also possesses the following advantages:

Table 5. Strain or loading rate on uniaxial compression tests used by several authors (Nunes, 1989).

Reference	Rock type	sample dimensions H/D	D(mm)	Rate of strain/ loading
Colback & Wiid, 1965	sandstones	2	25	0.7 MPa/sec
Handin et al., 1967	sandstones	2	13	0.1-100 microstrain/sec
Sangha & Dhir, 1972	sandstones	2.5	54	2.5 microstrain/sec
Preston, 1976	sedimentary rocks	3	13	0.5 millistrain/sec
Bell, 1978	Fell sandstone	2	38	0.07 MPa/sec
Ferreira et al., 1981	Caiua sandstone	2	54	0.07 mm/min
ISRM, 1978	rocks	2.5-3.0	54	0.5-1.0 MPa/sec
ASTM, 1980	rocks	2.0-2.5	47	failure in 5-10 min
Yoshinaka & Yamabe,1981	sedimentary rocks	2	50	0.0025 mm/min
Pells & Ferry,1983	sandstones	2	54	0.1-0.2 MPa/sec
Dobereiner, 1984	sandstones	2	38	0.114 mm/sec
Dyke, 1984	sandstones	2	54	0.01-0.02 mm/sec
Gunsallus & Kulhawy,1984	sedimentary rocks	2.0-2.5	54	12.5-25 microstrain/sec
Hosseini & Hayatdavoudi,1985	sandstones	2	25	0.075 MPa/sec
Faria santos, 1986	sandstones	2.0-2.5	54	416 KPa/sec
Nunes, 1989	sandstones	2	38	0.02 mm/sec

- it can be directly mounted on the sample with no special preparation,
- it allows the complete measurement of the stress-strain curve, even after failure,
- low cost of use.

Figure 9 presents details of the sample set-up for axial and radial clip-gauges. The excellent performance of these devices is well illustrated by the stress-strain curves of the sandstone and aluminium samples presented in Figure 10. The high resolution of the clip-gauge and its larger measurement area permit the measurement of global deformations within the sample, including the creation of microfissures at the onset of dilatancy.

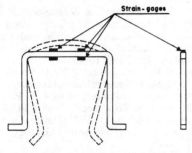

Figure 8. Clip-gauge for the determination of large deformations (Nunes, 1989).

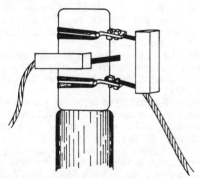

Figure 9. Axial and diametral transducers fitted on a rock sample (Nunes, 1989).

6 CONCLUSIONS

The use of international standards, such as the ISRM suggested rock testing methods, should be strongly encouraged by all engineers involved with rock testing, in order to obtain a broader understanding of rock behaviour, by allowing test results from different rock types and countries to be compared.

An improved understanding of rock deformation, including the accurate interpretation of rock texture and mineralogy is in practice not always obtainable during routine testing, however it should be included in research testing programmes, in order to enable understanding of rock deformability to develop. It is

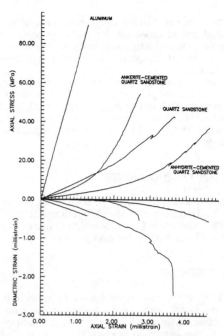

Figure 10. Stress-strain curves of sandstones and aluminium obtained with the clip-gauge device (Nunes, 1989).

observed, from the results described in this paper, that the deformability of sandstones is greatly dependant on grain contact relations, and that the development of microfissures within the cement or grain contacts is dependant on several testing conditions.

In practice, sandstone deformability testing should include the following procedures:
- the evaluation of the volumetric strain-stress curves, to allow increased understanding of rock deformability behaviour, and enable the correct interpretation of the rock's elastic constants. Young's modulus should be determined at stress levels below the onset of dilatancy,
- due to the sensitivity of rock deformation and strength, to sample moisture content, it is easier to test samples in the saturated state or at high values of moisture content. The accurate control of moisture contents in the lower range is difficult and time consuming. Submersion of the sample within water for 24 hours, should guarantee a moisture content sufficiently high to produce reproducable strength and deformability test results,
- Young's modulus measured at stress levels below the onset of dilatancy, is not influenced by sample moisture content, and can therefore be determined at any moisture condition,
- sample preparation requirements, for the weaker materials, can be less strict than recommended by the ISRM. Pells & Ferry (1983) outline that non parallel (but flat) specimen ends, by up to 2°, are sufficiently accurate for strength and deformability testing,

353

- the range of loading rates recommended by the ISRM are adequate for the testing of sandstones. However to facilitate comparison of data with other rock types, it is desirable to define a more precise strain rate,
- recently developed strain measuring devices, such as the clip-gauge, seem to be more practical for systematic rock testing than resistance strain gauges. The technical and economical advantages of these systems should make their use more frequent in the future.

ACKNOWLEDGEMENTS

The Authors acknowledge the financial support given to L. Dobereiner by the "Conselho Nacional de Desenvolvimento Cientifico e Tecnologico - CNPq" (Brazil) which made this work possible.

REFERENCES

ASTM, 1980. Standart test method for elastic moduli of intact rock core specimens in uniaxial compression. ASTM (American Society of Testing Materials). D 3148-80.

Atkinson, B.K. 1984. Subcritical crack growth in geological materials. J. Geophys. Res. 89: N.B6: 4077-4144.

Attinger, R.D. & Koppel, J. 1983. A new method to measure lateral strain in uniaxial and triaxial compression tests. RM & Rock Engng. 8:73-78.

Bell, F.G. 1978. The physical and mechanical properties of the Fell sandstones, Northumberland, England. Engineering Geology. 12: 1-29.

Colback, P.S.B. & Wiid, B.L. 1965. The influence of moisture content on the compressive strength of rocks. Proc. 3rd Canadian Rock Mechanics Symp. 65-83.

Denis, A. ; Durville, J.L.; Massieu, E. & Thorin, R. 1986. Problèmes posés par un calcaire très poreux dans l'étude de la stabilité de une carrière souterraine. Proc. 5th Congress Int. Assoc. of Eng. Geology, Buenos Aires, 2.3.2: 549-557.

Dobereiner, L. 1984. Engineering geology of weak sandstones. PhD Thesis, Imperial College, Univ. of London. 472p.

Dobereiner, L. & Dyke, C.G. 1986. Caracteristicas de deformabilidade de arenitos em função da variação de umidade na rocha. 2nd South American Rock Mech. Symp., Porto Alegre. 2: 57-66.

Dobereiner, L. & de Freitas, M.H. 1986. Geotechnical properties of weak sandstones. Geotechnique, 36: 79-94.

Dobereiner, L. & Oliveira, R. 1986. Site investigation on weak sandstones. Proc. 5th Congress Int. Assoc. of Eng. Geology, Buenos Aires, 2.1.2: 411-421.

Dyke, C.G. 1984. The pre-peak deformation characteristics of sandstone at varyng moisture contents. MSc Thesis, Imperial College, Univ. of London, 135p.

Faria Santos, C.A.F. 1986. Effect of laboratory-simulated weathering on the properties of Loyalhanna sandstones. MSc Thesis, Pennsylvania State University. 109p.

Ferreira, R.; Monteiro, L.C.C.; Peres, J.E. & Prado Jr., F.A.de A. 1981. Análise de alguns fatores que influem na resistencia a compressão do arenito Bauru. 3rd Brazilian Congress of Eng. Geology.(ABGE), Itapema. 3: 89-102.

Gunsallus, K.L. & Kulhawy, F.H. 1984. A comparative evaluation of rock strength measures. Int. J. Rock Mech. Min. Sci.& Geom. Abst. 21: 5: 233-248.

Handin, J.; Heard, H.C. & Magouirk, J.N. 1967. Effects of the intermediate principal stress on the failure of limestone, dolomite and glass at different temperatures and strain rates. J. Geophys. Res. 72: 611-640.

Hassani, F.P. ; Whittaker, B.N. & Scoble, M.J. 1979.Strength characteristics of rocks associated with opencast coal mining in the U.K. Proc; 20th U.S. Symp. on Rock Mech., Austin. 347-356.

Hawkes, I. & Mellor, M. 1970. Uniaxial testing in rock mechanics. Engineering Geology. 4:177-285.

Height, D.W. 1983. Laboratory investigations of sea-bed clays. PhD Thesis, Imperial College, Univ. of London.

Homand-Etienne, F. 1985. Comportement mécanique des roches en fonction de la temperature. PhD Thesis, Institute National Polytechnique de Lorraine.

Hosseini, M.S. & Hayatdavoudi, A. 1985. Reservoir characterization of tusaloosa sand by mineralogical and petrological data; Soc. of Petroleum Eng., SPE 14274

Howarth, D.F. 1987. The effect of pore-existing microcavities on mechanical rock performance in sedimentary and crystalline rocks. Int. J. Rock Mech. Min. Sci. & Geom. Abst. 24: 4: 223-233.

ISRM, 1978. Suggested methods for determining the uniaxial compressive strength and deformability of rock materials. Int. J. Rock Mech. Min. Sci. & Geom. Abst. 16: 2: 135-140.

Jardine, R.J.; Brooks, N.J. & Smith, P.R. 1985. The use of electrolevel transducers for strain measurements in triaxial tests on weak rocks. Int. J. Rock Mech. Min. Sci. & Geom. Abst. 22: 5: 331-337.

Kitaowa, M.; Endo, G. & Hoshino, K. 1977. Influence of moisture on the mechanical properties of soft rock. 5th National Symp. on Rock Mech., Japan.

Koshima, A.; Frota, R.G.Q.; Lozano, M.H. & Hoshisk, J.C.B. de F. 1983. Comportamento e propriedades geomechanicas do arenito Bauru. Simp. Geot. Bac. Alto Parana, ABGE-ABMS-CBMR, Sao Paulo, 2b: 173-189.

Nunes, A.L.L. 1989. Um estudo sobre as caracteristicas de resistencia e deformabilidade de arenitos. MSc Thesis , Pontificia Universidade Católica do Rio de Janeiro. 280p.

Okubo, S. & Nishimatsu, Y. 1985. Uniaxial compression testing using a linear combination of stress and strain as the control variable. Int. J. Rock Mech. Min. Sci. & Geom. Abst. 22: 5: 323-330.

Pells, P.J.N. & Ferry,M.J. 1983. Needless stringency in sample preparation standards for laboratory testing of weak rocks. Proc. 5th Cong. Int. Soc. of Rock Mech., Melbourne. sec. A:203-207.

Preston, D.A. 1976. Correlation of certain physical and chemical properties of sedimentary rocks. Proc. 17th U.S. Symp. Rock Mech..2AB-1 - 2AB-6.

Priest, S.D. & Selvakumar, S. 1982. The failure characteristics of selected British rocks. A Report to the Transport and Research Laboratory, Department of Environment and Transport, Imperial College, Univ. of London.

Sangha, C.M. & Dhir, R.K. 1972. Influence of time on the strength deformation and fracture properties of a Lower Devonian sandstone. Int. J. Rock Mech. Min. Sci. & Geom. Abst. 9: 343-354.

Singh, D.P.; Sastry, V.R. & Srinivas, P. 1989. Effect of strain rate on mechanical behaviour of rocks. Proc. Symp. Rock at Great Depth, Pau. 109-114.

Wiid, B.L. 1970. The influence of moisture content on the pre-rupture fracturing of two rock types. Proc. 2nd Cong. Int. Soc. Roch Mech., Belgrad. 3: 239-245.

Yoshinaka, R. & Yamabe, T. 1981. Deformation behaviour of soft rocks. Proc. Int. Symp. on Weak Rock, Tokyo. 87-92.

6th International IAEG Congress / 6ème Congrès International de AIGI, © 1990 Balkema, Rotterdam. ISBN 90 6191 130 3

Determination of residual bond strength by the pulling test method

La détermination de la force de liens résiduels par la méthode de l'essai de traction

E.J.Ebuk, S.R.Hencher & A.C.Lumsden
Department of Earth Sciences, University of Leeds, Leeds, UK

ABSTRACT: Shear test results on carefully sampled intensely weathered rocks often show an apparent cohesion intercept on the shear stress axis which may be attributed, in part, to relict structural bonds. Such materials may also exhibit "virtual preconsolidation" behaviour (in the voids ratio vs effective stress plots), which can also be attributed to collapse of relict fabric. This paper describes a tension test developed at Leeds University for the quantification of residual bond stength. The test is simple to carry out and results obtained in the laboratory are compared to data from field in-situ tests. Its applications to practical engineering geological situations are discussed.

RESUMÉ: Les resultats d'analyse de cisaillement sur l'echantillons bien preparé de roches intensivement altéres, montrent souvent, un cohesion apparent d'intercept sur l'axe d'effort de cisaillement qui peut être attribuée, en partie, aux veuve de liens structurale. Une telle terre peut aussi exposer un charactère de "preconsolidation virtuelle" (sur le diagramme du taux de porosité contre charge effective) qui peut être aussi attribuer à l'enfondrement de liens fabrique. Cet article decrives un test de tension dévèlopé à l'université de Leeds pour la quantification de la force de liens résiquat. Le test est simple à éxecuter et les resultats obtenu dans la laboratoire sont comparés à ceux du terrain. Ses application au traveux geotechnique practicale sont bien discutés.

1. INTRODUCTION

The contribution of relict bonding to engineering behaviour of chemically weathered granitic rocks and residual soils has been discussed by several workers particularly with respect to apparent quasi-preconsolidation behaviour in compression tests (Vargas 1953, Wallace 1973, Pender 1971, Vaughan et al. 1988, Ruddock 1967, Brink and Kantey 1961, Sowers 1963). Using Lumb's decomposition index (1962), as a means of quantifying degree of weathering, Baynes and Dearman (1978) suggested that such behaviour might be explained by the development of clay bonds as weathering progresses. Vaughan et al. (1988) explained the behaviour of leached residual soils in terms of the effect of relict fabric on the basis of model testing and theory. On review, it seems certain that bonding in weathered igneous rocks as shown both in virtual preconsolidation behaviour and in providing increased shear resistance, manifested in part as an apparent cohesion must be attributed generally to more than one mechanism namely:

1. Primary bonding - the contribution of crystalline textural bonds which will decrease with increasing degree of weathering, and
2. Secondary bonding - due to the development of clay minerals and cementation by silicates, iron oxides and in some circumstances, carbonates.

The contribution of each mechanism will depend on the type and degree of weathering and the parent rock type.

Existence of these bonds leads to differences in the engineering behaviour of chemically weathered rocks from that of typical uncemented sedimentary soils of similar grading. For example, in an oedometer test, cohesive bonds within the often open fabric of weathered rocks resist compression until a yield stress which marks the collapse of the relict fabric, following which with increased stress consolidation proceeds normally. Precise determination of the yield stress in compression, shear or tension is essential for accurate prediction of material behaviour.

The oedometer test is the most commonly used method for determining yield stress in compression. However, paradoxically, the voids ratio - log effective vertical stress plot can sometimes give similar values for materials which are clearly weathered to different degrees. For very weakly bonded soils, the bonds may be destroyed during the first level of normal loading, giving a voids ratio - pressure plot similar to that of normally consolidated soils. Despite these disadvantages and the possibility of an overestimation of results (Wilmer et al, 1982) this method still yields useful information for the prediction of behaviour of materials under foundation loading conditions. Because the bonds are broken in compression, the results are not immediately applicable to situations such as slope stability where the bonding is destroyed in shear or tension.

Vargas (1953), working in weathered granite terrain in Brazil, demonstrated that at depths greater than 7m below the ground surface the virtual pre-consolidation pressure was equivalent to the overburden pressure but he also stated that at lesser depths the value is independent of overburden pressure and may range from 0 to 400 kPa. However, it is unlikely that depth can be generally be used as an indicator of bond strength, because of the complexity and variability of weathering profiles.

Attempts at measuring, in shear, the yield stress due to relict bonding have met with limited success. The shape of the strength envelope achieved is dependent upon many factors, particularly degree of saturation, and with weak material sensitive to disturbance, bonds may be disrupted during sample preparation and application of confining pressure.

Maccarini (1987) used the Brazilian test to measure the bond strength of a weakly cemented sand by the indirect tensile test method, and this technique may be applicable to some weathered rocks. The major difficulty will arise in sample preparation, particularly in the case of coarse grained rocks. The test may also be inapplicable to very weak samples.

This paper describes an alternative laboratory method for determining the bond strength in tension and forms part of continuing research on the characterisation and engineering properties of weathered granitic materials.

2. MATERIALS

Tests have been carried out on highly and completely weathered and hydrothermally altered granites from South west England. Four different rock types have been tested as listed in Figure 1. Figure 2 shows the typical characteristics of the rock material weathering grades.

3. TEST DETAILS

A modifed version of a new shear box, designed at Leeds University for testing weak rocks, was used as illustrated in Figures 3 and 4. The design of the basic shear box is described by Ebuk, et al (1990, in press). Firstly, an undisturbed block sample of the weathered material is trimmed to fit into the box. Two versions of the test have been developed for the tension test. For an early series of tests a hole was drilled into samples and a serrated rod fixed into place using epoxy resin. More recently a brass plate is fixed onto one free surface of the sample, care being taken to ensure that first all loose particles are removed. For both methods, the free surface around the test area is covered by a restraining plate. When the resin is set, the whole assembly is placed into the direct shear carriage and held rigidly in place by means of brass blocks and carriage screws. A horse-shoe shaped loading ram is fitted with end to the serrated rod or a rod protruding from the plate bonded to the sample, and the other to the plunger connected to the proving ring. The force is applied through the proving ring. Tensile force, horizontal displacement and vertical distortion of the horse-shoe shaped loading ram are measured by transducers connected through an interface to a computer. The tension force is increased until either a cone of material (rod) or a thin film of soil (plate) is pulled from the sample. At the end of the test, the dimensions of the failed material are measured and used in determining the bond strength.

4. RESULTS AND DISCUSSION

The stress - strain curves from the plate technique show an increase of strength (Fig

Rock Name	Quarry	Nomenclature
I. Weathered		
Biotite granite	Hingston Down	HDG4, HDG5
II. Hydrothermally altered		
Topaz granite	Trethosa Pit	TPG4, TPG5
Globular Quartz granite	North Goonbarrow Pit	NGG4, NGG5
Megacrystic Lithium mica granite	Virginia Pit	VPG4, VPG5

Figure 1. Materials used in the study.

Grade and Description	Typical Characteristics
IV Highly Decomposed	N-Schmidt hammer rebound value 0-25. Does not slake readily in water. Penetrometer strength > 250 kPa. Large pieces broken by hand. Grains plucked from surface.
V Completely Decomposed	No rebound from N Schmidt hammer. Slakes readily in water. Rock texture preserved.

Figure 2 Decomposition grades for weathered granite and volcanic rocks. (modified from Hencher and Martin, 1982).

Figure 3. Photograph of the set up

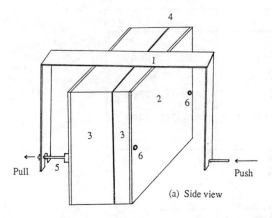

1. Horse shoe loading ram
2. Restraining plates
3. Brass box containing the sample
4. 2mm Aluminium plate
5. Extension rod from plate bonded
   to the sample surface
6. Screws
7. Exposed part of the sample
8. Plate bonded to the sample

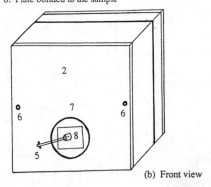

Figure 4. Schematic diagram of test
apparatus (Plate method)

5) with strain till a peak value is reached followed by abrupt failure. The strain at failure is usually small, and often less than 1 mm. Pulling is essentially horizontal until failure occurs.

Using the rod pull out method, the grade IV materials usually fail by sudden brittle fracture. For the weaker grade V material, failure is usually gradual, often with no sudden brittle failure, and occurs at relatively large strains. Occasionally, a few irregular microfractures develop prior to failure making it difficult to define the geometry of the failure surface precisely. Variables for the rod test include, depth of the hole in the sample and optimum radius of the hole on the top restraining plate to

avoid grain locking. Values of the peak strength for these materials from individual tests are presented in Table 1.

Table 1 Peak bond strength of the materials tested.

	Bond Strength (kPa)	
Sample	Plate	Rod
HDG4	241.90	–
VPG4	–	73.99
TPG4	–	129.24
NGG4	–	178.45
HDG5	29.11	27.05
VPG5	19.01	10.52
TPG5	14.45	26.47
NGG5	39.93	26.90

As can be seen from the table, where comparative tests have been conducted, the surface plate technique generally gives a higher bond strength than the hole method for grade V material. The exception is TPG5 which is a fairly friable material with low cohesion but densely packed. Its behaviour in triaxial test was similar to that of a densely packed clean sand. In this case, the higher value for the rod method may be partially due to grain locking caused perhaps by the small radius of the restraining plate used for the test.

Hencher et al (1990, in press) described a technique for determining the residual bond strength in the field using a hand penetrometer with load spreader to perform insitu direct shear tests on unconfined blocks of similar materials to those tested here (Figure 6). A comparison of test rssults shows that strengths measured in the field are similar despite the inherent difference of the two types of test. In detail the correlation is better for the grade V materials. The relatively high values for NGG5 given in table 1 as compared to the field test results perhaps reflect differences in moisture content (14% insitu, 8% laboratory). Grade IV materials generally give higher values by the pull out method than in the field and are considered more representative of the natural materials. In the field, relatively thin blocks had to be prepared for grade IV material due to the limited force that could be applied by the hand penetrometer.

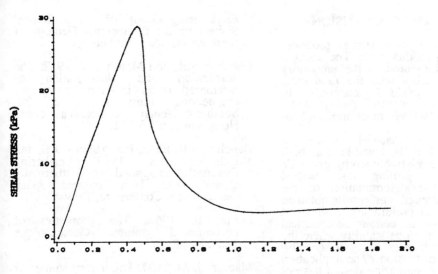

FIGURE 5. SHEAR STRESS VS HORIZONTAL DISPLACEMENT FOR NGG5

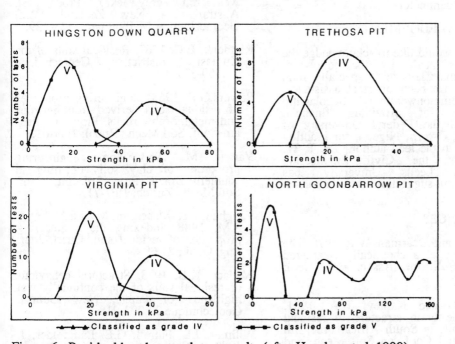

Figure 6. Residual bond strength test results (after Hencher et al, 1990)

5. SUMMARY AND CONCLUSION

Intense chemical weathering produces weakly bonded residual soils. The strength of the bonds measured in the laboratory compares favourably with the field data, particularly for grade V materials. In general the field method described by Hencher et al (1990) is most applicabl for testing
weaker grade V materials, while the laboratory method is suitable for both grade V and the relatively strong grade IV materials. The pulling test method facilitates precise determination of the residual bond strength and results obtained may be useful in preliminary design and stability analysis in certain geotechnical situations such as slope stability where the initial shear surface is likely to develop either in shear or tension. The application of such data for enginering design has not yet been established but it is expected that through correlation with other standard strength tests and field analysis the relatively simple tests described will provide a useful method for establishing engineering parameters.

6. ACKNOWLEDGEMENTS

The authors would like to acknowledge the contributions made by the following: English China Clays and especially John Howe who has been of great assistance. The Commonwealth Scholarship Commission for providing financial support and the Federal University of Technology, Owerri, Nigeria, for granting study leave to the lead author. Dr. T. W. Cousens of the Civil Engineering Department, Leeds University, has provided useful suggestions and comments.

7. REFERENCES

Baynes, F.J. and Dearman, W.R. 1978. The microfabric of a chemically weathered granite. Bull. Int. Assoc. Eng. Geol; 18:91-100.

Brink, A.B. and Kantey, B.A. 1961. Collapsible grain structure in residual granite soils in South Africa. Proc. 5th International Conf. on Soil Mech. and Foundation Engineering, Paris, France, pp 611-614.

Ebuk, E.J. Hencher, S.R. and Lumsden, A.C. 1990. The influence of structure on the shearing mechanism of weakly bonded soils. 26th Annual Conf. of the Engineering Group of the geological Society on the Engineering Geology of weak rocks, Leeds University.

Hencher, S.R. and Martin, R.P. 1982. The description and classification of weathered rocks in Hong Kong for engineering purposes. Proc. 7th Southeast Asian Geotechnical conf., Hong Kong; p125-142.

Hencher, S.R. Ebuk, E.J. Abrams, J.H. and Lumsden, A.C. 1990. Weathering Description as applied to hydrothermally altered rocks. 10th Southeast Asian Geotechnical Conference, Taiwan.

Lumb, P. 1962. The properties of decomposed granite. Geotechnique 12:226-243.

Maccarini, M. 1987. Laboratory studies of a weakly bonded artificial soil. Unpublished PhD Thesis, University of London.

Pender, M.J. 1971. Some properties of weathered greywacke. Proc. 1st. Australian - New Zealand Conf. Geomechanics, Melbourne, 1:423-429.

Ruddock, E.C. 1967. Residual soils of the Kumasi district. Geotechnique 17:359-377.

Sowers,G.F. 1963. Engineering properties of residual soils derived from igneous and metamorphic rocks. Proc. 2nd Pan-am Conf. Soil Mech., Brazil; 1:39-61.

Vargas, M. 1953. Some engineering properties of clay soils occuring in southern Brazil. Proc. 3rd. Int. Conf. Soil Mech., Zurich; 1:67-71.

Vaughan, P.R. Maccarini, M. and Mokhtar, S.M. 1988. Indexing the engineering properties of residual soil. Quart. Jour. Eng. Geol.; 21:69-84.

Wallace, K.B. 1973. Structural behaviour of residual soils of the continually wet highlands of Papua, New Guinea. Geotechnique 23:203-218.

Willmer, J.L. Futrell, G.E. Langfelder, J. 1982. Settlement predictions in Piedmont residual soils. Proc. ASCE specialist conf. on construction and engineering on tropical and residual soils, Honolulu, p629-646.

6th International IAEG Congress / 6ème Congrès International de AIGI, © 1990 Balkema, Rotterdam. ISBN 90 6191 130 3

Stress-history dependent behaviour of soft clay

L'influence de l'histoire de l'effort sur le comportement d'argile mou

M.K.El-Rayes
Suez Canal University, Egypt

F.A.K.Hassona, M.A.Hassan & A.M.Hassan
Civil Engineering Department, Minia University, Egypt

ABSTRACT: The behaviour of soft clay during undrained loading is significantly influenced by consolidation stress history. A simple approach was adopted in an attempt to examine the possible effect of stress history on the behaviour of soft clayey soil. Samples were consolidated to the same mean effective stress (P´) in the Bishop-Wesley triaxial cell along three different paths which were : isotropic, one-dimentional, and isotropic followed by drained compression. The results indicate that due to stress anisotropy, samples sustain higher undrained strength and smaller strains. Also, pore pressure generation was more likely for the isotropically consolidated samples. Stress history was found to have a fairly little effect on failure conditions (i.e. critical state line and and angle of internal friction).

RESUME: Le compotment de moulle argile pendant l´actions des charges sans drainage est plus gand affectué par l´histoire des efforts appliqués. Un approach simple a été effectué pour esseyer d´examiner L´influence de L´histoire d´un effort sur le comportment d´un mou argilé sol. Les echantillons sont fait solid toujours par l´urage de la même effort (p´), sur le principe de la Bishop-Wesley cell a´trois dimensions. Trois route differents ont été utilisé, ces sont: isotropique, uni-dimensiel ou isotropique suivi par comprêssion a´drainage.les resultant indiques l´acceptance des echantillons de plus grand valeurs de solidaité, a´cause de ll´anisotropie de l´effort, mais aussi de moins deformation sous charge. En meme temps, une generation de pression dans les pores a´eté trouvé pour les echantillions solidifiés par la route isotropie.

1 INTRODUCTION

The behaviour of soil as a composite material is generally influenced by several factors. Some of these factors are material dependent factors which come to play as a result of the conditionsof the material, its shape, its internal composition, and its size. Soil behaviour is also affected by stress dependent factors which are due to the stress existing in, and/or imposed on the material.

One of the stress dependent factors that affects the soil behaviour is the consolidation pressure. The influence of consolidation on the behaviour of soil has been investigated by many reasearch workers. El-Ruwayih 1975´tested identical specimens of sand under confining pressures ranging from 290 to 870 Kpa. His results indicated that the brittleness of the specimens and their tendency to dilate decrease with an increase in consolidation pressure. At high confining pressure the volumetric strains

become totally compressive.

Ilyas (1983) tested samples of clay under different isotropic consolidation pressures. His results indicted that samples which were consolidated to higher initial values of P sustained higher values of deviator stress at failure.

Period of consolidation is also a stress dependent factor that affects soil behaviour It has been suggested by Ohsaki (1969) that the age of a sand deposit can affect its strength due to creep effect. This view was also confirmed by the work of Seed (1976). Daramola (1978) tested typical sand samples which are consolidated to the same confining pressure for a time of 0, 10, 30, and 152 days. He concluded that the deformation behaviour of sand is very much dependent on the period of consolidation.

The stress strain behaviour of soil is also dependent on the stress path and the previous stress history because soil deformation during primary loading are largely irrecoverable. The influence of stress history on the behaviour of soil may be quite significant under some circumstances due to the inherently inelastic behaviour of soil. The work done by Gerrard (1967) illustrated the stress path dependency of the deformation behaviour of sand. Similar results were obtained by Lade and Duncan (1976) and El-Sohby (1969). The results of Bishop and El-Din (1961) indicated that, wiyhin the the range of stress used, the angle of internal friction is independent of the stress history. Hassona (1986) concluded that sand samples consolidated isotropically are more susceptable to liquefaction than those consolidated anisotropically or isotropically followed by drained shear.

A simple approach was adopted in the present investigation in an attempt to examine the possible effect of one of the stress dependent factors such as the consolidation history on the undrained behaviour of a soft clayey soil. Application of Cam-Clay model developed by Schofield and Wroth (1968) to the results of the present investigation to predict stress-strain and strain-pore water pressure behaviour are also presented.

2 MATERIAL AND TESTING PROCEDURE

The material used to form the samples was a silty clay. Blocks of soil were taken from north of Nile Delta, Egypt. The blocks were dried, pluverized and thoroughly mixed to produce a homogenous soil batch. Several tests were carried out to classify this soil and to determine its properties. The gradiation curve of the tested soil is shown in Fig. 1 and its constituents are: 57% clay, 38% silt, and 5% fine sand. Its physical properties are: liquid limit of 74.6%, plastic limit of 31.7%, and specific gravity of 2.73. According to Casagrande plasticity chart it is classified as a silty clay with very high plasticity (CH).

In order to obtain uniform samples of known stress history , the clay was mixed at a mosture content of about its liquid limit by means of electric mixer. The mixer was then transfered and consolidated one-dimensionally in a hydraulic consolid-ation cell (Rowe cell) similar to that developed by Rowe (1966) under a vertical stress of 50 Kpa. After the application of the vertical pressure, readings of time and displacement were taken. This process took about two weeks until the clay was completely consolidated.

The triaxial samples were then retrieved by means of thin-walled tubes of 38 mm internal diameter and 40.5 mm external

diameter following the recommendations of Atkinson and Kubba (1981). The tubes were then sealed with wax and stored in constant temperature room.

Samples were then set up in a stress-path triaxial cell similar to that developed by Bishop and Wesley (1975). At the end of sample set up, they were saturated and consolidated under different consolidation paths. Three series of consolidated undrained triaxial tests were carried out. Each series contained five identical samples During consolidation phase, Samples were consolidated to the same mean effective stress value , P'_0 following three different stress paths similar to those adopted by Hassona (1986) and shown in Fig. 2.

p'_0 is defined by the equation

$$p'_0 = \acute{\sigma_1} + 2\acute{\sigma_3} / 3$$

Values of p'_0 of 90, 140, 200, 300 and 400 Kpa was used. Undrained loading was then applied until failure. Measurements of load, displacement, pore water pressure, and volume change was done electronically using measuring transducers which were connected to 16 channels logging system. The data logger is also connected to an IBM micro-computer used to analize and process the results. Complete discription of test equipment and instrumentations are described in details in Hassan (1990).

An oedometer test was also performed in order to determine the critical state parameters of the tested soil needed for the cam-Clay model application.

3 TEST RESULTS

3.1 Stress-strain and pore water pressure

The three test series conducted in this investigation will be refered hereafter as A, B, C as indicated in Fig. 2. Results of all the tests carried out in series A, B, and C have indicated that, specimens which were consolidated to higher initial values of p'_0 typically sustained higher values of deviator stress at failure irrespective of the type of consolidation path. Similarly, higher values of pore water pressure were observed when samples were tested under higher values of p'_0 . Figrs. 3 and 4 shows an example of such results.

Normalization of stress-strain curves for each set of tests was done by plotting q/p'_0 against € as shown in Fig. 5. It can be seen that the normalized behaviour is fairly independent of the consolidation

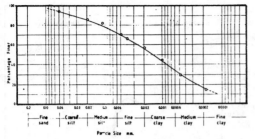

Fig.1 Grain size distribution curve for the tested clay

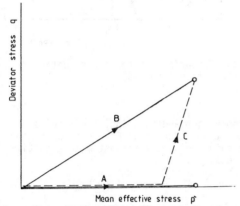

Fig.2 Schematic diagram fro the types of consolidation history used

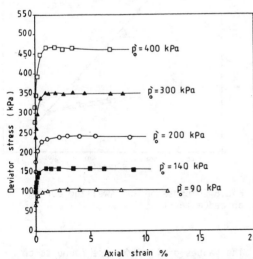

Fig.3 Axial strain-deviator stress curves for one-dimentionaly consolidated samples

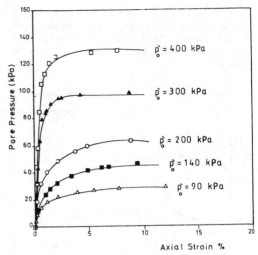

Fig.4 Axial strain-pore pressure relationships for K_0-consolidated samples

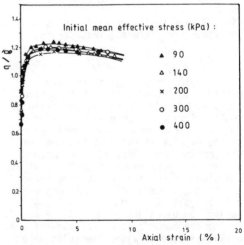

Fig.5 Nrmalized axial strain-deviator stress for samples consolidated following path C

pressure as expected from critical state theory, Schofield and Wroth (1968). Comparisons of the stress-strain behaviour for different sets of samples consolidated to the same value of p_0' indicated that there is no significant difference in the behaviour of samples which were consolidated following paths B and C. The ε - q curves of these samples are very similar in shape. However, samples which were isotropically

365

consolidated (i.e. series A) behaved in a quite different manner. These samples tend to get larger strains before comming to a peak value of deviator stress as shown in Fig. 6.

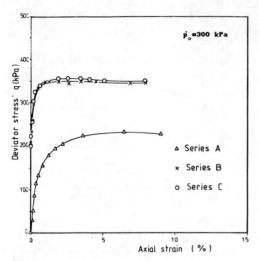

Fig.6 Axial strain-deviator stress behaviour for different test series

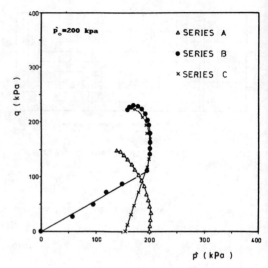

Fig.7 Comparison of the stress paths of different test series

compression and swelling. The results of this test is shown in Fig. 8 as aplot of specific volume, v against mean effective stress, p´.

A comparison of the stress paths of tests consolidated to the same $p_o^{'}$ along paths A, B, and C is shown, for example, in Fig. 7. One can observe that samples followed paths B and C sustained typical values of peak stresses at failure while the specimen consolidated along path A showed much less value of deviator stress and higher pore water pressure generation.

The above mentioned results may be attributed to the induced anisotropy during consolidation along paths B and C. During the consolidation process the clay particles tend to aline themselfs in a parallel array perpendicular to the loading direction. Thus, higher shear strength and lower pore water pressure generation are expected. Unlike the specimens consolidated along path A, the microstructure is expected to be flocculated and lower shear streangth and higher pore water pressures were observed.

3.2 Application of Cam-Clay model

In order to detrmine the slope of the virgin compression line ⋏ and the slope of the swelling line **x** , a one-dimentional cons-olidation test using the oedometer has been carried out. The test included both

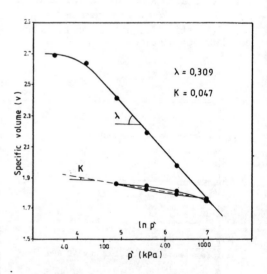

Fig.8 Compression and swelling during an oedometer test

The values of **x** and ⋏ were found to be 0.309 and 0.047 respectively. The parameter **Λ** is calculated as:

$\Lambda = 1 - \varkappa/\lambda$

substituting, we get the value of as 0.848. The critical state parameters for the clay used in the present investigation along with that for other clays are given in Table 1.

Table 1. Critical state parameters for the tested clay and other soils

Soil	λ	$\varkappa$	Λ	M	Γ
Tested clay	0.309	0.047	0.848	1.181	3.416
London clay [*]	0.161	0.062	0.614	0.888	2.759
Weald clay [*]	0.093	0.035	0.638	0.950	2.060
Kaolin clay [*]	0.260	0.050	0.807	1.020	3.767
Pottery clay [#]	0.110	0.025	0.773	1.200	2.265

[*] After Schofied and Wroth 1968, P.157

[#] After Hird,(1972)

The one-dimentional consolidation, the isotropic consolidation , and the critical state lines for the tested clay are all shown in Fig. 9. In this figure, the parallizm of the three lines is confirmed and in agreement with the critical state theory.

The relationship governing the stress-strain behaviour for Cam-Clay is given by:

$$\epsilon = -\frac{x}{Mv_0} \ln \left(-\frac{M}{M - q/p} \right)$$

where v_0 is the initial specific volume of sample and M is the slope of the critical state line in q-p space.

The above relationship is used to predict the stress-strain behaviour. Fig. 10 shows the predicted stress-strain along with the observed one for one of the tests performed in this investigation. It can be seen that actual and predicted curves are fairly in a good agreement.

Similarly, a prediction of pore water pressure at any stage of the test can be done using the relation provided by the critical state theory as follows:

$$u = (p_0' - p') + q/3$$

wher p_0' is the initial value of mean effective stress and p' is the current value of mean effective stress. The predicted and actual C-u curves for the same test are shown in Fig.11 which shows a very good agreement. Also, during an undrained test

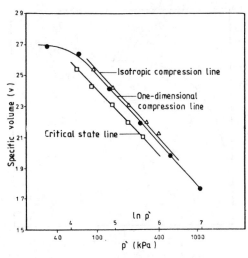

Fig.9 Comparison of the isotropic, one dimentional, and critical state lines

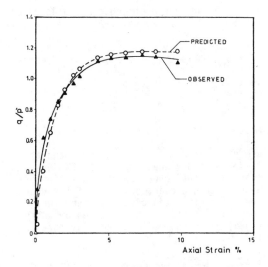

Fig.10 Observed and predicted axial strain-stress ratio q/p_0

for normally consolidated clay, the relationship between the initial mean effective pressure p_0' and its final value p_f' (at critical state) is given by:

$$p_0' / p_f' = \exp.\Lambda$$

Hence, the undrained shear strength is:

$$S_u = 1/2 \ q_f' = 1/2 \ M \ p_f' = 1/2 \ M \ p_0' \ \exp.(-\Lambda)$$

367

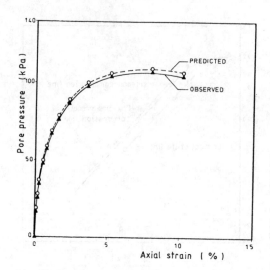

Fig.11 Observed and predicted axial strain - pore pressure relationship

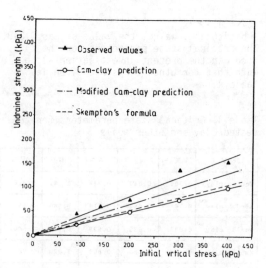

Fig.12 Observed and predicted undrained strength for isotropically consolidated saples

Thus, the undrained strength ratio is given by: $S_u / p_o' = 1/2\ M\ \exp.\ (-\Lambda)$

For isotropically consolidated samples, $p_o' = \sigma_{vo}'$ where σ_{vo}' is the initial effective vertical stress and thus,

$$S_u / \sigma_{vo}' = 1/2\ M\ \exp.\ (-\Lambda)$$

Measured and predicted values for S/σ_{vo}' are shown in Fig. 12. It may be noted that the predicted values are about 30% less than the actual ones. In the same figure also shown a predicted values of the undra- ined strength ratio given by other two formulae. The first one is given by Wroth (1984) based on modified Cam-clay model as follows:

$$S_u/\sigma_{vo}' = M/2\ (1/2)$$

This formula gives an underestimated strength ratio of about 14%. The second formula is provided by Skempton (1948) for actual deposits as follows:

$$S_u/\sigma_{vo}' = 0.11 + 0.37\ P.I.$$

This formula underestimates the undrained strength by about 26%. Thus it may be concluded that the modified Cam-clay model provides the nearest estimation of the undrained strength of isotropically conso- lidated samples in relation to the present results.

The Normalized K_o undrained strength given by Wroth (1984) is:

$$S_u/\sigma_{vo}' = (\sin \phi'/2a)[(1+a^2)/2]$$

where $a = (3-\sin \phi')/2(3-2 \sin \phi')$

For clays tested under K_o condition, Mayne (1988) suggested the following formula for the undrained strength:

$$S_u/\sigma_{vo}' = 2/3 \sin \phi'$$

The initial vertical stress σ_{vo}' against the undrained strength, S_u, for one- dimentionally consolidated samples are plotted along with those given by Skempton (1948), Wroth (1984) and Mayne (1988) are shown in Fig. 13. It can be concluded that the formula given by Mayne (1988) gives the least underesti- mation of the undrained shear strengh for one-dimensionally consolidated samples.

The stress path of a sample of virgin isotropically consolidated Cam-clay is given by

$$q = (M\ p'/\Lambda)\ln\ (p_o'/p')$$

Fig. 14 shows the stress path for a test carried out in the present investigation along with the predicted ones. It is obvious that the experimental points are far removed from the predicted ones. Similar results were obtained by Ilyas (1984). It is believed that this difference is due to that the specimen was theoriti- cally suposed to be under perfectly equal

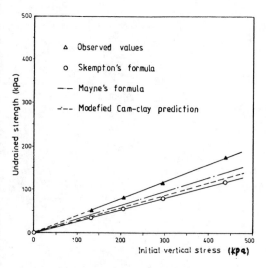

Fig.13 Observed and predicted undrained strength for K_o-consolidated samples

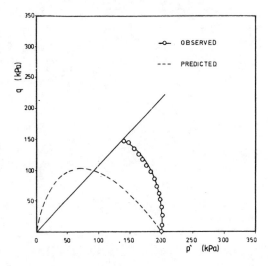

Fig.14 Stress path of a test compared with Cam-clay prediction

all-round compression. Experimentally this can not be achieved because there is shearing resistance on the top and bottom of the specimen.

CONCLUSIONS

As indicated from test results and within the range of stresses used one can conclud

that stress anisotropy during consolidation proves to have a distinct influence on the stress-strain behaviour of the tested soil. It increases the undrained strength and decreases the axial strain at peak deviator stress. Thus, a significant decrease in the built-up pore water pressure during undrained loading is observed.
Stress history is found to have a fairly little effect on failure conditions of the tested samples (i.e. critical state line and angle of internal friction). No significant difference is observed in the behaviour of samples whichare one-dimensionally consolidated and those consolidated with isotropic followed by drained compression. These two sets of samples have approximately tha same deviator stress at the end of consolidation phase. Hence, the value of deviator stress after consolidation is the major factor that affects the behaviour of soil. The normalized behaviour $(q/p_o' - \epsilon$ and $u/p_o' - \epsilon)$ is found to be fairly independent on the preconsolidation pressure as expected from the critical state theory. This is clearly noticed with anisotropic than with isotropic consolidation. It is possible to fit Cam-clay model to the tested soil. The predicted stress-strain and strain-pore water pressure behaviours predicted by Cam-clay show a good agreement with the observed behaviour.

REFERENCES

Atkinson, J.H. and Bransby, P.L. 1978. The mechanics of soils, An introduction to critical state soil mechanics, McGraw Hill Co., London.
Bishop, A.W. and Wesley, L.D. 1975. A hydraluic triaxial apparatus for controlled stress path testing. Geotechnique, 25, No.4
Bishop, A.W. and El-Din, G.A.K. 1975. The effect of stress history on the relation between ϕ and porosity in sand. Proc. 5th ICSMFE, Paris.
El-Ruwayih, A.A. 1975. Stress strain characteristics of rockfill and of clays under high pore water tension. Ph.D. Thesis, University of London.
El-Sohby, M.A. 1969. Deformation of sand under constant stress ratio. 7th ICSMFE, Mexico, vol.1
Hassan, A.M. 1990. Effect of stress history on the behaviour of soft clay. M. Sc. Thesis, Minia University, Egypt.
Hassona, F.A.K. 1986. Studies of the liquefaction behaviour of cohesionless materials. Ph.D. Theis, University of Sheffield, England.

Hird, C.C. 1972. On the strength of
    Pottery clay. Interm Report, The Univ.
    of Manchester Institute of Science and
    Technology, England.
Ilyas, T. 1983. Development and application
    of triaxial stress path soils testing,
    M. Sc. Thesis, Sheffield University,
    England.
Lade, P.V. and Duncan,J.M. 1976. Stress -
    path dependent behaviour of cohesionless
    soil, Journal of the Geotechnical Eng.
    Division, ASCE, vol. 103, No. GT 1.
Mayne P.W. and Stewart, H.E. 1988. Pore
    pressure behaviour of K_o-consolidated
    clays. ASCE, vol. 114, No. 11.
Ohsaki, Y. 1969. The effect of local soil
    conditions upon earthquake structures.
    Proc. 7th ICSMFE, Mexico.
Schofield, A.N. and Wroth, C.P. 1968.
    Critical state soil mechanics. McGraw
    Hill Book Co., London.
Skempton, A.W. 1954. The pore pressure
    coefficients A and B , Geotechnique 4.
Wroth, C.P. 1984. The interpretation of in
    situ soil tests. 24th Rankine Lecture,
    Geotechnique 34.
Rowe, P.W. and Barden, L. 1966. A new
    consolidation cell, Geotchnique, 16.

Pressuremeter, laboratory and in situ test correlations for soft cohesive soils

Corrélations entre essais pressiométriques, in situ et en laboratoire sur des sols mous cohérants

H.Giannaros & J.Christodoulias
Public Works Research Laboratory, Athens, Greece

ABSTRACT: In this article an attempt is made to evaluate and correlate the results of the pressuremeter test with conventional in situ and laboratory tests for soft cohesive soils. The above tests took place in several parts around Athens, and besides pressuremeter tests, standard penetration tests (SPT), static penetration tests (CPT) and vane tests were also carried out. From the boreholes, undisturbed samples were available for unconfined copression tests, quick triaxial tests (UU), and consolidation tests. From the above tests correlations between their results have been proposed and were compared to the corresponding correlations of the international literature .

RESUME: Dans cet article nous avons cherché à evaluer et à obtenir une corrélation entre lesrésultats des essais pressiométriques et des essais courants in situ et en laboratoire sur des sols cohésifs mous. Ces essais ont été réalisés sur des sites divers aux alentours d'Athènes et, mis à part les essais pressiométriques, nous avon en outre effectué des essais au pénétromètre statique (CPT), des essais standard penetration test (SPT), et des essais au scissomètre. Des echantillons intact provenant des forages ont été mis à notre disposition pour des essais en compression non confineé, des essais triaxiaux non consolidés-non drainés (UU), et des essais oedométriques. Nous avons proposé des corrélations entre les résultats des essais ci dessus, et nous les avons comparés aux corrélations correspondants dans la littérature internationale.

INTRODUCTION

During recent years the pressuremeter test has gained acceptance in Greece as a valuable in situ test in geotechnical investigation. As the method of interpretation of the pressuremeter test results became more elaborate and refined, it became obvious that this test could yield data on soil properties that could not be obtained by any other in situ test apparatus.

Since many engineers are not yet familiar with the pressuremeter test and its evaluation, an attempt was made to correlate the results of pressuremeter test with conventional in situ and laboratory tests, for soft cohesive soils.

In the pressumeter test, the soil around the probe is subjected to an increasing stress and the resulting deformations are assumed to be in a plane strain condition. The typical plot of a pressuremeter test performed in a borehole (figure 1) indicates that the volume-pressure curve contains three phases:

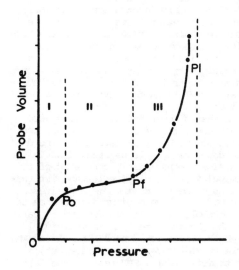

Figure 1: Typical pressuremeter curve

- The first phase (I) corresponds to the rebuilding of the pressure on the face of the borehole up to the initial horizontal stress which was acting at that point before the borehole was made.
- The second phase shows a pseudo-elastic relationship between the stress and strain which are approximately proportional, following the laws of elasticity.
- The third phase indicates a plastic behaviour of the soil around the probe as failure is gradualy reached.

In order to get the best out of the test, it is important to insure that all three phases are well defined on the plot, because each one of the three phases contributes to the definition of some of the soil parameters which can be obtained by the pressuremeter test.. If the test is performed in such a way that large errors are made on the value of P_o (horizontal earth pressure at rest), that no linear part of the curve can be defined and that no limit pressure can be evaluated, then the parameters obtained from that test will certainly be of a questionable value. The shape of the plot is mainly influenced by the properties of the soil and also by the quality of the prepared borehole for the pressuremeter test. Especially in the case of soft clays, where the remoulding of the clay around the hole produces very poor test results.

TESTING PROGRAM

An extensive investigation program took place in several parts around Athens, where soft clays are located, and besides pressuremeter tests (using a Menard's pressuremeter), standard penetration tests (SPT), static penetration tests (CPT) and vane tests were also carried out. From the boreholes undisturbed samples were available for unconfined compression tests, quick triaxial tests (UU) and conventional consolidation tests in order to establish empirical correlations for first order approximations.

The tested geotechnical profiles were mainly consisted of Quaternary deposites (Holocene and Upper Pleistocene). The dominate deposits were lacustrine, diluvial and coastal sediments mainly consisted of red loam, clays, grey clay, silts and intercalations of peat beds in some deposites.

In order to have minimum disturbance of the borehole wall, it was decided to carry out each separate test in an individual borehole. Thus, round each borehole in which a pressuremeter test was carried out, static penetrometer tests, SPT tests, vane tests were carried out, in a star-shaped arrange-

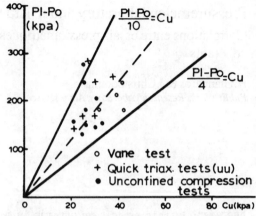

Figure 2. Correlation between pressuremeter and undrained shear strength.

ment, as well as one more borehole for undisturbed soil samples.

RELATIONSHIP BETWEEN LIMIT PRESSURE Pl AND UNDRAINED SHEAR STRENGTH Cu.

The undrained shear strength Cu of clayey soils is correlated to the pressiometric limit pressure Pl by the semi empirical formula:
Cu=Pl-Po/b   where:
Pl: the pressuremeter limit pressure
Po: horizontal earth pressure at rest at the test level.
b : factor, generally taken as 5.5
    Calhoun (1970), Amar et al.(1975).

The experimental relationship between Cu and Pl-Po is shown in figure 2. The values of the undrained shear strength Cu were obtained from quick triaxial tests (UU),unconfined compressive tests and vane tests.

A quite wide scatter is observed. Most of the points lie between Cu=Pl-Po/4   and Cu=Pl-Po /10 , with a mean value of the order Cu=Pl-Po /6 .

The scatter and the difference between the experimental results and the theoretical relation could be attributed to various causes. The test procedure influences the results of the measurement. In soft soils the preliminary hole may disturb considerably the soil and modifie certain characteristics obtained with the pressuremeter. On the other hand the triaxial or unconfined compression tests are performed on relatively small samples, whereas the pressuremeter tests a larger volume of soil. Also the failure patterns of these tests are very different with regard to distribution of stresses and planes of shear.

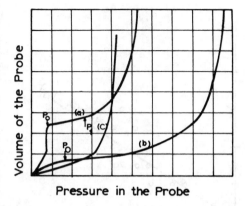

Figure 3. Typical pressuremeter test curves

If we could run a pressuremeter test without having borehole disturbance, the lateral earth pressure at rest Po could be accurately measured by the pressuremeter test. As shown in figure 3 this determination is more or less difficult and accurate depending on the quality of the test curve. If the borehole is larger than the probe a good curve (a) is obtained from which Po can easily be determined; if the borehole has about the same diameter as the probe a type (b) curve is generated and the determination of Po requires some guessing. Finally if the diameter of the hole is slightly less than that of the probe, the recompression stage of the test is eliminated and the phases I and II of the pressuremeter curve are missing (curve c). In this case Po cannot be determined. In soft clays due to borehole disturbance and consequent remoulding of the clay around the hole, (c) type curves are very often encountered.

An attempt was made to correlate the values of Po determined by the pressuremeter tests (where that was possible) to those derived from calculations, figure 4. The values of Po were calculated by the procedure proposed by Brooker & Ireland (1965).In the plot of computed versus measured values we observe some scatter and generally the pressuremeter values are greater than the computed values. Similar relationship was observed by Tavenas et al. (1975) and Lukas & De Bussy (1976).

RELATIONSHIP BETWEEN CREEP PRESSURE Pf AND PRECONSOLIDATION PRESSURE Pc.

The preconsolidation pressure Pc was correlated to the pressuremeter creep pressure Pf. The preconsolidation pressure was determined from conventional consolidation tests using the Cassagrande method. A plot of these data is presented in figure 5. A wide scatter is observed and the preconsolidation pressure lies between 0.8 Pf and 0.3 Pf.

A somewhat different relationship was observed by Lukas & De Bussy (1976), where Pc was almost equal to Pf. A wide scatter of the data was also observed. This difference is probably due to the different mechanism that gave rise to the preconsolidation pressure of the soils tested. The preconsolidation pressure of the glacial tills, investigated by Lukas & De Bussy,is due to the removal of the overburden stresses during the geological history, whereas the preconsolidation pressure of the soft cohesive soils of this study is attributed to various phenomena such as long term secondary compression, aging and cementing effects,Jamiolkowski et al. (1985), Bjerrum (1967).

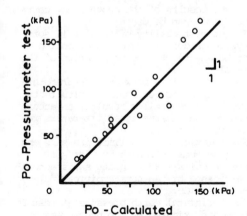

Figure 4. Horizontal pressure Po, calculated vs. measured values.

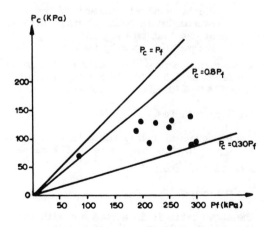

Figure 5. Preconsolidation pressure Pc vs. creep pressure Pf.

373

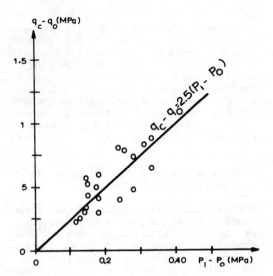

Figure 6. Correlation between pressuremeter
& static penetrometer.

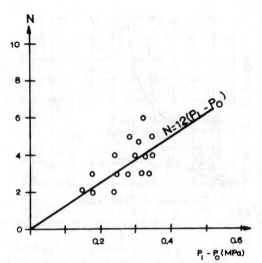

Figure 7. Correlation between pressuremeter
& Standard Penetration Test.

RELATIONSHIP BETWEEN STATIC PENETROMETER -
PRESSUREMETER

The undrained cohesion Cu is given directly
by the results of the pressuremeter and
static penetrometer tests using the formulae:

$Cu = Pl-Po/b$ for the pressuremeter

$Cu = q_c-q_o/Nc$ for static penetrometer

or $q_c-q_o/Pl-Po = Nc/b$ (1)

where, q_c:point resistance of static penetrometer
q_o:the vertical pressure of the
overburden in terms of total stress,
at the test level.
Nc:bearing capacity factor

As mentioned before the factor b is generally taken as 5.5 . The factor Nc lies between $10 \leqslant Nc \leqslant 20$. Thus (1) becomes:

$1.8 \leqslant q_c-q_o/Pl-Po \leqslant 3.6$ (2)

The experimental relationship between Pl-Po
and q_c-q_o is shown in figure 6. The regression line of the points is:

$q_c-q_o/Pl-Po = 2.50$

The above ratio is in accordance with the
theoretical relation (2) and also agrees
with the international literature, where

the above ratio lies between 2.5 and 3.5
Baguelin et al. (1978), for very soft to
soft clays.

RELATIONSHIP BETWEEN SPT - PRESSUREMETER
TEST

The experimental relationship between N and
Pl-Po is shown in figure 7.
The ratio N/Pl-Po=12 is in accordance
with the international literature, Cassan
(1978), where for the same type of soils
the ratio N/Pl-Po lied between 10-30.

CONCLUSIONS

From the results of the above study certain
conclusions can be drawn:
To get best quality of results in soft
clays, considerable care and attention
must be devoted to the preparation of the
borehole.
The correlations between pressuremeter
and the other conventional in situ tests
(SPT,CPT) gave similar results, in accordance to the international literature for
this type of soils.
The relation between limit pressure and
undrained shear strength Cu, for soft
cohesive soils, can be quite satisfactorily expressed by the semi empirical relation
Pl-Po = 5.5 Cu.
The horizontal earth pressure at rest Po
can be approximated by the pressuremeter
test, but because of the borehole distur-

bance the accuracy of this measurement is
limited.

Although the results are somewhat scattered the experimental relation between the
preconsolidation pressure and pressuremeter
creep pressure is $0.3 \ Pf \leqslant Pc \leqslant 0.8 \ Pf$

ACKNOWLEDGEMENTS

The laboratory and in situ   test results
were obtained from the Public Works Research
Laboratory.

The typing of the manuscript was produced
by Mrs V. Florou.

REFERENCES

Amar, S. et al. (1975). In situ shear resi-
    stance of clays. Proc. of the Conf. on
    In Situ Measurement of Soil Properties,
    ASCE, Vol. I, pp 22-44
Baguelin, F., Jézéquel, J.F. & Shields, D.H.
    (1978). The pressuremeter and foundation
    engineering.Trans. Tech. Publication.
Bjerrum, L. (1967). Engineering geology of
    Norwegian normally consolidated marine
    clays as related to settlements of buil-
    dings. Geotechnique, vol. 17, No 2,
    pp 81-118.
Brooker, E.W. & Ireland, H.O. (1965). Earth
    pressures at rest related to stress histo-
    ry. Canadian Geotech. Journal, vol. II,
    No 1, Feb., p 14.
Calhoun, M.L. (1970). Field load testing
    with the pressuremeter. Proc. of the 19th
    annual Soil Mech. & Found. Conf., Univ.
    of Kansas, Lawrence, Mar.
Cassan, M. (1978). Les essais in situ en
    mechanique des sols. Edition Eyrolles.
Jamiolkowski, M. et al. (1985). New develop-
    ments in field and laboratory testing of
    soils. 11th Int. Conf. SM & FE, San Fran-
    sisco, State-of-the-art, vol. 1, pp 57-153
Lukas, R.G. & De Bussy, B. (1976). Pressure-
    meter and laboratory test correlations for
    clays. Journal of the Geotech. Div., ASCE
    vol. 102, GT9, pp 945-962.
Tavenas, F.A. et al. (1975). Difficulties in
    the in situ determination of Ko in soft
    sensitive clays. Proc. of the Conf. on
    In Situ Measurement of Soil Properties,
    ASCE, vol. I, pp 450-476.

375

The effects of geological history on the behaviour of Miocene claystones on a regional scale

Les effects de l'histoire géologique sur le comportement d'argilolithes du Miocène à une échelle régionale

Jiří Herštus
Stavební geologie Praha Czechoslovakia

ABSTRACT: The paper discusses the effects of the long-term state of stress and diagenetic associations on the shear strength of claystones on a regional scale. Separately, the influence of overconsolidation and diagenetic lithification of claystones an its changes with the magnitude of the state of stress are dealt with. Attention is also paid to the influence of discontinuities on the strength of the sediment.

RESUME: Dans le travail on a suivi en échelle régionalle influence de l'histoire des contraintes et joints diagenétiques sur la résistance an cisaillement des argiles durs. On a suivi séparément l'influence de surconsolidation et durcissement diagenétique et ses variation avec le niveau des contraintes. On a preté attention an l'influence des fissures sur la résistance du sediment.

1. INTRODUCTION

In north-western Bohemia there occurs a brown-coal basin of Tertiary age, which covers an area of more than 600 km². In the NW it is limited by the Krušné hory mountain range (Fig.1) built up of gneissose rock. During the whole period of sedimentation the basin was exposed to tectonic movements on both longitudinal and transverse faults. The Tertiary fresh-water lacustrine sedimentation ends with a thick layer of Miocene claystones overlying the coal seams. Tectonic movements within the basin continued after the termination of Miocene sedimentation and resulted in markedly different values of denudation between the sectors studied (Fig.1).
The claystones were the object of extensive engineering-geological study, required with respect to intensive opencast mining of coal in the area. This study only concerns the properties of unweathered claystones.

2. DESCRIPTION OF MIOCENE CLAYSTONES OVERLYING THE COAL SEAM

Unweathered claystones are dark-grey bedded sediments interpenetrated by an irregular network of predominantly vertical discontinuous fissures (they usually do not run across more than 2 beds). Additionally, the claystones contain up to several metres long joints and sporadic shear zones associated with post-sedimentary tectonic movements. Statistical data on the plasticity limit of claystones are given in Table 1. At the locality No.5 the claystones were studied to a depth of 279 m. The W_L values vary with depth quite randomly and the mean value is constant. The IP value increases moderately with depth. The porosity of claystone decreases with depth and its bulk density increases simultaneously. The mean value of the bulk density of the dry mass matter was 1650kg m^{-3} within the whole thickness of 279 m (107 determinations); the mean deviation - 110 kg m^{-3} represents the difference between the base of the bed and its near-surface part.

Table 1.

| Loca-lity | W_L | | | I_P | | |
	n	$\bar{x}$	σ	n	$\bar{x}$	σ
1	103	96.5	12.0	103	45.1	10.1
2	27	70.2	18.7	27	40.2	18.4
4	170	89.3	12.4	171	53.1	12.1
5	107	74.0	8.8	160	43.6	8.3

$\underline{n}$ - number of determinations
$\bar{x}$ - mean value
σ - mean deviation

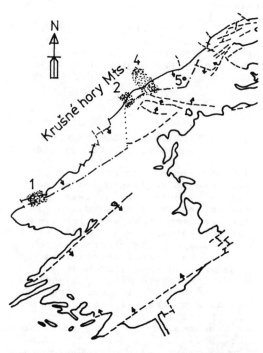

Fig.1. Tectonic sketch of the brown-coal basin with designated extent of the localities studied: solid line - margin of the basin, dashed line - faults, dot-and-dash line - axis of the basin.

The natural moisture content of unweathered claystones is near or slightly below the plasticity limit.

On the basis of palaeomagnetic dating of samples taken from a borehole located in the deepest part of the basin, the age of sediments ranges from 21.3 Ma at a depth of 315 m to 17.3 Ma at a depth of 18.0 m (M.Malkovský et al.,1989). The mean rate of sedimentation was 73.7 mm/ka.

The claystones contain on the average 26% montmorillonite, 30% illite, 15% kaolinite and 20% quartz; the remaining amount consists of 4% siderite, 2% carbonates and 3% organic matter (M.Malkovský el al., 1985). They do not show any signs of recrystallization of the clay minerals and no cementation by another substance at the contacts of clay particle aggeregates has been evidenced, although they behave as diagenetically lithified rock.

3. THE DEVELOPMENT OF THE STATE OF STRESS OF CLAYSTONES

Sedimentation in the basin occurred under the conditions of a feshwater lake. The deepest lying sediments are sandy, permeable. In places not affected by exploitation an artesian horizon was established in basal sandy layers by means of piezometers; the pressure head reached almost the level of the present-day ground surface. The pore pressure corresponding to this head was assessed using pore water pressuremeters installed in the claystones overlying the coal seams.

The claystones are diagenetically lithified and their moisture content is near the plasticity limit. Consequently, it is assumed that (1) diagenetic bonds formed as late as in the closing of sediment consolidation due to the weight of the overlying complex; (2) the porosity corresponding to the primary overlying complex (original effective stress) remains preserved in unweathered claystone also after denudation (at a degree of precision sufficient for regional considerations).

The determination of the bed thickness before denudation was based on the changes in claystone porosity with depth, which have been assessed by laboratory tests of undisturbed samples. For estimation of the thickness before denudation

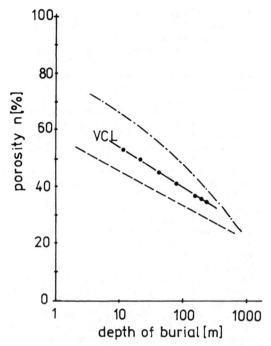

Fig.2. Baldwin s curves and oedome-
        ter test on slurry sample
dashed line - Baldwin s set, lower
boundary
dot-and-dashed line - Baldwin´s set
mean curve
solid line - oedometer test

we have used the curve showing the
dependence of porosity on the thickness of the overlying rock complex
(Fig.2) as published by B.Baldwin
(1971). The first determinations
(S.Hurník,1978) carried out with the
application of Baldwin´s mean curve
have shown disproportionately high
denudation values. Therefore laboratory oedometer tests on a slurry
sample prepared from the claystone
have been made. This test (Fig.2)
has shown that in considering the
secondary consolidation it will be
more appropriate to use rather the
lower limit than the medium values
of Baldwin´s results in apprecia-
ting the extent of denudation.
The values thus established for
individual localities are plotted
in Fig.3. At the locality No.1
(Fig.1) the denudation value determined by means of Baldwin´s curve was correlated with the results
of the laboratory tests of undisturbed samples; the results are
given in Table 2.

Table 2. Values of denudation established in the basin axis at the
locality No 1.

Mode of Baldwin´s curve	Denudation (in m)
modified Baldwins curve	50
laboratory test	60;60;53;38;36;50

The scater of values determined by
laboratory tests is obvously connec-

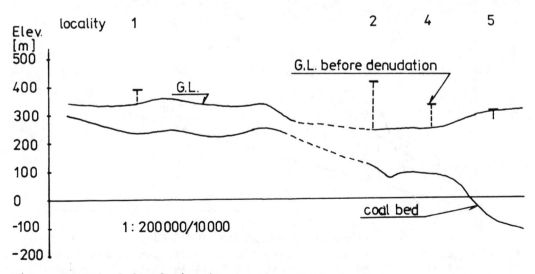

Fig. . Extend of denudation in the axis of the basin

ted with the existence of microcracks in the claystone.

4. SHEAR STRENGTH OF NORMALLY CON-
SOLIDATED, DIAGENETICALLY LITHI-
FIED CLAYSTONE

As is seen in Fig.3, the locality
No.5 was practically not affected
by denudation since the termination
of claystone sedimentation. We have
taken 107 undisturbed samples at
this locality from the whole depth
of 279 m in depth intervals of 3 to
5 m. (Kcnečný V.,1965.)
The tests of selected non-fissured
samples in a shear box apparatus
performed at a normal stress corresponding to the effective pressure
of the overlying rocks provided the
envelope of the shear strength of
the normally consolidated, diagenetically lithified claystone (Fig.
4). Also the results of the tests
of intact claystone samples from
other localities made at the normal stress corresponding to the
effective value of the overconsolidation pressure vary within the
range of experimental results from

the locality No.5. This envelope
can be compared with the envelope
of shear strength of a slurry sample prepared by crushing of the
claystone examined. The difference
between envelope A and envelope B
expresses the effect of diagenetic bonds on the sediment shear
strength. This difference may be
interpreted in two terms: as an
increment of $\Delta \tau$diag having
the character of cohesion or as
an increment of $\Delta \sigma$diag of
attraction between the clay particles or aggregates of clay particles. In both cases a logarithmic dependence on porosity is
evident (Fig.5).
If we accept the former variant,
i.e. the consolidation of the contacts between the aggregates of
clay particles by coagulation of
the solidifying substance from pore
water, the number of contacts should
increase logarithmically. Such cementation bonds, however, have not been
assessed. This variant is also con-
tradicted by intensive swelling of
claystones at a lower than overconsolidation stress, which is accompani-

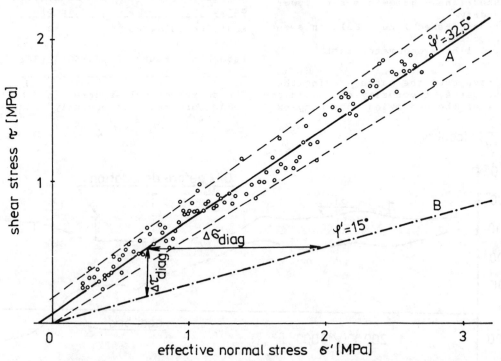

Fig.4. Comparison of the shear strength of normally consolidated
diagenetically lithified claystone (A) from the locality No.5 and
of normally consolidated clay (B).

ed by a decrease of strength.
The latter variant presumes that
physico-chemical attractive forces
originate at the contacts, which
depend only on porosity (i.e.proxi-
mity of one aggregate to another).

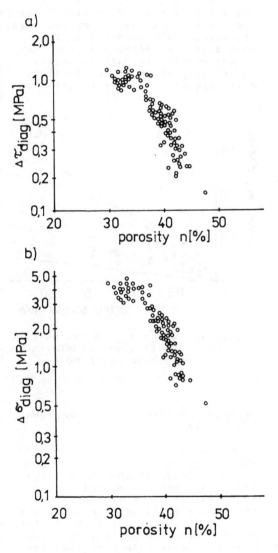

Fig.5. Dependence of diagenetic
       strength on porosity.

The resulting forces acting at the
contacts of clay particle aggregates would then be the sum of the
outer effective loading and these
inner forces.

This variant is to a certain degree
supported by the results of the experiments made by L.Bjerrum and K.
Y.Lo (1963) to verify the effect of
aging on the properties of normally
consolidated clay. These experiments indicated that clay becomes
more brittle with time (decrease of
deformation after peak strength was
attained, a higher decrease of
strength after surpassing the peak
value).

5. SHEAR STENGTH OF CLAYSTONES AT
   NORMAL STRESSES LOWER OR HIGHER
   THAN OVERCONSOLIDATION STRENGTH

At the localities Nos.1, 2 and 4
large numbers of direct shear and
triaxial tests of undisturbed claystone samples were performed. The
tests of intact samples gave the
values of shear strength envelopes
presented in Fig.6. The points of

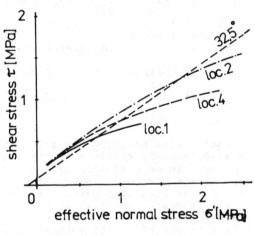

Fig.6. Comparison of the strength
       of claystones from the loca-
       lities Nos.1, 2, 4.

crossing of these curved envelopes with the envelope of the normally consolidated, diagenetically
lithified claystone ($\varphi' = 32.5°$)
correspond to the overconsolidation pressure. At the locality No.
2 a triaxial test of intact claystone samples was made additionally up to high normal stress values
(Fig.7). The test has shown that
with the application of normal effective stresses lower than the

overconsolidation ones (point a)
the envelope of shear strength is
above the envelope of shear strength
of the normally consolidated claystone. The dotted area (I) represents the effect of overconsolidation on the strength.After the over-

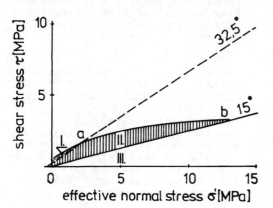

Fig.7. Triaxial test of claystone
from the locality No.2.
Delimitation of the zones
influence:
I.   overconsolidation in-
       fluence
II.   diagenetic lithifica-
       tion
III.   normally consolidated
       clay

consolidation pressure has been surpassed, the diagenetic bonds obviously begin to be gradually destroyed
by the application of normal stress.
The dashed area (II) between the curved envelope of strength of the diagenetically lithified claystone and
the envelope of of the normally consolidated clay demonstrates the influence of diagenetic bonds. At a
certain value of normal stress (point b in Fig.7) the strength of diagenetically lithified claystone
equals the strength of normally consolidated clay, i.e. the effect of
lithification disappears This stress
may be called the transitional
stress.

6. THE EFFECT OF DISCONTINUITIES ON THE STRENGTH OF CLAYSTONE

We have thus far discussed the shear strength of intact claystone, i.
e. not affected by the presence of
discontinuities. Similarly as in

rock massifs, the resulting strength
of claystones is also controlled by
the continuity, undulation, roughness and orientation of disconti-
nuitites. As mentioned above, the
claystones involve bedding surfaces,fissures joints and shear dislocations. The differences in the
character of these discontinuities
permits the oscillation of the shear
strength of the massif from the
peak strength (defined in the text
above) to the residual strength,
as is exemplified at the locality
No.4 (Fig.8). The same results ha-

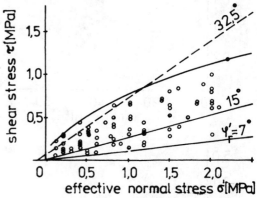

Fig.8. Influence of discontinuities
on the strength of claystone
from the locality No.4.

ve been obtained at the remaining
localities.

7. CONCLUSION

Regional study of the shear strength of Miocene claystones has shown
the possibility of distinguishing
the effect of diagenetic lithification of claystones from the influence of overconsolidation. With respect to the behaviour of claystones, three zones of normal stress
may be defined. With normal stresses that are lower than the overconsolidation pressure, the effects
of both the diagenetic bonds and
overconsolidation are apparent.
With normal stresses exceeding the
overconsolidation stress only the
diagenetic bonds are effective.
Their proportion in the shear

strength decreases with the increase of normal stress, disappearing after the transitional stress has been attained. If the values of normal stress are higher than the value of transitional stress, the strength of claystone is equal to the strength of normally consolidated clay.
The existence of discontinuities may, depending on their character and orientation, reduce the strength of the claystone massif to a whatever value between the peak and residual strength.

REFERENCES

Hurník S.(1978). Reconstruction of the thicknesses of the overlying formation in the North Bohemian brown-coal basin (in Czech).Čas. pro mineralogii a geologii, Vol. 23,No.3, pp 265-276

Malkovský M.edit (1985). Geology of the North Bohemian brown-coal basin and its neighbourhood (in Czech). Academia Praha.
Malkovský M., Bucha V., Horáček J. (1989). Velocity of Tertiary Sedimentation in the Most part of the North Bohemian brown-coal basin (in Czech). Geologický průzkum, Vol.34, No.1, pp.2-5
Konečný V.(1965). Unpublished internal report, VÚHU Most

Development and application of a hollow cylinder triaxial cell
Développement et application d'une cellule cylindrique triaxiale creuse

Jing-Wen Chen
National Cheng Kung University, Taiwan

ABSTRACT: The design of a new hollow cylinder triaxial cell is presented. Design criteria were based on sample preparation and sample dimension. The cell was used to investigate response of Monterey NO. 0/30 sand under two cyclic stress paths. The behavior of Monterey NO. 0/30 sand was found to differ substantially with these different stress paths.

Résumé: Présentation du dessin d'une nouvelle cellule cylindrique triaxiale creuse. Les critères du dessin ont été basés sur la préparation et la dimension des spécimens. L'appareil était destiné à étudier la réaction du sable Monterey NO.0/30 soumis aux deux trajets de pression cycliques. On a trouvé que le comportement du sable Monterey NO. 0/30 diffère substantiellement selon ces différents trajets de pression.

1 INTRODUCTION

Stress dependency of soils has been recognized for years. Conventional taiaxial tests were conducted for investigating effect of stress path (Habib 1952, Lorenz et al. 1965). Because of its inherent limitations, conventional triaxial apparatus can not be used for the extensive study of stress dependency of soils. True triaxial cell developed by Ko and Scott (1967) has been improved and adopted for the observation of stress path effect of soils (Ko and Masson 1976, Sture and Desai 1979). However, stress paths generated by the true triaxial cell are limited in static condition.

This paper describes the development of a new hollow cylinder triaxial cell. The cell is capable of producing uniform stress and strain distributions for soil specimen, and of simulating in-situ stress paths. The cell has been used for observation of Monterey NO. 0/30 sand under both static and cyclic stress paths (Chen 1988). Typical results of two cyclic stress path tests will be discussed in this paper.

2 THE HOLLOW CYLINDER TRIAXIAL CELL

The hollow cylinder triaxial cell consists of eleven detachable components and three accessories used for sample preparation. A general layout of the apparatus is shown in Fig.1.

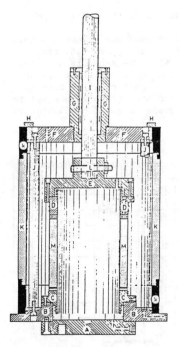

Fig.1 General layout of the Hollow Cylinder Triaxial Cell

The demensions of a specimen control the distributions of stress and strain in a hollow cylinder specimen. Therefore, selection of specimen dimension became an important task when the cell was designed. Dimensions with 25.4 and 20.3cm in outer and inner diameters, respectively; and with a height of 25.4 cm were finalized. It is believed that both stress and strain distributions would be maintained in a reasonable range when a specimen with these dimeneions is loaded (Hight et al. 1983).

The specimen is jacketed between an outer and an inner membranes and capped by ringshaped porous platens. The cell was also designed with potential to study the effect of principal stress direction rotation. Twelve stainless steel blades are set in the surface of the top and bottom ringshaped porous platens. These blades transfer torque applied to the specimen.

The inner chamber of the specimen is filled with water, being isolated from the outer chamber. Thus, the stress at the inner boundary of the specimen can be varied independently of the stress imposed at the outer boundary, but, if desired, both chambers can be connected each other, therefore permitting equal confining pressure to be applied to the specimen.

3 TEST MATERIAL AND SAMPLE PREPARATION

Monterey NO. 0/30 sand was chosen in this study. Specific gravity of the sand was found to be 2.65. The sand is a uniform clean sand with a coefficient of uniformity of 1.6, a cofficient of curvature equal to 1.0, and mean grain size equal to 0.45mm. The grain size distribution of the sand is shown in Fig.2. The maximum and minimum dry densities are equal to 16.6 and 14.4kN/m3, respecively.

The sand sample was prepared by dry pluviation. The desired relative density is obtained by pluviating a fixed amount of sand from a constantly rotating funnel through emperically determined nozzle opening at a fixed fall height. The relative densities of samples were about 38%.

4 STRESS PATH

Two cyclic stress paths, CYR-A and CYR-C, were used in this study. Stress paths presented in a p-q space is shown in Fig.3. The CYR-A stress path test was conducted with a stress ratio equal to 0.23. The stress ratio is defined as the single cyclic axial stress amplitude divided by two times of the effective initial consolidation pressure. Stress-controlled cyclic tests were conducted using invert sine wave with a frequency of 0.5 Hz. The invert sine wave function provided a compressive load which is followed by an extension load.

The CYR-C test was conducted under a stress ratio of 0.43, and a frequency of 0.5Hz. In order to follow the stress path shown in Fig.3, confining pressure needed to be changed when axial load was applied on the sample. Two pneumatic loaders were connected to the hollow cylinder cell to change the confining pressure. One provided confining pressure to the outer chamber, and another one to the inner chamber. Sinusoidal waves were used in both loades. They were synchronized to avoid any pressure different in two chambers.

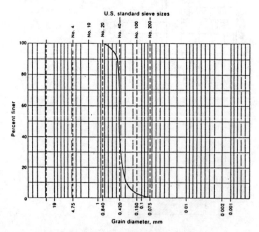

Fig.2 Grain Size Distribution of Monterey No. 0/30 Sand

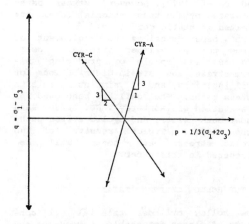

Fig.3 Stress Paths of Two Cyclic Tests

5 TEST RESULTS AND DISCUSSION

The effective stress path of the CYR-A test is shown in Fig. 4. In comparison to the total stress path of this test presented in Fig. 3, the pore pressure generation may be divided into three phases. The first cycle of loading generated a rather large excess pore pressure as compared to those generated by each subsequent cycle of loading. In the second phase of londing an equal amount of excess pore pressure increment was generated at every cycle of loading. The slope of the stress path is approximately equal. The amplitude of the cyclic deviator stress decreased slightly with increasing number of cycles. The last phase is the effective stress path touched the failure envelope. A large excess pore pressure increment was generated for each loading cycle. The stress path gradually flattened with increasing cycles of loading. The amplitude of the deviator stress also decreased gradually. Finally, the sample reached the initial liquefaction condition where the excess pore pressure was approximately equal to 207 kPa.

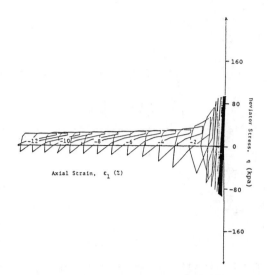

Fig.5 Stress-Strain Relationship of CYR-A Test

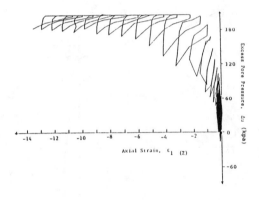

Fig.6 Excess Pore Pressure vs. Axial Strain for CYR-A Test

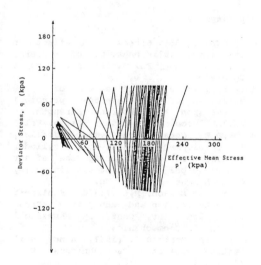

Fig.4 Effect Stress Paths of CYR-A Test

The stress-strain curve for test CYR-A is shown is Fig.5. The stress curve became flatter with increasing number of stress cycles. It means that the specimen was soften during the cyclic loading. The excess pore pressure and axial strain relationship for the CYR-C test is shown in Fig. 6. The excess pore pressure was finally equal to the effective confining pressure and the sample liqufied.

Fig.7 shows the effective stress path of the CYR-C test. The excess pore pressure generation increased with number of cycles, and the sample eventually liquefied. The relationship between deviator stress and strain is shown in Fig 8. It can be seen that the sample was softened and large flow deformation took place with increasing number of cycles. Fig. 9 shows the excess pore pressure generation versus axial strain. In the last few cycles, the sample was liquefied.

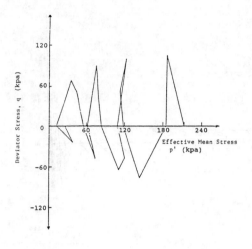

Fig.7 Effective Stress Path of CYR-C
Test

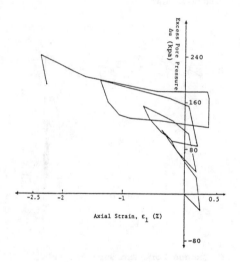

Fig.9 Excess Pore Pressure vs. Axial
Strain for CYR-C Test

6 CONCLUSION

The capabilities of the hollow cylinder
cell has been checked out by Monterey No.
0/30 sand under two cyclic stress paths.
The cell is capable of conducting tests
under combination of loading conditions.

Through the experimental study using
different stress paths, it is found that
the behavior of Monterey NO. 0/30 sand
is quite stress path dependent. The
strength, stress-strain relationship and
pore pressure generation are different
under different stress paths.

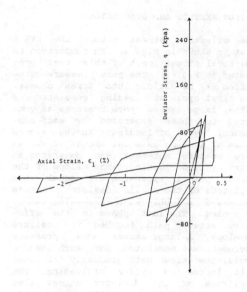

Fig.8 Stress-Strain Relationship of
CYR-C Test

7 REFERENCES

Chen,J.W. (1988). Stress path effect on
static and cyclic behavior of Monterey
No. 0/30 Sand , Ph. D. Thesis , Univ. of
Colorado

Harbib,p. (1952). Influence of the
Variation of the average principal
Stress upon the Shearing Strength of
soils, Proc. 3rd . ICSMFE, Paris, Vol.1

Hight,D.W., Gens,A., and Symes, M.J.(1983)
The development of a new hollow cylinder
apparment for investigating the effect
of principal stress rotation in soils,
Geotechnique, Vol.33 No. 4

Ko,H.Y. and Masson, R.M. (1976). Nonlinear
characterization and analysis of Sand.
Proc. 2nd Int. Conf. on Numerical
Methods in Geomechanics.

Ko, H.Y and Scott,R.S. (1967). A new soil
testing apparatus, Geotechnicque , Vol.
17 No. 1

Lorenz, H. Neumeuer, H., and Gudehus. G.
(1965). Tests concerning compaction and
displacements performed on sample of
sand in the state of plane deformation ,
Proc. 6th ICMFE , Montreal, Vol. 1.

6th International IAEG Congress / 6ème Congrès International de AIGI, © 1990 Balkema, Rotterdam. ISBN 90 6191 130 3

The novel rock fracture toughness testing method:
Cracked-Chevron-Notched Brazilian Disc method
Une nouvelle méthode d'essai de la ténacité de fracture de roche: Méthode de disque Brésilien de l'Encoche à Chevron craqué

Jun Fang Chen
Structural Analysis and Geotechnical Engineering Ltd, UK

ABSTRACT: This paper presents a novel test method using a chevron-notched Brazilian disc (CCNBD) specimen for mode I rock fracture toughness measurement. It is developed based on three dimensional Boundary-element and finite-element methods of calibration, minimum specimen size requirements investigation and experimental validation by comparison with the two recommended standard Chevron-Notched specimens by the testing commmission of the ISRM -- Chevron Bend specimen and Chevron Notched Short Rod specimen. It has been proved that the chevron-notched Brazilian disc specimen has good prospect to be utilized as the third chevron-notched rock fracture specimen.

RESUME: L'Article présente une nouvelle methode d'essai utilisant un specimen sous forme du disque Brésilien de l'encoche à Chevron (CCNBD) pour mesurer la ténacité de fracture de roche de mode I. Basée sur les méthodes de calibration à élément-borne at element-fini tridimensionels, de l'investigation de la nécessité de format des spécimens minimum et de la validation expérimentale, la méthode s'est développée en comparant avec deux specimens standards recommendes par la commission d'essai de ISRM - spécimen cintré à chevron-et spécimen de barreau court encoche a chevron. Il a été bien éprouvé que comme un troisième spécimen de fracture de roche de l'encoché à chevron. le spécimen sous forme du disque Brésilien de l'encoché à chevron présente one bonne prosperite de l'utilisation.

1 INTRODUCTION

Rock fracture toughness (mode I) is being gradually used by geotechnical Engineers in the areas such as: <1> Tunnelling machine performance prediction. <2> Rock cutting mechanism analysis. <3> New intact rock and rock mass classification index for rock mechanics and rock engineering etc. The testing commission of the ISRM has recommended two standard chevron-notched specimens for rock fracture toughness testing (Figure 1 and 2). Both recommended rock fracture specimens can be machined directly from rock cores and have chevron-notches at the center of specimen. Chevron-notched fracture specimens have unique features such as <1> extremely high stress concentration at the tip of chevron notch; <2> a crack can be initiated at a low applied load, eliminating the need to precrack a specimen, a costly and time-consuming procedure; <3> the fracture toughness can be evaluated from the maximum failed load, etc. The author here proposes a new chevron-notched Brazilian disc specimen for mode I rock fracture toughness testing. The geometry with basic notation of the CCNBD specimen is shown in Figure 3. It is a disc having two symmetrical curved notches at the center of the disc. For chevron-notched fracture specimen, rock fracture toughness can be evaluated using the following formula:

$$K_{ic} = F_c * P_{max} / (BD^{\frac{1}{2}}) \quad \ldots\ldots \quad (1)$$

Where:

K_{ic} -- mode I rock fracture toughness;
F_c -- critical dimensionless stress intensity factor;
P_{max} - maximum failure load;
B -- specimen thickness;
D -- specimen diameter.

The CCNBD specimen preparation is quite simple. A diamond saw with diameter less than 100mm and 1mm thickness is used for the curved chevron notches cutting. Both top and bottom surfaces of the CCNBD specimen are machined to the required thickness. The Chevron-notches are formed by both back and front cutting.

2 THREE DIMENSIONAL NUMERICAL CALIBRATION OF THE CCNBD SPECIMEN

Maximum confidence in rock fracture toughness testing methods can only be achieved after careful study and calibration by numerical techniques, experimental validation testing and specimen size requirement study. Three dimensional boundary element calibration for the CCNBD specimen was carried out using BEASY boundary element software. Stress intensity factor at the crack tip is calculated using displacement technique. The boundary element idealization of the CCNBD is shown in Figure 4. On account of its symmetry, only one eighth of the CCNBD specimen was modelled. The calibration results for different geometry of the CCNBD specimen are listed in Table 1.

3 EXPERIMENTAL APPARATUS AND PROCEDURES

A RDP testing machine with controlled loading line displacement rate was used. Tests were all conducted under constant loading line displacement rate of 0.1 mm/min. Loading line displacement and crack opening displacement was measured by AC-operated LVDTs. All the data was recorded by datalogger. The test rig is shown in Figure 5.

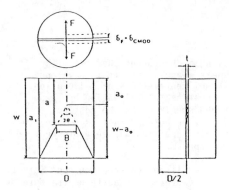

Figure 1 Geometry of Short Rod Specimen with Basic Notation

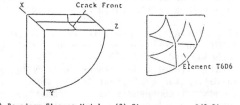

(1) Boundary Element Model   (2) Element at x = B/2 Plane

(3) Element at y = 0 Plane (4) Element at r = D/2 Surface

Figure 4 Bondary-Element Idealisation of CCNBD

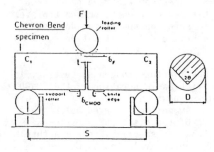

Figure 2 Geometry of Chevron Bend Specimen with Basic Notaion

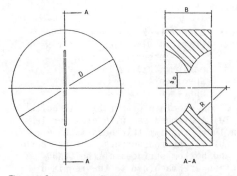

Figure 3 Geometry of Cracked-Chevron-Notched Brazilian Disc with Basic Notation

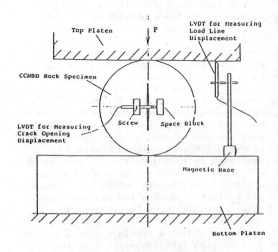

Figure 5 The Displacement Measurement Apparatus

390

Table 1 Summary of the Calibration Results of CCNBD

Disc-ID	D, mm	D', mm	B, mm	a₀, mm	C	a_1/R	Critical a_c	F_0 ($\xi=0$)	Critical F_{IC}
Da01	100	100	40.0	15	23.30	0.846	0.657	1.539	1.629
Da02	100	100	35.0	15	19.80	0.797	0.600	1.547	1.573
Da03	100	100	30.0	15	17.30	0.757	0.595	1.469	1.519
Da04	100	100	25.0	15	14.80	0.710	0.590	1.328	1.336
Da05	100	100	35.0	20	21.67	0.824	0.671	1.666	1.782
Da06	100	100	80.0	13	41.72	0.986	0.728	1.955	2.011
Da07	100	100	70.0	20	39.17	0.976	0.734	2.067	2.156
Da08	100	100	60.0	20	34.17	0.949	0.687	1.735	1.823
Da09	100	100	50.0	20	29.17	0.909	0.634	1.663	1.734
Da21	100	75	55.0	15	30.63	0.737	0.545	1.585	1.664
Da22	100	75	30.0	15	18.13	0.642	0.521	1.532	1.547
Da11	100	52	30.0	10	17.00	0.488	0.402	0.899	0.946
Da12	100	52	25.0	10	14.50	0.466	0.360	0.909	0.940
Da13	100	52	20.0	10	12.00	0.438	0.355	0.833	0.869
Da14	100	52	30.0	8	16.20	0.482	0.389	0.889	0.932
Db01	75	52	30.0	10	17.00	0.651	0.520	1.121	1.181
Db02	75	52	25.0	10	14.50	0.622	0.520	1.111	1.141
Db03	75	52	20.0	10	12.00	0.584	0.453	1.074	1.134
Db04	75	52	30.0	8	16.20	0.643	0.519	1.092	1.156
Db21	75	75	52.5	15	29.38	0.976	0.734	2.067	2.156
Db22	75	75	45.0	15	25.63	0.949	0.687	1.735	1.823
Db23	75	75	37.5	15	21.88	0.909	0.634	1.663	1.733
Dc01	50	52	20.0	10	12.00	0.876	0.585	1.814	1.926
Dc02	50	52	15.0	10	9.50	0.804	0.546	1.592	1.716

Table 2 Experimenal Results for the Size Requirements Study

Disc-ID	GN	a/R	F_{IC}	K_{IC}, $MN/m^{1.5}$	Average ± Std.
Da01	1	0.846	1.629	0.605, 0.607, 0.577, 0.604	0.598 ± 0.012
Da02	1	0.797	1.573	0.616, 0.624, 0.614, 0.608	0.615 ± 0.006
Da03	1	0.757	1.519	0.610, 0.603, 0.615, 0.612	0.610 ± 0.004
Da04	1	0.710	1.336	0.624, 0.616, 0.625, 0.613	0.619 ± 0.005
Da05	1	0.824	1.782	0.628, 0.597, 0.618, 0.620	0.616 ± 0.012
Da06	2	0.986	2.011	0.443, 0.456, 0.449, 0.455	0.451 ± 0.005
Da07	2	0.976	2.156	0.484, 0.428, 0.465, 0.469	0.462 ± 0.021
Da08	2	0.949	1.823	0.442, 0.426, 0.434, 0.440	0.435 ± 0.006
Da09	2	0.909	1.734	0.427, 0.430, 0.432, 0.433	0.430 ± 0.002
Da21	3	0.737	1.664	0.621, 0.619, 0.625, 0.616	0.620 ± 0.003
Da22	3	0.642	1.547	0.619, 0.603, 0.623, 0.621	0.617 ± 0.008
Da11	4	0.488	0.946	0.589, 0.599, 0.606, 0.600	0.599 ± 0.006
Da12	4	0.464	0.940	0.629, 0.637, 0.610, 0.623	0.625 ± 0.010
Da13	4	0.438	0.869	0.583, 0.588, 0.587, 0.589	0.587 ± 0.002
Da14	4	0.482	0.932	0.591, 0.619 0.610, 0.611	0.608 ± 0.010
Db01	5	0.651	1.181	0.599, 0.593, 0.617, 0.609	0.605 ± 0.010
Db02	5	0.622	1.141	0.597, 0.627, 0.628, 0.612	0.616 ± 0.013
Db03	5	0.584	1.134	0.623, 0.617, 0.625, 0.608	0.618 ± 0.007
Db04	5	0.643	1.156	0.598, 0.608, 0.598, 0.599	0.600 ± 0.004
Db21	6	0.976	2.156	0.395, 0.433, 0.432, 0.421	0.420 ± 0.015
Db22	6	0.949	1.823	0.312, 0.312, 0.342, 0.332	0.325 ± 0.013
Db23	6	0.909	1.734	0.339, 0.368, 0.367, 0.357	0.358 ± 0.012
Dc01	7	0.876	1.926	0.530, 0.586, 0.534, 0.581	0.558 ± 0.026
Dc02	7	0.804	1.716	0.501, 0.537, 0.527, 0.537	0.526 ± 0.015

4 SIZE DEPENDENCE INVESTIGATION OF KIC RESULTS USING THE CCNBD METHOD

As well known, the rock fracture toughness testing results depends on specimen size. In the Department of Mining Engineering, University of Newcastle upon Tyne, England, a series of experiemnts were conducted in order to find out minimum size of the CCNBD specimen which will generate reliable rock fracture toughness results. The dimensions of the CCNBD specimen used is listed in Table 1. A fine grained sandstone was used. Specimen dimaeter varies from 50 to 100 mm. The diameter of diamond saw used for chevron notches cutting are 52, 75 and 100 mm. Specimen thickness varies from 15 to 80 mm. The initial crack length varies from 8 to 20 mm. In total 24 geometrical varieties of specimen were tested. For the CCNBD specimen having the same geometry, usually at least 5 specimens were tested and results were averaged. The test results for minimum size investigation of the CCNBD method are listed in Table 2. The following conclusions could be reached: <1> rock fracture toughness test results using CCNBD method is insensitive to specimen thickness; <2> CCNBD specimens with 50 mm diameter generate lower value of Kic. Specimens with 75 and 100 mm can generate comparable results; <3> The ratio of maximum crack length (both top and bottom surface) vs specimen diameter should be less than 0.85; <4> Kic test result is insensitive to dimensionless initial crack length; <5> A standard geometry CCNBD specimen DB01 with 75 mm diameter, 30 mm thickness was used for experimental validation of the CCNBD method.

5 EXPERIMENTAL VALIDATION OF THE CCNBD METHOD

The validation testing of the CCNBD method was performed by comparing the kic testing using three Chevron-Notched rock fracture specimens, i.e. CB, SR and CCNBD specimes. The procedure of kic testing using CB, SR methods is basically the same as that recommended by Finn Ouchterlony. The kic testing using CCNBD method was performed under constant loading line displacement rate of 0.1 mm/minute. The test results using CB, SR and CCNBD methods are listed in Table 3. It shows that three chevron-notched rock fracture toughness testing method can generate comparable results. The repeatability of Kic results using the CCNBD method is best among these three chevron-notched rock fracture toughness testing methods.

391

6 DETAILED COMPARISON OF THREE CHEVRON-NOTCHED ROCK FRACTURE SPECIMENS

The comparison of three chevron-notched rock fracture toughness testing methods is listed in Table 4. It cab be seen that the CCNBD method has its obvious advantages over conventional checron-notched rock fracture specimens. Its major advantages can be listed as following: <1> easy specimen preparation; <2> experimental procedure is quite simple; <3> a relative small size of rock specimen is required; <4> it is easy for investgating the effect of rock anisotropy on Kic value; <5> The repeatability of the test results is quite good .

7 CONCLUSIONS

Based on the results obtained above, it can be seen that the Cracked-Chevron-Notched Brazilian Disc method has potential to be utilized as the third chevron-notched specimen for mode I rock fracture toughness measurement.

Table 3 Comparison of the Results by the CCNBD, CB and SR Methods

Rock Type	K_{IC} by CCNBD	K_{IC} by CB	K_{IC} by SR
Pennant Sandstone	1.872, 1.961	1.738,1.538	1.951, 2.131
Sandstone-Dos	0.466, 0.477	0.472,0.593	0.467, 0.522, 0.385, 0.467
Sandstone-Disc	0.612, 0.610,0.621,0.613	0.593,0.694	0.627, 0.760
Limestone-Eim	0.747, 0.765	0.771, 0.798	1.092, 0.798
Limestone-2	0.701,0.678, 0.781	0.956, 0.756	0.681, 0.921
Limestone-Hard	2.491, 2.467	2.399,2.724	1.787, 1.829
Gneiss-Eim5	1.922, 2.130, 2.210, 1.923	2.281, 2.342	Transverse tensile failure
Rhyolite-Eim4	1.971, 1.845, 1.769	2.091, 1.967,	2.271, 1.968
Gypsum-Dos	0.551, 0.565	0.54, 0.624	0.732, 0.657
Ore-Eim	1.395, 1.435	1.541, 1.498	1.623, 1.734
Sandstone-Fai	0.371, 0.398	0.369, 0.410	0.398, 0.423
Sandstone-7	0.561, 0.604,	0.484, 0.509	0.498, 0.512
Sandstone-9	0.694, 0.740,	0.847,0.811	0.752, 0.698
Sandstone-31	0.566,0.577,0.577,0.577	0.589, 0.623	0.715, 0.822
Sandstone-33	0.364,0.376	0.324, 0.357	0.364, 0.385, 0.330
Sandstone-34	0.586, 0.597	0.632, 0.598	0.550, 0.630, 0.630
Sandstone-Spr41	0.671, 0.682	0.721, 0.687	0.645, 0.714
Gypsum-Pink48	0.836, 0.920	0.931, 0.923	0.945, 0.834
Limestone-54	1.480, 1.267	1.369, 1.678	1.497, 1.298
Sandstone-S15	0.311, 0.354	0.367, 0.324	
Sandstone-S18	0.329, 0.387	0.412, 0.498	0.326, 0.398
Sandstone-S25	0.324, 0.345	0.310, 0.365	
Sandstone-S26	0.307, 0.324	0.345, 0.312	

ACKNOWLEDGEMENTS

Thanks to Dr R. J. Fowell, Mr. A. Szeki, University of Newcastle upon Tyne and Prof. Finn Ouchterlony, SveDeFo in Sweden for their invaluable help for this reserach programe.

REFERENCES AND BIBLIOGRAPHY

Barker L.M. "Theory for Determining Kic from small, non-linear specimens", Int. J. of Fracture 15, pp 513-536, 1979.
Barker L.M. "Specimen Size Effects in Short Rod Fracture Toughness Measurements", Chevron-Notched Specimens, Testing and Stress analysis, ASTM STP 855, pp.117-133. Philadelphia PA, 1984.
Chen Jun Fang "The Development of Cracked-Chevron-Notched Brazilian Disc method for Rock Fracture Toughness Measurement and Tunnelling Machine Performance Prediction' PhD Thesis, 1989, University of Newcastle upon Tyne, England, U.K.
Ouchterlony F. "Review of Fracture Toughness Testing of Rock", SM Archives No.7, pp.131-211, 1982.
Ouchterlony F. "Suggested Methods for Determining Fracture Toughness of Rock Material. International Society for Rock Mechanics Commission on Testing Method", J.A. Franklin (Canada); Sun ZongQi (China); B.K. Atkinson and P. Mere (England); W. Muller (Germany); Y. Nishimatsu and H. Takahashi (Japan) 1987.
Shetty D. K., Rosenfield A.R., and Duckworth W.H. "Fracture Toughness of Ceramics Measured by a Chevron-notched Diametral Compression Test", J. American Ceramics Society, 1985, C325-C327.

Table 4 Comparison of Three Chevron-Notched Methods

Item of Comparison	The CCNBD Method	The CB Method	The SR Method
Size of specimen	Small	Long	Small
Source of sample	Easy to obtain	difficult	easy
Preparation apparatus	Simple	Simple	Complex
Specimen preparation time	30 samples/day	20 samples/day	20 samples/day
Setup of testing rig	Simple	Complex	Complex
Loading machines	Compressive	Compressive	Tensile
Preload Requirement	Less than 2 kN	Zero	Zero
Auxiliary apparatus	Simple	Complex	Complex
Displacement measurement	Simple	Complex	Complex
Loading method	Compressive Loading	Three Point Bending	Tensile Loading
Failure load range (usually)	$\geq 1kN$	$\leq 1kN$	$\leq 1kN$
Experimental time (usually)	5 minutes	10 minutes	5 minutes
Repeatability of results	Excellent	Reasonable	Reasonable
Requirement of Testing Machine	Ordinary	High	High
Study of Rock Anisotropy	Easy	Difficult	Middle
Transverse Tensile Failure	No	No	Yes
Availability	Easy	Difficult	Difficult

Possibilities of true triaxial experiments in the laboratory for rock mechanics of the Delft University of Technology

Possibilités des expérimentations traxiales authentiques dans le laboratoire de mécanique des roches à l'Université de Technologie de Delft

W. Kamp & M.J. Cockram
Delft University of Technology, Netherlands

ABSTRACT: The possibilities of true triaxial testing in the Laboratory for Rock Mechanics recently have been improved significantly. This was achieved through the installation of a new force control system. Also a new set of sophisticated pressure platens has been added, equipped with spherical seats. A range of cubic sample sizes can be tested from 74 to 300 mm. Possibilities exist to carry out measurements of P-wave velocities in three perpendicular directions. It is also possible to measure acoustic emission.

After a description of the hardware of the true triaxial compression machine, an overview is given of the testing possibilities of this machine and the various measurements. Two tests are described. Firstly, the measurement of the elastic properties of an anisotropic shale sample. Secondly, the strength of a sandstone sample with support stresses including the P-wave velocities during failure.

RESUME: Les possibilitees des expériments triaxials authentiques dans le laboratoire de mécanique de roche sont récemment perfectionné significativement. C'était réalisé par l'installation d'un système nouveau pour controller les pressions. Aussi un nouveau assemblage des plaques pression est installé, équipé avec des sièges sphérique. Echantillons cubiques peuvent être testé avec dimensions entre 74 et 300 mm. Il y a des possibilitees pour mesurer les vitesses des ondes de compression en trois directions perpendiculaires. C'est aussi possible de mesurer l'émission acoustique.

Après la description de la structure de la machine de compression triaxial un sommaire est donné des expériments differents et les mesures variées. Deux expériments sont presenté. Premièrement, le mesure des propriétées élastiques d'un échantillon anisotropique de schiste. Secondement, la résistance d'un échantillon de grès avec des contraintes laterales, inclusivement les vitesses des ondes de compression pendant le creusement.

1. INTRODUCTION

Research is carried out in the framework of the project "Cataclastic-Plasto-Elastic equilibria in brittle rock under triaxial differential loading conditions" of the E.C. R&D programme Raw Materials and Recycling, contract MA1M-0018-NL.

In general a rock-mechanics problem consists of two parts. Firstly, there is the lay-out of the problem itself. The shape (tunnels, caverns, etc.) and the boundary conditions such as depth, in-situ stresses, temperatures and history.

Secondly, there are the material properties that play a role in the problem. This can be (linear) elasticity, anisotropy, strength, post-failure behaviour, etc. For those properties tests have to be carried out to obtain the required parameters.

There are two ways to approach the problem. One is to calculate and predict the material behaviour and/or the stability of the opening. Numerical methods such as the finite element/finite difference method nowadays provide strong possibilities to simulate complicated situations with complicated material properties. For the determination of material properties sophisticated possibilities now exist in the true triaxial compression machine.

Another way is to simulate the underground situation with model tests. The true triaxial compression machine can accommodate cubes of 300 mm. Combined with forces of maximum 3500 KN, simulations can be made of in-situ stresses of depths down to approx. 1500 m. Such cubes can be provided with a hole representing an underground opening (tunnel, etc.). Deformations and mechanisms of failure around such a hole can be studied during and after the test.

2. DESCRIPTION OF THE TRUE TRIAXIAL COMPRESSION MACHINE

2.1. General

The machine consists of three uniaxial compression systems assembled in perpendicular directions so that cubic samples can be compressed. Each of the uniaxial systems consists of a piston with a pressure platen at one side and a pressure platen at the other side, connected with four tension bars. The three uniaxial systems can slide relative to each other. Also vertical movements of the horizontal compression systems are possible as they are mounted on vertical springs. In this way shear forces on the sample surfaces are minimized.

On five of the six sides of a cubical sample spherical seats are mounted. The bottom side is fixed. In this way small rotations of the pressure platens are possible which are necessary in case the sample surfaces are not perfectly parallel.

2.2. Sample dimensions

Possible sample dimensions:
       74* 74* 74 mm.
      105*105*105 mm.
      115*115*115 mm.
      300*300*300 mm.
The sample sizes are always chosen a few millimetres larger than the endplatens (resp. 70, 100, 110 and 280 mm.) to avoid touching of the endplatens of perpendicular directions due to sample deformations. A disadvantage is that there are stress free edges of the samples, FEM calculations have shown that the influence is small as long as the deformations are elastic, nevertheless fracture initiation in the corners and the post-failure behaviour can be influenced by this effect.

2.3. Force control

The forces on the sample are determined from the oil pressures in the cylinders and the piston surfaces. The pistons are

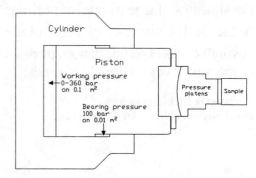

figure 1. Cylinder and piston (schematic).

equipped with hydrostatic bearings (see figure 1). The bearing pressure (100 bar) works against the main pressure but on a smaller area making it possible to retract the pistons. This reverse force is subtracted from the main force to obtain the actual force on the sample.

The main pressures are controlled with hydraulic servo valves. The actual pressures are compared with setpoints and the difference is amplified to obtain the steering signal for the servo valves. Since the hydraulic system has nonlinear characteristics, nonlinear amplification factors are used to obtain optimal response over the whole working range.

Force control is carried out by a personal computer equipped with A/D and D/A convertors. This computer also controls several logic functions such as display of pressures/forces and buttons presses. It also checks for several safety conditions.
       Force specifications:
       Maximum force: 3500 KN
       Minimum force: 5 KN
       Accuracy: 3 KN

3. MEASUREMENTS

3.1. Forces

Although the setpoints for the forces are usually followed very accurately it is possible that, due to failure, the sample is not capable to sustain the force. Therefore the actual forces are also available for registration.

3.2. Displacements

The displacement measurement system is mounted as near to the face of the endplatens as possible to exclude the strain of the compression system from the

394

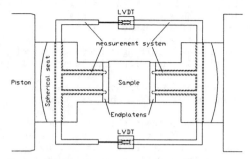

figure 2. Displacement measurement system.

measurements (see figure 2).

Using stress free arms and cross bars this displacement is transferred away from the compression system. Here linear voltage displacement transducers (LVDT's) measure the relative displacement of opposite sides of the sample. Two LVDT's are used to compensate for small rotations of the endplatens, resulting in a few microns of accuracy.

3.3. P-wave velocities and emission

To carry out P-wave velocity measurements and acoustic emission measurements, a stand alone system is used controlled by a personal computer. Each of the smaller endplatens (70–110 mm.) are equipped with a piezoelectric P-wave transducer. The equipment is shown in figure 3.

Using a custom made computer controlled six channel switchbox any transducer can be configured either as a transmitter or as a receiver. During a test a 300 V pulse is sent through the switchbox to the transmitter. The signal from the receiver goes through the switchbox to a broadband amplifier. After amplification the signal is digitized using a 100 MHz computer controlled digital scope. The digitized signal is then transferred to the control computer where in real time a sine is fitted to the signal for the calculation of the first arrival. In this way P-wave velocities can be determined with an

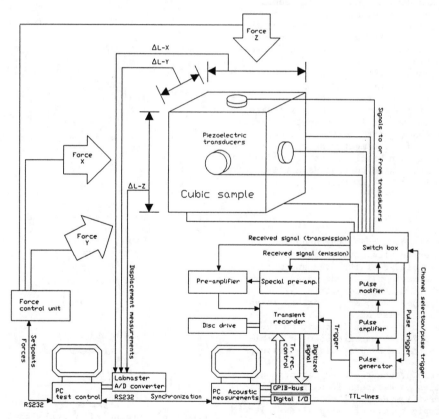

figure 3. Lay-out of equipment used during automatic testing.

accuracy of approximately 1 percent.

For the measurement of emission a second channel of the switchbox is used which is connected to a charge amplifier before being digitized.

A complete cycle of P-wave and emission measurements takes from 5 to 10 seconds.

3.4. Special measurements

Especially for the model tests, described below, special measurements are required. Temperatures can be measured by installing thermocouples in the models and are interesting when the model is (locally) heated.

Convergence of a tunnel (on scale) can be measured with specially designed displacement transducers.

4. TEST PROCEDURE

The lay-out of the equipment, operating during a test, is shown in figure 3.

For the control and measurements of forces and displacements two personal computers are required. The first computer (force control unit) is dedicated to the measurement of the various oil pressures and the control of valves of the compression machine.

The second computer generates the set-points for the forces during a test. It also stores and displays the measured forces and displacements. Using a software library it is relatively easy to generate complex loading paths either force controlled or displacement controlled. This computer is synchronized with the acoustic measurement system so that data files can be combined after the tests.

5. MATERIAL TESTS

In order to understand the behaviour of rock under in situ stress conditions, proper knowledge of its pre- and post-failure behaviour is required as well as the mechanisms that result in failure. Similar experiments are described in van Mier (5) and Reat et al. (6).

To obtain this type information the cubes of 74-115 mm. are used in combination with the P-wave velocity measurements.

The test procedures used for pre-failure properties is a stress loop in three directions (figure 4), to obtain data on the anisotropic behaviour using stress, strain and P-wave velocities. This stress loop is done six times for averaging purposes.

In figure 5 the stress path is shown for this procedure on a cube of shale. The sample is first loaded to a hydrostatic stress after which the stress loops are started. The resulting strains are shown in figure 6. Anisotropy can be observed clearly from the stronger deformations in the Z-direction (perpendicular to the layering). In table 1 the resulting Young's moduli and Poisson ratio's are shown. In this table the calculated velocities refer to theoretically determined values that can be obtained by applying the equations of motion to the anisotropic form of Hooke's law. The full theory can be found in Houwink (2).

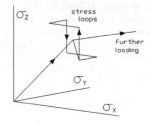

figure 4. Stress loops to determine anisotropic elastic properties.

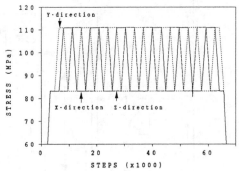

figure 5. Sequence of X-, Y- and Z-stresses during loops as shown in figure 4.

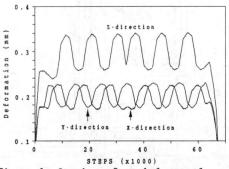

figure 6. Strains of a shale sample as a result of the stress loops described above.

396

E GPa	Poison ratio	calculated C_P m/s	measured C_P m/s
E_x=87	ν_{yz}=0.18	C_{Px}=6140	C_{Px}=6345
E_y=79	ν_{xz}=0.18	C_{Py}=5800	C_{Py}=6310
E_z=51	ν_{xy}=0.25	C_{Pz}=4630	C_{Pz}=5170

Table. 1. Anisotropic elastic properties of a shale and measured and calculated P–wave velocities.

As a second example the failure behaviour of a sandstone sample is shown. First, the sample was loaded hydrostatically to 3 MPa. In both X– and Y–directions the support pressure of 3 MPa was maintained and in the Z–direction the stress was increased until failure. The loading path is shown in figure 7 and the deformations in figure 8. Continuously P–wave velocities were measured and the results are shown in figure 9.

It can be observed from the velocities as well as from the deformations that there is an original anisotropy, which tends to disappear during hydrostatic loading while the velocities increase significantly. During deviatoric loading a further increase of C_{Pz} is observed and a decrease of C_{Px} and C_{Py}. This strong stress and/or plasticity induced anisotropy indicates the development of microfractures in the Z–direction.

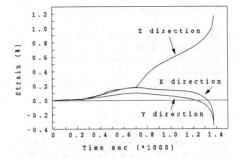

figure 8. Strains of the sandstone sample.

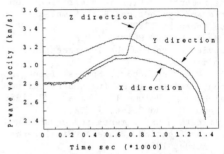

figure 9. Velocities measured in the sandstone sample during the test.

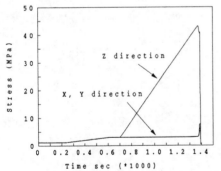

figure 7. Loading sequence of the test on the sandstone sample.

6. MODEL TESTS

Various possibilities exist to carry out model tests. Usually the maximum size cube (300 mm.) is used. It is possible to carry out tests on cubes with a model cavity. Such cavities can model an underground

gallery, a coal gasification chamber or a storage space of nuclear waste in rock salt. Such tests can include heaters, thermocouples or convergence measurements of the hole. Currently hydraulic fractures are studied using acoustic imaging techniques.

Projects have been carried out in the following research fields:
 – Underground coal gasification
 – Destressing techniques for tunnels
 – Hydraulic fracturing
 – Nuclear waste storage in rock salt
An overview is given in Kamp and Roest (3) and a detailed description of the application to underground coal gasification is given in Kamp and Roest (4).

7. CONCLUSIONS

The recent modification of the control unit of the true triaxial compression machine, the redesign of the pressure platen assembly and the P–wave velocity measurement system, have improved experimental possibilities significantly.

Pre– and post–yield behaviour of rock with complex loading paths can be studied in three independent directions.

8. ACKNOWLEDGEMENTS

This work has been supported by the
Commission of the European Community
as a part of the programme "Primary
raw Materials"

9. REFERENCES

1. Blok, P.
Das Hydrostatischer Keilspaltlager: Be
rechnung und Anwendung bei Hydrozylindern.
Dissertation, Delft, 1976.
2. Houwink, H. Static and Dynamic
Measurements of Elastic coefficients on
cubic and Cylindrical Rock Samples.
TUD, february 1990.
3. Kamp, W. and Roest, J.P.A.
Overview of rock-mechanics research for
deep mining.
Delft Progress Report. Special Issue on
Recent Research in Mining and Petroleum
Engineering. Volume 13, Number 1/2. 1988.
4. Kamp, W. and Roest, J.P.A.
Verification experiments for underground
coal gasification. Phase A. Final report.
TUD, August 1989.
5. Mier, J.G.M. van.
Strain-softening of concrete under
multiaxial loading conditions.
Dissertation, Eindhoven, 1984.
6. Reat, J.-F., Gooch, C.F., Hac, A. and
Cheatham, J.B.
Yield Surface and Strain-Hardening
Properties of Cordova Cream Limestone.
Int. J. Rock Mech. Min Sci. & Geomech.
Abstr. Vol 26, No. 5, pp 373-380, 1989.

On the shear strength parameters of some silty clays
Caractéristique de la résistance au cisaillement des argiles silteuses

U. Kołodziejczyk
Technical University, Zielona Góra, Poland

ABSTRACT: The basic shear strength parameters of soil are the shear strength and the strain. This paper presents the variation analysis of these parameters in silty clays. Three series of samples, which were collected from the geological structure in three directions perpendicular each other was object of investigation. These directions are in accordance with a-, b- and c-structural coordinate. Results analysis set in relation: stress deviator $/\delta_1-\delta_3/_{max}$, time of shearing $/t_z/$ and longitudinal strain $/\varepsilon/$. Obtained results shows that silty clays are characterized by anisotropy of shear strength parameters what is direction of acting stress as well as methodology of laboratory investigation.

RESUME: Dans ce travail on a présenté les effets des études sous contraintes triaxiales dans des argiles silteuses. La recherche est fondee sur trois series d'echantillons tailles dans trois directions perpendiculaires. La contrainte principale δ_1 a ete appliquee parallement aux coordonnees a, b et c qui indiquent les directions de la structure geologique. Chaque serie d'echantillons a ete etudiee a 3 vitesses. On a constate qu'il existe l'anisotropie des parametres de resistance.

1 INTRODUCTION

Many soils show shear strength anisotropy, as described by Bishop and Little /1965/, Duncan and Seed/1966/, Kowalski /1975/, Kaczyński /1981/ and others. In case of glacitectonically deformed this is due to their ability to memorize the history of successive loading-unloading cycles associated with the activity of continental glaciers - Kołodziejczyk and Kotowski /1985/. Applying appropriately oriented stress fields and adequately chosen axial strain rates can restore soil's memory, which makes possible studying the anisotropy of its strength parameters.

The present paper concentrates on showing the control exerted by the direction of the applied principal axial stress δ_1 and axial deformation speeds on the following strength parameters: longitidinal strain ε, time of shearing t_z and stress deviator $/\delta_1-\delta_3/_{max}$.

The samples studied were glacitectonically deformed silty clays from western Poland. Three series of samples have been investigated. The samples' axes were chosen relative to the structural coordinates a, b, c as defined by Sander /1970/ and to the primary vertical stress direction V. The samples within the successive series were oriented as follows: $V \wedge \delta_1 \| c$ /series 1/, $V \wedge \delta_1 \| b$ /series 2/ and $V \wedge \delta_1 \| a$ /series 3/, which means that when shearing the samples the principal stress δ_1 was oriented parallel to the respective structural coordinates and diagonally to the primary vertical stress V. This method of orientating samples was introduced here after Kotowski /1980/. Each series comprised 90 samples sheared at three speeds: 1.6, 2.4 and 4.8 mm/h. The shearing was carried out in a Norwegian type triaxial apparatus, after 24 hour primary consolidation, at the confining pres-

sure of $\bar{\sigma}_2=\bar{\sigma}_3=196.2$ kPa and with
simultaneous pore pressure recor-
ding.

2 TIME OF SHEARING

The assesment of shearing time was
performed on the basis of shearing
time measurement density curves
/see: Fig.1/, constituting curves
obtained at a constant axial deformation speed from differently oriented series of samples. The investigations showed that the shearing
time measurement density curves
displayed extremes clustering from
8% up to 40% measurements. For high
axial deformations speeds /v=4.8
mm/h: Fig.1c/ the curves show normal distribution; in all the series
the time of shearing was in most
cases /30-40% measurements/ 0.6-1.0
h. Lower deformation speeds resulted
in differentiation of the shearing
time measurement curves. This seems
to be the effect of restoring the
soil's memory which reflects its
geological history. For example,
at deformation speed of 1.6 mm/h
/Fig.1a/ soil samples oriented
$V\underset{30}{\triangle}\bar{\sigma}_1\|c$ revealed most /20% measu-
rements/ shearing times of 2.6 h,
those oriented $V\underset{60}{\triangle}\bar{\sigma}_1\|b$ revealed ti-
mes of 2.3 h /23% measurements/,
and those of $V\underset{85}{\triangle}\bar{\sigma}_1\|a$ orientation
were most often /21% measurements/
sheared during 2.5 h.

Axial deformation speed of
2.4 mm/h revealed the existence of
one distinct extreme /23% measure-
ments/ for samples of orientation
$V\underset{30}{\triangle}\bar{\sigma}_1\|c$ /Fig.1b/. On the other
hand, samples of another orienta-
tion displayed distinct bipartity
of extremes: for instance three
equivalent extremes /18% measurements each/ were obtained for times of shearing of 1.0, 1.3 and
1.75 h applied to samples oriented
$V\underset{60}{\triangle}o_1\|b$. It is only the series of
samples oriented $V\underset{30}{\triangle}\bar{\sigma}_1\|c$ that yiel-
ded one distinct extreme on shea-
ring time density curves /normal
distribution/, irrespective of axial deformation speed. This permits unequivocal determination of
shearing times for soil samples of
the latter orientation dependent on
different axial deformation speed.
The samples on another spatial
orientation show strength anisotropy which results in an abnormal

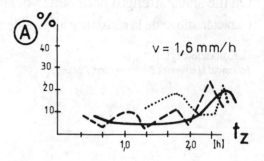

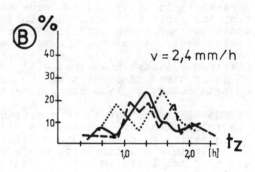

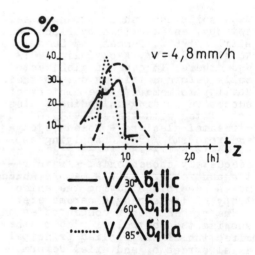

Fig.1 Shearing time t_z measurement
density curves for variously oriented cylindrical samples and various
axial deformation speeds

400

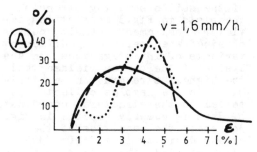

/multi-extreme/ distribution of shearing time density curves. The probability of a univocal determination of shearing time is in that
case not complete /two extremes in
series $V \wedge \delta_1 \| a$/ or only partial
/two or three extremes in series
$V \wedge \delta_1 \| b$/.

3 LONGITUDINAL STRAIN

The relative deformation density
curves obtained for various sample
orientations and various deformation speeds are presented in Figure 2. It illustrates longitudinal
strain observed at the moment of
shear failure of the samples, that
is at the maximum magnitudes of the
stress deviator. The samples of
orientation $V \wedge \delta_1 \| c$ display normal
distribution, irrespective of axial
deformation speeds; the longitudinal strain being 3-4% in the dominant number of cases /28-50% measu-
rements/. The remaining series show
bipartity of the extremes. The se-
ries of samples oriented $V \wedge \delta_1 \| b$ is
characterized by two extremes /25%
and 43% measurements/ for the shearing speed of 1.6 mm/h /Fig.2a/,
whereas the series of orientation
$V \wedge \delta_1 \| a$ shows bipartity of the extreme at the deformation speed of
4.8 mm/h /Fig.2c/. In series 2 and
3 a tendency of the extremes can be
perceived to be displaced towards
the higher strain magnitudes. For
example, in Fig.1a and 1c the extremes for series of samples orien-
ted $V \wedge \delta_1 \| b$ and $V \wedge \delta_1 \| a$ attain
their maximum values /30-45% measu-
rements/ for the longitudinal strain
of 5%. This attests to a higher deformability of the soil in the direction of a and b coordinates. On
the other hand an increase of stress
acting perpendicularly to the stra-
tification planes /$V \wedge \delta_1 \| c$/ shoud
result in a smaller longitudinal
strain /of the order of 3-4%/. The
described anisotropy is the effect
of the loading history of the soil,
which was doubly consolidated in
situ by continental ice sheet, in
the direction of the coordinate c.

4 SHEAR STRENGTH

It was assumed to consider the maximum magnitude of the stress de-
viator /$\delta_1 - \delta_3$/$_{max}$ as the measure

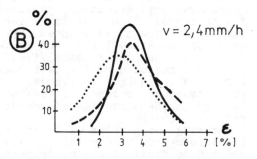

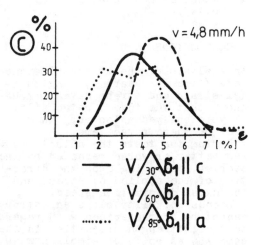

Fig.2 Longitudinal strain ε measurement density curves for variously
oriented cylindrical samples and
various deformation speeds

of the soil's shearing strength.
The plots in Fig.3 show stress deviator measurements density curves
obtained when performing different
variants of investigations. An analysis of the curves yielded that
the highest strength can be attributed to the samples collected parallel to the structural coordinate
b. This is clearly visible in Fig.
3c, where the magnitude of the
stress deviator is 110 kPa /30%
measurements/. The latter coordinate b corresponds to the axial trend
of the syncline from which the samples have been collected. It is,
therefore, the direction, along
which no intense compression ever
acted; since this was not the tectonic transport direction, parallel
to the coordinate a, nor the direction of the ice sheet-induced compression corresponding to the coordinate c. Much higher loads /i.e.
greater magnitudes of the deviator
stress/ are required to produce
soil structure collapse if they are
applied in the direction of coordinate b.

Shearing the samples at lower axial deformation speeds /Fig.3a and
3b/ led to distinct bi- or triparti-
ty of the extremes, most probably
reflecting the soil's deformational
history, including the first and
second loading by successive gla-
ciers /$V \wedge \delta_1 \| c$/ and the effects of
the glacier-induced drag force
/$V \wedge \delta_1 \| a$/.

5 CONCLUSIONS

An analysis of the strength parameters /time of shearing, longitudinal strain and stress deviator/ has
shown that the investigated silty
clays displayed anisotropy in respect of these parameters.

During the tests the anisotropy
was better or worse revealed by the
samples, depending upon the direction of the applied δ_1 stress and
the axial deformation speed.

Becouse of silty clays anisotropy
cannot be characterized with regard
to their strength properties in the
same way as soils of simple stress
history. The glacitectonically deformed media show distinct anisotropy of strength properties spatially
related to the symmetry axes of the
glacitectonic deformations, to the

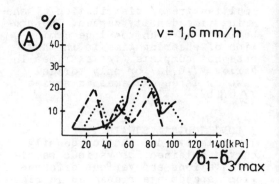

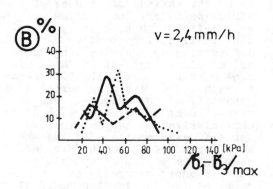

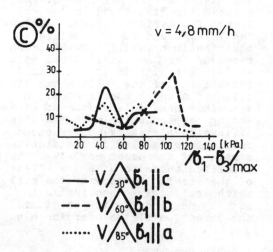

Fig.3 Stress deviator /$\delta_1 - \delta_3$/
measurement density curves for variously oriented cylindrical samples
and various axial deformation speeds

402

axes of the resultant geological structures and to the directions of tectonic transport. The general rule stading that soils show the greatest strength in the direction perpendicular to the sedimentary stratification is not always valid in the case of glacitectonically distorted soils. The decisive is here the role of local stress fields iducated by the glacier.

REFERENCES

Bishop A.W. Little A.L. 1965. The influence of the size and orientation of the sample on the apparent strength of London clay at Maldon. Essex.Proc.of the Geot. Conf. Oslo.
Duncan J.M. Seed H.B. 1966. Strength variation along failure surfaces in clay. Proc. ASCE, vol.92.
Kowalski W.C. 1975. Zróżnicowanie wytrzymałościowe skał. Ibidem. Warszawa.
Kaczyński R. 1981. Wytrzymałość i odkształcalność górnomioceńskich iłów zapadliska przedkarpackiego. Biul. Geol. t.29. Warszawa.
Kołodziejczyk U. Kotowski J. 1985. Anizotropia wybranych parametrów wytrzymałościowych iłu pylastego. V-th Glacitectonics Symposium. Zielona Góra.
Kotowski J. 1980. Wybrane zagadnienia z analizy strukturalnej iłów zaburzonych glacitektonicznie. III-rd Glacitectonics Symposium. Zielona Góra.
Sander B. 1970. An introduction to the study of fabrics of geological bodies. Pergamon. Oxford.

6th International IAEG Congress / 6ème Congrès International de AIGI, © 1990 Balkema, Rotterdam. ISBN 90 6191 130 3

Geotechnical properties of the Neogene sediments in the NW Peloponnesus, Greece
Caractéristiques géotechniques des sédiments du Neogene en Péloponnèse du N-O, Grèce

G. Koukis
Department of Geology, University of Patras, Patras, Greece

D. Rozos
Engineering Geology Department, Institute of Geology and Mineral Exploration, Athens, Greece

ABSTRACT: The lithostratigraphic development, physical state and geomechanical properties of the neogene sediments in the Achaia region (NW Peloponnesus) were examined after a detailed mapping, representative sampling and the execution of a series of laboratory testing. So, the evaluation of the geotechnical parameters for the various lithological phases was made and their corresponding engineering geological behaviour was assessed.

RESUME: Nous avons examiné le developement lithostratigraphique, la situation physique et les propriétés geomecaniques des sediments néogénes de la région d'Achaia (Peloponnése NW), d'aprés la réalisation d'une cartographie detaillée, d'un échantillonage représentatif et d'une série d'essais an laboratoire. Ainsi nous avons effectué une évaluation des parametres geotechniques des differents faciés lithologiques et nous avons aussi estime leur comportement geotechniques correspondant.

1 INTRODUCTION

The Neogene sediments of Achaia region (NW Peloponnesus), of Plio-pleistocene in age, are mainly lacustrine to marine-lacustrine deposits. They have been developed in three sedimentation basins, and are characterized by a variety of lithological horizons (Rozos, 1989). In the lower ones, the fine-grained facies (alternations of clayey marls, marls, silty sands and weak sandstones) prevail, with a progressive transition upwards to the coarser facies, which at least in the Corinthian basin give coherent conglomerates of great thickness.

Concerning the total thickness of the sediments, this ranges from 350 to 2000 m, explaining thus the great uplifting movements having taken place in the referred region.

Generally, the broader area (Tsoflias, 1970) is emplased in the Olonos-Pindos geotectonic zone, which is characterized by an "inherent weakness" connected with the alpine cycle of sedimentation and orogenesis. The induced stresses due to neotectonic movements (Plio-pleistocene) must be added (Doutsos et al,1987 & Doutsos et al,1988). This is particularly ob-

vious within this zone, through the presence of the still active Corinthian graben being, since the middle Pleistocene in an environment of extensional stresses. Here and mainly in the neogene sediments, a great instability is occurred due to their lithology, active tectonism, high seismicity, the abrupt morphology, and some other secondary exogene factors (Koukis,1988).

2 LITHOSTRATIGRAPHIC CONSIDERATIONS

The sedimentation basins are namely the Corinthian, Patraikos and Leondion ones (Fig.1). The stratigraphic sequence and the lithology of the sediments consisting them have as follows:

2.1 Corinthian basin

The sediments of this basin are developed up to an altitude of 1800 m and can be subdivided into two unities, Lower and Upper.

The Lower unity. The deeper horizons of this consist of gray coloured sandy clays, clayey marls and marls. These sediments are characterized by a low coherent and include lignite intercalations. A sedimentary series of about 400 to 500 meters in thickness covers the above horizons. It starts (lower beds) with alternations of clayey marls, fine to medium grained sands and sandy clays. The intermediate beds consist mainly of sandy clays in alternations with clayey marls, sands, marls and sandy marls, while gravels and pebbles also take part, with an increasing proportion from the bottom to the top beds. Among the beds of this series, humus intercalations with leaves and sprouts imprints as well as lignite bodies are present, while slumpings indicating the unstable character of the basin have also been recognized. In the upper beds, the sedimentation continues with alternations of clayey marls, sands, and sandy marls, with lignite bodies and increasing proportion of gravels and pebbles (30-40 %), which at some places form conglomerate beds. The above described lower series end with a transitional horizon of a thickness of about 30 meters. This horizon starts with alternations of sands, gravels, clayey marls, and conglomerates and terminates in conglomerate beds (1-4 meters thick).

The upper unity, consists of highly fractured conglomerates, with a thickness of more than 400 meters, usually in conformity with the underlying sediments.

. As far as the conditions of sedimentation are concerned the following remarks could be made:

-The unstable character of the sedimentation basin is revealed (quick alternations and lateral transitions of the lithologic units, presence of slumpings, etc.).

-The finding of vegetable fossils (leaves and sprouts) and the existence of lacustrine microfossils (Cypridinae and Candoninae) indicate the fluviolacustrine to lacustrine or locally brackish character of the lower unity's sediments (Rozos, 1989).

2.2 Patraikos basin

Its sediments, which are developed up to an altitude of 200 meters can be separated into two groups, as follows:

The lower horizons, with a confirmed thickness of 110 meters, consist of black-grey silty clays and sandy silts. They do not appear in the surface throughout the basin (Rozos,1987).

The upper horizons, show a thickness of approximately 100 meters and start from the bottom with clayey marls with sands,

gravels, pebbles, and clayey silts with thin lenses of conglomerates, in alternations and frequent lateral transitions. Clayey marls with sands as well as sands of weak to medium diagenesis with an increased proportion of gravels and pebbles at places, forming locally weak conglomerates of small thickness, dominate upwards.

According to bibliographic data (Frydas, 1987 & Zelilidis et al, 1988) the lower beds of the upper horizons are characterized as shallow marine Pliocene sediments, while stratigraphic data for the lower horizons are deficient.

2.3 Leondion basin

It has been filled by almost terrigenous coarse-grained deposits, which consist of three main horizons:

The lower one consists of conglomerates and has an approximate thickness of 200 meters. These are multi-mixed coarse-grained deposits of gravels and pebbles of various types and sizes, and red connecting material (clayey silt).

The intermediate, of a thickness of approximately 100 meters, is composed of conglomerates and red clayey silt alternations, with lateral transitions from one unit to another.

The upper, with a thickness of about 30 meters, consists of red clayey marls with sands.

The above formations, as it is assumed, are of fluvial or fluviolacustrine origin and Plio-pleistocene deposits in age.

3 SAMPLING

For the study of the mineralogical composition and mechanical characteristics of the various horizons of the Neogene sediments, a great number of samples were obtained from 26 positions throughout the area (Fig.1).

The collected samples can be grouped in seven unities, as follows:
(a)Clayey marls, (b)Marls, (c)Sandy silts, sands and weak sandstones, (d)Weak conglomerates of the fine-grained sediments, (e)Strong conglomerates of the Corinthian basin, (f)Fine-grained formations of Leondion basin, and (g)Coarse-grained formations of the latter.

4 GEOMECHANICAL PARAMETERS

These parameters of both fine-grained

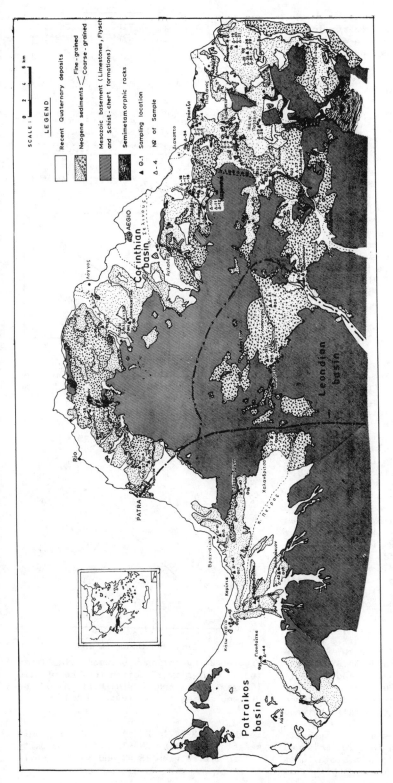

Figure 1. Geological map of the studied area

(clayey marls, marls, sandy silts, sands, and weak sandstones) and coarse-grained (conglomerates) sediments were examined by testing all the samples collected.

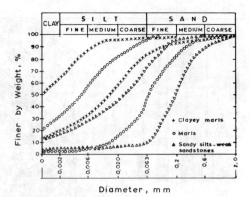

4.1 Fine-grained sediments

4.1.1 Physical properties

From <u>grain size analysis</u> the presence of silt in high percentages and the low participation of the clay fraction, is revealed (Fig.2).

In general, the clayey marls show Aggregation ratio, "Ar",(Davis, 1967) lower than unity, because of the additional presence of non clay minerals in their clay fraction. On the contrary, marls, silty sands, sands and sandstones, which have "Ar" values higher than unity, are characterized by the presence of clay mineral aggregates (Koukis,1977 & 1985).

<u>The Atterberg limits</u> of the clayey marls present a rather large range of values, while these of the marly horizons lie between the latter and those of the silty sands.

Classification of the different horizons, according to the above values, revealed that the c l a y e y m a r l s are characterized as inorganic clays of low to medium plasticity (CL) and rarely of high plasticity (CH), and as inorganic silts-clays of low plasticity (CL-ML).

The m a r l y h o r i z o n s are classified as inorganic silts and fine-grained sands or silty clays of low plasticity (ML) and rarely as inorganic clayey silts to inorganic silts, usually of low plasticity (CL-ML).

Also, the s a n d s - w e a k s a n d s t o n e s are considered as well-graded sands to silty sands (SW-SM) and rarely as silty or clayey sands (SM-SC) and/or poorly-graided sands (SP) (Fig.2).

Finally, with reference to the activity values, clayey marls and marls are characterized as non active soil materials with activity value 0.52 (Fig.3).

<u>The other physical characteristics</u> give the following values for the various phaces:

The clayey marls have specific gravity values ranging from 2.67 to 2.70, moisture content between 12.1% and 35.6%, total unit weight 18.5-23.4 KN/m³, and dry unit weight from 13.7 to 20.9 KN/m³. Also, their porosity was estimated between 21.9%-49.5%, and their void ratio between 0.28-0.98.

The marls present specific gravity values ranging from 2.65 to 2.70, moisture content between 0.5% and 21.2%, total unit

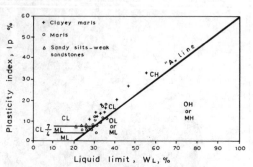

Figure 2. Particle size distribution envelopes and plasticity chart for the fine-grained Neogene sediments

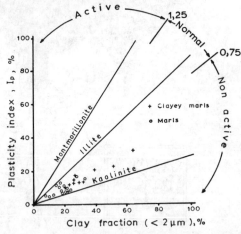

Figure 3. Relation between plasticity index and clay fraction for the plastic (Clayey marls and Marls) horizons of the Neogene sediments (Activity: 0.52)

weight 13.3-24.1 KN/m³, and dry unit weight 13.2-21.9 KN/m³. The porosity values lie between 18.0% and 50.5%, while void's ratio fluctuates from 0.22 to 1.02.

Silty sands and sandstones show specific gravity between 2.65-2.68, moisture content between 4.9% and 19.4%, total unit weight ranging from 14.1 to 24.9 KN/m^3, and dry unit weight 13.3-22.1 KN/m^3. Porosity has values from 16.7% to 49.7% and void's ratio from 0.20 to 0.99.

4.1.2 Mechanical properties

The investigation of the mechanical properties of the fine-grained neogene sediments included determination of uniaxial compressive strength, shear strength (by means of both direct shear and triaxial tests), and the compressibility characteristics of the soft clayey marls.

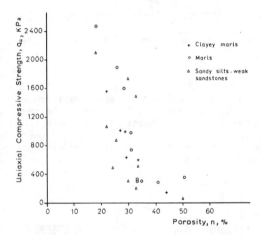

Figure 4. Relationship between uniaxial compressive strength and porosity for the fine-grained horizons (clayey marls, marls, sandy silts-weak sandstones) of the Neogene sediments

The calcitic marls and the sandstones show the highest values of the uniaxial-compressive strength (q_u=1900-2470 KPa), while the soft clayey marls and the loose silts the lowest (q_u=40-400 KPa). In general, from the range of the strength values of all the samples, it can be concluded that the Neogene sediments are characterized as very stiff to hard soil formations or weak rocks of very low strength, according to B.S.5930(1981). The good correlation between strength and porosity (Koukis,1974) reveals the meaning of the amount and arrangement of the accessory minerals, in relation to the percentage of the clay minerals (Fig.4).

The cohesion values of the samples examined by means of the direct shear test, from every horizon, do not show significant differences (mean range of values 30-110 KPa), with the only exception samples in which the percentage of sand is very high and the cohesion low (3-25 KPa). On the contrary, the values of the angle of friction present a progressive increase from those of clayey marls (23°-42°) to that of marls (29°-46°), and then to these of the sands and sandstones (33°-58°) (Fig.5).

The corresponding comparison of the results from the more reliable triaxial test shows an increase of the angle of friction and a corresponding decrease of cohesion, from clayey marls (φ_u=7°-33°, c_u=14-290 KPa) to marls (φ_u=27°-39°, c_u=65-140 KPa), and then to the coarser horizons, namely sandy silts, sands and sandstones (φ_u=30°-40°, c_u=20-180 KPa).

Generally, the cohesion values from the direct shear test results are always lower than those obtained from the triaxial test results, while the values of the angle of friction are always higher. These one-way deviations should be attributed to the disadvantages of the direct shear test.

The consolidation test, which was carried out using two samples from the loosest horizon of the Neogene sediments (black-grey clayey marls) shows that the compression index, C_c, has values of 0.160 and 0.240. These values correspond to Kaolinitic clay values and characterize materials without special problems, with regard to consolidation settlement.

4.2 Coarse-grained sediments

The conglomerates present specific gravity values between 2.63 and 2.65, dry unit weight from 24.3 to 26.2 KN/m^3, while porosity was found to have values from 1.1% to 7.9%. The highest values of dry unit weight and the lowest ones of porosity correspond to thick Corinthian conglomerates, while the lowest values of dry unit weight and the highest ones of porosity characterize the conglomerates in the upper horizons of the fine-grained sediments of the same basin. The conglomerates from the Leondion basin have values between those of the two former unities.

With regard to their uniaxial compressive strength examined by means of point load testing (Brook,1980), the highest values (37-42 MPa) correspond to samples from the thick Corinthian conglomerates, while conglomerates from the upper horizons of the fine-grained sediments of the same basin show the lowest values (12-23 MPa). Comparing the values of porosity

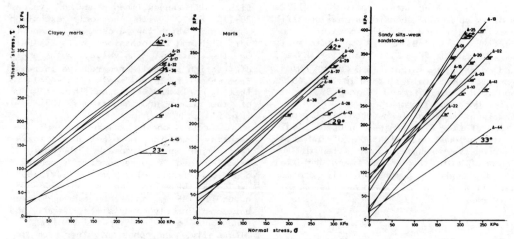

Figure 5. Shear strength envelopes of the samples from the three main horizons of the fine-grained sediments

and strength of the conglomerates, the influence of porosity on the mechanical characteristics of these rocky formations is revealed (Fig.6).

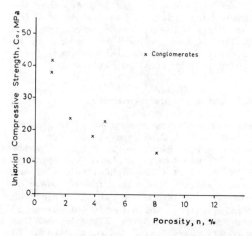

Figure 6. Relationship between uniaxial compressive strength and porosity for the coarse-grained horizons (conglomerates) of the Neogene sediments

Examination of the shear strength parameters of these coarse-grained sediments (Brook,1979),shows that the cohesion values increase from the conglomerates of the fine-grained Corinthian basin sediments (3000-5500 KPa) to those of Leondion basin (5000-6000 KPa), and then to the thick conglomerates of Corinthian basin

(9000-10,000 kPa), while the angle of friction values do not show differences (in all samples, "φ_u" was found between 37°and 44°). Triaxial test results of a characteristic sample (Λ-9) from the thick conglomeratic horizons of Corinthiakos basin is given in Fig.7.

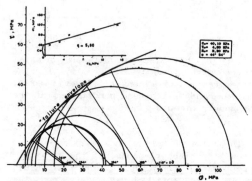

Figure 7. Triaxial test results of the Λ-9 sample from the thic conglomeratic horizons (Corinthiakos basin)

The shear parameters of characteristic discontinuities were also studied, by means of a portable shear box apparatus (Ross-Brown et al,1975, Barton,1976, Bandis,et al, 1981).
The study of the peak and residual angle of friction of such discontinuities, which determine the rockmass behaviour, shows that the thick conglomerates of Corinthian basin present higher values (φ_P=60°-61°,

$\varphi_r=57°$) than those of Leondion basin ($\varphi_p=52°-55°$, $\varphi_r=46°-49°$). The conglomerates of the fine-grained Corinthian basin sediments are of low coherence and they did not allow the formation of the necessary specimens.

5 CONCLUSIONS.

The results of the tests, which were carried out for the examination of the physical and mechanical characteristics of the Neogene sediments, lead to the following general remarks:

1. The comparatively low specific gravity values of marls and mainly those of sandy silts and weak sandstones – as compared with these of clayey marls– are attributed either to the relatively increased aggregation degree of clay particles (forming micropores), or to the presence of low crystallinity or amorphous calcitic grains (also forming micropores with air).

2. The moisture content values present a decrease, either to their range or to their limiting values, from the fine-grained to coarse-grained horizons. The low limiting values of two samples from marly horizons (0.5%), which diverge to the above rule, are attributed to the very high percentage of $CaCO_3$ and the corresponding limited clay fraction (3-9%).

3. The values of total unit and dry unit weights do not show considerable differentiations in all the samples from the three main unities of fine-grained sediments. The relatively lower values in samples from the marly horizons are ascribed to the abundance of amorphous or microcrystalline $CaCO_3$ and to its micropores.

4. The values of porosity and void ratio also present insignificant differentiation from one horizon to another, while the narrowest range is noticed in samples from clayey marls.

5. With regard to the mecanical properties of these sediments, calcitic marls and sandstones show the highest values of uniaxial compressive strength (1900-2470 KPa), while soft clayey marls and loose sandy silts the lowest (40-400 KPa). From the examination of the shear strength parameters, it is shown that there is a decrease of the cohesion values and a corresponding increase of the angle of friction values progressively from the clayey marls to the sandy silts-weak sandstones (c_u=290-20 KPa and φ_u=7°-40°). Also, the compression index values of the soft clayey marls has values between 0.16 and 0.24, showing that serious problems due to consolidation settlement, should not be expected.

6. Through the general consideration of the results from the physical and mechanical characteristics of the conglomerates, it is concluded that the thick conglomerates of the Corinthian basin are characterized by finer and more cohesive cementing material than both the conglomerates of Leondion basin and those of the upper horizons of fine-grained sediments of the Corinthian basin. Therefore, the thick Corinthian basin conglomerates show low porosity, high strength, high cohesion values as well as a better rockmass behaviour (higher values of peak and residual angle of friction of characteristic discontinuities). On the contrary, conglomerates from the fine-grained sediments in the same basin present the lowest values of geomechanical parameters, which clearly reveal the presence of coarser cementing material with low coherence. Finally, the Leondion basin conglomerates show intermediate values with regard to the above parameters.

6 REFERENCES

BANDIS,S, LUMSDEN,A, BARTON,N. (1981). Experimental Studies of Scale Effects on the Shear behaviour of Rock Joints. Int. J. Rock Mech. Min. Sci and Geomech. Abstr, Vol.18, pp 1-21. Pergamon press,London.

BARTON,N. (1976). Rock mechanics review: The shear strength of Rock and Rock joints. Int. J. Rock. Mech. Min. Sch. & Geomech. Abstr., Vol.13, pp.255-279. Pergamon Press, G. Britain.

BROOK,N. (1979). Technical note: Estimating the triaxial strength of rocks.Int.J. Rock Mech. Min. Sci. & Geomech. Abstr., Vol.16, pp 261-264. Pergamon press, G. Britain.

BROOK,N. (1980). Size correction for point load testing.Int. J. Rock Mech. Min. Sci., Vol.17, No4, pp231-235. Pergamon press, Oxford.

B.S.5930. (1981). Code of practice for site investigation. England

DAVIS,G.A. (1967). The mineralogy and phase equilibrium of Keuper Marl.Q. Jl. Engin. Geol., Vol.1. pp25-38. London

DOUTSOS,T, KONTOPOULOS,N & FRYDAS,D. (1987).Neotectonic evolution of north-western-continental Greece. Geologische Rundschau 76/2, pp 433-450. Stuttgart.

DOUTSOS,T, KONTOPOULOS,N & POULIMENOS,E. (1988). The Corinth-Patras rift as the initial stage of continental fragmenta

tion behind an active island arc
(Greece). Basin Research,1, pp 177-190.
Germany.

FRYDAS,D. (1987). Kalkiges Nannoplankton
aus dem Neogen der NW-Peloponnes,
Griechenland. N. Jb. Geol. Palaont. Mh,
H.5, pp 274-286. Stutgart.

KOUKIS,G. (1974). Physical , Mechanical
and chemical properties of the Trias-
sic sandstone aquifer of the Vale of
York. Ph.D.Thesis, Department of Earth
Sciences, University of Leeds.
England.

KOUKIS,G. (1977). Studies on the geological setting and physical-mechanical
characteristics of the Neogene sediments of Pyrgos - Helia area . Bullet.
of P.W.R.C., No.2/77, pp 69-79.
Athens, (in Greek, Eng. abs and fig.).

KOUKIS,G. (1985). Engineering-geological
conditions in the open Zante theater
founding area. Bullet. of P.W.R.C.,
No 3-4, pp 3-14. Athens, (in Greek,
eng. abs. and fig.).

KOUKIS,G. (1988). Slope deformation
phenomena related to the engineering
geological conditions in Greece. Proc.
of the Fifth Int. Symp. on landslides,
Vol.2. pp 1187-1192. Laussanne.

ROZOS,D. (1987). Geotechnical investigation for the foundation of labor es-
tates in Amaliada and Kato Achaia
areas. I.G.M.E., unpublished report,
pp 64. Athens, (in Greek).

ROZOS,D. (1989). Engineering-geological
conditions of Achaia province - Geomechanical characters of the Plio-
pleistocene sediments. Ph.D. thesis.
University of Patras, Dept of Geology,
453p. Patras, (in Greek, eng. summary).

ROSS-BROWN,D, WALTON,G. (1975). A portable
shear box for testing rock joints. Rock
Mechanics Ost., Vol.7, No3, pp 129-153.

TSOFLIAS,P. (1970). Geological constrac-
tion of the Northest part of Peloponnesus (Achaia province). Ann. Geol. d.
Pays Hell., vol.XXI, pp 554-651.
Athens, (in Greek).

ZELILIDIS,A, KOUKOUVELAS,I & DOUTSOS,T.
(1988). Neogene paleostress changes be-
hind the forearc fold belt in the
Patraikos Gulf area, western Greece. N.
Jb. Geol. Palaont. Mh, H.5, pp311-325.
Stuttgard.

6th International IAEG Congress / 6ème Congrès International de AIGI, © 1990 Balkema, Rotterdam. ISBN 90 6191 130 3

Application of acoustic emission parameters as a criterion of stress-strain behaviour of rocks

Les paramètres de l'émission acoustique comme un critère du comportement des roches du point de vue de la résistance et de la déformation

J. Krajewska-Pininska
Warsaw University, Poland

Z. Karska
Technical University of Bialystok, Poland

ABSTRACT: Results of laboratory testing of the Sudetic and Carpathian rocks, loaded till destruction are presented; strain value, sonic, ultrasonic and total acoustic emission, an attenuation of longitudinal acoustic waves in their transmission throughout rock material, and an impact of lithology on acoustic emission pattern are recorded and discussed. Of three phases of acoustic emission the last one occurring at about 80% of critical stress, with acoustic signals of specific frequency, can be used as an indicator of oncoming failure.

RESUME: Dans ce travail furent presentes les resultats des recherches sur les roches provenant des Sudetes et des Carpathes soumises a la charge jusqu'au moment de leur destruction. Durant la charge furent enregistrees les valeurs des deformations, l'emission acoustiques (AE) dans les limites sonores et ultrasonores, de meme que le changement du temps de passage des ondes ultrasonores. L'influence de la lithologie sur le deroulement de l'emission se fit remarquee. On identifier a ussi le moment de l'emission qui prouvaient le rapprochement de destruction. De tels signaux appraissaient le plus souvent quand la valeur de la tension atteignait 80% de la tension critique et se caracterisaient par une frequence determinee.

1 INTRODUCTION

Progress in research on acoustic emission (AE) of deformed rocks has been closely related to the development of the theories of the brittle microcracking in solids (Brace, Bombolakis 1963, Sholz 1968a,b). Evolution of microcracks releases elastic energy of deformed rocks, measurable as acoustic emission, what can be used for monitoring of fracuring and failure of rocks. Since Duval and Obert (1942) first noted a radiation of small noises attributed to loading of rock samples, the recording technics and processing and further interpretation of AE data much improved (Dunegan et al 1970, Shimada et al 1983, Horii & Nasser 1985, Zuberek 1988). Research on the AE is aimed at the determination of critical stress value, e.g. prediction of the highest admissible stress, what is an essential factor in various geological hazards. Basing on the AE testing of magmatic, metamorphic and sedimentary rocks an attempt to formulate a reliable prediciton system on rock failure has been made.

2 EXPERIMENTAL TECHNICS

The experiments were carried out following scheme by Karska (1987). In the uniaxial compressive tests the AE events were registered by known method of detection, using piezoelectric transducers (Sholz 1968a). Amplified acoustic effects of microfracturing were tape recorded, pen-recorded and oscillograph-monitored. Simultaneously were recorded values of the ultrasonic longitudinal waves velocities and their changes due to the progress in deformation (Pininska & Karska 1986, 1987).

Eleven types of variable unhomogeneity rocks have been selected for testing. Diabases: 1- massive, undisturbed, 2- disturbed, with quartz inclusions, 3- highly disturbed (fractured) and then cemented with quartz; metamorphic shists: 4- slaty, undisturbed, 5- disturbed, with quartz inclusions, 6- highly disturbed and then cemented with quartz; 7- breccias of crystalline rocks; sandstones: 8- cracked, thin-bedded, 9- laminated, thin-bedded, 10- compact, corrugated, 11- porous, thick bedded. All have been cut into cylindrical samples of height = diameter = 5cm. Their averge geotechnical parameters are presented in Table 1.

3 EVALUATION OF ACOUSTIC EMISSION

The following AE performance parameters were obtained from the recorded data:

1. Occurrence pattern of AE. Three-phase development of AE appears on the oscillographs for the most of rocks. A moderate emission activity is characteristic for the first phase; an acoustic stillness characterizes the second phase; and the third phase is characterized by a remarkably strong acoustic emission, rapidly accellerating towards the failure (Sholz 1968, Khair 1977, Pininska & Karska 1987); Such a general pattern of rock behaviour, irrespective of geological properties disparity, seems to be accurate for

Table 1. Average geotechnical properties

Rocks type	Volume density T/m³	Longitudinal wave velocity C_L (m/s)	Transversal wave velocity C_T (m/s)	Poisson ratio ν	Dynamic modulus E (MPa) $\cdot 10^3$	Compressive strength (MPa)	Total axial strain ε_{TOT} (%)
1	2	3	4	5	6	7	8
1	2.80	3480 – 6100	1650 – 2217	0.33	33.7	91.79	1.21
2	2.77			0.37	28.8	69.37	1.41
3	2.77			0.32	25.9	69.38	1.75
4	2.68	3404 – 5746	2018 – 2717	0.29	27.9	65.0	1.08
5	2.68			0.29	29.1	66.26	1.90
6	2.68			0.32	33.4	67.02	1.58
7	2.71	3410 – 5456	1714	0.32	29.3	48.19	1.43
8	2.33	1783 – 2080		0.22	6.6	12.0	0.40
9	2.51	2060 – 4000	3200	0.14	11.6	32.8	1.00
10	2.62	2670 – 6250	up to	0.30	47.5	80.0	0.25
11	2.40	1275 – 3370		0.18	34.7	20.3	0.60

homogeneous, undisturbed rocks only. With the increase of unhomogeneity of rock sample this pattern becomes perturbed, e.g. in breccias and in strongly disturbed rocks distinct signals are being emitted during the whole process of deformation, so that the phase boudaries are not obvious. Various AE patterns are presented in Fig. 1.

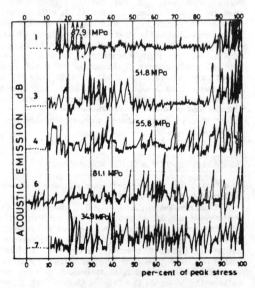

Fig.1. Acoustic emission pattern for various rocks 1, 3, 4, 6, 7 - rock types, see chapter 2; 51.8MPa - σ_{cr}

2. Frequency of acoustic activity events. A progress in individual crack growth follows an idealized cracking profile, similar to that presented by Horii and Nasser (1985) (Fig. 2) in whose opinion the individual cracks are nucleated into larger defects eventually leading to the brittle crack failure while approaching peak load.

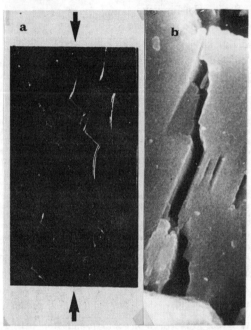

Fig. 2. Growth of microcracks: a) in polimers (after Horii & Nemat-Nasser 1985); b) in the Sudetic meamorphic shist (magn. ~5000x);

A concentration of cracks is manifested by density of emited signals. Registered increase and decrease in number of events produced by microcracking in a unit of time is an important factor, which can be used for determination of the stage of deformation. The frequency of events (N) represented by number of events per second increases strongly before failure for all kind of rocks. The increase in emission activity prior to failure is, however, less distinctive when rocks are disturbed, fractured and unhomogeneous, and their histograms of AE become bi- or multimodal (F ig. 3).

3. Sonic and ultrasonic emissions relationships. The sonic/ultrasonic emission ratio is another important parameter of acoustic emission. The energy of sonic and ultrasonic emissions produced by loaded sample were simultaneously registered by an ultrasonic sounder KD-10 of frequency range up to 180 kHz, and by an accellerometer KD-31 within frequency range up to 30 kHz.

It has been observed that ultrasonic signals appeared first, and their frequency of events (N) was much higher than those of the following sonic signals. Thus despite a fact that the

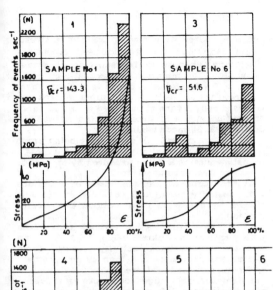

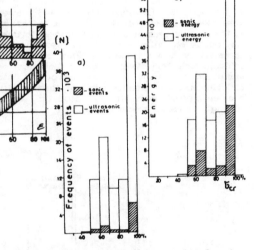

The sonic/ultrasonic emission events ratio (N_s/N_u) varies between 0 and 0.15 for various stages of deformation. The sonic/ultrasonic emission energy ratio ranges from 0.12 to 0.34; The energy of AE increases with the increase of both sonic and ultrasonic activity. At the final stage of deformation an ultrasonic emission increases faster than that of sonic, thus the resulting ultrasonic emission ratio decreases. The cause of faster increase of ultrasonic emission has not been satisfactorily explained, yet.

4. Acoustic spectrum of individual signals. The shapes of individual signals depend on a stage of deformation, (Fig. 5) and their acoustic spectrum varies accordingly. The characteristic stages of deformation were identified after the analog and numerical analyses of individual signal spectra. In the spectrum of individual signals the most common frequencies were discriminated. To assess an energy level of the frequency discriminated from the spectrum of individual signal, against a total energy of signal, a coefficient E_A has been introduced and then analysed on the background of various stress values (Fig. 6).

Fig. 3. Diagrams of events frequency (N) versus percentage of total axial strain (ε); 1, 3-6 - rock types, see chapter 2.

sonic signals are stronger, and their relative energy (E_s) conventionally accepted as signal amplitude (A) with constant factor (k) is higher, the increase of sonic energy at segments of σ_{cr} is lower than that of ultrasonic. Besides, both sonic and ultrasonic histograms show similar distribution of events, and show significant increase of activity towards a critical state of stress, (Table 2, Fig. 4.).

Fig. 4. Diagrams showing: a) frequency of sonic and ultrasonic events (N) versus percentage of critical stress (σ_{cr}); b) total energy of signals (E) versus critical stress (σ_{cr}).

Table 2. Energy and frequency of sonic and ultrasonic events, sample 40: metamorphic shists.

	Number of events (N) energy (E)	% of peak stress σ_{cr}					
		40-50	50-60	60-70	70-80	80-90	90-100
Ultra-sonic	Nu	50	16 900	29 750	13 850	16 550	54 950
	Eu	80	17 750	32 000	17 750	20 100	58 500
Sonic	Ns	0	1 300	2 090	550	580	8 450
	Es	10	2 900	5260	850	1400	19 750
Sonic/ultra-sonic	$I_N = N_s/Nu$	0	0.077	0.070	0.040	0.035	0.154
	$I_E = Es/Eu$	0.125	0.163	0.164	0.048	0.070	0.338

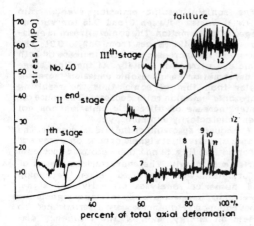

Fig. 5. Shape of individual signals at various stages of deformation.

4. CHANGES IN ULTRASONIC WAVE VELOCITY

A velocity of ultrasonic longitudinal waves, generated and received by 500 kHz sound boxes, at the early stages of loading increases to over 20% above an initial value (C_{Lo}) then it drops successively down towards 0 long before the critical state of stress. This effect can be explained by development of microcracs into dislocation zones attenuating, and then totally preventing transmission of acoustic waves through the rock material (Fig. 7).

5. DISCUSSION

The close correlation among prefailure stress-strain behaviour, the development of micro-fracturing, changes in transmission of waves throughout the material, and acoustic emission activity for the tested rocks is very clear when the results of testing are superimposed on idealised rock deformation curve with the three pre-critical stages of stress (Hoshino, Koide 1970, Costin 1983, Shimada 1983).

At the priminary (I) stage of closing pores and fissures with elasticity modulus $E_L \rightarrow 0 < E_L < \frac{d\sigma}{d\varepsilon}$, acoustic emission is being produced by process of pore and fissure compaction.

At the second (II) stage of linear deformations with elasticity modulus $E_L = d\bar{\sigma}/d\varepsilon$, an acoustic silence prevails as deformations are mostly elastic.

At the final (III) stage of critical peak stress with elasticity modulus $E_L \rightarrow d\bar{\sigma}/d\varepsilon > E_L > 0$, strong increase of emission activity results from microcrack nucleation and then by development of fractures prior to failure.

In disturbed, unhomogeneous rocks e.g. in breccias the stages of deformations may be somewhat less obvious, because of variable strength of chips of rocks and their quartz cementation. The stage III normally begins long before the failure. In tests of Sholz (1963 a, b, Khair 1977, Costin 1983, (as various, however, as granitic rocks and tuffs) emission activity arises sharply at the 80 % of critical stress.

For most of tested Sudetic and Carpathian rocks it arises when the load between 40 % and 90% of critical stress (σ_{cr}) is being applied. The more

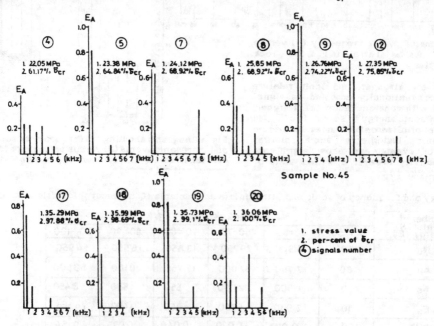

Fig. 6. Dominant frequencies in individual signals at various stress values; sample No. 45, rock type (3)-diabase, highly disturbed, cemented with quartz.

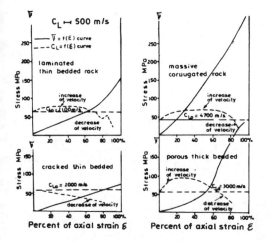

Fig. 7. Longitudinal wave velocity (C_L) versus strain ($\mathcal{E}$), and stress versus strain diagrams for the Carpathian sandstones.

unhomogeneous is rock, the earlier comes the acoustic "warning" (40 % for metamorphic shists indicates that there is still 60 % of load needed to reach a failure), and vice versa: the more homogeneous is rock, the later comes warning (e.g. 10 % before the failure for diabases).

An increase in frequency of events (N) before the failure is more readible while presented on a background of a strain development diagram (Fig.3). With the decrease in homogeneity of loaded rocks, as caused by tectonic disturbances, weathering etc. the number of events increases even at relatively small strain level. But despite of their homogeneity, or unhomogeneity, for all of tested rocks the most dramatic increase of (N) takes place at around 80 % of critical strain.

An analysis of spectra of singular signals, recorded for characteristic points of emission occurrence pattern, shows the following regularities:

a) at end below the 50 % of critical stress, impulses with acceleration of 1 to 2 m/s² appear at intervals of 120 us;

b) between the 50 % and 70 % of critical stress impulses with acceleration of 2 to 6 m/s² appear at intervals of 120 us;

c) between the 70 % and 90 % of critical stress impulses with acceleration of 3 to 8 m/s² appear at intervals of several us;

d) at end above the 70 % of critical stress impulses with acceleration of 3 to 10 m/s² appear at intervals of 5 us.

To distinguish between signals of the first (I) and the third (III) stages of deformation, an emission indicator of dominant frequency of signals (C_e) has been introduced:

$$C_e = f_i / \Sigma f_i ;$$

where f_i - energy of dominant frequency of individul signal, Σf_i -total energy of singular signal.

Typical for stage I is almost complete lack of dominant frequencies and C_e is low, while at the stage III the C_e indicator is over 0,75 high.

6. CONCLUSIONS

1. An evidence of failure, as taken from an analysis of acoustic emission bases on microcracking theory of failure in the brittle rocks. Due to development of microcracks an acoustic emission intensifies at the early (I) and the late (III) stages of precritical deformation; the emissions begins with predominate ultrasonic signals, then a share of sonic signals arises.

2. The occurrence pattern of acoustic emission depends on the lithology and fabric of the rock material. The most distinct separate stages of emission were obtained from tested diabase and shists.

3. As typical warning signals of the approach to a critical state before the failure are the attenuatiion of waves tremsmited through the material, the strong increase in number of acoustic emission events and an apearance of characteristic frequency of individual signal spectrum.,

4. The tested rocks of undisturbed homogeneity were approaching a stage of intensive deformation at 80 % of destructive deformation. At this level of deformation an increase of acoustic emission was recorded. For the less resistant rocks an increase in accoustic emission may begin much earlier, at less advanced stage of deformation e.g. at 40% of destructive deformation for metamorphic shists. For much unhomogeneus rocks (quertz cemented shists, diabases) even two extremes of increase in acoustic emission were noted.

5. For practice purposes an empiric safety limit, when the indicator of dominant frequency of individual signal (C_e) reaches 0.75, has been determined as a warning signal of approaching failure.

REFERENCES

Brace, W.F. & E.G. Bombolakis 1963. A note on brittle crack growth in compression. J.Geoph.Res. 68:3709-3713.

Costin, L.S. 1983. A microcrack model for deformation and failure of brittle rocks. J.Geoph. Res. 88:B 9485-9492.

Dunegan, .L., D.O. Harris & D.Tetelman 1970. Detection of fatigue crack growth by acoustic emission techniques. Material Eval. 23.10.

Horii, H. & S. Nemat-Nasser 1985. Compression induced microcrack growth in brittle solids: axial splitting and shear failure. J. Geoph. Res. 90:B 3105-3125.

Hoshini, K. & H. Koide 1970. Process of deformation of the sedimentary rocks. Proc. Congr. Int. Soc. Rock Mech. 2nd. I:353-359.

Ishido, T. & O. Nishizawa 1984. Effects of zeta potential on microcrack growth in rock under relatively low uniaxial compression. J. Geoph Res. 89:B:4153-4159.

Study on mechanical property of fault gouge under ground stress environment

Etude des propriétés méchaniques des failles dans la roche broyée, dans l'environnement de la contrainte du sol

Li Rongqiang, Kong Defang & Nie Dexing
Chengdu College of Geology, Sichuan, People's Republic of China

ABSTRACT: The paper studies systematically the physical and mechanical property of fault gouge at Ankang Hydro-electrical Station on the basis of the ground stress environment. In addition, a change tendency of the physical properties of the gouge and several causes leading to shear strength parameter decreased are researched by a simulation test of two mechanical processions of compression and relaxion. The results have showed that shear strength of the fault gouge will greatly increase when it is kept in an in-situ closed stress environment.

RESUME: Le present article a systematiquement recherche la caracteristique mecanique et physique pour la boue de la couche brisee de la station hydro-electrique de Ankang dans l'environnement de la contrainte de terrain. Et puis, avec des essais de maquette de deux phases mecaniques comprime et detendu, on a aussi recherche la tendance de changement de la caracteristique physique pour la boue de la couche brisee et quelques facteurs qui conduisent a la reduit des parameters de la contraine de cisaillement. Le resultat de recherche a montre que la resistance de cisaillment pourra beaucoup plus elever quand la boue de la couche brisee reste dans l'environnement de la contrainte fermee sur place.

The shear strength of fault gouge is of considerable importance in evaluating the stability of the rock mass in dam foundation. Usually, the test in-situ and lab. all change the natural status of the fault clay at some degree and make the mechanic property of the fault gouge worsen. In this paper, the mechanic feature of the fault gouge in nature geo-stress field was studied through a practical example of Ankang hydro-electric station. The result shows that the shear strength of the fault clay increase obviously if it is kept at natural status.

1 GEOLOGICAL BACKGROUND

1.1 Geology characteristic

Ankang hydro-electric station is one of large power projects which are

being built in China. The exploration showed that the stability of the dam is under control of the gentle dip fault (f_1^b) within the dam foundation (Fig1). The fault, dipped at north-

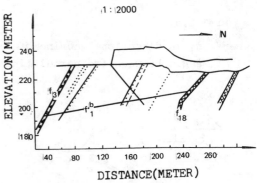

Figure 1. Geological profile map

Table 1.　　　Measured result of geo-stress

Measuring place	Measuring content	Main stress (KPa)	β (°)	α (°)
Left bank	3-D stress	5200 4100 1300	246 291 194	62 -21 -19
Right bank	plane stress	5300 3100	297 21	0 0
River bed	plane stress	4000 2600	307 27	0 0

Remark:　is dip angle which is plus and minus above and under sea
　　　　level separately.
　　　　is azimuth.

eastern 10 degrees with dip of 20
degree and undulating plane, is
situated at 180--240 m level underneath the dam, stretching about 200
m. The fault zone is 2--5 cm in width, in which the tectonite is consisted of mylonite, breccia and the
fault clay. The thickness of clay is
30 mm on an average and 40 mm at its
maximum. The clay near the gallery
wall presented soft-plastic state.

1.2 Geo-stress field characteristic

The geo-stress in the dam foundation, measured by Chinese Earthquack
Bureau using stress-relief method,
is arranged in Tab1. It is found
that geo-stresses are compressive
stresses. It is easy to obtain the
calculation formular for the stress
on any inclined plane in the condition of the main stress field in
light of elactic theory (Fig2.).

$$\sigma_n = n_i^2 \, \sigma_{ii}$$

$$\tau_n = (n_i^2 \, \sigma_{ii} - \sigma_n^2)^{\frac{1}{2}}$$

(i=1,2,3)

where: n_i is directional cosine of
　　　　angle between the normal line

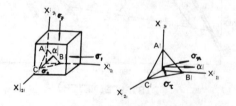

Figure 2. Main stress field and the
stress state on any inclined plane

on inclined section and main
　　stress.
The main stress on the fault is calculated through formular (1) to be
2100 KPa. In such geo-stress surrounding, the fault clay is in a compress-
ion-densitied state and in addition
to 600 KPa of engineering load, normal stress on the fault plane will be
2700 KPa so that the gouge will probably be denser under larger compression stress in future.

2 PHYSICAL AND MECHANICAL PROPERTY
OF THE FAULT CLAY

2.1 Physical feature of the clay

To study the influence of the excavation disturbance on the mecha
nical feature of the clay, sampling in varial depth of surrounding
rock and measuring at once water
content and natural density etc.
in situ were carried out. Comprehensive measured result is listed
on Tab.2.from which we can find
out follow laws:
　　1. The natural water content and
dry density of the clay gradually
decrease and increase separately
from the gallery wall toward rock
mass. Fig3 showes that the water
content of sample at depth of 80-
100 cm basically tend to constant
12%, but dry density gradually increase, which is consistent with
the decrease of the water content
and when the water content is basically constant, the dry density also
stablized about 1900 Kg/m , and the
pore ratio of soil varies between
0.4 and 0.87.

Table 2. Grain composition and physical property of fault clay

No. of sample & depth of sampling (cm)	grain size distribution				physical property							
	gravel $0.5 \sim 0.25$ (mm)	sand $0.25 \sim 0.050$ (mm)	silt $0.05 \sim 0.005$ (mm)	clay < 0.005 (mm)	W	γ	ρ	ρ_s	e	W_l	W_p	S_r
						$\times 10^4$	$\times 10^3$	$\times 10^3$				
	%	%	%	%	%	KN/m³	Kg/m³	Kg/m³		%	%	%
A-10	2.47	43.0	32.0	22.5	20.0	1.83	1.52	2.84	0.87	25.8	21.2	65.3
A-15	1.89	42 6	35.0	20.5	16.1	1.98	1.71	2.81	0.64	29.5	18.8	70.5
A-20	2.10	42.9	34.0	21.0	14.3	2.07	1.81	2.52	0.39	29.5	19.7	92.2
A-50	1.17	54.0	30.0	14.0	15.3	1.93	1.54	2.77	0.80	36.2	24.4	87.4
B-0					23.1	1.94	1.58					
B-10					20.7	1.97	1.63					
B-15					14.7	2.08	1.81					
B-20					14.6	2.12	1.82					
B-25					14.9	2.14	1.82					
B-30	2.93	49.1	31.5	16.5	16.1	2.14	1.84	2.83	0.54	34.3	20.6	86.7
B-50					15.0	2.08	1.81					
B-60					12.4	2.21	1.96					
B-70					11.2	2.16	1.94					
B-75	5.18	40.8	31.0	23.0	11.9	2.16	1.93	2.84	0.47	31.2	20.3	71.6
B-80					10.1	2.23	2.01					
B-90					11.1	2.17	1.97					
B-95					11.2	2.12	1.90					
C-25	1.17	32.8	36.0	30.0	13.3	2.20	1.94	2.77	0.43	33.2	19.4	85.4
C-60	2.70	45.3	38.0	14.0	16.0	2.14	1.85	2.85	0.54	34.6	21.3	84.2
C-70	0.23	50.8	31.0	18.0	16.4	2.21	1.90	2.88	0.52	35.8	18.8	80.6
C-110	2.50	47.5	33.0	17.0	15.5	2.13	1.84	2.87	0.56	32.1	20.9	79.5
D-60	1.13	38.9	39.0	21.0	15.3	2.06	1.79	2.85	0.59	37.9	16.8	73.7
D-80					10.8	2.23	1.94					
D-100					10.7	2.25	1.99					
A-70					138	2.07	1.82					
A-100					12.9	2.08	1.87					
A-110					12.0	2.11	1.93					
C-100					13.9	2.17	1.88					

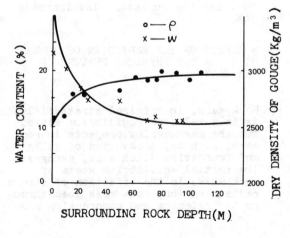

Figure 3. Variation tendency of the water content and dry density of the clay at various depth of surrounding rock at gallery wall

2. In light of the particle sub-clay with 10 to 30 percent of clay particle and the contend of sand particleiss more than one of silt particle, meanwhile, the particle substance has no variation from the gallery wall toward surrounding rock mass.

3.The liquified and plastic limitation of the clay is 25 to 36 and 17 to 24 percent separately,and the clay presents in solid state in the depths of the surrounding rock. The excavation of gallery cause the stress within the surrounding rock to redistribute, and the features of the clay change relevantly. The water content of the clay reaches 20 to 30%, so that it takes on soft -plastic state .

2.2 Mechanical feature of the clay

The analyses result for the shear test of the clay sample at different depth shows its characteristic as follow.

1. When normal stress keeps constant, the peak shear strenth of the clay continuously increases, with the variation of water content of the clay from gallery wall toward the depth of surrounding rock mass (Fig4.). The curve of shear stress to shear displacement displaies a plastic deformation characteristic.

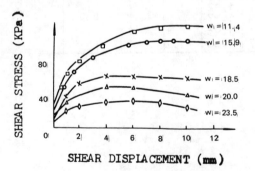

Figure 4. Relation betweem shear stress and shear displacement of the fault clay at varial depth of rock when normal stress is 300 KPa

2. The interrelated relatiom between frictional coefficient and water content of the clay is ob-

tained by statistic method. The statistic formular and curve (Fig 5.) is as follows.

$$f=1.2 -0.157w^{0.6}$$
$$(r=0.92)$$

where: wis water content((percent)
where: w is water content (percent). r is interrelated coefficient.

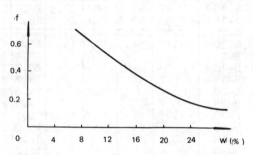

Figure 5. Relation between the frictional coefficient and the water content of the gouge

It is clear that water content will worsen mechanical property of gouge.

3. Tab 3. listed out the prediction value of the shear strength under condition of constant water content (12%) as well as the experimemtal value by means of medium-sized shear tests. It is known from Tab 3. that the stress relaxion of the surrounding rock is greatly influence on the shear strength of the gouge. When the gouge is kept in the natural state, the frictional coefficient is able to increase 80%, and the cohesion also increase 24%.

3 EFFECT OF THE VARIATION OF STRESS STATE ON THE PHYSICAL FEATURE OF THE CLAY

Rock mass, in original stress field, is in a relative equilibrium state, but the engineering projects in rock mass, such as, excavation of gallery amd foundatiom ditch etc., changes the initial equilibrium state of rock mass. In the effection of stress redistribution, rock mass goes through the relaxiom and compression pro-

Table 3. Physical and mechanical coefficient of fault gouge in disturbed sample and non-disturbed sample

physical index	ρ_s (kg/m^3)	W (%)	ρ (kg/m^3)	e	W_l (%)	W_p (%)	f	c (KPa)
disturbed sample	2010	19.97	1675	0.66	30.89	18.90	0.22	130
non-disturbed sample	2160	11.85	1930	0.47	31.0	20.31	0.44	170

cession respectively. After excavation of the gallery, f_1^b fault produced relaxion deformation, so that the mechanical property deterioration was occured in the clay which was originally in a high dense state, with the stress relaxion and the effect of the groundwater, the decrease of density, the increase of water content. The phynonmena was verified in the investigation in the field and the research in Lab. and imitated test.

3.1 Compression test

The disturbed sample of soil was made up to one whose water content is close to the liquid limit of the clay, tested at different normal stress, so that constant pore ratio and water content etc. were obtained at various normal stress (Fig 6, Fig7 and Tab4). It is known by analyses that relative constant water content

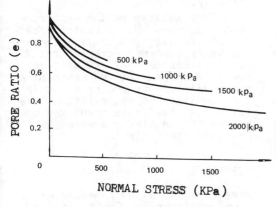

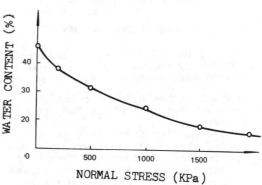

Figure 6. Compression curve of the clay in different condition of normal stress (initial water content is about 48%)

Figure 7. Relative curve between normal stress and stable water content

Table 4. Physical index of the clay before and after compression test

No. of sample	maximum compression stress (kPa)	physic index before compression				physic index after compression			
		ρ_s (kg/m^3)	ρ (kg/m^3)	W (%)	e	ρ_s (kg/m^3)	ρ (kg/m^3)	W (%)	e
J-1	200	2750	1830	30.8	0.97	2750	1715	29.1	0.81
J-2	500	2750	1740	29.3	0.90	2750	1960	22.4	0.72
J-3	1000	2750	1950	31.2	0.85	2750	1980	15.1	0.59
J-4	1500	2750	1950	33.1	0.88	2750	2000	12.4	0.53
J-5	2000	2750	1950	36.0	0.91	2750	2000	12.0	0.50

and the density of the gouge are coresponding with the test result in situ, when the normal stress is 2.0 MPa.

3.2 Unload test

After the test under 2.0 MPa of normal stress enter a stage of stable consolidation, the samples were unloaded level by level. The lower the normal stress after unloading is, the more the water content increased. Comparision of the result with one in Fig8. shows the water

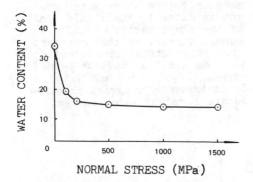

Figure 8. Relative curve between unloading normal stress and coresponding stable water content

content of the clay at relaxion is lower one at compression, because:

1. Unloading procession is relevented with time, andthere only was the recover of elastic deformation during the test. When normal stress is 200 kPa, for example, the pore ratio of sample is 0.57 which is far lower one in the process of compression;

2. Hydrodynamic condition is not similar. Water pressure during the test is close to zero, so that the water is not easy to enter the sample. The saturation of the sample, which is 105 to 118 after compression, but is 50 to 90 after unloading, confirm the point of view mentioned above.

The simulated test verified that there is better relerant relation between geo-stress and physical index of the fault clay, so that the geo-stress field about the fault has to be thought over when evaluating the shear strength of the fault clay.

4 CONCLUSION

1. The geo-stress is one of the important factors which controls the basic characteristic of rock mass, so the mechanical test may be taken basing on the research of stress environment in which rock mass exsited.

2. The compression and the unloading are two different mechanic process which is of important controling effect to the physical feature of the fault clay. The compression test in long time shows that certain normal stress is better corresponded with certain dense degree of the clay and the continuous increase of water content as the clay dis turbed after unloading is mean reason for the deterioratation of the mechanical feature of the clay.

3. From gallery wall toward surrounding rock, the mechanical feature of the clay gradually increase The water content decrease from 25 to 30 percent at gallery wall to about 12 percent at the depth of a meter within surrounding rock. In the condition of non-relaxion, the frictional coefficient and the cohesion of the clay increase 50% and 24% separately.

REFERENCES

Jeager, C. 1979. Rock Mechanics and engineering. Cambridge Uni. Press.
Curran, J. H. 1983. Influence of shear velocity on rock joint strength. Proceedings of the 5th inter. confer. rock mech.
Lama, R. D. & Vutukuri, V. S. 1978. Handbook on mechanical Properties of rock. Term Tech. Publications.

6th International IAEG Congress / 6ème Congrès International de AIGI, © 1990 Balkema, Rotterdam. ISBN 90 6191 130 3

Observations on different procedures for the oedometer test
Observations sur différentes exécutions de l'essai oedométrique

R. Mortari & T. Gerardi
Rome University 'La Sapienza', Italy

L. Budassi
Rome, Italy

ABSTRACT: Several oedometer tests have been carried out with different procedures on a clay sample having uniform properties. Time by time the following parameters have been varied: a) the consolidation degree reached for each loading, b) waiting time between consolidations of two consecutive loadings, c) specimen shape, d) section area of the specimens. Compression index, preconsolidation pressure and the variation of the coefficient of consolidation have been observed. Preconsolidation pressure and compression index underwent variations in relation to sample shape and consolidation degree. If only section area was changed instead, we observed that all parameter maintained constant values. This proves that smaller section than normal can be employed to carry out oedometer tests on clays similar to sample. This gives the advantage reaching higher pressure values and of shortening testing time.

RESUME: On a exécuté quatre séries d'essais oedométriques avec differents procédés sur des éprouvettes d'argile de moyenne consistence et de baisse sensibilité, qui possédaient caractéristiques très uniformes. Les paramètres qui ont été changés sont respectivement: le degré de consolidation atteint par l'argile pour chaque pression, le temps attendu entre la fin d'une charge et le début de la charge suivante, la forme des éprouvettes, la grandeur de la section. On a controlé les variations de l'indice de compression, de la pression de préconsolidation et du coefficient de consolidation. Les deux premiers paramètres ont des considérables variations lorsque on change la forme des éprouvettes ou le degré de consolidation. Mais tous les trois paramètres ne montrent pas des variations lorsque on change seulement les dimensions des éprouvettes. Sur argiles de moyenne consistence on peut donc éxecuter des essais oedométriques avec sections et hauteurs réduites, avec les avantages de terminer l'essai pendant deux ou trois jours et d'attendre pressions plus élévées.

1 INTRODUCTION

To carry out a standard oedometer test, a specimen 2 cm high and having a section area of at least 20 cm² is routinely loaded, and the load values are redoubled every 24 hours. This implies that every single test takes from two to three weeks to be completed (from start of loadings phase to expiring of unloading phase).

Attempts have be made to vary testing procedures, particularly in order to reduce performing times: as with constant strain-rate (CSR) tests, or constant consolidation (CC) tests (Crawford 1988). All these procedures require more complex equipment and do not allow the assessment of the coefficient of consolidation value (c_v) (Kolisoja et al. 1987).

The present work was finalized to observe changes in the results of the oedometer test when change of standard procedures had been made. The following parameters were observed in particular: compression index C_c (determined on the virgin part of the compression curve); preconsolidation pressure p_c (as determined by Casagrande's 1936 graphic method); in a few cases, coefficient of consolidation (c_v).

To attain this aim, a 45 cm thick block of clay was employed, belonging to a continental formation of the middle Pleistocene in the nearabouts of Rome. In this block the values of water content and preconsolidation pressure were assessed at regular intervals of 2 cm each, in order to prove the uniformity of the material.

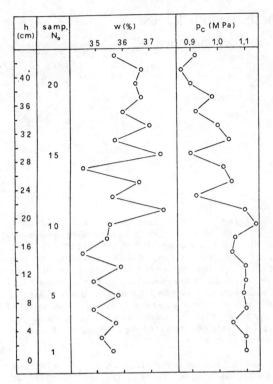

h (cm)	samp. N_o	w (%)			p_c (M Pa)		
		35	36	37	0.9	1.0	1.1

Figure 1. Variation of water content and preconsolidation pressure within the clay block analyzed. In spite of apparent uniformity the sample shows constant characteristics only between 6 and 14 cm above the bottom. For the following tests all specimens except one were taken at middle level where greater uniformity is shown.

The results are shown in figure 1. In particular, a great constancy of the preconsolidation values can be observed from 6 to 14 cm above the bottom. In the other parts of the block, uniformity is much less, or practically nonexistent.

Within the interval showing a higher degree of uniformity a particular level was chosen at about 10 cm from the bottom, showing the following physical properties: water content w = 35.8 %, unit weight γ = 18.2 kN/m^3, unit weight of solid particles γ_s = 26.3 kN/m^3, liquid limit w_L = 63.8 %, plasticity index IP = 40.0 %, clay fraction <2μm = 64 %. Values of sensitivity were very closed to 3.

In a first series of tests, the specimens were loaded with loading-time periods (t_1) ranging from 24 hours to 1 minute respectively. Next, with same section area, height was varied from 2.90 to 0.41 cm while loading times were made propor

tional to the square of height values.

Successively the height/diameter ratio was constantly kept equal to 0.40, while section areas (A) were modified from 20 to 2.5 cm^2, on parallel lines with the loading-time periods, in order to get the same consolidation effects.

2 RESULTS

2.1 Loading times

First we compared the effects produced by different loading-time periods (t_1) with the effect usually obtained by redoubling the loads each 24 hours. Eleven specimens, all 1.41 cm high and with a section area of 10 cm^2 were loaded with different loading times, ranging from 24 hours to 1 minute.

Since loading-time periods ranging from about 240 to 1440 minutes are equally valid from a practical point of view, only shortest times offer the advantage of more loadings in the same day. Figure 2 shows the results attained: both the compression index and the preconsolidation pressure display two different trends for loading-time periods shorter and longer than about 40 minutes respectively. This is about the time when completion of primary consolidation occurs, as determined on the consolidation curves in a representation of the deformation readings versus log of time. It can be deduced therefore that the two recorded trends, both for the C_c parameter and p_c parameter, as shown in figure 2, are due to the occurrence of pore pressures when loading times are shorter than about

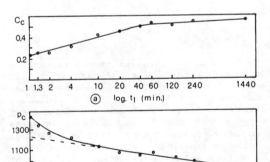

Figure 2. Variation of compression index C_c (a) and preconsolidation pressure p_c (b) for different loading times t_1, on specimens 1.41 cm height and having a section area of 10 cm^2.

40 minutes. Such pressures had no time to dissipate between two consecutive loadings.

On the other hand, for t1 longer then 40 minutes, the trends are governed by the secondary consolidation reached by the specimens.

2.2 Waiting times between two consecutive loadings

One of the above tests - the one with one-minute loading time - was repeated varying the time intervals between two consecutive loads, when loading is stopped and no further settlements take place. This time interval was called waiting time (t_w) and was varied from 1 minute to 24 hours.

Clearly only the longest times allow the pore pressure arising during the loadings to disappear completely.

Figure 3 shows the results of such tests. Compression index C_c never changes, while preconsolidation pressure p_c varies for waiting times shorter than 40 minutes, emphasizing the previous hypothesis made at different loading times: that for specimens 1.41 cm high pore pressures disappear after about 40 minutes.

Notably too, the constant preconsolidation value obtained for waiting times t_w longer than 40 minutes fits rather well with the data shown in figure 2b for t_w longer than 40 minutes.

On the other hand, the fact that the compression index C_c shows no change when t_w varies, means that pore pressures influence only the parallel shifting of the virgin part of the compression curve.

2.3 Height/diameter ratio

Eight specimens having different heights ranging between 0.41 and 2.90 cm and a constant section area of 10 cm^2 were prepared, so that the height/diameter ratio varied from 0.115 and 0.81. Usually, oedometer tests are carried out with specimens having a height no lower than 2 cm, occasionally no lower than 1.3 cm, because of the possibly disturbing effect of the specimen shaping (Van Zelst 1948; ASTM 1988).

Unfortunately, higher heights cause a quadratic increase of the consolidation times and stress the lateral friction effects.

In order to compare the results coming from each test, the loading times t_1 were quadratically increased with linear heights increasing. Moreover, the loading times were varied from 126 to 3 minutes in order to remain in proximity of the end of primary consolidation, which in this case, for specimens 2.0 cm height, is, in a large range of loads, next to 1 hour. Loadings maintained as far as the end of primary consolidation are recommended by Ladd (1971), Lowe (1974), Brumund et al. (1976), Leonards (1976), Jamiolkowski et al. (1985).

The results are shown in figure 4. It is noteworthy to say that compression index C_c varies sharply, though slightly, in a way inversely proportional to the logarithm of the height/diameter ratio, with values ranging between 0.461 and 0.505, while preconsolidation pressure p_c varies in a more important way in opposite directions, between the final values of 885 and 1150 kPa.

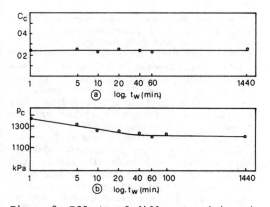

Figure 3. Effects of different waiting times t_w between two consecutive loadings applied on specimens alike to the previous ones after loading times of 1 minute.

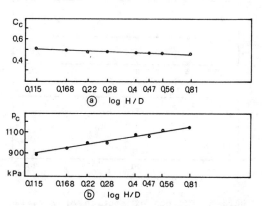

Figure 4. Shape influence on specimens with section area of 10 cm^2 and heights ranging from 0.41 to 2.90 cm. Loading times are as long as primary consolidation.

2.4 Section area

Four tests were carried out on specimens of different section areas of respectively 2.5, 5, 10 and 20 cm^2.

The specimen shape was kept constant with a height/diameter ratio of 0.40, while loading times were ordered as in the former series of tests so that at the end of each load the end of the primary consolidation was reached: respectively these times were 7.5, 15, 30 and 60 minutes.

The results obtained are shown in figure 5. No considerable variations were observed, either of compression index or of preconsolidation pressure; and this should demonstrated that the disturbing effect was totally negligible.

Another, analogous series of tests was carried out on four specimen of different section areas, of respectively 2.5, 5, 10 and 20 cm^2, but with greater loading times, ranging from 3 to 24 hours, so that it was possible to asses the value of the coefficient of consolidation c_v from the representation of the deformation readings versus log of time.

Another test was carried out with the section area of 20 cm^2 on a specimen coming from a level at half height in the clay block.

The coefficients of consolidation at the various loads are plotted in the figure 6 versus log of pressure. In all graphic displays a sharp lowering of the coefficient in the nearabouts of the values of the preconsolidation pressure is to be observed, in conformity to the data of Lambe and Whitman (1969).

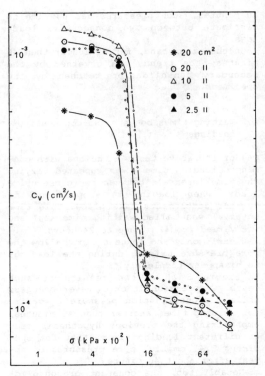

Figure 6. Coefficient of consolidation c_v versus log of pressure for five specimens, four of which were coming from the same level and were tested with different section areas, but same shape and same consolidation degree. These show that specimen dimension is uninfluent in respect to c_v values, while stratigraphic position is clearly a determiner.

3 DISCUSSION

The present work as pointed out that there is no variation of preconsolidation pressure p_c nor of the compression index C_c nor of the coefficient of consolidation c_v in similar experimental conditions. It was previously pointed out (Mortari 1976) that variations in section areas cause no change in preconsolidation pressures. This is true only when sample shapes are not varied and if the loading times are modified accounting for the change of height.

But the two first parameters considerably change, with opposite signs, when the loading times are varied. This behavior can be explained by the fact that longer loading times cause greater settlements (and therefore a higher C_c) and a wider structural disturbance (also a lower p_c value). This two parameters are subject to wider variations when the loading times

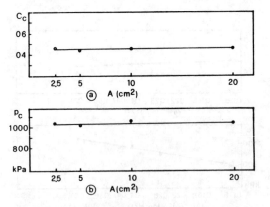

Figure 5. Effects of section area variations, while specimen shape and degree of consolidation are maintained constant. Neither compression index nor preconsolidation pressure are influenced.

t_l decrease below the values corresponding to the and of the primary consolidation.

It is noteworthy to say however, that the changes of the same two parameters are attributable to different reasons, as can be deduced from the fact that variations in the waiting times t_w eventually affect only the p_c trends, as shown in figure 3.

The change in preconsolidation, along with the loading times t_l, is in agreement with Bjerrum's experience (1967): according to that, in fact, when during an oedometer test a soil is loaded for shorter periods, its preconsolidation pressure increases.

Changes in the height/diameter ratio cause variations in the two parameters C_c and p_c considered; and it can easily be deduced that such modifications are closely dependent on lateral friction, which causes the leak of a share the vertical pressure applied.

On the other hand, it can be excluded that a decrease of the height/diameter ratio is joined by a wider remolding effect along the base at least for soils that are analogous to the one considered here.

If that were the case, we would obtain different compression index and preconsolidation pressure values for the specimens having smaller sizes.

4 CONCLUSIONS

The results attained by a standardized oedometer test (i.e. section area A = 20 cm^2, height H = 2.0 cm and loading time t_l = 24 hours) on a clay characterized by middle consistency and low sensitivity could altogether be obtained by carrying out a test on specimens of smaller sizes. A test carried out in t_l = 45 minutes, on specimen 0.35 cm high, with A = 10 cm^2, and therefore a lower height/diameter ratio, reaches the same degree of consolidation as above but is much closer to natural conditions, since lateral friction effects are higher for less flattened shapes.

Standardized oedometer tests generally give compressibility and preconsolidation values considerably different from real behavior, because experimental conditions cannot account for "in situ" consolidation times.

Specimen height reduction not only favors the experimental reconstruction of actual soil behavior, but also allows the test to be completed in much shorter times (one-two days only, as opposite to the two-three weeks needed for a standard test) without affecting the consolidation degrees usually attained.

REFERENCES

ASTM (1988).Annual book of ASTM standards.Section 4. 953 pp. Philadelphia

Bjerrum, L. (1967). Engineering geology of Norwegian normally consolidated marine clays as related to settlement of buildings. Géotecnique 17,81-118.

Brumund W.F., Jonas, E. & Ladd, C.C. (1976). Estimating in situ maximum past preconsolidation pressure of saturated clays from results of laboratory consolidation test. Special Report 163, Transportation Research Board, 4-12.

Casagrande, A. (1936). The determination of the preconsolidation load and its practical significance. Proc. 1st Int. Conf. SMFE, 3., 60-64.

Crawford, C.B. (1988). On the importance of rate of strain in the consolidation test. Geotechnical Testing Journal, GTJODJ, 11 (1), March, 60-62.

Jamiolkowski, M., Ladd, C.C., Germaine, J.T. & Lancellotta, R. (1985). New developments in field and laboratory testing of soils. Theme Lecture. 11th Int. Conf. SMFE, S. Francisco.

Kolisoja P., Sahi K. & Hartikainen J. (1987). Automated oedometer device. Proc. 9th Europ. Conf. SMFE, Dublin, 1, 67-70.

Ladd, C.C. (1971). Settlement analyses for cohesive soils. Dep. of Civil Engr. Research Report R71-2, Soils Publ. N.272, MIT.

Lambe, T.W., Whitman, R.V. (1969). Soil Mechanics. John Wiley & S., New York.

Leonards, G.A. (1976). Estimating consolidation settlements of shallow foundations on overconsolidated clay. Special Report 163, Transportation Research Board.

Leonards, G.A. and Altschaeffl A.G. (1964). Compressibility of clay. Journal of the Soil Mechanics and Foundation Division, ASCE, 90 SM5, 135-155.

Leonards, G.A. & Girault, P. (1961). A study of the one-dimensional consolidation test. Proc. 5th Int. Conf. SMFE, Paris, 1, 213-218.

Lowe, J. (1974). New concepts in consolidation and settlement analysis. Journal of the Geotech. Engng. Div., June.

Mortari, R. (1976). Elementi per una nuova interpretazione della preconsolidazione delle argille. Geologia Applicata e Idrogeologia, 12 (2), 189-200.

Schmertmann, J.M. (1955). The undisturbed consolidation of clay. Trans. ASCE, 120, 1201-1227.

Van Zelst, T.W. (1948). An investigation of the factors affecting laboratory consolidation of clay. Proc. 2nd Int. Conf. SMFE, 7, 52-61.

Characterisation of tropical weathering profile derived from Precambrian gneiss-migmatite complex in parts of South Western Nigeria

Caractéristiques du profil de l'altération tropicale, provenant de complexes gneiss-migmatite précambriens, dans quelques régions du Sud-Ouest nigérien

S.F.Oni
Department of Environmental Sciences, The Polytechnic, Ibadan, Nigeria

ABSTRACT: The weathered profile of the migmatite-gneiss complex of parts of South Western Nigeria was studied. The basic index properties were determined according to BS 1377 of 1975. Selected fractions passing No. 200 sieve were analysed geochemically and mineralogically using AAS model 188-70 and X-R-D machine with FeK_α radiation respectively.

The parameters obtained were examined and compared in line with the weathering profile defined by Adekoya (1987). No appreciable pattern of variation laterally or vertically was observed in the index properties. This might be due to mechanical and thermal instability manifested in the breakdown of the samples in the insitu state or chemical and mineralogical changes accompanying drying.

The mineralogy in each zone is the same but varies in abundance within the locations of the same zone. Nevertheless, there is appreciable difference in their mineralogy and their abundance down the profile.

There is high SiO_2 content, low Al_2O_3 and Fe_2O_3+FeO but very little K_2O, CaO, Na_2O, MgO and MnO. The presence of K and Mg including illite is indicative of incomplete weathering in all the profiles.

1. INTRODUCTION

Studies have shown that weathering is generally intensive in the tropics to sub-tropical regions of South Western Nigerian basement complex. The borehole records (Adekoya (1987), Teme, and Oni (1989)) in this area revealed that the thickness of the weathered soil varies from 0.0cm (bare rock outcrop) to an average of about 12.0m. Adekoya (1987) attempted the classification of the weathered profiles in this area and defined six weathering zones as Top soil, Stone layer, laterite zone, clay zone, Saprolite zone and partly weathered rock zone. Most authors, Du Preez (1956), Malomo et-al (1983), Mesida (1986), Akpokodje (1986), Ogunsanwo (1986, 1988) who have worked on weathering profiles focused their attention more on laterite or clay zones than others,

thereby lost sight of the inter-relationship of the properties of the six zones.

This Paper is, therefore, intended to examine the possible relationship between the geotechnical, chemical and mineralogical characteristics of these zones in the weathered soils derived from the precambrian gneiss-migmatite complex of the Nigerian basement.

2. GEOLOGICAL SETTING

South-Western Nigeria is underlain by rocks of the crystalline basement complex. (Fig. 1). Earlier workers, Oyawoye (1972), Rahaman (1988) to mention a few have described and or classified the rocks into different rock groups.

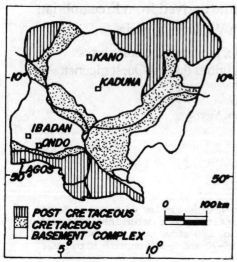

Fig.l: Generalised Geological Map of Nigeria.

The gneiss-migmatite complex consists of: (i) Banded, migmatatic and granite gneisses; and (ii) The massive, pseudo-conglomerate, and schistose quartzites. The migmatites are heterogeneous rocks in which rocks of different composition and origin co-exist. They have similar mineralogy with granite gneiss except in the feldspar constituent. The banded gneisses exhibit alternating leucocratic quartz-oligoclase rich bands and melanocratic hornblende-biotite rich bands (Burke, 1972). The quartzite, which contains essentially quartz (90%) could either be massive, pseudo-conglomerate or schistose. They are very difficult to separate as units on the map. The gneiss-migmatites complex is often intruded by pegmatites and dolerite dykes. It was described to have its most pronounced deformation and remobilisation during the Pan African orogency (Odeyemi, 1981).

The South-Western Nigerian basement complex is underlain by two types of soil profiles. The in-situ and transported weathered soil profiles. The in-situ weathered soil profile occupies the area of discussion. It is formed as a result of the tropical weathering of the migmatites, gneisses, schists and quartzites in the study area.

3. MATERIALS AND METHODS

Soil samples derived from the precambrian rocks in some selected places within the region were collected from the existing zones in each profile. Average of five profiles was used as representative sample for each location (Fig. 2).

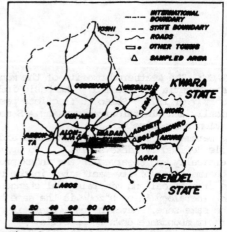

Fig. 2: Location Map of Study Area.

The basic index properties were determined according to BS 1377 of 1975. Atomic Absorption spectrophotometer model 188-70 was used in geochemical analysis of the selected samples. The mineralogy of the clay size fractions ($2\mu m$) was determined by the use of X-Ray Diffractometer with Fe-K radiation generated at 30Kv and 10mA.

4. GEOTECHNICAL RESULTS AND ANALYSIS

Visual observation of the profiles revealed that the top soil is essentially a dark humus soil with evidence of organic matter within the gravelly to silty-clayey sandy soils. It is reddish brown in colour where the terrain is not thickly vegetated. The loose gravelly sandy soil with thin organic cover at times dominates the stream channels. Whereas the stone-layer has angular to sub-angular gravelly grains with silty

432

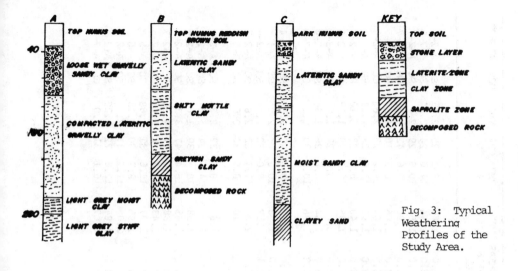

Fig. 3: Typical Weathering Profiles of the Study Area.

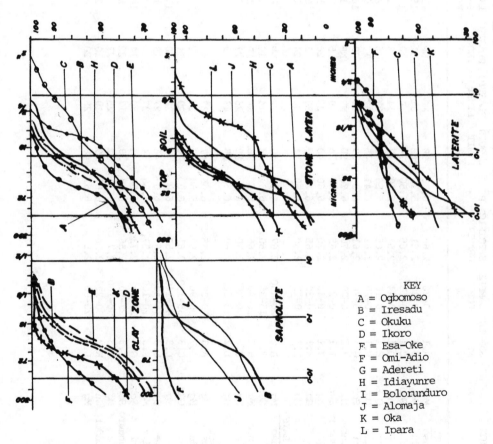

KEY

A = Ogbomoso
B = Iresadu
C = Okuku
D = Ikoro
E = Esa-Oke
F = Omi-Adio
G = Adereti
H = Idiayunre
I = Bolorunduro
J = Alomaja
K = Oka
L = Ipara

Fig. 4: Typical Particle size curves of the study area.

TABLE 1.0   PHYSICAL AND GEOTECHNICAL RESULTS OF THE ZONES IN THE STUDY AREA

LOCATION	PARENT ROCK	SAMPLE NO	SAMPLE DEPTH (cm)	SPECIFIC GRAVITY	pH	LIQUID LIMIT %	PLASTIC LIMIT %	PLASTIC INDEX %	LINEAR SHRIN-KAGE %	MDD gm/cc	OMC %	CBR %	NMC %	UCS	AASHTO
TOP SOIL															
Ogbomoso	Mg	A.1	0-20	2.65	6.70	18	NP	NP	1.0	1.60	8.2	25	2.25	SM	A-2-4
Iresadu	Mg	B.1	0-30	2.61	5.50	62	31	31	15.0	1.94	14.0	24	2.24	SC	A-7-5
Okuku	GGn	C.1	0-15	2.65	6.50	24	14	10	11.0	1.52	10.9	29	2.0	SC	A-2-7
Ikoro	Mg	D.1	0-44	2.57	5.50	17	14	3	1.0	1.96	9.6	42	2.86	SM	A-2-7
Esa-Oke	Q.Sch	E.1	0-16	2.65	5.50	43	27	16	8.0	1.86	8.8	10	4.30	SC	A-7-5
Omi-Adio	Mg	F.1	0-20	2.61	6.41	45	32	13	6.0	1.64	19.2	23	2.63	CL	A-2-4
Adereti	Sch	G.1	0-20	2.61	6.5	39	30	9	10.0	1.97	11.6	28	1.69	SM	A-2-4
Idiayunre	Mg	H.1	0-20	2.70	6.20	20	NP	NP	1.0	1.40	8.0	23	2.10	SM	A-2-6
Bolorunduro	Sch	I.1	0-27	2.65	6.68	35	30	5	8.0	1.96	10.2	28	1.69	SM	A-2-6
Alomaja	Mg	J.1	0-40	2.70	6.60	40	20	20	1.0	2.02	9.6	27	1.8	SC	A-2-6
Oka	Gn	K.1	0-30	2.65	6.21	39	36	3	6.0	1.84	13.6	26	1.36	SM	A-2-4
Ipara	Mg	L.1	0-15	2.60	6.50	54	30	24	1.2	1.96	10.2	30	0.5	SC	A-7-6
STONE LAYER															
Ogbomoso	Mg	A.2	20-100	2.67	6.6	10	NP	NP	1.0	1.57	8.0	20	2.0	SM	A-2-4
Okuku	GGn	C.2	15-35	2.60	6.0	24	14	10	1.0	1.58	10.0	23	2.0	SC	A-4
Idiayunre	Ng	H.2	20-40	2.66	6.6	16	NP	NP	1.1	1.57	8.0	32	2.1	SM	A-2-4
Alomaja	Mg	J.2	40-80	2.70	6.5	52	25	27	1.3	2.0	11.2	22	2.8	GC	A-2-7
Ipara	Mg	L.2	15-20	2.60	6.5	58	49	9	1.4	2.10	8.5	30	0.5	SC	A-5
LATERITE															
Ogbomoso	Mg	A.3	100-290	2.65	6.65	47	33	14	1	1.6	9.6	32	2.3	SC	A-2-7
Iresadu	Mg	B.3	30-90	2.60	5.50	59	27	32	13	1.7	14	34	5.54	SC	A-2-7
Okuku	GGn	C.3	35-130	2.61	6.0	40	29	11	1	1.58	10	30	1.6	SC	A-2-6
Ikoro	Mg	D.3	44-140	2.57	6.0	42	22	20	11	1.65	15.4	47	3.3	SC	A-2-7
Esa-Oke	Q.Sch	E.3	16-40	2.67	6.0	46	27	19	11	1.88	12.0	41	4.31	SC	A-2-7
Omi-Adio	Mg	F.3	20-100	2.57	6.50	54	37	17	10	1.43	14.2	6	6.47	CH	A-7-5
Adereti	Sch.	G.3	20-83	2.73	6.54	44	30	14	10	1.97	11.6	31	2.68	CL	A-5
Bolorunduro	Sch.	I.3	27-50	2.56	6.38	39	17	22	5	1.85	8.7	30	3.15	SC	A-6
Alomaja	Mg	J.3	80-160	2.7	6.5	52	25	27	8	2.0	11.2	44	2.8	SC	A-2-4
Oka (Ondo)	Gn	K.3	30-220	2.56	6.64	29	19	10	2	1.84	12.6	22	1.25	SC	A-2-4
Ipara	Mg	L.3	20-60	2.5	6.52	48	28	20	8	1.67	16.1	31	1.0	CL	A-7-6

table 1 cont.

LOCATION	PARENT ROCK	SAMPLE NO	SAMPLE DEPTH (cm)	SPECIFIC GRAVITY	pH	LIQUID LIMIT %	PLASTIC LIMIT %	PLASTIC INDEX %	LINEAR SHRINKAGE %	MDD gm/cc	OMC	CBR	NMC	UCS	AASHTO
CLAY ZONE															
Ogbomoso	Mg	A.4	290–315	2.65	6.7	52	30	22	1.0	1.5	12	34	4.8	CH	A-7-5
Iresadu	Mg	B.4	90–210	2.60	6.0	57	30	27	7.0	1.68	16.3	29	6.67	SC	A-7-6
Okuku	GGn	C.4	130–360	2.63	6.0	22	14	8	1.0	1.52	10.4	27	0.9	CL	A-4
Ikoro	Mg	D.4	140–300	2.64	6.0	61	40	21	12	1.98	9.2	47	4.59	SC	A-6
Esa-Oke	Q.Sch	E.4	40–90	2.65	6.64	45	22	23	6	1.51	16.2	38	5.51	SC	A-2-7
Omi-Adio	Mg	F.4	100–360	2.57	6.50	46	23	23	11	1.40	17.1	17	5.7	CH	A-7-6
Adereti	Sch	G.4	83–230	2.66	6.5	45	20	25	10	1.98	11.1	17	5.75	CL	A-5
Idiayunre	Mg	H.4	40–150	2.65	6.5	35	25	10	1	1.59	9.5	23	0.6	SC	A-2-4
Bolorunduro	Sch	I.4	50–240	2.66	6.25	37	34	3	5	1.81	8.9	44	1.86	SM	A-2-4
Alomaja	Mg	J.4	160–240	2.7	6.0	65	30	35	1.5	1.90	14.0	818	2.8	SC	A-7-5
Oka	Gn	K.4	4220–340	2.65	6.64	48	43	5	10	1.86	13.1	126	2.75	CL	A-5
Ipara	Mg	L.4	60–100	2.60	6.5	47	23	24	1.0	1.97	11.8	35	1.2	CH	A-7-6
SAPROLITE ZONE															
Iresadu	Mg	B.5	210–240	2.61	6.0	49	28	21	7.0	1.97	12	25	6.89	SC	A-2-7
Omi-Adio	Mg	F.5	360–400	2.56	6.21	45	26	19	6.0	1.86	13	18	14.5	CH	A-7-6
Alomaja	Mg	J.5	240–300	1.70	6.0	61	29	31	5.0	1.74	16.8	29	4.8	SC	A-7-6
Ipara	Mg	L.5	100–170	2.60	6.0	39	29	10	6.0	1.98	11.1	30	0.6	SC	A-4

Mg = Migmatite gneiss,   GGn = Granite Gneiss,   Q.Sch = Quartz Schist,   Sch. = Schist,   UCS = Unified Classification System.

435

to clayey sand matrix. Where this zone is not absent, it often appears laterally with the profile.

The laterite zone showed reddish to brownish silty to sandy clay or clayey sand, with occasional pebbles or gravels. Its unique boundary is not practically easy to define on the field as the reddish colour often extends from the top soil to saprolite zone of a profile, particularly in the quartz schist areas.

The clay zone is richer in the fines than other zones. It is sometimes grey in colour but mottled colour predominates. It may be dry and compact or soft. It usually contains silty or sandy clay.

The saprolite zone was not investigated in most of the cases but the few sampled showed a sandy clay texture with micaceous and feldsparthic mineral grains not fully decomposed. (Fig. 3).

The particle size distribution curves of the top-soil are fair to well graded but the samples are coarser in one relation than the other. But in the stone layer, the soil is fair to poorly graded with varying coarseness, whereas the laterite, clay and saprolite zones can be said to be generally poorly graded to well graded. (Fig. 4).

According to unified classification system, the top soil ranges from loose silty to slightly clayey sand, whereas the stone layer is classified as loose gravelly to silty sand. The laterite and saprolite zones are predominantly sandy clay and the clay zone belongs to various plasticity range of clay.

AASHTO classification soils put the top soil predominantly in the A-2-4 while others are in the A-7 category. The stone layer soils are in A-2-4, A-2-7 and A-5 groups while the laterite zone has A-2 group predominating over A-5 and A-7 groups. The clay and saprolite zones are in the A-2-7, A-7-6 and A-4 groups.

In each zone, the maximum dry density (MDD) does not considerably vary but tends to have higher values southwards of the study area. Whereas the California Bearing Ratio (CBR), Optimum Moisture Content (OMC), Natural Moisture Content (NMC), and the linear shrinkage limit (L.S) vary widely

laterally in each zone as well as down the profile, no specific pattern of variation is observed.

The pH and the specific gravity of the soils are uniform relatively for the profiles. (Table 1).

Cassagrande's chart revealed that the top soils have about a 50% plot above and below the A-line in the low to medium plasticity range. An indication of the presence of organic and inorganic clay. The stone layer showed scattered points on the chart, whereas most of the plots of the laterite soils are above the A-line with ones scattering below the A-line. Those below the A-line are not organic clay but inorganic clays with higher content of silts. Most of the soils of the clay zone plotted above the A-line within the medium and high plasticity range. But the few data of the saprolite zone are dispersed within the chart. This could be associated with different weathering stages.

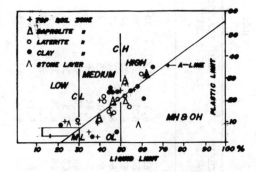

Fig. 5: Cassagrande's Chart for the Zones in the Study Area.

5. MINERALOGICAL RESULTS AND ANALYSIS

Hand specimen showed that quartz in the predominant inorganic mineral that is present. The clayey character must have been introduced by the plant decay in the respective areas.

The stone layer is mostly absent hence not well represented for a good catalogue. Quartz as well as iron concretions are common but the quartz or feldspathic minerals are often cemented by oxides of iron.

TABLE 2: GEOCHEMICAL RANGE OF VALUE FOR THE STUDY AREA

TOP SOIL	SiO₂	Al₂O₃	*Fe₂O₃	K₂O	CaO	Na₂O	MgO	MnO
Top Soil	67.82-71.02	13.1-16.83	7.09-8.72	3.72-5.5	0.3-1.04	0.61-1.08	0.30-0.83	0.07-0.12
Stone Layer	68.61-69.05	14.64-16.37	7.64-9.56	3.80-5.52	0.20-0.76	0.25-0.50	0.10-0.60	0.03-0.07
Laterite Zone	68.33-70.10	14.20-16.59	7.60-9.01	3.07-5.42	0.04-1.0	0.54-1.89	0.11-0.71	0.02-0.09
Clay Zone	67.88-72.76	13.26-20.31	0.06-9.07	3.07-6.22	0.31-0.86	0.32-1.89	0.11-0.82	0.02-0.11
Sarolite Zone	67.32-71.16	13.66-16.57	6.07-9.11	4.73-5.96	0.37-0.59	0.62-0.78	0.38-0.68	0.03-0.09

* Fe₂O₃+FeO

437

In the laterite zone, quartz, kaolinite, feldspar, illite, smectite and chlorite are the minerals identified from the X-R-D analysis of the fines. The coarser fractions often contained partially weathered feldspar and quartz in all the profiles. But in the clay zone, kaolinite and plagioclase are prominent at very high intensity while quartz, illite, smectite anddlepidolite occured at very low intensity on the diffractograph in all profiles. The plagioclase only appeared in the gneisses and migmatites but absent in the quartz-schist, instead lepidolite is present.

Quartz and kaolinite are predominant in the saprolite zone while illite, Heamatite, montimorillonite and vermiculite are the secondary minerals in this zone.

6. GEOCHEMICAL RESULTS AND ANALYSIS

The results of the geochemical analysis indicated that the samples in each profile or profile zone contained very high percentage of SiO_2, low percentage of Al_2O_3, Fe_2O_3+FeO and K_2O and CaO, Na_2O, MgO and MnO appeared in trace amount. There is no appreciable increase in the elemental oxides from place to place within the zones nor along each profile.

The high silica in the clay zone is in line with Holmes' (1949) report that clay minerals resulting from large in-situ disintegration and decomposition of pre-existing rock, do generally contain much of the "Soluble Silica gel" and some of the potash set free during weathering process.

The low percentage of CaO in each zone is indicative of the leaching effectiveness, however, the presence of K and Mg in the soil implied that leaching is incomplete in any of the zones laterally or vertically. (Table 2).

7. DISCUSSION AND CONCLUSION

The particle size distribution patterns are closely similar in all zones even though the coarseness varies. The exception to this rule are the soils in the lateritic zone with varying distribution pattern and soils of the saprolite zones that varied a lot both in pattern and coarseness of the particles. The soils show various modifications of sand to clay soils without a particular pattern of distribution within each zone. The variation is pronounced along the profile rather than within in the locations of a zone.

The irregular pattern in the lateral variation of the optimum moisture content and the maximum dry density might be due to thermal instability resulting from chemical and mineralogical changes that accompany drying or promoted by it. The soil can be stable in this new state depending on the initial conditions (Malomo 1979). The higher values of the optimum moisture contents than the natural moisture contents is in the line with Brand and Hognsnvi's (1969) result that, ovened samples have higher values than unovened samples. Since maximum dry density and the California Bearing Ratio are directly related (De Graft-Johnson et-al 1969), the CBR, therefore, shows no pattern like the maximum dry density. The irregularity in the distribution patterns laterally within the zones of Atterberg limits, moisture and linear shrinkage might be due to mechanical instability caused by remoulding and manipulation of the soil in its in-situ state. Thus, breaking down cementation and structure of the soil Newill (1962), Townsend et-al (1969) and Matyas (1969).

The relatively uniform values of the pH and specific gravity may be due to felsic character of the parent rocks combined with the leaching and groundwater characteristics.

Each zone has a unique mineral assemblage from place to place in the zone but vary in abundance. This is a result of the degree of leaching that might have taken place in the parent rock. The low percentage of CaO shows that leaching was effective while the presence of K, Mg and illite is an indication that leaching was not excessive.

Both field and experimental results indicated that the physical, geotechnical and geochemical properties of each zone similarly vary as they do down the profiles.

The coarseness of the weathered profile varies not only down the profiles but also within each zone. The percentage of fines increases down the profile.

The weathering profile of quartzite/quartz-schist contained coarser fractions than migmatites and the older granites.

The inconsistencies observed in the engineering properties of the soil might be due to thermal or mechanical instability of the soils.

Most of the clay in this area are inorganic in origin and are predominantly medium to high in plasticity.

Leaching is very effective in this area but not excessive. Some of the soils in the first three zones are either fair to good foundation or fill materials. Some are good sub-base materials.

REFERENCES

Adekoya, J. A. 1977. A note on jointing in the basement complex of the Ibadan area, Oyo State, Nigeria. J. Min. Geol. 14(1): 48-52).

Adekoya, J. A. 1987. A preliminary geological and geotechnical study of lateritic weathering profiles derived from SW Nigeria. 9th Reg. Conf. for Afr. on soil Mech. and found. Eng. pp 21-30.

Akpokodje, E. G. 1986. The geotechnical properties of lateritic and non-lateritic soils of S.W. Nigeria and their evaluation for road construction. Bull. IAEG No. 33 pp 115-121.

Du Preez, J. W. 1956. Origin classification and distribution of Nigerian laterites "Proc. 3rd Int. West Africa Conference, Ibadan.

Holmes, A. 1914 Investigation of laterite in portuguese East Africa, Geolo. Mag. Decade VT, 1 529-537.

Malomo, S. 1979. Engineering problems peculiar to laterite soils - A new approach. J. Min. Geol. 16(1): 55-61.

Malomo, S. et-al 1983. An investigation of the peculiar characteristics of laterite soils from S. Nigeria. Bul. IAEG 28: 197-206.

Malomo, S. and Ogunsanwo, O. 1983. The pre-consolidation pressure of a laterite soil. Bul. IAEG 28: 261-265.

Matayas E. L. 1969. Some Engineering problems peculia to laterites: A new approach. Jn. Min. and Geol. 16: 55-63.

Mesida, E. A. 1986. Some geotechnical properties of residual mica schist derived subgrade and fill materials in the Ilesha area, Nigeria. Bul. IAEG 33: 13-17.

Newill, D. 1961. A laboratory investigation of two red clays from Kenya. Geotechnique II, pp. 302-318.

Ogunsanwo, O. 1986. Basic index properties, mineralogy and microstructure of an amphibolite derived laterite soil. Bul. IAEG 33: 19-25.

Ogunsanwo, O. 1988. Basic geotechnical properties chemistry and mineralogy of some laterite soils from S.W. Nigeria. Bul. IAEG 37: 131-135.

Oyawoye, M. O. 1972. The basement complex of Nigeria. In African Geology pp. 67-99. Edited by Dessauvagie and Whiteman, U. I. Press.

Rahaman, M. O. 1988. Recent advances in the study of the Basement complex of Nigeria. GG.S.N. Pub.: 11-41.

Teme, S. C. and Oni, S. F. 1988. Detection of Groundwater in the fractured media - Ibadan, A Case Study. Journal of African Earth Sciences (in press).

Townsend et-al, 1969. Effect of remoulding on properties of a lateritic soil. Highway Res. Board Rec. 284: 76-84.

On oedometric and shear moduli of sands
Sur les modules oedométriques et de cisaillement des sables

P. Previatello & G. Rossato
Istituto di Costruzioni Marittime e di Geotecnica, University of Padova, Italy

ABSTRACT: The design problems of geotechnical engineering are essentially related to two main features: deformations in safe conditions and safety itself with respect of the shear strength of soils. This paper reports the oedometric modulus during consolidation and the shear modulus during shear in the Direct Simple Shear test, and includes an experimental programme carried out by the authors on sand from the river Adige. Both the oedometric and shear moduli are of considerable design interest and are related to each other, depending on the relative density of the sand and on the confining pressure applied.

RESUME: Les problèmes de projet dans le domaine de l'ingégnerie géotechnique sont liés a deux caractéristiques principales: c'est-à-dire, la déformation en condition de sécurité et de la sécurité proprement dite en fonction de la résistance au cisaillement du sol. On considère le module oedomètrique pendant la phase de consolidation et le module de cisaillement pendant l'exécution des essais de cisaillement simple direct. L'article illustre un programme expérimental utilisé par les sables du fleuve Adige. Les deux modules, oedomètrique et de cisaillement, sont particulièrement importants dans le domaine du génie civil et sont reliés entre eux en fonction de la densité relative et de la pression de confinement appliquée.

1 INTRODUCTION

The stress-strain behaviour and shear strength of soils may be studied by means of the standard triaxial apparatus, in which cylindrical specimens are subject to an axial-symmetrical state of stress. In standard compression triaxial tests, major principal stress (σ_1) is vertical (axial); while minor (σ_3) and intermediate principal stresses (σ_2) are equal to each other and horizontal (radial). Therefore, as both intensity and direction of the principal stresses are known, the state of stress inside the specimen can be determined by using Mohr's circles of stresses. However, in the standard triaxial test, the principal stresses have fixed directions. Moreover, intermediate principal stress

(σ_2) is equal to minor principal stress (σ_3). Instead in situ, states of stresses with principal stresses different from each other may occur ($\sigma_1 \neq \sigma_2 \neq \sigma_3$), also with rotation of the principal axes.

Starting from some cases of instability and failure occurring with plane strain conditions and rotation of the principal stresses (e.g., fig. 1), several types of direct shear apparatus have been designed. The simplest type is the well-known Casagrande box, which subjects the specimen to a state of plane strains with failure occurring in a predetermined horizontal plane (fig. 2a). Then the direct simple shear device was designed with the aim of overcoming the main drawbacks of the Casagrande box (failure occurs along a predetermined plane without measurement of

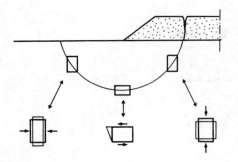

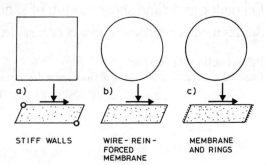

EXTENSION　　SIMPLE SHEAR　　COMPRESSION

Fig.1 Example of shear in plane strain conditions with rotation of principal stresses

STIFF WALLS　　WIRE - REIN - FORCED MEMBRANE　　MEMBRANE AND RINGS

Fig.3 Types of direct simple shear device: Cambridge (a), NGI (b), SGI (c)

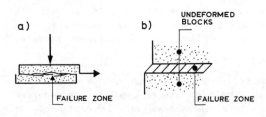

Fig.2 Schematic view of Casagrande box (a) and failure zone (b)

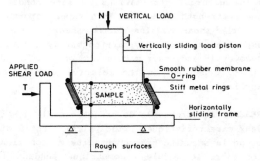

Fig.4 Schematic description of direct simple shear device used

pore pressure, correct estimation of strains is difficult, stresses are not uniform) and of reproducing strain conditions in the shear zone (Hvorslev, 1937 - Fig. 2b).

Direct simple shear devices are essentially of:

a) Cambridge type (Roscoe, 1953), in which square cross-section specimens are placed between rough horizontal plates, connected to each other by means of smooth, stiff, parallel surfaces linked at their ends (fig. 3a);

b) NGI type (Bjerrum and Landva, 1966), in which circular cross-section specimens are placed between rough horizontal plates and a smooth wire-reinforced rubber membrane (fig. 3b);

c) SGI type (Kjellman, 1951), in which circular cross-section specimens are placed between rough horizontal plates and a smooth rubber membrane externally reinforced by a column of stiff metal rings (fig. 3c).

This paper refers to a direct simple shear device of the SGI type (fig. 4). The specimen is placed between rough horizontal plates, through which shear is applied, and a smooth membrane externally reinforced by stiff metal rings.

2 EXPERIMENTAL PROGRAMME

The soil tested is a medium-fine uniform sand coming from the mouth of the Adige river (Italy). Figure 5 shows the grain size distribution. Characteristic values are: mean grain size D_{50} = 0.42 mm and coefficient of uniformity C_u = 2.0; density of solid particles γ_s = 26.58 kN/m^3; minimum and maximum dry unit weights are 13.58 kN/m^3 and 16.51 kN/m^3 respectively, according to ASTM D4253 and D4254 standard procedure. The mean sphericity index of grains, according to Rittenhouse's (1943)

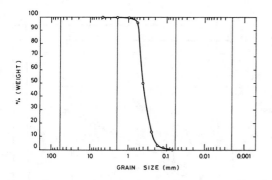

Fig.5 Grain size distribution of Adige river sand

ε_v = vertical deformation = $(h_o-h)/h_o \neq 0$

ε_h = horizontal deformation = 0

γ = shearing deformation = s/h=0

σ_v = vertical (effective) stress = $N/A \neq 0$

τ = shear stress = T/A=0.

During shear (fig. 6b) :

$\varepsilon_v \neq 0$

$\varepsilon_h = 0$

$\gamma \neq 0$

$\sigma_v \neq 0$

$\tau \neq 0.$

classification, is 0.8; particles are angular or subangular, but a few are platy, according to Powers's (1953) classification. Carbonatic elements are 20% weight and quartz is about 15%.

The diameter of the specimens was 70 mm with an initial height of about 20 mm. They were obtained by pouring the sand into a rubber membrane supported by a column of stiff metal rings and filled with water, and vibrating the base of the system during sedimentation. Figure 4 gives a schematic representation of a direct simple shear device of the SGI type (Kjellman, 1951); figure 6 shows the stress and strain conditions applied to the specimen during consolidation (a) and shear (b).

During consolidation (fig. 6a) :

The vertical load, shear load, axial external deformation and shearing external deformation have been measured during tests by means of an electronic data acquisition system. All tests were drained, with constant velocity of shearing deformation (during shear) of 0.2%/min. During shear vertical load was kept constant, so that the sand was free to dilate.

Figure 7 shows the testing programme carried out in the present study. Numbered as 11-13 and 21-23 is the first series of tests with initial density before consolidation D_i constant, varying the vertical load. Numbered as 31-33 and 41-43 is the second series of tests with applied vertical load σ_v x A, varying the initial

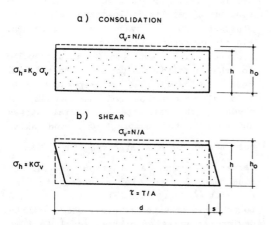

Fig.6 Stress and strain conditions applied to specimen during consolidation (a) and shear (b)

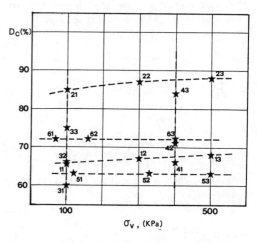

Fig.7 Testing programme: density before shear (D_c) versus vertical consolidation stress (σ_v)

density. Numbers 51-53 and 61-63 represent
the third series of tests, with density
after consolidation D_c constant, varying
the vertical load. In figure 7 the points
refer to conditions after consolidation.

3 RESULTS

The first phase of a direct simple shear
test is consolidation of the specimen under
a K_o ratio between horizontal and vertical
stresses. Consolidation is reached by
successively increasing the vertical load
applied and measuring the vertical
displacement of the load piston. This is in
fact an oedometric test, except that, in
the direct simple shear test, the specimen
is confined by a smooth rubber membrane,
while in the oedometric test it is confined
by a smooth metal ring. So we consider the
consolidation phase of the direct simple
shear test to be equivalent to the
oedometric test. Note that this is true at
sufficiently high stresses, while at low
stresses up to 50 kN/m,^{2} it appears to be
"apparent" compressibility of the specimen
as an effect of the confining elastic
membrane, together with assessment of the
plate-specimen interface.

By relating the vertical displacement of
the load piston to the load applied during
consolidation, the oedometric modulus is
given by

$$M = \frac{\sigma_v}{\varepsilon_v} \qquad (1)$$

(where σ_v is vertical stress applied and
ε_v is vertical deformation of the
specimen). Now we consider the well-known
equation:

$$M = m\sigma_a \left(\frac{\sigma_v}{\sigma_a}\right)^{1-a} \qquad (2)$$

first proposed by Janbu (1963). It
expresses tangent modulus M in the

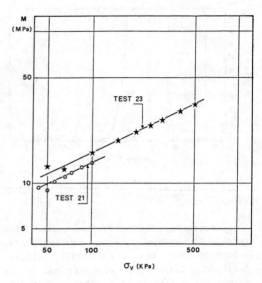

Fig.8 Secant modulus M during consolidation
versus vertical stress applied σ_v

oedometer test as a function of vertical
stress applied, σ_v, by means of modulus
number m and experimental constant a ($\sigma_a =$
1 atm). Furthermore, equation (2) may
properly be extended to the secant modulus.
See, for example, figure 8 which
schematically reports direct simple shear
tests 21 and 23 in log-log scale (M is the
secant modulus given by the ratio between
vertical stress applied and vertical
deformation). Note in figure 8 evident the
scatter of results at low stresses, due to
assessment of the system (plate-specimen,
elastic membrane, interface).

Janbu's (1963) equation (2) is also valid
for the consolidation phase of the triaxial
test. In that case M represents the
triaxial compression modulus and may be
given by the ratio between axial stress
applied to the sample (σ_v) and axial
deformation of the specimen (ε_v). As a
matter of fact, a relation similar to
equation (2) may be extended to the
triaxial moduli calculated at the same
point during the tests (e.g., secant at the
deviatoric stress mobilized equal to some
percentages of the maximum deviatoric
stress: see fig. 9) (Simonini, 1988) and
vertical consolidation stress. It may also

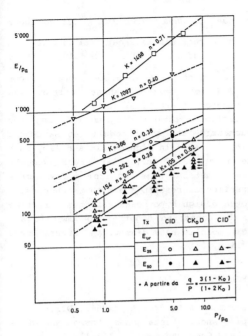

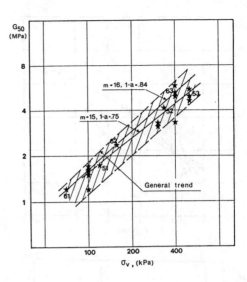

Fig.9 Secant triaxial compression moduli versus mean stress applied, both normalized by atmospheric pressure (from Simonini, 1988)

be extended to the shear moduli calculated at the same point during direct simple shear tests and vertical consolidation stress: see fig. 10 (log-log scale) (Rossato, 1988), which gives as an example the modulus number and exponent of equation (2) extended only for the tests 51 -53 and 61-63, but the general trend is evident too. G_{50} indicates the secant modulus calculated at a shear stress of 50% maximum shear stress.

Note that this paper refers to a shear modulus calculated as secant to the stress-strain curve at a mobilized shear stress of 50% maximum shear stress only for simplicity. It is related to a generic safety factor SF = 2 in design practice. However, the equations valid for G_{50} are also valid for the tangent modulus or other secant moduli different from G_{50}.

Both shear strength and stress strain behaviour of sands are strictly related to sample density and the stress level applied (see in extension Ladd et al., 1977). So we

Fig.10 Shear modulus G_{50} versus vertical stress applied in direct simple shear tests (from Rossato, 1988)

can also expect oedometric modulus M of the consolidation phase and shear modulus G of the shear phase of a direct simple shear test to be related to each other too. This relation will depend on relative density and stress level.

In fact, the oedometric modulus increases with both increasing density and vertical stress applied (fig. 8) and the shear modulus also increases with both increasing density and vertical stress applied (fig. 10). If we complete a diagram M-G, we find a clear pattern of trends as in figure 11a. In figure 11 only tests numbered 11-43 are reported, while tests 51-63, carried out at a constant density D_c after consolidation, did not show well-defined ratios between oedometric modulus and shear modulus, as a consequence of the different sample preparation for each test that a similar series of tests implies. Figure 11b emphasizes the general trend of the whole test programme in log-log scale. In figures 11a and 11b, the combination of effects on moduli of both density and stress applied (and also different sample preparation) are interesting.

445

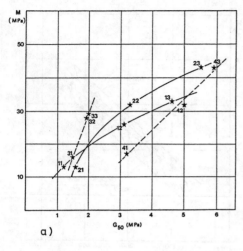

a)

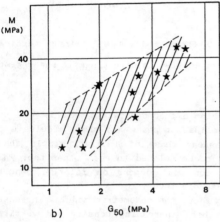

b)

Fig.11 Oedometric modulus M versus shear modulus G_{50} in direct simple shear tèsts

extended to shear modulus G_{50} during shear in the direct simple shear test and it also fitted (fig. 10). Modulus G_{50}, corresponding to a mobilized shear stress of 50% of maximum shear stress was considered of practical design interest.

Lastly a simple diagram relating the oedometric and shear moduli was sketched. As can be seen in figures 11a and 11b, both the general and particular trends can be drawn. That is, starting from a few simple experimental results and with the necessary proven design hypotheses, all the variables can be connected together: oedometric and shear moduli, sand density and stress level applied.

4 CONCLUSIONS

Adige river sand was investigated by means of the direct simple shear apparatus within a meaningful range of initial testing conditions, with the aim of establishing if a relation between the one-dimensional compression modulus during consolidation and the shear modulus during shear could be drawn.

Equation (2) (see Janbu, 1963) was extended to the secant oedometric modulus calculated during consolidation in the direct simple shear test, and equation (2) fits (fig. 8). Equation (2) was then also

REFERENCES

Bjerrum, L. & Landva, A. (1966). Direct simple shear tests on a Norwegian quick clay. Geotechnique 16, 113-116.

Hvorslev, M.J. (1937). Uber die Fesigkits-eigenschaften Gestorter Bindiger Boden. Ingeniorvidenskabelige Skrifter A, 45, Kobenhavn.

Janbu, N. J. (1963). Soil compressibility as determined by oedometer and triaxial tests. Proc. ECSMFE, 1, 19-25, Wiesbaden.

Kjellman, W. (1951). Testing the shear strength of clay in Sweden. Geotechnique 2, 225-232.

Ladd, C.C., Foott, R., Ishihara, K., Schlosser, F. & Poulos, H.G. (1977). Stress-deformation and strength characte-ristics. Proc. 9th ICSMFE, 2, 421-494, Tokyo.

Powers, M.C. (1953). A new roundness scale for sedimentary particles. J. Sed. Petr. 23, 117-119.

Rittenhouse, G.A. (1943). Measuring inter-cept sphericity of sand grains. Am. Jour. Sc. 241, 109-115.

Roscoe, K.H. (1953). An apparatus for the application of simple shear to soil sam-ples. Proc.3rd ICSMFE,1, 186-191, Zurich.

Rossato, G. (1988). Parametri di resistenza e di deformabilità di un mezzo granulare con prove di taglio semplice. Specializa-tion Thesis, Padova University.

Simonini, P. (1988). Parametri di deforma-bilità in formazioni sabbiose. C.N.R. Congress 173-192, Padova.

Study of the effect of some factors on shear strength of expansive soils in Egypt

Etude des effets de quelques facteurs sur la résistance au cisaillement de sols expansifs en Egypte

M. M. Reyad
General Organization for Housing, Building and Planning Research, Egypt

ABSTRACT: The research inhand aims at studying the effect of both the initial moisture content and the initial density of expansive soils on their shear strength. Two different soils from two different locations in Egypt were used. Two groups of tests were carried out. In the first group the initial moisture content of the soil was taken as a variable while in the second group the initial dry density of the soil was variable. The different properties of the used soils were determined. The conventional triaxial set up was used to carry out the required tests. It was found that shear strength increases with the increase of the initial moisture content and reaches its ultimate value when the moisture content is slightly higher than the shrinkage limit of the soil. Any slight increase in moisture content results in a great loss in shear strength. It was also found that the higher the soil density the higher the shear strength of the soil.

RESUME: Cet article décrit les recherches faites pour étudier les effets de la teneur en eau initiale et la densité initiale sur la résistance au cisaillement des sols expansifs. Deux sols differents enlevés de deux chantiers en Egypte ont été utilisés. Deux séries d'essais ont été effectuées. Dans la première série on a choisi la teneur en eau comme variable, tandis que dans la deuxième, la densité a été variée. Les propriétés différentes de sols utilisés ont été determinées. On a employé l'appareil triaxiale convenventtionale pour effectuer les expériénces. On a trouvé que la résistance au cisaillement croit avec la croissement de la teneur en eau initiale et atteint une valeur maximale à une valeur de la teneur en eau légèrment supérieure de la limite de retrait du sol. Une petite augmentation de la teneur en eau produit une chute rapide de la résistance au cisaillement. On ce qui concerne l'effet de la densité, les résultats des expériénces ont démontré que la résistance au cisaillement est plus grande lorsque les valeurs initiale de la densité sont plus elevées.

1 EXPERIMENTAL WORK

The soils used to carry out the experimental work were obtained from Fayoum at south of Cairo and from Aswan at upper Egypt. In order to control the initial conditions of the different specimens, remoulded samples of 0.0381 m diameter were prepared. The used soil was air dried and pulverised. The fraction passing through the ASTM sieve No. 40, (0.42 mm) was used for preparing the remoulded samples. The conventional triaxial set up was used. Shear parameters were determined using consolidated drained triaxial tests on the basis of maximum values of stresses observed. The different properties of the used soils were determined and given briefly in table 1.

To study the effect of initial moisture content, Fayoum soil samples were prepared at reasonable constant initial dry density γ_{di} of 1670 kg/m^3 and variable initial moisture contents W_j of 3.8, 7.8, 11.4, 14.5, 16.8, 19.0 and 21.0 %. These values were chosen such that they lie around the shrinkage limit of the soil. It was not easy to remould samples with higher moisture contents. Aswan soil samples were prepared at initial dry density of 1650 kg/m^3 and initial moisture contents of 6.2, 9.1, 12.2, 14.9, 16.7, 19.0 and 22.0 %.

To study the effect of initial density, Fayoum soil was used only for this study. Samples were prepared at constant moisture content of 11.4 % and variable dry densities of 1570, 1670, 1740, and 1800 kg/m^3.

Table 1. Properties of the tested soils

Item		Fayoum	Aswan
Liquid limit	%	89.00	59.50
Plastic limit	%	35.00	28.80
Plasticity index	%	54.00	30.70
Shrinkage limit	%	12.00	13.30
Specific gravity		2.89	2.83
Sand (fine)	%	zero	3.5
Silt	%	4- 6.5	56.0
Clay (< 2 mic.)	%	96-93.5	40.5
Classification		clay high plasticity	clay high plasticity

Chemical composition		
$Si\ O_2$	46.13	53.60
$Al_2\ O_3$	17.54	20.20
$Fe_2\ O_3$	7.75	3.95
$Ti\ O_2$	1.31	zero
$Ca\ O$	3.74	3.30
$Mg\ O$	3.25	1.95
$Na_2\ O$	1.79	0.32
$K_2\ O$	1.05	0.16
$S\ O_3$	5.87	1.52
Loss on ignition	11.35	14.50

Mineralogical composition	Montmor-illonite Kaolinite	Montmor-illonite Kaolinite Feldspars Calcite Hematite Barite Quartz

2 RESULTS AND CONCLUSIONS

The test results indicate that Fayoum expansive soil behaves almost similar to Aswan soil in spite of the clear distinction in their properties. This behaviour can be summarized in the following.

2.1 Effect of initial moisture content

The test results showed that the effect of initial moisture content on shear strength of the tested soils is as follows:

1. Increasing the initial moisture content results in an increase in the cohesion c up to a certain value after which it begins to decrease rapidly as shown in

Figures 1 and 2. The moisture content at which the cohesion reaches its maximum value was found to be slightly higher than the shrinkage limit of the soil.

2. Cohesion increases slowly whenever the initial moisture content is still lower than the shrinkage limit. The rate of increase begins to be noticeable and more rapid when the moisture content approaches the shrinkage limit.

3. The higher the initial moisture content the lower the value of the angle of internal friction ϕ as shown in Figures 1 and 2.

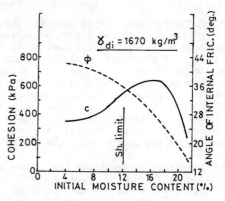

Fig.1 Effect of initial moisture content on shear parameters of Fayoum clay

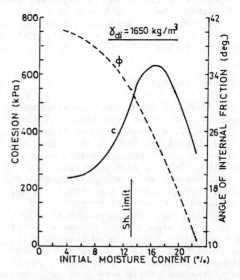

Fig.2 Effect of initial moisture content on shear parameters of Aswan clay

448

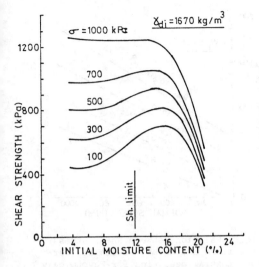

Fig.3 Effect of initial moisture content on shear strength of Fayoum clay

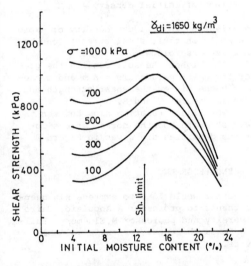

Fig.4 Effect of initial moisture content on shear strength of Aswan clay

4. The rate of decrease of the angle of internal friction becomes more noticeable whenever the moisture content is higher than the shrinkage limit of the soil.

5. Applying Coulomb's formula $\tau_f = c + \sigma \tan \phi$ to calculate the shear strength τ_f for different values of normal stress σ, Figures 3 and 4 show that:

- At the lower values of moisture content, the shear strength increases slowly with the increase of moisture content.
- The rate of increase of the shear strength becomes more obvious when the moist-

ure content of the soil is within its shrinkage limit.
- Shear strength reaches its ultimate value at an initial moisture content slightly higher than the shrinkage limit.
- When the shear strength reaches its maximum value, any slight increase in moisture content results in a great loss in shear strength.
- The linear relationships between shear strength and normal stress for various moisture contents are given in Figures 5 and 6.

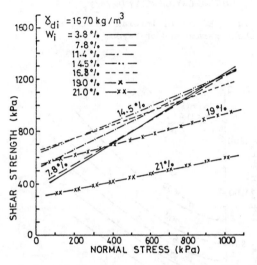

Fig.5 Relation between shear stress at failure and normal stress for various initial moisture contents for Fayoum clay

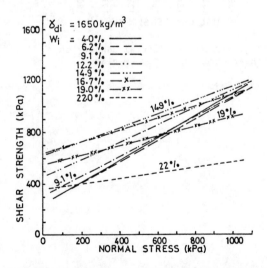

Fig.6 Relation between shear stress at failure and normal stress for various initial moisture contents for Aswan clay

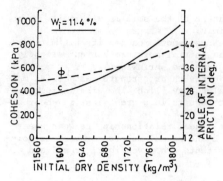

Fig.7 Effect of initial density on shear parameters of Fayoum clay

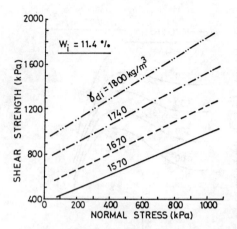

Fig.9 Relation between shear stress at failure and normal stress for various initial densities for Fayoum caly

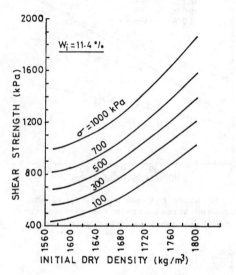

Fig.8 Effect of initial density on shear strength of Fayoum clay

2.2 Effect of initial density

The effect of initial dry density of expansive soils on their shear strength can be summarized as follows:

1. The higher the soil density the greater the shear parameters c & ϕ and consequently the higher the shear strength as shown in Figures 7 and 8.

2. The linear relationships between the shear strength and the normal stress for various densities are given in Figure 9.

3 ACKNOWLEDGEMENT

The author would like to express his deepest thanks to professor A.Abouleid, Cairo University and peofessor M.Eldemery, Building Research Center for their valuable advice.

6th International IAEG Congress / 6ème Congrès International de AIGI, © 1990 Balkema, Rotterdam. ISBN 90 6191 130 3

Dielectric behavior of rocks under uniaxial compressive stress

Le changement de la constante diélectrique de rochers sous une compression uniaxiale

Rakesh Sarman & Donald F. Palmer
Department of Geology and the Water Resources Research Institute, Kent State University, Kent, Ohio, USA

ABSTRACT: The influence of stress on the transmission of electromagnetic waves through limestone and sandstone has been measured for radio waves in the 19-22 GHz range under uniaxial compression. The results show a gradual increase in the radio wave transmission followed by a rapid increase near the microfractures level. Within the elastic limit the transmission properties are independent of the loading sequence. Data indicate a gradual phase shift along with the amplitude, caused by the changes in stress conditions.

Changes in the transmission characteristics of EM radiation measured here suggest a significant change in dielectric constant at 50 to 70% of the peak strength. These results offer real promise for the development of a ground penetrating radar system that can be used to measure changes in in-situ stress conditions in near surface materials in underground structures, slopes, and embankments.

RÉSUMÉ: L'influence de l'effort de compression sur la transmission des ondes électromagnétiques à travers le calcaire et le grès a été mesurée pour des fréquences radios entre 19 et 22 GHz sous compression uniaxe. Le résultat démontre une augmentation graduelle de la transmission des ondes radios suivi d'une augmentation rapide au voisinage du niveau des microfissures. Dans la limite élastique, les propriétés de transmission ne dépendent pas de la séquence de chargement. Le data indique un déplacement graduel de la phase et de l'amplitude, à cause des changements de force.

Les mesures suggèrent un changement important de la constante diélectrique a 50-70% de la résistance a l'écrasement. Les résultats font espérer le développement d'un radar sonde pour mesurer les changements des conditions de force dans les matériaux près de la surface, les structures souterraines, les pents, et les remblais.

1. INTRODUCTION

The measurement and monitoring of in-situ stress in rocks is of major importance in geotechnical engineering where the design of safe and stable tunnels, foundations, or embankments are concerned. An understanding of the stress distribution is very critical for evaluating the stability of underground openings (Oliveira and Garca, 1987; Goodman, 1980). In-situ stress may develop or change as a result of numerous factors including overburden pressure, pore water pressure, tectonic stresses, thermal stresses, blasting, grouting, or excavation of material from the surface and underground (Hanna, 1985).

Estimation of these stresses has been tried using numerical analytical methods and by attempting direct measurement with different types of instruments. The numerical methods treat rocks as elastic to elasto-plastic materials to calculate the stress regime in space. Instrumentation has been developed to measure in-situ stress and changes in stress caused by geotechnical activities by both direct or indirect methods. Direct methods involve the counterbalancing of stresses by the use of pressure cells, whereas indirect methods measure some physical quantity, such as deformation, which can be related to stress (Kovari et al, 1977). Extensiometers, inclinometers, deflectometers and shear strips are the most common instruments for measuring deformation (Johnson and DeGraff, 1988). These instruments are installed in drill holes along the excavation at several points and information is interpolated for the intermediate locations. The results obtained by these methods usually yield inconsistent values with a large scatter, and they often do not give an accurate picture of the larger rock mass (Oliveira and Garca, 1987).

Efforts to develop geophysical methods of determining in-situ stress have included studies of the basic electrical properties of rocks. Issacson (1962) studied the effect of pressure on the electrical resistivity in the Kolar gold fields and found a non-linear inverse relationship. This method was found to have difficulties in rock- transducer contact and in the calibration of the instrument for in-situ tectonic stress (Farmer, 1968).

Alternative methods have looked at the dielectric properties as well as resistivity. While most solids behave as conductors at low frequency alternating currents, they behave as dielectrics at high frequencies (Houck, 1984). The dielectric constant decreases with increasing frequency until about 1 MHz, beyond which it becomes constant (Hipp, 1974; Sen, 1981; Zheluder, 1971). The relationship between dielectric constant of rocks and pressure has been studied by Decker and Zhao (1983) and by Parkhamenko (1967) who showed that the ratio of the dielectric constant at pressure P (Ep) to that at atmospheric pressure (Eo) increases more rapidly with pressure at low pressure than at higher pressures. While these studies were conducted under the very high pressure and temperature conditions similar to those of the earth's mantle, the results do suggest that a detectable variation may exist at the low to moderate pressures and temperatures of interest to the geotechnical engineer.

2. EXPERIMENTAL INVESTIGATIONS

The pressure dependence of the dielectric constant at low pressures and at room temperature has been evaluated by measuring changes in the amplitude of radio frequency radiation transmitted through rocks at various loads. The results are preliminary because the equipment available for this study could only measure the amplitude and not the phase shift of the signal. Since the dielectric constant is a vector quantity (Alvarez, 1973), the phase component as well as the amplitude of the electromagnetic

signal is necessary to calculate its value (Hallikainen et al., 1985).

2.1 Sample Description and Preparation

Testing was done on samples of thinly bedded Berea sandstone, Graften sandstone, and Indiana limestone. While shale was also tested, the absorption of microwave energy by the clay minerals was so great that the amplitude of the transmitted wave could not be measured dependably. Therefore only the results from experiments on sandstones and limestone are discussed here.

The Berea sandstone samples were thinly (2 mm) bedded, well-sorted, fine to medium grained, and were composed primarily of subrounded quartz grains small amounts of feldspar, and iron oxide stain. The bedding is defined by concentrations of oxide minerals. The rock is moderately well cemented but has a high porosity of 15 to 20 percent.

Two types of Graften sandstone were used for testing. Both were obtained from the same area in West Virginia but differed in the amount of oxidation shown in red staining throughout the matrix. The Graften Sandstone is an immature sandstone composed primarily of subrounded to subangular quartz and feldspar grains with a few percent mica. The rock is generally massive with subtle bedding defined by variation in mica content. The samples did not break preferentially along the bedding planes. The sandstone is poorly cemented and has a high porosity.

The Indiana Limestone is a dark grey to black massive crystalline biosparite. Faint bedding planes on the scale of 4 mm are defined by very slight changes in color. The rock breaks along irregular fracture surfaces with only a small preference for fracture on bedding planes.

Rock samples were NX size cores as well as rectangular prisms ranging from 50 to 199 mm in size. All samples used for measurement were prepared with approximately a 2:1 height to width ratio. The samples were shielded with aluminum foil to limit external sources of radiation. Two slits in the foil on opposite sides of the sample where aligned with the wave guides. Samples showed no anisotropy in the transmission of microwaves but for the tests under compression, transmission was measured in the plane of bedding with compression at right angles to the bedding.

2.2 Experimental Setup

The experiment was performed using Hewlett-Packard (HP) electronics and wave guides, a Forney press, and specially manufactured parts for wave alignment. Square-wave modulated signals were generated using an HP 83508B/83570A sweep oscillator operating between 19 and 22 GHz. During the first stage of the experiment both horn antennae and wave guides were tested as transmitters and receivers. For the set up and sample sizes chosen, wave guides were superior to antennae in reducing signal scattering and providing better reproducibility in the observations. The initial, transmitted, and reflected signals were monitored with an HP 8756A scalar network analyzer. The experiment was controlled using an interactive program on an HP 9816 computer which also stored and processed the initial, transmitted, and reflected signals.

The experiment first measured the general electromagnetic wave transmission characteristics of rocks and then changes in these characteristics as a function of internal stress. The reproducibility of measurements of the transmitted and reflected signals through same sample was tested for different samples and loading sequences at a range of frequencies. The effect of sample dimensions on the recorded signals was tested by making measurements across different sides of the same prismatic sample and by using different size samples of the same rock. The amplitude of the initial, transmitted, and reflected signals were recorded in decibels at every predetermined load level and therefore measurements are logarithmic values. The aluminum foil around the sample caused multiple reflections within the sample which generated standing waves. The peaks observed are the results of constructive and destructive resonance between the standing waves within the sample.

2.3 Compression of Samples

The load was applied using a Forney press which is capable of applying up to 400,000

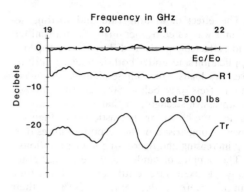

Figure 1. Transmitted (Tr), reflected (R1) and Ec/Eo curves for Berea Sandstone at 500 lbs load.

pounds (180,000 Kg) load and of recording the applied load to nearest pound. The increments in load level for measurement were determined based on the unconfined compressive strength of rocks as determined by the point load test. The test gave an approximate strength value and it was close enough to decide upon the load intervals for measurement. The stress levels for data analysis were calculated from the load levels measured from the Forney press and the area of cross section of the sample. The load intervals ranged from 5000 lbs (2268 Kg) at low load conditions to 1000 lbs (about 454 Kg) for the upper quarter of the compressive strength. The rate of load increase was less than the 200 lbs/min recommended by ASTM (1987). The initial signal levels which are used for comparison purposes were recorded at a negligible load level of 100 to 500 lbs (about 45 to 227 Kg) to ensure good contact between the rock samples and the platens of the Forney press. The transmitted and reflected signals were measured and stored and became the baseline against which all other measurements for that sample were compared (Figure 1). The initial transmitted signal is compared with the transmitted signals at various loads by calculating the logarithmic difference between the two signals. If the dielectric constant is the only electrical quantity that changes as a result of loading, the logarithmic difference between the signals will give the ratio (Ep/Eo) of the dielectric constants at those loading levels. The reflected signal is not used in the analysis since no significant changes were observed in this quantity during increase in load.

The effect of loading rate and loading sequence was tested by measuring the transmitted and reflected signals at different rates of load application and under both increasing and decreasing loads. In both sequences, the maximum stress was below 50% of the uniaxial compressive strength, so that the rock can be assumed to be within its elastic limit. Figure 2 gives a comparison of the response curves during increasing and decreasing load conditions.

The samples of sandstone were loaded gradually at fixed rate until failure. Limestone samples were loaded only to 50% of their compressive strengths as determined from the point load test in order to avoid any damage to the instruments at failure. The loading was stopped for about one minute for measuring the transmitted and reflected signals and then it was continued again at the same fixed rate. Measurements taken in this way showed no drift as a function of the interruption of loading rate and were reversible. In all cases the rate of loading decreased suddenly a little before the failure point although the load kept on increasing. This slight drop in loading rate is interpreted as due to development of microfractures.

The computer program used to control the experiment was designed for recording changes in the amplitude of the three signals at every 7.5 MHz frequency interval. Therefore amplitude variation was recorded at four hundred points within the 3 GHz frequency range. The program also displayed the three signals on the scalar network analyzer oscilloscope at every load increment so that signal quality could be checked at every stage.

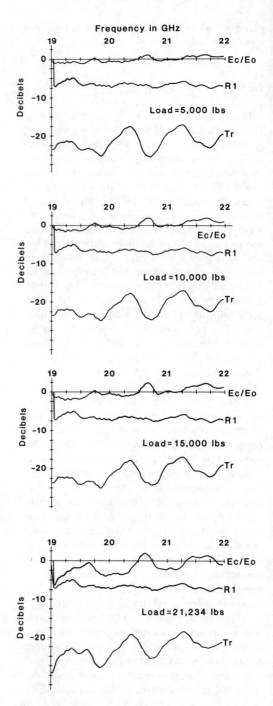

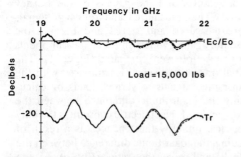

Figure 2. Comparison of the transmitted and Ec/Eo curves during increasing and decreasing loads. Solid line represents data for conditions of increasing load and dashed line represents data for decreasing load conditions.

Figure 3. Changes in the transmitted, reflected and Ec/Eo curves with increase in load for Berea Sandstone. The reflected wave curve does not change as a function of load although a large change does occur for transmitted and Ec/Eo curves.

3. RESULTS AND DISCUSSION

During the initial investigations it was observed that the response of transmitted waves changed considerably with stress. On the other hand, the reflected signal remained within a few percent (+5%) even when the stress was increased close to the failure level. Comparison of Figures 1 and 3 show the lack of change in the reflected wave with increase in pressure for Berea Sandstone.

3.1 Berea Sandstone

Berea sandstone was loaded to a maximum of slightly above 25000 lbs (about 25 MPa) where it failed. Changes in the Ep/Eo for Berea sandstone with increasing stress are shown in Figure 3. The loading rate was fixed for each experiment but it suddenly dropped at about 21234 lbs (about 17.4 MPa), interpreted as the initiation of microfracturing. At all four stages, three peaks of constructive resonance can be identified. The magnitude of these peaks increases with the increase in load levels. Figure 5a shows the change in the Ep/Eo with stress at these peaks. The ratio increases with stress at two higher frequency peaks until it reaches the microfracturing stage. The ratio increases by almost a factor of two and then drops a little. This trend is not so clear for the changes at the lower frequency peak.

3.2 Graften Sandstone

The first Graften sandstone sample failed at a little before 27000 lbs (59 MPa) and the microfracturing was developed at 25845 lbs (55 MPa). The changes in Ep/Eo with stress for Graften sandstone I are shown in Figure 4a. It is observed that the changes in the ratio are very little for an increase in load from 15000 to 25000 lbs but the development of microfractures considerably affect the ratio. Here we can recognize four peaks of constructive resonance on all plots. Changes in the ratio of Ep/Eo at

Figure 4. Changes in the Ec/Eo curves for (a) Graften Sandstone I, (b) Graften Sandstone II, and (c) Indiana Limestone at selected load levels.

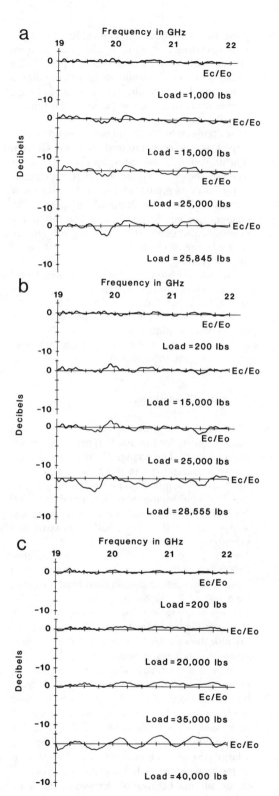

455

these peaks is shown in Figure 5b. Here the ratio decreases at the first loading and then increases with load for 2nd and 3rd peak. The changes are not as uniform as for the Berea sandstone, however, there is a sudden change in the behavior of all the peaks at the time of microfracturing.

Graften sandstone II showed weaker transmitted signals compared to those in the Graften sandstone I but the ratio Ep/Eo has similar trend as shown in Figure 4b. The four peaks in this case, as plotted in Figure 5c, also show creases with load for 2nd and 3rd peak. The changes are not as uniform as for the Berea sandstone, however, there is a sudden change in the behavior of all the peaks at the time of microfracturing.

Graften sandstone II showed weaker transmitted signals compared to those in the Graften sandstone I but the ratio Ep/Eo has similar trend as shown in Figure 4b. The four peaks in this case, as plotted in Figure 5c, also show a marked change in the ratio at the development of microfracturing.

3.3 Indiana Limestone

The results of transmission characteristics measurements for Indiana limestone are shown in Figure 4c. The changes in the ratio Ep/Eo are very pronounced in this case compared to the previous samples. The microfracturing in the limestone sample was developed at 37200 lbs (about 41 MPa) and the measurement was stopped at 44 MPa, about half of its estimated compressive strength. The ratio Ep/Eo increases gradually until the loading reaches 35000 lbs and then the microfracturing caused a sudden increase in the magnitude of the ratio, especially at the higher frequency signals. The changes in the ratio at four peaks with load are shown in Figure 5d for the limestone. The figure also shows a similar trend as in the previous samples, only more pronounced.

There is a systematic shift in the location of peaks with load in all samples. This is perhaps due to the effect of phase change which could not be measured. The most important feature in all samples was the change in the behavior of the Ep/Eo with stress. We are suggesting that a large part of these changes in the amplitude of the transmitted signal are due to stress increase and not because of changes in the sam-

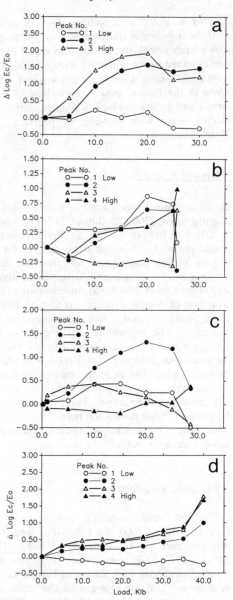

Load vs Δ Log Ec/Eo at Peak Values

Figure 5. Variation in the peak intensity as a function of load for (a) Berea Sandstone, (b) Graften Sandstone I, (c) Graften Sandstone II, and (d) Indiana Limestone. The peak numbers 1 (low) through 3 or 4 (high) refer to low to high frequency peaks on the Ec/Eo curves in Figures 3 and 4. Accelerated variation and occasional reversals in the sense of change are seen close to the microfracture level.

ple dimensions. This is based on the observations we made on the samples of different dimensions from the same rock. We did not observe any systematic increase or decrease in the amplitude of the transmitted signal as a function of small changes in the path length while we do see such changes with stress increase. The ratio E_p/E_o increased by a factor of 2 to 10 depending upon the rock type when the stress increased to about 50 to 80 percent of its compressive strength. This suggests that these kinds of measurements may be used to develop an early warning system for rock failure.

4. CONCLUSIONS

The experimental results show that the electrical properties of rocks, in the radio frequency range, are dependent on the prevailing stress regime. Within the elastic limit of the rocks, the electrical properties depend on the applied stress and are independent of the loading sequence. The changes in phase and amplitude of the transmitted signals, in the 19-22 GHz range, show that the dielectric constant of rocks can be used to study the changes in stress concentration resulting from natural effects or geotechnical disturbances caused by excavation or additional loading. The logarithmic difference in the amplitude of transmitted signals at the current load and that at the initial negligible load gives the ratio of dielectric constants E_c/E_o at the two load levels. A substantial increase in E_c/E_o at 50 to 70% of the compressive strength of rocks suggests that a measurement of changes in dielectric constant can be used to detect stress increase before the failure occurs. The adaptation of the laboratory experiment to equipment that operates in the field may be used to develop a stress estimating radar which might provide an early warning system before rock or soil failure. Such a testing method would be non-destructive and could also provide a continuous picture of the stress regime over a wide area between sites where other tests have been done. These data would be useful in a number of practical applications and in theoretical modelling.

5. ACKNOWLEDGEMENTS

The authors thank the BP America Corporation for their loan of equipment to conduct these experiments. The help given by Mr. Timothy Miller in operating the compression machine is also acknowledged. Dr. Abdul Shakoor gave many constructive suggestions during the experiment.

6. REFERENCES

ASTM 1987. Annual book of ASTM standards; Soil and rock, building stones and geotextiles. Am Soc. Testing and Materials. 408: 1188 pp., Philadelphia.

Alvarez, R. 1973. Complex dielectric permittivity in rocks: a method for its measurement and analysis. Geophysics 38(5): 920-940.

Decker, D.L. and Zhao, Y.X. 1983. Changes in the BaTiO3 phase transition to 40 KBar. High Pressure in Science and Technology. 1: p179-182, North-Holland, New York.

Farmer, I.W. 1968. Engineering properties of rocks. E. & F.N. Spon Ltd.,180 p, London.

Goodman, R.E. 1980. Introduction to rock mechanics. John Wiley and Sons, 478 p, New York.

Hallikainen, M.T., Ulaby, F.T., Dobson, M.C., El-Rayes, M.A. and Wu, L-K 1985. Microwave dielectric behavior of wet soil- Part 1: Empirical models and experimental observations. IEEE Trans. Geoscience and Remote Sensing. GE 23(1): 25-34.

Hanna, T.H. 1985. Instrumentation in geotechnical engineering. Trans Tech Publ., 843 p., Clausthal-Zellerfeld, F.R.G..

Hipp, J.E. 1974. Soil electromagnetic parameters as function of frequency, soil density and soil moisture. Proc. IEEE. 62(1): 98-103.

Houck, R.T. 1984. Subsurface imaging with ground penetrating radar. Unpublished Ph.D. Dissertation, 172 p, Pennsylvania State University.

Issacson, E.S.Q. 1962. Rock pressure in mines. Mining Publications, 260 p. London.

Johnson, R.B. and DeGraff, J.V. 1988. Principles of engineering geology. John Wiley and Sons, 497 p, New York.

Kovari, K., Amsted, Ch. and Fritz, P. 1977. Integrated measuring technique for rock pressure determination. Proc. Symp. Field Measurements in Rock Mechanics 1, p. 289-316, Zurich.

Oliveira, R. and Garca, J.C. 1987. In-situ testing of rocks. Ground Engineer's Reference Book, F.G. Bell (ed)., p. 26.1-26.28, Butterworth, London, U.K..

Parkhamenko, E.I. 1967. Electrical properties of rocks. Plenum Press, 314 p., New York.

Sen, P.N. 1981. Relation of certain geometrical fractures to the dielectric constants of rocks. Geophysics. 46(12): 1714-1720.

Zheluder, I.S. 1971. Physics of crystalline dielectrics. Plenum Press 2, 620 p., New York.

Prediction of volumetric increase of selected mudrocks

La prédiction de l'augmentation volumétrique des roches argileuses

Rakesh Sarman & Abdul Shakoor
Department of Geology and Water Resources Research Institute, Kent State University, Kent, Ohio, USA

ABSTRACT: Mudrocks (shales, claystones, siltstones) from 28 different sites in the United States were studied to investigate the relationship between their volumetric increase caused by immersion in water and their engineering properties. The properties investigated included the natural water content, Atterberg limits, dry density, absorption, adsorption, swelling pressure, and volumetric increase. X-ray diffraction analysis and scanning electron microscopy were used to determine the clay mineralogy and fabric of the mudrocks studied.

Preliminary results indicate that absorption, adsorption, and slake durability index show fairly strong correlations with volumetric increase measured on 2.5-cm cubical samples. X-ray diffraction analysis indicates that volumetric increase is not restricted to mudrocks containing swelling type clay minerals. Scanning electron microscopy reveals that in many cases mudrock fabric is more closely related to their volumetric increase than mineral composition.

RÉSUMÉ: Les roches argiliques (les argiles, l'argillite, les roches limoneux) de 28 sites differents aux États Unis étaient étudiées pour trouver la rapport entre de l'augmentation volumétrique et des propriétés géotechniques (telle que le teneur en eau, les limites d'Atterberg, la densité sèche, l'absorption, l'adsorption, et la pression de gonflement). Les techniques de la diffraction aux rayons X et au microscope électronique à balayage ont été utilisées pour déterminer la minéralogie des argiles et l'étoffe des roches argiliques.

Les resultats préliminaires, measuré sur les echantillons cubicale de 2.5 cm indiquent que l'absorption, l'adsorption et l'indice de la durabileté en immersion montre une correlation assez forte à l'augmentation volumétrique. La diffraction des rayons X indique que l'augmentation volumétrique n'est pas limitée a les roches argiliques contenant les minéraux de argile de type gonflant. Le microscope électronique à balayage révèle que dans plusieurs cas l'étoffe est plus relatif à l'augmentation volumétrique qu'à la composition minéralogique.

INTRODUCTION

Mudrocks are the most commonly occurring sedimentary rocks. The term mudrock was first proposed by Ingram (1953) and since then different researchers have defined it differently. Presently, a fine to very fine-grained siliciclastic sedimentary rock with more than 35 percent of the grains smaller than 60 µm is considered to be the generally acceptable definition (Grainger, 1984; Press and Siever, 1985).

Because of their widespread occurrence, mudrocks are the most frequently encountered rock type in engineering construction. They are also the most problematic rocks because of their low strength, high compressibility, and high swell potential. The engineering problems associated with the swelling of mudrocks are well-recognized all over the world (Chen, 1975). The expansion of mudrocks is known to have caused damage to tunnels (Madsen and Müller-Vonmoos, 1985), building foundations (Penner et al., 1970; Ola, 1982; Berube et al., 1986), dam foundations (Zeng, 1981; Ramachandran et al., 1981), underground structures (Wittke, 1982; Lee and Klym, 1978), slopes (Flemming, 1970), and highway embankments (Strohm et al., 1978).

Different researchers have used different properties to predict the swelling potential of mudrocks. The most commonly studied properties include mineral composition, texture, natural water content, dry density, and compressive strength (Lee and Klym, 1978; Preber, 1984; Madsen and Muller- Vonmoos, 1985; Berbue et al., 1986; and Seedsman, 1986). Olivier (1979) developed a geodurability classification system for mudrocks based on correlation of the uniaxial compressive strength and deformation modulus with the swelling coefficient. Kojima and others (1981) used the dry density and the water absorption capacity of pulverized mudrock samples to predict swell potential. Huang and others (1986) proposed a more general classification based on adsorption under varying conditions of relative humidity, water content, and ion concentration of the pore fluid.

The swelling behavior of mudrocks remains poorly understood because of their highly variable nature. A few classification systems that have been developed to predict the swelling behavior of mudrocks are of limited application because they were based on a small number of mudrocks tested and considered only a few of the properties related to swelling. The study presented in this paper is part of an ongoing research project intended to develop a more comprehensive classification that could be used to predict the swelling potential of a wide variety of mudrocks using a number of different properties. Only a few of the properties investigated, and their relationship to swell potential, will be discussed in this paper.

RESEARCH METHODS

Sampling

Mudrock samples were collected from 28 sites which are spread over 9 states and represent 25 different formations. Figure 1 shows the geographic distribution of the sampling sites. Most of the samples were collected from highway cuts using hand tools. An attempt was made to collect as fresh samples as possible, and sufficiently large in size and quantity to enable different lab tests to be performed.

Laboratory Testing

Samples collected from the field were subjected to a series of laboratory tests in order to determine their natural water content, Atterberg limits, density, absorption and adsorption characteristics, slake durability, swelling pressure, and 3-D volumetric increase. In addition, x-ray diffraction analysis and scanning electron microscopy were performed to determine clay mineralogy and textural characteristics of the mudrocks involved.

The natural water content of the samples was determined in accordance with ASTM procedure D 2216. The Atterberg limits were determined using ASTM procedures D423 and D424. The powdered samples for Atterberg limits test were prepared by subjecting the mudrock samples to alternate cycles of wetting and drying. This procedure was followed to prevent any crushing of the clay minerals (Diaz, 1973). Several methods were used to determine the volume of irregular samples required for calculating the dry density of mudrocks. The method that was finally chosen consisted of submerging the sample in water and measuring the volume of the water displaced using the Archimedes principle. However, in order to eliminate absorption of water by the samples, the samples were coated by a thin film of spray lacquer. The film was thick enough to retard the water absorption but thin enough to weigh less than 0.02 g compared to sample weights of 10 to 50 g. Percent absorption was measured according to ASTM method C 97 whereas adsorption was measured at 25°C under conditions of 100 percent humidity. The slake durability test was performed according to the procedure outlined by Franklin and Chandra (1972) and approved by ISRM (1979).

Figure 1. Geographic distribution of the sampling sites.

Measurements of 3-D volumetric increase were made on cubical samples, the cube sides ranging from 2.4 to 3.5 cm in length. The apparatus used for measuring the 3-D volumetric increase is shown in Figure 2 and was designed by the Structural Behavior Engineering Laboratories of Arizona, U.S.A., in accordance with the ISRM specifications (1979). The cubical samples of mudrocks were loaded in the apparatus with their bedding planes (laminations) perpendicular to the vertical axis, and were subsequently immersed in water. The expansion in three mutually perpendicular directions was measured for a period of 24 hours, using gages capable of measuring to 0.0001 inch. The cumulative linear expansion in the three directions was potted against time as shown in Figure 3. Three cubes of each mudrock were tested in this manner to obtain the average values of 3-D volumetric increase.

Data Analysis

The average values of 3-D volumetric increase for each mudrock were correlated with the corresponding results of other tests to identify those properties which best correlate with the swell potential as indicated by the respective values of correlation coefficient. The correlations were then evaluated in the light of mineral composition and textural characteristics of mudrocks involved.

RESULTS

Table 1 provides a summary of the mudrock properties used in this paper. The natural water content varies from a low of 1.09 percent to a high of 22.17 percent and does not appear to be related to the swell potential. This is because all of our samples were obtained from near surface where natural water content tends to be more a function of local climatic conditions than a reflection of the intrinsic properties of mudrocks, especially when the samples are collected from regions of very different climates. A plot of Atterberg limits on Casagrande's plasticity chart (Figure 4) reveals that mudrocks studied consist of material that can be classified as silt of low plasticity (ML) to clay of low plasticity (CL). The dry density values range from 1.46 g/cm3 to 2.75 g/cm3 with a mean value of 2.31 g/cm3. Absorption values vary from 1.20 percent to

44.00 percent whereas adsorption values range between 0.20 percent and 8.69 percent. The second cycle slake durability indices range from 0.2 percent to 99.0 percent indicating that the mudrocks included in this study cover a wide range (very low to high) of durability. The 3-D volumetric increase varies from a low of 0.22 percent to a high of 54.15 percent.

Figure 2. Apparatus used for measuring volumetric increase.

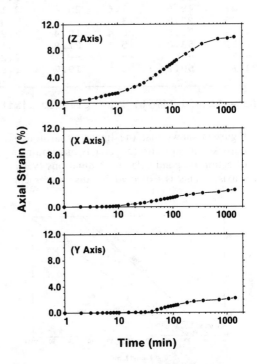

Figure 3. Axial strain vs time for a typical mudrock sample.

461

Table 1. Selected engineering properties of the mudrocks studied.

Natural Water Content (%)	Atterberg Limits			Dry Density (g/cm³)	Absorption (%)	Adsorption (%)	Slake Dur. Index (%)	3-D Volumetric Increase (%)
	LL	PL	PI					
3.29	27.7	17.5	10.2	2.59	0.20	5.50	91.00	0.24
3.43	29.2	18.7	10.5	2.54	3.05	44.00	4.00	53.99
3.40	36.0	19.8	16.2	2.37	0.98	19.00	51.00	15.33
10.60	35.4	26.6	8.8	2.22	1.49	20.50	4.80	11.35
2.20	25.2	18.3	6.9	2.44	0.77	8.30	86.10	0.80
4.00	35.5	22.1	13.4	2.26	2.25	15.20	66.70	4.29
2.80	40.0	28.8	11.2	2.49	0.37	11.70	92.80	2.80
1.50	25.8	19.7	6.1	2.45	0.33	4.40	98.30	0.59
1.80	31.4	23.7	7.7	2.57	0.65	5.10	98.70	1.83
1.20	31.8	24.6	7.2	2.51	0.22	6.80	97.80	2.50
1.80	--	--	--	2.33	0.15	2.50	97.80	1.11
14.76	--	--	--	2.33	1.60	8.10	90.30	1.64
5.89	--	--	--	2.58	0.46	4.80	92.50	4.81
1.09	--	--	--	2.57	0.11	3.40	98.90	0.22
4.28	64.4	40.2	24.2	1.46	5.31	39.00	79.00	10.19
2.19	--	--	--	2.75	0.09	1.20	99.00	0.27
3.36	37.0	25.5	11.5	2.08	1.12	17.80	19.00	5.34
5.97	39.5	27.7	11.8	2.04	2.99	17.00	78.00	10.64
14.14	35.8	22.1	1.37	1.97	1.89	21.00	0.20	12.97
3.32	26.2	20.1	6.1	2.43	1.93	21.60	16.10	17.82
5.36	37.0	21.6	15.4	2.38	4.28	36.80	0.30	11.62
2.30	34.0	19.7	14.3	2.39	0.68	24.10	90.50	8.37
10.21	48.6	22.6	26.0	2.15	6.58	34.00	0.20	54.15
4.89	25.9	18.3	7.6	2.36	0.54	6.30	90.40	0.87
13.15	36.8	26.4	10.4	2.22	8.69	13.70	7.50	25.82
19.51	53.0	25.1	27.9	2.10	6.80	20.70	2.90	30.13
6.37	--	--	--	2.24	3.13	9.80	95.70	1.04
22.17	50.2	30.5	19.7	1.80	7.80	29.70	5.90	32.62

LL = liquid limit; PL = plastic limit; PI = plasticity index

Figure 3 shows that the linear expansion of mudrocks on exposure to water is maximum in a direction perpendicular to laminations (vertical axis). This is expected because water can enter more easily along the horizontal laminations causing greater expansion in the vertical direction. Figure 5 shows a plot of 3-D volumetric increase versus time for a typical mudrock sample. It is clear from the figure that volumetric increase is slow in the beginning, becomes faster, and then slows down again. The slow rate of swelling in the beginning indicates that during the initial stage of immersion the sample absorbs water without much increase in volume. Later, the absorbed water starts opening the pores and microfractures, causing an increase in the rate of expansion. After about 2 hours, most of the fractures have been opened and filled with water and, therefore, the rate of swelling slows down. Thus, most of the swelling occurs during the first two hours.

The relationship between 3-D volumetric increase versus the dry density, absorption,

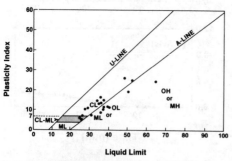

Figure 4. Classification of mudrock material based on Atterberg limits.

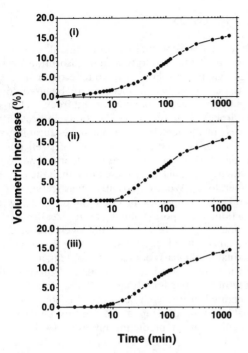

Figure 5. Volumetric increase vs time for a typical mudrock sample.

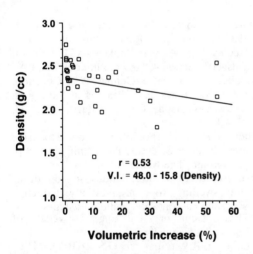

Figure 6. Relationship between volumetric increase and dry density.

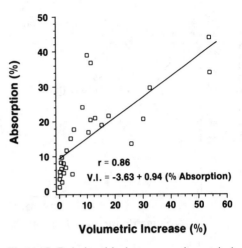

Figure 7. Relationship between volumetric increase and absorption.

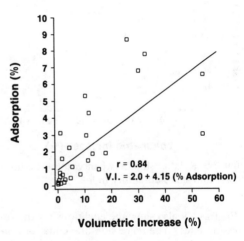

Figure 8. Relationship between volumetric increase and adsorption.

adsorption, and slake durability is shown in Figures 6-9, respectively. The dry density does not appear to be related to volumetric increase as indicated by the very low (0.53) value of correlation coefficient (Figure 6). Generally, however, the swell potential decreases with increasing density. Percent absorption shows a moderately significant correlation (r = 0.86) with volumetric increase (Figure 7) and may be used to predict swell potential. Adsorption also shows a moderate correlation (r = 0.84) with volumetric increase (Figure 8). The relationship between slake durability index and volumetric increase shows two separate trends (Figure 9). Mudrocks with slake durability indices less than 20 percent tend to follow a more gentle trend (r = 0.63) than those with slake durability indices more than 20 percent (r = 0.89).

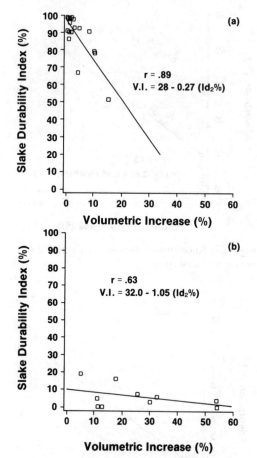

Figure 9. Relationship between volumetric increase and slake durability index.

Figures 6-9 also show the equations for the regression lines. The regression line equations involving absorption, adsorption, and slake durability can be used to estimate swell potential with a reasonable degree of confidence.

X-ray diffraction analysis of the mudrocks studied showed illite and kaolinite to be the most common clay minerals. In spite of this more or less similar clay mineralogy, a wide range (0.24 to 54.15 percent) in volumetric increase was observed. An examination of selected samples under the scanning electron microscope revealed that mudrocks which exhibited large increase in volume were characterized by the presence of a large number of open, inter-connected, voids and microfractures. This suggests that mudrock fabric and structure are probably more important in controlling swell potential than the clay mineralogy.

CONCLUSION

Based on our preliminary findings, it can be concluded that absorption and adsorption are significantly correlated with volumetric increase and that these two properties can be used to predict the swell potential of mudrocks. For mudrocks exhibiting slake durability indices in excess of 20 percent, even the slake durability index can be used as an indicator of swell potential. Also, the structure and fabric tend to influence volumetric increase more than the clay mineralogy. We are currently investigating several other properties including the percent clay fraction, total pore volume, pore size distribution, compressive strength, and swell pressure. We also intend to subdivide the mudrocks studied on the basis of their clay content, and re-evaluate the corretions in the light of these subdivisions. We hope that the large data base generated on the completion of this study will enable us to develop a more comprehensive classification for predicting mudrock swell potential.

REFERENCES

American Society for Testing and Materials 1987. Soil and Rock, Building Stones, Geotextiles. Annual Book of ASTM Standards 4.08 Section 4, 1189 p. Philadelphia, PA.

Berube, M.A., Locat, J. Gelinas, P. & Chagnon, J.Y. 1986. Black shale heaving at Saint-Foy, Quebec, Canada. Can. J. Earth Sci. 23: 1774-1781.

Chen, F.H. 1975. Foundations on expansive soils. Elsevier Scientific Publishing Co., 28 p. Amsterdam, The Netherlands.

Diaz, A.F.C. 1973. Interrelations of slake durability, swelling properties, plasticity characteristics, and natural water content of shales. M.S. Thesis, Dept. of Civil Eng., University of Louisville, Kentucky.

Flemming, R.W., Spender, G.S. & Banks, D.C. 1970. Empirical behavior of clay shale slopes. Technical Report 15(1), 93 p. U.S. Army Corps of Engineers Nuclear Cratering Group, Wicksburgh, Mississippi.

Franklin, J.A. & Chandra, R. 1972. The slake durability test. Int. J. Rock Mech. and Min. Sci. 9: 325-341.

Grainger, P. 1984. The classification of mudrocks for engineering purposes. Quat. J. Eng. Geol. 17(4): 381-387.

Huang, S.C. Aughenbaugh, N.D. & Rockaway, J.D. 1986. Swelling pressure studies of shales. Int. J. Rock Mech. Min. Sci. and Geomech. Abstr. 23(5): 371-377.

Ingram, R.L. 1953. Fissility of Mudrocks. Bull. Geol. Soc. Am. 64(8): 869-878.

International Society of Rock Mechanics 1979. Suggested Methods for determining water content, porosity, density, absorption and related properties, and swelling and slake durability index properties. Int. J. Rock Mech. Min. Sci. and Geomech. Abstr. 16(2): 143-156.

Kojima, K., Saito, Y. & Yokokura, M. 1981. Quantitative estimation of swelling and slaking characteristics. Proc. Symp. Weak Rock 1, p.292-223. Tokyo, Japan.

Lee, C.F. & Klym, T.W. 1978. Determination of rock squeeze potential for underground power projects. Eng. Geol. 12(2): 181-192.

Madsen, F.T. & Müller-Vonmoos, M. 1985. Swelling pressure calculated from mineralogical properties of a Jurassic opalinum shale, Switzerland. Clay and Clay Minerals. 33(6): 501-509.

Ola, S.A. 1982. Geotechnical properties of an attapulgite clay shale in Northwestern Nigeria. Eng. Geol. 19(1): 1-13.

Olivier, H.J. 1979. A new engineering-geological rock durability classification. Eng. Geol. 14(2): 255-279.

Penner, E., Gillot, J.E. & Eden, W.J. 1970. Investigation of heave in Billings Shale by mineralogical and biogeochemical methods. Can. Geotech. J. 7: 333-338.

Preber, T. 1984. Engineering properties of the Maquoketa Shale in Northeastern Illinois. Proc. 25th Symp. Rock Mech., 381-389. Evanston, Illinois.

Press, F. & Siever, R. 1985. Earth, 656 p. W.H. Freeman and Co., New York.

Ramachandran, B., Majumdar, N. Bandose, R.B. & Rawat, J.S. 1981. Influence of weak pyritiferous shale on the design of Dudhganga dam, India. Proc. Int. Symp. Weak Rock 3, p. 1472-1475. Tokyo, Japan.

Seedsman, R. 1986. The behavior of clay shales in water. Can. Geotech. J. 23: 18-22.

Strohm, W.B., Bragg, G.H. & Ziegler, T.W. 1978. Design and construction of compacted shale embankments, 142 p. Federal Highway Administration, U.S. Department of Transportation.

Wittke, W. 1982. Fundamentals for the design and construction of tunnels in swelling rocks. Proc. 23rd Symp. Rock Mech., p. 710-729. Berkely, California.

Zeng, Z. 1981. An engineering geological study of argillaceous soft rocks at the Gezhouna Project on the Yangtze river. Proc. Int. Symp. Weak Rock 2, p. 1127-1133. Tokyo, Japan.

465

6th International IAEG Congress / 6ème Congrès International de AIGI, © 1990 Balkema, Rotterdam. ISBN 90 6191 130 3

Shear behaviour and strength prediction studies on an Indian quartzite and sandstone

Une étude qui prédit l'allure et la résistence de cissaillement sur le grès et le quartzite indiens

R.K.Srivastava, A.V.Jalota & Ahmad A.A.Amir
M.N.R.Engineering College, Allahabad, India

ABSTRACT : In the present study strength behaviour of two types of Indian rocks, a quartzite and a sandstone is reported. Confining pressure upto 100 Kg/cm^2 have been used. Assessment of applicability of strength criteria proposed by Bieniawaski, Hoek-Brown and Rao et. al. indicates that the criteria proposed by Rao et. al. is most suitable.

RÉSUMÉ : Dans cet étude on a rendu compte deux types de conduit de force sur les roches indiennes, une quartzite et l'autre le grès. Une pression enfermée jusqu'à 100 Kg/cm^2 était utilisée. L'évaluation de l'applicalilite du critère de résistance proposé par Bieniawaski, Hoek-Brown et Rao nous indique que le critere proposé par Rao est le plus conrenable.

1 INTRODUCTION

From the engineers point of view, the increasing magnitude of projects and the resulting responsibility demands, quantitative information and an insight into strength and deformation behaviour of rock and rock masses. Natural geological conditions are usually complex. Especially in India, Himalayan geology is very complicated. India being a developing country with limited resources, the responsibility of engineers is considerably increased to come up with designs which are techno-economically viable. Thus understanding, appraisal and prediction of strength behaviour in a given geologic environment are of paramount importance.

Recent advances in testing techniques and equipment and also computational methods are proving to be very helpful in providing a greater insight into the stress-strain behaviour of rock and rock masses. Stability of any rock mass is essentially a function of induced stresses in relations to its inherent strength. Usually a rock formation occurs in triaxial state of stress. A study of their strength response under simulated laboratory conditions is less costly and essential in developing any understanding of their behaviour.

Many strength criteria have been proposed to predict the behaviour of rock and rock masses. But none of the several theories proposed have been observed to be universally true for all types of rocks. Because of scale effects, strength and deformation behaviour determined from the tests on small laboratory specimen are not normally directly applicable to large rock mass and it is necessary to relate observed and predicted behaviour from experience in similar type of rocks. Thus there is a need to generate large amount of data of strength behaviour of various types of rocks in different geologic environment. With this is view, present investigations have been carried out to study the strength behaviour of two Indian rocks viz. quartzite and sandstone. Further a brief review of some important strength criteria have been carried out along with a comparison of experimentally observed variation of σ_1 with σ_3 and theoretically predictable variation of σ_1 with σ_3 using Bieniawaski (1974), Hoek and Brown (1980) and Rao et. al. (1985) strength criteria.

2 EXPERIMENTAL INVESTIGATIONS

Following experimental investigations have been carried out :
1. Determination of physical properties viz., Water absorption, density, specific gravity and porosity.
2. Determination of strength indices viz., point load strength, Brazilian strength and uniaxial compressive strength.
3. Determination of triaxial behaviour of the two rocks using confining pressure σ_3 = 25, 50, 75 and 100 kg/cm^2.

The samples of quartzite and sandstone used in this investigation are from state of Uttar Pradesh, India. and they have been designated as Vindhyachal sandstone and Baraundha quartzite.

3 STRENGTH CRITERION

Franklin (1971) has very aptly defined strength criterion and its main functions. As per Franklin (1971), a strength criterion is an algebric expression used to describe the locus of a stress point as it travels in the strength surface. Its main functions are :

1. To allow strength prediction for the purpose of rock structure design. The algebric expression allows interpolation, sometimes extrapolation from observed strength values.

2. To allow rock classification according to strength; parameters (constant terms) of the criterion are characteristics of the shape of strength surface for a given rock sample.

After the pioneering experiments on specimens of marble (under triaxial conditions) by Von Karman (1911), many researchers have studied the various aspects of rock under triaxial conditions (e.g. Hoffmann 1958, Handin and Haeger 1958, Heard 1960, Schwartz 1964, Waversik and Brace 1971 etc.). One of the major conclusions of their experimental studies is that peak strength (σ_1) varies non-linearly with confining pressure (σ_3). Based on these observations, several non-linear failure criteria were proposed by various researchers.

The well known Navier-Coulomb theory based upon the maximum shear strength criterion predicts a linear behaviour. The classical Griffith's criterion based upon failure in tension in the rocks, predicts to some extent a non-linear behaviour. However, these classical theories, though simple in concept and in use, fail to correctly predict rock behaviour universally.

Realising the inadequacy of the basic theories in predicting intact rock strength, an empirical 'power law' was proposed by Murrell (1965). The original power law proposed is in the form.

$$\sigma_1 = \sigma_c + B \; (\sigma_3)^A$$

where σ_1 = major principal stress

σ_3 = minor principal stress

σ_c = Unconfined compressive strength

B and A are constants

In the normalised form the equations can be writen as

$$\sigma_1 / \sigma_c = 1 + K \; (\sigma_3 / \sigma_c)^A$$

where K is a constant

Franklin (1971) has listed a number of empirical yield criteria suggested by various researchers using the normalised form of Murrell's equation. Among the various yield criteria proposed, the criterion in terms of principal stresses as suggested by Bieniawaski (1974) has been widely used in practice in recent years. It is of the form

$$\sigma_1 / \sigma_c = 1 + B \; (\sigma_3 / \sigma_c)^\alpha$$

where α is the slope of the plot between ($\sigma_1 / \sigma_3 - 1$) Vs. σ_3 / σ_c on log-log scale and B is a material constant.

Bieniawaski suggested for a range of South African rocks that α is a constant equal to 0.75 for all rock types and B = 3 for siltstone and mudstone, 4 for sandstone, 4.5 for quartzite and 5 for norite.

Hoek and Brown (1980) proposed another empirical yield criterion based on the non-linear failure envelopes predicted by classical Griffith's theory. This criterion for intact rock and rock mass is expressed as

$$\sigma_1 = \sigma_3 + (m \sigma_c \sigma_3 + s \sigma_c)^{\frac{1}{2}}$$

where m and s are constants.

For intact rock s=1 and for completely broken rock s = 0. The range of variation of m is very wide and is believed to be dependant on rock type and quality

A semi-empirical strength criterion has been proposed by Rao (1984), Rao et. al. (1985). This has been developed on the basis of Mohr-Coulomb yield criterion and is applicable for both intact and jointed rocks. This strength criterion has been expressed as,

$$\sigma_d / \sigma_3 = B \; (\sigma_c / \sigma_3)^\alpha$$

(applicable for all values of $\sigma_3 > 0$)

where,

σ_d = deviator stress ($\sigma_1 - \sigma_3$)

B = a constant, which is a function of rock type and quality

α = slope of the plot between σ_d / σ_3 and σ_c / σ_3 on log-log scale.

A value of 0.8 has been proposed for α for all rock types and a mean value of parameter 'B' has been proposed for different rock types e.g. for Argillaceous rocks it is 2.0, for Arenaceous rocks it is 2.4 (2.2 for sandstones and 2.6 for quartzite) for chemical rocks it is 2.6 and for Igneous rocks it is 2.8.

In the present study, the applicability of Bieniawaski (1974), Hoek and Brown (1980) and Rao et. al. (1985) strength criteria has been evaluated and reported with the help of data generated from the experiments carried out.

4 RESULTS AND DISCUSSIONS

4.1 Physical properties
Average value of the physical properties of the Vindhyachal sandstone and Baraundha quartzite are presented in table 1.

Table 1 Physical properties of sandstone and quartzite

Property	Sandstone	Quartzite
Water absorption	2.95 %	0.727 %
Sp. Gravity	2.65	2.66
Density (a) dry	2.45 g/cm^3	2.78 g/cm^3
(b) Saturated	2.52 g/cm^3	2.80 g/cm^3
Porosity	7.36 %	2.02 %

4.2 Strength indices

The mean value of results of uniaxial compressive strength, Brazilian and point load strength test are presented in table 2. It is evident that quartzite exhibits higher uniaxial compressive strength than sandstone. The Brazilian and point load test results also indicate that strength is higher for quartzite.

As per Deere and Miller's (1966) classification the Baraundha quartizte can be classified as 'BM' and Vindhyachal sandstone as 'CM'. The values of strength indices i.e. uniaxial compressive strength, Brazilian strength and point load strength index for Vindhyachal sandstone and Baraundha quartizte also lie within the range of test data published by various researchers (e.g. Kulhawy 1975).

4.3 Stress-strain behaviour

The stress-strain variation of Vindhyachal sandstone and Baraundha quartzite is shown

Table 2 Strength index properties of sandstone and quartzite tested.

Property	Sanstone	Quartzite
UCS (kg/cm^2)	698.6	1034.5
Point load Strength index (kg/cm^2)	52.50	119.85
Brazilian Strength (kg/cm^2)	49.7	102.6
Deere & Miller (1966) classification	CM	BM

in fig 1 and 2. The values of tangent modulus E_t (determined at 50 % of peak strength) as obtained from uniaxial compression strength test and values of cohesion and friction angle as obtained from triaxial test data analysis for both the rocks are presented in table 3.

Table 3 Values of E_t, c and ϕ

Property	Sandstone	Quartzite
E_t x 10^5 (kg/cm^2)	1.49	4.19
c (kg/cm^2)	105.47	129.29
ϕ	36.3^0	31.7^0

4.4 Strength criteria for rocks

The strength and deformation behaviour of the sandstone and quartzite is non-linear under the applied stress combinations. Three strength criteria Viz., Bieniawaski (1974), Hoek and Brown (1980) and Rao et. al. (1985) have been considered for prediction and comparison with the experimentally observed values. The values of σ_1 has been obtained for σ_3 = 25, 50, 75 and 100 kg/cm^2 confining pressures. For the purpose of prediction, for various strength criteria, the values of constants have been choosen for sandstone and quartzite as suggested by the authors of the strength criteria. The general values

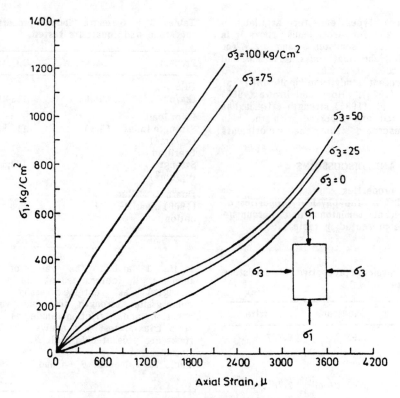

Fig. 1 Stress–strain variation plot – sandstone.

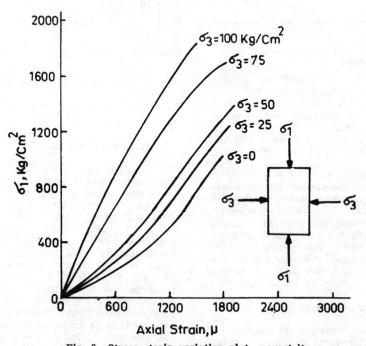

Fig. 2 Stress–strain variation plot – quartzite

suggested by Bieniawaski for sandstones is $\alpha = 0.75$ and $B = 4$ and for quartzite is $\alpha = 0.75$ and $B = 4.5$. In case of Hoek-Brown, the values suggested are $m = 15$ and $s = 1$ for both types of intact rocks. In case of Rao et. al. criterion the values suggested for sandstone are $\alpha = 0.8$ and $B = 2.2$ and for quartzite are $\alpha = 0.8$ and $B = 2.6$. Using the above mentioned very general values for prediction of σ_1 for various confining pressures, variation of σ_1 with σ_3 has been plotted (figures 3 and 4) for Vindhyachal sandstone and Baraundha quartzite. A comparison of experimental and predicted values clearly indicates that in general Hoek and Brown (1980) and Rao et. al. (1985) strength criteria predict intact rock behaviour very close to experimentally observed behaviour in both the cases. Bieniawaski (1974) strength criteria is observed to slighty over predict the peak stress values.

5 CONCLUSIONS

The present study attempts to characterise two types of rocks by studying their physical and engineering properties. Further applicability of various strength criteria proposed for rocks and rock masses have been verified for the two Indian rocks tested. Followi main conclusions are listed from the study carried out in the present work.

1. The sandstone and quartzite tested have typical properties of their group as reported in published literature.

2. The sandstone can be classified is 'CM' and the quartzite as 'BM' from Deere and Miller's (1966) classification charts.

3. The stress-strain variation in case of both the rocks is non-linear. The peak strength varies non-linearly with confining pressure .

4. The failure of rock is brittle is both the cases. The shear strength parameters for sandstone are $c = 105.47$ kg/cm^2 and ϕ 36.3° and for quertzite are $c = 129.29$ kg/cm^2 and $\phi = 31.7°$

5. A comparison of experimentally observed variation of σ_1 for various values of confining pressure with theoretically obtained variation using Bieniawaski (1974), Hoek and Brown (1980) and Rao et. al. (1985) strength criteria shows that in general all the three criteria can be very useful for strength prediction . However amongst the three, Hoek-Brown and Rao et. al. yield criteria predict value very close to actually obtaind ones. Especially Rao et. al. criteria prediction are very close at lower confining stress ranges

which are encountered more generally. The prediction at higher confining stress ranges is comparatively more non-linear and tends to slightly underpredict the peak stress values leading to slightly oversafe designs.

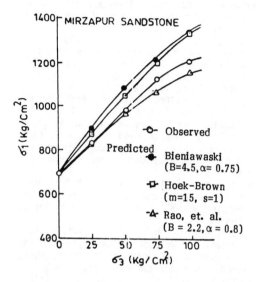

Fig. 3 Comparison of different strength criteria - sandstone.

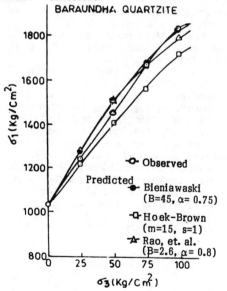

Fig. 4 Comparison of different strength criteria - quartzite.

REFERENCES

Bieniawaski, Z.T. 1974. Estimating the strength of rock materials. J.S. Afr. Inst. Min. Metall. Vol 74, No. 8 : 312-320

Deere, D.V. and Miller, R.P. 1966. Engineering classification and index properties for intact rock. Tech. report, A.F.W. lab. New Mexico.

Franklin, J.A. 1971. Triaxial strength of rock materials. Rock mechanics, Vol.3, 86-98.

Handin, J. and Haeger, R.V. 1958. Experiment deformation of sedimentery rocks under high confining pressure : tests at high temperature. Bull, Am. Soc. Petrol. Geol., Vol. 42 : 2894-2934.

Heard, H.C. 1960. Trassition from brittle fracture to ductile flow in Solenhofen limestone as a function of temperature, confining pressure, and intestitial fluid pressure. Rock deformation. Griggs and Handin, eds. Geol. Soc. Am. Memoir 79 : 193 - 226.

Hoek, E. and Brown, E.T. 1980. Underground excavations in rock. Institution of Mining and Metallurgy, London.

Hoffmann, H. 1958. Investigations into carbonic rocks under triaxial pressure for the purpose of rock stress computation, Proc. Int. Strata control Cong. Leipzig.

Kulhawy, F.H. 1975. Stress deformation properties of rock and rock discontinuties. Eng., Goel. Vol. 9, No. 4 : 327-350.

Murrell, S.A.F. 1965. The effect of triaxial stress systems on the strength of rock at atmospheric temperatures. Int. J. Rock Mech. Min. Sci. Vol. 3 : 11-43.

Rao. K.S. 1984. Strength and deformation behaviour of sandstones. Ph.D. thesis, I.I.T. Delhi, India.

Rao, K.S., Rao, G.V. and Ramamurthy, T. 1985. Rock mass strength from classification. Paper no. 3, Proc. Workshops on Engineering classification of rocks, CBIP, New Delhi, India : 27-50 .

Schwartz, A.E. 1964. Failure of rock in the triaxial shear test. Proc. of the 6th symp. on rock mech., Roll, Mo., 109 - 151.

Von Karman, 1911. The Festigskeitsversuche unter Auseitigem druck, Zeits. Verein Deutsch. Ing., Vol. 55 : 1749 - 1757 .

Waversik, W.R. and Brace, W.F. 1971. Post failure behaviour of a granite and diabase. Rock Mech., Vol. 3 : 61 - 85.

Correlation of slaking with the unconfined compressive strength of cement stabilized clayey admixtures

Corrélation du pourcentage de l'érosion avec la résistance à la compression simple des mixtures d'argile stabilisés avec le ciment

E. I. Stavridakis, K. A. Demiris & T. N. Hatzigogos
Department of Civil Engineering, Geotechnical Engineering Division, Aristotelian University, Thessaloniki, Greece

ABSTRACT:The existence of some difficult soils (non-durable) has motivated an experimental investigation of the correlation of slaking with the unconfined compressive strength of artificial soils stabilized with cement. The good correlation between the experimental results of the tested clayey admixtures, which consisted of less than 75% sand (grain size >74μm), indicate that an empirical prediction of slaking=(100-Id$_2$) (2) may be possible. This empirical relationship is defined by the upper (q$_u$=110Kg/cm², 100-Id$_2$=5%) and lower (q$_u$=20Kg/cm², 100-Id$_2$=100%) values of unconfined compressive strength and slaking. The empirical prediction valid between these limits.

RESUME: L'existence de quelques sols difficiles (non durables) donne lieu à une recherche de la correlation du pourcentage de l'érosion avec la resistance à la compression simple des sols artificiels stabilises avec le ciment. Le bonne correlation entre les resultats experimentals de ces mixtures d'argile qui contiennent de la sable en pourcentage moins de 75% (diamètre des grains >74μm), indique qu' une prévision empirique de l' érosion (100-Id$_2$) (2) est possible. La limite superieure de cette correlation empirique est (q$_u$=110Kg/cm², 100-Id$_2$=5%) et la limite inferieure (q$_u$=20Kg/cm², 100-Id$_2$=100%). La prévision empirique se base sur ces limites.

INTRODUCTION

Erosion problems may result when the dispersive properties of clays are not suitably accounted for the design of earth dams. These problems may become evident immediately upon impounding of water in the reservoir. Soil stabilization techniques may be used against such problems by improving the physical properties of the soil. Cement-Stabilization of soils has been succesfully used in facing or lining of highway embankments and drainage ditches to improve long-term resistance to mechanical stresses and particularly to reduce the risk of erosion. However, very litlle information regarding its erodibility (slaking) is available. In an effort to characterize the erodibility of compacted cement-stabilized soil, both slaking and unconfined compressive strength tests were carried out on artificial soils mixed with various amounts of cement (4%,8%,12%), in various percentages of compaction (90%,95%,100%) and in 3,7,14,28 days of curing (Table 1). Based on the results, a correlation was derived between the unconfined compressive strength and the slaking (100-Id$_2$).

2. DESCRIPTION OF MATERIALS

The tests were carried out on clayey admixtures which consisted mainly of two clayey soils, namely Bentonite and Kaolin. The behaviour of bentonite is characterized by Bell (1976),Croft (1967) and others as active and that of kaolinite as inert. The behaviour of weathered minerals such as "degraded" Illites, Chlorites, Vermiculites can be similar to that of montmorillonite and require large amounts of cement for stabilization. Kaolinite and illitic soils are largely inert, and generally develop satisfactory strengths with economical amounts of cement after short curing periods.

Commercially available kaolin and sand were used while bentonite was natural.

The basic index properties of kaolin, bentonite and sand are given in Table 2.

The qualitative characteristics are given by the X-rays diagrams in Fig. 1, 2

Table 1. Clay-sand mixtures designs.

Mixtures	Mixtures Conditions
20M12.5K67.5A	
20M50K30A	
20M80K	
5M25K70A	
12.5M25K62.5A	
25M25K50A	
30M25K45A	
100K	4%/95%/28
20M25K55A	
10M90K	
10M62.5K27.5A	
25M75K	
35M65K	
20M62.5K17.5A	
30M62.5K7.5A	
82.5K17.5A	
62.5K37.5A	
50K50A	

25M25K50A	
30M62.5K7.5A	
25M75K	
20M50K30A	12%/95%/28
20M25K55A	
35M35K30A	
45M25K30A	

35M35K30A	
20M25K55A	8%/95%/28
100K	

20M25K55A	
10M62.5K27.5A	4%/100%/28

20M25K55A	8%/100%/28

20M25K55A	12%/100%/28

20M25K55A	4%/90%/28

20M25K55A	8%/90%/28

20M25K55A	12%/90%/28
- - - - - - - - - - -	
20M25K55A	4%/100%/7

20M25K55A	8%/100%/7

20M25K55A	12%/100%/7

20M25K55A	4%/95%/7
10M62.5K27.5A	

20M25K55A	8%/95%/7

20M25K55A	12%/95%/7

20M25K55A	4%/90%/7

20M25K55A	8%/90%/7

20M25K55A	12%/90%/7
- - - - - - - - - - -	
20M25K55A	4%/95%/3
- - - - - - - - - - -	
20M25K55A	4%/95%/14

N.B. Where M=%Bentonite, K=%Kaolin,
A=%Sand
4%,8%,12%=%cement
100%,95%,90%=%compaction
3,7,14,28=days of curing

and 3.

The uniform gradation of medium to fine sand with Hazen coefficient 2.19<5 is shown in Fig. 4.

Table 2. Index properties of clays.

Soil property	Kaolin	Bentonite	Sand
Liquid limit (%)	34	111.50	—
Plastic limit (%)	29.61	42.19	—
Plasticity index (%)	4.39	69.31	—
Moisture content (%)	0.6	12.37	0.20
% finer than 74μm (%)	100	100	0.48
% Montmorillonite (%)	0	36	—

3. TESTING PROCEDURE

Two tests were performed for comparison of unconfined compressive strength with slaking:

a) The strength (q_u) was measured using a commercially available device named Versa Tester/Soil test Inc.

The dimensions of the cylindrical specimens tested were 35.5mmD by 71mmL. The rate of strain was 0.6604mm/min.

b) The slaking (100−Id_2) (2) was measured using the device developed by Franklin (Franklin and Chandra 1972).

The apparatus consists of a drum comprising a 2.00mm standard mesh, a trough to contain the drum with the slaking fluid (tap or distilled water,HCl,e.t.c.) and a motor drive capable of rotating the drum at a speed of 20 rpm (Fig. 5).

A representative "sample" comprising of eleven specimens of each admixture weighing 40–50gr each (total sample weight 450–550gr) was placed in the clean drum.

The dimensions of the cylindrical specimens tested in slake durability tests were 35.5mmD by 23.7mmL.

The testing procedure was as follows:

a. The sample was placed in the drum and dried for 6 hours to constant weight at a temperature of 105°C, in a ventilated oven. The weight (A) of the drum plus sample was recorded and the sample was immediately tested.

b. The lid was replaced, the drum mounted in the trough and coupled to the motor.

c. The trough was filled with slaking fluid (tap water, 20°C) to a level 20mm below the drum axis, and the drum was rotated at 20rpm for 10 min.

d. The drum was removed from the trough, the lid removed from the drum and the drum plus retained portion of the sample was dried to constant weight at 105°C. The weight (B) of the drum plus retained portion of the sample was recorded.

e. Steps (b–d) were repeated and the weight (C) of the drum plus retained por—

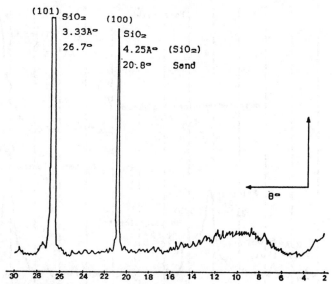

Fig. 1. X-ray diffraction traces of sand.

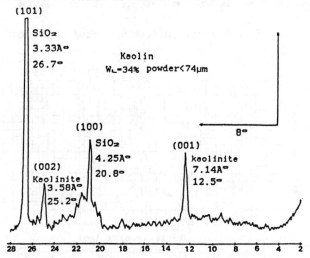

Fig. 2. X-ray diffraction traces of Kaolin.

tion of the sample was recorded.

f. The drum was brushed clean and its weight (D) was recorded.

g. The slake-durability index (second cycle) was calculated as the percentage ratio of final to initial dry sample weight as follows:

$$\text{slake-durability index } Id_2 = \left(\frac{C-D}{A-D}\right)\% \quad (1)$$

$$\text{slaking}\% = (100 - Id_2) \quad (2).$$

4. SAMPLE PREPARATION

The samples were prepared at the optimum moisture contents and maximum dry densities (Proctor test according to B.S.I. 1377: 1972 B2 test 12 and B.S.I. 1924:1975).

The cylindrical samples were prepared according to A.S.T.M. 1632–63/1979.

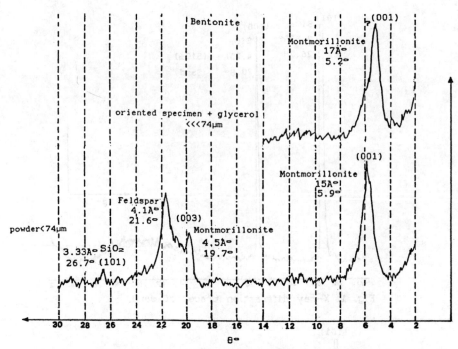

Fig. 3. X-ray diffraction traces of Bentonite.

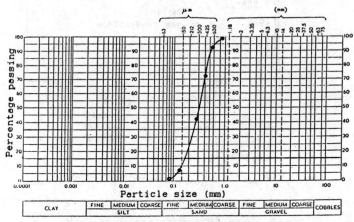

Fig. 4. Particle size distribution of sand.

The test programme involved 141 (47x3) specimens for 141 unconfined compression tests and 517 (47x11) specimens for 47 slake durability index tests.

The clay–sand–cement mixtures were cured at approximately 95.5% relative humidity and temperature 21°C for 3,7,14,28 days.

5. RANGE OF PARAMETERS

The range of proportions of clays in the admixtures and the percentages of cement, compaction and the curing time are shown in Table 1.

The limits in pecentages of cement were selected to give noticeable changes in strength.

The Portland Cement Association reported that 85% of soils need ⪡14% cement to give noticeable changes in strength. As much as 5% addition of cement has been shown by Bell (1976) to be necessary if noticeable changes are to be brought in the strength of montmorillonitic clays, whereas at least 2% was required to modify the kaolinitic clays.

The range of compaction (100%,95%,90%) refer to the degree of compaction obtained in field projects.

The curing time was related to the minimum acceptable compressive strength, for base coarse material in road construction, approximately 17.5 Kg/cm2 after 7 days of curing.

6. PRESENTATION OF RESULTS

Durability under atmospheric conditions of wetting and drying is an aspect of cement-stabilized soils, behaviour that has been neglected in favour of other properties such as strength. However it is an important feature of many commonly encountered engineering problems.

The slake–durability tests may be used

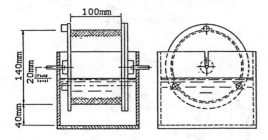

Fig. 5. Critical dimensions of Slake–
Durability test equipment.

to predict the potential deterioration due to climatic wetting and drying. The durability tests are used in combination with other index tests such as the unconfined compression test, as an aid for classification and quality control of soils stabilized by cement.

The results summarized in Table 3 and plotted in Fig. 6 show a negative power relationship y=ax$^\beta$ between the decay in unconfined compressive strength with increasing slaking, as the percentage of cement (12%,8%,4%) decreases. The reason is that in a soil–cement mixture any calcium ions liberated from the cement during hydration and hydrolysis react rapidly with the silicates and aluminates from the soil minerals to form insoluble products. The letter coat the grains to form a skeletal structure of considerable strength. Therefore, the amount of liberated calcium ions are related to the amount of cement, strength and slaking.

The results summarised in Table 4 and

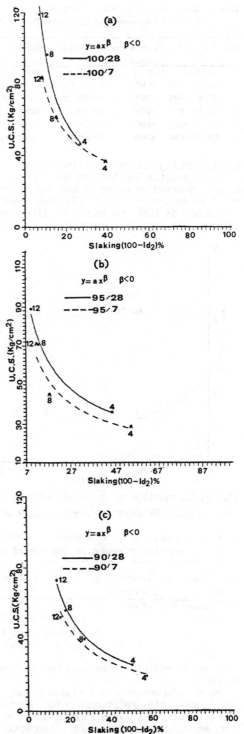

Fig. 6. Correlation of q$_u$ with slaking
in 4%,8%,12% cement Table 3.

Table 3. Correlation of q_u with slaking in 100%,95%,90% compaction for 4%,8%,12% cement.

A/A	a	β	Correlation coefficient	Correlation	Mixture Conditions
1	526.10	−0.73	−0.994	$y=ax^a$	100%/28days
2	303.10	−0.56	−0.997	$y=q_u(Kg/cm^2)$	95%/28days
3	615.47	−0.81	−0.998	$x=\%(100-Id_2)$	90%/28days
4	263.60	−0.54	−0.999	for 4%,8%,12%	100%/7days
5	251.80	−0.55	−0.962	cement	95%/7days
6	482.53	−0.78	−0.985	20M25K55A	90%/7days

plotted in Fig.7 show a similar to the previous relatioship between the unconfined compressive strength and slaking as the time of curing (28,14,7,3) decreases. The reason is that the amount of liberated calcium ions from the cement during hydra-

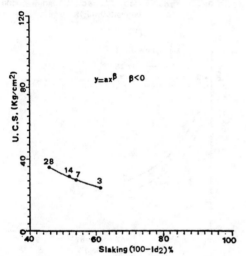

Fig. 7. Correlation of q_u with slaking in 3,7,14,28 days of curing Table 4.

Table 4. Correlation of q_u with slaking in 3,7,14,28 days for 4% cement and 95% compaction.

A/A	a	β	Correlation coefficient	Correlation	Mixture Conditions
1	5212.42	−1.3	−0.996	$y=ax^a$ $y=q_u(Kg/cm^2)$ $x=\%(100-Id_2)$ 20M25K55A	4%/95% 3,7,14,28 days of curing

tion and hydrolysis is related to the time of curing.

The results summarised in Table 5 and in Fig.8 show the relationship between the unconfined compressive strength with slaking as the percentage of compaction (100%,95%,90%) decreases. That is because poor compaction, results in corresponding

loose structure of the soil–cement mixture which in turn affect the unconfined compressive and slaking.

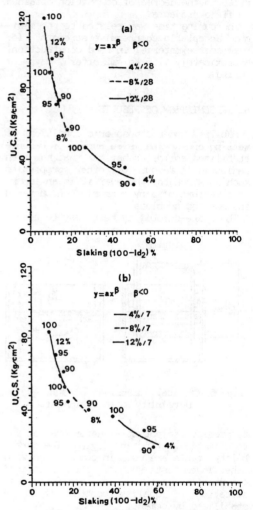

Fig. 8. Correlation of q_u with slaking in 90%,95%,100% compaction Table 5.

Table 5. Correlation of q_u with slaking in 4%,8%,12% cement for 100%,95%,90% compaction.

A/A	a	β	Correlation coefficient	Correlation	Mixture Conditions
1	612.5	−0.78	−0.90	$y=ax^a$	4(100,95,90)28
2	900.3	−0.96	−0.94	$y=q_u(Kg/cm^2)$	8(100,95,90)28
3	445.39	−0.69	−0.97	$x=\%(100-Id_2)$	12(100,95,90)28
4	4882.33	−1.33	−0.89		4(100,95,90)7
5	288.60	−0.60	−0.84		8(100,95,90)7
6	390.20	−0.71	−0.98	20M25K55A	12(100,95,90)7

Finally a general curve (Fig.9 and Table 6) was derived involving all 47 clayey admixtures prepared with 4%,8%,12% cement, 90%,95%,100% compaction and 3,7,14,28 days curing time. The curve was fitted with a function of the type y=ax$^\beta$ where

(2) for the type of soil mixtures tested, which contained less than 75% sand (grain size >74μm). The range of validity of this relationship is defined by the upper Limit [q_u=110kg/cm^2, (100−Id_2)%=5%] where according to Franklin the material behaves

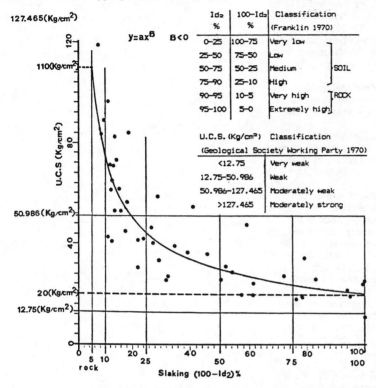

Id_2 %	100−Id_2 %	Classification (Franklin 1970)	
0–25	100–75	Very low	
25–50	75–50	Low	
50–75	50–25	Medium	SOIL
75–90	25–10	High	
90–95	10–5	Very high	ROCK
95–100	5–0	Extremely high	

U.C.S. (Kg/cm^2)	Classification (Geological Society Working Party 1970)
<12.75	Very weak
12.75–50.986	Weak
50.986–127.465	Moderately weak
>127.465	Moderately strong

Fig. 9. An empirical prediction and classification of q_u and slaking (100−Id_2).

y=q_u(Kg/cm^2) and x=%(100−Id_2) slaking (2).

In Winterkorn and Hsai 1975, a maximum is obtained, of U.C.S.(qu), in soil cement mixtures after 7 days of curing when they contain 75% material with grain size larger than 50μm. For this reason the test programme dealt with a group of admixtures which contained <75% sand (grain size >74μm).

7. CONCLUSIONS

The experimental results indicate good correlation between unconfined compressive strength and slaking in all cases.

The general curve (Fig.9) gives an empirical prediction of slaking (100−Id_2)

Table 6. Empirical relationship for prediction of slaking.

A/A	a	β	Correlation coefficient	Correlation	Mixture Conditions
1	277.64	−0.57	−0.86	y=ax^a	4%,8%,12% cement
				y=q_u(Kg/cm^2)	100%,95%,90% compaction
				x=%(100−Id_2)	7,28 days of curing

as a rock and a lower Limit [q_u=20Kg/cm^2, (100−Id_2)%=100%] where the material is beginning to behave as durable.

The Fig.9 could be used for the classification of soil-cement mixtures and can be applied for the design of relevant engineering projects.

Further research on natural soils stabi-

479

lized by cement is currently underway.

REFERENCES

Akpokodjie, E.G.(1986). A method of reducing the cement content of two stabilized Niger delta soils. Q.J.Eng. Geology. London, Vol.19, pp.359–363.

Akpokodjie, E.G.(1985). The stabilization
of some arid zone soils with cement and
lime. Q.J.Eng. Geol. London, Vol.18,
pp.173–180.

A.S.T.M. D.1632–63. Standard Method of
Making and curing soil–cement compression and flexure test specimens in the
laboratory.

Bell, F.G.(1976). The influence of the
Mineral Contents of clays on their stabilization by cement. Bulletin of the
association of Engineering Geologists
Vol.XIII No4.

B.S.I.(1975).1924. Methods of test for
stabilized soils. The Institution.

B.S.I.(1975).1377. Methods of test for
soils for Civil Engineering purposes.

Croft, J.B.(1967). The influence of mineralogical composition on cement stabilization. Geotechnique,17,119–135.

Croft, J.B.(1968). The problem in predicting the suitability of soils for cemen-
titious stabilization. Engineering Geology. Elsevier Publising Company,
Amsterdam, Printed in the Netherlands.

Franklin, J.A. & Chandra, R.(1972). The
slake durability test. Int. J. Rock
Mech. Min.Sc. Vol.9.

Ingles, D.G. & Metcalf, J.B.(1972). Soil
stabilization. Butterworths, Sidney,
Australia.

Oswell, J.M. and Joshi, R.C.(1986).
Development of an Erosion Test for
soil cement. American Society for Testing and Materials.

Rizkallah, V.(1989). Stabilization of difficult soils in developing countries.
Proceedings of the Twelfth International
Conference on Soil Mechanics and Foundation Engineering.

Winterkorn, F.H. & Hsai-Yang F. (1975).
Foundation Engineering Handbook. Van
Nostrand Reinhold·Company.

An approach to the determination of geostress using the Kaiser effect
Recherche en détermination de géostress au moyen de l'effet de Kaiser

Tianbin Li & Lansheng Wang
Chengdu College of Geology, Chengdu, People's Republic of China

ABSTRACT: The determination of geostress by means of the Kaiser effect of rock allows the 3-dimensional stress of rock masses to be understood. On the basis of acoustic emission tests and geostress calculations for basalt samples from the damsite of a hydroelectric project in China, the acoustical properties of the rock and the geological factors affecting the Kaiser effect of the rock are analyzed. It is concluded that the maximum geostress previously or currently acting on the rock during modern geological history, can be determined according to the Kaiser effect of rocks. Combining geological analysis with back analysis of a stress field, We can determine the maximum historical or modern geostress and also the environmental conditions under which the historical geostress existed.

RESUME: La détermination du géostress au moyen de l'effet de Kaiser de la roche permet de comprendre l'efforts à trois dimenssions des masses de la roche. Sur la base d'essai de l'émission acoustique et le calcul du géostress d'échantillons basatique partant de la base de barrage à la centrale hydro-électrique en Chine, la propriété acoustique de la roche et les facteurs géologique affectant l'effet de Kaiser de la roche sont analysée. Il est conclu que le géostress maximum auparavant et couramment agissant la roche au cours de la histoire géologique est capable d'être déterminé d'après l'essai d'effet de Kaiser de la roche. En combinant l'analyse géologique avec l'analyse du revers d'un champ d'efforts, on peut déterminer maximum du géostress historique et moderne et aussi les conditions environnantes où le géostress historique aexiste.

1 INTRODUCTION

The determination of geostress by means of the Kaiser effect of rock is a new, economical laboratory technique which allows the 3-dimentional stress of rockmasses to be understood. At present, many scholars are interested in this technique, a lot of reseach have been taken such as the independence of the Kaiser effect, the elimination of the noise from sample ends and so on (Hayashi 1979, Yoshikawa & Mogi 1981, Lin 1982, Lu 1987, etc.). The debate continues, however, about whether the geostress determined by the Kaiser effect represents paleo-conditions or modern conditions. According to the Kaiser effect principles, the test results should be the maximum geostress that acts on a rock during geological history, but parts of them conform to the present geostress of rock (Huang 1988, Yao 1989).

In order to approach the subject mentioned above, Tongjiezi damsite in China was chosen to test geostress by way of the Kaiser effect, where the evolution history of the geostress fields has been studied in detail. It has been shown that in this area the geologic body had undergone not only the tectonic movement, but also the epigenetic and superficial geological procecces, subjected to E-W and NW directional crushing stress fields during the paleo-geohistory and the deformation movement called time-dependent deformation from Pliocene to early Pleistocene epoch, even when the river downcutting. The epigenetic time-dependent deformation structures are the main fracured phenomina in the subhorizontal strata of Permian basalt of the damsite(Fig.1). They were originated as the result of regional erosion of overburden and consequent release of stored residual stress acting in NWW direction of 282° ~ 300° (Li 1988). The epigenetic discontinuities had further displacement during or after the river downcutting.

On the basis of acoustic emission tests and geostress determination for a basalt from Tongjiezi damsite, the acoustical properties of the rock and the geological factors that affect the Kaiser effect as well as the memory features of the Kaiser effect are analyzed and discussed in this paper.

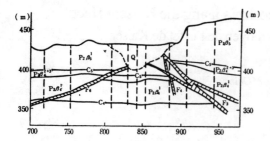

Fig.1 A geologic section of Tongjiezi damsite, where F_3, F_6, C_4, C_5 are some of the epigenetic discontinuities, $P_2\beta_4^1$, $P_2\beta_4^{2+3}$, $P_2\beta_5^1$ are Permian basalts

2 TEST AND RESULTS

2.1 Test equipments

The block diagram of test equipments is showed in Fig.2. The loading instrument is the hydraulic stress test machine. The YJ-16 strain apparatus is used for receiving the loading signals

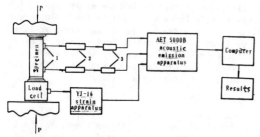

Fig.2 The block diagram of test equipments
   1---AE transducer 2---Pre-amplifier
   3---Wave filter

from load cell and converting them into the simulation voltage. Finally, the loading which acts on specimen can be recorded automatically in the form of voltage by the acoustic emission apparatus. The AET 5000B acoustic emission apparatus is a key block in the tests. At first, AE signals are received by micro-transducer with 300 KHz natural frequencies and amplified to 1000 times in the pre-amplifier(60db). After being filtered, they are amplified again to 50 times in the sub-amplifier, at least, recorded, processed and analyzed by AE apparatus.

2.2 Test methods

A oriented basaltic rock from Tongjiezi dam

foundation under 17 meters was sampled in lab in accordance with six directions as shown in Fig.3, 5 specimens each direction. The specimen was a rectangular parallelopiped 20 mm×20 mm ×70 mm, whose unparalell degree of ends is less than 0.02 mm.

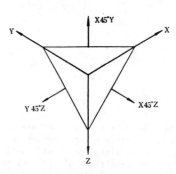

Fig.3 The directions of the specimens for AE test. In our test, the X-, Y-direction represent separately east and north.

In order to eliminate the allochthonous noise and the untrue AE signals caused by ends effect, two pieces of the teflon film and a rubber gasket were put at the ends of specimens. These two kinds of materials can absorb the acoustic waves well and almost don't produce AE, and the teflon film has very small friction coefficient. Therefore, they can effectly reduce the ends friction and absorb the allochthonous noise. In addition, the AET 5000B acoustic emission apparatus has a fine function of linear positioning, so the acoustic emission events in the middle part of specimen were only recorded and counted for above-mentioned purpose.

After all of this procedure, the uniaxial compressive deformation experiments of the specimens were conducted. In the meantime, the acoustic emission test system received and recorded the AE signals of the specimens, and the data processing system outputed the various experimental curves.

2.3 Test results

A series of test results were obtained after the acoustic emission tests of all specimens. The typical AE activity curves of the specimens are showed in Fig.4. The experimental results show that the Kaiser effect is very clear for the basalt. Without a definite value of load, no AE events appear, and if the load comes up to or exceeds a critical value, AE events will come out and increase quickly, but the critical value is not the same for the specimens of different directions(Fig.4)

482

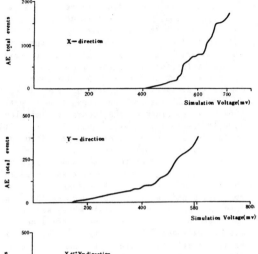

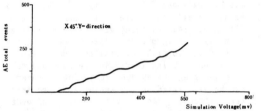

Fig.4 Some results of AE tests of the basalt,
AE total events---simulation voltage
curves

By the relation between simulation voltage
and stress, the stress value that corresponds
with the point at which AE events increase
markedly was determined for every test curves.
In other words, the normal stresses for 6 directions of the basalt were obtained (Table 1).

Table 1 The average normal stress of each
direction after tests

Directions	XX	YY	ZZ	X45Y	Y45Z	Z45X
Magnitude (MPa)	29.82	12.86	10.32	8.84	7.63	10.76

3 DETERMINATION OF GEOSTRESS

According to the independance of the Kaiser
effect and the theory of elasticity mechanics,
the genenal state of stresses for measuring
point can be calculated so long as the normal
stresses of 6-directions, σ_x, σ_y, σ_z, $\sigma_{x45^\circ y}$,
$\sigma_{y45^\circ z}$ and $\sigma_{z45^\circ x}$ can be determined.

Suppose the normal stress is σ_k, then the
relation between σ_k and stress components is
expressed.

$$\sigma_k = A_{k1}\sigma_x + A_{k2}\sigma_y + A_{k3}\sigma_z + A_{k4}\tau_{xy} + A_{k5}\tau_{yz} + A_{k6}\tau_{zx} \quad (1)$$

in which,
$A_{k1} = l_{ij}^2$, $A_{k2} = m_{ij}^2$, $A_{k3} = n_{ij}^2$, $A_{k4} = 2l_{ij}\,m_{ij}$,
$A_{k5} = 2m_{ij}\,n_{ij}$, $A_{k6} = 2n_{ij}\,l_{ij}$, $K = i \times j$,
i is the numbers of specimen directions,
j is the numbers of specimens for
each direction,
l, m, n are the direction cosine.
changing (1) into matrix formulation

$$[\sigma] = [A] * [\sigma^*] \quad (2)$$

where
$[\sigma]$ is a column matrix composed of σ_k,
$[A]$ is a coefficient matrix with $K \times 6$ orders,
$[\sigma^*]$ is a column matrix for stress components.
By using the least square method, the optimum
stress components can be obtained form(2). Then
the magnitude of the principal stresses σ_1,
σ_2, σ_3 is determined through solving the
following third equation.

$$\sigma^3 - J_1 \sigma^2 + J_2 \sigma - J_3 = 0 \quad (3)$$

Where J_1, J_2 and J_3 represent separately the
three invariants of stress components.
The direction of the principal stresses is
also decided by (4) and (5) as follows.

$$\begin{bmatrix} \sigma_x - \sigma_i & \tau_{xy} & \tau_{zx} \\ \tau_{xy} & \sigma_y - \sigma_i & \tau_{yz} \\ \tau_{zx} & \tau_{yz} & \sigma_z - \sigma_i \end{bmatrix} \begin{bmatrix} l_i \\ m_i \\ n_i \end{bmatrix} = 0 \quad (4)$$

$$l_i^2 + m_i^2 + n_i^2 = 1 \quad (5)$$

where i = 1, 2, 3.

For the convenience of geological analysis,
the direction of the principal stresses is expressed by way of their projection direction
β_i on horizontal plane and dip angle α_i
(Fig.5). The conversion formula (6) is given.

$$\begin{cases} \alpha_i = \text{arc } \sin(n_i) \\ \beta_i = \text{arc } \sin(m_i / \cos \alpha_i) \end{cases} \quad (6)$$

in which the positive directions of α_i and
β_i are showed in figure 5.
The above-mentioned method was programmed to
process the test results. The 3-dimensional geostress obtained at the measuring point were
showed in table 2. According to the oriented
position of the sample and table 2, it is concluded that the direction of maximum principal
stress at the measuring point is NWW (294.4°),
coinciding with that of stored residual stress
which resulted in forming the epigenetic time-
dependent deformation structures.

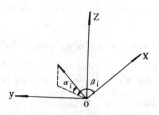

Fig.5　A way of expressing principal stress
　　　　directions

Table 2　The principal stresses at the
　　　　measuring point

| Stress | Magnitude(MPa) | Direction(°) | |
		β	α
$\sigma 1$	38.02	-24.4	-13.9
$\sigma 2$	15.25	81.1	-47.3
$\sigma 3$	-0.28	53.9	39.3

The back analysis of a stress field by Finite Element Methed (FEM) was adopted to estimate the environmental conditions under which the geostress existed. A sectional model for the back analysis including the point of determining geostress was established along the direction of maximum principal stress. During the calculation, the magnitude of structural stress at the left and right boundarys of the model as well as the position and shape of the top boundary was repeatedly changed until the calculating stress fitted in well with the testing geostress. The results show that in the case of the present topography the stresses can not be fitted at any rate, only if 30 MPa structural stress and 318.3 m overburden (not 17m) above the measuring point of geostress. Therefore, the geostress determined by the Kaiser effect at Tongjiezi damsite represents historical condition under which there was 318.3m of overburden above the measuring point. this condition corresponds with that of the formation of the epigenetic time-dependent deformation structures.

It can be seen from above-mentioned results that the geostress level determined in our tests is the stored residual stress leading to form the epigenetic time-dependent deformation structures during modern geologic times.

4 DISCUSSION

The principle of the Kaiser effect shows that the geostress determined by means of this method should represent the maximum one to which

rocks have been subjected during geological history. Obviously, our test result is in contradiction with this principle.

In fact, the acoustic emission signals of rock materials are a kind of elastic wave caused by the failure of the micro-cracks or the micro-defects in rocks. As the reloading stress in a specimen equals or exceeds the fracture toughness for the propagation of previous micro-cracks, the acoustic emission signals come out in a great quantity. Thus, it is the features of micro-cracks that claim the results of the Kaiser effect tests.

The occurence of micro-cracks and their development and change were greatly complicated in the course of geologic processes, which were controlled and affected by many factors. In Tongjiezi area, the rockmasses which are located near surface at present were buried deeply during paleogeologic history. So they could hardly develop the micro-cracks under the condition of high confining pressure; even if they actually took form , the micro-cracks could be healed due to high temperature and pressure and so on, losing memory to the relevant geostress. During the epigenetic geologic process, the basalts rised to the subsurface and fractured strongly due to the unloading of overburden and the release of stored residual stress, the relevant micro-cracks being kept in rocks. During the superficial geologic process, the residual stress was further released, and only the epigenetic discontinuities had further displacement. It is thus clear that the microcracks in basalt mainly resulted from the epigenetic time-dependent deformation. This is why the result of the Kaiser effect tests represents the residual geostress. In some areas not intensely subjected to the epigenetic geologic process, the residual geostress in rockmasses have almost been remained, and in river areas they have been probably increased during the superficial geologic process, owing to river downcutting deeply, such as the situation in the damsites of Ertan and Laxiwa hydroelectric projects in China. For this reson, the microcracks in rockmasses mainly resulted from the present geostress. This is why that some results of the Kaiser effect tests conformed to the present geostress measured in-situ, for example, in Ertan and Laxiwa damsites.

It can be seen, therefore, that the micro-cracks produced during modern geological history could be Kept in rocks because of relatively constant environmental conditions. the authors infer that the geostress determined by the Kaiser effect represents commonly the maximum grestress that previously or currently acts on a rock during modern geologic histony, rather than all geologic period.

To Sum up, some viewpoints on the Kaiser effect of rock are as follows:

1. It is necessary to know the evolution of

whole geologic processes in research area in the course of the determination of geostress by means of the Kasser effect.

2. In general, the geostress determined by this method represents the maximum one previously or currently acting on a rock during modern geological history.

3. It is very useful to combine geological analysis with back analysis of a stress field with FEM for distinguishing the historical or present geostress.

REFERENCES

Huang, R.Q.(1988). A new view for the determination of geostress by the Kaiser effect. Proc. 3rd national symp. of Eng. Geol., 1110 -1117. Chengdu, China.

Li, T. (1988) The mechanism of the epigenetic time-dependent deformation and failure in rockmasses. A paper for M.S., Chengdu college of geology.

Lin Q.X.(1982). The measurement of stresses in rocks by the acoustic emission of rocks. Underground engineering, 40-46.

Lu X.Y. (1987). A preliminary study on the relation between the Kaiser effect and stress components. Joural of Chongqing college of architecture and engineering, 83-91.

Yoshikawa, S. & Mogi, K.(1981). A new method for estimating of the crust stress from cored rock samples: Laboratory study in the case of uniaxial compression. Tectonophysics, Vol.74, 323-339.

Zhang, J.(1987). Measuring the in-situ stresses in rockmasses by the Kaiser effect of AE. A paper for B.S., Chengdu university of sci. and tech.

Les caractéristiques géotechniques des terrains résiduels
The geotechnical characteristics of the residual soils

L.E.Torres
Escuela Politécnica Nacional, Quito, Ecuador

RESUME: A partir du profil Géologique - Géotechnique des terrains résiduels de la zone du fleuve "Luis" au sud de l'Equateur, ont été identifiées six couches bien differenciées qui vont de la roche saine à la couche de sol residuelle évoluée. Les essais de laboratoire sur sols et roches nous ont permis d'établir les paramètres physiques et mécaniques sur lesquels on a pu determiner les caractéristiques de chacune des couches des terrains d'altération telles que: la plasticité des matériaux diminue avec la profondeur, tandis que la taille des grains augmente, ce qui a rélation avec le degré d'altération du sol. Les paramètres de resistance du sol (φ,c) en contraintes totales comme effectives sont en général bas. La compressibilité des matériaux diminue avec l'augmentation de l'altération du sol. Dans le cas de la roche altérée, s'il y a des valeurs differentes, elles ne sont pas très marquées pour nous permettre de différencier clairement d'une couche à l'autre, ce que l'on a obtenu uniquement avec la géophysique.

ABSTRACT: Here is presented the Geophysical - Geological profile of residual soils of the Rio Luis zone of the southern Ecuador. It has been differenciated 6 layers, well characterized, from fresh rock until mature residual soil. Lab tests of soil and rocks have permited to stablish the mechanical and physical parameters, in order to characterized each layer, such as: plasticity which diminished with depth, whereas grain size increased due to meteorization stages of material, resistency (φ, c) which in general are low total and effective values, compresibity which decreased with alteration of soils. In case of altered rocks, although there a range of values, these are not typical to permite a clear cut off among the layers, only obtained with geophysical complementary work.

1 GENERALITES

La plupart des études qui ont eté realisées dans le pays sur les sols résiduels font référence, en général à la partie superficielle. Dans le cadre des études de faisabilité du projet hydroeléctrique de moyenne capacité, denommé "Rio Luis", pour les travaux souterrains fondations spéciales et grands terrassements, qui furent réalisées dans leur grande totalité par l'auteur de ce rapport, on a disposé de moyens suffisants pour mener un étude approfondie des terrains résiduels en dessous de la couche superficielle, avec l'aide de la géophysique, pour identifier toutes les couches d'altération, à partir de la roche saine, jusqu'au niveau du sol résiduel évolué.

Les essais de mécanique de roches n'ont pas eté exécutés pendant cette étude, mais plustard par des essais au laboratoire de l'Ecole Polytéchnique National, à titre de recherche, faisant partie d'un thème de travail de fin d'études pour l'obtention du titre d'ingénieur en Géotechnique de l'élève M. Byron HEREDIA sous la direction de l'auteur de ce rapport.

La temperature moyenne annuelle de ce bassin varie en forme linéaire selon l'altitude oscillant entre une valeur absolue maximun de 35°C et une valeur minimum de 11°C. L'humidité relative est de 81%. La précipitation moyenne en 1983 fut de 1530m.m. L'altitude est comprise entre 900 m. et 1400 m. au dessus du niveau de la mer.

2 CONTENU DES ETUDES EFFECTUEES

En plus du levé géologique régional détaillé, des travaux superficiels avec une densité acceptable ont eté effectués: puits, tranchées et sondages mécaniques sur le terrain, sismique de refraction, résistivité électrique, etc. En même temps des échantillons alterés, ainsi que des échantillons inaltérées ont été prélevés, pour des essais de Laboratoire.

3 LE CONTEXTE GEOLOGIQUE

La zone d'étude est couverte en majeur partie par des roches de la "Formation Celica" faisant partie d'un complexe volcanique constitué principalement par des laves et andesites porphyritiques avec une grande variété de textures telles que les tufs et brèches consolidées de semblable composition. L'étude comprend également les terrains résiduels qui sont le fruit de l'altération des laves et andesites porphyritiques qui se trouvent prophylitisées dans leur majeur partie.

Du point de vue hydrographique, la zone est controlée structuralement par la nature des roches trouvées. Dans les zones où la roche se trouve immédiatement en dessous des terrains superficiels, la topographie est du type falaise, mais, lorsque l'épaisseur du sol résiduel augmente, les talus deviennent plus douces, mais dans ce cas là, se présentent des mouvements de terrain et des phénomènes d'érosion.

La Figure 1 présente un profil transversal Géologique - Géophysique sur lequel on peut voir comme l'épaisseur des terrains résiduels augmente en direction de la partie haute de la montagne et disparaît pratiquement vers le lit des rivières.

Les laves montrent une texture aphanitique avec phénocristaux d'hornblende de couleur gris-verdâtre. Les roches de type andesite-porphyritique par contre, sont constituées par quartz, plagioclase et hornblende, de texture porphyritique avec une matrice granulaire fine et prophylitisée.

Du point de vue structural on a pu déterminer trois systèmes principaux de discontinuités, bien définis sur le terrain, tant sur le plan macro que sur le plan microstructural.

4 LE PROFIL D'ALTERATION

Bien que les talus du chemin, ainsi que les puits ont permis la description en détaille des niveaux supérieures d'altération, ce sont les études de géophysique et les sondages mécaniques qui ont montré de façon compléte la variation du profil transversal et longitudinal d'altération de la zone en question (voir Fig. No. 1).

Trois couches principales présentes dans le profil d'altération ont été parfaitement définies, c'est à dire la couche rocheuse, la "saprolitique" et celle du sol résiduel proprement dit. Les caractéristiques de chacune de ces couches sont définies dans la partie suivante.

4.1 La couche de roche

Formée par des laves et andesites porfiritiques, constitue le substratum rocheux de la zone et par conséquence la "roche mère", d'où proviennent les sols résiduels. Ces couches ont suivi un procesus d'altération chimique intensif qui ont donné naissances aux différentes couches du sol résiduel. Cette couche a été divisée en:

Sous-couche A: qui correspond à la roche saine qui affleure dans certains et peu d'endroits, qui fut detectée lors de sondages et par les profils géophysiques, dont les valeurs de résistivité (ρ) et de vitesse sismique (V) sont les suivantes:
(ρ): infinite
V = plus grande que 4700 m/s.

Sous-couche B: Roche avec fissures espacées et altération précaire incluant l'altération minéralogique comme chloritisation et propilitisation. Sa couleur varie du gris-verdâtre à vert clair. Le carbonate de calcium rempli les fissures de la roche. Les valeurs de résistivité et vitesse sismique sont:
ρ = Infinite V = 4400 - 4700 m/s.

Sous-couche C: Roche propilitisée avec une grande densité de fissures d'oxydation moderée à fortement oxydée à partir des plans de fractures.
ρ = 1100 - 1300 Ω/m; V = 3000-4050 m/s.

Sous-couche D: Roche peu consistant, fortement fissurée, altérée et oxydée de couleur vert-jaunâtre.
Présence d'oxydes et argile dans les plans des fractures.
ρ = 500 - 1300 Ω/m;
V = 2000 - 3000 m/s.

4.2 La couche saprolitique (E):

Constituée par les terrains de transition entre la roche saine et le sol présentant les caractéristiques prédominantes de ce dernier. Elle est constituée fondamentalement par des matériaux mous (sol saprolitique) comme matrice, avec des blocs de roches décomposées par méteorization sphéroidale. On note la décomposition chimique presque totale des minéraux ainsi que la texture de la roche originale, avec la présence de kaolin, argiles, grains de feldesphate et quartz. La couleur génèrale est jaune-marron

ρ = 65 - 90 Ω/m; V = 1630 - 2120 m/s.

4.3 La couche de sol résiduel (F):

Constituée par minéraux d'argile et oxydes hydratés feraluminiques d'origine secon - daire donnant origine à un sol mou de couleur rouge-brun-jaunâtre.

Ce sol garde sa texture originelle dans la partie inférieure, tandis que dans la partie supérieure le sol est un sol résiduel terminé (mur).

ρ = 230 - 600 Ω/m   V = 620 - 1320 m/s

5. CARACTERISTIQUES GEOTECHNIQUES

5.1 Essais sur le terrain

En plus des profils sismiques et des sondages électriques, avec les valeurs donnés ci-dessus, ont été effectué des essais de pénétration standard (S.P.T.) et des essais de perméabilité du type Lugeon nonnant les résultats suivants.

Tableau 1. Resultats des essais de pénétration S.P.T.

Couche	SPT (n)
Sol résiduel	7 - 13
Saprolite	13 - 40

Tableau 2. Résultats des essais de perméabilité

Couche	Perméabilité (Unités Lugeon)
D	9
B et C	1

5.2 Essais de Laboratoire

Les échantillons de sols proviennent des excavations à ciel ouvert (remaniés et intacts(, ainsi que des sondages à rotation (SPT et shelby). Les échantillons de roche ont été prélevés des forages avec le carrotier à double paroi.

Le carrotage du sol résiduel mou n'a pas présenté des difficultés lorsqu'il a été executé dans la partie supérieur, cependant la difficulté a augmenté avec la profondeur. Par contre, le sol du type saprolite n'a pas permis un carrotage continu à cause de la presence de cailloux, blocs et des zones de caolins plus ou moins consolidé, donc un échantillon continu n'a pu être prélevé ni pour le sol ni pour la roche.

Granulométrie. Le pourcentage de fins (qui passent le tamis No. 200) atteint les 90% en surface et diminue avec la profondeur.

Plasticité. On peut observer que la plasticité augmente avec le degré d'altération du sol avec une variation général des MH jusqu'à ML, c'est-à-dire que les sols sont plus plastiques à la surface.

La Fig. No. 2 montre la carte de plasticité des sols résiduels d'après Casagrande;

Poids spécifique. On a trouvé une valeur moyenne de 2.78 entre les valeurs extrêmes qui vont de 2,5 à 3,1 sans relation apparente avec les couches de méteorization.

Résistance au cisaillement direct. Les valeurs de la cohesion (C') obtenues à partir des essais de cisaillement direct sur les échantillons de sol résiduel varient de 0.05 Kg/cm² à 0.4 Kg/cm², tandis que varie entre 32.6° et 26.9°. Dans le sol saprolitique par contre on a c' = 1.5 Kg/cm² et $\emptyset$ = 30° (voir Fig. No. 3).

Essai triaxial C.U. La Fig. No. 4 nous donne les résultats des essais triaxiaux C.U. éffectués sur les échantillons de sol résiduel terminé (mur), lesquels dû a leur homogeneité nous ont permis d'obtenir des résultats parfaits. Les éprouvettes de ces échantillons montrent une rupture plastique, typique des argiles. Les résultats sont donnés dans le tableau suivant:

Tableau 3. Résultats des essais triaxiaux C.U. pour le sol résiduel terminé (mur)

Ech. No.	Prof. (m)	φ' (o)	φ (o)	c' (Kg/cm²)	c (Kg/cm²)
MC-71	5	36.5	11	0	0.6
MTP-6	4	24.0	26	0	0
MTP-5	5	35	12	0	0.3

489

SHG	8	38	16	0	1.0
MTP-2	3.5	26	12.3	0	0.6

Sur la Figure No. 4 a été tracé la droite de Coulomb. Les essais traxiaux C.U., effectués sur les échantillons de sol résiduel, qui gardent les traits de texture de la roche originelle, sont montrés sur la Figure No. 5.

Les résultats montrent une grande héterogeneité, dûe à la nature des matériaux même, comme on peut le voir dans le tableau suivant.

Tableau 4. Résultats des essais triaxiaux C.U. pour le sol résiduel qui garde les traits texturales.

Echant. No.	Prof (m)	φ' (o)	φ (o)	c' (Kg/cm²)	c (Kg/cm²)
55-8	8	36.5	-	0	-
58-SH3	8.5	31	-	0	-
MI-1	1.0	28	9.6	0.15	0.35
57-SH16	22	39	28.5	0.25	0.6

Compressibilité. A partir de quelques essais de consolidation qui ont été effec - tués, on peut noter que la variation des caractéristiques de consolidation va en fonction de la nature du sol (saprolitique, résiduel avec traits de texture antérieure, et résiduel terminé (mur). Dans le tableau NO.5 on peut voir que le sol résiduel terminé (mur) est le plus compressible, et le moins compressible correspond au "saprolitique".

Tableau 5. Caractéristiques de la compressibilité des différentes couches

Sol	Echant. No.	Prof (m)	Eo	Cc	Cv (m²/s)
Fb	MTP-5	3	1.68	0.47	10E-7
	ST-SH4	4.5	1.61	0.3	10E-6
Fa	57-SH27	33	1.22	0.17	10E-7
	59-SH1	6.0	0.88	0.17	10E-7
E	MI-1	1.5	0.78	0.19	10E-8

Fb= Sol résiduel terminé (mur);
Fa= Sol résiduel de textureantérieure;
E = "saprolite".

5.3 Essai de laboratoire sur roche

Pour arriver à connaître les propriétés physiques et mécaniques des différentes couches définies par la géophysique, l'élève Byron Heredia a effectué quelques essais de mécanique de roches sous la

direction de l'auteur du présent travail, cela faisant partie de la thèse pour obtenir le titre d'ingénieur en géotechnique.

Les propriétés physiques de la roche. Le tableau suivant résume les resultats de quelques essais de mécanique de la roche:

Tableau 6. Quelques propriétés physiques de la roche.

Echantillon	Couche	Absorption W %	Porosité n %
1	B	0.11	0.32
2	B	0.17	0.47
3	C	0.29	0.78
4	C	0.29	0.80
5	D	0.40	1.10
6	D	0.30	0.78
7	E	0.27	0.73
8	E	0.21	0.60

Vitesse sonique. La vitesse longitudinale de l'onde a été mesurée sur plusieurs échantillons pris de différentes couches d'altération; on a obtenu des valeurs qui varient entre 4800 m/s et 5520 m/s, ceux-ci ne gardent aucune relation avec la couche dont ils proviennent apparemment, mais on peut admettre une valeur moyenne pour la roche sans fissures de 5400 m/s, tandis que pour la roche fissurée et cimentée, la valeur correspond à 4800 m/s.

Résistance à la charge ponctuelle.
On a effectué un total de 37 essais sur les échantillons prélevés des différentes couches B, C, et D, ainsi comme dans les blocs de la couche E. Les résultats obtenus ne sont pas en accord avec ceux qu'on attendait pour chaque couche; celà est dû au niveau d'altération de chaque échantillon essayé et aux effets d'échelle, exception fait des échantillons de la couche E qui sont clairement différents des autres. La roche saine ou légèrement alterée et sans microfissures a des valeurs moyennes plus hautes que Is (50) = 0.54 MPa; tandis que la roche moyennement alterée, propilitisée et avec microfissures remplies de ciment trouve sa résistance diminuée et donne des valeurs moyennes de Is (50) = 0.44 MPa, finalement pour la roche très alterée et propilitisée Is (50) = 0.34 Mpa.

Resistance à la compression simple.
Dans ce cas il n'y a pas, non plus, une corrélation directe entre la couche et la valeur de la résistance à la compression simple comme nous pouvons le voir sur le

tableau No. 7.

Tableau 7. Résistance à la compression simple de la roche

Echantillon	Couche	Résistance à la compression simple (MPa)
S - 1	E	80.26
S6- 1	D	137.93
S6- 2	D	55.09
S3- 1	C	69.14
S3- 2	C	65.82
S6- 3	C	58.29
S3- 1	B	86.86
S3- 2	B	83.71

Conclusions Pratiques:

- On peut noter un certaine relation entre le degré d'altération et la plasticité des matériaux a capacité d'absobtion (WL) des liquides augment dans les sols résiduels terminés (murs) (voir Fig. 6).

- La granulométrie des matériaux montre de façon plus objective le degré de méteorisation des materiaux, car elle permet de connaître le degré de décomposition des cristaux et la formation de nouveaux minéraux (voir la Fig. 6).

- Le poids spécifique montre une diminution relative selon le degré d'altération, tendance qui n'est pas commune à ce type de materiaux.

Tableau 8. Rélation entre les propriétés physiques et le degré d'altération (meteorisation)

Sol	WL	% fins	S.U.C.S	γ T/m³	(météorisation)
Résiduel Terminé (mur) Fb	60	80	MH	1.5-2.0	debut de lateritisation + lixiviation
Résiduel de texture reçue	40 à 60	50 à 80	ML MH	1.5-2.0	Kaolinisation + lixiviation
Saprolitique	NP à 40	50	NP à ML	1.7_2.2	Kaolinisation

- La classification de Bieniawsky coincide parfaitement avec les différentes couches de sols résiduels (voir Fig. 7).

- L'indice de qualité du massif IQ, fut définie en tenant compte de la valeur de la matrice rocheuse (5400 m/s) de chaque couche, tel comme on peut le voir au tableau 9.

Tableau 9. Valeurs de l'indice IQ pour chaque couche

Couche	Vitesse sismique m/s	Vitesse sonique m/s	IQ	
E1	1630-2120	5400	0.3 -	0.39
D	2000-3000	5400	0.37 -	0.55
C	3000-4050	5400	0.55 -	0.75
B	4400-4700	5400	0.81 -	0.87

- Les valeurs obtenus à partir des essais de pénétration SPT (N coups) sont des indicateurs des couches stratigraphiques.

- Les essais triaxiaux du type U.U. et C.U. ont été réalisés sous les normes internationales; ils ont donné des résultats de paramètres en corelation. Les paramètres de résistance au cisaillement en conditions de contraintes efectives (c' et φ) ne montrent pas une dépendance avec les donées de texture, donc ils sont plus confiables. Par contre, ceux obtenus à partir de contraintes totales (c et φ) sont affectés par les donées de la texture originelle, donc il faut utiliser les valeurs minimums.

Le tableau No. 10 montre les valeurs de cet obtenues à partir des essais de résistance au cisaillement.

- Pour la compressibilité, on a déjà vu que la couche de sol résiduel terminé (mur) montre des caractéristiques plus compressibles que les autres, résultat principal de sa grande homogeneité.

- Bien que, du à l'effet d'échelle, il n'existe pas une différence claire entre les résultats obtenus des essais de mécanique de roches sur les échantillons prelevés des différentes couches; la géophysique nous donne un excellent contraste entre une couche et l'autre, de telle façon que nous pouvons obtenir le profil d'altération complet (voir Fig. 1).

REFERENCES

Annold, W. Essai de description géotechnique des caractéristiques qui determinent les comportement des sols résiduels, tropicaux en vue de leur classification, in VV.AA. ISSMFE, Brasilia, 1985.

De Sola, O. Weathering in the Guri Area - Venezuela, in VV.AA., First Conference on Geomechanics in tropical lateritic and saprolitic soils, ISSMFE, Vol. II, Brasilia, 1985.

Fernandez, A.I. Problemas encontrados en los taludes en suelo residual existentes en la carretera Duarte de la Republica Dominicana, in VV.AA., Revista Tecniberia, Espana, 1984.

Fredlung, D.G.; Rahardjo, H; Theorethical context for understanding unsatured Residual Soils Behavior, in VV.AA., First Conference on Geomechanics in tropical lateritic and saprolitic soils, ISSMFE, Vol. II, Brasilia, 1985.

Heredia, B. Contribucion al Conocimiento de las caracteristicas geomecanicas de los horizontes de meteorizacion del material existe te en la zona del Proyecto Rio Luis (EPN, Quito).

Karunakaran, C;; Sinharoy, S; Laterité profile development linked with policyclic geomorphic in South Kerala, in VV. AA., &st. Seminar on Laterisation Processes, India, 1981.

Nogami, J. Mechanical and hidraulic properties of tropical soils - particularly as related to their structure and mineral components, topic 2.4.2, Committee on tropical soils of the ISSMFE, Brasilian Society for soil Mechanics, Brasil, 1985.

Nogami, J.S. Caracterization, identifi - cation and classification of tropical soils, Progress Report, Theme 1, First Conference on Geomechanics in tropical lateritic and saprolitic soils, ISSMFE, Vol. II, Brasilia, 1985.

Smith, D.M. Geology, Geotechnology and investigation methodology for lateritic soils, near worsley, in VV.AA., First Int. Conference on Geomechanics in tropical lateritic and saprolitic soils. Technical sessions (ISSMFE), Brasilia, 1985.

Torres L.E., INECEL-CEC. Estudio de Factibilidad avanzada del Proyecto Centrales Hidroeléctricas de Mediana Capacidad Grupo Cuatro-Aprovechamiento Luis Chorrera, Tomo No. 3, Vol. No. 1, Quito, 1985.

TABLEAU Nº 10 RESULTATS DES ESSAIS DE RESISTANCE AU CISSAILLEMENT

COUCHE	ECHANT	ESSAI	CONDITIONES TOTALES		CONDITIONES EFFECTIVES	
			C(KPa)	φ' (O)	C'(KPa)	φ (O)
Fb	MR 11 a	C.D.			4.0	32.6
	MR 11 b	C.D.			2.6	27.0
	MR22a	C.D.			0.5	32.62
	SH-12	U.U.	21	0		
	SH-20	U.U.	6	18		
	MC-7'	C.U.	6	11	0	36.5
	MPT-6	C.U.	0	26	0	34.0
	MPT-5	C.U.	3	12	0	35.0
	SH-6	C.U.	10	16	0	38.0
	MTP2b	C.U.			0	26.0
Fa	MI-2	C.D.			2.4	26.2
	MI-5	C.S.			0.75	-
	SH-8	C.U.	-	-	0	36.5
	SH-3	C.U.	-	-	0	31
	M 1-2	C.U.	3.5	9.6	1.5	28
	SH-16	C.U.	6	28.5	2.5	38
E	MI-1	C.D.			15	30

C.D.= Cissaillement Direct ; C.S. = Compresion Simple

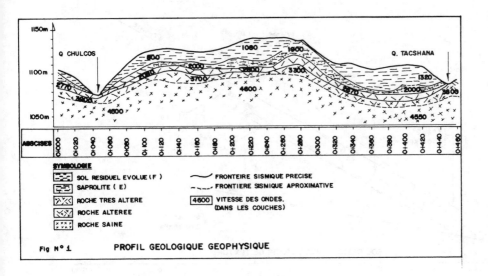

Fig Nº 1   PROFIL GEOLOGIQUE GEOPHYSIQUE

SYMBOLOGIE

SOL RESIDUEL EVOLUE (F)
SAPROLITE (E)
ROCHE TRES ALTERE
ROCHE ALTEREE
ROCHE SAINE

FRONTIERE SISMIQUE PRECISE
FRONTIERE SISMIQUE APROXIMATIVE
4600 VITESSE DES ONDES.
(DANS LES COUCHES)

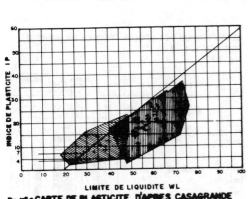

Fig Nº2 CARTE DE PLASTICITE D'APRES CASAGRANDE

Fig Nº3   RESULTATS OBTENUES A PARTIR
DES ESSAIS DE CISAILLEMENT DIRECT

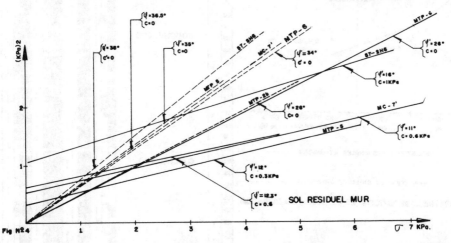

Fig Nº4

DROITES DE COULOMB A LA RUPTURE EN CONTRAINTES TOTAUX ET EFECTIVES OBTENUES DES ESSAIS TRIAXIAUX CU.

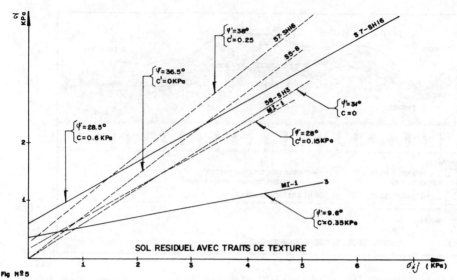

σ₁ KPa

$\psi'=38°$
$C'=0.25$

S7-SH16

S7-SH16

$\psi'=36.5°$
$C'=0KPa$

S5-6

58-SH3

MI-1

$\psi'=31°$
$C=0$

$\psi'=28.5°$
$C=0.6$ KPa

$\psi'=28°$
$C'=0.15KPa$

MI-1 3

$\psi'=9.6°$
$C=0.35KPa$

SOL RESIDUEL AVEC TRAITS DE TEXTURE

2

1

1   2   3   4   5   6

σ'_{xj} (KPa)

Fig N°5

DROITES DE COULOMB A LA RUPTURE EN CONTRAINTES TOTAUX ET EFECTIVES OBTENVES DES ESSAIS TRIAXIAUX C.U.

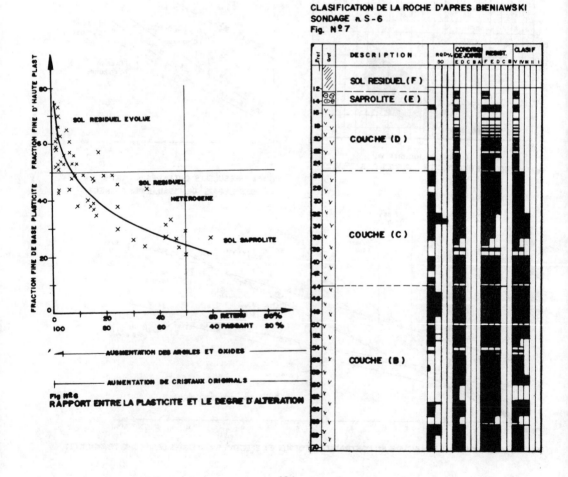

CLASIFICATION DE LA ROCHE D'APRES BIENIAWSKI
SONDAGE n S-6
Fig. N°7

FRACTION FINE D' HAUTE PLAST
FRACTION FINE DE BASE PLASTICITE

SOL RESIDUEL EVOLUE

SOL RESIDUEL
HETEROGENE

SOL SAPROLITE

80

60

40

20

0

0   20   40   60 RETENU   80%
100   80   60   40 PASSANT   20 %

AUGMENTATION DES ARGILES ET OXIDES

AUMENTATION DE CRISTAUX ORIGINALS

Fig N°6

RAPPORT ENTRE LA PLASTICITE ET LE DEGRE D'ALTERATION

494

Influence of microscopic structure on the abrasivity of rock as determined by the pin-on-disc test

L'influence de la texture microscopique sur l'abrasivité des roches, déterminée avec l'essai 'pin-on-disc'

P.N.W.Verhoef, H.J.van den Bold & Th.W.M.Vermeer
Delft University of Technology, Faculty of Mining and Petroleum Engineering, Section of Engineering Geology, Delft, Netherlands

ABSTRACT: Pin-on-disc experiments have been performed on artificial glass concrete, to determine the influence of angularity of the glass grains on abrasive wear. The measured weight loss of the metal pins was compared with the wear coëfficient "F" developed by Schimazek. It was found that the "round" and "angular" concrete had different physical and mechanical properties and were not completely identical but for the roundness of the glass grains. This drew attention to the influence of microscopic structure on abrasivity. Also data on natural rock indicate that the factors contained in Schimazek's coëfficient "F" (mineral hardness, grainsize and strength) are not all the factors involved, although the F-value has a linear relationship with abrasive wear.

RESUME: Des expériments "pin-on-disc" ont été executé sur béton avec perles de verre comme abrasives pour étudier l'influence de l'angularité sur l'usure abrasive. Malgré que les résultats ont indiqué que les propriétés physiques et méchaniques de béton angular et sphérique étaient incomparables, on a trouvé une influence de la texture microscopique sur l'abrasivité. Une même influence peut être exister dans des roches naturels, comme indiquer par les résultats des essais "pin-on-disc" de Schimazek. Le coëfficient d'usure "F" de Schimazek contiens la dureté des mineraux, la granulométrie et la résistance en traction. Les expériments ont indiqué que "F" a une relation linéaire avec l'usure abrasive, mais pas tous les facteurs importants, comme la texture microscopique, sont y compris.

1 INTRODUCTION

It is difficult to quantify the factors causing abrasive wear of rock-cutting tools. Wear is dependent on the "wearing system" of the rock-cutting process, which is described by the mechanical properties and geometry of the tool, the petrographical-mechanical properties of the rock and the physical-chemical properties of the engineering environment (Verhoef, 1989).

One of the important contributing factors is the nature of the rock. Various attempts have been made to assess those properties of the rock which are important in the determination of its abrasivity. It has been suggested that strength, deformability, failure mechanism, mineralogical content (especially quartz content) and microscopic structure (grain size, angularity of the grains, cementation factor) have influence on the potential abrasivity (or "abrasive capacity") of the rock. (See Verhoef, op. cit., for a review).

Most assessments of abrasivity of rock are undertaken using some sort of abrasivity test (Schimazek and Knatz, 1970; Suana and Peters, 1982; West, 1989), or by determining the rock hardness by one or another method thought to be indicative of abrasivity (Abrasion Test (ASTM C 241-85, 1988); Brown, 1981), or by using a combination of index tests to evaluate abrasive capacity (Roxborough and Phillips, 1974; Tarkoy, 1973). As is known from tribological engineering, whatever tests or assessments are being used, results should be calibrated against the wear experienced during the full-scale cutting process. The usefulness of a certain test or assessment procedure must be evaluated in rock-cutting practice, comparing the assessment with the actual wear experienced.

One of the methods which has proved to be of value (at least in the coal-mining industry) was developed by Schimazek and co-workers (Schimazek and Knatz, 1970, 1976; Paschen, 1980). Schimazek found that the wear determined from a pin-on-disc test on rock could be linearly related to the product of some petrographic and strength parameters. Schimazek performed tests on artificial rock (concrete with a variation in quartz content and grain size) and found a high correlation between the weight-loss of the steel pins and a wear factor "F", where F is defined as follows:

$$F = 10*Q*D_q*\sigma_t \quad [N/m] \quad , \text{ where}$$

Q = quartz content (in vol.%)
D_q = diameter of quartz grains (in mm)
σ_t = Brazilian tensile strength (in MPa)

Although no statistical information is given in the original papers (Schimazek and Knatz, op. cit.), the fit of the data for the artificial samples seems particularly good and holds at least for the quartz grain sizes used (average Ø 0.02 - 0.45 mm). In order to gain insight into the effect of angularity on abrasive capacity and to gain an insight into the factors influencing pin-on-disc type wear experiments, experiments on artificial concrete samples have been repeated by the authors. Instead of quartz, glass was chosen as the abrading mineral, because both perfectly round and angular grains could be used to construct the samples. The experimental set-up, however, differed substantially from the original tests performed by Schimazek.

2 PIN-ON-DISC EXPERIMENTS

2.1 Preparation of the glass-concrete samples

The glass-concrete samples were constructed by adding a certain volume of glass grains to Portland-B cement. Spherical glass pearls could be obtained commercially. The angular grains were prepared by crushing glass pearls. The roundness of the latter was determined using Powers' roundness classification, which classifies roundness (or angularity) of grains into 6 classes (Powers, 1953). A value for the roundness has been obtained by classifying at least 100 grains in these classes which have been numbered, in steps of 20, from zero (very angular) to 100 (well-rounded). The average of this classification is defined as the "roundness number".

To maintain a proper concrete also at higher volume percentages of glass, a mixture of size fractions had to be chosen which could assure an optimum bond between

glass grains and cement. A gap-graded aggregate was used to obtain a minimum amount of void space between the grains. Two size gradings were used to obtain two different average grain sizes and two types of aggregate were used; round glass pearls and crushed glass pearls. The four types of glass aggregate are given in Table 1.

Eight concrete cores were constructed using each type of glass aggregate with a glass volume percentage increasing from 10% to 80%. The cement mixtures were poured into PVC tubes of a diameter of 46 mm and a length of 300 mm. After at least 28 days of hardening the cores were sawn-up into 5 discs of 30 - 40 mm length and cores of 80 - 90 mm length. This material was used for the wear tests and the determination of the physical and mechanical properties of the cores. For each concrete type one determination of the unconfined compressive strength, porosity and density was made, and 5 determinations of Brazilian tensile strength (Table 2).

Both unconfined compressive- and tensile strength increased with increasing glass content up to a maximum value at about 40 - 50 percent by volume, above which the strength dropped (Fig. 1 and 2). This behaviour was also found by Schimazek with quartz concrete and is clearly related to the cement-glass ratio for at high glass percentages the binding capacity of the cement diminished. It is interesting to note that the angular glass concrete had higher compressive strengths but lower tensile strengths than the glass-pearl concrete specimens. The porosity values (Table 2) indicate the cause for this behaviour for they are much higher for the angular concrete samples. The bond surfaces and thus the grain-cement adhesion should be higher for the glass spheres. This is because the glass sphere cement mixes had a higher workability, giving a higher contact area with the cement. In compression the higher frictional resistance of the angular grains contributed to the higher compressive strength of the angular glass concrete.

2.2 Apparatus and Experimental programme

The experimental set-up differed considerably from that used by Schimazek and co-workers (op. cit.). Schimazek used 10 mm diameter rod-shaped pins with a 90° conical point, but flattened to 0.3 mm diameter. These pins (Steel 50; tensile strength 706 MPa) were placed on a rock disc, horizontally positioned on a rotating table. A weight of 4.5 kg was placed on the pin. The pin moved outward from the centre, and the

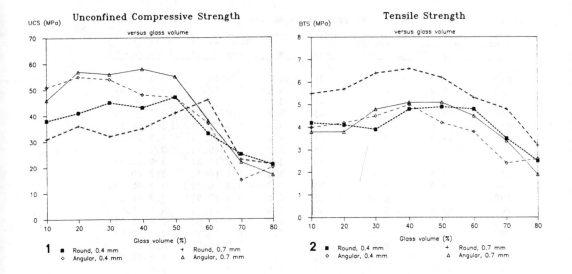

Table 1. Types glass aggregate used for the concrete

Sample nr. [1]	Smallest fraction (39 vol.%: mm)	Largest fraction (61 vol.%: mm)	Average Diameter (mm)	Roundness Number
E	0.055 - 0.095	0.490 - 0.700	0.392	100
F	0.110 - 0.180	0.850 - 1.230	0.690	100
I	0.055 - 0.095	0.490 - 0.700	0.392	6.4
II	0.110 - 0.180	0.850 - 1.230	0.690	7.6

[1] E,F glass "pearls"; I,II broken glass "pearls"

resulting Archimedal spiral path had a length of 16 m. The rotating table revolved at 25 rpm. The disadvantage of a conical point is the increasing contact area which developes with continuing wear, thus giving decreasing contact pressure at constant normal load. Another drawback is the variation in velocity under the pin, due to the constant rotation velocity of the turning table.

The new experiments were performed on a lathe. The 46 mm diameter specimens of about 40 mm length were centred in the clamping device of the lathe. The pin could be placed in a specially devised holder, which ensured that the flat-end of the core was at right angles to the axis of the lathe rotation. A constant load could be applied to the pin via a dead-weight construction. The pin used was made of a soft steel (Steel 37, tensile strength 360 MPa) and had a square (3.95 x 3.95 mm) prismatic shape. The flat end of the concrete cylinder was sawn using a sawing device which could be placed in the chisel holder of the lathe. This was repeated before each experiment so that more than one test could

be done on one concrete cylinder. The pin was prepared for each test by carefully square polishing the end face using a diamond grindstone. Before each test the pin was weighed with an accuracy of ± 0.1 mg. The pin was positioned square on the concrete flat surface, 9.00 mm away from the centre of the core cylinder. The desired normal pressure was developed by placing a weight on the loading device. The pin was moved outward from the centre of the core with a particular translation velocity. Since the lathe had a constant rotating velocity, during translation from the inside to the outside the linear velocity of the core under the pin increased. This, together with the limited diameter of the sample, restricted the amount of translation and the preferred situation of sliding of the pin only on a fresh concrete surface could not be met. Movement was restricted between 9.00 and 18.00 mm away from the centre of the core. The linear velocity given in the text and tables is calculated at 13.5 mm away from the centre of the core. The variation in linear velocity with respect to the average velocity at 13.5 mm

Table 2. Physical and mechanical properties of the glass concrete cores.

Sample nr.[1]	porosity %	density Mg/m^3	UCS Mpa	BTS MPa	Wear at 145 rpm (mg)[2]	Wear at 275 rpm (mg)[2]
E10	12.3	2.57	38	4.2	1.6 (0.4)	2.1 (0.8)
E20	9.0	2.68	41	4.1	2.4 (0.6)	1.8 (0.4)
E30	10.5	2.35	45	3.9	3.1 (0.4)	2.6 (0.7)
E40	12.4	2.23	43	4.8	5.0 (0.4)	3.1 (0.7)
E50	13.2	2.36	47	4.9	5.3 (0.5)	3.5 (0.8)
E60	11.3	2.79	33	4.8	4.8 (0.5)	4.2 (0.5)
E70	9.7	2.77	25	3.5	4.0 (0.2)	3.7 (0.6)
E80	8.0	2.78	21	2.5	2.8 (0.3)	3.4 (0.5)
F10	9.9	2.37	31	5.5	1.9 (0.7)	1.9 (0.7)
F20	9.2	2.39	36	5.7	3.3 (0.7)	3.3 (0.7)
F30	10.5	2.41	32	6.4	4.8 (0.5)	3.7 (0.5)
F40	13.4	2.55	35	6.6	6.8 (1.0)	4.6 (1.0)
F50	12.6	2.45	41	6.2	7.8 (0.8)	5.6 (1.0)
F60	13.1	2.57	46	5.3	8.8 (0.4)	5.6 (0.9)
F70	10.3	2.79	23	4.8	9.0 (0.8)	6.0 (1.2)
F80	11.0	2.66	21	3.2	6.4 (0.5)	6.4 (0.5)
I10	32.0	1.73	51	4.0	0.8	0.6
I20	28.3	1.88	55	4.2	1.4	1.1
I30	23.4	1.98	54	4.5	2.5	1.7
I40	16.0	1.97	48	5.0	3.4	1.7
I50	17.8	2.07	47	4.2	4.6	3.7
I60	23.1	3.83	37	3.8	3.6	--
I70	26.8	2.44	15	2.4	--	--
I80	28.2	1.89	20	2.6	--	--
II10	33.2	1.72	46	3.8	0.6	0.6
II20	27.8	1.87	57	3.8	1.5	1.1
II30	25.2	1.98	56	4.8	3.6	1.7
II40	21.4	2.09	58	5.1	6.3	3.2
II50	18.6	2.16	55	5.1	7.4	3.9
II60	15.4	2.22	38	4.5	5.3	5.2
II70	20.1	2.05	22	3.4	--	--
II80	22.9	2.07	17	1.9	--	--

[1] Number following sample type (Table 1) gives volume % of glass
[2] Experiments carried out under a contact pressure of 2.5 MPa, standard deviation is given between brackets.

is ± 33 % under these conditions. To maintain a constant wear-path length the product of the rotation velocity and translation time ($\omega.t$) was kept constant for each experiment, to obtain a total wear path length of about 35.5 m. After each test the pin was weighed. Each abrasion test was repeated 5 times on each core. The following variables were investigated:

- grain size of particles
- volume percentage of glass particles
- roundness of particles
- velocity of pin against the abrading surface
- applied normal pressure of the pin against the sample

With regard to the influence of normal pressure and velocity on the abrasive wear it was found that, when increasing the contact pressure, the wear of the pin increased linearly, and the wear of the sample exponentially (Fig. 3). At high contact pressures (about 1 - 2 times the tensile strength of the concrete) irregular wear processes or failure of the specimen occurred. Thus for most of the experiments a contact pressure of 2.5 MPa was chosen. When increasing the rotation velocity an exponential decrease of wear occurred. At velocities above 0.5 - 1 m/s a polishing effect occurred due to the adhesion of metal particles from the pin onto the concrete surface. At lower velocities a stable wear situation was present. Most experiments were carried out with two velocities; 145 rpm (average linear velocity 0.205 m/s, variation from 0.137 - 0.273 m/s) and 275 rpm (average linear velocity 0.389 m/s, variation from 0.259 - 0.419 m/s).

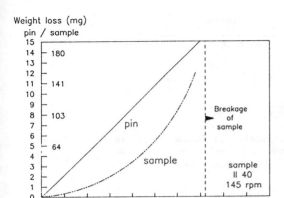

Weight loss (mg)

3

2.3 Calculation of the F-value

To be able to compare the results of the pin-on-disc test with the results of Schimazek and Paschen (op. cit.), the F-value for the concrete specimens had to be calculated. To do this, the "hardness" value of glass had to be determined using the hardness scale of Rosiwal. Schimazek (pers. comm. 1989) used this hardness scale, simply because this gave better results than other hardness values such as Vickers Hardness. A comparison of Rosiwal and Mohs hardness is given in Table 3. The Rosiwal hardness for the glass used (Mohs H = 5.5) was 15, while the Rosiwal hardness for the carbonate cement was 3. The concrete specimens consisted of glass, cement, and pore space, and in the F-value the porosity had to be considered because this does not contribute to the abrasive wear. For the grain size the average grain size of the glass spheres was chosen and for the cement a grain size of 0.01 mm was used. The wear factor (F) was thus determined as follows:

$$F = 10*((X*15/100*D_g) + ((1-X) * (3/100*0.01)) * (100-n)/100*\sigma_t \quad [N/m]$$

where X = glass content (%)
D_g = average diameter glass (mm)
n = porosity (%)
σ_t = tensile strength (MPa)

F-values determined this way can be compared with the F-values determined on rock by Schimazek and Paschen, using the Rosiwal hardness for minerals to obtain an "equivalent quartz hardness" (Schimazek and Knatz, 1970).

2.4 Results of the pin-on-disc tests

Schimazek & Knatz (1970) compared the results of their pin-on-disc test with tensile strength, equivalent quartz hardness and grain size and came to the conclusion that, although no linear correlation could be found with either tensile strength or hardness, the product of the above factors gave a linear correlation with the weight-loss of the steel pins. Fig. 4 - 7 show the relationship between Schimazek's F-factor and the wear due to angular and round glass concrete (two grain sizes) using two velocities. The tests, although done with quite a different experimental set-up than Schimazek's, showed a significant linear correlation between the F-value and the weight-loss of the pin. Examing figures 4 - 7 it is clear that at lower rotation velocities a higher abrasive wear is experienced, the correlation line does not pass through the origin, and the angular glass concrete seems to give lower values of wear compared with round glass concrete.

The last two observations must be examined further, because they are related. Firstly it is clear from Table 2, examining the porosity and density data of the round (E+F) and angular (I+II) glass concrete, that the properties of these materials differ not only in grain angularity, but also in other ways. Corresponding glass concretes with similar glass volumes have not identical physical and mechanical properties (Fig. 1 and 2). The F-value corrects this somewhat by the incorporation of tensile strength and porosity into the equation, but probably the test results are also influenced by other factors, not incorporated in the F factor.

The good linear distribution of the data points along the regression lines of Figures 4 - 7, shows that the regression lines can be used to predict wear-loss under exactly identical conditions of testing, that is testing at the same velocity, with the same contact pressure etc. The regression coëfficient of the lines may be dependent on the type of wear processes actually occurring during the pin-on-disc test.

2.5 Discussion

The first conclusion that may be obtained from the above results is that the F-value, as suggested by Schimazek and also with the modified test procedure, is linearly related to abrasive wear. Furthermore this linear relation seems also dependent on other factors not included in the F-value

499

Table 3. Comparison of Mohs, Rosiwal and Vickers Hardness

mineral[1]	talc	gypsum		calc.	fluor.	apat.		orth.	quartz	topas	corundum
Mohs[2]	1	2		3	4	5		6	7	8	9
Rosiwal[3]	0.3	1.25	2.0	4.5	5.0	6.5	18	37	120	175	1000
	0.25	1.0	1.7	3.8	4.2	5.4	15	31	100	146	833
Vickers[4]	20	50		125	130	550		750	1000	1850	2300

[1] calc.=calcite, fluor.=fluorite, apat.=apatite, orth.=orthoclase
[2] relative scale of mineral scratch hardness
[3] reciprocal of volume loss of grinding test scaled to corundum=1000 (Rosiwal, 1916); the second row gives the hardness relative to quartz=100
[4] indentation hardness [kg/mm²], large variation in reported values (Uetz, 1986)

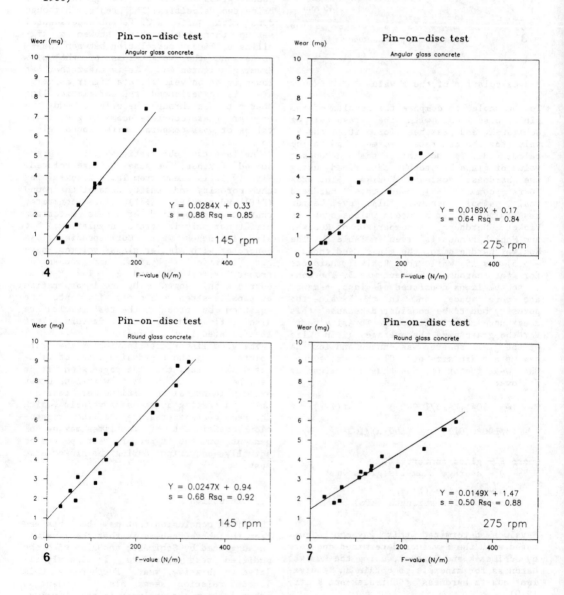

It is thought that in particular the direction coëfficient of the regression lines (m) is an indication of relative abrasivity, because the major factor influencing the F-value is the volume content of glass particles. The influence of angularity has been assessed as follows. On samples E and I, wear was measured at 5 different velocities. The ratio of the direction coëfficients of angular to round glass concrete varied around 1.4. If the assumption holds that the regression coëfficient m is a measure for the relative abrasivity then angularity may be included in the F-value as follows:
F = α F (Schimazek), were α is about 1.4 for the tests performed here.

3 DATA ON NATURAL ROCKS

The only data on natural rocks available at present were kindly provided by Dr. Paschen (Paschen, 1980). Paschen performed a multivariable linear regression analysis and examined the statistical value of Schimazek's F-factor. Paschen found that the F-value correlated well with the wear measured with Schimazek's pin-on-disc experimental set-up. Fig. 8 gives the data

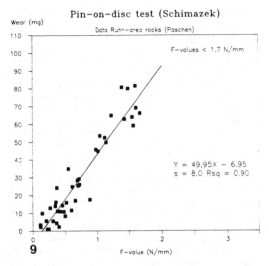

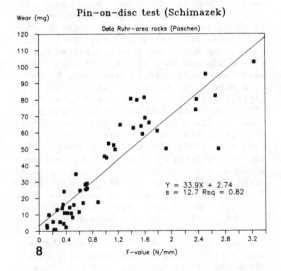

available. The rocks were Carboniferous sandstones and shales from the Ruhr-area. Paschen (1980) notes that the deviation from linearity at high F-values is caused by a group of strong sandstone rocks. If these are left out, what actually means leaving out the F-values > 1700 N/m, a much better linearity and a higher correlation is found (Fig. 9). The multivariable linear

regression analysis lead to a formulation of a wear factor in terms of summations of products of mineral content, grain size, tensile and compressive strength, but it was concluded by Paschen that the abrasivity of the natural rocks cannot be described completely by the mineralogical and mechanical parameters used.

4 INTERPRETATION OF PIN-ON-DISC TEST RESULTS

Although the pin-on-disc tests on glass concrete were performed with a very different apparatus and procedure, it appears that Schimazek's F-value gives a reasonable linear correlation with the abrasive wear of pins, either flat or pointed. The tests on glass concrete show that the result is very dependent on test velocity, but also on factors other than the ones considered in the F-value (i.e. hardness, grain size and tensile strength). The angular glass concrete gave, unexpectedly, less wear than round glass concrete tested under similar circumstances. In general, the direction coëfficient of the regression line for angular glass concrete was higher than that for round, which suggests that if the angular concrete would be similar on microscopic scale it would be more abrasive. The result is not satisfactory in this respect and further testing on better artificial materials should be performed. The test results, however, have shown that the abrasive processes operating are also dependent on the type of material being tested, and that this material is not completely described by mineral hardness, grain size and strength. The remaining, unknown, factor may be called "microstructure". If it were possible to construct angular glass concrete with a microscopic structure completely similar to that of the round glass concrete, except for the angularity of the glass particles, the factor α would describe this microstructural difference.

The data on natural rocks of Paschen (1980) are very interesting in this respect. Apparently a group of similar rocks - with probably a similar type of abrasive wear occurring during the pin-on-disc test - shows a good linear correlation with wear, but this relationship cannot be used to predict the amount of wear occurring of another rock type with other microstructural, mineralogical and rock mechanical characteristics. This result shows that the determination of the F-value according to Schimazek alone can never be used to predict amounts of wear. However, the results (Fig. 4-7, 8, 9) show that a combination of mineralogical and mechanical data is a sensitive parameter that can be used in abrasive wear assessment.

Fig. 10 shows some data on pick point wear of different types of suction cutter dredgers in different projects which has been related to F-values determined on rock samples obtained from the dredging sites. Although these data are very crude in the sense that highly different dredgers and working conditions existed, the F-value seems to give a fair indication of the wear experienced.

Pick wear data

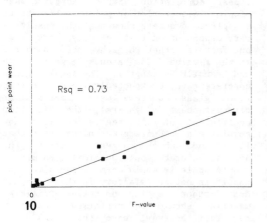

5 DISCUSSION

The pilot study described here to assess the applicability of a pin-on-disc type of test to measure abrasivity of rock has given information about the many factors involved in abrasion. Pin-on-disc type tests are very common in tribological engineering (e.g. Uetz, 1986) and have been used to assess digger teeth steel quality (Eyre and Mashloos, 1986). The experiments on glass concrete have shown that the test is sensitive to more factors than are involved in the wear value F of Schimazek and Knatz (1970). Paschen (1980) has reached similar conclusions on this subject. The data on natural rock (fig. 8 and 9) suggest that it is necessary to investigate the influence of other factors on the abrasive processes occurring during the test. One of these factors is very probably microscopic structure. Schimazek (1989, pers. comm.) recommends that tests be done on more rock types than the sedimentary rocks tested by him from the Ruhr-area.

The authors have determined the F-value of many rock types which have been obtained from dredging contractors. These rocks come from all rock groups (sedimentary, metamorphic, igneous). F-values for some of these are illustrated in Fig. 10. Due to the high competition in dredging, wear data are highly confidential and it was not possible to construct a reliable data base of pick point wear and characteristic rock data. One of the questions that was raised during this study was what type of mineral hardness should be used to assess the hardness of a mineral relative to quartz. For example, both the Vickers Hardness (Uetz, 1986) and the CERCHAR index (Suana & Peters, 1982) of feldspar are about 70% of that of quartz in contrast to Rosiwal's hardness (33%). The data of Paschen (1980) not really favor either of the methods, because the content of feldspar of these rocks is relatively low (Fig. 11). Since feldspar is a mineral with cleavage, it is doubtful that Vickers indentation Hardness can be a good index in cases of sliding abrasion and the "grinding hardness" of Rosiwal (1916) intuitively seems better. This merits further research, especially since in tribology the usage of Vickers Hardness is very common (Uetz, 1986).

6 CONCLUSION

This study has shown that a pin-on-disc type of test is a good method of comparing the abrasivity of different rock types. The F-value of Schimazek and Knatz (1970) gives a good indication of abrasive capacity of

Pin-on-disc test (Schimazek)

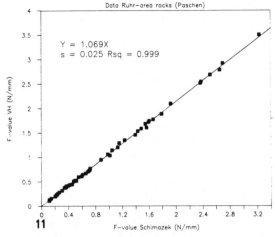

Data Ruhr-area rocks (Paschen)

Y = 1.069X
s = 0.025 Rsq = 0.999

F-value VH (N/mm)

F-value Schimazek (N/mm)

11

rock, but the resulting wear is dependent on the actual abrasive mechanism occurring during the test. This mechanism is dependent on such factors as contact pressure, rotation velocity and also on rock factors not yet contained in the wear value "F". Angularity of abrasive grains is one of these factors but generally "microstructure" seems important in this respect. This study has only generally indicated the factors involved. Further study is under way to investigate these, using a better instrumented pin-on-disc set-up on a lathe with provision to measure and maintain constant normal pressure and constant velocity under the pin.

ACKNOWLEDGEMENTS

The experimental program was performed by H.J.van den Bold and Th.W.M.Vermeer under supervision of the first author. The data have been analysed by the first author, who is also responsible for the contents of this paper. This study is supported by the Technology Foundation (STW).

REFERENCES

ASTM C 241-85 (1988): "Standard method for abrasion resitance of stone subjected to foot traffic". Annual book of ASTM standards, Vol. 04.08: Soil and rock, building stones, geotextiles.

Bold, H.J. van den (1986): "De abrasieve slijtage van snijwerktuigen door gesteenten". Memoirs of the Centre for Engineering Geology, Delft; no. 38; pp.98.

Brown E.T. (1981): "Rock characterization, testing and monitoring". ISRM Suggested Methods for determining hardness and abrasiveness of rocks. Permagon Press, Oxford, 95-103.

Eyre T.S. and Mashloos, K.M. (1986): "Abrasive wear and its application to digger teeth". Tribology International, vol. 85, 259 - 266.

Paschen, D. (1980): "Petrographische und geomechanische Charakterisierung von Ruhrkarbongesteinen zur Bestimmung ihres Verschleissverhaltens". Dissertation, Technischen Universität Clausthal. pp. 202.

Powers M.C. (1953): "A new roundness scale for sedimentary particles". Journ. of Sed. Petrology, vol. 23, 117 - 119.

Rosiwal, A. (1916): "Neuere Ergebnisse der Härtebestimmung von Mineraliën und Gesteinen, ein absoluter Mass für die Härte spröder Körper". Verh. d.k.u.k. Geol. Reichsanst. Wien, 5/6; 117-147.

Roxborough,F.F.; Phillips,H.W. (1974): "Experimental studies on the excavation of rock using picks". ISRM Proceedings of 3rd Congress,Denver. Vol.II.B; 1407-1412.

Schimazek,J. (1981): "Construction underground - present and future". Vortrage der Stuva-tagung Berlin, Stuva, Köln, 42-45.

Schimazek,J.; Knatz,H. (1970):"Der Einfluss des Gesteinsaufbaus auf die Schnittgeschwindigkeit und den Meisselverschleiss von Streckenvortriebsmaschinen." Glückauf, vol.106.6. 274-278.

Schimazek,J.; Knatz,H. (1976): "Die Beurteilung der Bearbeitbarkeit von Gesteinen durch Schneid- und Rollenbohrwerkzeuge". Erzmetall. Vol.Bd. 29; 113-119.

Suana,M.; Peters,Tj. (1982): "The CERCHAR abrasivity index and its relation to rock mineralogy and petrography". Rock Mechanics, vol. 51/1, 1-8.

Tarkoy P.J. (1973): "A study of rock properties and tunnel boring machine advance rates in two mica schist formations". 15th Symposium Rock Mech., Custer State Park, South Dakota.

Uetz, H. (1986): "Abrasion und Erosion". Carl Hanser Verlag, München. pp. 830.

Verhoef, P.N.W. (1989): "Towards a prediction of the abrasive wear of cutting tools in rock dredging". Delft Progress Report, Vol13 nr 1/2; 307-320.

West, G. (1989): "Rock abrasiveness testing for tunneling". Technical Note. Int. J. Rock Mech. Min. Sci. & Geomech. Abstr. Vol. 26, No. 2; 151-160.

Quantitative approach on micro-structure of engineering clay
Etude quantitative de la microstructure de sols cohérents liée aux travaux

Y.X.Wu
Xiamen Development & Research Centre of Environmental Geology, CAGS

Z.H.Zhang & Z.M.Ling
The Institute of Hydrogeology and Engineering Geology, CAGS

ABSTRACT, This present paper introduced a temperature gradient vacuum sublimation instrument designed by the author to dry the quickly frozen samples of undisturbed soil and a computer system set up by the author and HEFEI polytechnic university for quantitatively studying of the images of soil micro-structure. New concepts are put forward about the structure state of soil and the entropy of structure state from the viewpoints of system theory. Based on quantitative analysis of structure states of NINGBO clay under different pressure stress and time, the change laws of the entropy of soil structure state, the mean area and mean shape parameter of micro-components of clay are pointed out. The relationships between structure change and special properties of clay are analyzed.

RESUME, A partir de la théorie du système de sols liés aux travaux, on propose, dans cet article, de nouveaux concepts sur l'état de structure du sol et l'entropie de l'état de structure, et présente un instrument de sublimation à vide et sous la différence de température, conçu pour sécher les échantillons du sol gelés rapidement et non remaniés, et un système d'analyse quantitative par ordinateur des images de la microstructure de sols cohérents. Fondées sur l'analyse quantitative de l'état de structure de sols argileux de Ningbo sous la différente pression et au différent moment, la loi de variation de cette entropie a été découverte, et la relation entre l'évolution de structure et les propriétés particulières physico-mécaniques de ces sols a été traitée.

1. INTRODUCTION

Engineering soil is a natural or an artificial deposit interacting with engineering constructions. Clay is one of the most popular type of it . This kind of engineering soil is of special properties quite different from other engineering materials. For instance, high un-homogeneousness , high anisotropy, non-linear constitutive relationship , and unconventional strain softening and hardening, etc.

Because of these special properties, it is quite difficult for us to appraise justly on its engineering behaviors by means of conventional continuous medium mechanics. For recent years, the researchists have more and more realized that they cannot regard the clay as iron and steel-like materials or as the general geological body, instead, they should dismantle them and look into its structure laws to make clear why they show such special properties, then to install them according to its structure laws. meanwhile, they expect to look for the method to change the clay's structure to improve its properties by analyzing the clay structure. Such, how to evaluate, especially how to appraise quantitatively on the structure of clay is to probe into the structure laws of clay and is

the first step to probe into the engineering quality of clay.

The main purpose of the essay lies in discussing how to study clay structure by what instruments and techniques to get this purpose and trying to look for a method to comment quantitatively on structure and trying to seek the law of behaviors of engineering soil with this method.

2. TECHNIQUES AND EQUIPMENTS

(1). Developing of a temperature gradient vacuum drying equipment

Drying of clay sample is the first step to analyze its structure. Freeze drying method is the one that can better keep the clay sample undisturbed in drying. However, the duration of sample preparation of the equipments reported nowadays takes about 15 hours, which makes it inconvenient to popularizes it. Improving the principles, the author designed a temperature gradient vacuum sublimation system. Its principle sketch shows in figure 1.

The obvious advantage of this system is that a temperature gradient sublimation system has been

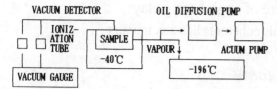

Fig. 1 The principle sketch of temperature gradient vacuum sublimation system

added into a vacuum sublimation device and this makes the speed of sample preparation raise in treble compared with the same instruments reported.

(2). Developing of a computer system for quantitative processing of soil micro-photos

In Nov.1987, together with the Computer and Information Department in Hefei Polytechnic University, the author had finished re-building an image processing system for analysis of soil micro-structure with micro-computer. The principle sketch of this system is shown in fig.2.

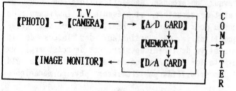

Fig.2 Principle sketch of the computer system for processing of soil structure-images

The structure image of soil is put into in form of imitating signal with the speed of 30 frames per second through a TV. camera with standard light guide pick up, then in 6 bits precision the input video frequency signal is transformed into digital signal through A/D transforming card. Then these data are stored in frame memory.

The stored image can be pre-processed through micro-computer. It includes, image time domain processing, image enhancement, image smoothing, image calculating and so on. The aim of the process is to get rid of vagueness and disturbance, to produce binary image and to take the outline of the main elements (particles and voids) in the photo. From the outline figure of soil components, the perimeter, area and the length and the position of the longest hypotenuse , as well as the shape coefficient of each elements of soil can be calculated automatically. By statistical analysis, these data are arranged into the porosity, void ratio, average particle diameter and average shape parameter of soil and into the parameter of soil structure state i.e. structure state entropy (this will be introduced later).Moreover, the components (particles and voids) appeared in this photo can be classified

in shape and area according to the classifying standard inputted in. The processed result can be output according to the request through printer and plotter.

3. CONCEPT OF THE SYSTEM STRUCTURE OF ENGINEERING CLAY AND ITS QUANTITATIVE EXPRESSION

(1). Systematic characteristics of engineering clay

As all known, the clay substance formed in natural geological circumstances consists of clay mineral of different shapes and sizes, non-qualitative substances, organism and small amount of other minerals and void fluid (void gas and liquid). We can call the complexity of these substances as the forming elements (or components) of clay. The engineering practices and experiments show that all these elements cannot represent clay body individually. Their own properties & functions differ greatly from that of clay,e.g., The intensity of mineral particles is much greater than that of clay. Obviously, the simple addition of the properties and functions of these elements doesn't equal that of clay. So we not only should analyze the formation elements but also cannot expect to study mechanically the individual element to understand the properties of clay, instead of studying the correlative relation and effect of all elements formed this organic body, that is to say to study the clay as a system.

The clay in natural environments is an open system. It exchanges substance and energy at times with the surrounding circumstances. When the main circumstance of this system is engineering structures, we call it engineering soil system.

(2). Structure levels of clay system and the definitions of micro-structure and its state

The so-called level of structure refers to sub-system divided according to the different degree of connection of system elements. To be concretely, concerning the engineering soil system we discussed, there is a great difference between the degree of connection of mineral particles and that of aggregates formed by them. They belongs to different levels of structure (sub-systems) and the low level of mineral particles is the forming element of the rather high level (system) of the aggregate of mineral particles. Similarly, the level of aggregate is the forming element of a higher system. This is the levels of system structure.

The theory of system says that the levels of substance system have the following properties,

a. A specified levels of substance structure suit a specified energy state. The levels of system will have broken when the energy of system has surpassed certain threshold and a new levels would come to exposed while the old ones disappear.

b. Following relation exist in a specific level of structure, that is,

$$L \times E = 10^{-7} \quad cm. ev......(1)$$

in the formula,
L = The measurement of the specified level of substance system (cm)
E = The combining energy between the elements of this level (ev)

c. The higher the substance level is, the less the abundance in the universe is, and the bigger the variety of its structure function is.

By taking the above properties as the standard to divide levels, combining with the practical observation and measurements, it is considered that three main levels can be divided of the engineering clay.

MACRO_LEVEL, The elements of this level is the blocks separated by cracks, feeble belts. These Blocks together with the crammed fluid between them have formed macro-levels by the connection of capillary power, friction force, and cementing power. The so-called macro refers to that the element-blocks of this level are rather big in measurement and can be seen with naked eyes. The connection of the elements of this level is very week it can be broken and expose the next level whenever a bit amount of energy is brought to bear on it.

MICRO_LEVEL, The micro-level is soil blocks. The elements of this level are the aggregates of mineral particles. These aggregates connected with each other through void liquid or cementing substance in a certain degree to form the micro-levels. And the combination of this level is obviously stronger than that of macro-levels. It cannot be separated unless more energy is supplied. The elements in this level are the aggregates of mineral particles, and the measurement is too small to be differentiated with naked eye.

SUPER-MICRO_LEVEL, The super-micro level is the aggregate of mineral particles, its elements are mineral particles. The particles combine each other with cement, electric double layer, and ionic band to form aggregate . The combination of this level is stronger, and the general engineering load cannot damage it.

As all know, the structure levels of substance have the indefinite separability. This principle also suits the engineering clay system. According to the above dividing methods, we can also further divide the levels into mineral level, molecule level, atom level and so on. But it is meaningless to divide such levels in terms of the system of engineering clay. We considered, for the general engineering load cannot affect these levels, that is to say, cannot deform and damage the mineral particles or molecules. The relative bits shift and disposition between the aggregates of mineral particles are the main internal cause of the deforming and damaging of clay. So it should be considered that the micro-level of clay structure is to be the key level to study. The structure the author refers to in the following is the structure of this level.

After making sure the systematic characteristics and the key level of structure of engineering clay, we can define the micro-structure of this substance as the interaction and interconnection order and relation between soil's micro-components (elements).

Although we have defined the structure of clay system , a lot of observations and measurements show that there are hardly two samples completely the same in elements arrangement. Not to speak of the interactive intensity between elements. Even in the same sample, the arrangement is quite different in different parts. Even if the same part of the same sample, cases also differ in different time. That means that the engineering clay does not have the fixed structure pattern as crystal structure and element structure have. There are uncertainty and confusion existing in the connection between its (key) elements.
However, the author has noticed that the confused connection between clay's components is just the characteristics of its structure. Although the confusion degree varies at the different time acted by the circumstances, the general confusion degree is definite at a certain time. So we can call this confused connection corresponding this relatively stable time as a structure state. The characteristics of the state can be overall generalized when making sure the degree of confusion. It may be considered that the degree of confusion is a measurement reflecting structure state.

(3). Quantitative expression of micro-structure state of clay

For engineering clay, the confusion degree of structure can be divided into three parts,

$$E = Ea \& Ed \& Ee \qquad ......(2)$$

where,
Ea=confusion degree of the arrangement of aggregates.
Ed=confusion degree of size distribution of aggregates.
Ee=confusion degree of the level of connecting energy between aggregates.

507

Because of the limitation of time and techniques, only the arrangement confusion is discussed in this study.

For the two-dimensional case, we can divide the direction of particles orientation into n parts. Suppose the probability of orientation of aggregates at direction part k is p (k), Ea can be expressed by the famous ENTROPY FUNCTION :

$$Ea = -\sum_{k=1}^{n} P(k) \, Log n P(k) \quad \ldots \ldots (3)$$

So, we can also call it the arrangement entropy that the confusion degree of aggregates.

It might be proven that 0<=Ea>=1. From the above definitions and properties we know that Ea is a statistic parameter, its value is between the range of [0,1]. The smaller the entropy value is, the smaller the arrangement confusion of the aggregates of clay minerals is. Oppositely, it will be bigger.

4 CHANGES OF STRUCTURE STATE OF CLAY WITH PRESSURE STRESS AND TIME

In order to make clear the connections between clay' behaviors and its structure when interaction with engineering construction, two kinds of tests were made on Ningbo clay and structure analysis were done by employing the instruments mention in episode 1.

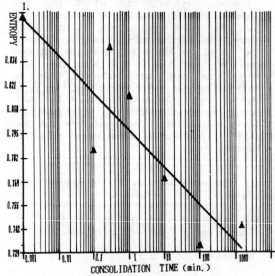

Fig. 3 variation of Ea with increase of time

Figure 3 shows the variation of Ea of micro-structure of Ningbo clay loaded with 100kpa pressure stress when time increases. Figure 4 gives the relationship between the entropy and

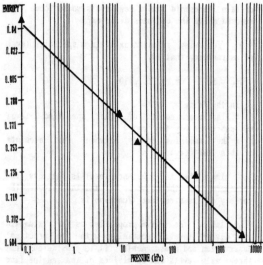

Fig. 4 Variation of Ea with pressure stress

pressure stress. These two figures show the tendency of linear decrease of the entropy in half logarithm coordinate system.

From the definition in episode 3, we know that what the entropy reduction reflects is the reduction of the confusion degree of structure arrangement. That is, the clay changes into better order from confused order with the increasing of pressure or time . Also, it reflects the rotation of particles of clay which causes the deformation of clay macro-cosmically.

It should be pointed out that the above mentioned laws only refer to the used pressure range. Because equipment and time are limited, the author has not made high-pressure test in this study, however, the engineering practice shows that when pressure has reached certain degree, the soil will be damaged, we might suppose that the entropy of clay at this time will have a sudden change from entropy reducing to entropy increasing. If this supposition is later proved right, the study of the critical entropy of clay will have great engineering significance.

What Fig.5 shows is the measuring data points and the regression curve of average area of the aggregates and Ea. In the figure, except one point that has more departure, the general tendency is that the average area appears the phenomenon of increasing as the pressure logarithm becomes bigger after the pressure get over certain value, and the bigger the pressure is, the quicker the increasing rate of the area is.

This is a quite special phenomenon, for it shows that the bigger the pressure is, the bigger the soil particles in statistic concept is. That is to say after beyond a certain pressure stress, the effect of pressure can make the aggregates

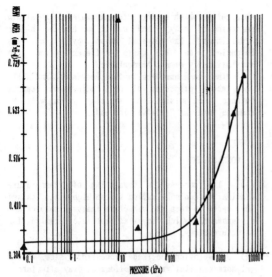

Fig.5 Change of average area of aggregates with increase of pressure stress

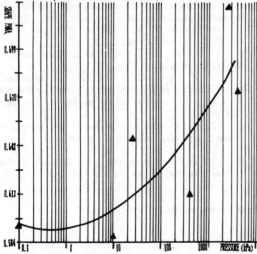

Fig.6. The relation between pressure stress and mean shape parameter. of clay particles

Fig.6 shows measuring points and a regression curve of the average shape parameter and pressure. It looks rather confused, but from the effect of regression, the shape parameter tends to become bigger with the increasing of pressure.

It is clear that the changing law reflects that the shape of soil aggregate becomes longer and longer in the direction vertical to pressure. From this, we know, with the increasing of pressure, the clay not only has changes in the degree of structure confusion between its element but also in its element, the bigger the pressure is, the more obviously the elements change.

5. THE MECHANISM OF CHANGES OF THE STRUCTURE STATE OF ENGINEERING CLAY AND ITS RELATIONSHIP WITH THE SPECIAL PROPERTIES OF CLAY

As is known, the newly deposited clay is similar to the deposited colloidal substance. Its elements are connected with each other by water glue. Because there exists repulsion wall (Fig. 7). The micro-elements stop at point "b", reaching the balancing of power. By this time, the arrangement of the elements is rather random, which makes the soil have certain isotropism on properties.

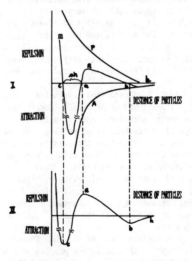

Fig.7 The interaction curves of colloidal particles
I. Power curve II. Potential energy curve
A. attraction curve. B. repulsion curve.
R. sum of power   a. repulsion wall.
b. No.2 gulf of potential energy.
c. NO.1 gulf of potential energy.

polymerization to form a bigger one. Obviously, this phenomenon is not obtained by other engineering materials, including sand soil. So sand soil is closer than clay to the linear elastic material. And because the clay has such strange polymerization property, it dissipates a great amount of external energy to adjust the structure and once more polymerization its elements, making it depart from linear elasticity and the bigger the pressure is, the more it departs.

Similarly, this relation is only suitable to be applied to the range of pressure stress used, the change of the area of particles under high pressure is to be experimented and studied.

When load increment (here refers to the pressure stress, the follows are same) is charged on clay system, it would assign and process this affecting load at the connecting points of its elements and pass the load to the surrounding circumstances through these points. At the same time, it gives a negative feedback to the engineering building (Fig.8) to keep its original structure state.

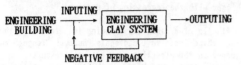

Fig.8 relationship between clay system and its circumstance

If this negative feedback cannot balance the input load, the soil system will begin adjusting the relation between its elements , making them reduce the contradictory confusion between themselves by non-linear synergy. It will increase its orderliness to dissipate the input entropy from outside in the direction vertical to the pressure stress By this way, the connection mode between clay's elements develop from points connection to surface connection, to increase resisting force, and output the energy and substances (void liquid) to the surroundings. By now the soil system reproduces a negative feedback to compare with engineering load. If this negative feedback still cannot balance this invader from outside, the soil system will repeat the above process...... again and again until the negative feedback balances with input. The soil system has then formed a structure state of higher orderliness. The author calls the process that the clay has rearranged itself to suit engineering load as evolution process.

It is just because of the evolution under the pressure stress and time, which makes the clay appear orderly structure and results in 7 the anisotropy of its mechanical properties. We can also see at the same time that the clay's deformation under pressure stress is shown mainly due to the re_orientation of its elements, that is, the elements turn round to the direction vertical to the pressure stress . The macro-deformation in statistic formed by the turning round of aggregates doesn't present the linear relation with affecting stress.

After a structure state of clay has formed, the clay will repeat this evolution process on a higher stage if a new increase of load has taken on . However, if only by reducing the confusion degree of micro-levels cannot resist the outside load. the soil elements will damage their originally connecting mode, breaking through the repulsive wall to polymerize and begin the evolution on the next structure level. This has led to the phenomenon mentioned before that the aggregates are getting bigger and their shapes are getting flatter. Because the element polymerization accompanied by the evolution under such bigger stress needs consume a large amount of energy, this makes the soil appear great resistant force macro-

cosmically, which is the internal cause of the strain hardening of clay. And now it has had the higher orderliness (macrospical anisotropy), so its hardening law is quite different from the conventionally strain hardening .

If the load has been further increased, and all the above processes of self adjustment cannot balance the outside load, the clay system will be damaged, and one "life" period ends.

6. CONCLUSIONS

(1). The success in developing of temperature gradient vacuum sublimation system highly has raised the degree to keep the structure analysis samples undisturbance, which supplied reliable premise for structure study.

(2). By employing the image processing system, we can get more qantitative informations of clay structure.

(3). Three structure levels of clay has been divided from the system points of view. Micro-structure is considered as the key level of engineering clay.

(4). It is analyzed that clay has not fixed structure but a series of structure state which may be relatively stable at certain time. Confusion degree of the connection between clay aggregates is taken as a quantitative parameter of clay's structure, and it can be expressed in following ,

$$E = Ea \& Eg \& Ee$$

in which , each item can be written in Entropy function,

$$Ei = -\sum_{k=1}^{n} p(k) Log_n P(k)$$

(5). From mechanic tests and structure analysis on Ningbo clay, following facts have been found,

a. Ea reduces by logarithm with time under charge of certain pressure.

b. Ea goes down by logarithm with the increase of pressure stress.

c. In average value, the area of aggregates become bigger when pressure get to certain value, and the bigger the pressure is, the quicker the increase rate is.

d. The average shape parameter of aggregates become bigger with increase of pressure.

(6). The structure evolution is the internal factor of clay's special properties and functions.

Reference(omitted).

Use of Schmidt hammer for estimating mechanical properties of weak rock

Utilisation du Schmidt hammer pour l'estimation des propriétés méchaniques des roches faibles

S. Xu, P. Grasso & A. Mahtab
Geodata, Torino, Italy

ABSTRACT: Weak rocks often contain healed fractures and deformation fabrics (foliation, schistosity) whose response to the impact of the Schmidt hammer is different from that of strong rocks which are usually devoid of such weaknesses. The relative direction of the impact and the proximity of the impact point to a fracture are significant influences on the Schmidt hammer index, R. Most of the existing empirical relationships tend to considerably overestimate the strength and deformation parameters of weak rock. The procedure described in this paper accounts for the rock fabrics during the actual Schmidt hammer test and in establishing correlations between the hammer index, R, and the mechanical properties. In all cases examined using regression analysis, close correlations were obtained between R and strength Co and modulus Et.

RESUME: Les roches présentent souvent des fractures remplie et des deformations (feuillation, scistosité) qui sous l'effet du Marteau de Schmidt donnent des différent réponses par rapport à celles qui sont données par des roches résistantes dépourvues d'alterations. La variation sur R, donnée par le marteau de Schmidt, est très grande lorsque nous avons différentes directions d'impact et la présence de fractures en proximité du point d'impact du marteau. La majeure partie des relations de type empyrique connues on la tendance à surextimer les paramètres de résistance et de déformations des roches alterées. Le procédé qu'on propose dans cet article tient compte de la structure de la roche pendant les épreuves avec le marteau de Schmidt et établi des corrélations entre l'index R du marteau et les proprietées méchaniques. Dans les cas examinés en utilisant une analyse régréssive, on a obtenu des coéfficients de corrélations significatifs entre R et les propriertées méchaniques (Co, Et).

1. INTRODUCTION

The Schmidt hammer, as a non-destructive and index testing method, has been used since the early 60's in engineering practice for rapid estimation of mechanical properties of rock. A standardized method of rebound testing using this hammer has been suggested by the International Society for Rock Mechanics (ISRM, 1978). Many empirical relations have been proposed (e.g., Deere & Miller, 1966, Haramy & DeMarco, 1985) for correlating the Schmidt-hammer index, R (obtained in situ or in the laboratory) with the uniaxial compressive strength (Co) and/or the tangent modulus of elasticity (Et) of intact rock.

However, the usual experience in correlating R with the mechanical properties of rock is generally not applicable to weak rocks. As will be demonstrated in the next section, most of the existing empirical relationships tend to considerably overestimate the strength and deformation properties of weak rock. This is due to the fact that the available procedures and correlation schemes do not account for the healed fractures and linear fabric (lamination and schistosity) associated with weak rocks.

Consequently, a research program was initiated in Geodata, with the following objectives:

1) to examine the applicability of the Schmidt hammer to weak rocks, such as the metamorphic rocks encountered during tunneling in Northern Italy;

2) to establish appropriate procedures for recording rebound values and for establishing correlations between the rebound value and the mechanical properties of weak rocks;

3) to derive reliable correlations for frequently encountered weak rocks.

The preliminary results of the research are reported in this paper. The following sections of the paper contain: a critical review of the existing correlations and procedures; a discussion of the influence of the fabric on R and the mechanical properties; a description of the proposed procedures for using the Schmidt hammer for characterizing weak rocks; and correlations between R and the strength and modulus of some weak rocks.

2. CRITICAL REVIEW OF EXISTING CORRELATIONS AND PROCEDURES

A survey of the literature shows that several empirical relationships are available for correlation the uniaxial compressive strength (Co) and tangent modulus (Et) of intact rock with the average Schmidt hammer rebound value. A summary of these correlations is provided in Table 1. The following observations can be made with respect to the correlations shown in this table.

1. The correlations between R and Co or Et are nonlinear and generally of an exponential form.

2. Density of the rock, dn, has been incorporated in various forms to improve the correlation, suggesting an implicit relation between R and dn. However, it is not clear why dn was never used as a denominator to "normalize" R. In

fact, this normalization gave a better fit to the data in some cases (see sec.5).

3. The correlations (e.g. ISRM's suggested method based on Deere & Miller, 1966) involved measurement of Co with the axial loading directed either parallel or normal to the linear fabric. This invariably results in the highest value of Co (see sec.3). These correlations are, therefore, not applicable to the general case where the fabric is inclined to the core axis or the direction of loading.

4. The use of these correlations may result in an unacceptably wide range of Co and Et as shown in Table 2 for five different types of weak rock.

All correlations, except that of Haramy & DeMarco (1985), overestimate considerably the mechanical properties. Incidently, the excepted correlation was based on tests made on coal which is a material that may be classified as a weak rock.

The available procedures for the application of Schmidt hammer have two main aspects: the technique for recording R and the method for correlating R with Co and Et.

The various recording techniques (7 reviewed by Poole & Farmer, 1980, and one of Shorey et al., 1984), vary widely in their specifications of the grid for impact testing and the total number of tests per sample. None of these techniques takes into account the effect of fabric. In fact the ISRM suggested method (ISRM, 1978) requires that the rock below the impact area "to a depth of 6cm shall be free from cracks, or any localized discontinuity...".

The objectives of the correlation procedure are (1) to calculate a representative value of R and (2) to correlate R with Co and Et.

In general the recorded values of R (often those belonging to the higher range) are averaged. As pointed out by Poole and Farmer (1980), none of these procedures is based on a detailed statistical analysis of a test program.

It appears reasonable to average the higher values in the case of laboratory tests where the samples are free of fabric and

Table 1 Summary of available correlations for intact rock.

References	Proposed Correlations
Deere & Miller, 1966	$C_o = 10^{(0.00014 \times dn \times R + 3.16)}$ in p. s. i. $E_t = 6.95 \times dn^2 \times R - 1.14 \times 10^6$ units
Aufmuth, 1973	$C_o = 6.9 \times 10^{[1.348 \times \log(dn \times R) + 3.16]}$ $C_o = 6.9 \times 10^{[1.061 \times \log(dn \times R) + 1.861]}$
Beverly et al, 1979	$C_o = 12.74 \times EXP(0.0185 \times dn \times R)$ $E_t = 192 \times (R \times dn^2) - 12710$ (as quoted by Haramy & DeMarco, 1985)
Kidybinski, 1980	$C_o = 0.447 \times EXP[0.045 \times (R + 3.5) + dn]$ (as quoted by Haramy & DeMarco, 1985)
Haramy & DeMarco, 1985	$C_o = 0.994 \times R - 0.383$ with $\gamma = 0.84$
Reanalysis of data from Haramy & DeMarco, 1985	Linear-fit : $C_o = 0.9437 \times R - 0.3833$ with $\gamma = 0.7005$ $\sigma = 8.9048$ Best-fit : $C_o = 0.2869 \times R^{1.3252}$ with $\gamma = 0.8535$ $\sigma = 0.3090$

Notes:
1) dn = air-dried unit weight in ton/m^3, except in the case of Deere and Miller, 1966.
2) R = the Schmidt hammer rebound index.
3) C_o = air-dried uniaxial compressive strength in MN/m^2.
4) E_t = tangent Young's modulus defined at the axial stress corresponding to 50% of C_o, in MN/m^2.
5) γ = correlation coefficient, σ = standard deviation.

discontinuities. For in-situ tests, as found by Shorey et al. (1984), "the lower mean of rebound values", that is, the mean of those values of R which fall below the arithmetic mean, is more appropriate.

In correlating R with Co and Et, it is desirable to make the measurements for R and the other properties on the same specimen. However, this is generally not the case. For example, Deere and Miller (1966), whose work forms the background of most other investigators, have consistently used different specimens for R and Co. This procedure is clearly not appropriate for correlating R and Co for weak rocks.

3. INFLUENCE OF FABRIC

The influence of fabric on the strength of a specimen of weak rock may be compared to the strength of a material containing weakness planes, as illustrated by Jaeger & Cook (1979) and summarized below.

Table 2. A comparison between predicted and measured values of C_o and E_t for various rock types containing weakness planes.

Rock Type		Predicted Values using the correlations listed in Table 1					Measured Values		
		Deere & Miller 1966	Aufmuth 1973	Beverly et al 1979	Kidybinski 1980	Haramy & DeMarco 1985	C_o (MPa) E_t (GPa)	Dry Unit Weight dn (tons/m^3)	Rebound Index R
Mica-schist	C_o (MPa)	88.50	181.03	94.87	46.91	39.09	27.30	2.66	40.8
	E_t (GPa)	46.04	72.37	42.72	—	—	8.23		
Prasinite	C_o (MPa)	93.22	186.86	99.51	49.76	39.35	30.22	2.71	41.0
	E_t (GPa)	48.36	74.20	45.10	—	—	9.57		
Serpentinite	C_o (MPa)	110.43	206.19	116.28	56.79	43.99	28.52	2.68	44.6
	E_t (GPa)	51.95	80.18	48.79	—	—	11.00		
Gabbro containing healed fractures	C_o (MPa)	172.8	259.55	175.50	85.23	50.37	46.86	2.87	49.4
	E_t (GPa)	68.12	96.10	65.42	—	—	20.92		
Mudstone	C_o (MPa)	19.96	38.63	24.12	10.04	10.56	10.06	2.27	15.2
	E_t (GPa)	6.76	21.45	2.33	—	—	0.69		

Let the weakness plane be inclined at β to the major principal stress, σ_1, (Fig. 1a). The region BQR shown in Fig. 1b includes all values of β for which shear failure is possible along the weakness plane according to the Coulomb criterion

$$\tau = c + \sigma_n \tan\phi \tag{1}$$

where c and φ are the cohesion and friction angle associated with the weakness plane. The failure conditions is expressed by:

$$\sigma_1 \geq \sigma_3 + \frac{2(c + \sigma_3 \tan\phi)}{(1 - \tan\phi \cot\beta) \sin 2\beta} \tag{2}$$

For an unconfined compression test, the above equation reduces to

$$\sigma_1 \geq \frac{2c}{(1 - \tan\phi \cot\beta) \sin 2\beta} \tag{3}$$

Referring once again to Fig. 1b, failure of the specimen for values of β falling outside the region BQR involves fracture across the weakness plane, through the intact rock. The resulting rock strength is higher as shown in Fig. 1c which also illustrates the shear and fracturing modes of failure.

It follows that the Schmidt hammer index, R, should relate to the range of β (or at least to the mode of failure) as observed in our laboratory experiments. A treatment of this aspect is, however, noticeably absent from the literature.

In our correlation procedure, based on the results of laboratory experiments, we have taken into account the mode of failure. A clear expression of the influence of β will be presented sometime in the future.

4. PROPOSED PROCEDURE

The procedure for the use of Schmidt hammer for estimating the mechanical properties of weak rocks consists of the following three steps:
(1) Technique for recording the rebound values.
A total of 20 individual tests (impacts) are made on each cylindrical specimen. [Note that ISRM, 1978, also suggested 20 tests]:
(a) 5 tests on each flat end, one at the center and 4 around the center,
(b) 5 tests parallel to the fabric (while the cylinder is lying horizontally on a steel base), and
(c) 5 tests normal to the axis of the cylinder when it is turned 90° from its position in (b).
The positioning of the impact points, and the type of hammer used, conform with the ISRM (1978) suggestions.
The impact is always directed vertically downward. We added the tests on the two ends of the specimens in order to better evaluate the influence of fabric or weakness.
(2) Method for processing the rebound values.
Experience suggests (e.g. Shorey et al., 1984) that, in the presence of weakness planes, the rebound values measured in-situ exhibit a wide scatter.
It is, therefore, necessary to include the lower range of the rebound values in calculating the rebound index, R. This investigation used laboratory specimens containing weakness planes. R was calculated according to the method suggested by Shorey et al. (1984), which was referred to in Sec. 2, that is, R=the lower mean of the rebound values.
It should be noted that the above procedure applies to the range of β in which the weakness plane controls the failure (by sliding) as illustrated by the theoretical example of Fig. 1. However, in practice the range of β for sliding failure may be from 15^0-70^0.
The mode of failure outside this range of β involves fracturing through rock. The ISRM (1984) procedure, using the higher mean of rebound values, is more appropriate in this case.
(3) Correlation procedure
Reference was made earlier to the fact that the same specimen should be tested for R and for Co in order to obtain meaningful correlations. This was, in fact, the case for all specimens tested in this investigation. In addition, only those tests in which the failure of

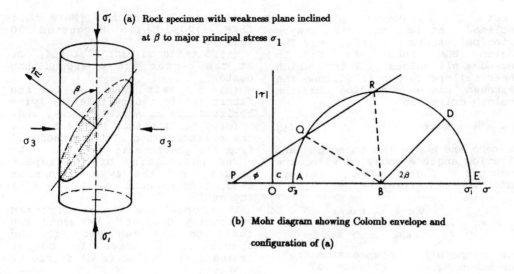

(a) Rock specimen with weakness plane inclined
at β to major principal stress σ_1

(b) Mohr diagram showing Colomb envelope and
configuration of (a)

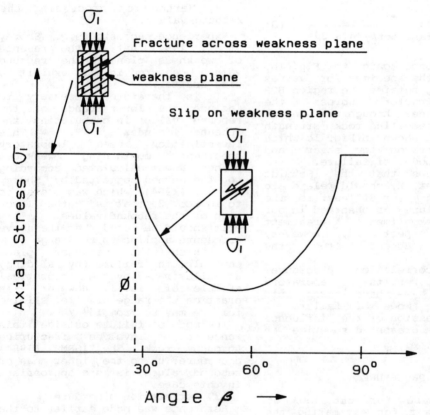

(c) Mode of failure and strength of specimen as a function of β

Fig. 1   Influence of fabric inclination on strength of weak rock

516

the sample was due to sliding along the weakness plane (foliation, healed fractures) were used for the correlation.

5. RESULTS

In this investigation, five rock types were examined. At least 15 specimens of each rock type were tested using the Schmidt hammer and under unconfined compression (for Co and Et).

Correlations between R and Co (and Et) were performed for each rock type using simple regression analyses incorporating the linear, the power, and the exponential model. For all rock types, the exponential model provided the best fit to the experimental data as measured by the correlation coefficient.

We have selected two rock types, Prasinite and Schist, to illustrate (Figs. 2 and 3, respectively) the correlation between R and the properties Co and Et.

The scatter in the data plotted in Figs. 2 and 3, is, in our opinion, largely due to the actual inclination, β, of the weakness planes (see Fig. 1c).

As mentioned earlier, the correlation procedure used here accounted for only the mode of failure and not β.

The scatter in the data should reduce, when β is incorporated in the correlation scheme (this is one of the aims of the research which is in progress).

The results of the correlation for all rock types are listed in Table 3, in terms of the constants and the coefficient of the exponential correlation.

No general trend can be observed among the correlation constants even though the correlation coefficient has a high value for each rock type. In fact, the wide range of the correlation constants of Table 3, as well as of Table 1, suggests that a general application of an equation (developed for a given rock type) is not warranted.

The above statement also recognize the inherent variability in different rocks and at different locations.

6. CONCLUSIONS

Good correlations are obtained between R and the properties, Co and Et, for five types of rock containing weakness planes, using new procedures directed to weak rock. It is suggested that the empirical relationship based on tests on a given rock are not acceptable for a general application.

Table 3   Matrix of correlation parameters.

Rock Type	C_O = EXP (a × R + b) (in MPa)				E_t = EXP (c × R + d) (in GPa)			
	a	b	γ	σ	c	d	γ	σ
Mica-schist	0.0556	1.0910	0.95	0.1850	0.0735	-0.9488	0.96	0.1890
Prasinite	0.0565	1.0950	0.91	0.3996	0.0395	0.9980	0.91	0.2645
Serpentinite	0.0272	2.0175	0.94	0.2236	0.0337	0.9441	0.88	0.4249
Gabbro	0.0504	1.3286	0.93	0.2926	0.0466	0.5609	0.95	0.2119
Mudstone	0.5227	0.2304	0.92	0.2659	0.3069	-2.6501	0.89	0.4227

Notes 1) When developing correlations for the weak mudstone, R values were normalized by corresponding values of dry-unit weight, dn.
2) All symbles used in this table were previously explained in the footnotes of Table 1.

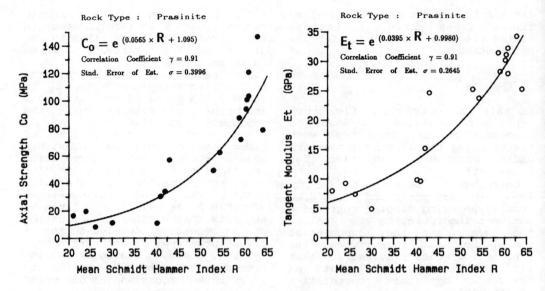

Fig. 2 The empirical relationship between R and the properties of
C_o and E_t for a dark-grey, weak to strong, fresh to
slightly weathered prasinite.

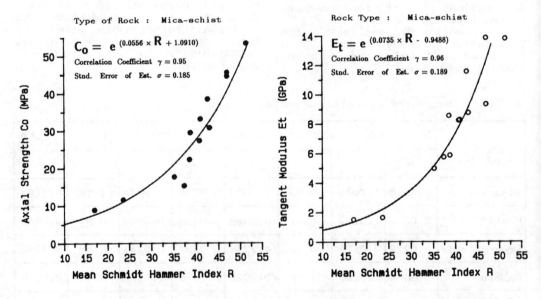

Fig. 3 The correlation between R and the properties of C_o and E_t
for a light-grey, weak to medium strong, slightly weathered
mica-schist.

From an engineering point of view, it is important for each project to develop its own data base for deriving specific relationship to be used in the site.

REFERENCES

Aufmuth, R.E. 1973. A systematic determination of engineering criteria for rock. Bulletin of Associate of Engineering Geologists. 11:3, 235-245.

Deere, D.V. & R.P. Miller 1966. Engineering classification and index properties of intact rock. Tech. Report NO. AFWL-TR 65-116. pp.300.

Haramy, K.Y. & DeMarco M.J. 1985. Use of Schmidt hammer for rock and coal testing. 26th U.S. symp. on Rock Mechanics, p. 549-555. Balkema (?).

Int. Soc. Rock Mechanics 1978. Suggested methods for determining hardness and abrasiveness of rocks. Int. J. Rock Mech. Min. Sci. & Geomech. Abstr. 15:89-98.

Jaeger, J.C. & Cook N.G.N. 1979. Fundamentals of rock mechanics, p. 65-67. London: Chapman & Hall.

Poole, R. & Farmer, I. 1980. Consistancy and repeatability of Schmidt hammer rebound data during field testing. 17:167-171.

Shorey, P.R., D.Barat, M.N.Das, K.P. Mukherjee & B.Singh 1984. Schmidt hammer rebound data for estimation of large in situ coal strength. Int. J. Rock Mechanics Min. Sci. & Geomech. Abstr. 21:39-42.

Physico-chemical and mechanical properties of peats and peaty ground
Propriétés physico-chimique et mécanique de tourbes et terres tourbeuses

H. Yamaguchi
Department of Civil Engineering, The National Defense Academy, Kanagawa, Japan

ABSTRACT: A series of physico-chemical and mechanical tests were performed on the various soils sampled from peaty ground, and the geotechnical properties of peats and peaty ground were investigated. The results of various physico-chemical tests were mainly summarized as a function of organic matter content involved. Also, by using the undisturbed samples of peat which were consolidated under different consolidation loads, the changes in pore volume and pore size distribution due to the one-dimensional compression were investigated with the porosimeter apparatus of mercury intrusion type.

1 INTRODUCTION

Peat is representative of soft soils and is classified into highly organic soils. In general, the peat is mainly composed of fibrous organic matter, i.e., partly decomposed plants such as leaves and stems. Therefore, it has been said that the peat shows the unique geotechnical properties in comparison with those of inorganic soils such as clay and sandy soils which are made up of soil particles. Then, in order to understand the physico-chemical properties of peats, it is very important to estimate both the quantity and the quality of organic matter involved. Also, the peaty ground with high compressibility is a typical example of soft ground. Then, in order to deal with the geotechnical problems in plactical such as consolidation settlement, slope stability and bearing capacity of peaty grounds, the mechanical investigations on consolidation behavior, permeability and shear properties of peats must be carried out. Hence, the geotechnical properties of peaty grounds must be discussed from the relationships between the physico-chemical properties and mechanical properties.

In this investigation, the various physico-chemical and mechanical soil tests were carried out on the many samples of several soils obtained from the direction of the depth in a peaty ground. Based on the various test results, the geotechnical properties of peaty ground were mainly discussed from the point of view of amount of organic matter involved.

2 EXPERIMENTS

2.1 Peaty ground investigated

In Japan, a great deal of peaty grounds are mainly distributed in the Tohoku and Hokkaido districts. These areas are located in the north of the Japan Islands. In this investigation, the peaty ground which is distributed in the Ishikari riverside near Sapphoro city, Hokkaido district, was investigated. The undisturbed samples of various soils were obtained from the peaty ground by using the piston sampler with the thin-walled tubes, 75 mm in diameter and 1000 mm in height. The soil sampling was carried out to a depth of about 12 m below the ground surface. The soil profile of peaty ground is shown in Fig. 1. The peaty ground has upper peat layer at the depth (Z) of 0.2 - 2.8 m below the ground surface, subpeat layers (I and II) at the depths of 7.3 - 7.9 m and 8.7 - 10.7 m and clay layer containing peat at the depth of 7.9 - 8.7 m. The ground water level was about 0.8 m below the ground surface. In particular, the samples of upper peat layer contained a considerable amount of vegetal fibers and showed very high natural water content.

2.2 Soil tests

The various physico-chemical and mechanical soil tests were carried out on the many samples of various soils until the depth of about 10 m. In this investigation,

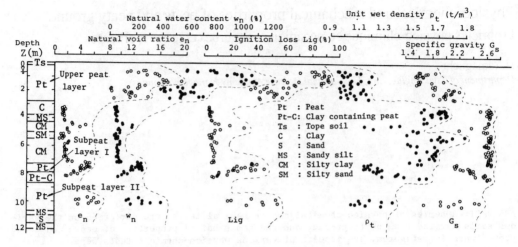

Fig. 1. Soil profile and soil property charts of peaty ground.

the physico-chemical soil tests were composed of the measurements of water content,
unit wet density, ignition loss, organic
matter content, humus content, pH and
element contents of carbon, nitrogen and
hydrogen. The ignition loss was defined as
the ratio of the mass lost by burning completely at a temperature of 800 °C to the
total dry mass at 110 °C of the soil sample. In case of soil such a peat which contains many amounts of vegetal fibers, its
ignition loss approximately corresponds to
the organic matter content. On the other
hand, as for the mechanical tests, oedometer, permeability and direct shear tests
were carried out. In particular, in the
permeability and direct shear tests, the
anisotropies of shear strength and permeability were investigated. Also, after the
completion of oedometer tests on the samples of upper peat layer, the consolidated
samples of peat were immediately dried on
the vacuum freezing for preventing the
samples from volume shrinkage due to drying, and then the measurements of pore size
distribution were performed by using the
mercury intrusion porosimeter apparatus.
The other measurements except ones of element contents and pore size distribution
mentioned above were upon the methods of
soil tests by Japanese Society of Soil Mechanics and Foundation Engineering. The
element contents of carbon, nitrogen and
hydrogen were obtained by using CN and CH
coder apparatuses. The availability of
vacuum freeze drying method has been already reported by Ahmed et al. (1974) and
Yamaguchi et al. (1986). Also, the principle of measurement of pore size distribution of soils has been described by

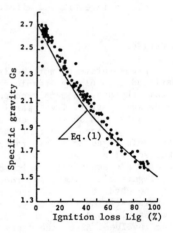

Fig. 2. Gs vs. Lig relationships.

Shridharan et al. (1971), Ahmed et al.
(1974), Bengochea et al. (1979) and Yamaguchi et al. (1989).

3 PHYSICO-CHEMICAL PROPERTIES

Based on the results of a series of physico-chemical soil tests, the distributions
of typical physico-chemical properties in
the direction of the depth are shown in
Fig. 1. Symbol of Lig in this figure signifies ignition loss. The upper peat layer
and subpeat layers have a very high natural water content (w_n) on comparison with
other soil layers. In particular, the soil
samples of upper peat layer had the natural water contents in the range between

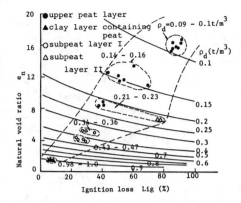

Fig. 3. e_n vs. Lig relationships.

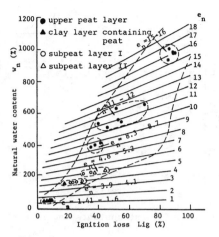

Fig. 4. w_n vs. Lig relationships.

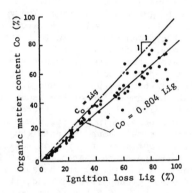

Fig. 5. Co vs. Lig relationships.

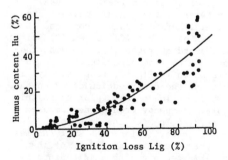

Fig. 6. Hu vs. Lig relationships.

400 % and 1100 % and also showed the values of Lig in the range of 40 % - 90 %. Hence, the values of wet density (ρ_t) and specific gravity (Gs) were very small on comparing with ones of clay and sandy soil layers.

The phase of peat soil is composed of organic matters, soil particles and void spaces. Then, from the phase consititution, the fundamental physical parameters can be represented by the following equations;

$$G_s = \frac{G_{so}G_{ss}}{G_{so}\left(1-\frac{C_o}{100}\right)+\frac{C_o}{100}G_{ss}} \qquad (1)$$

$$\rho_d = \frac{\rho_w}{1+e}\left\{\frac{G_{so}G_{ss}}{G_{so}\left(1-\frac{C_o}{100}\right)+\frac{C_o}{100}G_{ss}}\right\} \qquad (2)$$

$$\rho_t = \frac{\rho_w}{1+e}\left[\left\{\frac{G_{so}G_{ss}}{G_{so}\left(1-\frac{C_o}{100}\right)+\frac{C_o}{100}G_{ss}}\right\}+\frac{S_r e}{100}\right] \qquad (3)$$

$$e = \frac{w}{S_r}\left\{\frac{G_{so}G_{ss}}{G_{so}\left(1-\frac{C_o}{100}\right)+\frac{C_o}{100}G_{ss}}\right\} \qquad (4)$$

where Gs, Gso and Gss represent specific gravities of peat soil, organic matter itself and soil particle, respectively, Co is organic matter content, ρ_d and ρ_w are dry density of peat soil and density of water, e and w are void ratio and water content and Sr is degree of saturation. Then, based on the results shown in Fig. 1, the relationships between Gs and Lig obtained from various soil layers are plotted in Fig. 2, and the calculated results by Eq.(1) are also shown with a solid curve. The calculation was carried out as Gso=1.5, Gss=2.7 and Co=Lig. Furthermore, by assuming as Sr=100 % and ρ_w=1 and using Eqs. (1) through (4), the relationships between physical parameters (e_n, w_n) and Lig are compared for the soil samples of upper peat layer, subpeat layers of I and II and clay layer containing peat in Figs. 3 and 4. In the two figures, the calcula-

ted relationships are shown with a family of solid curves. It was found from these figures that the calculated values almost agree with the experimental values. Thus, it can be said that the fundamental physical parameters of peaty ground are remerkably controlled by the amount of organic matters.

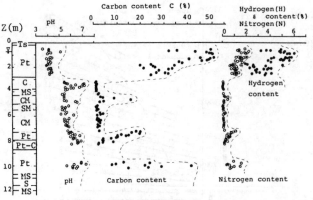

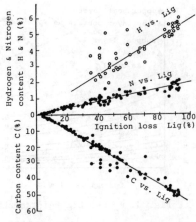

Fig. 7. Element contents and pH distributions.

Fig. 8. Element contents vs. Lig relationships.

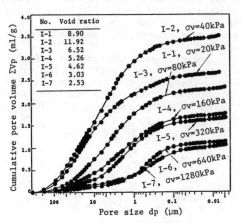

Fig. 9. ΣV_p vs. d_p relationships.

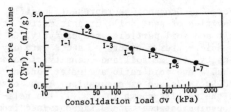

Fig. 10. $(\Sigma V_p)_T$ vs. σ_v relationships.

In Figs. 5 and 6, the organic matter content (Co) and humus content (Hu) are plotted against ignition loss (Lig). The values of Co do not coincide with the values of Lig but the proportional relations are approximately recognized between Co and Lig. However, the amount of scatter shows an increasing tendency with the increase in Lig. It will be thought that the difference in Co at the same value of Lig is dependent upon the quality of organic matters involved. As can be seen in Fig. 6, the amount of scatter in the humus contents (Hu), which represent humidity of peat soils, nearly corresponds to that of Co vs. Lig relationships in Fig. 5. In the soil samples of peaty ground investigated, it can be said that 20 - 30 % humic organic matters are roughly contained in the total amount of organic matters involved.

In general, the organic matters such as vegetal fibers are mainly composed of the elements of carbon (C), oxgen (O), nitrogen (N) and hydrogen (H). The distributions of the element contents of C, N and H are shown in Fig. 7, respectively. In addition, the distribution of pH in the depth direction is also plotted in this figure. The peaty ground was composed of the acid soils with the value of pH less than 7. In particular, the soil samples of upper peat layer showed pH of about 4 and exhibited a strong acidic property. Based on the results shown in Fig. 7, the relationships between three element contents (C, N and H) and Lig are indicated in Fig. 8. These relationships can be approximated by the straight lines, respectively. Thus, it can be said that three element contents almost show the proportional increase with increase in Lig. The element contents of C, N and H approximately correspond to 50 %, 2 % and 6 % of Lig, respectively.

4 MECHANICAL PROPERTIES

By using the undisturbed samples obtained from the peaty ground, one-dimensional consolidation, permeability and direct shear tests were carried out. The measurements of pore size distributions were performed on the specimens of upper peat lay-

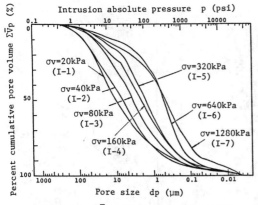

Fig. 11. $\Sigma \bar{V}p$ vs. dp relationships.

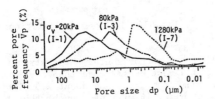

Fig. 12. $\bar{V}p$ vs. dp relationships.

er which were consolidated under different consolidation loads (σv). The pore sizes of peat specimens were calculated by the following equation (Ahmed et al., 1974 and Bengochea et al., 1979);

$$dp = - \frac{4 \sigma \ ccs \ \theta}{p} \qquad (5)$$

where dp and p are pore size and mercury intrusion pressure, respectively, σ is surface tension of mercury, θ is an angle of contact between mercury and specimen. The measured results of pore size distributions are shown in Figs. 9 through 12. Fig. 9 shows the relationships between pore size (dp) and cumulative pore volume per unit dry mass (ΣVp). the values of ΣVp gradually decrease with increasing consolidation loads (σv) because the amounts of consolidation settlements are increased with the increase in σv. Thus, the slopes of ΣVp vs. dp curves decrease in the range of large pore sizes more than about 0.1 μm and the volumes of pores tend to decrease remarkably. Then, the total pore volumes $(\Sigma Vp)_T$ were plotted against σv in Fig. 10. The relationships between (ΣVp) and σv can be approximated by a straight line on log-log plane. Also, can be seen in Figs. 11 and 12, the larger consolidation loads increase, the more the curves of percent cumulative pore volume ($\Sigma \bar{V}p$) vs. dp relationships sift to the right direction.

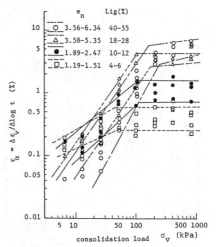

Fig. 13. ε_{α} vs. σv relationships.

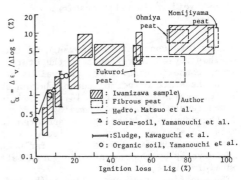

Fig. 14. ε_{α} vs. Lig relationships.

Thus, from the pore frequency curves of percent pore volume ($\bar{V}p$) corresponding to respective pore sizes, the locations of peak pore frequency gradually change into the range of small pore sizes with the increase in σv. Hence, it can be said that the pore size distribution of peat soil is influenced considerably by the consolidation.

Based on the oedometer test results obtained from the undisturbed samples of various soil layers, the coefficient of secondary consolidation (ε_{α}) are plotted against σv in Fig. 13. The curves of vertical strain (ε_v) vs. logarithmic time (log t) in the region of secondary consolidation were approximated by the straight lines. Thus, the value of ε_{α} corresponds to the slope of ε_v vs. log t straight line. As is evident from Fig. 13, the values of ε_{α} are dependent upon σv and Lig. However, in the cases of the samples with almost the same values of Lig, the values of ε_{α} are

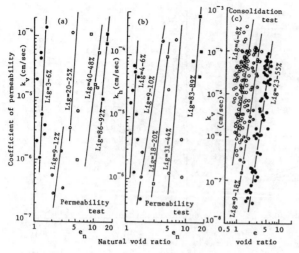

Fig. 15. Coefficients of permeability vs. void ratio relationships.

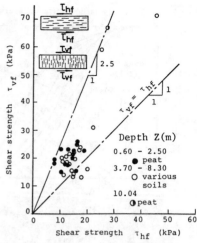

Fig. 16. τ_{vf} vs. τ_{hf} relationships.

roughly held constant in the range of large σv. This region corresponds to the state of normal consolidation. Then, the values of ε_α in the state of normal consolidation are plotted against Lig in Fig. 14. In addition to the results (Iwamizawa sample) in Fig. 13, the results of other peats and organic soils are also showed in this figure. It is clear from this figure that ε_α is influenced considerably by the amount of organic matter containing in soils.

The coefficient of permeability (k_v and k_h) obtained from permeability and oedometer tests are plotted against void ratios in Fig. 15. The suffixes of v and h signify the vertical and horizontal directions, respectively. In general, it is said that the permeamility of soils is mainly controlled by the void ratio. In the cases of peat and organic soils, however, the permeability is affected remerkably by not only the void ratio but the amount of organic matter. As can be seen in Fig. 15, in the case of soil samples with almost the same values of Lig, the relationships between coefficient of permeability and void ratio can be approximated by the straight lines on log-log plane. Thus, the soil samples with large values of e and Lig exhibit a very high permeability.

The anisotropy of shear strength is shown in Fig. 16. The results were obtained from direct shear tests on vertical and horizontal samples of peats and other soils. There is remarkable difference in shear strength between vertical and horizontal direction. The ratios of vertical shear stren-

gth (τ_{vf}) to horizontal shear strength (τ_{hf}) almost became more than one, and the maximum of τ_{vf}/τ_{hf} was about 2.5. Thus, in particular, it will be thought that the shear strength of fibrous peats is influenced considerably by the anisotropic structure formed during accumulation.

5 CONCLUSIONS

From the results of a series of physico-chemical and mechanical tests, it was found that the geotechnical properties of peaty ground were strongly controlled by the amount of organic matter involved. In particular, the physico-chemical and mechanical properties of peat soils can be systematically represented by the function of ignition loss.

REFERENCES

Ahmed,S., Lovell,C.W. & Diamond,S.D. 1974. Pore size and strength of compacted clay. ASCE, Vol.100, No.GT4 : 407 - 425.

Bengochea,I.G., Lovell,C.W. & Altschaeffl, A.G. 1979. Pore size distribution and permeability of silty clays. ASCE, Vol. 105, No.GT7 : 839 - 856.

Shridharan,A., Altshaeffl,A.G. & Diamond,S. 1971. Pore size distribution studies. ASCE, Vol.100, No.GT4 : 407 - 425.

Yamaguchi,H. & Ohira,Y. 1986. Drying shrinkage and pore diameter distribution of peats. JSSMFE, Symposium on thermal engineering properties of soils : 53 - 60.

Yamaguchi,H. & Nakayama,M. 1989. Change in pore structure of peat due to consolidation. JSSMFE, Symposium on engineering properties of peats : 33 - 42.

6th International IAEG Congress / 6ème Congrès International de AIGI, © 1990 Balkema, Rotterdam. ISBN 90 6191 130 3

Geotechnical properties of tertiary mudstone ground
Propriétés géotechniques de terre pélite de l'ère tertiaire

H. Yamaguchi & Y. Mori
Department of Civil Engineering, The National Defense Academy, Kanagawa, Japan

I. Kuroshima & M. Fukuda
Technical Research Laboratory, Mitsui Construction Company, Chiba, Japan

ABSTRACT: The slaking tests were carried out on the blocks of mudstones cutted from tertiary mudstone ground. The crushing properties due to slaking of tertiary mudstones were mainly investigated. In particular, the influences of degree of saturation and ultraviolent radiation were discussed. Furthermore, by using the crushed samples of mudstones produced by the effect of slaking, x-ray analyses, measurements of pore size distributions and swelling tests were also carried out and investigated on the changes in geotechnical qualities of tertiary mudstones with slaking histories.

1 INTRODUCTION

The mudstone is classified into the sedimentary soft rocks. It is generally crushed by the effect of slaking. Then, in order to deal with the geotechnical problems in practical such as the settlement, slope stabilty and bearing capacity of mudstone ground, it is a very important to perform the fundamental studies on the crushing properties of mudstone due to slaking.

In this paper, in the slaking tests on tertiary mudstones, the influences of degree of saturation and amount of ultraviolent radiation on crushing properties were paticulary investigated. Also, after slaking tests, by using crushed mudstones, the measurements of mineral composition, ignition loss and pore size distribution were performed with x-ray analysis and porosimeter apparatuses, and the changes in the geotechnical qualities of tertiary mudstones with the slaking histories were discussed. Furthermore, by using the clastic mudstone produced by the effect of slaking in the field, the influence of initial gradation of packed samples on the behavior of swelling and settlement was also investigated.

2 TERTIARY MUDSTONES TESTED

The samples tested in this investigation are three kinds of tertiary mudstones with physical properties shown in Table 1. The sample A and B are sound rocks of mudstones.

Table 1. Physical properties of mudstones

Properties		Sample A	Sample B	Sample C
Natural water content	(%)	23.4	20.9	4.25
Saturation	(%)	91.7	92.7	1.65
Specific gravity		2.45	2.53	2.49
Ignition loss	(%)	9.85	12.1	12.9
Liquid limit	(%)	63.8	65.2	65.5
Plastic limit	(%)	31.5	32.1	32.3

Photo. 1. State of slaking in the field

As shown in Photo. 1, the sample C is clastic mudstone, which was produced by the effect of slaking in the field. These samples were obtained from the region of tertiary mudstone stratum at a place called Hayama, about 10 km south from the center of Yokosuka city, Kanagawa, Japan. The blocks of sound rocks of sample A and B were taken from the depth of 0.5 - 1.0 m below the ground surface by using rock drill. The

527

sound rocks showed a very high water content more than 20 %. The sample C of clastic mudstone existed at air-dried state in the ground surface.

3 TESTING PROCEDURE

By using the sound blocks of tertiary mudstones, the influences of saturation and ultraviolent radiation on crushing properties were investigated. The several soil tests for investigating the changes in mineral composition, pore size distribution due to slaking and the behavior of swelling and settlement were also carried out on the clastic mudstones produced by slaking.

3.1 Slaking tests

Firstly, the blocks of sample B were dried for different durations at constant temperature of 20 °C, and then were submerged. Based on the test results, the relationships between crushing states after submergence and immediate degree of saturation before submergence were investigated. Also, after the different amounts of ultraviolent radiation were applied to the blocks of sample B which were oven-dried for one day at constant temperature of 110 °C, the blocks were submerged and the influence of ultraviolent radiation on the degree of crushing was discussed. The accumulative amount of ultraviolent radiation applied for six hours corresponds to that per about one year in the field. The application of ultraviolent radiation was carried out with sunshine weather meter apparatus shown in Fig. 1. Furthermore, cyclic tests of oven-drying (temperature 110 °C) and submergence were also carried out on the blocks of sample A and B. The respective durations of oven-drying and submergence were specified to be one day. In all the tests mentioned above, the states of crushing were observed by grain size analyses.

3.2 X-ray analyses, measurements of pore size distribution and ignition loss

By using the crushed mudstones of sample A and B which were produced in the cyclic slaking tests of oven-drying (temperature 110 °C) and submergence, in order to estimate the changes in mineral composition, pore size distribution and crystal water content involved with slaking histories, x-ray analyses, measurements of pore size distribution and ignition loss were carried out. The measurements of pore size distribution for crushed particles were performed with mercury intrusion porosimeter apparatus. In ignition loss tests for obtaining the crystal

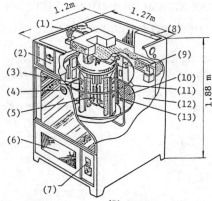

(1) Ark cooler
(2) Control panel
(3) Filter glass
(4) Door
(5) Carbon rod
(6) Humidity control tank
(7) Humidity control meter
(8) Exhaust duct
(9) Thermometer
(10) Sample rack
(11) Sample rotating disk
(12) Test box
(13) Water spray

Fig. 1. Sunshine weather meter

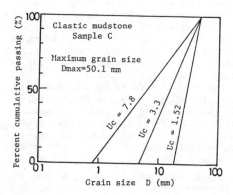

Fig. 2. Gradations of clastic mudstone

water content, the crushed particles were again burned at the temperature of 800 °C after oven-dried at 110 °C. The value of ignition loss (L_i) was defined as the ratio of the lost mass by burning at the temperature of 800 °C to the total dry mass at 110 °C.

3.3 Swelling and settlement tests

The clastic mudstone of sample C, which was produced by the effect of slaking in the field, was sieved by using sieves in the three different gradations shown in Fig. 2. The samples were packed in CBR mold (150 mm in diameter and 450 mm in height) and then constant dead loads were applied to the upper surface of samples. The behavior of swelling and settlement with elapsed time was observed under the repetitions of submergence and drying processes. In the method of drying, electroheater was used.

4 CRUSHABILITY OF MUDSTONE DUE TO SLAKING

The present authors (Yamaguchi et al. 1988) have already reported in detail concerning the mechanism of slaking and the mechanical properties of tertiary mudstones. In this section, the influences of saturation and ultraviolent radiation on the crushing properties due to slaking were mainly discussed on the mudstone blocks of sample B. Firstly, the blocks of sound rocks, which had degree of saturation (Sr) more than 90 % in the field, were satisfactory submerged in a water tank. The saturated blocks were secondly air-dried for different durations at constant temperature of 20 °C, and the degree of saturation was kept under control. Then, the unsaturated blocks of which the values of Sr were in the range of about 20 % - 100 % were obtained. Also, by oven-drying at 110 °C, the dry block of Sr=0 % was prepared. Then, the blocks with different degrees of saturation were again submerged and the states of crushing were observed during submergence.

Fig. 3 shows the relationships between crushing zone and Sr. In cases of tertiary mudstone tested, the crushing phenomena due to slaking are induced under the degrees of saturation less than about 80 %. The Sr of 80 % corresponds to water content (w) of about 16 %. The results of grain size analyses obtained from crushed samples are shown in Figs. 4 and 5, respectively. As can be seen in these figures, the degrees of crushing are dependent upon Sr at submergence. Thus, it can be seen that the remarkable crushing pheno-

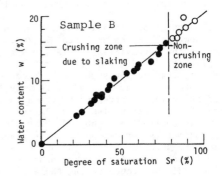

Fig. 3. Relationships between crushing zone and Sr

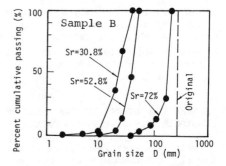

Fig. 4. Effect of degree of saturation

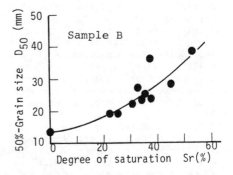

Fig. 5. Dgree of crushing with changes in Sr

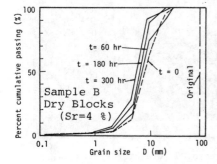

Fig. 6 Effect of ultraviolent radiation

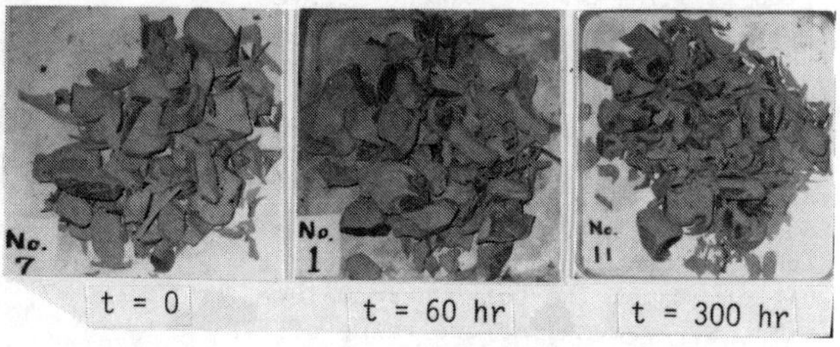

Photo. 2. Comparison of crushing states

mena are observed with the reduction in Sr.

The effect of ultraviolent radiation is shown in Fig. 6 and Photo. 2. In the tests, the ultraviolent radiation was applied to the dry blocks of Sr=4 %. After the application, they were submerged in a water tank. The amounts of ultraviolent radiation were controlled by the duration (t) of application. The accumulative amounts applied for t=60 and 300 hours approximately correspond to those given from the sunshine for 10 and 50 years in the field, respectively. The crushing of blocks did not cause during the applications of ultraviolent radiation. The crushing phenomena shown in Photo. 2 caused at the submergence. However, as shown in Fig. 6, the effect of ultraviolent radiation was only slightly recognized. Hence, it can be said that the degree of crushing due to slaking is influenced considerably by the degree of saturation rather than the amount of ultraviolent radiation.

5 CHANGES IN GEOTECHNICAL QUALITIES OF MUDSTONES WITH SLAKING HISTORIES

The cyclic tests of oven-drying (temperature 110 °C) and submergence were performed on the sound blocks of sample A and B. The number of cycle (N) was taken to be ten. The grain size analyses were performed at the ends of each cyclic stage. In order to investigate the changes in geotechnical qualities with slaking histories, the measurements of pore size distribution, ignition loss (crystal water content involved) and x-ray analyses were performed on the crushed particles which were produced at each cyclic stage.

The changes in the states of crushing in cyclic slaking tests are shown with grain size frequency curves in Fig. 7. In case of sample A, the states of crushing at the stages of the one and the ten cycle are also indicated in Photo. 3. From sifting to the left direction of grain size frequency curves, it is very clear that the degree of crushing is gradually accelerated with increase in number of cycle (N).

The measured results of pore size distributions for crushed particles obtained from typical cyclic stages are indicated in Fig. 8. Symbol of ΣVp signifies cumulative pore volume per unit dry mass. The total pore volume $(\Sigma Vp)_T$ was assumed as the value of ΣVp till pore size (dp) 0.0035 μm. The re-

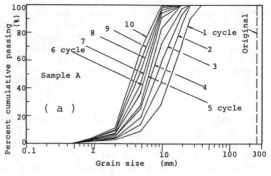

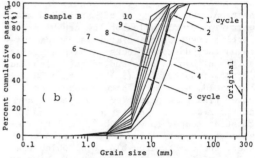

Fig. 7 Changes in gradation in cyclic slaking tests

Photo. 3. Crushing states in cyclic slaking tests

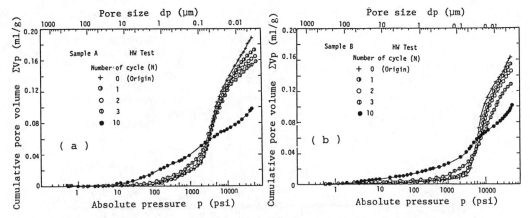

Fig. 8. Changes in pore size distribution in cyclic slaking tests

lationships between $(\Sigma Vp)_T$ and $N + 1$ are also plotted in Fig. 9. It can be said from these results that the pore size distributions of mudstones are affected considerably by the slaking histories and the crushed particles with small pore volumes are formed by the developments of slaking. Moreover, the changes in ignition loss (Li) with number of cycle (N) are shown in Fig. 10. The value of Li must signify crystal water content which is contained in mudstone. As well as the measured results of pore size distributions, it will be thought that the decrease in Li is closely related with the developments of slaking.

Figs. 11 (a) and (b) indicate the changes in mineral compositions with the developments of slaking. Based on these results, for making easy to understand, the intensities of main minerals are again compared in Fig. 12. The existence of montmorillonite was recognized from the results of x-ray analyses on the same samples treated with ethylene glycol. The minerals of tertiary mudstones of sample A and B are mainly composed of montmorillonite, illite, chlorite, quartz, anorthite and calcite. However, as can be seen in Figs. 11 and 12, the changes in mineral composition with repetitions of oven-drying and submergence were hardly observed. Thus, as shown in Fig. 10, it will be said that the developments of slaking only bring about the reduction in crystal water contents rather than the changes in mineral composition.

6 SWELLING AND SETTLEMENT BEHAVIOR OF CLASTIC MUDSTONE

The clastic mudstone of sample C, which was

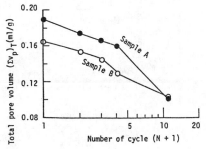

Fig. 9. $(\Sigma Vp)_T$ vs. N relationships

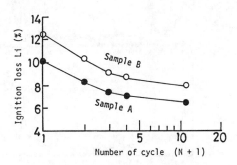

Fig. 10. Li vs. N relationships

produced by slaking in the field, was packed in CBR mold, and then the different magnitudes of constant dead loads were applied on the sample surface. The behavior of swelling and settlement was observed at the cyclic stages of submergence and drying. In the tests, the packed samples with three different gradations shown in Fig. 2 were used. Fig. 13 shows the behavior of swelling and settlement under different constant loads (P) for the samples with uniformity coefficient (Uc) of 3.3. The packed samples

531

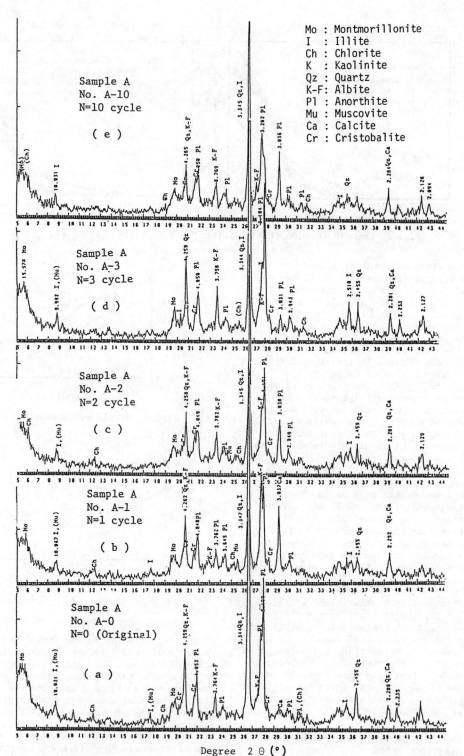

Mo : Montmorillonite
I : Illite
Ch : Chlorite
K : Kaolinite
Qz : Quartz
K-F: Albite
Pl : Anorthite
Mu : Muscovite
Ca : Calcite
Cr : Cristobalite

Sample A
No. A-10
N=10 cycle
(e)

Sample A
No. A-3
N=3 cycle
(d)

Sample A
No. A-2
N=2 cycle
(c)

Sample A
No. A-1
N=1 cycle
(b)

Sample A
No. A-0
N=0 (Original)
(a)

Degree 2 Θ (°)
Fig. 11(a). Changes in mineral composition with developments of slaking (Sample A)

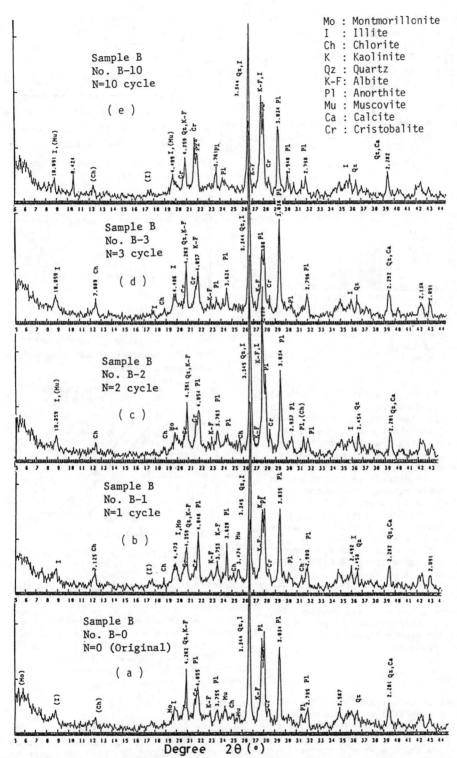

Fig. 11(b). Changes in mineral composition with developments of slaking (Sample B)

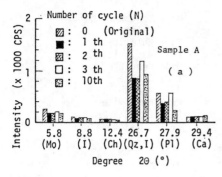

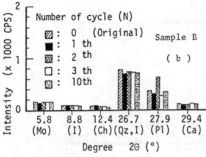

Fig. 12. Changes in main minerals
due to slaking

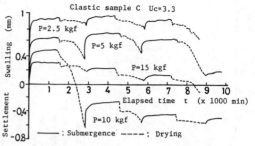

Fig. 13. Behavior of swelling and
settlement (Uc=3.3)

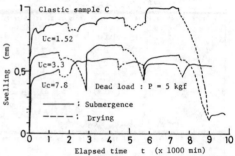

Fig. 14. Behavior of swelling and
settlement (P=5kgf)

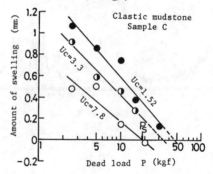

Fig. 15. Swelling vs. P relationships

showed the behavior of swelling at each stage
of submergence. On the other hand, at each
stage of drying, the behavior of settlement
due to drying shrinkage was observed. The
behavior of swelling and settlement is influenced considerably by the magnitudes of
P. As shown in Fig. 14, the behavior is also
controlled by the gradations of packed samples. The amounts of swelling become very
large in the case of packed sample with
uniform gradation. Then, based on the test
results, the swelling pressure (Ps) can be
estimated. The amounts of swelling at the
first stages of submergence were plotted
against surface loads (P) on semi-log plane
in Fig. 15. It will be thought that Ps
corresponds to P at zero swelling. In case
of Uc=7.8, the value of Ps is equal to about
20 kgf. However, the swelling pressure is
dependent upon the gradation of packed
sample, and may become large on the sample
with uniform gradation.

7 CONCLUSIONS

(1) Crushability of tertiary mudstone due
to slaking is remarkably dependent upon
degree of saturation but is hardly affected by the amount of ultraviolent radiation.
(2) The changes in pore size distribution
and crystal water content with the deve-
lopements of slaking bring about but the
changes in mineral composition are hardly
observed.
(3) In case of clastic mudstone which was
produced by slaking, the behavior of swelling and settlement with repetitions of
submergence and drying is influenced
considerably by the gradations of packed
samples.

REFERENCE

Yamaguchi,H., Yoshida,K., Kuroshima,I. and
Fukuda,M. 1988. Slaking and shear proper-
ties of mudstone. ISRM, Rock Mechanics and
Power Plants, Vol.1: 133 - 144.

Triaxial testing of rock/soil under cryogenic conditions

Test triaxial de roches/sols sous conditions cryogéniques

N. F. Zorn & P. E. L. Schaminée
Delft Geotechnics, Delft, Netherlands

ABSTRACT: A triaxial facility and test procedure to determine the mechanical properties and thermal strain as function of the temperature are presented. The facility allows to simulate the in situ conditions of rock/soil in the surrounding of an unlined cryogenic storage cavern for a range of conditions. Results achieved during tests with rock and overconsolidated clay are presented.

RESUME: Un appareil triaxial et le procedé a suivre pour determiner les proprietés mechaniques et les deformations thermals en function de la temperature sont presentés. Le system permet de simuler les conditions in-situ de roches et sols, dans les alentours d'une caverne de storage cryogenique sans séperation entre soil et fluid, pour toute une gamme de conditions. Les résultats obtenus durant des tests effectuer avec des roches et des argiles surconsolidés sont présentes.

1. INTRODUCTION

The storage of liquefied hydrocarbons poses engineering problems that require knowledge on material behaviour at temperatures commonly not investigated in geotechnical and geological engineering (Homer et al, 1989). Extensive studies have been performed to investigate the properties of steel and concrete (see Van der Veen for a survey, 1988) which are used for construction of tank installations on land.

Unlined underground storage has many advantages when compared to conventional insulated storage in tanks, especially with regard to environmental and safety aspects. It could even become commercially attractive provided that the geological conditions are favourable and that design methods are further developed so that the advantages can be taken into account.

This paper deals with laboratory equipment and experiments that have been developed (Horvat et al, 1988) to determine rock and soil properties under conditions that simulate the mechanical and temperature loading that the material will be exposed to in situ, surrounding an unlined underground crygenic storage cavern.

Experience and facilities available from triaxial testing of soil samples were used and combined with new elements to cool the samples and monitor its behaviour during cooling.

2. TEST AIM

The aim of the tests is twofold:

a) to determine mechanical soil properties as function of the temperature under conditions that resemble the mechanical and temperature flow conditions as can be found in situ (i.e. a low temperature front originating in the center of a volume and not at the surface) and

b) to gain insight into material behaviour and phenomenons during a special type of scale model test.

The first aim, the determination of the mechanical properties is subdivided into properties that directly describe the mechanical behaviour, i.e. Youngs modulus and Poissons´ ratio and strength parameters, and the temperature strain relation, i.e. the volumetric change of the sample due to the temperature loading. The latter was chosen as integral indicator and may become a very important characteristic for design of unlined caverns, since contraction due to cooling will cause tension and may lead to cracking. Cracks around a storage cavern increase the contact surface between the cryogenic liquid and the surrounding and this is unfavourable with respect to the temperature field and the corresponding energy losses (boil off rate).

An integral result of the tests therefore is to determine the temperature-strain relation and verify the proposed characteristics of rock and clay as shown in figure 1.

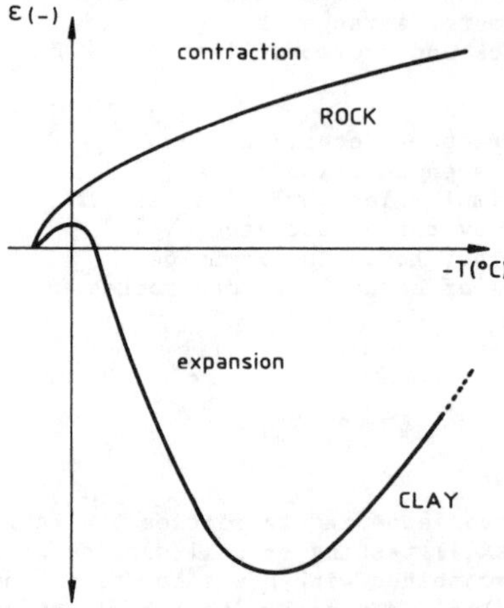

Figure 1: Proposed temprature strain behaviour of rock and clay

Once a design procedure is developed that takes the volumetric strain behaviour into account this test provides an objective criterion which makes it possible to qualify different soil and rock formations with respect to their suitability for unlined underground storage of liquefied hydrocarbons under different mechanical (in situ) loading conditions.

The second aim, analysis of phenomena that occur during the cooling process, i.e. the interpretation of the triaxial tests as a scale model test applies to different materials and phenomena to varying extents.

Tests in rock where the geometry of the cooling hole diameter is in the range of two to ten times the maximum diameter of the size of the minerals, depending on the type of rock, are limited with respect to

simulation of fracture mechanic effects on scale. The characteristic tensile strength becomes the strength of the mineral and not the strength of the composite rock.

Scale tests for this type of loading should consider parameters as a characteristic volume and stress redistribution due to plastic material behaviour which also counts for "brittle" materials as granite and concrete (see Walraven, 1990 and Hillerborg et al 1976). Both materials show tension softening in the post peak part of the uniaxial stress strain relation. It however may be possible to analyse the opening and stress relaxation effect of an existing joint which crosses the cooling hole.

When analysing clay samples the phenomenon of cracking can be analysed to a larger extent, since the characteristic volume is related to the particle size. The short coming of the stress distribution which is influenced by the sample size however stil exists. The favourable expansion of the material during part of the freezing branch however will lead to compressive stresses as shown in figure 1.

3. TEST FACILITY

The test facility used for the cryogenic temperature loading tests is a modified high precision triaxial test apparatus for cylindical samples with a diameter of 75 mm and a heigth of 140 mm. High precision referring to the load control and deformation measurement devices.

In triaxial testing the sample, which is sealed in a latex membrane, is exposed to mechanical loading in radial and axial direction while deformations and pore water behaviour are monitored during predescribed loading cycles. The radial stress, corresponding to the cell pressure is applied with silicon oil, the vertical surcharge with a plunger.

The utilization of a standard test facility as starting point for a new testing device makes it possible to use the "standard" options and experience gained as deformation measurements, mechanical boundary conditions [isotropic and vertical surcharge stresses], drainage or pressure measurement of the pore water and data acquisition and control. The dimensions of the facility are also sufficient to test a wide range of standard cores drilled in different parts of the world.

Additional installations were added to cool the sample, to control the cooling process and to measure the temperature in the sample at different locations. Figure 2 presents a schematic sketch of the facility.

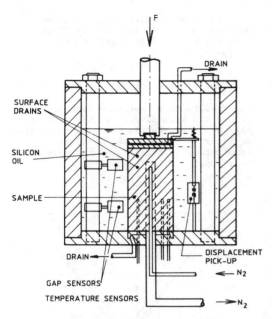

Figure 2: Modified triaxial cell for cryogenic testing

The mechanical boundary conditions that can be achieved in the currently used cell are: isotropic pressure max. 10 bar and a surcharge load which at the mentioned diameter results to additional 10 bar in vertical direction. The samples thus can be exposed to stress conditions equivalent to in situ stresses at a depth of aprox. 100 m.

The combination of pressure and temperature loading made it necessary to redesign the cell wall. In spite of the sensitivity to cracking a perspex wall was given the preference to a steel wall, allowing the technician performing the tests to visually follow the tests.

The deformation measurement devices which were used to monitor the sample are contactless gap sensors bases on the eddy current principle with an accuracy of 5 μm in radial direction. For axial deformations transducers (Sony magnesensors) with an accuracy of 5 μm were used.

Using the drainage system, which consists of two filters one in the base and one in the top plate, and a set of paper drains connecting the sample surface with these filters the pore water flow (for drained conditions) or pressure (for undrained conditions) can be monitored by either measuring the volume flow and the pressure in the saturated closed water circuit. This system ofcourse will only function in the first phase of a temperature loading test, until the water in the filters turns to ice.

The additional installations that were added are: the cooling circuit, making it possible to cool the sample in the central axis, temperature sensors, to monitor the sample temperature and a software PID controller to steer the cooling and warming up process, heating devices, used to heat the silicon oil surrounding the sample in order to warm the sample up as controlled as it was froozen.

It is the intention to include acoustic emission sensors located on the sample to monitor acoustic activity possibly resulting from micro and macro cracking during the cooling process.

4. TEST PROCEDURE

The preparation of a cored sample basically consists of drilling holes for the cooling in the center and temperature measurements at different distances. Clay samples shaped from larger lumps require trimming to a cylindrical shape. Clay and other saturated samples are generally consolidated prior to thermal testing, i.e. the sample is mechanically loaded in radial and axial direction according to the in situ stress state that the test is performed for. The duration of this consolidation phase depends on the permeability of the sample, it is monitored by control of the pore water flow.

After the sample has been mounted and the mechanical boundary conditions have been met (see figure 3) the control units are programmed to maintain these conditions during the entire duration of the tests in order to measure deformations resulting from temperature effects only.

The samples are cooled with liquid nitrogen that is injected into the central cooling hole in an open unpressurized circuit. The temperature sensor nearest to the cooling hole is chosen as reference temperature for the cooling rate. This so far has been varied between shock cooling, 20 °C and 1 °C per hour. A micro computer HP 310 steers the valve that controls the amount of nitrogen injected into the sample by comparing the measured reference temperature with a calculated value as function of time. The cooling and the warming up phases are both controlled during the test. Warming up basically means the supply of less cooling energy due to the heat input from the silicon oil. When reapproaching ambient temperature after long tests the silicon oil has to be heated to warm the sample up according to the predescribed rate.

The measurements that are performed during the test are deformation measurements in radial and axial direction to determine the volumetric strain and temperature measurements at different locations.

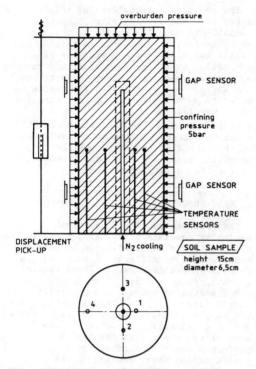

Figure 3: Sample mounted in triaxial
testing facility

The change of the outer dimensions are
monitored with two sets of three gapsensors
monitoring the diameter change at different
heights and three displacement transducers
which measure the deformation in axial
direction. The volumetric strain ε_v is
determined according to:

$$\varepsilon_v \approx \varepsilon_z + 2 \varepsilon_h$$

ε_v and ε_h being the vertical and
horizontal or radial strain.
   Elastic properties as Youngs' modulus and
Poissons' ratio can be derived from
incremental deformations resulting from
additional surcharge loadings at
predetermined temperatures.
   After a test the sample can be inspected
for cracks with different techniques.

5. TEST RESULTS FOR ROCK AND CLAY

A number of tests have been carried out
following the outlined procedure. This
chapter will present results for rock and
clay tests which have been performed for
different stress boundary conditions and
different cooling rates. The more recent
rock tests which also include the warming
up phase show that the control of the
temperature has been significantly
improved.

5.1 Test on rock sample

The results of a test on a rock sample
cooled and warmed up at a rate of 1 °C per
hour are presented in the figures 4 through
6. Figure 4 presents the measured
temperature as function of the time for the
four locations in the sample.

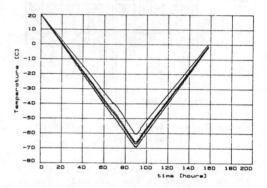

Figure 4: Temperature-time history at four
locations in rock sample

A good or rather high thermal conductivity
of the rock can be concluded when
examining the temperature gradient in the
sample, it is less than 10 °C for aprox.
22 mm.

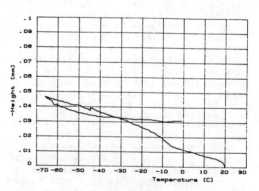

Figure 5: Axial deformation as function of
the reference temperature

The deformation measurements in radial
and axial direction are presented in
figures 5 and 6. The radial deformation is
presented for both locations while the
axial deformation presented is the avarage

of the three measured values. All three
curves show contraction during cooling and
expansion during warming up. The hysteresis
in the curves can be explained by the
reversal of the cooling front and the
offset at the end may be due to the effect
of some internal micro cracking that may
have occured (Wiedemann, 1982). The
difference in radial deformation for the
two locations indicates inhomogenity, this
may result from the sample material or the
cooling.

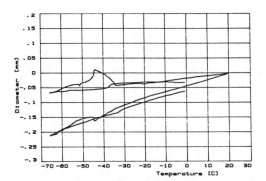

Figure 6: Radial deformation as function of
the reference temperature for two
locations

The volumetric strain that can be
calculated based on these results shows
that cooling the sample down to -65 °C
results in a strain of 0.6%. This value,
scaled to the dimensions of a full scale
cavern gives an indication of the tensile
stresses that may be expected or the
opening of existing joints.

5.2 Test on clay sample

The results of a test on a sample of
overconsolidated clay (found near Iperen)
are presented in the figures 7 through 9.
Figure 7 presents the temperature time
history, which when compared to figure 4
shows more fluctuation and a temperature
gradient in radial direction of 22 °C. In
addition the warming up of the sample was
not monitored because the sample was
inspected for cracks in its froozen state.
The larger fluctuations can be explained
by the not optimised cooling control
circuit, nozzle and valve have been
significantly improved since, and the
larger gradient indicates a lower
conductivity of the clay.

The results of the deformation measurements
in radial and axial direction are presented
in figures 8 and 9. The radial deformation
is presented for both locations while the
axial deformation presented is the avarage
of the three measured values. In contrast

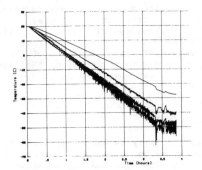

Figure 7: Temperature-time history for four
locations in clay sample

to rock the deformations do not only show
contraction. All three curves show
contraction during a short cooling period,
until aprox. 0 °C is reached and then
expansion. The difference in radial
deformation for the two locations again
indicates inhomogenity, resulting from the
sample material or the cooling.

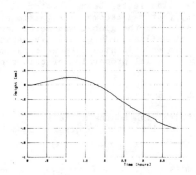

Figure 8: Axial deformation as function of
the test time elapsed

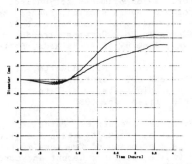

Figure 9: Radial deformation as function of
the test time elapsed for two
locations

The inspection of the froozen sample did
not reveal any significant cracking.
However a test in which the sample was

confined with a nominal pressure only, lead to multiple cracking of the clay.

If the three curves presented in figures 7 through 9 are combined to determine the volumetric strain as function of the temperature a curve verifying the proposed curve for clay presented in figure 2 would result.

6. CONCLUSIONS

A test facility in which the mechanical and temperature boundary conditions as they occur in situ surrounding an unlined underground cryogenic storage has been developed. The available facility is capable of covering a temperature range to -65 °C measured at the surface of the sample, further extensions of the temperature range pending.

Following the test procedure as it is presented results in quantitative information on mechanical material properties as function of the stress boundary conditions and the temperature, and more integral information about the temperature-strain behaviour of the material. The latter being a material property that may become of significant importance for the global and detailed design of unlined underground cryogenic storage in which finite element codes may require material properties of this type.

AKNOWLEDGEMENT

The authors wish to thank Shell International for the permission to publish the results.

REFERENCES

Homer,J.B., Horvat,E., Unsworth, J.F., Laboratory experiments for unlined cryogenic cavern storage, Proceedings of the International Conference on Storage of Gases in Rock Caverns, Trondheim, Norway, June 89, A.A. Balkema, Rotterdam

Horvat,E., Zorn, N.F., Unsworth, J.F., A method and cell for freezing formation layer sample tests, U.K. Patent No. 8828205 filed Dec. 1988

Hillerborg,A., Modeer,M., Petersson, P.E., Analysis of crack formation and growth in concrete by means of fracture mechanics and finite elements, Cement and Concrete Research, Vol.6 No.6, Nov. 1976, pp773-782.

Van der Veen, C., Properties of concrete at very low temperatures, a survey of literature, CUR, STW report 88-3, Gouda, Holland

Walraven, J.C., Scale effects in beams with unreinforced webs, loaded in shear, Progress in Concrete Research, Annual Report, Vol. 1, 1990, Delft University of Technology

Wiedemann, G., Zum Einfluss tiefer Temperaturen auf Festigkeit und Verformung von Beton, Dissertation Technische Universität Braunschweig, 1982

1.4 Procedures, classification and interpretation for engineering design
     Procédures, classification et interprétation pour des plans techniques

6th International IAEG Congress / 6ème Congrès International de AIGI, © 1990 Balkema, Rotterdam. ISBN 90 6191 130 3

Use of remote sensing to estimate earthwork volumes for road construction

Utilisation de la télédétection pour estimer le volume de travail lors de la construction d'une route

Joseph Olusola Akinyede
Nigerian Building and Road Research Institute, Lagos, Nigeria

Keith Turner
Colorado School of Mines, Department of Geology and Geological Engineering, Golden, Colo., USA

Niek Rengers
International Institute for Aerospace Survey and Earth Sciences (ITC), Enschede, Netherlands

ABSTRACT: Quantities of earthwork must be evaluated and optimized to meet geometric specifications for road alignments and to reduce road construction costs. This requires quantification of the internal relief and valley density. These variables were assessed using remote sensing data to model in a GIS the terrain roughness and grading volumes along potential road routes in northeastern Nigeria.

RESUME: L'ampleur du travail de sol doit être évaluée et optimisée de manière à rencontrer les spécifications géométriques découlant des alignements de la route et de manière à réduire les coûts de construction. Ceci demande une quantification du relief interne et de la densité des vallées. Par le biais de la télédétection, ces variables sont gérées de manière à modéliser dans un SIG la rugosité du terrain et l'effet topographique le long de profils routiers potentiels dans le nordest du Nigéria.

1 INTRODUCTION

The cost of excavating soil and rock materials is one of the largest single costs in road construction projects. It is directly related to the total volume of materials excavated and moved.

The most ideal economic situation is to balance the cut-and-fill quantities, which requires longitudinal profiles to determine the total volume of materials excavated and used for fill by assessing the difference between the maximum road gradient and the terrain gradient. A proper estimation of the volume of earthwork involved in road construction thus requires knowledge of the topography (or terrain roughness). This information was used to optimize the volume and cost of earthworks for potential road routes in northeastern Nigeria (Akinyede 1990).

A quantitative evaluation of the terrain form is usually based on the description and determination of two important characteristics: "internal relief" and "valley density". A dimensionless "ruggedness number" can be generated from the product of these two (Rengers 1981). The ruggedness number expresses the overall configuration of the terrain (or terrain roughness) which can be used in assessing the earthwork quantity. Classes of topographic forms defined in terms of internal relief and valley density can be related to terrain mapping units. These data can be stored in a geographic information system (GIS), and used to determine the valley widths, valley side gradients and potential quantities of cut-and-fill.

2 DEVELOPMENT OF THE TERRAIN MAPPING UNIT

A test area of approximately 15,000 km^2 in the Gongola valley area, NE Nigeria (figure 1), was selected for study. The terrain was classified in mapping units on the basis of a predefined system of land classification (Meijerink 1990), and the relevant cut-and-fill values were determined for each mapping unit.

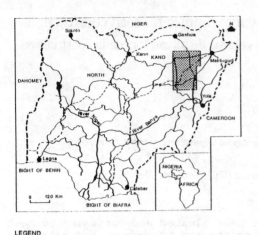

Figure 1. Map of Nigeria showing some of the existing road network development.

The terrain in the test area is characterized by rugged hilly regions underlain by granites and sandstones, volcanic plateaus, isolated steep volcanic plugs and low swampy clay plains.

A Precambrian basement complex, dominated by granites and gneisses, underlies the area. These rocks are overlain by Cretaceous sedimentary rocks, predominantly sandstones and shales of both marine and continental origin. The older sedimentary rocks are complexly folded and create long ridges. The younger rocks

are more gently folded. All are faulted and covered by basaltic lava flows and pyroclastic materials.

This variety of conditions creates potential problems and opportunities for road construction. Highly plastic "black-cotton" clay soils are found in areas of basalt flows and expansive shales. A monotonous occurrence of both massive and dissected sandstone hills and ridges and complex rugged granite hills constitute a topographic constraint which affects the road network development in the area.

2.1 Classification of the land system mapping units

The basic components of the land classification system are land system mapping units (LSMU). The terrain is assessed on the basis of geologic, geomorphologic and geotechnical characteristics which are spatially related by defining areas with essentially uniform characteristics (Akinyede & Rengers 1990). Internal relief and valley density provide useful information about the average steepness of slopes, which can be used to distinguish between mapping units. Figure 2 illustrates two mapping units in the test area with similar internal relief but with different valley densities and slope steepnesses.

The land system mapping units were delineated by visual photo interpretation of Landsat thematic mapper (TM) satellite images and side-looking airborne radar (SLAR)

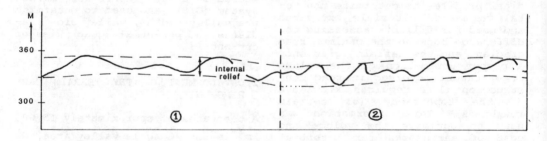

Figure 2. Average sampled cross-section showing two land system mapping units, Gulani sandstone hills (1) and dissected Bima sandstone hills (2) with similar internal relief (50 m) but different valley density (Vd1/Vd2 = 1/2), Gongola test area, NE Nigeria.

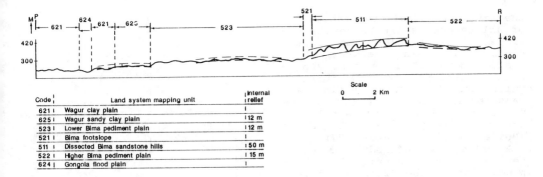

Figure 3. Cross-section illustrating land system mapping units with corresponding internal reliefs (IR), Gongola test area, NE Nigeria.

images (scale 1:250,000). Further divisions of the land system units into sub-units, with characteristic relief components, were made from the interpretation of SPOT images (scale 1:100,000), topographic maps (scale 1:50,000) and aerial photographs (scale 1:25,000).

2.2 Determination of internal relief

The internal relief (IR, also known as relief amplitude or local relief) is the average of the height differences (in metres) between local drainage divides or hill tops and the valley bottoms within a land system mapping unit. The internal relief is determined for each land system mapping unit. A typical example of a 2-dimensional cross-section, showing some land system mapping units and their internal reliefs, is shown in Figure 3 (see fig. 4).

The minimum number of required observations is based on the internal variation within each mapping unit. Logarithmic class intervals can be specified for both hilly and low-lying terrains, as shown in table 1, or the minimum and maximum internal relief per mapping unit can be used direcly in the calculation of earthwork volumes.

2.3 Determination of valley density

In many road engineering studies, drainage density (DD) is used as an important indicator of the linear

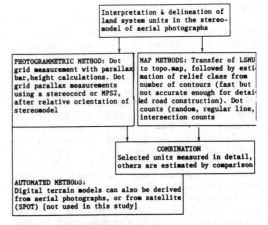

Figure 4. Scheme of methods for obtaining the internal relief (modified from Meijerink 1988).

scale of landform elements in stream-eroded topography. Drainage density is defined and measured as

Table 1. Possible classes of internal relief (IR) for the terrain in Nigeria.

symbol (relief)	IR class (in m.)
F - Flat	0-5
L - Low	5-10
I - Intermediate	10-20
M - Moderate	20-40
H - High	40-80
VH- Very high	80-160
EH- Extremely high	160-320

the ratio of cumulative length (L)
of all channel segments to the basin area (DD = ΣL/A). Recent studies have shown, however, that
"valley density" is a more appropriate descriptor of the terrain.
Valley density (VD) is defined and
measured as the length (Lv) of the
valley bottoms (or drainage lines)
per unit area (A), i.e., VD = ΣLv/A
(Meijerink 1988, Schumm 1977).

A valley is defined as a slope
reversal along a section line perpendicular to the drainage or valley bottom, with walls having a
given length. Valley density is
usually determined from maps and
stereo aerial photos using photogrammetric measurement techniques
(Meijerink 1988).

In this study, however, valley
density was estimated for each mapping unit using a counting circle
template, as shown in figure 5a.
The circle's diameter depends upon
the mapping scale, such that at
least eight values of the ridge/
valley crossings should be obtained
for statistical analysis. A circle
diameter of 8.3 cm was used for the
1:25,000 scale photos. The essential information on valley geometry
for estimating cut slope gradient
and earthwork quantities is the
valley width.

The circle, on transparent overlay, is placed on the stereo aerial
photographs and the number of ridge
or valley crossings is counted for
each sample area. Many locations
are sampled for each mapping unit
and the mean number of ridge/valley
crossings and standard deviations
are calculated.

Ridge/valley crossings were
counted in four directions (N-S,
NE-SW, E-W and SE-NW; figure 5a),
to evaluate the influence of direction on the earthwork volume. The
valley frequency count was estimated in terms of fractional valleys
per circle diameter. For example,
figure 5b shows a typical profile
along a diameter. The distance 0-1
is calculated as a half ridge/val-
ley crossing and the distances 1-2,
2-3, 3-4, etc. are calculated as
approximately one ridge/valley
crossing each. The total number of
ridge/valley crossings from 0 to 6
is thus 5.5. This implies that a
valley is approximately equal to
two ridge/valley crossings, i.e.,
from ridge to ridge.

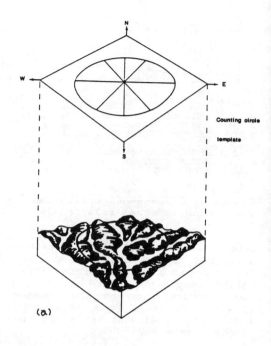

(a)

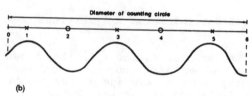

(b)

Figure 5. Schematic illustration of
the determination of number of
ridge/valley crossings. (a) Block
diagram of part of the Gongola
granite plain (NE Nigeria) with a
counting circle template for determining the number of ridge/valley
crossings. (b) Generalized profile
of the terrain with the ridge/val-
ley crossings.

If W = valley width (in m),

$$W = \frac{X \text{ (in cm)} \times \text{scale number}/100}{N/2}$$

The approximate gradient (in %)
of the valley side slope can be determined as follows:

$$\frac{\text{Height difference, h (in m)} \times 100}{\text{valley width}/2}$$

The numbers of ridge/valley
crossings were stored in the database of a geographic information
system (GIS) and used in the manipulation of the relevant equations

to determine valley widths, valley side gradients and earthwork volumes in a road construction project.

3 COMPUTATION OF EARTHWORK VOLUMES

Estimation of earthwork volumes was based on a number of simplifying assumptions, which include:
- local mass balance (cut-and-fill equalities) can be readily achieved;
- side slope gradients in cuts and fills are roughly equal;
- material bulking and shrinkage ratios are either very small or offsetting, and can be neglected;
- the amount of excavated material that is unsuitable for placement in fills, and thus must be wasted, can be determined from data defining the lithologic properties and the percentage soil/lithologic composition in the soil/lithologic complex, as illustrated in figure 7.

Correction factors are later included in the database for each land system mapping unit (LSMU) which partially compensate for these simplifications.

3.1 Determination of actual volumes of fill (or cut)

The determination of actual volume of fill (or cut) is based on a geometric representation of an idealized roadway fill (fig. 6b), width L, crossing a V-shaped valley with a typical symmetric geometry. The roadway may be supported by a fill volume having a maximum height of h, a side slope gradient of g and a length of fill D. To balance cut-and-fill volumes for economic purposes, longitudinal profiles are prepared and the volumes of cut are computed in a similar fashion. Thus, the total volume of fill (or cut) =

$$1/2 \ (DhL) + 1/3 \ (Dhh)/g) \, 100$$

where: D = typical valley width at the fill heigth; h = typical local relief or depth of valley below road level; L = roadway width; g = fill side slope gradient.

L and g are governed by roadway design standards, and g in cut (or

fill) is also related to the local soil and rock conditions, and can be defined for each mapping unit. The values of D and h are related to the terrain mean valley width, W, and local relief, Z (fig. 6a), and vary with the terrain roughness for each land system mapping unit.

For any desired principal direction, a valley side slope gradient (vsg) can be computed from the typical valley width and local relief data, as shown in figure 6c. The vsg value can be compared to the maximum allowable longitudinal roadway gradient (rg) (which is defined in the Nigerian Road Design Manual, FMWH 1973). As shown in figure 6c, the maximum cut/fill height (h) can vary between maximum (hmax) values and minimum (hmin) values, where:
hmax = (local relief)/2 = Z/2;
hmin = hmax - Δh;
h = { (rg) . (W/4) }/100={ (rg) . W}/400;
and average typical h = hmax - Δh/2
Z/2 - { (rg) . W}/800,
where: Z = local relief; h = typical maximum height of fill (or cut); hmax = largest maximum height of fill (or cut); hmin = smallest maximum height of fill (or cut); rg = standard longitudinal roadway gradient; W = typical valley width.

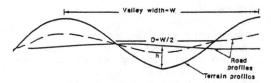

Figure 6a. Sketches of typical road profiles on rough or hilly terrain

The volume of cut-and-fill depends on the difference between the valley side gradient (vsg) and the maximum road gradient (rg). In a flat terrain, the values of vsg < = rg(max), the volume of earthwork will approach some minimum value; a subgrade height is then determined which can be used to calculate the volumes of earthwork for raising the road level above the surrounding plain. In an undulating to hilly terrain, the earthwork volume increases as vsg becomes increasingly larger than rg.

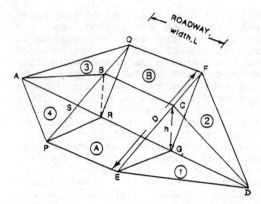

Figure 6b. Geometrical representation of an idealized roadway fill.

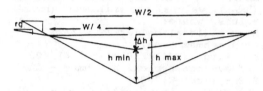

Figure 6c. Sketches showing geometrical idealized relationships between valley side gradient (vsg), allowable longitudinal road gradient (rg) and typical fill parameters: D (= W/2) and maximum height (h).

3.2 Volume adjustments

Adjustments must be made to compensate for changes in the volume of earthwork caused by shrinkage and bulking, unsuitable excavated materials and excess cut or fill materials which result from different cut-and-fill slopes. The assumptions used in this study were based on the engineer's experience in earthwork operations. Correction factors must be introduced, however, based on the available terrain data, to account for the necessary volume changes, and thus meeting the requirements for cut-and-fill balances. For example, the mass haul diagram in figure 7a illustrates a deficiency of fill materials caused by an unsuitable layer in the lithologic complex. If this layer is exported as spoil during earthwork operations and cannot be used as fill material, then an equivalent volume of suit-

able material must be imported to balance the fill. Similarly, figures 7c and 7d describe other conditions where cut materials are either not sufficient for fill and material has to be imported, or they are in excess of the volume of fill and the excess cut material has to be exported.

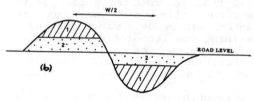

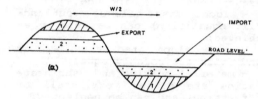

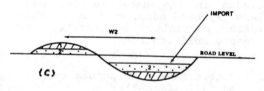

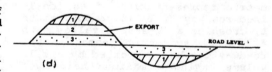

Figure 7. Mass haul diagrams for earthwork balances.

The computation of earthwork volume is based on a typical symmetrical V-shaped valley, but the volume can be adjusted from the valley density by 1 or 2 standard deviations to compensate for the natural terrain. Such adjustments must conform with local or field practices. This aspect is not considered in this study.

3.3 Example of evaluation of typical earthwork volumes

Figure 8 shows an example of typical terrain profiles for two different mapping units in the test area, a granite plain and a clay plain. In the Gongola granite plain, the average numbers of ridge/valley crossings per 2 km for the N-S and E-W directions are 5.6 and 9.1 respectively.

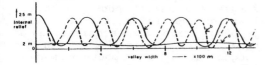

Figure 8. Exaggerated terrain profiles of the Gongola granite plain (a + b) and the Wagur clay plain (c), NE Nigeria.

The approximate valley widths are 750 m for the N-S direction (profile a) and 450 m for the E-W direction (profile b).

The Wagur clay plain has fewer ridge/valley crossings of 1.4 per 2 km and gives a larger valley width of 3,000 m for the N-S direction (profile c).

The Gongola granite plain and Wagur clay plain have average local reliefs of 25 m and 2 m respectively. Thus a steeply rising or falling curve indicates heavy cut or fill in the Gongola granite plain. In addition, the E-W direction shows a steeper valley gradient of 11%, which may lead to a larger earthwork volume in that direction than in the N-S direction with a valley gradient of 7%.

The flat curve of the Wagur clay plain shows that the earthwork quantity is small compared with the other mapping unit. The valley side gradient is less than 1%.

In general, the higher the number of ridge/valley crossings, the steeper the valley gradient and the larger the earthwork volume. The SE-NW and NE-SW directions show the least and highest values of earthwork volumes which correspond to the least and highest valley side gradients respectively for the land system mapping units in the Gongola test area (fig. 9). Similarly, the

highly undulating to hilly terrains, such as the Bima sandstone hills (LSMU 10 and 11) and the Gongola granite plains and hills (LSMU 21, 22 and 23), show comparatively high valley gradients and high earthwork volumes because of high average internal reliefs. In smooth terrains, however, the earthwork volumes show general similarities for all directions; the volumes are related to the requirement of raising the road bed above the surrounding terrain to ensure adequate drainage.

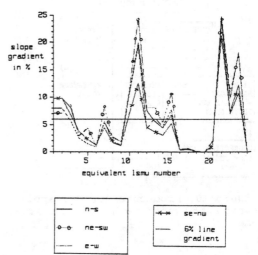

Figure 9. Plots of valley slope gradients per mapping unit for N-S, NE-SW, E-W and SE-NW directions.

The determination of the total volumes of earthwork to achieve suitable road alignments thus requires systematic procedures in which many variables, such as lithology and lithologic complexity, valley density, internal relief (coded per land system mapping unit number), are fully integrated in a geotechical information system (GIS), as illustrated in figure 10. The resulting database can be queried to generate the relevant information about valley widths, valley gradients and typical heights of fills. The analysis of both spatial and attribute databases can be used to generate earthwork volumes which can be translated into the potential earthwork costs per metre of road length (Akinyede 1990).

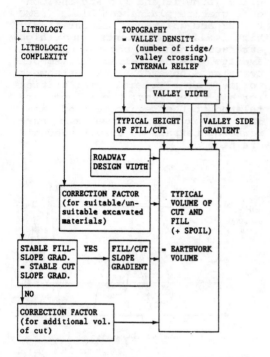

Figure 10. Prodecure for GIS application to generate earthwork volume.

4 CONCLUSION

The determination of earthwork volumes for roads in northeastern Nigeria was based on the quantitative evaluation of the topographic form, measured in terms of local relief and valley density. These variables can be used to generate information on valley geometry and to assess the general terrain roughness and earthwork volumes along potential route corridors.

In regions with high relief and marked variations in the number of ridge/valley crossings, some directional differences can be seen. The SE-NW and NE-SW directions show the least and highest grading volumes which correspond to the least and highest valley side gradients and number of ridge/valley crossings respectively, particularly for the rolling to hilly terrain mapping units in the test area.

The total volume of earthwork was determined from a geometrical representation of an idealized roadway fill, with the necessary volume adjustments to compensate for local mass-balance equality factors which are related to unsuitable excavated materials and differences between cut-and-fill slopes. The volume estimate procedures allow for the integration and analysis of cut-and-fill variables in a geotechnical information system (GIS).

The interpretation of satellite images and aerial photographs allows for the definition of terrain mapping units and acquisition of data which define the terrain roughness. The photogrammetric methods are relatively simple and can be used to optimize earthwork quantities (and grading costs) when planning a road route.

REFERENCES

Akinyede, J.O. 1990. Highway cost modelling and route selection using a geotechnical information system. PhD thesis Technical University Delft, 221 p.
Akinyede J.O. & Rengers N. 1990. Highway cost modelling using a geotechnical information system. Proc. 6th Eng. Geol. Congress, Amsterdam 1990.
FMWH (Federal Ministry of Works and Housing) 1973. Highway Manual, part 1, Design. FMWH, Lagos, Nigeria.
Meijerink, A.M.J. 1988. Data acquisition and data capture through terrain mapping units. ITC Journal 1988-1, pp. 23-44.
Remeijn J.M. 1973. Forest road planning from aerial photographs. ITC Journal 1973-3, pp. 429-443.
Rengers, N. 1979. Remote sensing for engineering geology: possibilities and limitations. ITC Journal 1979-1, pp. 44-67.
Rengers, N. 1981. Special application of photographs in engineering geology. ITC Lecture Notes on course h.30, pp. 51-70.

Hydric resources generated by snow melting on the Spanish mountain
Ressources hydriques produites par la fonte de la neige dans les montagnes espagnoles

M.Arenillas
Universidad Politécnica, Madrid, Spain

I.Cantarino & R.Martinez
Universidad Politécnica, Valencia, Spain

A.Pedrero
Ministerio de Obras Públicas, Madrid, Spain

ABSTRACT: Since 1984 the Spanish Ministry of Public Works, in collaboration with the Engineering Geology Departments of Madrid and Valencia Politechnical Universities, launch a proyect to quantify the hydric resources generated by snow melting in high mountain areas.

The purpose of this study has been to reach a better knowledge of the Spanish rivers with nivo-pluvial and pluvio-nival regimes, as well for floods' prevention and fluvial regulation, as for optimizing hydroelectrical production, water supply, land irrigation, etc.

In this paper we expose the methods and techniques used during the development of the proyect, the results reached by the moment and other future activities.

RESUME: Dés 1984 le Ministére des Travaux Publics espagnol avec la collaboration des départements de Géologie de l'ingénieur des Universités Polytecniques de Madrid et Valencia, a commencé un projet sur la quantification des ressources hydriques produites par la fusion de la niege dans les zones de haute montagne.

L'objet de ce travail a été atteindre une meilleure connoisance des rivières espagnoles avec regimes nivo-pluviales et pluvio-nivales, ainsi que la prevention des crues et la regulation fluviale, afin d'optimizer la production hydroélèctrique, approvisionnement en eau, l'irrigation, etc.

Dans cet article nous exposons les méthodes et les tecniques que nous avons utilisé pendant le developpement du projet, les résultats que nous avons atteint jusqu'au moment present, et les activités futures.

1 INTRODUCTION

In Spanish territory the snow only reaches important volumes on the principle mountain ranges: the Pyrenees, the Cantabrian Range, the Iberian Range, the Central System and the Sierra Nevada (Fig. 1). In these areas the melting of the snow, concentrated in the last months of spring and the first months of summer, sharply characterize the states of the rivers which have their origins in these mountains (Fig. 2, 3 and 4). Although this fact has been well known for a long time, only in the last few years has the quantification of the hydraulic resources of Spain, whose origin is water from the melting snows, been commenced. In determined zones the understanding of this process is of fundamental importance both in optimizing the use of these resources and in preventing or correcting the effect of flooding. Precisely with these ends in view, in 1984 the Dirección General de Obras Hidráulicas (DGOH) of the Ministry of Public Works and Urbanism (MOPU) initiated the study of the quantification of the hydric resources which result from the melting of the snow in the spanish mountains (EHRIN programme). The Departments of Engineering Geology from the Polytechnical Universities of Madrid and Valencia advise the DGOH on this study. Moreover, in 1986 the DGOH promoted an agreement to collaborate with the National Institute of Meteorology (INM),

Fig. 1. Areas of interest

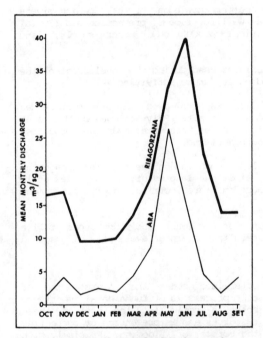

Fig. 2. Pyrenees: Standard hydrographs

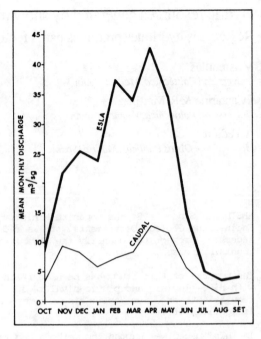

Fig. 3. Cantabrian Range: Standard hydrographs

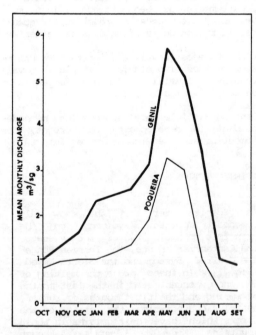

Fig. 4. Sierra Nevada: Standard hydrographs

the hydroelectric companies through their research organisation (ASINEL) and the association of ski stations (ATUDEM). In accordance with this agreement the DGOH is developing the study using an hydrological-statistical method. The INM is in the way of testing a phenomenological model and ASINEL is working on the use of remote sensing. ATUDEM provides the logistic support for the field work. In the following paragraphs we shall primarily refer to the work being carried out by the DGOH with the consultancies of the Polytechnical Universities of Madrid and Valencia (Pedrero et al. 1987). The work is still in hand and few years will be required to complete a general methodology and to obtain the final results.

2 AREA OF INTEREST

The initial interest of the DGOH was also justified by the launch in 1983 of a series of automatic hydrological information systems (SAIH), which should cover the whole of the Spanish territory. With these systems (distributed throughout the large Spanish hydrographic basins), real-time information is being obtained on the hydrological and hydraulic variables of the respective river networks, which allows an obvious improvement in the management systems, by means of the opportune computerization of the data received. In order to complete the information indicated, the DGOH considered it necessary to start up the studies we are now referring. With these it is hoped to include the data related to the hydric resources directly produced by the melting snow in the SAIH networks and also establish adequate systems for the prediction of the yearly flow and volume which may be expected from this process.

The Spanish Pyrenees were chosen (Fig. 5) for beginning the study as this is the mountain range where not only is the snow most abundant but also most persistent. Within the Pyrenees the central section was selected, where the abundance and persistence of the snow is a maximum. The surface area of this zone is some 5,500 km^2 and mainly corresponds to the basin of the River Ebro. (A small part -10%- drains towards the Garonne and another, corresponding to the Ter basin, directly towards the Mediterranean. The first quantifications obtained in this sector have demonstrated that the melting snow directed towards the Ebro produced an annual volume of water of some 2,200 hm^3 which is equivalent to approximately 50% of the total hydric resources of this sector. Furthermore, this volume supposes more than 10% of the total water which circulates in the Ebro basin, whose area is 86,000 km^2. An important data is that the 2,200 hm^3 derived from the melting snow each year are concentrated in some

three months (Table 1).

Once the working methodology was defined and the first quantifications in the Pyrenees obtained, the study was extended, for the time being, to the Cantabrian Range (Fig. 6 and Table 2) and to the Sierra Nevada (Fig. 7 and Table 3), where the first field work has been carried out and the observation network of the snowpack installed.

3 METHODOLOGY

In essence, all the lines of investigation on the melting of the snow and its hydrological consequences can be put into two broad groups, according to whether they are based on phenomeno-logical analysis or on facts of a hydrological and statistical nature (Guillot 1982, Gray & Male 1981, WMO 1986).

In our case we are working along these two lines of investigation as well as the use of remote sensing, in accordance with the distribution of activities which have been indicated above (Pedrero 1988).

The hydrological-statistical method which we are developing in essence consists of correlating two basic observation networks: the snowpack and the water flow generated by the melting snow. Once such a correlation is defined, the series of data proceeding from the variable snow serves to predict the flows generated by the thawing of the snow.

The method has been applied in the Pyrenees (Arenillas and Martínez, 1988), which was the first area to be studied. For this, 140 poles, 4 m high, were installed and these constitute the network for measuring the snowpack. These poles are placed after careful selection of the points of insertion, in such a way that each one in principle represents the average conditions of the snow level in its own environment. This representati-veness has been proved for the majority of the points during four field work campaigns (1986 to 1989). On the other hand, after various previous tests, 14 poles have been selected on which the average snowpack density is systematically measured. This reduced number of snow density controlled points is justified by the small proven spacial variability of this factor.

553

Table 1. Pyrenees

No	BASIN	SURFACE AREA (km²) >1000 m	SURFACE AREA (km²) >1500 m	POLES	GAUGE STATION	SNOWPACK WATER EQUIVALENT (hm³) 1988	SNOWPACK WATER EQUIVALENT (hm³) 1989
1	Irati	186	4	2	1	0.4	0.5
2	Salazar	191	1	1	1	0.2	0.4
3	Esca	330	60	2	1	10.0	9.9
4	Veral	144	52	2	1	11.5	10.8
5	Subordan	272	138	3	2	34.9	31.4
6	Aragon	342	158	6	1	61.5	55.2
7	Gallego	882	309	14	1	134.5	90.1
8	Ara	491	262	8	1	143.9	91.8
9	Cinca	764	500	12	3	262.0	176.8
10	Esera	812	486	12	1	247.2	197.2
11	Ribagorzana	554	412	24	8	165.8	102.8
12	Pallaresa	1823	1177	22	7	246.0	158.4
13	Segre	1753	1123	25	4	(1)	(1)
14	Ter	498	283	3	3	(1)	(1)
15	Garona	566	439	4	2	211.2	143.2
	TOTAL	9608	5364	140	37		

(1) Not calculated

Table 2. Cantabrian Range

No	BASIN	SURFACE AREA (km²) >1000 m	SURFACE AREA (km²) >1500 m	POLES	No	BASIN	SURFACE AREA (km²) >1000 m	SURFACE AREA (km²) >1500 m	POLES
1	Navia	580	22	2	19	Sil	1097	296	23
2	Ibias	285	29	2	20	Navea	127	12	6
3	Narcea	527	88	3	21	Bibey	127	108	7
4	Piqueña	270	82	4	22	Jares	248	33	3
5	Trubia	375	49	4	23	Cabrera	558	140	3
6	Pajares	503	129	7	24	Tera	298	148	4
7	Nalón	270	40	2	25	Eria	560	135	4
8	Sella	486	66	4	26	Omañas	402	116	3
9	Cares	455	76	4	27	Luna	500	249	10
10	Deva	644	80	6	28	Bernesga	712	315	9
11	Nansa	214	18	2	29	Porma	407	166	6
12	Saja	202	15	2	30	Esla	607	260	9
13	Besaya	308	0	0	31	Carrion	228	194	6
14	Pas	250	1	1	32	Pisuerga	301	54	6
15	Miera	110	0	0	33	Ebro	200	50	4
16	Ason	143	1	1	34	Nela	50	0	1
17	Burbia	492	21	5	35	Trueba	50	2	2
18	Cua	482	54	2		TOTAL	13068	3049	157

Three series of snow thickness and density measurements are carried out annually at the points indicated (January, March and April) (Fig. 8 and 9). With this the evolution of the snowpack is analyzed and in each case the corresponding "snowcourses" are established. These snowcourses are at present defined as an exclusive function of the variable altitude, once the study area has been divided up and the effects of the other variables -orientation, valley morphology, etc.- are eliminated. This has been achieved by applying linear regression techniques to the series of data which has been collected.

The integration of these laws with the corresponding topographic conditions, provides the value of the total volume of

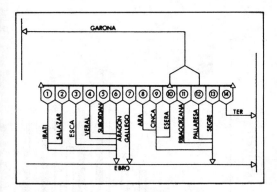

Fig. 5. Pyrenees: Diagram of principal basins and fluvial network

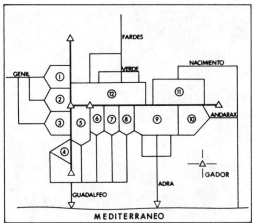

Fig. 7. Sierra Nevada: Diagram of principal basins and fluvial network

snow stored in each one of the river sub-basins considered. By applying the average density to these volumes, the value of the volume of water stored in the form of snow is obtained (Fig. 10).

These figures are compared later with the volumes actually controlled in the network of outflow gauge stations. These have been situated in the various water courses, as near as possible to the respective snow areas.

In this way the volume of water measured in the gauge stations and that calculated in the snowpack are recorded by means of a coefficient, that we call a transfer coefficient, which must be defined in each case.

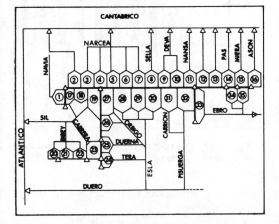

Fig. 6. Cantabrian Range: Diagram of principal basins and fluvial network

This process, evidently simpler than any other used in phenomenological methods, has given acceptable results to date in the Pyrenees. Through it forecasts can be made on the total volume of water generated by melting snow.

4 RESULTS

Table 3. Sierra Nevada

No	BASIN	SURFACE AREA (km²)	
		> 1600 m	POLES
1	Genil	155	4
2	Monachil	30	3
3	Dilar	32	1
4	Durcal	28	1
5	Lanjaron	27	1
6	Poqueira	72	3
7	Trevelez	80	4
8	Cadiar	54	1
9	Adra	122	2
12	Fardes	88	3
	TOTAL	688	23

The results that have been obtained to date are sufficiently homogeneous to assure the definitive start-up of the method used within few years. In order to obtain results which refer to volumes, action has been taken as indicated below (Arenillas and Martínez, 1988).

Fig. 8. Density measurements in Sierra Nevada (Jan., 1990)

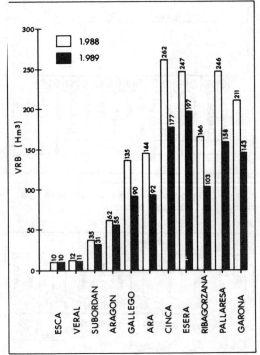

Fig. 10. Pyrenees. Water ressources generated by snowmelting. 1988 and 1989.

E and S : thickness and surface in terms of altitude.

The volume (gross) of water in the snowpack is obtained by multiplying by the average density value (Table 1).

$$V_b = V_N \cdot d_m$$

In order to correlate these gross volumes with the net ones measured in the gauge stations, it is necessary to determine previously the remaining factors which form part of the process. For this the periods corresponding to the circulation of the runoff derived from the melting snow are identified in the various hydrograms. Below are obtained the total volumes gauged during these dates and, through meteorological information, the rain received in each basin in the same periods is calculated.

Fig. 9. Meteorological measurements and snowpack control in Pyrenees. Jan. 1990

From the snowcourses E(z), volumes of snow in the different basins are easily obtained by multiplying by the corresponding hypsometric curves. That is to say:

$$V_N = \int_{H_1}^{H_2} E(z) \cdot s(z) \cdot dz \qquad (1)$$

Calling V_T the total volume gauged, V_p the rain received and k_e the runoff coefficient corresponding to this rainfall, in each basin the volume of water obtained from melting snow, or the

556

net resources (V_n), may be defined by:

$$V_n = V_T - k_e \cdot V_p \qquad (2)$$

The correlation between the resources calculated in the snowpack, or the gross resources (V_b), and the resources obtained from melting snow, or net resources (V_n), may be expressed generally, for each basin, introducing a transfer coefficient (k_n) which collects all losses as a result of sublimation, infiltration and runoff. In this way:

$$V_n = k_n \cdot V_b \qquad (3)$$

From (2) and (3) the general equation for the process is obtained:

$$V_T = k_e \cdot V_p + k_n \cdot V_b \qquad (4)$$

From the checks that are carried out each year, the values of V_T, V_p, and V_b in this equation are known either by calculation or measurement and this permits the corresponding values of k_e and k_n to be adjusted. It is evident that the precision of this adjustment will be greater when a greater number of checks are available. However, even with little data some determined orders of magnitude can be given, taking into account, on the one hand, the figures by which the runoff coefficient k_e would normally vary and, on the other hand, whether it is considered that a fraction of the transfer coefficient k_n also represents a runoff which, as a rule, must be lower than this corresponding to the liquid precipitation. That is to say:

$$k_n = k'_e \cdot k'_n \qquad \text{and}$$

$$k'_e = a \cdot k_e \qquad (a \leq 1)$$

In this way, equation (4) has the following form:

$$V_T = k_e \cdot V_p + k'_e \cdot k'_n \cdot V_b = k_e (V_p + a \cdot k'_n \cdot V_b) \qquad (5)$$

where k'_n is a new transfer coefficient which fundamentally covers losses from sublimation.

The results obtained up to now indicate variation ranges in the coefficients considered of the following order:

$$0.70 \leq k_e \leq 0.90$$

$$0.50 \leq k'_e \leq 0.70$$

$$0.65 \leq k'_n \leq 0.85$$

The definition of the hydrograms for melting (outflows) is considerably more complex and laborious than the corresponding volumes. Nevertheless, in the various basins studied in the Pyrenees it has been possible to prove that the main melting is always produced within very concrete periods, over a constant period of time, in each basin. Moreover, once this main melting has started, the eventual detentions in the process (normally from a lowering of temperatures), tend to only have a slight effect, since they are practically always phenomena lasting a short time. For general explanation of these sequences and their changes a well calibrated phenomenological model would have to be available. Nevertheless, the data which has been obtained up to the present indicate the possibility of using simpler models (based, fundamentally, on thermal variations) which allow sufficient approximations to achieve the desired ends. Along this line we have already achieved acceptable adjustments in some basins and we are working with these criteria at present.

6 CONCLUSIONS

A hydrological–statistical method has been used in the Spanish Pyrenees since 1984 with the objective of quantifying the hydric resources proceeding from the snowfalling. The method is based on establishing simple correlations between the snowpack and the flows which are the result of it melting. Different parameters of a territorial nature are made to influence these correlations and these explain the process of the melting snow in the area studied. With these premises it has been possible to define an operative method, which has proved to be efficient in achieving the objectives proposed. In fact the results obtained up to now have proved the validity of the method and define some very clear lines for future action.

As it is logical, the methodology proposed admit certain improvements, some of which are evident. On the one hand, those relating to the meteorological observations should be mentioned (Almarza & Peinado 1984), since a large number of them allow a more exact quantification of the rain factor and also a better adjustment of the snowcourses in the higher altitude sectors. On the other hand, those which require the support of otner study methods, should also be mentioned. Such is the case of the seasonal melting sequences, which are not sufficiently explained except by conceptual models (Ríosalido 1985); these do not have to be excessively complex, given the degree of precision demanded.

In any case, the hydrological-statistical method, such as been initiated in the Pyrenees, permits sufficient results to be achieved for the proposed objectives, both with regard to the final correlations between the gross and net resources and in that relating to some partial aspects which are of special interest. Thus, for example, the calculation of snowcourses leads to the establishment of a level of representativeness for each one of the sites of the sample network. This fact will allow the selection of the most suitable points for installing automatic remote controlled snow gauges which will be connected to the SAIH networks. With these the methodology which has been developed is directly applicable to the elaboration of the data supplied by the said stations and, in this way, it is possible to consider the incidence of melting snow on hydrological management and planning of the basins affected by this process.

In the other mountain ranges (Cantabrian and the Sierra Nevada) the hydrological-statistical method is being adapted to the peculiarities of each one. With this increase in the areas studied it is hoped that, in the near future, the necessary comparison between the established models will be achieved, distinguishing the factors which admit a clear generalization of those which reflect particular points proper to each territory studied.

REFERENCES

Almarza, C. and Peinado, A. (1984). Notas para una climatología de la nieve y bases para un estudio de la cobertura nivosa invernal en España. Instituto Nacional de Meteorología. Serie A-88.

Arenillas, M. and Martínez, R. (1988). El método hidrológico-estadístico. En La Nieve en el Pirineo español. 109-126. MOPU, Madrid.

Gray, D.M. and Male, D.H. (1981). Handbook of snow. Pergamon Press.

Guillot, P. (1982). Mesure de l'equivalent en eau de la couche de neige. ANENA.

Pedrero, A. (1988). El programa ERHIN. En La nieve en el Pirineo Español. 9-28. MOPU, Madrid.

Pedrero, A., Arenillas, M. and Martínez, R. (1987). Cuantificación de recursos hídricos derivados de la fusión nival en el Pirineo español en Avenidas. Sistemas de previsión y Alarma. Barcelona.

Pedrero, A., Arenillas, M. and Martínez, R. (1987). El programa ERHIN: Evaluación de recursos hídricos procedentes de la innivación en la alta montaña española. Conferencia Ibero-Americana sobre Aprovechamientos Hidráulicos. Lisboa

Ríosalido, R. (1985). Estimación de la fusión nival mediante un modelo de balance energético. Revista Meteorológica. AME. nº 5, junio, 27-35.

W.M.O (1986). Intercomparision of models of snowmelt runoff. Secretariat of the W.M.O. Geneva.

A holistic approach to detailed geotechnical data gathering in southern Africa

Une approche holistique pour détailler des données géotechniques recueillies dans le sud de l'Afrique

Oliver B. Barker
Davies, Lynn & Partners (Formerly of Barker & Associates), Johannesburg, South Africa

ABSTRACT:The philosophy in support of and the mechanism for achieving an integrated, comprehensive approach to detailed geotechnical and geological investigations in the civil and mining engineering industry in Southern Africa is presented. In order to achieve this goal there is an urgent need for geologists to be in executive positions in organisations responsible for the conception, planning and execution of such projects

1.0 INTRODUCTION

There are not more than 19 published geological map sheets at a scale of 1:50 000 or larger in South Africa. This constitutes about one percent of the land surface of the country. Thus for the foreseeable future, descriptive geology will remain an essential geological activity in Southern Africa (Kuhn, 1970 and de Waal, 1988). Consequently engineering geology will similarly rely on primary data gathering for most projects.

Most technically advanced countries and professional associations publish guidelines for data collection and presentation. Conferences are held and numerous books and papers are published annually on mapping and geological data collection. Despite this, many government and private companies and corporations continue to ignore the need to provide for the collection of detailed information.

It is axiomatic that analysis based on fundamental sources of data can never better the precision or accuracy of the field mapping and sampling. There are many examples of maps, books and reports where prejudice and preconceived notions and inadequate planning and management have influenced the data gathering and thus the analysis and conclusions.

Unlike research or field mapping, data gathering in many civil engineering and mining situations is a unique and momentary opportunity to evaluate record and sample a geological entity. Once completed the rock or soil material is removed or covered and the evidence is obliterated, seldom to be seen again in the form it was mapped. This reality places a grave responsibility on the geologist, for his records will remain the authoritative record of the rock or soil surface or layer investigated.

This paper will review and analyse the philosophy and practice of detailed data collection in situations where the information is exposed for a brief period in time as described above. The subject is vast. Thus by necessity we will focus on: project planning; in situ mapping; and photography.

It is self evident that much of what is said applies equally to aerial photographic interpretation, geophysics, borehole drilling and logging and other primary forms of data gathering.

Space precludes the presentation of case studies in this text. A poster display will provide these and should be seen as an integral part of this paper.

2.0 PROJECT PLANNING

2.1 The scientific process

Planning must remain the key to a successful project. Those who have written a research paper will have come to terms with the scientific, planned approach required by refereed or examined works. This approach entails the careful separation of source data from analysis, discussion (model postulation) and finally conclusions and where relevant, recommendations.

This process includes:
1. time to review the project, background data and the philosophy behind the research proposed;
2 opportunity to precisely and accurately research and collect data:
3 time and facilities to thoroughly analyse the information discuss the implications and to draw thoughtful and well founded conclusions.

2.2 The commercial process

By contrast the commercial, consulting or production line seldom operates on such a rigorous basis. Time is money and swift decisions and productivity are the objectives. Seldom can time be afforded to the geologist to produce a method statement. The need to research facts, theories and conclusions often become secondary to production.

It is often only when problems are encountered and sometimes only once they have become so serious that lives are lost or financial losses are being incurred, that the need for and precedence is given to, geological data gathering. At this stage it is often too late as evidence is obliterated. Back analysis is then impossible or at best full of uncertainty. In fact very few civil engineering contract documents allow for the planning of geological mapping as an essential element of the production process.

2.3 The need for a system

The problem in most civil and mining engineering situations revolves around the absence of time for adequate project planning. Consequently there is insufficient time to think about the implications of actions and their influence on logistics of the project as a whole.

The nuclear industry, owing to its inherent dangers and sociopolitical sensitivities world-wide, very early in its history, adopted management and planning techniques designed to reduce the possibility of errors to an absolute minimum.

One of these methods involved the concept of the Quality Assurance Programme which originated in the aircraft industry. QA, as it has become known, was developed to assure management that the quality controls necessary to the safe construction and operation of aircraft were carefully researched and implemented.

These same concepts have been successfully applied in the nuclear industries overseas and in South Africa. However, other than SA Bureau of Standards document No. 0157, the South African civil engineering an mining literature is remarkably silent on this subject, except in relation to the construction of the Koeberg Nuclear Power Station (Barker 1980).

2.4 Application of QA management

The specifications developed under a QA programme lead to the rigorous research and development of comprehensive and appropriate quality control (QC) specifications. QA also lays down rules for the documentation of deviations from or need to alter, QC specifications, which arise out of production-related variations, rigidities, problems or changes in logistics, design or demand criteria. QA defines documentation and review procedures for quality control programmes at all stages in an engineering operation from concept to operation. In the nuclear industry the process extends to the final decommissioning and dismantling of the facility.

In the civil, geotechnical engineering aspects of the construction of a nuclear power station, QA implements an auditing system which ensures QC specifications are maintained in relation to data collection, whether it be core logging, mapping or sampling; standardisation of equipment, calibration or recalibration, analytical procedures, or the appointment of contractors and their activities. It is understood that QA was first applied in South Africa in 1976 in the design and construction of the Koeberg Nuclear Power Station. It was here that South African geologists were first introduced to the system when planning and preparation began for the foundation mapping programme. QA required the preparation of documents which defined all aspects of our data gathering programme. These documents had then to be reviewed by a competent, acknowledged expert as were the activities during project execution to report completion.

2.5 Discussion

The application of QA led to the successful completion of the geotechnical and civil projects at Koeberg. The projects:

1) commenced with all parties aware and in agreement as to what was required, objectives to be met and the time frame in which the work had to be completed;
2) were within budget and on time;
3) were comprehensively planned and programmed and once started, were completed with few deviations and thus few revisions.

All personnel involved in the project were comprehensively briefed and "indoctrinated" as to the QC requirements. Consequently the geotechnical staff in the field and office were fully informed of the expected geology and prepared with mechanisms to rapidly respond to unexpected conditions or field requirements. The period of precedence given to geological and geotechnical investigations and mapping, photography and sampling was sufficient to allow for adjustments to the work plans to be revised and approved within the QA/QC system.

The maps and data produced were accurate and precise records of the geology and geotechnical parameters of the rocks encountered. It is generally agreed that the geotechnical studies at Koeberg rank as one of the most detailed and accurate in South Africa. Project planning and management, through the application of QA was certainly the major contributor to the success of the Koeberg projects.

3.0 IN SITU MAPPING

3.1 General comments

In this paper we are dealing with detailed data collection which will allow rigorous analysis and form the basis for detailed engineering design.

In such a project situation and where a QA/QC programme is in operation the fundamentals and basis of the mapping programme will be carefully planned. Not only does this include the mapping and data collection itself but also logistics such as accessibility and clarity of the rock or soil surface. In order to be brief the necessary steps are taken as regards these aspects of a project are assumed and even though not discussed further remain vital to the accuracy and precision attained in mapping, sampling and data collection.

3.2 The human factor

The management of a project may have provided the best possible working environment and background support. There is, however an essential ingredient without which even the best planned projects will fail to meet expectations. This is the geologist at the rock face.

The selection of the geologist must be as rigorous as the planning of any other aspect of the work. The choice as to competency will obviously depend initially on the needs of the project and pertinent training and experience of the geologists available.

Data collection in civil and mining projects is hazardous, and requires a combination of physical and mental talents not present in every geological graduate or technician.

Thus training and experience are not the only criteria which should be used in selection. There can be little doubt that less easily quantifiable attributes such as diligence, physical fitness, mental agility and an enquiring mind may be as critical to success as a high IQ.

Finally a dedicated person with the appropriate, optimum blend of the aforementioned attributes is an ideal which should be sought.

3.2 Choice of scale

3.2.1 Some concepts

The scale of map ultimately defines the level of detail which can be achieved. This choice is thus an important element of the QC stage of planning.

3.3.2 The alternatives

Ideally more than one alternative should be provided to the field staff. This also implies that appropriate base plans and mapping sheets are then designed and printed prior to start up.

This approach then allows the mapper at the face to make scale changes which meet the project needs without disruption to the overall plan of action. In this way detailing of significant features becomes routine and will not constitute a deviation. It will then be automatically incorporated as an integral element of the data base.

The implication of such a system is that more time may be required at the face and this must be accepted by the contrator. As such it must also be part of the project documents, either as an acceptable delay to the contractor or as an acceptable extra over claim item by the client.

Unfortunately the scale factor is often not considered important as a planning item and is left to the data collector to decide. This decision is then left to be made either at the rock face or at best at the last instance before commencement of the project. Both options will lead to delays in compilation and analysis.

3.3.3 Alternative Techniques

Prior to the development of modern photographic and subsequently, computer graphic technology, scale changes were logistically more difficult to cope with during data compilation and also analysis. With these new techniques, data at different scales is now easily combined or incorporated by one means or another.

The mapping techniques to be used in adapting the level of detail to the circumstances rather than vice versa are numerous and can be addressed in several ways:

   1. certain elements can be ignored when they become too complex to draw at the scale being used;
   2. the complexities can be simplified;
   3. a change of scale can be applied and the details recorded. A two stage process is useful here: a) stage one is to draw a simplified version of the main elements at the present scale; b) stage two will then follow at the smaller scale factor;

   4. cartoon it showing the basic elements but with key structural measurements eg. dip, strike, azimuths.

3.4 Discussion

The implications of the above conceptual approach are vital to an understanding of what, to most geologists is the hardest and to others, also the most exciting aspect of mapping or data collection. The four alternatives are discussed in order below.

3.4.1 The cost of ignorance

It is tempting to ignore complex situations when collecting data or mapping a rock surface which may also be poorly exposed or partly concealed by shotcrete, mud or dust. Who will ever know (as the surface will be blasted or shotcreted immediately the work is done), is a thought that must have passed through the minds of many geologists at one time or another when faced with this situation.

Such an attitude may be disasterous and may have severe consequences or presupposes certain knowledge about future developments which may not be justified.

3.4.2 To be or not be

The issue of simplification evokes much arguement. The process implies interpretation and analysis. The question is; to what extent should analysis be applied at the rock face. should it not be conducted in the calm of the site office with all the data to hand. Carried out at the work face in the mine, tunnel or excavation, it may lead to the loss of vital information.

There is necessarily a counter arguement, which calls on the professional judgement and experience of the geologist. The answer to this would lie in his level of involvement in and knowledge of, the initial project planning. It must needs also imply a level of knowledge of the encountered and expected geology, as well as all the potential geotechnical implications of the feature. Given these aforementioned implications, is the rock face the place to make these decisions when it is only time and effort which is required to obtain the data".

3.4.3 The need for change

There are thus strong arguments in support of the detailed approach. This implies a change of scale of mapping in order to detail a complex feature difficult to map at the larger scale. The diligence and accelerated mapping effort must be complemented by a contractor's patience and logistical assistence and by an understanding client. With this planned and well managed approach, the results will generally outweigh the effort, time and money expended.

Finally the last option lies somewhere between the two extremes discussed. The use of cartoons and actual measurements will result in a largely conceptual record not fixed in space and thus of limited value to engineering design.

3.4 Presentation of mapping

The representation of a three dimensional vertical or subvertical surface on a map under the rigorous conditions at a stope or tunnel face or sidewall of an excavation requires the special combination of skills already discussed as well as a degree of artistic talent.

Mapping may vary from three dimensional representations to simple line drawings. The latter is less demanding than the former. However if one is recording data from a very uneven rock surface the simplification to representative simplified lines may result in a significant loss of detail. It may also lead to difficulties in correlation with photographs of the face.

Both approaches have their place and should be used selectively. Three dimensional representation is very important when photography is an integral element of the data base for the reason already stated. Three dimensional mapping allows for rapid comparison with the photography. As already mentioned this is not so immediate with line mapping especially in the case of irregular or curved surfaces in blocky rock conditions.

4.0 PHOTOGRAPHY

4.1 General statement

Photography in geology is both an art and a science and should form an integral part of the mapping process. The skillful use of this medium can produce a data base as important as mapping. It can also be used to underscore certain issues and can reflect bias in much the same way as mapping. Whatever the argument, it is undeniably valuable.

It is frequently said that photography never lies. This may be so to the photographer but photographs can equally reflect a specific point of view. This is an aspect that can be used to advantage in engineering related disputes.

The orientation, scale, use of light, all influence the result. It is quite possible to photograph the same rock mass from two orientations each giving a completely different picture. The difference can be as great as that between a massive and a closely jointed body of rock (See poster display).

4.2 Photographic techniques

The choice of the optimum blend of scales when taking photographs is very important. The final product must be supportive of the mapping detail but must also reflect the overall strategy of the investigation. The latter will determine what is required in the final analysis, a combination of overviews at a small scale combined with details is an optimum objective. The addition of stereo photography is an added extra which, in certain circumstances can be very valuable if detailed back analysis is ever required (See poster display).

The level of detail to be captured will once again depend on the interpretive judgement of the geologist. this will be in relation to the overall project objectives, and the engineering circumstances at the face. In general, however it is good practice to take too many photographs than too few.

The photography of large surfaces can be achieved in two ways, a) with the use of wide angled lenses and b) by the application of mosaiicing techniques to a carefully composed suite of overlapping photographs.

The first technique results in gross radial distortion at the edges of the photograph.

The second produces a composite mosaiic which can be rephotographed and is relatively free of scale and angular distortion. It is taken for granted that all photography is scale-related by the use of mapping frames (poster display) or to surveyed or scaled marks on the rock surface.

Photography that will meet the low levels of radial and angular distortion in the mine workings or at the work face or sidewall in a tunnel or any other production situation requires a good understanding of photographic techniques and equipment.

The photographs need to be kept as close to a common scale as possible. This is achieved by taking photographs from a stable base of constant distance while maintaining angular distortion to a minimum between successive photographs.

Without very sophisticated stands and given the limitations which exist in the production situation, considerable overlap between individual and rows of photographs should be obtained. This allows the mosaiic compiler the chance to select a combination which produces the lowest overall distortion.

4.3 Compilation of photography

All photographic data should be compiled as meticulously as mapping. Specifically formatted sheets are best for this purpose just as are map sheets for mapping. The photos are then compiled and annotated and filed with the relevant map sheets for analysis. At a very minimum, all photography should be annotated with time, location, orientation and photographer's name.

4.4 Discussion

A photographic data base so constructed (See poster display) is without doubt a valuable and in some instances (eg. in a dispute situation) invaluable. It will remain a comprehensive time and place record of conditions and events. At best the data can be used for analysis, or supplementary mapping. It could conceivably be used for design purposes where scaling is sufficiently accurate (eg. in military engineering strategic planning). Photography of very briefly exposed rock masses, or those which are unstable and thus hazardous to map at close range may prove to be the only record possible. The methodical photographic mapping in such environments could prove vital to the success of a project or the definitive record in a claim related situation.

Finally it is axiomatic that if such a data base is to be representative the more photographs available from which to choose the fewer chances there are for omissions to occur. Likewise the opportunity to make one's point is infinitely greater. It is reiterated: rather take too many photographs than too few, there is seldom a second chance.

5.0 DISCUSSION

The philosophy and some important specific aspects of the planning, management and operation of detailed data gathering and mapping projects in the civil and mining industries have been reviewed.

The need for detailed and unbiased data collection is emphasised. However, this must be within a framework of an integrated, well planned and managed environment for the interpretive skills of the data collector to be fully realised. Briefing and indoctrinating the project team with the project's objectives and philosophy is essential.

Having said this, it is extremely difficult to predict the encountered conditions in a geological investigation, even under the best conditions and availability of investigative expenditure.

It is therefore very important to provide the field staff with alternatives within a carefully structured system. Such a system would ideally also provide for an iterative process of reassessment both at the rock face and in the office.

The need for an iterative process involving the stages of researching, planning, management programming in the early stages is vital. As important is a similar process of observation, analysis, synthesis and deductive testing during the project. It is this aspect which is possibly the most difficult to apply as it requires constant re-evaluation and self criticism.
The ultimate level of detail can never be prescribed prior to the event. For example it may be found that the crystal structure in intrusive dolerites have significant geotechnical implications. The data may be found to be unimportant in an existing project. If published, it may find application in academic research. The

result of the research may eventually be of importance to subsequent engineering projects.

Practising geologists have an important responsibility in this respect. Data gathered in a project, be it large or small are generally unique in South Africa. The availability of such detailed geotechnical and geological data to both the practising and academic earth scientists and engineers is very limited.

Consequently the responsibility lies with these technologists to publish the data in a form useful to the geological and engineering profession at large. A corresponding responsibility must also lie with the owners of data to allow it to be published within its useful life. Only in this way will the detailed data base in South Africa grow and in time reduce repetitive and unnecessary expenditure in civil and mining geotechnical investigations.

In publishing their findings, professional geologists will supply the research scientists with information which can be used to the ultimate benefit of the profession (de Waal, 1988).

6.0 CONCLUSIONS

This paper has demonstrated the advantages of integrated planning and management procedures to the attainment of high quality geological and geotechnical data collection. Whereas QA has a role to play -- it may too costly to apply fully in most small projects. However it has been argued that that had QA or a similar system been in operation in certain major civil projects or mining operations in South Africa disasters may have been avoided or mitigated. It is further concluded that the philosophy behind QA may be found to be cost effective even in smaller projects in the long run.

Many projects have failed or run into serious cost overruns or tragic human losses due to incomplete or limited initial geotechnical and geological investigations and research. It is suggested that the answer may lie in the absence of geologists in key positions in the mangement systems of these projects. The frequent lack of understanding of the need for more integrated geological and geotechnical research and planning at the inception of a project may also be a contributing factor.

In order to change this the geologist or engineering geologist must have executive powers in the project team or even at a higher level in a client, contractor or professional consulting engineering organisation.

Is it prejudice or training which has led to the frequent absence of geologists in key positions in relation to geotechnical projects. Recent developments do indicate this to be changing. However very few seem to be present in the management structures of the major corporations and none in the public sector. Whatever the cause it is vital that this position changes.

Geological underperformance is not only potentially disasterous but continues to hold the profession back in attaining a key position in the conceptual planning stages of civil and mining projects. While this situation remains unchanged numerous professional opportunities are being lost to geologists as are vast amounts of unique data which remains unpublished.

ACKNOWLEDGEMENTS

The opportunities afforded the author to work on some major civil engineering projects from which material for this paper and the poster displays is derived is greatfully acknowledged. The editorial assistence from my wife and critical comments from colleagues have assisted in achieving whatever excellence is reflected in these pages.

REFERENCES

Barker, O.B. 1980. Palaeo proof from the Piddock. Nulear Activ. 22:2-8

De Waal, S.A. 1988. Of barons and barriers, Presidential address, Geol. Soc. S.Africa. 91(3):305-315

Kuhn, T.S. 1970. The structure of scientific revolution, Second edition, Univ. Chicago Press, Chicago. 210pp.

6th International IAEG Congress / 6ème Congrès International de AIGI, © 1990 Balkema, Rotterdam. ISBN 90 6191 130 3

Special purpose engineering geological methods for mapping and interpretation of rock mechanical phenomena

Méthodes de géotechnique spécialisée pour la cartographie et l'interprétation de phénomènes de mécanique des roches

V. Bräuer & A. Pahl
Bundesanstalt für Geowissenschaften und Rohstoffe, Hannover, FR Germany

ABSTRACT: Basic methods of engineering geological investigations for rock mass characterization were improved, and tested at the special structural and geological conditions of the Aare Granite, Switzerland. In a German-Swiss cooperation (NAGRA, BGR, GSF), the research projects "Stress Measurements" and "Fracture System Flow Test" were carried out by BGR. A prerequisite for the planning and specification of the rock mechanical and hydraulic tests was the engineering geological mapping of the test site. Fracture and microcrack analyses of a test borehole allowed for a detailed interpretation of the rock stress measurements. Relations exist between frequency of fractures and absolute amount of stress, and between orientation of fractures and of stresses, respectively. In zones of high fracture density, relatively small amounts of stress were measured, whereas in zones of compact rock, high values of stress were observed. A rotation of the horizontal stress direction, in an almost unfractured region, from mainly ESE/WNW to SSE/NNW, correlates with an identical change in orientation of the microcracks. Due to this geologic site characterization, we could perform a subdivision into various, structurally homogeneous zones. Thus, the results obtained in connection with the stress measurement project, could be applied to larger rock masses and site areas.

RESUME: Lors d'essais dans la région d'Aare Granite, Suisse, des méthodes de caractérisation des roches cristallins au moyen de technique de la géologie de l'ingénieur ont été utilisées et ameliorées. Dans le cadre d'une coopération entre les instituts allemands et suisse suivants: CEDRA, BGR, GSF, le BGR a effectué des mesures de l'état de contraintes et de perméabilité des fissures préexistantes. Le relevé géologique du site a d'abord été nécessaire en vue d'établir le planning et les characteristiques des essais hydrauliques et de mécanique des roches. Les analyses des fractures et des microfissures permettent l'interprétation détaillée des mesures de contraintes in situ. Il existe en effet des relations entre la fréquence des fractures et les contraintes in situ: Dans les zones très fracturées, les contraintes mesurées sont relativement peu elevées. Au contraire, dans des zones compactes, les contraintes mesurées sont très elevées. Nous avons observé une rotation de la contrainte horizontale, dans une zone peu fracturée, de ESE/WNW a SSE/NNW, correlée avec une changement identique de l'orientation des microfissures. Grâce à cette caractérisation géologique du site, nous pouvons realisé une subdivision du site en differentes zones de structure homogène. En appliquant cette méthode, il est alors possible de prévoir les caractéristiques géologiques des massifs rocheux ou des volumes plus importants.

1 INTRODUCTION

Under the terms of a German-Swiss cooperation project the Grimsel Test Site, Switzerland, was the site of the testing and further development of engineering geological, rock mechanical, rock hydraulic and geophysical investigative methods. The partners at the test site are the National Cooperative for the Storage of Radioactive Waste (NAGRA), Baden, the Federal Institute for Geosciences and Natural Resources (BGR), Hannover, and the Research Centre for Environmental Sciences (GSF), Braunschweig (BREWITZ & PAHL 1986).

The Grimsel Test Site, operated by NAGRA, is the location of tests to gain practical experience and know how in the planning, execution and assessment of final disposal storages for radioactive wastes in hard rock, as for example in crystalline rock. The laboratory galleries and the test areas are situated within the Aare granite of the Swiss Central Alpine Massif in the vicinity of the Grimsel Pass, and were cut in 1983/84 with a full face tunneling machine. The test site in the granite and granodiorite was selected by NAGRA (NAGRA 1985) on the basis of exploratory holes drilled from the main access tunnel to the Central Grimsel II Oberhasli AG power station at a depth of around 450-500 m below the Juchlistock (Fig. 1).

taminated substances as functions of hydraulic gradients and rock mass structure.

Knowledge of the size and direction of the principal stresses in rock masses is of considerable importance in the engineering planning and assessment of the long-term deformation behaviour of a final storage facility. The rock stresses are entered as parameters into various stability calculations. A different research project, the "Rock Stress Measurement" project, also undertaken by the BGR at the Grimsel Test Site, concerned the testing of various overcoring procedures designed to determine the load-deformation behaviour of the rock, and also their further development for use in deep drilling.

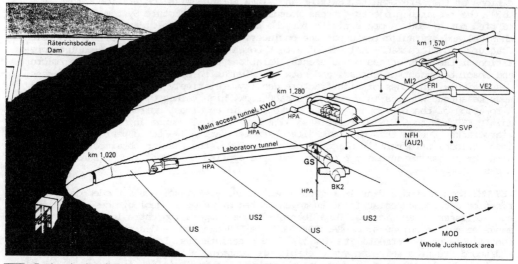

US	Exploration borehole	FRI	Fracture zone investigation	NFH	Near-field hydraulics
	US-borehole	HPA	Hydraulic parameters	SVP	Predict. ahead of tunnel face
AU2	Excavation effects	MI2	Migration test	US2	Underground seismics
BK2	Fracture system flow test	MOD	Hydrodynamic modelling	VE2	Ventilation test

Fig. 1: Perspective of the Grimsel Test Site after LIEB 1988 with the Locations of "Stress Measurements" (GS) and "Fracture System Flow Test" (BK2)

It is very important in assessing the long-term stability of a repository for radioactive wastes to be in a position to determine the influence of individual fractures and joint systems on the rock mechanical and hydraulic situation in the vicinity of a storage facility. For this reason the BGR has been undertaking permeability tests in fractured granite at the Grimsel Test Site since 1984. The purpose of the BGR project "Fracture System Flow Test" is, amongst others, to determine directional permeability dependencies and also the rates of migration of con-

The two research projects, "Fracture System Flow Test" and "Rock Stress Measurement" (which was completed in 1988) were preceded by comprehensive geological preparatory work, and accompanied by detailed engineering geological data recording. These engineering geological studies, and some extracts of the test results, are described and detailed in the following.

2. ENGINEERING GEOLOGICAL METHODS

Geological special mappings and core recordings provided the information base for
determining the individual test configurations, locations and direction of drilling, and interpretation of the rock
stress and flow data results. An additional objective was to indicate the existence
of dependencies between the fracture system
and the direction and level of stress, and
to extend the stress measurement results
to cover more extensive rock areas.

2.1 Mapping of GS and BK test galleries and boreholes

The test galleries of the GS and BK test
areas were mapped in the period late 1983
to early 1984. These detailed mappings,
targeted especially at the recording of
fracture surfaces, provided the essential
data needed for the subsequent determination of the arrangement of boreholes and
the planning of the injection and measurement configurations (BRÄUER, KILGER & PAHL,
1989).

The mapping of the test areas showed
that the apparently homogeneous granite
could be differentiated into zones having
different structural geological and petrographic characteristics. At the GS test
area, in the front part of the cavern, the
typical light Central Aare granite has
zones with a biotite-rich variety (Fig. 2).
Similarly, the fracture surface distribution in both areas shows in part pronounced differences in the direction and
dip of the systems. The darker biotite-
rich granite of the GS area has a relatively uniform fracture system in the direction of the primary schistosity in the
Grimsel area, whereas in the BK area, in
exclusively light granite, fracture systems
of various orientations are to be found.
The pattern of fractures on the cavern
floor also indicates that the surface of
the primary schistosity in the GS area
are, for the most part, continuous, whereas the fracture surfaces in the BK area
are often disjointed and wedge out.

The NAGRA-developed method of core recording was used for geological recording
of the cores from the test drillings. In
this method, the structural and textural
rock characteristics and discontinuity
planes (e.g. fractures) are recorded on a
film. In the case of continuously oriented
cores it was possible to determine, on a
1:1 scale, the location of fractures with

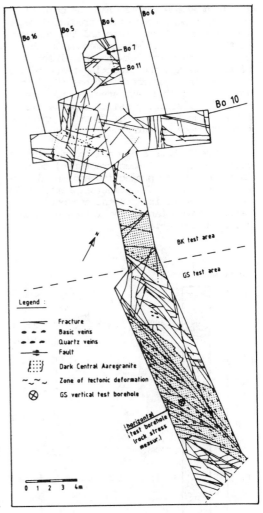

Fig. 2: Geological Map of GS and BK-Test
Areas

respect to the borehole axis from the
bearing of the outcrop. This mechanical
method of core orientation was checked
after completion of the GS test borehole
with the help of an acoustic down-hole
video system, whereby gaps in core orientation, due to missing cores, were filled
in. The true spatial location fracture
data was then input into a computer and
finally subjected to an overall statistical evaluation.

2.2 Rock stress and geological structure

The detailed geological recordings of the
cores from the GS test borehole described

in the previous chapter had the objective of clarifying a connection predicted between geological structure and the results of the rock stress measurements. To this end the connection between the direction of fracture and the direction measured for the principal horizontal stress (S_{H1}) was investigated. The next step was to determine the existence possible dependencies between the frequency of fracture and the rock stresses measured.

Figure 3 shows a comparison of joint direction (plot of open fractures) on a lower hemisphere projection and the directional values of the principal horizontal stress as determined in the overcoring tests. In this manner the following zones can be differentiated.

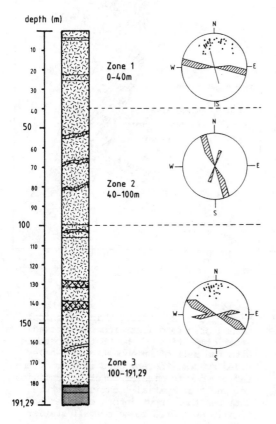

Fig. 3: Geological Cross Section of GS Test Borehole with Direction of Horizontal Stress (Rosette) and Orientation of Open Fractures (points in Schmidtnet, Lower Hemisphere Projection)

Zone I (0 – 40 m)
The direction of S_{H1} has a maximum at around WNW/ESE. This maximum is also that of the main direction of the open fractures.

Zone II (40 – 100 m)
The direction of S_{H1} shifts to approximately NNW/SSE – N/S. Open fracture surface were not mappable in this zone.

Zone III (100 – total depth)
The direction of horizontal stress and the maximum of the open fractures are comparable with the results in Zone I and lie in a WNW/ESE direction.

An obvious connection exists between the direction of the open fractures and the horizontal stress direction, and in Zone II the stress levels measured display definite departures from the borehole overall. The levels for S_{H1}, which are on average around 30 MPa, increase in this borehole section to approximately 40 MPa and more. It may then be surmised, that the lower frequency of fractures, primarily the absence of open fractures in Zone II, results not only in a change in the direction of stress, but also explains the higher levels of stress measured. This becomes clearer in Fig. 4, in which, for the measurement depth in question, the horizontal stress values are compared with the number of fractures determined and averaged in the near zone. The plot shows not only the correlation curve but also the upper and lower 95 % confidence limits. The two relatively low stress levels of 10 and 15 MPa were measured at a depth of 180 m at the contact to the lamprophyre/schistous zones.

2.3 Microcracks

In additon to the fracture analysis, 31 thin sections taken from 14 different depths of the GS test borehole were subjected to microcracks analysis. The objective here was to determine any connections and possible parallels between the fracture characteristics and fracture orientation and the microcrack direction. An additional target was to determine whether or not the results of the stress measurements were correlated with the results of the microcracks studies. This would allow statements on the stress field in a rock mass zone to be made at relatively little expense, without the need to undertake stress measurements.

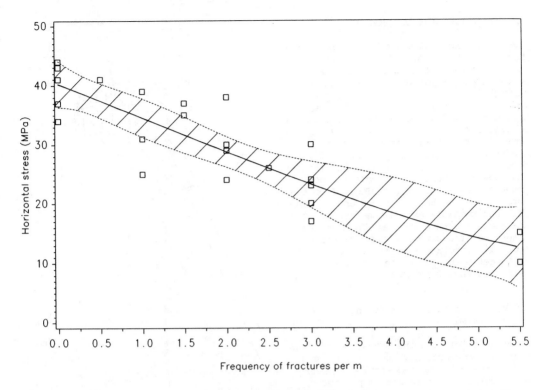

Fig. 4: Correlation of Rock Stress and Frequency of Fractures

The oriented cores recovered from the GS test borehole made it possible to undertake a direct comparison of macroscopically recorded fractures with microcracks. The overall evaluation of the thin sections resulted in a clustering of the primarily vertical microcracks in a EW to NNW/SSE direction. The results of a comparison between the fractures and the closed microcracks showed a clear agreement in direction in three sections of the test borehole (Fig. 5). And again, similarly to the open fractures, the open microcracks in Zone II deviate compared with the upper and lower borehole sections. The cracks, which otherwise orient in a ESE/WNW direction, turn here to an NNW/SSE direction. A comparison with the measured horizontal stress directions brings to light clear parallels in all three zones. Taking the results of the microstructural analysis into account, it is concluded that the open cracks are an indicator of the recent stress field in the Central Aare granite.

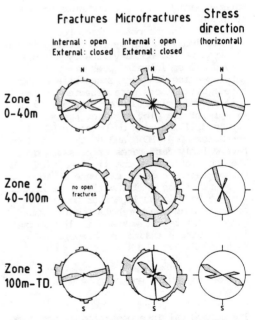

Fig. 5: Relationship between Direction of Fractures and Microfractures and Stress Orientation

3. ROCK HYDRAULIC INVESTIGATIONS

The BGR is currently involved in permeability tests being carried out in the area adjacent to the GS test field and targeted at developing equipment and methods for permeability investigations in fractured media. To date the first phase has involved the drilling of 12 boreholes in the BK test area. Packers have been set in the boreholes to shut-in various injection and observation sections for a number of individual tests (PAHL et al., 1986) (Fig. 6). The objective here is to increase knowledge of the structural geological configuration of the BK test field and to confirm and possibly extend the details of the known fracture systems. In order to determine effective and connected rock hydraulic fracture systems, the plan was to target open, and hence potentially water-bearing, fractures.

3.1 Water pressure observations in the fractured rock

In order to clarify the situation of rock hydraulic connections, the drilling work during phase 2 involved setting packers and pressure cells to create pressure measuring points at certain depths in all observation boreholes present in the near zone of the BK test field. Thus, when water-bearing zones are encountered the resultant pressure reduction would be detectable in the observation boreholes. The sections shut-in by the packers in the observation boreholes virtually all displayed a drop in water pressure when water-bearing fracture zones were encountered. Differences were detected in the rate of fall of pressure. Figure 7 is an example of the plot of water pressure during drilling work on borehole BO 12. Boreholes BO 4, BO 5 and BO 6 resulted in particularly fast rates of pressure drop. The oscillations in the first part of the curves indicate some influence by the pressure of the drilling fluid, of around 4.5 - 5 bar, in BO 12. After completion of borehole BO12 the borehole opening was packed. The result was an immediate rise in the water pressure in boreholes BO 4, BO 5 and BO 6.

4. DEMARCATION OF HOMOGENEOUS ROCK MASS ZONES

The geological interpretation of the rock stress measurements and the results of the analysis of the flow tests resulted in the

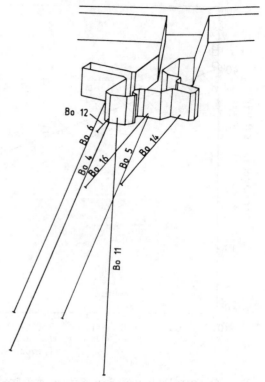

Fig. 6: Boreholes for Permeability Tests in BK-Test Area

subdivision of the rock mass into rock mechanically and rock hydraulically homogeneous zones. Figure 8 is a cross section of the geological subdivisions. The following zones were differentiated.

Zones 1 and 3 have relatively heavily fractured granite, displaying considerable dislocation at the contact to the adjacent lamprophyre. The frequency of the fractures has resulted in a partial compensation of rock stress, such that levels of horizontal stresses are predicted to be lower here, with maximum levels of 30 MPa. The direction of S_{H1} has also shifted to approximately ESE/WNW.

The maximum permeability in these homogeneous zones is determined by the numerous and, for the most part, hydraulically connected open fractures. The permeability values in these heavily fractured areas are approximately 1×10^{-7} m/s.

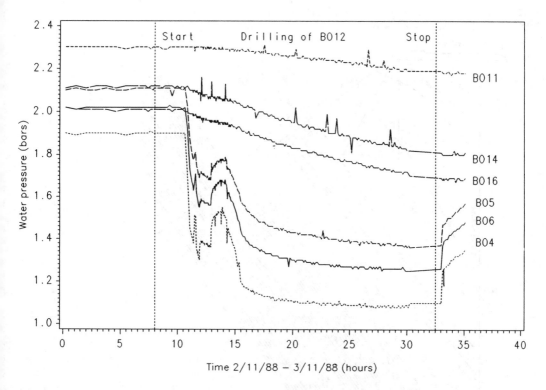

Fig. 7: Response of Water Pressure in BK-Test Boreholes while Drilling Borehole BO 12

Zone 2 is located for the most part in compact granite. Due to the low number of fractures the horizontal stress is expected to be relatively high at around 40 MPa in an SSE/NNW direction.

The permeability of the rock mass in this zone is determined by isolated hydro-thermally altered zones and by the very dense granite matrix. Average values for permeability here are around 1×10^{-11} to 1×10^{-12} m/s.

5 CONCLUSIONS

The analysis and interpretation of rock stress measurements at the Grimsel Test Site indicate a direct connection between geological structure and the measurement results. Similarly, the microstructural studies show dependencies between the microcrack direction and the direction of horizontal stress. The details of rock stresses generally consist of values determined as point values, then extended to cover larger areas. In conjunction with the geological recordings it is possible

to translate homogeneous zones with certain characteristic structures over more extensive rock zones. The number of measurements required may therefore be reduced. By comparing zones of similar rock mechanical characteristics with zones of similar hyraulic characteristics it is possible to determine the parameters necessary to allow an assesment of the barrier effect of the rock mass in the storage of waste materials.

REFERENCES

Bräuer, V.; Kilger, B. & A. Pahl 1989. Ingenieurgeologische Untersuchungen zur Interpretation von Gebirgsspannungsmessungen und Durchströmungsversuchen. - NAGRA Techn. Bericht, NTB 88-37, Baden/Schweiz.
Brewitz, W. & A. Pahl. 1986. German Participation in the Grimsel Underground Rock Laboratory in Switzerland - Targets and Methods on In-Situ Experiments in Granite for Radioactive Waste Disposal. - Int. Symp. on the Siting Design and Construction of Underground Repo-

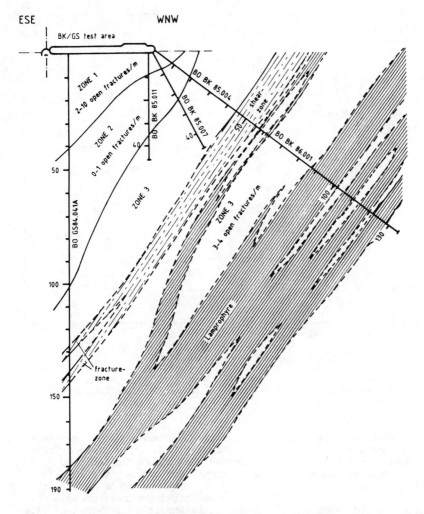

Fig. 8: Cross Section of GS/BK-Test Area with Structurally Homogeneous Zones

sitories for Radioactive Wastes, IAEA-
SM 289, Hannover, F.R. Germany.
Lieb, R.W. 1988. The Grimsel Test Site
from 1983 to 1990 - An Overview. -
Nagra Bulletin, Baden, Switzerland.
Nagra 1985. Felslabor Grimsel, Rahmen-
programm und Statusbericht. - NAGRA
Techn. Bericht, NTB 85-34, Baden/
Schweiz.
Pahl, A.; Bräuer, V.; Heusermann, S.;
Kilger, B. & L. Liedtke 1986. Results of
Engineering Geological Research in
Granite.   Bull. IAEG, 34: 59-65,
Paris.

Engineering performance evaluation of the foundation rocks of Narmada Sagar Dam Project, Central India

Evaluation des performances techniques des fondations rocheuses du projet de barrage de Narmada Sagar, Inde Centrale

Vishnu D.Choubey & Shailendra Chaudhari
Engineering Geology Division, Indian School of Mines, Dhanbad, India

ABSTRACT : A comprehensive laboratory testing programme was carried out to evaluate and correlate the physicomechanical properties of foundation rocks whereas the strength and elastic properties are basic and important parameters for the consideration of design and functional stability of the sturcture. Results of the study lead to conclusion that the foundation rocks of Narmada Sagar Dam Project are competent, having high strength and of low modulus ratio.

1 INTRODUCTION

The Narmada Sagar Dam, a 92 m high and 635 m long concrete gravity structure is proposed to be constructed across the river Narmada near Punasa in Central India. In case of gravity dams , the distribution of stresses depend largely on the strength and elastic properties of the dam and foundation. The foundation deforms under the load imposed by the dam and headwater and this deformation which is due to the weight of dam plays an important role in the formation of the state of stress at the dam to foundation contact. The foundation of Narmada Sagar Dam is mainly composed of quartzite with thin intercalations of siltstone. Physicomechanical properties of these foundation rocks were determined in the laboratory as these properties are of prime importance in design part of rock related components and stability evaluation.

2 PHYSICAL PROPERTIES

The observed values of physical properties of the rocks tested have been presented in Table 1. An average bulk dry density of quartzite is 2.56 gm/cc, average water absorption determined is 0.44% and specific gravity is 2.75 gm/cc. The average bulk dry density of siltstone was found to be 2.61 gm/cc, average water absorption which is of much higher value than quartzite, is 1.26% and an average specific gravity observed was 2.68 gm/cc which is slightly lower than quartzite.

Table 1. Physical properties of the tested rocks

No. of tests	Rock type	BDD gm/cc Max.	BDD gm/cc Min.	Water ab. % Max.	Water ab. % Min.	Sp. gr. gm/cc Max.	Sp. gr. gm/cc Min.
4	Qzt	2.67	2.49	0.60	0.21	2.84	2.66
4	Slt	2.69	2.54	1.63	0.47	2.74	2.62

Qzt - Quartzite (refer in all tables and text)
Slt - Siltstone (refer in all tables and text)

3 STRENGTH AND ELASTIC PROPERTIES

The study of reaction of rocks to applied stress is very useful for predicting the behaviour of foundation rocks. Strength and elastic properties were determined on intact rock core samples of Nx size (54 mm diameter). The cores were first cleaned of dirt and clay by washing with water and dried by putting in an oven at a temperature of 110°C for about 8 hours. The specimens were cut to the required length by Lapro Slab Saw 24 rock cutting machine using oil as a cooling medium. The procedure adopted to flatten and polish the surfaces consisted the usage of carborendum powder with water followed by rubbing the surface against an iron flat rotating plate of polisher Ecomet II grinder.

Uniaxial compressive strength (σ_c) tests were performed on the Servo control, Stiff testing machine MTS 413 keeping the slenderness ratio of 2:1. Rock cores were loaded axially in between the platens. The samples of both the rock types failed violently in brittle manner resulting in typical conical ends. Determined σ_c of quartzite and siltstone is 1920.44 kg/cm^2 and 1348.14 kg/cm^2 respectively. Test results are given in table 2.

Table 2. Uniaxial compressive strength of rocks

Sample No.	Rock type	Length cm	Failure load kg	σ_c kg/cm^2
SC1	Qzt	10.245	38880	1657.97
SC2	Qzt	10.432	57100	2458.02
SC3	Qzt	10.474	45200	1961.80
SC4	Qzt	10.260	63500	2737.06
SC5	Qzt	10.376	31400	1358.71
SC6	Qzt	10.544	31300	1349.13
SS1	Slt	10.262	29600	1277.51
SS2	Slt	10.494	33100	1418.77

Mean σ_c of quartzite	-	1920.44 kg/cm^2	
Standard deviation	-	527.31	
Mean σ_c of siltstone	-	1348.14 kg/cm^2	
Standard deviation	-	70.63	

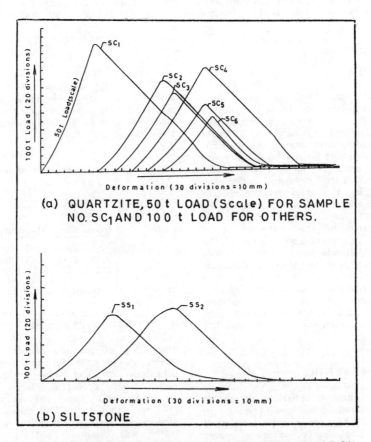

(a) QUARTZITE, 50 t LOAD (Scale) FOR SAMPLE NO. SC$_1$ AND 100 t LOAD FOR OTHERS.

(b) SILTSTONE

FIG. 1   MACHINE PLOT FOR LOAD − DEFORMATION CHARACTERISTICS OF FOUNDATION ROCKS. TESTS CONDUCTED IN MATERIAL TESTING SYSTEM, 413, MASTER CONTROL PANEL.

Modulus of elasticity (stress at 80% peak strength/strain), secant modulus (stress at 50% peak strength/strain at 50% peak strength) and tangent modulus (Δ stress at 50% peak strength/Δ strain) were calculated with the help of machine plots obtained from stiff testing machine. Under static loading conditions, quartzite approached ideal elastic behaviour with increase in load from zero condition. The slope of stress-strain curve increases as the pores and interstices in the material close with increasing pressure. Further the rock material shows almost a linear stress strain relationship. Calculated elastic properties are given in table 3.

Table 3. Elastic modulus parameters of rocks

Sample No.	Rock type	E kg/cm^2 $\times 10^4$	E_s kg/cm^2 $\times 10^4$	E_t kg/cm^2 $\times 10^4$
SC1	Qzt	7.3	6.3	4.9
SC2	Qzt	8.7	8.2	5.5
SC3	Qzt	8.4	7.5	5.4
SC4	Qzt	4.6	7.8	6.2
SC5	Qzt	7.4	6.6	4.2
SC6	Qzt	6.9	6.3	3.6
SS1	Slt	6.3	5.2	3.6
SS2	Slt	6.3	5.4	3.4

Modulus ratio (E_t/σ_c) of the foundation has been determined from the plots of σ_c and E_t values using the figure given by Deere and Miller (1966). It is inferred from the figure that the rocks fall under the class of "low modulus ratio" (Fig.2).

Tensile stress and tensile fractures confined mainly to the rock foundation at the heel of dam attributed to overstressing of rock under tensile stress. Tensile strength determination becomes important in this regard. Brazilian tests were performed on the Universal testing machine of 20 ton load capacity to carry out indirect tensile strength (σ_t) of rocks in which the rocks in biaxial stress field fail in tension when one principal stress is compressive. Slenderness ratio was kept between 0.5 to 1.0 and the diametrical loading of the disc shaped rock sample caused most of them to fail along the diameter joining the loading points/platens. Six tests were performed on quartzite and σ_t was found to be 164.26 kg/cm^2 whereas it was 144.52 kg/cm^2 for siltstone by performing five tests. Test results are given in table 4.

The ratio of uniaxial compressive strength and indirect tensile strength (σ_c/σ_t) of of quartzite is around 11 and the same ratio

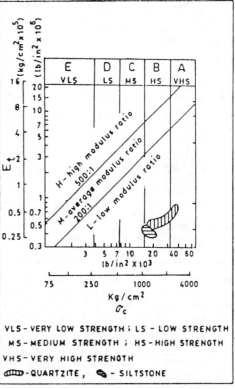

FIG.2. PLOT OF MODULUS RATIO FOR QUARTZITE AND SILTSTONE

VLS - VERY LOW STRENGTH ; LS - LOW STRENGTH
MS - MEDIUM STRENGTH ; HS - HIGH STRENGTH
VHS - VERY HIGH STRENGTH
▥ - QUARTZITE , ▧ - SILTSTONE

of siltstone is around 9. Although this ratio depends basically on the nature of rock and environmental factors and no unique value should be adopted for all the rocks.

Brittleness is an important mechanical property of rocks. Higher the angle of internal friction, higher will be the brittleness in the rocks with fracture failure. It is observed that the difference between σ_c and σ_t increases with increasing brittleness. Brittleness coefficient (B) (Hucka and Das 1974) is determined from σ_c and σ_t of quartzite [see (i)] and siltstone [see (ii)].

$$B(Qzt) = \frac{1920.44 - 164.26}{1920.44 + 164.26} \times 100$$

$$= 84.24 \qquad \text{.... (i)}$$

$$B(Slt) = \frac{1348.14 - 144.52}{1348.14 + 144.52} \times 100$$

$$= 80.63 \qquad \text{.... (ii)}$$

Shear strength of a rock can be defined as the particular shear stress level at which maximum

Table 4. Indirect tensile strength of rocks

Rock type	Length cm	Failure load kg	σ_t kg/cm^2
Qzt	2.532	3960	182.74
Qzt	2.702	2700	116.94
Qzt	2.612	3240	146.00
Qzt	2.232	3960	207.26
Qzt	2.412	3780	182.90
Qzt	2.912	3740	149.76
Slt	2.648	3760	168.26
Slt	2.234	2600	136.68
Slt	2.246	1320	69.00
Slt	2.234	3480	183.68
Slt	2.254	3180	165.02

Mean σ_t of Quartzite	-	164.26 kg/cm^2
Standard deviation	-	29.77
Mean σ_t of siltstone	-	144.52 kg/cm^2
Standard deviation	-	44.69

Table 5. Indirect shear strength of rocks

Rock type	Length cm	Dia.of puncher cm	Failure load kg	S_o kg/cm^2
Qzt	0.604		2240	488.01
Qzt	0.838		2980	467.81
Qzt	0.772		2020	344.12
Qzt	0.954		3760	517.90
Qzt	0.706		3500	651.76
Qzt	0.648		3100	628.00
Qzt	0.984		4680	625.66
Qzt	0.662	2.424	2000	397.61
Qzt	0.600		2120	464.91
Slt	1.035		2170	275.45
Slt	1.190		2266	250.18
Slt	0.958		2260	285.51
Slt	1.281		2295	235.38
Slt	1.202		1975	215.87
Slt	0.996		2260	298.18
Slt	1.154		2190	249.33
Slt	1.074		2070	253.22

Mean S_o of Quartzite	-	509.62 kg/cm^2
Standard deviation	-	101.15
Mean S_o of siltstone	-	257.89 kg/cm^2
Standard deviation	-	25.30

Table 6. Point load strength index of rocks

Rock type	Length cm	Failure load kg	I_s kg/cm^2
Qzt	8.312	165	80.19
Qzt	7.028	200	97.21
Qzt	7.028	175	85.05
Qzt	6.852	200	97.57
Qzt	6.852	220	107.32
Qzt	7.256	175	85.43
Qzt	7.256	130	63.46
Qzt	7.334	240	116.84
Qzt	7.334	220	107.11
Qzt	7.334	160	77.89
Qzt	7.246	200	97.57
Qzt	7.029	190	92.69
Slt	6.908	100	48.81
Slt	6.908	80	39.05
Slt	6.732	98	47.84
Slt	6.854	118	57.43
Slt	7.336	76	37.10
Slt	7.224	84	40.97
Slt	8.324	110	53.70
Slt	8.028	90	43.84
Slt	7.234	80	38.94
Slt	8.334	94	46.13

Mean of I_s of Quartzite	-	92.36 kg/cm^2
Standard deviation	-	14.18
Mean I_s of siltstone	-	45.38 kg/cm^2
Standard deviation	-	6.36

load bearing capacity has reached and beyond which major or structural changes take place in the material of a body. Mathematically it is the ratio of the tangential breaking force to the sheared surface area. It is measured to obtain information fully about the shearing resistance of the intact rock and that of large scale discontinuity planes occurring in the rock mass. The term shear strength is used for rocks and the term cohesion for soil. The greater the porosity of rock, lower the value of shear strength. Shear strength of pure soil is zero i.e. the zero cohesive force. Punch shear method was adopted to determine the indirect shear strength (S_o) using core samples of small thickness. In this method very high concentrations are generated along the circumference of the punch which cause the rock disk to break at a low stress value. Diameter of the puncher used was 2.424 cm. S_o value (Table 5) determined was 509.62 kg/cm^2 for quartzite and 257.89 kg/cm^2 for siltstone.

The point load strength index (I_s) test requires relatively a simple loading frame and can be carried out virtually on any shape and size of rock sample. The samples were placed between opposing cone shape platens and subjected to compression. The test generates tensile stresses normal to the axis of loading. The core samples of Nx size, having length of 0.7 times the diameter, were used. The cores over the diametrical area were placed between the two conical points. I_s value determined for quartzite was found to be 92.36 kg/cm^2 and it was 45.38 kg/cm^2 for siltstone. Test results are presented in table 6.

The ratio of uniaxial compressive strength

and diametrical point load (σ_c/I_s) of quartzite was found to be 20. In case of siltstone, the above ratio was around 29. Uniaxial compressive strength and point load strength index properties have a definite correlation which can be shown as

$$\sigma_c = C\ I_s \qquad \ldots\text{(iii)}$$

where C is a constant.

However, a constant value of 24 has been suggested by Broch and Franklin (1972) for all rocks and the correlation proposed by ISRM (1985) is

$$\sigma_c = 22\ I_s \qquad \ldots\text{(iv)}$$

in the case of foundation rocks of Narmada Sagar Dam Project, the above correlation was found to be

$$\sigma_c = 22.95\ I_s \qquad \ldots\text{(v)}$$

The plotted graph is shown in Fig. 3.

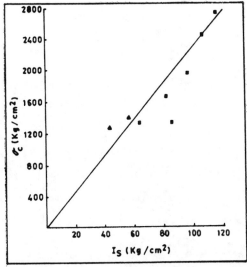

FIG.3. CORRELATION BETWEEN σ_c AND I_s OF ROCKS
■ QUARTZITE ▲ SILTSTONE

4 DISCUSSION AND CONCLUSIONS

Physical properties of rocks indicate that quartzite has low bulk dry density and water absorption, and a high specific gravity in comparision to siltstone. Quartzite is of high compressive strength, falling in the class 'VS' of Stapledon (1968), class 'B' of Deere and Miller (1966) and 'Very strong' in the classifica-suggested by Anon (1979,a). The compressive strength of siltstone is also high according to the classifications mentioned above. Siltstone fall in the class 'S', class 'B' and class 'Very

strong' respectively. It is evident that quartzite exhibits higher uniaxial compressive strength than siltstone. It has been observed that the variation in modulus of elasticity, secant modulus and tangent modulus for these rocks is similar to that observed in strength i.e. quartzite has greater value of E, E_s, E_t than of siltstone. The relation between tangent modulus and uniaxial compressive strength has been presented and on the basis of this relation the foundation rocks are classified as BL to AL for quartzite and BL for siltstone. Point load strength index of rocks show that both the rock type belong to a group of 'Very high strength' in the classification suggested by Franklin and Broch (1972). The brazilian and point load index test show that the strength is higher for quartzite than siltstone.

In summary, the quartzite and siltstone which are the main foundation members of dam are brittle and hard of high strength and of low modulus ratio.

ACKNOWLEDGEMENTS

The help and cooperation received from Department of Irrigation, Govt. of M.P. during field visits and financial assistance from University Grants Commission are thankfully acknowledged. Shri P K Litoria and Shri Gopal Dhawan have generously cooperated during finalisation of manuscript.

REFERENCES

Anon (1979,a). Classification of rocks and soils for engineering geological mapping. Pt I, Rock and soil materials, Bull. Int. Ass. Engg. Geol., No.19, pp.364-371.

Deer, D.U. and Miller, R.F. (1966). Engineering classification and index properties for intact rock, technical report No.AFWL-TR-65-116, Air Force Weapon Laboratory, Kirtland Airbase, New Mexico.

Franklin, J.A. and Broch, E. (1972). The point load strength test. Int. Jour. Rock Mech. and Min. Sci., vol.9, pp.669-698.

Hucka, V. and Das, A. (1974). Brittleness determination of rock by different methods. Int. Jour. Rock Mech. and Min. Sci., vol.11, pp.389-392.

ISRM (1985). Suggested methods for determining point load strength. Int. Jour. Rock Mech. and Min. Sci. and Geomech. Abstr. vol.22, No.2, pp.51-60.

Stapledon, D.H. (1968). discussion of D.F. Coate's rock classification. Int. Jour. Rock Mech. and Min. Sci., vol.5, pp.371-373.

Stages of geotechnical investigation for townships

Etapes d'investigations géotechniques pour communes

J.H.de Beer
Steffen, Robertson & Kirsten Inc., Johannesburg, South Africa

ABSTRACT: The paper describes the different stages of geotechnical investigation that should accompany the phases of planning associated with township development. Adherence to these stages of work will reduce the geotechnical risks associated with development.

RESUME: Le papier decrit les differents stages d'investigations geotechniques qui devraient accompagner les differentes phases de plannification associees au development de communes. L'adherence a une telle planification reduit les risques geotechniques associes au development.

1 INTRODUCTION

Recent socio-political changes in South Africa have led to an acute shortage of housing in urban areas. This is as a result of the migration of the population from rural areas to the apparently prosperous urban centres. Much of this housing shortage is in the lower income groups of the population. New housing programmes aim to provide suitable First-World type homes on a freehold ownership basis which will be financed by commercial institutions.

In view of the urgency to provide housing, there is often the danger that essential stages of geotechnical investigation are missed or curtailed.

The main theme of this paper is not the identification of the type of unstable soils, nor a suitable foundation solution, but the timing and extent of geotechnical investigations associated with the different phases of township planning and development.

2 BACKGROUND TO THE PROBLEM

The geology of the Transvaal Highveld region is extremely variable and it may have a significant effect on the development of residential townships. Unstable transported or residual soils may give rise to heave or collapse settlement of structures or, even worse, the formation of sinkholes.

To avoid these problem soils or to become aware of their extent and degree of severity, as early as possible, is essential for the successful development of a township. However, care is also required to ensure that unnecessary large sums of money are not expended during the early phases of the project when its viability is still uncertain.

3 GEOTECHNICAL CONSIDERATIONS

The geotechnical characteristics of a site may play a major role in the planning and development of a township. These characteristics are identified by dividing the site into geotechnical zones wherein the conditions are broadly similar. This information is used by town planners during the various phases of township development.

On the Highveld region of the Transvaal, many soils are moisture-sensitive and may undergo changes in volume following development of an area.

A problem that also affects large parts of the Highveld is the occurrence of cavernous dolomitic materials which may give rise to sinkholes and subsidences.

3.1 Collapsible Soils

Collapsible soils such as unconsolidated hillwash or windblown sands, and deeply weathered residual granitic soils cover a major part of the Highveld. They will have a major impact on the cost of house foundations and building maintenance if no construction precautions are taken to minimise the effects of collapse settlement. The presence of collapsible soils will, however, seldom make the development of a particular township inviable, nor will these areas be excluded from development.

Collapsible soils will have some effect on the cost of road construction, necessitating treatment and a possible loss of material on compaction.

3.2 Expansive Soils

Expansive soils such as alluvium and gullywash, or residual soils derived from mudstones and certain basic igneous rocks, also occur in large parts of the Transvaal. The consequences of building on expansive soils without precautions are more severe and more costly to repair than on collapsible soils, especially if the soil has become desiccated by a previous vegetative cover. Areas of highly expansive soils may be zoned for development other than residential such as parkland or industrial. Where these areas are extensive it may make a township inviable.

3.3 Sinkholes and Subsidences

Sinkholes and doline subsidences are prevalent over areas that are underlain by dolomite of the Transvaal Sequence. Investigations should identify the main factors affecting sinkhole formation such as, position of the water level and its relationship to dolomite bedrock, the nature of the dolomite residuum and, most important, the type and thickness of blanketing layer over the potentially cavernous dolomite. Sinkhole-prone areas are identified in terms of high, medium and low risk of sinkhole formation. High risk areas are unsuitable for the development of residential townships. Their boundaries need accurate definition and if these areas are extensive, the viability of the township may be in question. Medium risk zones require the implementation of water management precautions to reduce the destabilising effects of township development. Low risk areas require no precautions.

3.4 General

Other geological factors that require identification at an early stage is the presence of outcropping or shallow bedrock, swampy or filled areas, areas of potential slope instability as well as construction materials.

Areas affected by steep slope or flooding will usually be identified by other parties responsible for the development of a township.

Over dolomitic areas, two separate investigations are required. Firstly, a geotechnical investigation to assess the near surface conditions that may affect houses and roads, and, secondly, a more intensive investigation to evaluate the dolomitic stability of the upper 30 m to 40 m of the profile. These investigations involve the execution of gravity surveys, rotary percussion drilling and monitoring of water levels. They are generally very costly.

4 PLANNING PHASES IN TOWNSHIP DEVELOPMENT

Planning of a high standard and the adherence to acceptable procedures are essential to facilitate the provision of serviced land and affordable housing for the lower income groups of the community. In developing low cost townships, planning may have to be undertaken with limited resources or without structure or service master plans which increases the risk of making untimely decisions.

Working to proven planning procedures will ensure that decisions regarding the expenditure of finances are made timeously and that crucial information is provided at strategic milestones in the planning process.

The following main phases of planning are generally adopted by township developers. These phases, together with the associated stages of geotechnical investigations, are described below.

Each stage of geotechnical investigation should commence with a desk study of available information.

While economic geotechnical investigations should be aimed at, they will not achieve their objectives if geotechnical consultants are required to price investigations on a competitive basis. This will place the risk of not having sufficient information on the developer.

4.1 Planning Phase 1: Feasibility

The main decision-makers during this phase of work are politicians and local authority policy-makers. Broad issues such as loss of high grade agricultural land or disturbance of environmentally sensitive areas are considered. Areas of land considerably larger than may be considered by individual developers are dealt with. Recognition of geological constraints are essential at this phase of planning and will be identified in the associated Geotechnical Feasibility Investigation.

During this planning phase, areas suitable for the different types of land use, such as residential, industrial, commercial, agricultural, mining, etc. should be identified. A broad zoning of the area delineating, and even ranking land use options, should be produced.

4.1.1 Geotechnical Stage 1: Feasibility Investigation

During this stage of work, extensive use will be made of available geological maps, aerial photographs, and reconnaissance site visits to inspect site features such as outcrops, erosion features, borrow areas, etc.

It is advisable to carry out some investigations by means of digging of test pits or drilling of boreholes. Investigation points are likely to be between 500 m to 2 km apart, but will depend on the complexity of the geology and the detail of information required. Collapsible soils will only be identified from visual inspection, while expansive soil may require limited laboratory testing.

In dolomitic areas the drilling of a number of boreholes is considered essential at this stage to obtain an indication of the factors affecting sinkhole formation.

It is imperative at this stage to attempt to identify ground water compartments, mainly from previous work, and to assess the likelihood of future changes of the water level as a result of possible mining or withdrawal of ground water for emergency water supplies. Extensive lowering of the water level will give rise to settlement of the ground surface.

4.2 Planning Phase 2: Reconnaissance

This phase of work is carried out to assess the feasibility of one or more particular township sites that may be considered by a developer. The work is carried out prior to purchasing of the land and the information presented in broad terms for the approval of the Local Government.
A provisional layout may be provided, but may change significantly during the ensuing phases.

The Geotechnical Reconnaissance Investigation, which will precede this phase of planning, is described below.

4.2.1 Geotechnical Stage 2: Geotechnical Reconnaissance

During this stage the least expenditure possible should be incurred as the property has not yet been purchased and there is a risk that the project may not proceed.
Test pits, using a backhoe, should be excavated to determine ·the soil profile and material properties in the upper 3 m of the profile. The greatest effort should go into assessing the geotechnical conditions at each investigation point as accurately as possible. This should be done with the aid of laboratory testing.
Boundaries between geotechnical zones will be very approximate. In designing a reconnaissance investigation it should be remembered that backhoe hire charges are on a daily basis and that between 15-20 test pits may be excavated in one day. At this stage of work it is advisable to have investigation points at 200 m to 300 m intervals. On this basis, townships up to 100 ha in size may be investigated with 15-20 test pits to provide the minimum feasible information. On townships of 100-200 ha, 30-40 test pits may be sufficient and so on.
The amount of work should provide sufficient information to allow purchasing of the land with confidence. It should ensure that there are no large uninvestigated areas of the site which, if found to be unsuitable, may make a township inviable.

4.2.2 Geotechnical Stage 2: Dolomitic Reconnaissance

Dolomitic stability investigations are very costly because of the effort required to model the geological and ground water conditions accurately.
If sufficient effort is not provided at this stage to identify the extent of high risk zones, prior to purchasing of the site, there is a danger that unfavourable conditions discovered too late may make the township inviable.
Nevertheless, many developers choose to do all dolomitic stability investigation during the next stage of planning, that is, after the property has been purchased.

During this stage of work, boreholes are usually drilled to determine broadly the factors affecting sinkhole formation. This stage is often used as a preparatory stage to design the next stage of investigation, especially as far as deciding on the details or need of a gravity survey.
As stated earlier, the position of the water level is critical in determining the dolomitic stability of a site. A separate geohydrological investigation, often extending beyond the boundaries of the site in question, may be required if the site lies within a recognised mining area or if there is any possibility of future lowering of the water level. This work should not be carried out later than during this stage and should preferably be done during the Feasibility Stage.

4.3 Planning Phase 3: Planning

This is the most important phase of work and it is carried out once a site has been purchased by a developer and he begins with the engineering investigation and design.
Sufficient information should be available to allow the optimisation of land use. Preliminary construction cost estimates are prepared by the developer. A township layout is prepared for final approval by Local Government. Sufficient geotechnical information should be available during this phase to ensure that the layout, in terms of boundary lines between geotechnical zones, does not require any future amendment as this will give rise to delays in approval.
The achievement of the optimal layout of a township is crucial if capital and maintenance costs are to be controlled.

4.3.1 Geotechnical Stage 3: Geotechnical Planning Investigation

The investigation will be similar to that described under Geotechnical Reconnaissance Investigation, except that boundaries of critical geotechnical zones need to be identified to an accuracy of approximately 30-50 m. A critical geotechnical zone, for example, is one where a thick and highly expansive clay horizon has been zoned as unsuitable for development.
It is important to provide some soil parameters for roads design at this stage.

4.3.2 Geotechnical Stage 3: Dolomitic Stability Investigation

This stage of work begins with a geophysical gravity survey, with gravity stations on a 30 m grid. The purpose of this survey is to identify gravity anomalies, represented by variations in bedrock depth, for later investigation by percussion drilling.
With the aid of drilling, the risk zones, especially the boundary of the high risk zone, should be defined with boreholes no further than 30-50 m apart. The financial gain in having more ground available for development by defining boundary lines to this accuracy is far greater than the cost of drilling the boreholes.

4.4 Planning Phase 4: Detailed Design

During this phase of planning, tender documents for the installation of services are prepared. Further investigation may be required on the extent of rock occurrences to refine quantities of possible blasting quality rock to be removed.

Roads over the whole township and foundations in poor areas will be designed during this phase utilising information from the associated geotechnical work described below.

4.4.1 Geotechnical Stage 4: Detailed Geotechnical Investigation

This investigation will address two aspects, namely, the design of roads and the design of foundations in areas where normal foundations cannot be installed.

A centre-line soil survey will be carried out for the design of roads. This will involve the excavation of test pits at various intervals related to the type of road. Extensive laboratory testing will be undertaken.

Further testing, both laboratory and in situ, may be initiated in zones where special foundations are required. This will be followed by the preparation of design drawings for foundations.

4.4.2 Geotechnical Stage 4: Detailed Dolomitic Stability Investigation

It is seldom that additional work is required at this stage to define dolomitic risk zones. In certain circumstances, however, it may be advisable to refine the boundary between a low and a medium risk zone to determine the extent of implementation of water management precautions.

4.5 Geotechnical Stage 5: Construction Geotechnical Investigation

During the construction phase of township development, particular queries may arise regarding unforeseen soil conditions. These will be investigated with specifically designed investigations, normally covering small areas.

Extensive lengths of services trenches will also be open at various times of construction and will provide ideal opportunities to up-date boundary lines between geotechnical zones and review foundation designs.

5 CONCLUSIONS

Diligent implementation of the planning and investigation stages outlined above will allow the successful development of a township.

The possible financial consequences of planning with inadequate geotechnical information or missing out some of the recommended stages far outweigh the costs of the actual investigations. It should, however, also be remembered that it is not always financially feasible to eliminate the risks associated with geotechnical conditions entirely.

6 ACKNOWLEDGEMENT

The author wishes to acknowledge with appreciation discussions held and unpublished information provided by Mr B Orlin (South African Housing Trust) and Mr S Foster (FHA Homes).

584

Recent developments in Urban Geology
Développements récents dans la géologie urbaine

Ed F.J.de Mulder
Geological Survey of the Netherlands, Haarlem, Netherlands

ABSTRACT: Urban geology is a relatively new branch of the earth sciences. This paper presents a state-of-the-art in Urban Geology and deals more specifically with the availability of data in urban areas, site investigations, and the correlation of engineering geological data in cities. Recent developments in the field of urban geology are: the growing awareness of the discipline among geoscientists, the developments in project acquisition and the overall automation in municipal administration, in data processing, and in data presentation.

RESUME: La Géologie Urbaine est une branche relativement nouvelle des Sciences de la Terre. Cet article donne un "state-of-the-art" dans la Géologie Urbaine et traite de manière plus spécifiée la disponibilité de données relatives au sol dans les régions urbaines, les méthodes d'analyse du sol et la corrélation entre les données de la géologie de l'ingenieur dans les villes. Les développements récents dans cette spécialité sont: la conscience croissante des scientifiques de la terre vis à vis de cette spécialité, les développements dans la domaine de l'acquisition de projects et l'essor de l'informatique dans l'administration locale et dans l'incorporation et la présentation de données relatives au sol.

1 INTRODUCTION

Most cities owe their origin and present existence to geological and geo-graphical factors that control the position of the coastline, the river courses, the situation of corridors through mountain ranges, the foundation conditions, the vulnerability to natural hazards, and the distribution of natural resources. In the course of time, man has modified the natural conditions in these places tremendously. The rapid increase in human popula-tion in the past decades took place primarily in urban centers. This popula-tion explosion caused a growing demand for natural resources such as fresh groundwater and construction materials (Schneider & Spieker, 1973; Lüttig, 1987). Furthermore, the urban population became exposed to a number of serious threats such as flooding, slope failure, seismic catastrophes, construction failures because of foundation instability, groundwater pollution and soil pollution.

City planners and decision makers now begin to realize that earth sciences play a key role in providing the desired resources and in mitigating natural and man-induced hazards (Prinya Nutalaya, 1988). Gradually, a new branch of geological science came into existence: Urban Geology.

Urban geology can be considered as the field of Applied Geology that deals with major population centres. This discipline combines those branches of the earth sciences that can assist in urban management and development. Urban geology is one of the most interdisciplinary fields of the earth sciences, it covers parts of engineering geology, environmental geology, and land management. In addition to conventional geological disciplines such as stratigraphy and tectonics, geotechnics (rock-/soil mechanics) and geohydro-logy are of major importance in urban geology.

The first urban geological activities that can be traced in literature date from the end of the nineteentwenties. In Germany special soil maps were made to support urban planning (Hoyningen-Huene, 1931; Stremme, 1932). Towards the end of the thirties the PreuBischen Geologischen Landesanstalt in Germany (Brüning, 1940) combined detailed maps on a scale of 1:10,000 and 1:5,000 with maps indicating the suitability for various kinds of land-use and compiled these in a "Bodenatlas" for urban expansion. As a result of the population explosion and economic revival inmediately after World War II the amount of urban geological activities increased dramatically in many countr-ies especially in Europe and in North America. New systematic geological mapping programmes were set up in a.o Germany, Czechoslavakia, and in The Netherlands. These were intended to support physical planning and pay attention to physical properties and to the succesion of strata in the subsurface (Hageman, 1963). An excellent example of such urban geological activities was the detailed (scale 1:5,000) mapping of foundation conditions in the city of Prague. The numerous data and maps of this city are constantly updated (Legget, 1973). For more than a dozen German cities atlasses with a number of thematic land-use maps were prepared (Müchenhausen & Müller, 1951). Although those maps were obviously intended for urban planning their readibility was generally poor. To economize on the high (colour)printing costs, too much information was crammed into too few maps. Information about soil properties and suitability for various kinds of land-use was presented on the maps and in the explanatory notes in qualitative terms only. This was undoubtly due to the very limited availability of geotechnical and geohydro-logical in-situ and laboratory test results.

As a result of the explosive economic growth in the United States of America after World War II and the subsequent urban expansion the number of geologists concerned with Urban geology grew tremendously. For example, at the end of the sixties about 150 geologists were employed in this field in the city of Los Angeles alone (McGill, 1973). In the same period a breakthrough in the use of geological data for urban planning and management was achieved in Canada mainly because of the publications of Legget (Refs. 1973; White, 1989). In the industrial countries concern about our natural environment grew and the dangers of pollution caused by large-scale waste disposal predominantly in and around urban centers, became apparent in the seventies. Detection, immobilization and restoration of polluted areas, and selection of appropriate sites for waste disposal opened up a new field of interest and created a new challenge for urban geologists. Geochemistry was added to their expertise and Environmental Geology became the subject of a fast growing number of new studies. This resulted in new ways of informing planners about the potentials and the limitations of the soil. "Geopotential maps", in which preferential land-use based on earth-scientific and related disciplines is indicated, were first introduced in the Federal Republic of Germany (Lüttig, 1978). This mapping system was later applied in several other countries. Many urban geological maps were made and printed by the U.S. Geological Survey during this decade (e.g., McGill, 1973; Baskerville, 1981; Merguerian & Baskerville, 1987). In various European countries special studies were carried out in order to investigate the most appropriate way to present (processed) geological data for urbanized areas on maps (e.g., Monroe & Hull, 1987; Forster et al., 1987; de Mulder, 1986).

In Spain, geotechnical maps for urban planning on a scale of 1:25,000 were made by the Instituto Geologico y Minero de España for the cities of Huelva, Granada, Palma de Mallorca, Almeria, Malaga, Cordoba, Alcoy, Valladolid (Refs. Instituto Geologico y Minero de España, 1984; Cendrero et al, 1987) and for the Madrid region on scales of 1:400,000 to 1:100,000 by the Instituto Technológico GEOMINERO de España (Ayala Carcedo, 1988).

The increased use of geohydrological and geotechnical models at the end of the seventies and at the onset of the next decade made it possible to predict and to quantify the effects of human interference in the geosphere. A striking example was the prediction of the harmful effects of land subsidence resulting from land reclamation in The Netherlands (Claessen et al., 1988). This study showed not only that the construction of a new polder would cause great damage to buildings in the surrounding land area but indicated also the amounts of forthcoming damage claims. The use of electronic data files facilitated the presentation of data on maps substantially. Thematic maps became more accessible to planners, decision makers, and engineers by deleting all kinds of marginal information which would not be understood directly by the target group and by presenting these data in a more quantified manner (Geological Survey of The Netherlands et al., 1986; de Mulder, 1986). Recently, in The Netherlands some more fundamental studies were carried out concerning the relationships between the geological, geotechnical and geohydrological conditions of the subsurface and the costs of construction and maintenance of infrastructural work, such as roads and sewerage systems. These studies show that a certain relationship exists between the costs and the vulnerability to settlement of the compressible Holocene beds under the cities in the west of The Netherlands. Apart from the vulnerability to settlement the change in groundwater level necessitated by these construction activities proves to be related to the above-mentioned costs (de Mulder & Hillen, in press [b]). Considerable attention is now given to the development of a criterium by which these relationships may be expressed with respect to the costs to the cities' municipal councils. After development of such a criterium, the subsurface of many Dutch cities will have to be mapped in detail to characterise these in terms of vulnerability to settlement and change in groundwater level.

Outside Europe and North America urban geology began to develop recently in Southeast Asia and the Pacific mainly because of the stimulating activities of the Economic and Social Commission for Asia and the Pacific (ESCAP, United Nations) since the mid eighties. Three volumes of the Atlas of Urban Geology compiled by J. Rau and published by ESCAP in Bangkok comprise the results of special urban geological studies and state-of-the-art reports from the Peoples Republic of China, Bangladesh, Fiji, Indonesia, Malaysia, Nepal, Pakistan, the Republic of Korea, Sri Lanka, Thailand and Viet Nam. The Association of Geoscientists for International Development (AGID) has organized several LANDPLAN seminars about the role of geology in the planing and development of urban centres in Southeast Asia (e.g. Tan & Rau, 1986). A few decades earlier, urban geological studies for the

city of Calcutta (India) were carried out and published (Dastidar & Ghosh, 1967) followed by a number of urban geotechnical studies scattered all over this subcontinent (Raju, 1987). In Africa no major urban geological studies or geological studies for land management apart from Togo and Morocco (Allaglo et al., 1987; Hafdi, 1987) are known to the author.

It is obvious that urban geological studies bring about very specific problems. These are discussed in section 2, while some of the present trends in this field are presented in section 3.

2. PROBLEMS IN URBAN GEOLOGY

In conducting geological studies in urbanised areas, certain technical, legal, and political problems are often encountered. Some of these problems are summarised below.

Availability of observation points
The concentration of construction activities in urbanised areas generally requires a greater number of data points concerning subsurface conditions than in rural areas. However, the extent that the results of these site investigation are available to (engineering) geologists depends largely on the way that site investigation, data collection and urban management are organised in that city. In countries like The Netherlands, most data for urban geological studies can be obtained relatively easily from the municipial departments of Public Works or Housing. Municipal administrations of the larger cities generally have organised, computerised and maintained their data files very well, whereas in some smaller cities the data files consist of a loose bunch of bore-hole descriptions and Cone Penetration Test (CPT) graphs in a drawer. The Geological Survey of The Netherlands also has access to data files of the principal private or semi-private geotechnical organisations and companies.

Although many site-investigation studies have been performed in cities in many countries, it may take a considerable effort to collect such data because the municipalities may have no responsibility for data collection or certain city departments might even fear adverse technical or political reactions to their willingness to deliver subsurface data to (engineering) geologists.

Without municipal cooperation, the urban geologist has to rely on data from geotechnical advisory companies, drilling companies, insurance companies, tunnel, railway and road construction companies, real estate bureaus, archi-tects, etc. for subsurface data.

The number of actual data for urban geological studies depends upon the scale at which maps should be produced. Because land-use in the urban areas often changes within geographically short distances, maps based upon interpreted subsurface data should, preferably, be made on detailed scales such as 1:10,000 to 1:5,000. The number of observation points necessary in geolo-gically less complicated areas is (on empirical grounds) assumed to be bet-ween some 25 - 50 for maps at a scale of 1:5,000 and 10 - 25 for maps at a scale of 1:10,000. Besides these relatively detailed maps it might be wise to produce some small scaled maps in order to draw attention to i.e. certain hazards which may threaten the city.

Site investigation in urban areas

When the number of data points available is not sufficient, additional data should be collected by site investigation. If exposures are lacking, which is normally the case in urbanised areas, site investigation will generally consist of drilling bore holes, or in-situ tests, such as CPTs or SPTs. Most surface geophysical measurement methods cannot be applied in cities because of the presence of electrically conductive materials or contaminations in the topmost metres of the ground or because of the inconvenience to the public, as will often be the case when using seismic methods. The only geophysical methods, which might provide good results under these circumstances, are georadar and geoelectrical measurements with a floating cable in canals or in other fresh-water courses. Older air photographs of the area made prior to urban expansion often provide a very helpful tool in mapping the subsurface conditions here.

For the execution of borings or in-situ tests, permission has to be granted by the municipal authorities. Furthermore, execution will often be hampered by limitations to the choice of suitable locations. Finally, in urbanised areas the topmost metres of the subsurface often contain fill material, such as rock fills, wood, demolition debris etc., which may hamper penetration.

Engineering geological correlation problems in urbanised areas

Specific correlation and mapping problems occur in urbanised areas. The topmost bed is often man-made and frequently consists of fill material. The distribution of this "bed" may be very irregular both in horizontal and vertical senses. Sometimes, the engineering geologist may rely on data about the bed's distribution and thickness from municipal sources (e.g. the Public Works Department), but often, especially in the older parts of the city, no data are available at all. Geological mapping in urbanised areas might be done in a very indirect way as well. This might be the case when a geological boundary line has to be traced in a built-up area. Then, the presence or absence of construction failures visible in buildings could be a good indicator for the situation of such a boundary line.

Another correlation and mapping problem often occurs in cities because of the variation in surface load exerted on the underlaying beds and because of the results of local groundwater extraction. Both may lead to sudden lateral changes in thickness of compressible beds such as peat and clay, and, conse-quently, also to dramatic lateral changes to the geotechnical and geohydrolo-gical properties of these beds. This creates an extra uncertainty factor to engineering-geological mapping of urban areas.

The reliability of boundary lines on (engineering) geological maps is a matter of great concern for geoscientists, the city municipality and the citizens. From such lines conclusions are drawn with respect to areas of potential risks, future maintenance works or property values which might in turn, cause speculation activities. At present no real satisfactory solutions have been found to avoid justifiable or injustifiable concern of the public and still indicate engineering-geological boundaries on large scaled maps of urbanised areas.

3. TRENDS IN URBAN GEOLOGY

Growing awareness among geoscientists

During the past decade many geoscientists climbed down from their ivory towers and became involved in geological aspects of environmental development and management. The rapid growth of Engineering Geology as a new discipline in universities in many countries, and the large number of members and publications on this subject in for example the Journals of the IAEG (International Association of Engineering Geologists) and AAEG (American Association of Engineering Geologists), and the recent creation of a Steering Committee on Environmental Geology in IUGS (international Union of Geological Sciences) demonstrate the growing awareness of the present problems in society, especially in densely populated areas. Furthermore, geoscientists increasingly succeed in a clear and straight foreward presentation of applied geoscientific information, for instance by preparing thematic maps. Such presentations appear to be of great relevance to bridge communication gaps between earth scientists and planners and decision-makers (de Mulder, in press).

Developments in project acquisition
Initiatives for urban-geological activities may be taken either by the municipal authorities or geoscientists. In some cases, as in The Netherlands, provincial or even national government agencies or ministries might generate such activities. Unless natural disasters or catastrophic shortages of natural resources have occurred, geoscientists generally face problems in convincing the (municipal) authorities to sanction urban-geological activities. Occasionally, municipal authorities may be convinced by pointing out the financial benefits. An attempt to make a cost/benefit analysis of urban geological activities has been made for the city of Amsterdam (de Mulder, 1988b).

The rapid development in urban-geological activities in South-east Asia can be considered as a major trend in project acquisition. In this region, both the Committee for Co-ordination of joint Prospecting for Mineral Resources in Asian Offshore Areas (CCOP) and the Economic and Social Commission for Asia and the Pacific (ESCAP) of the United Nations, and the Association of Geoscientists for International Development (AGID) have organised various workshops and seminars with geoscientists, planners and decision makers on geological boundary conditions for urban development. At present, eleven countries in South-east Asia participate in the ESCAP programme "Geology for Urban Development". In addition to these activities, an explosive growth in

geological studies of SE Asian urban centres took place in the eighties. In a number of publications (a.o. ESCAP, 1985; ESCAP, 1987; ESCAP, 1988a; ESCAP, 1988b) urban geological information is presented from a.o. Bangkok, Kuala Lumpur, Kuching, Cebu, Manila, Sydney (Australia), Bombay, Hong Kong, Jakarta, Surabaya, Karachi, Singapore, Rabaul (PNG), Katmandu, Shanghai, Tianjin and many other cities in the Peoples Republic of China. In some countries in this area special sections in the Geological Surveys have been installed in order to pay continuous attention to the field of Urban Geology. Furthermore, an international Working Group on Urban Geology is coming into being.

Automation in municipal administration
In the past few years, all kinds of data-files from municipal administrations have been stored in automated databases, especially in developed countries. Taking advantage of the lessons learned in developed countries automation has now started in various developing countries. After storage of population data and other geologically irrelevant data, many cities have now reached the stage that hydrological, hydrogeological and geological data have been or are about to be stored in databases. Advantageous effects of this process, and of vital importance to the development of urban geological activities, are
1) the need for the municipality to collect as many data as possible and to expose files that were previously not accessible;
2) the necessity to store data in a standardised format;
3) the ability to process such data rapidly and to produce all kinds of single-value maps (for example distribution maps or isopach maps) or to combine groundwater, geotechnical and geological data in thematic and special maps (for example on the vulnerability to subsidence); and
4) the simplification of new data entry and data correction, so that file maintenance is guaranteed much better.

The rapid availability of various kinds of fancy coloured but not printed maps might facilitate the process of convincing city authorities of the value of this type of information to their physical planning and decision-making. This, in turn, might elicit more political and financial support for an increased effort in the expansion or improvement of the data

589

files. The introduction of Geographical
Information Systems (GIS) in Geological
Surveys, Provincial and Planning
Organizations and even in municipal bodies
of some major cities have stimulated and
facilitated this process considerably.

Due to the development of geohydrological
and geotechnical modelling predic-
tion/prevision studies on the impact of
man's activities in and above the ground
can be conducted quite well nowadays. The
results of such studies prove to be of
major importance for planners and decision
makers (Claessen et al., 1987). In order
to feed the models with reliable geodata
such data should preferably be expressed
quantitatively and go together with
reliability intervals.

4. CONCLUSIONS
In this chapter some conclusions are drawn
concerning recent developments in Urban
Geology.
1. Although the first steps in this field
   date back some 60 years, the number of
   activities in Urban Geology increased
   rapidly in the sixties in the United
   States of America with an explosive
   growth on a global scale in the last
   decade.
2. In many countries the initial problems
   of communication between geologists
   planners and decision-makers are
   becoming eroded. This is due to a much
   more straight foreward presentation of
   geological information on the one hand
   and a growing awareness by the public
   about the relevance of earth scientific
   information for hazard mitigation.
3. The rapid developments in automatic
   data storage and data handling may
   provide excellent new opportunities to
   improve the remnants of the above
   described communication gap.
4. Geoscientific data should preferably be
   expressed in quantitative terms in
   order to provide reliable input for
   geotechnical and geohydrological
   models.

References
ALLAGLO, L.K. et al. (1987): Togo, its
geopotential and attempts for land-use
planning- A case study. In: [Arndt P. &
Lüttig G.W.]: Mineral resources'
extraction, environmental protection and
land-use planning in the industrial and
developing countries: 243-270.
Stuttgart, (Schweizerbart).

AYALA CARCEDO, F.J. et al. (1988): Atlas
Geocientifico del Medio Natural de
la Comunidad de Madrid. 83 p. Madrid
(Instituto Technológico GEOMINERO de
España).
BASKERVILLE, C.A. (1981): The foundation
geology of New York City. In: [Legget,
R.F.]: Geology under cities. Geol. Soc.
Amer. Rev. eng. Geol., 5: 95-117.
BRüNING, K. (1940): Bodenatlas von
Niedersachsen. Göttingen (Wirtschafts-
wis. Ges. Stud. Nieders.).
CENDRERO, A. et al. (1987): Detailed
geological hazards mapping for urban and
rural planning in Viscaya (Northern
Spain). Norg geol. Unders., spec. Publ.
2: 25-41, Trondheim.
CLAESSEN, F.A.M. et al. (1987): Secondary
effect of the reclamation of the
Markerwaard Polder. Geol. & Mijnb. 67:
238-291, Dordrecht.
DASTIDAR, A.G. & P.K. GHOSH (1967):
Subsoil conditions of Calcutta. J. Inst.
Eng. (India), 48, 3, CI 2: 692.
ESCAP (1985): Geology for Urban Planning;
selected papers on the Asian and
Pacific Region. ST/ESCAP/394, 41 p.,
Bangkok.
ESCAP (1987): Geology and Urban
Development; Atlas of Urban Geology, 1.
Bangkok.
ESCAP (1988a): Urban Geology in Asia and
the Pacific; Atlas of Urban
Geology, 2. ST/ESCAP/586, 228 p.,
Bangkok.
ESCAP (1988b): Urban Geology of coastal
lowlands in China; Atlas of Urban
Geology, 3. ST/ESCAP/624, 168 p.,
Bangkok.
FORSTER, A. et al. (1987): Environmental
Geology maps of Bath and the surrounding
area for engineers and planners. In:
[CULSHAW, M.G. et al.]: Planning and
Engineering Geology. Geol Soc. eng.
Geol. spec. Publ. 4: 221-235. London.
GEOLOGICAL SURVEY OF THE NETHERLANDS et
al. (1986): Subsoil Uncovered. 36 p.
Haarlem, Delft, Wageningen.
HAFDI, A. (1987): Approach of a
methodology for drawing up a
habitability map. In: [Arndt P. &
Lüttig G.W.]: Mineral resources'
extraction, environmental protection and
land-use planning in the industrial and
developing countries; 271-278,
Stuttgart, (Schweizerbart).
HAGEMAN, B.P. (1963). A new method of
representation in mapping alluvial
areas. Verh. kon. ned. geol. mijnb.
Gen., geol. Serie 21-2, Jub. Conv. 2:
211-219, Amsterdam.

HOYNINGEN-HUENE, P.F. von (1931):
Übersichstkartierung im Gebiet der
MeBtischblätter Kempen, Krefeld,
Viersen, Willich nebst Randgebieten.
"Briefe" des Landesplanungsverbandes
Düsseldorf, 21, Berlin.

INSTITUTO GEOLOGICO Y MINERO DE ESPAñA
(1984): Mapa Geotechnico par
ordenacion territorial y urbana de
Valladolid. Madrid.

LEGGET, R.F., (1973): Cities and Geology.
624 p. (McGraw-Hill).

LüTTIG, G. (1972): Naturräumliches
Potential I, II und III. In:
Niedersachsen, Industrieland mit
Zukunft: 9-10 Hannover. Nds. Min.
Wirtsch. öff. Arb.

LüTTIG, G.W. (1978): Geoscientific maps of
the environment as an essential tool in
planning. Geol. & Mijnb. 57, 4: 527-532,
Amsterdam.

LüTTIG, G.W. (1987): Approach to the
problems of mineral resources'
extraction. In: [Arndt P. & Lüttig
G.W.]: Mineral resources' extraction,
environmental protection and land-use
planning in the industrial and
developing countries; 7-13, Stuttgart,
(Schweizerbart).

MCGILL, J.T. (1973): Growing Importance of
urban Geology. In: [Tank, R.W.]: Focus
on Environmental Geology; 378-385. New
York (Oxford University Press, Inc.).

MERGUERIAN, C. & C.A. BASKERVILLE (1987):
Geology of Manhattan Island and the
Bronx, New York City, New York. Geol.
Soc. Amer. Cent. Field Guide- NE
Section.

MONROE, S.K. & J.H. HULL (1987):
Environmental Geology in Great Britain.
Norg geol. Unders., spec. Publ. 2: 111,
Trondheim.

MüCKENHAUSEN, E. & E.H. MüLLER (1951):
Geologisch-bodenkundliche Kartierung
des Stadtkreises Bottrop i. W. für
Zwecke der Stadtplanung. Geol. Jahrb.
66: 179-202. Hannover.

MULDER, E.F.J. DE (1986): Applied and
Engineering Geological Mapping in
The Netherlands- Proc. V int. Congr.
Int. Assoc. eng. Geol., 6: 1755-1759,
Buenos Aires.

MULDER, E.F.J. DE (1988a): Thematic
geological maps for urban management and
planning. Proc. int. Symp. Urban Geol.;
46-59. Shanghai. Bangkok (ESCAP).

MULDER, E.F.J. DE (1988b): Engineering
Geological maps: a cost-benefit
analysis. In: [Marinos P.G. & Koukis
G.C.]: The Engineering Geology of

Ancient Works, Monuments and Historical
Sites: preservation and protection.
1347-1357, Rotterdam (Balkema).

MULDER, E.F.J. DE (1989): Thematic Applied
Quaternary maps: a profitable Investment
or expensive wallpaper? In: [De Mulder
E.F.J. & Hageman B.P.]: Applied
Quaternary Research. 105-117, Rotterdam
(Balkema).

MULDER, E.F.J. DE (in press): Engineering
Geology for Urban Planning -
Proc. Sem. on Geology & Land-Use
Planning, Kuching. Bangkok (ESCAP).

MULDER, E.F.J. DE & R. HILLEN (in press
[a]): Preparation and Application of
Thematic Engineering and Environmental
geological Maps in The Netherlands.

MULDER, E.F.J. DE & R. HILLEN (in press
[b]): Recent developments in Engineering
Quaternary Geology in the Geological
Survey of The Netherlands.

PRINYA NUTALAYA (1988): Geologic and
hydrologic hazards and their effects on
urban development in Southeast Asia.
Proc. int. Symp. Urban Geol.; 60-71,
Shanghai. Bangkok (ESCAP).

PRINYA NUTALAYA & J.L. RAU (1981): The
sinking Metropolis. Episodes 4: 3-8.

RAJU, K.C.C. (1987): Geoscientific maps in
land-use planning in India. In: [Arndt
P. & Lüttig G.W.]: Mineral resources'
extraction, environmental protection and
land-use planning in the industrial and
developing countries; 281-298,
Stuttgart, (Schweizerbart).

SCHNEIDER, W.J. & A.M. SPIEKER (1973):
Water for the Cities - the Outlook.
In: [Tank, R.W.]: Focus on Environmental
Geology; 385-392. New York (Oxford
University Press).

STREMME, H. (1932): Die Bodenkartierung
als wichtigste Vorarbeit der
Generalplanung. In: [Mauesmann, A.]: Die
Umst. im Siedlungswesen, Berlin.

TAN, B.K., & J.L. RAU (1986): Landplan II;
Role of Geology in Planning and
Development of Urban Centres in
Southeast Asia. Rep. Ser., 12. 92 p.
Bangkok (AGID).

WHITE, O.L. (1989): Quaternary Geology and
urban planning in Canada. In: [De Mulder
E.F.J. & Hageman B.P.]: Applied
Quaternary Research. 165-175, Rotterdam
(Balkema).

Site characterisation for tunnels in jointed rock masses
Caractéristiques de site, pour tunnels, dans des masses rocheuses compactes

Y.R.Dhar & V.D.Choubey
Indian School of Mines, Dhanbad, India

H.S.Pandalai
Indian Institute of Technology, Bombay, India

ABSTRACT : In this paper an attempt has been made to present an analysis for stability and site-characterisation of the rock mass surrounding an excavation on the basis of kinematic feasibility of the blocks whose geometry is defined by the orientation parameters of the joint sets. This method quantifies the rock condition in terms of rock load factors (RLF) for a given excavation orientation. It can also be used to work out reasonably stuitable orientation within a given rock mass.

1 INTRODUCTION

The general rule for sit ing of tunnels in bedded strata is that the alignment or the axis of a tunnel should not be parallel to the strike of the rock formations. The basic rationale behind this judgement is that the weakness planes in the form of shear seams, bedding shears etc., if present, are encountered for long lengths, and in case of high tangential stresses at periphery of the excavation, overbreaks increase two to three times of what is considered as normal. Further, in folded mountain belts, it is generally presumed that if tectonic stresses are operative, the azimuth of the maximum principal stress vector acts in the direction perpendicular to the strike of the rock formation. This situation is considered undesirable in case the plane containing major and intermediate axis of principal stress is tangential to some parts of the periphery of the opening. Such parts of the opening are at times exposed to stress problems like spalling or even rock bursting.

For jointed rock masses in which primary planar features are absent, decision on alignment orientation becomes a difficult job. The problem is generally tackled on the basis of intuition, judgement and experience. However, the general rule in such cases is to have the alignment oriented at an angle of at least 25 degrees to the strike of the steeply dipping smooth or clay filled joint planes.

Many a time, situations arise where one has to deviate from the general thumb rules. In such situations it is not clear that if one has to deviate, how much relative cost increase is there going to be. This calls for a quantitative assessment of tunnelling conditions in different directions. With the advent of classification systems namely RMR (Bieniawski, 1973) and RSR (Wickham et al, 1974) such situations have been quantified although subjectively. However, in both these methods consideration has been mainly given to orientation parameters of only one planar feature which obviously refers to the primary planar feature (bedding) of the sedimentary rocks or the prominent secondary planar feature (schistosity or foliation) of metamorphic rocks.

2 JOINT CONTROLLED STABILITY OF ROCK MASSES

A typical rock mass, however, contains several sets of discontinuities that it can be considered to be a mosaic of systematically interlocked blocks whose geometry is controlled by orientation of several discontinuities. When shallow depth excavations are penetrated through, the rock mass surrounding the excavation is loosened. Therefore, the kinematic characteristics of the blocks in the rock mass should necessarily influence support deformations.

For a given excavation, some of the blocks, contained in the rock mass mosaic in the sourroundings of the opening, may be critically situated to trigger of movements of larger masses of rock. Such blocks have been termed variously as critical blocks, kinematically feasible blocks (Priest, 1980) and Keyblocks (Shi and Goodman, 1981). The identification of the critical blocks requires the assessment of whether the discontinuities could form a block which could fall or slide into the cavity. Once this is established then identification of the sliding (plane(s), direction of the resultant force, and maximum volume of the critical blocks bounded by planar joint planes and exposed free surface provide a rational way for site characterisation of underground openings in jointed rock masses.

Hoek (1977), Hoek and Brown (1982), Priest (1980), Den-Yeming (1982) have used stereographic projections to ascertain kinematics and failure modes of the critical blocks. Of late, Goodman and Shi (1985) have given an elegant theoretical technique in addition to stereographic techniques for identification of keyblocks. The main theme of all these works is to find the critical blocks created by intersection of discontinuities in a rock mass excavated along the defined surfaces. In application all these approaches lead to the same conclusion i.e. identification of critical blocks. Also fundamental assumptions of all these methods appear to be the same.

In this work, the relationships visualised from stereographic projections (lower hemisphere) have been computerised for an easy quick and correct solution and have been extended for quantitative estimation of rock load factors in different alignment directions for an overall characterisation of rock mass. Initially, the present concept shall be explained for a rock mass with three discontinuities.

3 IDENTIFICATION OF CRITICAL BLOCKS

For determination of critical blocks in roof several logical tools have been explained by Hoek and Brown (1982), Goodman and Shi (1985). Dhar (1986, 1988). From these tools it can be concluded that with three planar (nonparallel) discontinuities in rock mass only one critical block (tetrahedral block) is formed at the roof of the excavation. In this situation all the three lines of intersection between different planes defined by points on the stereographic projection (Fig. 1) lie inside of the outer great circle (reference circle). If the centre of the projection (centre of the net) is enclosed within the triangle formed by cyclographic traces of joint planes, the critically disposed block may fall into the excavation; and if the centre of projection lies outside the triangle the block has tendency to slip into the cavity. It can slip only when the resisting frictional

forces are exceeded by sliding forces along the preexisting joint plane(s) on which weight of the block rests.

For computational purpose, it means that if the three lines of intersections formed by the joint planes intersect the excavation surface, the block is critical or potentially critical.

4 CRITERIA FOR SLIP OR FALL

To determine whether the block is going to slip or fall into the cavity the following criteria can be visualised from these projections (Dhar, 1986, 1988).

1. If the base area of the wedge at the roof is equal to the sum of the subareas defined by the projections of the intersecting planes on the roof surface, the wedge will fall (Fig. 1b).

2. If the base area defined by the wedge at roof is less than sum of the subareas defined by the projections of lines of intersections of planes, the wedge can slide (Fig. 1c).

5 CRITERIA FOR DETERMINATION OF NATURE OF SLIP ON PLANE

While both types of movements i.e. fall or slip can be encountered on the roof, for walls only slips can be anticipated. For determining the mode of slip failure various criteria have been suggested by Panet (1969), Goodman (1976), Goodman and Shi (1985). The criteria which have emerged from model studies conducted by Dhar (1986) are summarized in the following :

5.1 Criteria for Roof

1. For sliding to occur on a single plane, the azimuth of the dip vector of any dipping plane should lie between the azimuths of the two lines of intersections plunging towards free surface of the opening and its dip should be steeper than all the intersection lines formed by the intersecting planes.

2. Sliding will occur along the line of intersection of two planes if condition 1 is not satisfied, and in such a case, the slip direction shall be the direction of the steepest intersection line formed by the planes.

5.2 Criteria for Walls

In the case of walls the criteria for finding slip direction of the blocks are more or less the same as those described for roof

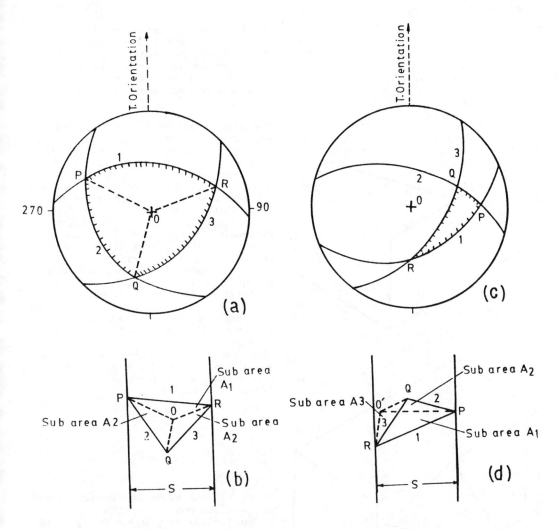

Fig. 1 Case of a falling and a slipping wedge. (a) Stereographic projection for a falling wedge, (b) horizontal projection of the falling wedge, (c) stereographic projection of a slipping wedge at roof, (d) horizontal projection of the slipping wedge.

except that all the intersection lines need not plunge towards the excavation. Even only one or two intersection lines between the planes plunging towards excavations can make the block slip towards cavity. Thus the following criteria emerge for the walls.

1. If only one intersection line between planes plunges towards the cavity, the block can slip on that intersection line.

2. If two intersection lines between planes plunge towards the cavity the blocks can either slip on intersection line or on a single plane.

3. If the azimuth of the true dip of a plane, dipping towards the cavity, lies between

the azimuths of two plunge vectors of the two intersection lines formed by the intersecting joint planes, the block will slip on the plane. If this condition is not satisfied it will slip on steepest intersection line dipping towards cavity.

6 COMPUTATIONS OF FACE AREAS, SUB-AREAS AND VOLUME OF THE BLOCKS

A simple computational method has been developed for determination of the areas of the planes of the wedge and its volume. In this method, the distance of the apex of the wedge from the base is assumed

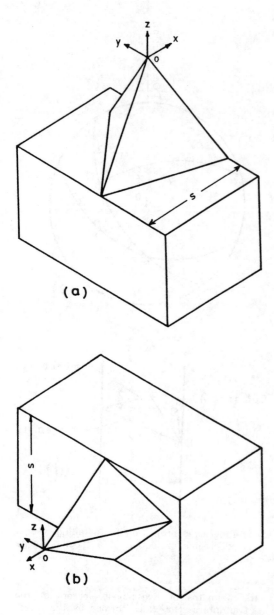

(a)

(b)

Fig. 2 Wedge fitted to the coordinate system oxyz (a) isometric view of the wedge at the roof (b) isometric view of the wedge at the wall.

as one unit, and with this assumption the length and inclination of ridges of the "unit wedge" can be easily computed. Further, if the equations of the planes bounding the "unit wedge" are determined from their dip amount and dip direction, and if these

planes are assumed to pass through a coordinate system OXYZ, the coordinates of the vertices of the wedge can be established; and thereby the area of joint planes bounding the wedge and its base area at the free surfaces of the excavation can be determined (Fig. 2).

For roof, coordinates of vertices of the triangular surface of the wedge in contact with the free surface are computed by assuming the roof as a horizontal plane situated at a distance of one unit below the wedge apex. Similarly, in the case of walls the coordinates of wedge at wall are generated on a vertical plane aligned along the axis of the excavation (Fig. 2b). For inclined tunnels the coordinates of the wedge are established by rotation along X-axis after testing mode of the movement of the wedge on the roof is established.

For determination of the real volume, the areas and the apex height for the wedge limited to the excavation of desired dimension, the values of the volume, the areas and the height of the unit wedge have to be multiplied by the respective factors as discussed below.

For example, the volume and the base area of the wedge at the roof of the excavation of the desired width is given by the following expression :

$$\text{Volume (V)} = v\,(B/S)^3$$

$$\text{Base area (A)} = a\,(B/S)^2$$

Where, V is the volume of the wedge limited to the desired excavation width, v the volume of the "unit wedge", B is the given width of the excavation, S (limiting dimension) of the excavation width of the "unit wedge" (Fig. 2a).

Similarly, the volume and the base area, i.e. area of the wedge in contact with the wall of the desired height can be obtained from the following expressions :

$$\text{Volume (V')} = v'\,(H/S')^3$$

$$\text{Base area (A')} = a'\,(H/S')^2$$

Where, H is the height of the wall of the desired excavation, S' is the limiting height of the "unit wedge" (Fig. 2b), v' is the volume of the unit wedge and a' is the base area of the unit wedge at the wall surface.

7 DETERMINATION OF ROCK LOAD FACTORS

After finding volume of the unit wedge block, its mode of movement, direction of slip, sliding plane(s), the rock load factors can be computed. The rock pressure at roof and the walls can be determined by the following equations :

For falling wedge :

$$\text{Av. P (roof)} = V. \gamma/A = v \, (B/S)^3 \gamma/a \, (B/S)^2$$
$$= v.B. \gamma/aS = N.B.\gamma$$

For slipping wedge :

$$\text{Av. P (roof)} = W \, (\text{Sin}\,\alpha - R)/A$$
$$= V\gamma \, (\text{Sin}\,\alpha - R)/A$$
$$= v \, (\text{sin}\,\alpha - R) \, B \, \gamma/aS$$
$$= N \, B \, \gamma$$

Similarly, the net rock pressure at the wall can be calculated from the following expression :

$$\text{Net P (wall)} = v' \, (\text{Sin}\,\alpha - R) \, H \, \gamma/a'S'$$
$$= M \, H \, \gamma$$

Where, V = volume of wedge limited to desirable width/height of excavation surface, A = base area of the wedge in contact with the excavation surface, γ = unit weight of the rock, W = weight of the block, v = volume of the "unit wedge", a = base area of the "unit wedge", H = height of the excavation, B = width of the excavation, S,S' = limiting width of the roof/wall, α = dip of slip vector, R = n tan φ/W in case of single sliding plane and (ni tangent φ i + nj tangent φ j)/W in the case of double plane sliding; n is the normal reaction for single plane sliding, ni and nj are the normal reaction on planes (i, j) for double plane sliding, φ is the friction of the single sliding plane, φ i, φ j are frictional angles of sliding planes in case of double plane sliding, N and M are the rock load factors.

8 SITE CHARACTERISATION

A computer programme, using the principles stated in the preceeding, has been prepared to compute rock load factors of the rock mass in the surroundings of the openings. In this programme facilities have been developed for computation of rock load factors (RLF) for the roof and the walls of openings along different geographic orientations and inclinations. The basic input parameters used in the analysis include orientation features of joint planes, frictional parameters of the joint planes, and alignment directions of the openings. At this stage, the programme does not take into account water pressures, fractures induced during excavation and rotational movement of the blocks. For safety, the friction parameters can be neglected if desired. It is intended that such an information generated from the analysis would be complementary to the data generated through other conventional methods.

A case of site characterisation for horizontal openings in the rock mass whose joint parameters are shown in Table 1 is explained in the following paragraphs.

Table 1 Joint parameters of the rock mass

Joint	Dip Amount	Dip Direction	Friction Angle
J1	50	010	20
J2	40	240	20
J3	70	180	20

The rock load factors generated for the rock mass in different directions have been plotted on polar graphs specially designed for the purpose (Fig. 3). These show as to how the loads due to the falling/slipping wedge(s) can vary in different directions. From this graph it is possible to know the adverse and the optimum direction for sit ing the opening. It can be seen in Fig. 3a that rock load factors for roof are maximum in 280 direction, while the rock load factors are maximum for the right wall oriented in the northerly direction. Therefore, the optimum tunnel direction should be along 70° or 300° direction. In other words, this direction should make an angle of about 20-25 degrees with the strike of the steepest joint plane provided its friction angle is equal to or less than the frictional angle of the other two planes. If the frictional angle of the steepest plane is considerably greater than the other two planes situation for optimum orientation can change.

9 CONCLUSIONS

The analysis for characterisation of site for tunnels in the rock mass with three joint sets has revealed that the optimal tunnel orientation should be fixed at an

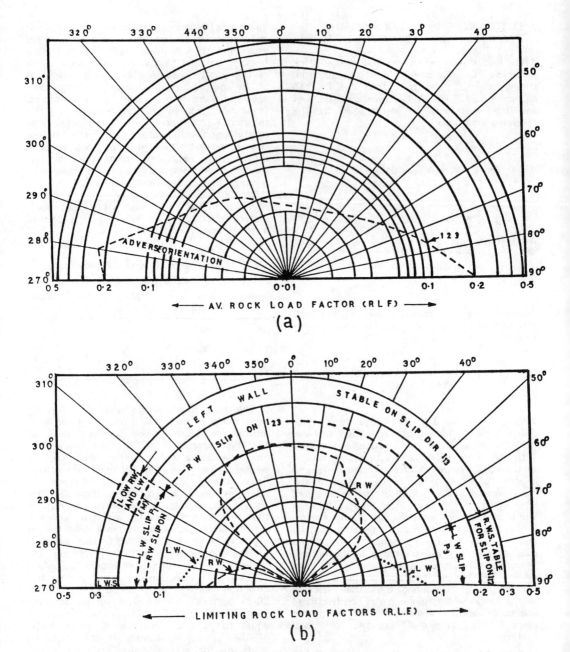

Fig. 3 (a) Variation of rock load factors (roof) along different alignment directions, (b) variation of rock load factors (wall) along different alignment directions.

angle of 20-25 degrees with the strike of the steepest joint plane and the worst direction is nearly along the strike of this joint. This concept can be applied for rocks with more than three joint sets also. Although, in this work the roof has been assumed horizontal, the error that is caused due to this assumption will be on the side of safety because the curved roofs or the curved walls will reduce volume of the wedges at the roof and the walls.

598

10 REFERENCES

Bieniawski, Z.T. (1973). Engineering Classification of Rock Masses, Transactions South African Institute of Civil Engineers, Vol.15, No.2 : 355-344.

Den-Yemining (1982). Stability Analysis of Rock Masses Around Tunnel Through Mount Da-Yao and Suggestions for Support with Shotcrete and Rock Bolts, Proc. IV Int. Cong. IAEG, Vol.5, New Delhi:143-149.

Dhar, Y.R. (1986). Role of discontinuities in stability of underground openings (unpublished).

Dhar, Y.R. (1988). Tectonic and geotechnical evaluation for stability analysis and assessment of rock loads for tunnels of Uri Hydroelectirc Project, Kashmir Himalaya (unpublished).

Goodman, R.E. and Shi, Gen Hua (1981). A new concept of support of underground and surface excavations in Discontinuous Rock Masses on a Key stone Principle, Proc. 22nd Symp. on Rock Mech., MIT, Cambridge, Mass, : 290-296.

Goodman, R.E. and Shi, Gen Hua (1985). Block Theory and its Application to Rock Engineering, Prentice Hall Inc. New Jersy : 338p.

Hocking, G. (1976). A method for distinguishing between single and double plane sliding for tertrahedral wedges, Int. J. Rock Mech. Min. Sci. & Geomech. Abst.13:225-226.

Hoek, E. and Brown, E.T. (1982). Underground Excavations in Rock, Inst. Min. Metal. London.

Lucas, J. M. (1980). A general stereographic method for determining the possible mode of failure of any tetrahedral rock wedge, Int. J. Rock Mech. Min. Sci. & Geomech. Abst., Vol. 17 : 57-61.

Panet, M. (1969). Discussion of graphical stability analysis of slopes in jointed rock. J. Soil Mech. Found Div. Proc. ASCE, 95 : 685-686.

Priest, S.D. (1980). The use of the inclined hemisphere projection methods for determination of kinematic feasibility, slide direction and volume of rock blocks. Int. J. Rock Mech. and Min. Sci. & Geomech. Abst., Vol. 17 : 1-23.

Wickham, G.E., Tiedman, R.R. and Skinner, E.H. (1974). Ground support prediction model - RSR concept. Proc. Rapid Excavation and Tunnelling conference, AIME, New York: 43-64.

Geological and geotechnical characterization of the morainic 'Amphitheater of Rivoli' in NW Italy

Caractéristiques géologique et géotechnique de 'l'Amphithéâtre de Rivoli' dans le N-O de l'Italie

A. Eusebio, E. Rabbi & P. Grasso
Geodata, Torino, Italy

ABSTRACT: A thick glacial sequence including a major morainic structure, is found in the lower part of the Susa valley, 10km West of Turin Italy. This complex is also called "Amphitheatre of Rivoli" and is composed dominantly of gravel, sand and silt with clay, pebbles and boulders. Because of the inherent inhomogeneity of the materials, tunnel excavation in this complex encounters a host of construction and support problems. The investigation described in this paper was designed to assess the geological setting of the complex and to perform a geotechnical characterization of the deposits for providing input parameters for engineering design . This is our first attempt to characterize such a complex. The investigation included geological mapping, drilling and logging boreholes, static and dynamic penetration tests, and an extensive laboratory test program to obtain the physical and mechanical properties of the various facies.
The procedure used in this investigation may be adapted for geotechnical characterization of similar complex deposits.

RESUME: Une structure moranique épaisse et complèxe appelée "Amphithéâtre de Rivoli" se situe à environ 10 km de Turin vers Ouest en Italie. Ce complèxe est composé de depôts glaciaires (gravier avec cailloux et gros blocs, sable, argile et vase). A cause de l'inhérente non-homogénéité des materiaux en question, le creusement de tunnel dans ce complèxe rencontre nombreux problèmses pour la construction et le soutènement. L'investigation qu'on décrit dans cet article a été faite pour mieux évaluer la situation géologique de ce complèxe et pour effectuer une caractérisation géotecnique des depôts pour avoir les paramètres nécessaires des plans de creusement. Celui-ci est le premier tentatif de caractérisation du complèxe moraniques pour un travail d'ingénieur. Parmis les investigation faites nous avons le positionnement de sondages avec perforation, éprouves de pénétration statiques et dynamiques sur le terrain, et nombreuses éprouves de laboratoire pour obtenir les proprietées physiques et méchaniques des différents types de terrain. La procedure utilisée dans cette enquête peut être utilisée pour une caractérisation géotechnique pour depôts semblables comme données de depart pour un projet d'excavation.

1. INTRODUCTION

The Susa valley lies East-West and perpendicular to the Alpine arch. It is generally flat with a mean gradient of 0.4%. Its morhology varies in the last part and 10km from Turin, three concentric and successive hills are present in the valley with a maximum altitude of 500m, above sea level.

These hills represent one of the best preserved moraine "Amphitheatres" of the North-

Western Alps.

A motorway along Susa valley linking the French-Italian border and Turin, an industrial city, and it is corrently being constructed by SITAF (Societa' Italiana Traforo Autostradale del Frejus), a well-known contractor in Italy. For this reason special attention was paid during the project that envisaged the construction of a tunnel through the innermost (and recent) morainic hill, a viaduct and a series of embankments on different kinds of alluvial and glacial deposits.

2. GEOLOGICAL HISTORY OF THE VALLEY

Some attempts at reconstructing the more recent history of the area - during the Quaternary period - were previously made by various authors. In particular, for historical reasons, we wish to remember the paper published by Sacco (1887). In more recent publications of Gabert (1962), Pietrucci (1970), and Bortolami-Dal Piaz (1970), a distinction was made between the different morphological units (Fig.1).

According to these studies,

during the Pleistocene period climatic conditions changed and a glacier expanded through the valley (1.5My).

The maximum lenght of the glacier was about 90km, the average width was 3km and the thickness about 500-600m.

Three important glacial stages were identified: the Mindel period (400000y) during which the glacier reached its maximum expansion; the Riss period (200000y) during which a retreat of the glacier occurred and, at the end, the Wurm period (15000-20000y) with a further retreat (2-4km) and a complete melting of the glacier during the Holocene period (10000y).

The present morphology of the valley is directly connected with the action of the glacier: the three successive hills correspond to the maximum expansion and the subsequent deposition of morainic material of each glacial events. Only in the upper part of the sequence it is possible to find fluvial and lacustrine deposits.

An attempt was made to classify the materials found in the "Amphitheatre of Rivoli" and in the nearby area, through a comprehensive geological

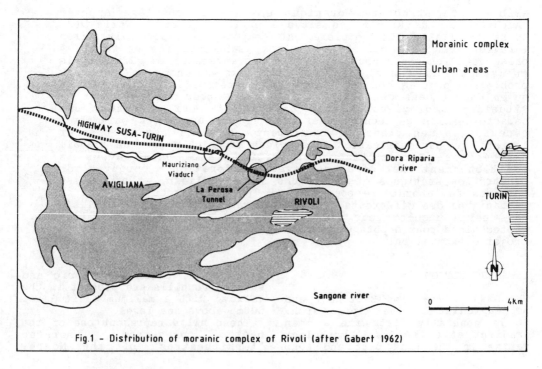

Fig.1 - Distribution of morainic complex of Rivoli (after Gabert 1962)

602

investigation including mapping, drilling boreholes and extensive in situ cone penetration tests (CPT) and standard penetration tests (SPT).

The results from this exploration, as summarized in Fig.2, will be described in the next paragraphs and generally agree with the findings of the previous investigators.

The morainic material is made up of gravel and sand with pebbles and muddy and/or clay layers.

The depositional model is chaotic and completly heterogeneous, as illustrated by various authors (AA.VV. 1983).

The erratic distribution of big boulders of different kinds of rocks is very important because it represents one of the main problems of excavation. In general the materials do not show any sign of weathering, however one may occasionally finds very thin layers of paleosoils which represent the fluctuation of the glacier even during the same period.

Because of separate events of expansion and retrait of the glacier, the deposition model for the transition area immediatly in front of the morainic hills (about 2km) is even more complex.

This transition area is all covered by recent fluvial materials consisting mainly of gravel and sand. Its thickness varies from 2m to 8m, and the top of this material is weathered to soil (see Fig.2).

Different kinds of glacial deposits were found at various depths in this part (Fig.2). The presence of these materials is connected with the existence of a big paleo-lake which was formed in the eastern part of the valley due to the damming of the valley by the morainic hills (Sacco 1887) and subsequent impounding of the so-formed reservoir by the melting water of the glacier. These deposits generally belongs to two facies, namely glacio-fluvial and glacio-lacustrine facies.

3. GEOTECHNICAL CHARACTERIZATION PROGRAM

For providing parameters to engineering design like excavation and foundations, it is necessary to characterize the geotechnical properties of the complex materials described in the previous section.

For this purpose, the initial geological investigation was complemented by sampling of boreholes and subsequent laboratory tests on the samples obtained. The laboratory tests included seiving tests for particle size distribution, direct shear tests, oedometer tests and measurements of density and Atterberg limits.

As many as 12 boreholes were sited both in the morainic hills and in the adjacent transition area, to various depth with a maximum of about 50m (see Fig.2) and continuous sampling was made for each borehole. These boreholes are all distributed along the alignement of the projected motorway.

In many cases CPT tests were made very close to the borehole so as to calibrate the results from such test, since conventional experience is generally not applicable for such heterogeneous materials.

All this allow us to define a geological profile along the motorway as already been shown in Fig.2, and a corresponding geotechnical section on which are shown the results of SPT, CPT and Pocket penetrometer together with the stratigraphic column associated with typical boreholes.

In next section the geotechnical properties of the complex deposits will be described.

4. GEOTECHNICAL CHARACTERIZATION

The complex deposits along the motorway have been classified into various units as shown in Fig.2 and Tab.1, according to their characteristics.

In this section we describe the geotechnical properties of each unit.

UNIT 1--Cover soil.

It consists mainly of clayed silts and is of weathering and transportated origin. The particle size of this material at the bottom of this unit is coarser (sand) and represents the transition to the underlying fluvial deposits.

The thickness of this unit varies from centimeters to about 1 meter. This material appears to be cohesive. No tests were made to characterize this unit since it will be removed during construction.

UNIT 2--Braided river deposits.

They are composed of sand and gravel and the thickness is in the range of 6-20m.

This unit may be divided into two subunits one of which is coarse and thick (2a), whereas the other is fine and thin (2b). Occasionally within this unit one can find cohesive layers which are a few meter thick, whose nature and consistency is similar to unit 3 that will be described next.

Both cone penetration test (Qc) and the standard penetration test values (Nspt) are high and not constant (see Tab.1). In fact it was sometimes impossible to go through this unit using cone penetrometer due to the high resistence offered by the gravels. In particular the friction sleeve (Fs) value is relatively high for unit 2a (see Tab.1).

UNIT 3--Fluvio-glacial deposits.

It consists of clayey silts and silty sands. In addition, both sand and clay lenses are also present in this unit. Based on the value of shear strengh, cone resistence (Qc) and Spt numbers, this unit have been further divided in 2 subunits: 3a and 3b (Fig.2 and Tab.1) which correspond to the upper cohesive part and the lower sandy part, respectively . The occurence of division is irregular as illustrated in Fig.4 and shown in Fig.2.

The thickness of the whole unit varies considerably ranging from 4 to 14m.

UNIT 4--Glacio-lacustrine deposit

This formation is composed dominantly of clay with some very thin lenses of sand. Its thickness exceeds 40m since its bottom was not met in all boreholes sited in this area. A thin layer at the top part of this cohesive unit, appears to be slightly overconsolidated as can be seen from the cone penetration test results shown in Fig.3. This unit, having a very low value of both Qc and Fs from the CPT test (Fig.3 and Tab.1), will be the main source of settlement or deformation for the embankment foundations and for the piled foundation (45m deep) of the viaduct.

UNIT 5--Morain deposits.

This unit is very inhomogeneous (Fig.5) consisting of materials ranging in particle size practically from clay to gravel and boulders. In fact the SPT and CPT tests carried out in this unit were often interrupted when pebbles and boulders were encountered.

Some time, as mentioned before, thin layers (less than 10cm), of paleosoils were also observed during the tunnel excavation.

As can be seen from Fig.2 this unit has also been divided in 2 subunits, 5a and 5b. Subunit 5a corresponds to the concentric hills, whereas 5b lies to the West of these hills and it is above the cohesive layers (unit 4).

The main difference between these two divisions is that division 5b generally does not contain boulders and it has a higher content of sand and silt.

LEGEND (see figg. 2 -3)

UNIT 1 - Soil

UNIT 2a - Gravels and sands, with locally silts

UNIT 2b - Sands

UNIT 3a - Silt, organic silts

UNIT 2a - Gravels and sands

UNIT 3b - Silts and silty sands

UNIT 4 - Clays

UNIT 5a - Silty sands with gravelly sands

UNIT 5b - Gravels and sand with glacial boulder and silt

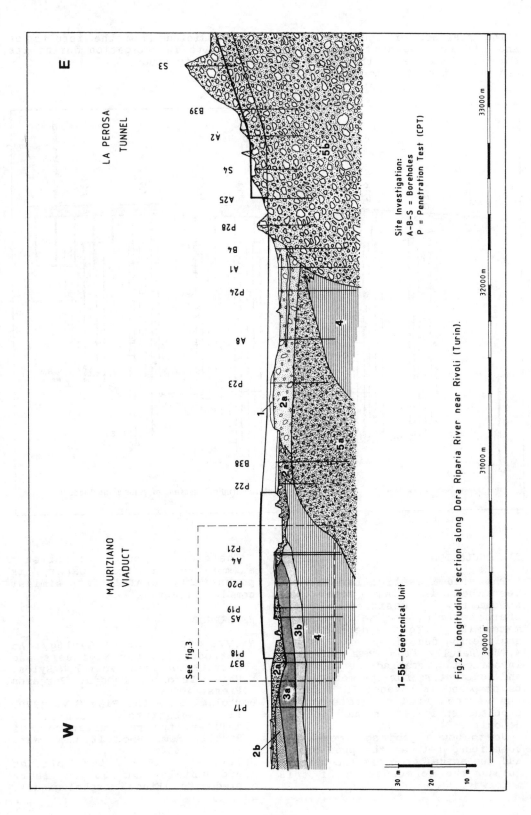

W

MAURIZIANO
VIADUCT

LA PEROSA
TUNNEL

E

Site Investigation:
A-B-S = Boreholes
P = Penetration Test (CPT)

1–5b – Geotecnical Unit

Fig. 2 – Longitudinal section along Dora Riparia River near Rivoli (Turin).

The presence of boulders in 5a and its heterogeneity has posed some difficulties to the excavation and support of the motorway tunnel as predicted from the results of this site investigation during its feasibility study.

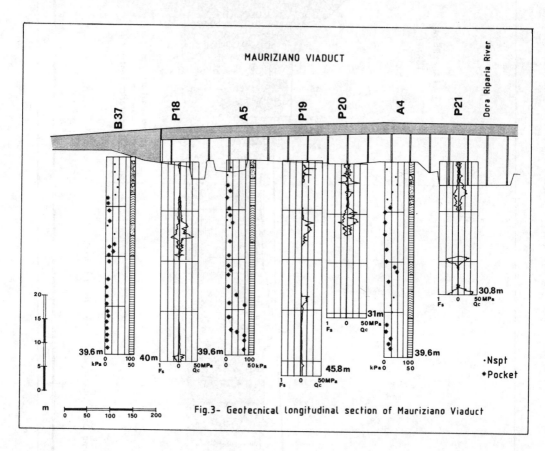

Fig.3- Geotecnical longitudinal section of Mauriziano Viaduct

5. CONCLUSION

The site investigation program described in this paper was specially designed for characterizing heterogeneous materials for geotechnical engineering design.

The results from this extensive investigation enabled the designers and contractors for the motorway to be prepared in advance in treating the various kind of geotechnical problems envisaged in such complex deposits.

Up to now a good agreement have been found between the predictions and the actual problems encountered during the construction of this motorway.

Therefore our experience summerized in this paper, is potentially useful for similar complex situation.

REFERENCES

AA.VV. 1983. Glacial Geology. An introduction for Engineers and Earth Scientists. Editor N.Eyles. Univ. Toronto. Canada. Pergamon Press: 409.

Bortolami G. & Dal Piaz G.V. 1970. Il substrato cristallino dell'Anfiteatro morenico di Rivoli. Mem. Soc. It. Sc. Nat. 18: 125-169.

Gabert P. 1962. Les plaines occidentales du Pô et leurs piedmonts (Etude morphologique).

Louis-Jean, Gap: 531.
Petrucci F. 1970. Ricerche sull'Anfiteatro morenico di Rivoli-Avigliana (Prov. Torino). Mem. Soc. It. Sc. Nat. 18: 95-124.

Sacco F. 1887. L'Anfiteatro morenico di Rivoli. Boll. R. Com. Geol. It. 18: 141-180.

Table 1: Syntesis of the geotechnical characteristics of the different units (from in situ tests and laboratory tests).

Unit	Nspt	CPT Qc MPa	CPT Fs kPa	c'	φ'	cu kPa	Pi %	Density kN/m3
2a	20-40	10-30	200	0	38	-	-	18
2b	40-60	10	100	0	34-36	-	<5	18
3a	-	10	50		18-20	50	10-15	19
3b	20-60	10-20	200		26-28	30-100	-	19
4	<20	<10	<10		18-20	30-70	10-25	19
5a	40	10	100-200	0	32	-	-	19
5b	20-60	40-100	100-400	-	30*	50-200	-	19

* in the silty layers

Fig. 4 - Artificial cut found in the innermost part of the morainic hills. For the description of different units see legend.

Fig. 5 - Excavation of La Perosa tunnel. The material found
belongs to unit 5 and is composed by coarse material
(gravel and erratic boulders) in a silty matrix.

Fig. 6 - The West portal of La Perosa tunnel during the first
stage of excavation.

Recent soils of the Po River: Statistical analysis of geotechnical properties
Les terrains récents du Pô: Analyse statistique des propriétés géotechniques

M. Favaretti & P. Previatello
Istituto di Costruzioni Marittime e di Geotecnica, University of Padua, Italy

ABSTRACT: During a study on salt intrusion in the branches of the Po River, several boreholes were drilled over a relatively small area. Several samples of soil deposited in same sedimentary environment meant that an extensive series of laboratory data could be studied using a statistical approach. In particular, frequency distribution curves relative to 11 geotechnical properties and possible linear regressions between pairs of the above variables were investigated. Other properties observed were: high scatter of the data from means values, the general tendency of frequency distribution curves to fit Beta distribution curves, and a strong and statistically significant influence of the liquid limit on other geotechnical properties.

RESUME: Pour une étude de la remontée de la salinité dans les bras du fleuve Po ont été effectués de nombreux sondages sur une petite surface. Etant donné que l'on s'est trouvé en presence d'un grand nombre d'echantillons de terrain qui s'étaient déposé dans un meme milieu de sédimentation, on a pensé d'analyser, en utilisant la statistique, les résultats des expériences de laboratoire. On a examiné en particulier les courbes de distribution des fréquences des 11 propriétés géotechniques et les possibles regressions linéaires de ces propriétés entre les différents couples. On a pu observer une dispersion élevée des données par rapport aux valeurs moyennes, une tendance générale des courbes de distribution des fréquences à se disposer selon les courbes de distribution du type Béta, une forte influence significative du point de vue statistique de la limite de liquidité sur les autres propriétés géotechniques considérées.

1 INTRODUCTION

The most important river in Italy is the Po, 652 km long, whose hydrographic basin occupies an area of more than 70,000 sq. km or about two-thirds the surface area of Northern Italy. The mouth of the river is of the delta type and the delta itself, about 500 sq. km in area, has undergone complex geological and anthropic events.

In the geological reconstractions of the Po delta, eight littoral bars of successive age and ten old cuspate deltas have been recognized (Ciabatti, 1966). The latter formed when the single mouth advanced, gradually accompanied by banking of sand in littoral bars, due to sea action, on both sides of the river. More recently, the further advance of the Po, with strong sedimentation along the river branches, no longer prevented by sea action, gave rise to two lobate deltas, of which the southern one is the more recent.

On historical times, man's actions and activities have most greatly influenced the morphology of the Po delta. In particular, the largest operation was the southward diversion of the northern branches of the river, carried out by the Republic of Venice in 1604 with the aim of preventing the Lagoon from silting up. The river responded by changing the style of its delta structures; in particular, it has been calculated that delta advance went

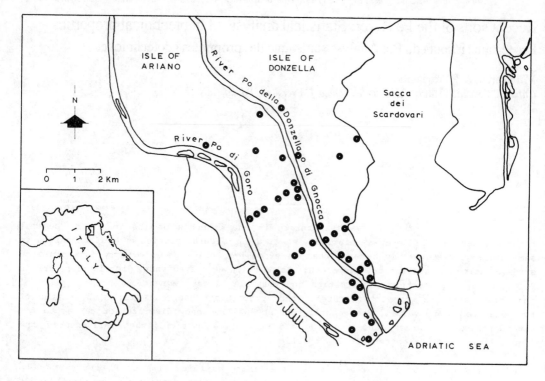

Fig. 1. Plan of the Po Delta and location of the boreholes.

from about 450 m/century to about 7 km/century. This speed of advance was arrested and the delta even began to retreat between 1950 and 1960, when methane-bearing waters were extracted, causing widespread subsidence over enourmous areas of the delta. The territory substantially entered a stationaty phase when the wells were closed down in the early 1960s.

The Po delta has recently been subject to salt intrusion through its branches, jeopardizing some of the crops in this area, one of the most fertile in the whole of Italy. In order to study the phenomenon of salt intrusion, a sample area (fig.1) was chosen in which to test the efficiency of a mobile barrage on one of the southern branches of the Po, at Po di Gnocca.

However, before examining the consequences of the barrage on groundwater regime, it was necessary to obtain detailed information on the characteristics of the superficial soils in the sample area. A total of 40 rotary drillings was thus

carried out with continual sampling, over a surface area of about 32 sq. km. Average boring depth was 8 m. All samples were tested in the laboratory of the Istituto di Costruzioni Marittime e di Geotecnica of the University of Padova.

The results of many laboratory tests therefore exist on samples coming from a relatively restricted area. The examined soils are also of very recent origin; they were all deposited in the same sedimentary environment: the delta, i.e., a transitional environment between continental and strictly marine. All these facts suggest that the results of laboratory tests and the geotechnical characteristics found should be analyzed with statistical methods.

2 STATISTICAL ANALYSIS

Soil properties characterizing a specific geological deposit show a significant variability around their mean values, and

property X_i	COHESIVE SOILS							GRANULAR SOILS			
	W_l	W_p	I_p	W	G_s	c_u	C_c	d_{10}	d_{60}	C_U	C
units	(%)	(%)	(%)	(%)		(kPa)		(mm)	(mm)		
N	90	90	90	41	17	24	18	56	56	56	56
mean	47.6	30.8	16.8	41.8	2.76	25	.39	80E-3	15E-2	2.39	1.28
median	48	30.5	16	42	2.76	24	.37	81E-3	15E-2	2	1.20
mode	48	30	15	40	2.78	24	.34	.12	22E-2	1.4	1.10
variance	68	24	28	52	8E-4	12E+1	4E-3	2E-3	4E-3	1.4	.1
std.dev.	8.3	4.9	5.3	7.2	3E-2	10.8	7E-2	45E-3	6E-2	1.2	.3
var.coeff.	.17	.16	.32	.17	.01	.43	.17	.56	.42	.50	.25
minimum	29	21	6	28	2.71	10	.31	4E-3	3E-2	1.2	.8
maximum	73	48	34	55	2.80	46	.58	.18	.35	7.5	2.1
range	44	27	28	27	0.09	36	.27	.176	.32	6.3	1.3
skewness	.27	.64	.49	-.10	-.29	.58	1.89	.24	.41	2.19	1.08
kurtosis	.62	1.22	.35	-.78	-1.11	-.40	4.15	-.88	.28	6.03	.39
95% confid.	45.9	29.8	15.7	39.5	2.74	20.5	.35	68E-3	14E-2	2.1	1.2
interv.mean	49.4	31.8	17.9	44.0	2.77	29.5	.42	92E-3	17E-2	2.7	1.4

Tab. 1. Summary of Statistical Parameters for Different Soil Properties.

the degree and manner of this variability are always of concern in engineering design and evaluation of laboratory data. There is significant scatter in the measurement of soil properties. This is due to natural heterogeneities of soils and inconsistencies in testing procedures. A single-point value is usually chosen from variable results for deterministic analysis. Nowadays the trend is for a probabilistic approach. In order to apply the latter, it is essential to know the mean and dispersion magnitudes of pertinent soil properties. Statistical analysis of soil test data provides this information.

We had the opportunity to perform an extensive statistical study of geotechnical properties of several soil samples, obtained from a relatively small area where soils were deposited in similar environments. This study has consisted of measures of dispersion tendencies and departures from normal frequency distribution curves. Central tendencies were measured by the arithmetic mean, the median, i.e., corresponding to 50% of frequency, and the mode, i.e., occuring with the greatest frequency.

Dispersion tendencies were measured by standard deviation, variation coefficient and 95% confidence intervals for the mean. On a normal standardized distribution curve plus or minus 1.96 times the standard deviations account for 95% of measurements.

In order to study the frequency distribution of the experimental data, the following coefficients were considered:

1. Skewness is the degree of asymmetry or the measure of departure from the symmetry of a normal distribution. For perfectly symmetrical curves, skewness is equal to zero; it is positive or negative when the frequency distribution curve presents a longer tail to the right or left, respectively, of the mode.

2. Kurtosis represents a measure of the peakedness of the frequency distribution relative to a normal distribution. It is equal to zero for a normal distribution, and assumes a positive or negative values when the peak is higher or lower, respectively, than the normal distribution one.

Statistical analysis was carried out on 112 disturbed and undisturbed soil samples taken from forty boreholes drilled in the Islands of Ariano and Donzella in the Delta

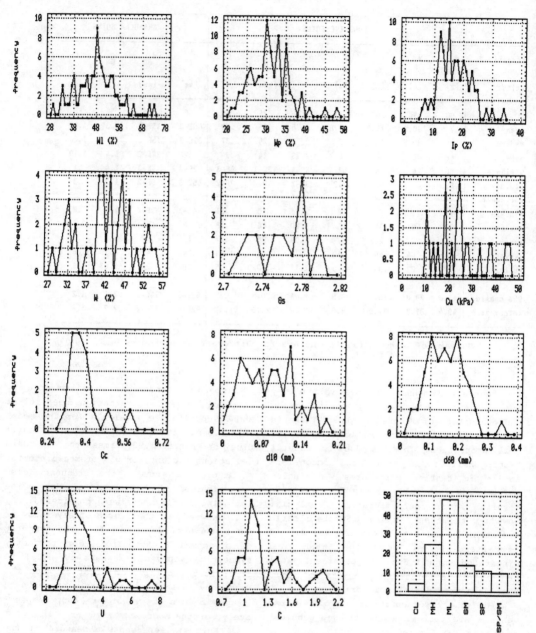

Fig. 2. Frequency distribution curves.

of the Po River.

The shallowest layers, to a depth of -10 m, do not show a planimetric continuity: this fact suggested that depth should be excluded from the variables considered in the linear regression analysis. It was observed that the nature of the

investigated layers shows many similarities in both islands, with a greater presence of ML (or A-7-5) cohesive soils and of SP, SM (A-3, A-2-4) sandy and silty soils.

Statistical analysis was therefore performed considering all available soil samples as coming from a place instead of

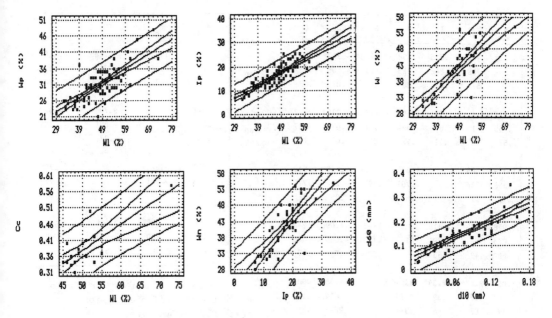

Fig. 3. Estimated regression lines with the confidence and prediction limits

from two smaller different sites.

The area was described, from a geotechnical point of view, by means of N_i observations of different soil properties X_i (i=1,...,11): liquid limit w_1, plastic limit w_p, plasticity index I_p, natural water content w, specific gravity of soil solids G_s, undrained shear strength c_u, and compression index C_c were considered for cohesive soils. Diameters d_{10} and d_{60}, uniformity coefficient C_U, and curvature coefficient C for granular soils. The frequency distribution of each variable X_i is unknown.

The computed statistical parameters (mean, median, mode, variance, standard deviation, variation coefficient, range, skewness, kurtosis, and 95% confidence interval for the mean) are summarized in Table 1.

The number of experimental observations ranged from a minimum of 17 for G_s to a maximum of 90 for w_1, w_p, I_p and w. All variables showed high values of variance and variation coefficient; in particular c_u, d_{10}, d_{60} and C_U showed the highest scatter from the mean. Only specific gravity G_s showed limited scatter, because the Po Delta clays have the same alluvial origin and a similar mineralogic composition.

Pearson's diagram was used in order to study the frequency distribution of each variable by means of skewness and kurtosis coefficients. Most of the frequency distributions showed a good fit with the Beta distribution curve, asymmetry to the righ,t and a high peaked curve.

Some possible relationships between pairs of different variables were investigated, asssuming a linear law between a dependent variable X_j and an independent one X_i of the following kind:

$$X_j = a + b X_i$$

where a is the intercept on the X_j axis, and b is the linear regression coefficient which represents the effect caused on X_j by a unitary variation of X_i.

613

	W_l	W_p	I_p	W	G_s	c_u	C_c
W_l	1	.446 / .755 / .57	.545 / .848 / .72	.698 / .826 / .68	2E-3* / .401 / .16	-.187* / -.105 / .01	8E-3 / .787 / .62
W_p	1.277 / .755 / .57	1	.365 / .336 / .11	1.143 / .644 / .42	2E-3* / .171 / .03	.215* / .072 / .01	7E-3* / .337 / .11
I_p	1.319 / .848 / .72	.309 / .336 / .11	1	1.017 / .791 / .63	2E-2* / .397 / .16	-.539* / -.212 / .05	.010 / .779 / .61
W	.978 / .826 / .68	.363 / .644 / .42	.615 / .791 / .63	1	1E-3* / .173 / .03	-1.086 / -.566 / .32	1E-3 / .688 / .47
G_s	91.03* / .401 / .17	18.63* / .171 / .03	72.39* / .397 / .16	33.45* / .173 / .03	1	4.76* / .014 / 2E-4	.901* / .383 / .15
c_u	-.059* / -.105 / .01	.024* / .072 / .01	-.084* / -.212 / .05	-.295 / -.566 / .32	4E-5* / .014 / 2E-4	1	2E-4* / .053 / .003
C_c	75.92 / .787 / .62	15.61* / .337 / .11	60.32 / .779 / .61	42.22 / .688 / .47	.163* / .383 / .15	14.71* / .053 / .003	1

Tab. 2. Regression, Linear Correlation and Determination Coefficients for pairs of cohesive soil properties.

	d_{10}	d_{60}	C_U	C
d_{10}	1	1.232 / .867 / .75	-18.865 / -.713 / .51	-3.067 / -.428 / .18
d_{60}	.611 / .867 / .75	1	-9.706 / -.521 / .27	-.959* / -.190 / .04
C_U	-.027 / -.713 / .51	-.028 / -.521 / .27	1	.194 / .717 / .51
C	-.060 / -.428 / .18	-.038* / -.190 / .04	2.648 / .717 / .51	1

Tab. 3. Regression, Linear Correlation and Determination Coefficients for pairs of Granular Soil Properties.

Tabs. 2 and 3 show the estimations of linear regression, correlation and determination coefficients, for each pair of variables X_i (columns) and X_j, (rows).

Significativity analysis of the linear regression coefficient was carried out assuming a significant level of 0.05. The regression coefficients marked with an asterisk in Tabs. 2 and 3 indicate that the effect produced by X_i on X_j is not statistically significant (probability P greater than 0.05). In these cases, it is not possible absolutely to exclude that any relationship exists between the considered variables, but it means that a simple linear equation cannot represent such a relationship.

In particular, it can be observed how specific gravity G_s never causes significant effects on the other variables. This may be due to the limited range of the G_s experimental values. Liquid limit w_l seems to be the most influential property: this fact, well-known in soil mechanics, was also founded in the statistical analysis.

Tabs. 2 and 3 show some cases of statistically significant regression coefficents associated to not good correlation and determination coefficients; this fact may be explained by the high variability of the variables in question.

Fig. 3 shows estimated regression lines, characterized by determination coefficients greater than 0.57. The plots are scatterplots of the original values, and the estimated regression lines and two pairs of dotted lines represent 95% confidence and prediction limits. The size of the confidence interval, nearer the estimated regression line, than the prediction interval increase proportionally with departure from the mean.

614

3 CONCLUSIONS

Statistical analysis was performed considering 112 soil samples taken from 40 boreholes drilled in the delta of the Po River. The following considerations can be done:

1. All geotechnical properties showed high values of variance and variation coefficients.

2. Most of frequency distribution showed a good fit with the Beta distribution curve.

3. Liquid limit seems to be the most influencial property, while specific gravity of soil solids never causes significant effects.

REFERENCES

Azzouz, A.S., Krizek, R.J. & Corotis, R.B. (1976). Regression analysis of soil com-compressibility. Soils and Foundations 16, 19-29.

Ciabatti, M. (1966). Ricerche sull'evoluzione del Delta Padano. Giornale di Geologia 34, 381-406.

Lumb, P. (1970). The variability of natural soils. Canadian Geotechnical Journal 7, 74-97.

Magnan, J. & Baghery, S. (1982). Statistique et probabilités en mécanique des sols Etat des conaissances. Laboratoire central des Ponts et Chaussées, Rapport de Recherche LPC 109, 192 pp.

Geology and pedology related to collapsible soils in Pernambuco-Brazil
Géologie et pédologie rapportées aux sols à structure ouverte de Pernambouc-Brésil

S.R.M. Ferreira
Federal University of Pernambuco, Brazil

ABSTRACT: This paper presents the geological and pedological characterization of some areas in the Northeast of Brazil, where many geotechnical
problems related to soil collapse were observed during the implantation
of housing and irrigation projects. The geotechnical investigation/pro -
gram includes both field and laboratory tests. The soil analysed
which presented most severe collapse problems was the Red-Yellowish Lato
soil in Santa Maria da Boa Vista. It was observed a tendency of reduction
of the collapse potential that's inversely proportional to the distance
from the sites to the coast.

RÉSUMÉ: Cet article presente la caracterization géologique et pedológique
de quelques regions du Nordest du Brésil ou plusieurs problémes relatifs
aux sols colapsibles ont été obsérvés pendant l' implantation de projets
d'irrigation et de batiments. Des essais de laboratoire et in situ ant été
realizés. On observe un tendence de réduction du potentiel de colapse dans
la mésure ou on s'eloigne de la côte. Des sols analysés, celui de Santa
Maria da Boa Vista, un latosol rouge-jaunâtre, présente les plus graves
problémes de colapse.

1. INTRODUCTION

The processes of lixiviation of fi
nes in the superficial horizons and
the moisture defficiency in arid and
semi-arid regions are important fac
tors contributing to the formation
of collapsible soils.
The concentration of heavy rains
over short periods of time, in most
areas of the Northeast of Brazil, is
one important factor contributing to
the formation of collapsible soils.
In the presente decade, many public
projects have been developed in areas
where the occurrence of collapsible
soils has been associated to geotechnical problems-as Aragão and Melo
(1982), Ferreira and Teixeira (1989),
Signer et alii(1989) describe them.
The sites investigated where collap
sible soil were identified are presented in Figure 1.Letters "H" and
"I" stand for "Housing Construction"
and "Irrigation" respectively.

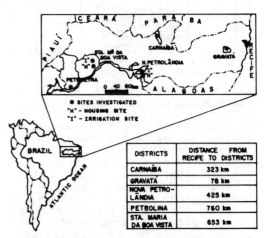

● SITES INVESTIGATED
"H" - HOUSING SITE
"I" - IRRIGATION SITE

DISTRICTS	DISTANCE FROM RECIFE TO DISTRICTS
CARNAÍBA	323 km
GRAVATÁ	78 km
NOVA PETRO-LÂNDIA	425 km
PETROLINA	760 km
STA. MARIA DA BOA VISTA	653 km

Figure 1. Location of collapsible
in the State of Pernambuco-Brazil.

2. GEOTECHNICAL CHARACTERISTICS

The geotechnical investigation program in the areas of study inclu̲ded field and laboratory observation. The field investigations consisted of percussion drilling, block sampling, specific weight and natural humidity tests. Oedometric tests with inundation at specific pressures, double consolidation and mine̲ralogic analysis were performed in̲ the laboratory.

In the site "H" of Santa Maria da Boa Vista, the percussion drilling and the observation wells showed that the soil profile had two layers before reaching a hard layer impenetrable to percussion. The superficial layer had a thickness varying from 0.50m to 2.0m, cons - tituted of red and yellowish little compact silty sand; the SPT varies from 5 to 10 strokes; the second layer has a thickness that varying from 0.0m to 2.0m, of a compact to medium compactness yeallowish silty sandy gravel, and the SPT varies from 10 strokes to the impenetrable. (See Figure 2).

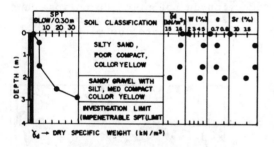

Figure 2 - Typical profile of soil Santa Maria da Boa Vista.

In the site of Nova Petrolândia, the sediments show a thickness that varies from 0.0m to 1.4m. It is constituted of a silty little clayish light brown an yellowish fine-grained and medium sand. Table 1 shows the results of charac̲terization tests in typical samples of each site. The absence of clay is characteristic of lixiviation of fine-grained soils on more super̲ficial horizons, in areas of alter-nate dry and intense rain seasons. For all soils analyzed, the Atter-berg limits fall above line "A" in the plasticity chart,with liquidity

Table 1. Geotechnical characteri-zation.

TEST	DISTRICTS	CARNA-ÍBA	GRAVA-TÁ	N. PETROLÂNDIA "H"	N. PETROLÂNDIA "I"	STA. Mª B.V. "H"	STA. Mª B.V. "I"	PETRO-LINA*
IN SITU CONDITION	e_0	0.823	0.750	0.689	0.558	0.704	0.857	0.592
	ρ_d (kN/m³)	15.00	14.97	15.90	17.79	16.20	14.47	16.70
	SR (%)	22.56	6.82	13.37	27.78	19.30	11.64	23.60
	γ_s (kN/m³)	27.34	26.20	26.86	26.18	26.14	26.86	26.59
	W (%)	5.94	1.96	3.40	5.92	5.20	3.72	5.23
GRADING (%)	GRAVEL	01	–	01	–	01	01	03
	SAND	63	–	91	77	70	72	70
	SILT	07	–	05	15	22	7	23
	CLAY	29	–	03	08	07	20	04
CONSIS-TENCY	WL (%)	40.40	–	NL	23	23	19.4	15
	WP (%)	22.20	–	NP	18	14	12.5	NP
	IP (%)	18.20	–	–	5	9	7	–
COM-PAC-TION	$P_{s\,máx}$ (kN/m³)	16.92	–	19.80	18.97	19.65	–	20.60
	Wot (%)	17.51	–	10.80	12.82	11.10	–	7.20

* ARAGÃO & MELO (1982).

limits that vary from 14.0 to 35.0%.

The mineralogic X-Ray diffraction analysis and the differential term, in the fraction % < 2 mm, showed that the material is constituted of quartz. The predominant clay mineral found in samples from Nova Petrolândia and Carnaíba was Illite, whereas in and others sites the predominant clay mineral was kaolinite.

In order to characterize the beha-vior of soil collapse in laboratory, double and simple consolidation tests were made in undisturbed and shaped-in-laboratory samples, in different conditions. Some typical results are presented in Figure 3.

Using the criteria of identifica-tion of collapsible soils sugested by Jenning and Knight (1957, Apud Clement and Finbarr (1981), it was observed that 35.8% of all undistur̲bed samples did not present collapse̲ problems; 34% presented moderated problems and 11.3% presented serious problems.

Figure 4 shows the potential col-lapse values for undisturbed soil samples. It was observed that the soils that presented the most serious collapse problems are in the site of Santa Maria da Boa Vista. There is a tendency of reduction of the poten̲tial collapse for the sites nearing the coast.

The identification criteria of collapsible soils proposed by Reginatto and Ferrero (1973), with collapse pressures determined by Pacheco and Silva's method, demons-trated that the soils of Nova Petro̲lândia "H" presented collapse pres̲sure values, in inundated tests , inferior to the ground pressure,and the collapsibility coefficient was

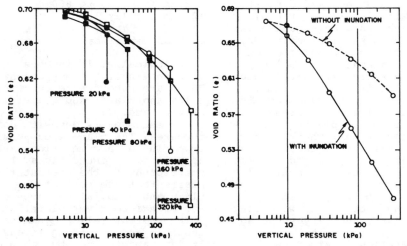

Figure 3. Void ratio change vs vertical pressure-Santa Maria da Boa Vista

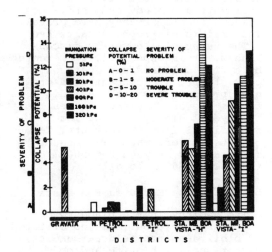

Figure 4. Collapse potential (%),
severity of problem versus inundation pressure and districts
(undisturbed samples).

negative. This is characteristic of
soils that crush down under its own
weight when inundated. These soils
are called the real collapsible
soils". The soils of Santa Maria da
Boa Vista presented collapse pressu
re values in the inundation test
larger than the ground pressure and
collapsibility coefficient between
zero and one. These soils are called
"collapsible-conditional-soils", and
it is necessary to apply an external
charge in order to have a collapse
during the moisturing process.

3. PEDOLOGICAL, PHYSIOGRAFICAL, GEOMORFOLOGICAL, AND GEOLOGICAL CHARACTERISTICS.

Based on studies made for the Radam
Brasil Project, Brazil (1981,1983),
the places where samples were
collected to make tests showed the
following characteristics:

Site of Carnaiba

Geologically, there are different
gneisses. The hornblend-gneiss,
garnet-gneiss, schistous gneiss,
migmatitic zones, calcisilicatic
intercalations, amphibolites,
quartzites and metamorphic limesto
nes. Monteiro Complex. Early to
middle Pre-cambrian.

The soils are brown calcic and
lithic eutrophic, which in the
studied area were the ones that
presented a greater percentage of
clay (29%). The relief is plane -
topped, with different heights and
drainage depths, generally separated
by valleys of deep bottom. The drai
nage depth intensity is very low.
It is a Savanna Region, with warm
semi-arid climate. The rainy season
is short and poorly distributed.
The average temperature is bellow
24ºC.

Site of Gravatá

It is geologically placed in the
remobilized Carnaíba Complex.

It presents a reworked denudation surface, with inclined, irregular barren planes as a consequence of succesive attacks, indicating predominance of the areolar erosion processes, truncating fresh or little altered rocks. Central pedi plane of Planalto da Borborema.

The present relief dynamics is unstable and of strong intensity. It shows irregular planes with residual elevations. There are moderate declivity degrees (5 to 10º) and is locally sloped (>45º). There are discontinuous detrital deposits.

The soil is podzolic, red - yellowish and eutrophic, with clay of low activity and clayey average texture.

The vegatation is arboreous steppe, open to pasture areas. The drainage is superficial and produces a generalized ablation with granular disaggregation and fragmentation.

The climate is tropical, with dry periods in the summer. The annual precipitation rate goes from 250 to 800mm, reaching 1200mm locally.

Site of Petrolândia

Geologically there occur paracon glomerates and green-dark-gray shales and siltstones. Occasionally, there occur calcareous lenses. Marizal formation-bahia supergroup-cretaceous-mesozoic.

Basins and sedimentary covers dominate the area. Levelling is the dominating genetic process. There are inclined planes, made uniform through different covers that result from reworking and sucessive reorganization that indicate predominance of areolar erosion processes.

The present relief dynamics is of average and weak intensity and transition. It shows barren table lands and colluvial slopes, with weak declivity degrees (2 to 5º). The colluvial deposits are continuous; they vary in thickness, sandy and sandy-silty texture.

The soil is made of allochtonous quartz sand. To the inland, the vegetation cover is continuous made of arboreous steppe. The climate is warm and the temperature goes above 18ºC. The annual rain rate varies from 250 to 800mm, reaching 1200mm locally.

Site of Petrolina

Geologically, it is constituted of predominating sandy deposits. There are subordinate quantities of silts and clays, and gravel levels at the base.

The sedimentary deposits dominate the area. The plane area results from fluvial accumulation, subject to periodical inundations. It corresponds to the present flood - plains that are part of the flood-plain and alluvial terraces of the São Francisco river plain.

The present relief dynamics is unstable and of strong intensity. It is characterized by foodplain and fluvial terraces with very weak declivity (0 to 2º). The sediments are sandy and sandy-clayey. Sometimes they have gravel. The soil is distrophic quartz sand.

The vegetation cover is discontinous. It is especially represented by cyclic cultures and spots of park steppe. The concentrated drai nage predominates with periodic floods provoked by the changes in the hydrologic system and local superficial flow in torrents. The climate is very warm. The annual rain rate is between 250-500mm; it reaches 800mm locallv.

Site of Santa Maria da Boa Vista

The geology in the places where the houses were built is constituted of biotite-quartz-feldspar gneiss, garnet-gneiss, biotite and/or hornblende-gneisses, tonalitic gneisses, clastic migmatic and granulitized rocks, early precambrian, regional metamorphism.

Where the irrigation project was installed, the geology is constituted of different granitoids, porphyroid granites with biotite and/or hornblende with remaining migmatized gneisses and frequent basic dioritic spots. There are granitoid silts, stonelike and abundant vegetation. Late pre - cambrian.

The exhumed reworked surface has inclined planes. It is made uniform by covers of different origin that resulted from sucessive reorganizations and reworkings. This indicates the predominance of areolar erosion processes.

The present relief dynamics is unstable and of strong intensity. It is characterized by irregular planes, with residual elevations. The declivity degrees are moderate (5 to 10º) and locally strongly sloped (45º). Descontinuous detrital deposits occur or not. The soils are eutrophic red-yellowish latosolic and red-yellowish podzolic.

The vegetation is open arboreous steppe with pasture areas.Dominating processes: superficial flow, producing generalized ablation; granular disaggregation and local fragmentation.

The climate is quite warm. The temperature is above 18ºC. The annual rain rate is between 250 and 800mm, reaching 1200mm locally.

4. CONCLUSIONS

Among the soils analized, the one that presented greater collapse problems is that located in the site of Santa Maria da Boa Vista; it is red-yellowish latosoil and conditionally collapsible. The soil of the site of Nova Petrolândia is truly collapsible. The soils of Petrolina and of Nova Petrolândia showed the same pedologic classifi cations, but different geotechnical behaviors, regarding collapsibility. A tendency of reduction of the collapse potential inversely pro - portional to the distance of the sites to the coas was observed.

The influence that the erosion processes have on the present relief is important and the geology is quite diversified between the different places.

ACKNOWLEDGEMENT

The author gratitude to the staff of LSI-UFPE, to Construtora Norber- to Odebrecht, Christianne Lyra No - gueira and Amaro Henrique Pessoa Lins, by their precious cooperation in the work and are also grateful to CNPq - Conselho Nacional de De- senvolvimento Científico of finan- cial supports.

REFERENCES

Aragão, C. J. G e Melo, A. C.(1982) Fundações rasas em solos colapsí veis. Um caso no semi-árido de Pernambuco - VIII Congresso Bra- sileiro de Mecânica dos Solos e Engenharia de Fundações. Vol. 2. P.19 - 40 - Recife - PE - Brasil.

Brasil. Ministério das Minas e Energia. Secretaria Geral. Proje to Radambrasil folhas SC 24/25 Aracaju/Recife; geologia geomor- fologia, pedologia, vegetação e uso potencial da terra. Rio de Janeiro, 1983. 856P.

Brasil. Ministério das Minas e Energia. Secretaria Geral. Proje to Radambrasil folhas SB 24/25 Jaguaribe/NAtal; geologia geomor fologia, pedologia, vegetação e uso potencial da terra. Rio de Janeiro, 1981. 744P.

Clemence, S. P. and Finbarr, A. O. (1981). Design considerations for collapsible soils journal of the geotechnical engineering di vision, USA (107):305-317. GT 3. 1981.

Ferreira, S. R. M. and Teixeira,D. C. L. (1989). Collapsible soil - a practical case in construction Pernambuco - Brasil. Proc. 12th ICSMFE, 603-606. Balkema, Rio de Janeiro.

Jennings, J. E. and Knight (1957) The additional settlement of foundations due to a colapse of strutture of sandy. Subsoils wetting. Proc. 4 th ICSMFE. 316- 319.

Reginato, A. R. and Ferrero J. C. (1973). Collapse potential of soils and soil-water chemistry. Proc. 8 th ICSMFE. 77-183. Moscou.

Singer, S; Marinho, F. A. M; San- tos, N. B. and Andrade, C.M.(1989) Expansive and collapsible soils in semi-arid region. Proc. 12th ICSMFE. 647-650, Balkema, Rio de Janeiro.

An experimental-statistical approach to study the shear strength of rock joints

Une approche statistique-expérimentale à l'étude du comportement au cisaillement des discontinuités naturelles

P.G. Froldi & S. Mantovani
Rock-Soil, Geological Survey, Milan, Italy

ABSTRACT: The shear behaviour of natural discontinuities is frequently observed only from a purely mechanical point of view by a deterministic approach and the general recognition of the geological context where the problem occurs, is often neglected. In this survey a statistical analysis about different kinds of stratification joints has been dealt; one of the main aims whas to obtain probabilistic information, in the local environment, generally utilizable. The tilt-test method as a source of data and the Barton's theory to process them, has been adopted.

RESUME': Le comportement au cisaillement des discontinuitées naturelles sont fréquemment observée sous un point de vue purement mécanique avec des approches déterministes; ces souvent ne prend pas in consideration les observations relatif au conteste geologique general. Dans cet étude on a developée une analyse statistique sur differents typologiés des joints de stratification avec les objectifs de obtenir informations probabilistes de bonne validité dans l'ambiant considéré. On a adoptè le methode du tilt-test et pour l'elaboration des donnés experimentales on a utilisée la théorie de Barton.

1 PREFACE

The shear behaviour of stratification joints has been analyzed. The rock samples to try come from "Scaglia Cinerea" formation (Paleocene) which appears in the S. Croce Valley (Belluno, North Italy) (Fig. 1); the litological nature of

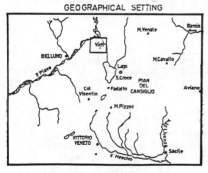

GEOGRAPHICAL SETTING

FIG. 1

rock change from marl to clayey marl with good stiffness. The surfaces of sampled discontinuity (19 in number) has been selected with geostatistical criteria.

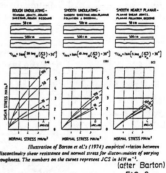

Illustration of Barton et al.'s (1974) empirical relation between discontinuity shear resistance and normal stress for discontinuities of varying roughness. The numbers on the curves represent JCS in MN m⁻².

(after Barton)

FIG. 2

The number of test carried-out in this survey is 150. Regarding the discontinuity, Barton (1973) and Barton & Choubey (1977) define one yield criterion with a curved envelope (Fig. 2) as Mohr's criterion. Here's Barton's law:

$$\tau_{mob} = \delta n \cdot tg\, \emptyset_{mob} =$$

$$\delta n \cdot tg\, (JRC \, \log_{10}(JCS/\delta n) + \emptyset\, b)$$

These symbols are supposed to be known by the readers.

2 EXPERIMENTAL SURVEY

The survey main aims are:
- Identification of behavioural characteristics of joints under shear strength, after their classification;
- Analysis of JRC obtained from back-analysis and its comparison with the values visually estimated (Fig. 3).
In order to allow one better and quicklier understanding, the operative criterion, through this flow diagram, has been shown.

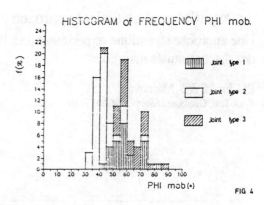

FIG. 4

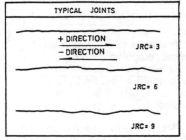

FIG. 3

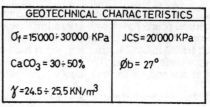

FIG. 3

<u>FLOW DIAGRAM</u>

	Description of geometry and characteristics of samples	
Statistical observations ←	Determination of $\emptyset$ mob, by the tilt-test (Fig. 4)	→ geomechanic observations
	Determination of i eff. by the law i eff = $\emptyset$ mob - $\emptyset$b	observations on data population
Statistical observations ←	Determination of JRC by $JRC = \dfrac{i\ eff}{\log_{10}(JCS/\delta n)}$	opening analysis about validity of this relation
Correlative observations ←	Comparative analysis of JRC(estimated) and JRC(back-calculated)	Observations on particular phenomena
Correlative observations ←	Determination of $\bar{a}°$ for different step sizes (a°= angle of roughness)	$\bar{a}° = \left[\sum_{i=0}^{m} arctg\dfrac{(Yi - Yi_{-1})}{(Xi - Xi_{-1})}\right]\cdot\dfrac{1}{n}$ Observation about parameter variations with increasing step size(s.s.)
Observations about realistic extrapolation of experimental data ←	Estimation of JRC nat with Barton's suggestion	final observations on Barton's metodology
statistical observations ←	Elaboration of frequency curve of parameter a°	observation about effect of roughness asymmetry on experimental data.

3 EXPERIMENTAL OBSERVATIONS

The sampled joints has been classified into three main groups:
Group 1: joints with sound surfaces and/or poor weathering.
Group 2: joints with weathered surfaces and/or thin clayed - silt infilling.
Group 3: sound, rough joints generated by recent fracturing with strong exfoliation.
The main characteristics about shear behaviour observed during the trial are:
Groups 1 and 3: besides the roughness the yield depends strongly on mating of joint sides.
Groups 2: subseguent failures has been observed at rather close tilt angles.
The bibliografic value of Øb has been assumed.

4 DATA PROCESSING AND INTERPRETATION

Three different kinds of graphics has been made from data source: statistical, constitutive, correlative
statistical: histograms   Phi mob (1, 2, 3 groups) (Fig. 4)
Constitutive: JRC (back-calculated)-JRC (estimated) (Fig. 5)
Correlative: JRC (b.c.) - $\bar{a}°$ (Fig. 6)
$\bar{a}°$ - ss (Fig. 7)

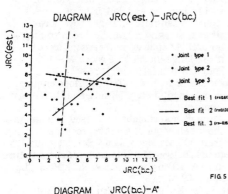

DIAGRAM   JRC(est.)–JRC(b.c.)

• Joint type 1
• Joint type 2
+ Joint type 3

— Best fit 1 (r=0.61)
— Best fit 2 (r=0.13)
— Best fit 3 (r=0.15)

FIG 5

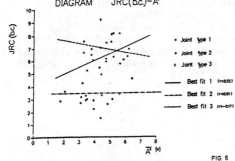

DIAGRAM   JRC(b.c.)–A°

• Joint type 1
• Joint type 2
+ Joint type 3

— Best fit 1 (r=0.92)
— Best fit 2 (r=0.01)
— Best fit 3 (r=0.17)

FIG. 6

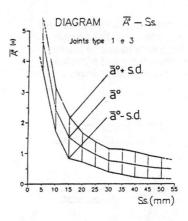

DIAGRAM $\bar{A}°$ – Ss.

Joints type 1 e 3

$\bar{a}°$+ s.d.
$\bar{a}°$
$\bar{a}°$- s.d.

Ss.(mm)

FIG. 7

From histograms it has been observed that the groups 1 and 3 show a greater frequency at high Ø mob while, in group 2, the unimodal distribution around medium Ø mob show a reduced roughness. However, this last group also show high Ø mob, withnessing localized roughness.

From graphics JRC (b.c.) - JRC (est) it has been recognized, only for group 1, the existence of linear law (positive) between roughness and friction angle. The JRC (b.c.) values of joints "2" are costant in trend, showing independance from roughness, because of thin infilling and weathering. The joints "3" show an appreciable decrease of JRC (b.c) at high JRC (est), because of progressive exfoliation failures: these failures play animportant role in damaging sliding surface.

Finally, only for group 1, the good rappresentativeness of $\bar{a}°$ parameter, compared with JRC is confirmed. It's possible, by the $\bar{a}°$-ss diagram, to extrapolate $\bar{a}°$ values for rock masses with different levels of subdivision (more or less closely jointed rock masses); this means different ss values for natural blocks as suggested from experiences of Rengers (1970) and Barton (1971). In this geologic-technical context has been found the follow relation(empirical):
$$\bar{a}° = 3.58 \cdot e^{-0.044 \cdot 55} \quad (e = Neper\ number)$$

5 JOINTS ASYMMETRY

The frequency curve of $\bar{a}°$ + and $\bar{a}°$ - has been processed on the basis of measurement developed above joint surfaces.(Fig. 3) The processing of statistical parameters such a Skewness and Kurtosis, showing the asymmetry and sharpening of frequency curve, has not found strong actual link between their values and sliding angles.

625

Developing an opening analysis about the curves shapes it's believed that a simple qualitative check of them can be interesting for the characterisation of geometry and geomechanic joints features (figg. 8, 9).

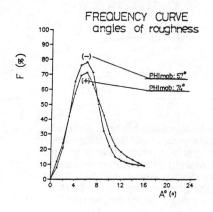

FIG. 8

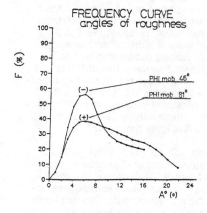

FIG. 9

It seems to be very important the centre of gravity of statistical distribution to affect the asymmetry of experimental data.

6 CONCLUSION

The main answers of this study regard the affecting of geological features on geomechanics behaviour of natural discontinuities. Starting from geostatistical sampling and analysis, it has been recognized the geotechnical parameters of structures. Looking at the obtained information it appear the possibility of setting-up regional data bank; the last one can allow a statistical - probabilistic analysis to neglet an improbable deterministic approach.

In this case observing the geometric and geomechanic characteristics of joints it's possible to state that in the shear resistence many factors of control exist, often neglected or underestimated in the study with a purely geotechnical approach.

REFERENCES

Bandis, S., Lumsden, A.C., Barton, N. 1981. Experimental studies of scale effects on the shear behaviour of rock joints. Int. J. Rock Mech. Min. Sci & Geomech. Abstr. 18: 1-21.

Barton, N. 1971. A relationship between joint roughness and joint shear strength. Proc. Int. Symp on Rock Mechanics. Nancy 1-8.

Barton, N. 1973. Review of a new shear strenght criterion for rock joints. Eng. Geol. 7: 287-332.

Barton, N. 1988. Some aspects of rock joint behaviour under dynamic conditions. 2 Ciclo di conferenze di meccanica ed ingegneria delle rocce, MIR 88, Politecnico di Torino 17: 1-14.

Barton, N. & Bandis, S. 1980. Technical note: some effects of scale on the shear strenght of joints. Int. J. Rock Mech. Min. Sci & Geomech. Abstr. 17: 69-73.

Barton N., & Choubey, V. 1977. The shear strenght of rock joints in theory and practice. Rock Mech 10: 1-65.

Farmer, I. 1983. Engineering behaviour of rocks. Chapman and Hall, Londra.

Forlati, F. & Zaninetti, A. 1988. Prove di resistenza al taglio di giunti e discontinuità naturali. 2° Ciclo di conferenze di meccanica ed ingegneria delle rocce, MIR 88, Politecnico di Torino, 3: 1-27.

Hoek, E. & Bray, J.W. 1977. Rock slope engineering. Institution of Mining and Metallurgy, Londra.

Ladanyi, B. & Archambault, G. 1969. Simulation of shear behaviour of a jointed rock mass. 11 congress US Rock Mech. Symp. Berkeley, 105-125.

Patton, F.D. 1966. Multiple modes of shear failure in rock. 1 Congress Int. Soc. Rock Mech., Lisbona 1 : 509-513.

Tse, R. Cruden, D.M. 1979. Estimating joint roughness coefficients. Int. J. Rock Mech. Min. Sci. & Geomech. Abstr. 16: 303-307.

Rengers, N. 1970. Influence of surface roughness on the friction properties of rock planes. Proc. 2.nd Int. Soc. for Rock mecchanics. Belgrade 1: 1-31.

Characteristic features of engineering geological survey on landslide prone slopes

Les particularités de la reconnaissance du sol de fondation des pentes, subissant des glissements

L.K.Ginzburg

Ukrspetsstrojproekt, Dnepropetrovsk, USSR

ABSTRACT: The characteristic feature of engineering geological survey on slopes stem from the unstable condition of the soil; the necessity to determine such strength characteristics of rocks which enable to estimate the degree of instability.The characteristics of soils are refined by reverse calculations. Estimates of the slope stability are finalized only according to repeatedly verified parameters.

RESUME:Les particularités de la reconnaissance du sol de fondation des pentes sont conditionnées par l'état instable des sols étudiés, par la nécessité de déterminer les caractéristiques demandées de résistance du sol et assurant la possibilité de détermination du dégré d'instabilité. Les caractéristiques du terrain sont précisées par calcul inverse. Les caractéristiques du terrain, maintes fois vérifiées, doivent servir de base pour le calcul de stabilité de la pente.

1 INTRODUCTION

The engineering geology studying a slope possesses a number of unique features.We can define the key features which differentiate these surveys from the common ones:the necessity to estimate the slope safety factor;specific and stringent requirements to the determination of the strength characteristics of soils (angle of internal friction and cohesion)which have to be studied under varying conditions of soils including preparation of their shear surface and humidity;the necessity to determine the actual and extreme surfaces of slide(their location and geometry).

It should be noted that in the USSR, when engineering geological survey is carried out on slopes,the required parameters are established by several methods, and for design purposes those characteristics are selected which are most closely correspond to the actual condition of the slope. To ensure the correct selection of characteristics, they are additionally checked by means of the so-called reverse calculations,the procedure of which is described in this paper.

The determination of the actual position of the sliding surfaces in a slope is the most difficult task in the process of engineering geological survey.To enable control,various methods are usually used.However, the actual sliding surfaces determined in the process of survey are additionally verified through computations. Performance of such calculations presents certain difficulties, first,due to the complexity of modifying various sliding surfaces, and, second, due to the abundance of methods for slope stability calculation, among which the true one has not yet been selected up to now. It should be emphasized that performance of such calculations is another characteristic feature of the engineering geological surveys on landslide-prone surfaces.

It is highly improbable that such survey methods for unstable slopes which would enable to unambiguously determine all the necessary parameters, will be developed within the nearest future.Therefore, we consider it necessary, when carrying out work on slopes, to continue the utilization of the adequate methods for engineering geological survey, laboratory studies and mathematical computations.

2 GENERAL SURVEY PROCEDURE

The engineering surveys on territories located on slopes generally include all kinds of geological activities carried out on ordinary sites.Special works carried out on inclined surfaces are specified by various specification documents developed in the USSR(Instructions, 1966;Instructions,1978;Construction Norms,1988, etc).

The order and the volumes of geodetic work, as well as types and quantities of engineering geological surveys are envisaged in the survey program which is developed in cooperation by engineers geologists and designers who formulate the technical specifications for carrying out the surveys. The program also includes all other details of the envisaged engineering geological activities.Among other things, for unstable slopes the characteristics of tectonics, neotectonics and seismicity of the territory being studied, are often critical, including seismicity scaling according to the microseismic zoning principle. Forecasting of the groundwater level changes is performed during the survey, taking into consideration the influence both of natural phenomena and of the envisaged construction activities which effect the conditions of the surface run-off and evaporation,and foster leakages from the underground service lines into ground.

If landslide displacements are clearly exhibited on a slope,then during engineering geological surveys on such landslide-prone slopes, in addition to the commonly used and standard survey procedures aimed at studying the geological structure,hydrogeological conditions and properties of soils,the following specific problems are usually solved for landslide-prone regions:
a)history of the studied sliding slope formation;
b)types of landslides which developed on the slope, their structure,

mechanism of displacement,and causes of occurence;
c)presence and intensity of other physico-geological processes which take place in the slope and influence the sliding processes;
d)condition of the sliding slope over the survey period, and forecasts of its future behaviour;
e)condition of the buildings and structures on the sliding slope(including the underground service lines).

It should be noted that in the process of engineering geological survey on sliding slopes the most critical activities are the following:determination of the strength characteristics of soils by laboratory and field methods;determination of the properties of soils within the undeformed texture directly within deep boreholes;determination of actual sliding surfaces in a sliding slope;determination of the stability degree of a slope;location of the most hazardous sliding surface by computation;finalization of strengths characteristics of soils by reverse calculations.

The procedures for performing these activities are described below.

3 FIELD AND LABORATORY STUDIES

The field methods for studying the engineering geological properties of soils during the survey on sliding slopes, in addition to those which are commonly used, include:
a)testing by the surface penetration method;
b)in-depth static and dynamic probing;
c)rotational cut through boreholes;
d)cut of pillars in pit holes and open pits;
e)determination of deformation properties of soils by in-depth die stamps;
f)location of the existing sliding surfaces in a slope.

Testing of rocks by the surface penetration method is performed to: divide the stratigraphically and lithologically homogeneous ground mass into layers of differing strength (engineering geological elements);finilize the points of sampling for laboratory studies.Penetration is performed using a variety of instruments available at a surveying organization(micropenetrometer, hydraulic penetrometer,"DORNII" striker, etc.).

The testing of rock by static and dynamic probing is used to reduce the volume of boring and laboratory work to determine the depth of occurence and thickness of the weakened zones(the landslide displacement zone,soft and viscous clay soils,water-saturated loess soils,etc.)located at a depth down to 15-20m.In engineering geological sections the areas of reduced resistance against the probe in individual probing core-points are connected by a line which characterizes the respective weakened zone.

Rock testing by rotational cut in boreholes is used to estimate strength,usually, in weak sand-clay rocks located at a depth down to 15m.The rotational cut method is usually not used in landslide accumulations containing coarse fragmental inclusions.⅃The studies of the shear strength by the pillar-cut method on sliding slopes are used to estimate the strength of inhomogeneous laminated or cleaved clay or medium-hard rocks and also to ascertain the possibility of displacements along the laminae cont-acts. A direct shear within the specified plane is usually effected according to a procedure which enables to simulate the shear conditions similar to those observed during landslide phenomena.

To determine the value of the deformation module, the deformative properties of soils are studied.The value of the deformation module is determined by testing the soils under vertical static loads on die stamps usually installed in special vertical shafts. Die-stamp testing in special shafts and pits during studying deeply located soils is difficult.Therefore, in such cases die stamp testing of soils in bore holes reinforced by casing pipes is employed. However testing in boreholes often does not yield the required effect, since when die stamps are installed in boreholes,it hampers control of the soil cleaning quality on the face, and taking into account the deformation of the pipe column and its friction against the borehole wall.

"Ukrspetsstrojproekt" has developed and used during the recent years die stamp testing of deep-seated soil layers by means of screw-type stamp/blades of 600 cm^2 area.The commercialized method enables to take into account the value of the lateral friction of the die stamp stem against the soil. With this in view, the die stamp is designed so that when the die stamp blade is immovable, its stem is free to move axially.The die stamp is driven into the soil from the surface,into a preliminarily bored leading borehole of a small diameter. To determine the die stamp stem lateral friction against the soil,the stem is tested for pull-out. The obtained friction value is corrected while the load value at each loading step is being determined. In such case, all parameters(subsidence load) are set according to the instruments located on the ground surface (Fig.1),and the vertical press-in load applied by the jack

Fig.1.Screw-type die stamp testing of soil

to the screw-type stamp through the stem is determined for each loading step by the formula

$$p_1 = p_i - p_2 + t, \qquad (1)$$

where p_1 is force transmitted by the jack, p_i is required load at the loading step, p_2 is weight of

the rod column (stamp stem), t is
force of the rod column friction
against the soil.

The use of the screw-type die
stamp of novel design has proved
that die stamp testing of soils at
a depth of 30m and more is feasible.

Soil samples for laboratory tests
are mostly taken from the weakened
zones of a slope. Such weakened zones are:contacts between varying
lithological differences; contacts ·
between clay rock and aquifers;zo-
nes of landsliding; zones of tectonic disturbances;clay filler of large fissures; etc.

In addition to the standard schemes for laboratory rock shear testing, special schemes for rock shear
testing are used which enable to simulate the state and conditions of
soil work in a slope, particularly,
within the landslide movement zone.
For instance,tests according to the
scheme of repeated shear along a
prepared humidified surface("die-
on-die") are used. This test scheme
enables to simulate sliding shifts,
when the soil is heavily humidified,
along the sliding surfaces existing
in the slope.The calculated rock
shear strength characteristics obtained according to such a scheme
are best suited to the tasks dictated by the necessity to estimate
the stability of landslide-prone
slopes,since they characterize the
stability of soil within the mostly
weakened zones. According to that
scheme the soil samples are initially tested according to the method-
"nonconsolidated accelerated shear
under complete water saturation".
Then the shear surfaces in the soil,
which were formed as a result of
the shear, are again wetted with
water and brought into register
whereupon the shear test is repeated again by the same method.

The laboratory shear studies are
carried out by means an automatic
instrument which enables to simultaneously perform shear of three samples under the specified vertical
loads. The plotted test results are
desplayed on the oscillograph screen
and registered on the oscillograph
paper.The instrument is based on
the method of shearing at a constant rate which varies as a function
of the soil types.

The instrumentation enables to increase production rate and reduce
labour consumption by simultaneously carrying out three tests and eliminating constant observation and
control by the operator.In addition,
it enables to perform tests at a
qualitatively new level,since the
factor of operator's subjectivism
is excluded.

4 METHODS FOR DETERMINING SLIDING SURFACES IN A SLOPE

The methods used in the USSR during
engineering geological surveys on
landslide-prone slopes are described below.

4.1 Visual observation methods

A sliding surface is determined visually. For instance,while boring
a borehole, cases of the tool"fall-
through"are registered which indicate the appearance of weakened zones.
Direct observation of the passed
through rocks is also carried out
during boring boreholes or driving
pits. The sliding surfaces are indirectly characterized by the visible shear planes,increased humidity
of soils, occurence of repeatedly
crumpled rocks,gliding planes and
scores, etc.(Ginzburg, 1979).Rather
often the sliding surface is also
revealed by the appearance of the
rocks passed through, e.g.,when
deluvial soils flow down along the
bedrock (hard,medium-hard,compact
clay, etc.).

4.2 In-depth hose-type bench mark

For this method, a thick-wall rubber hose of 35-50mm diameter is used
which is lowered into a borehole,
usually to 10m deep (Kuntzel and
Novikov,1964).

Observations are carried out using
a measuring reel weighted at the
end,or a steel rod (Fig.2).If the
hose is bent by the sliding movement,the rod comes up against an
obstacle on its passage,where the
hose is bent.This enables to determine the upper boundary of the most
severe landslide deformation zone,
and the very fact of the initiated
landsliding as well. The latter is
of utmost interest when very slow,
invisible for a human eye,sliding
movements occur.

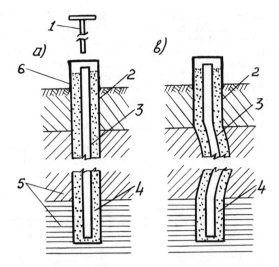

Fig.2. In-depth hose-type bench mark

a-observation well before landslide movement;b-ditto after movement; 1-rod-type probe; 2-borehole;3-rubber hose;4-sand feel;5-geological section; 6-branch pipe

4.3 In depth bench mark made of tubes

Into a borehole an in-depth bench mark consisting of lengths of metal or concrete tubes is lowered. The tubes of 1m lengths are installed vertically one above the other by means of a guiding column of smaller diameter tubes. Displacements of tube lengths relative to each other are measured by means of a column of bore rods, rod feeler or measuring reel weighted at one end.Periodical measurements enable to determine the direction and velocity of sliding movements at various depths.

4.4 Electrical displacement indicator

This method recommended by Prof. I.J.Baranov is used at some landslide areas of the Caucasus Coast of the Black Sea. An electrical bench mark consists of a closed circuit of thin electrical wires.One of the wires is axial. Lead-ins are connected in series to it with a 0.5m interval. The ends of the wires are run to the ground surface and fixed to a panel. All the wires are lowered into a borehole. The borehole is grouted to add rigidity to the bench mark. If sliding deformation occurs,the ce-

ment pole breaks,and the enclosed electric wires break too.Monitoring is carried out by means of a special meter,consisting of a milliammeter, switching scale and power unit.When the wires are broken by a landslide the closed circuit gets open at those intervals,where the soil masses had been displaced, and this is determined by the ammeter readings.

4.5 Determination of relative displacement of layers

The method enables to determine relative displacement of soil layers over a long period of time(Rubanik, 1972). A pipe made of elastic material is placed into a borehole,while the space around is filled with sand. Measurements are taken using meters containing a set of rods of different diameters which are lowered into the pipe.

4.6 Reverse float plumb

A reverse plumb of Fundamentprojekt consists of a wire (Fig.3),the lower end of which is anchored in the

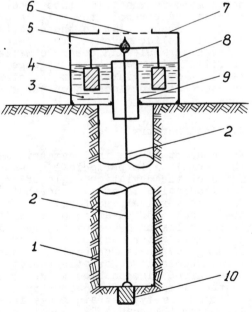

Fig.3 Scheme of a reverse float plumb: 1-borehole wall; 2-wire; 3-water in a tubular vessel;4-annular float; 5-clamp;6-glass with a coordinate grid;7-cover;8-pipe of 900-1000mm diameter;9-pipe of 300mm diameter;

631

10 -anchor in the bedrock.

stable bedrock,while the upper end
is fixed to the float freely buoyant
in a special tubular water-filled
vessel. When surface soils are displaced,they carry the tubular vessel
there with, while the float remains
in its place.Hence,the clamp marks
the displacement value of the surface soil layers on the upper glass
with a grid.

4.7 In-depth strain gauge bench mark

The sensing element of the strain
gauge bench mark(Petrenko,e.a.,1972)
which measures horizontal displacements of soils,consists of a resili-
ent steel rule 4cm wide and 0.1cm
thick. To transform mechanical movements (bending deformations of the
rule) into electric signals,resis-
tance strain gauges are used which
are glued to the rule surface at a
50 cm distance between the measuring
points(pairs of strain gauges).Befo-
re the rule is assembled into an integral strain gauge bench mark,each
strain gauge is calibrated on a
stand. The in-depth strain gauge
bench mark is installed into a bore-
hole.In the borehole the rule is arranged perpendicular to the direction of landslide movement.The space
between the rule and borehole walls
is filled with sand. Strain gauge
readings are registered at the strain gauge station located on the surface.

4.8 Inclinometry

The concept of the inclinometric method consists in periodical measurements of the deviation angle of the
longitudinal axis of each borehole
from the vertical,and formation of
the actual configuration of a series
of boreholes on a geological section
of a slope after manifestation of
sliding movements.Connection of the
lowest borehole deformation points
by a curved line enables to determine the sliding surface geometry.
 A pendulum-type strain gauge inclinometer (Blehman,1988) consists of
the inclinometer proper with a measuring element lowering device and
a field electronic digital recorder.
The measuring element of the inclinometer comprises a pendulum suspended
in bearings,with an elastic steel

beam to which tensoresistors are
glued.The measuring element is arranged inside a cylindrical metallic
housing which is connected to a distance piece. The distance piece comprised of the upper and lower identical clamping hinged systems serves
to install the measuring element in
a borehole coaxially with its longitudinal axis. Thus, the measuring
element axis is coaxial with the
axis of the borehole section and has
the same angle to the vertical,while
the pendulum of the measuring element is positioned vertically. As
this takes place,the steel beam is
bent to a certain angle,and the strain gauges glued to it change their
resistance which is registered on
the surface by the electronic digital recorder to which a wired cable
is run from the resistance transducer.

5 DETERMINATION OF THE EXTREMAL AND OF STRENGTH CHARACTERISTICS BY REVERSE CALCULATIONS

Methods for slope stability calculation are plentiful (Ginzburg,1986,
1988).Therefore, in the present discussion only the most critical relations are used:

$$K_u = \frac{\Sigma R}{\Sigma Q} ; \qquad (2)$$

$$E_{lp} = K_{uz} \Sigma Q - \Sigma R, \qquad (3)$$

where K_u is slope safety factor,
E_{lp} is landslide pressure, K_{uz} is
assumed safety factor, ΣR is the
sum of retaining forces along the
slope (in the calculated cross-section), ΣQ is the sum of the shear
forces.

 Determination of the most hazardous sliding surface (extremal) and
finalization of the soil strength
characteristics by reverse calculations are carried out in the sequence described below (Ginzburg,1984).
 Consider case - the slope is unstable,but,according to the sliding
surface and strength characteristics
of soils under natural conditions
determined during the survey(φ_f,C_f)
we obtain $K_u > 1$. It means,that during
engineering geological survey some
of the parameters were determined in-
accurately and should be finalized
by reverse calculations, based on

the assumption $K_u \approx 1$, because this condition refers to the ultimate equilibrium state. First, according to the actual (under natural condition of soils)strength characteristics φ_f and C_f the search for the most hazardous sliding surface is done. For this purpose, a series of possible sliding surfaces is plotted. It is necessary to find such forms of the sliding surface (Fig.4) which, for instance, correspond to relations: $K_{IV} > K_{III} < K_{II} < K_I$ (while $K_{IV} > K_I$); $K_I > K_V > K_{VI} < K_{VII}$(while $K_{VII} > K_I$), where K_I, K_{II}, K_{III} ... are slope safety factors for the

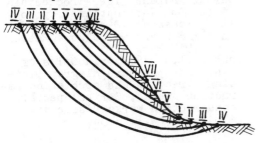

Fig.4 Determination of the extremal sliding surface

I - the sliding surface determined during engineering geological survey; II-VII- the sliding surfaces assumed while determining the extremal by the selective method.

respective sliding surfaces.After a family of such curves has been plotted,for subsequent calculations only that sliding surface is to be left for which the slope safety factor is minimal(K_u min). Such a slope sliding surface is termed extremal. Assuming that for this sliding surface $K_u \approx 1$, the values of φ and C are finalized by reverse calculations. In such case the strength characteristics are selected in the range

$$\varphi min < \varphi < \varphi f; \quad C min < C < C f, \quad (4)$$

where φ min, Cmin are the minimal strength characteristics of soils at the sliding surface level,which are determined on soil samples during laboratory shear tests,usually according to the "die-on-die" scheme,on a wetted surface;φ, C are the strength characteristics of soils(internal friction angle and specific cohesion) assumed for subsequent calculations

(finalized by reverse calculations); φ_f, C_f are the strength characteristics of soils under natural conditions (determined by laboratory or field methods). On the basis of the known values of φ and C the landslide pressure values are then determined for the extremal surface per each slice of the calculated cross-section,under which the landslide pressure diagram is plotted,based on these values.

When the soil strength characteristics are refined by reverse calculations,both strength characteristics are being varied - φ and C. In such case they depend on the following functional relations(specified by the Ukrspetsstrojproekt studies):

for clay soils $\varphi = 14 \lg C + 22.8;$ (5)

for loams $\varphi = 15.1 \lg C + 29;$ (6)

for loess-like loams
$$\varphi = 4.4 \lg C + 19; \quad (7)$$

for sandy loams $\varphi = 10.2 \lg C + 34.$ (8)

In formulae (5)-(8) the internal friction angle φ is given in degrees, specific cohesion C - in MPa.

All the above calculations are performed by computers, using specially developed programs.Finalization of sliding surfaces and strength characteristics of soils by the above methods guarantees an adequate degree of reliability in designing antilandslide facilities.

6 CONCLUSIONS

The engineering geological surveys on landslide-prone slopes have their own characteristic features as compared to the survey operations on ordinary sites. Therefore it is reasonable to accumulate the experience of field and laboratory studies under those conditions. It should be noted, that all types of surveying operations on unstable slopes are far from ideal. Site geological investigations, geodetic observations of the displacing rocks,laboratory soil tests, methods for determining sliding surfaces, methods for calculating slope stability -all these elements of engineering geological survey on slopes need further research and development. Therefore, these aspects deserve special attention. As various methods are used at different stages of survey, the required parameters should be defined by various ways. Such an approach to engineering geological survey on landslides will guarantee reliable data for design, reliable stabilization of unstable slopes and reliable construction of buildings and structures thereon.

REFERENCES

Instructions(1966) on engineering geological survey in landslide-prone regions."Fundamentprojekt" Institute.-Moscow,Strojizdat.79pp.

Instructions(1978) on engineering geological survey on landslide-prone slopes of the South Coast in the Crimea."PNIIIS"Institute, Gosstroj,USSR.-Moscow,Strojizdat. 74pp.

Constructions norms and rules(1988). Engineering survey for construction.SNIP 1.08.07-87.Gosstroj, USSR, GUGK,USSR.-Moscow,CITP,Gosstroj,USSR.104pp.

Ginzburg,L.K. 1989.A review of landslide processes on the territory of the Soviet Union.In:Landslides: Extent and Economic Significance. Edited by Earl.E.Brabb and Betty L.Harrod.-Balkema.Rotterdam.P.213-220.

Ginzburg,L.K. 1979.Antilandslide retaining structures.- Moscow,Strojizdat.81pp.

Kuntzel,V.V. and Novikov,P.A.1964. Methods for monitoring sliding movements by means of in-depth bench marks.In:"Landslides and their prevention. Proceedings of the North-Caucasian seminar on landslide studies and prevention experience".-Stavropol.P.420-426.

Rubanik,M.N. 1972. Method for determining relative displacements of soil layers. Inventor's Certificate No.337466.-Bull."Otkritiya, izobreteniya,promishlennije obraztsi, tovarnije znaki",1972, No.15,Moscow.

Petrenko, V.B.,Kozelsky,D.D. and Burdin, A.G.1972.Experience of using strain gauge bench marks for studying horizontal displacements of a soil massif by Ukrvostokgiintiz Institute. In:"Problems of engineering technological survey,design and construction in seismic and landslide-prone regions of the Southern Coast in the Crimea".-Yalta.P.64-65.

Blehman, D.A.,Labzov, J.V. and Pildysh,S.A.1988. Experience of using inclinometry to determine some parameters of landslides."Erection and special construction works.Series: Special construction works".Issue 4, 1988.-Moscow.P.15-20.

Ginzburg,L.K. 1986.Recommendations on the selection of methods for calculating slope safety factor and landslide pressure.-Moscow, CBNTI Minmontazhspetsstroy,USSR. 124pp.

Ginzburg,L.K. 1988. Stabilization of landslide slopes by pile structures. In:Landslides-Proceedings of the 5th International Symposium, Lausanne 10-15 July 1988. Editor Bonnard,Ch.Balkema.Rotterdam.P.915-919.

Ginzburg,L.K. 1984.Principles of landslide pressure diagram plotting."Erection and special construction works. Series:Special construction works."Issue 11, 1984.-Moscow.P.26-34.

A case history of the problems of determining the ground parameters of 300 sites in 6 months in Indonesia

L'histoire des problèmes entraînés en déterminant les paramètres du sol sur 300 endroits, en 6 mois, en Indonésie

H.R.G.K. Hack
ITC (International Institute for Aerospace Survey and Earth Sciences), Delft, Netherlands

C.S. Kleinman
Grabowsky & Poort B.V., Consulting Engineers, Maastricht, Netherlands

G.J. de Koo
Grabowsky & Poort B.V., Consulting Engineers, Hoorn, Netherlands

ABSTRACT: A survey for determining the ground parameters of 300 different sites along approximately 500 kilometre of railway track, was executed in Indonesia. The number of sites and different ground types opened an excellent opportunity to assess and compare the potentials of field and laboratory testing procedures. A lower bound relation between plasticity index and the ground parameters cohesion and angle of internal friction is presented.

RESUMÉ: On a fait des analyses du sol, en Indonésie, pour déterminer les paramètres du sol sur 300 endroits différents le long de 500 kilomètres de voie ferrée. Le nombre des endroits et leurs différents types de sol offrirent un occasion excellent pour justifier les problèmes et les possibilités des méthodes de recherche aux sites et dans les laboratoires de procédures d'essai. Une relation de limite basse entre l'Indice de Plasticité et les paramètres du sol, comme la cohésion et l'angle de frottement interne, est présentée.

1 INTRODUCTION

The region of South-East Asia shows a rapidly increasing economic activity. One of the countries in the region which is particularly affected by this economic revival is Indonesia. To be able to fully profit of the economic developments the infrastructure of roads and railways has to be modernized and expanded. The Indonesian Government designed a major schedule to upgrade and renovate the railway lines in Indonesia. The Netherlands Government was prepared to finance part of this upgrading as well as the survey to establish the scope of work. This paper describes the investigation for the upgrading of the bridges of five railway lines on the islands Java and Sumatra, Indonesia (fig. 1). The lines were built starting at the turn of the century up to the worldwide recession between the two world-wars.

The railway lines have to be made suitable for heavier train loads and higher train speeds. This embraces in addition to the upgrade of embankments also the upgrade and renovation of bridges. The five railway lines incorporate 289 bridges crossing mainly rivers and canals. Span lengths are up to 85 metres.

The existing steel super-structures of the bridges are not suitable because the 'life' loads (train loads) will be increased by 62 to 116 %, and 'fatigue' had lowered the strength of the steel structures. Further, corrosion of the steel had severely weakened the structures of some of the bridges. The steel structures have to be replaced by new and heavier designed structures and have to be installed on existing abutments and piers to minimize costs.

To assess the state of maintenance and the suitability of the existing abutments and piers for the new loads a survey was carried out during 1988 and 1989. The actual field survey took approximately 6 months while the subsequent interpretation of the field data, calculation of safety factors, formulation of renovation schemes and design of steel super-structures took another 6 months.

2 BRIDGE SURVEY

The five railway lines incorporate 289 bridges along railway lines with a length of 498 km. Investigating 289 bridge locations in a reasonable amount of time requires a very proper and efficient set up of the investigation scheme. To minimize logistics a permanent project office was

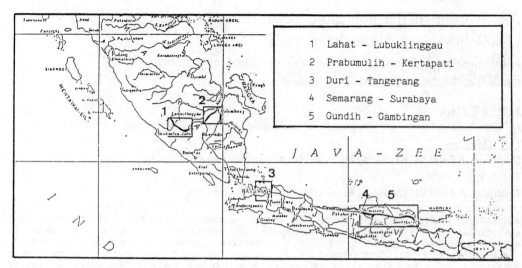

Figure 1. Locations of railway lines

established in Bandung to maintain the contacts with the Indonesian State Railways (PJKA) and testing laboratories while temporary site offices with site laboratories were erected at various locations along the railway lines.

The following aspects were included in the survey:

1. A survey of all existing historical and present-day data, eg. as-built drawings of the existing bridges, river levels and rainfall data, records of the Indonesian State Railways, etc..

2. A ground-investigation survey to assess the stability of the foundations of the existing abutments and piers.

3. A hydrological survey to determine maximum river water levels and to assess maximum water pressures on abutments.

4. A survey of the strength and state of maintenance of the existing abutments and piers.

5. A survey of the steel super-structures.

This paper describes the ground-investigation survey only. The other aspects of the survey will be reported upon in different papers.

3 GEOLOGY

The islands Java and Sumatra belong to an island arc of volcanic and orogenic origin.

They form the boundary of the Euro-Asian plate in the north and the Indo-Australian plate in the south. Present and past tectonic movements and abundant volcanic activity govern present day geology. The results of this geological setting are encountered during the survey. Two of the five railway lines are located in rock areas where the bridges are founded on weak tufaceous rock. The three remaining railway lines are located in coastal areas where the formations consist of marine and alluvial clays, marine and alluvial volcanic ash clays and limestone muds and sands. The marine and alluvial clays contain varying amounts of silt and carbonate.

4 CALCULATION SAFETY FACTORS

The aim of the survey was the determination of the safety of each bridge after renewal of the steel super-structure and with the new loading scheme applied. A major aspect in the safety of a bridge is the stability of its foundations. The stability depends on: the loads on the foundation level, the type of foundation and the ground parameters of the foundation layers.

4.1 Loads on the foundation level

The loads on the foundation of an abutment or pier can be divided into dead and life loads. The dead loads are the loads caused by the weight of the abutment or pier, steel super-structure and passive and active ground pressures. The live loads are loads caused by a passing train, their influence on the water pressures at the active side of the abutment, wind and river-water pressures and earthquake loads. A train which comes to a stand-still on a bridge is regarded as a permanent load although the time the train will be on the bridge is small compared to the other permanent loads.

4.2 Type of foundation

As-built drawings of the foundations and revision data with drawings were available for the majority of the bridges. During the survey the as-built and revised drawings of abutments and piers proved to be accurate so that details about the foundations (depth, area, etc.) and type of foundation (raft, piles, ground improvement, etc.) were known.

4.3 Ground-parameters

The foundation analyses of the existing bridges consisted of analyses of ultimate bearing capacity according to Brinch-Hansen, overturning, sliding and slope stability. To calculate these failure mechanisms the following ground parameters have to be determined: dry and wet unit density, angle of internal friction, cohesion, layer thickness, and ground-water tables.

Under the new loading scheme only the load of the train will be increased. (The weight of the new steel super-structures is only marginally higher than the weight of the old structures.) Settlement of the bridges should have been completed in the 50 years the bridges were already in position and additional settlement due to the increase of the train load will be minimal. Extra settlement has, therefore, not been evaluated.

For the evaluation of ground parameters the available ground information from the time the bridges were built has also been used. This was available for a few of the bridges. The data have been compared with the information obtained during the survey and no significant differences were found.

Revised bridge drawings and data indicated whether problems with particular bridges on a particular foundation had been experienced in the past and probably could be expected in the future for other similar bridges with similar foundations.

5 GROUND-INVESTIGATION SET-UP

To determine the ground parameters the following two approaches are possible:

1. Extensive ground survey
An extensive ground-investigation for each bridge. This would have resulted in very high costs and would have taken a considerable amount of time to execute.

2. Cost-limited ground survey
The determination of ground parameters for a particular bridge by correlating and interpolating ground parameters determined along the railway lines, without executing a whole set of tests at each bridge. This lowers the accuracy, but speeds up the investigation and therefore lowers the costs considerable.

The cost-limited approach (2) was chosen because of the following reasons:

1. Available data
The type of foundation was known and the available ground data from the time the bridges were built was reasonably accurate.

2. Required accuracy
The accuracy of the survey did not have to comply with the standards necessary for building completely new bridges.

Secondary arguments for a cost-limited ground survey were the available time and the budget. The budget for the survey was based upon a field investigation with a duration of approximately six months. An extensive ground survey would have severely extended the time necessary and thus would have required a larger budget.

Although the cost-limited ground survey was selected, all incoming data from the survey were carefully checked, so that it was discovered during the course of the work whether the cost-limited ground survey was unsuitable for a particular bridge and the survey had to be extended. If this happened for many bridges the time and budget restrictions would obviously have had to be re-assessed.

During the survey it was observed that:

1. Field data obtained during the survey indicated that correlation of ground-parameters over long distances was possible.

2. The majority of the bridges are still standing-up without any signs which could indicate failure of the foundations. This implies that the safety factor for these bridges under the old loading scheme had to be at least 1 at the time of the investigation. Abutments which showed possible failure of the foundations, for example: tilting, should have a safety factor of less than 1 at the time of investigation. Safety factors

calculated under the old loading scheme with ground-parameters obtained by correlation and interpolation according to the cost-limited ground survey gave results which confirmed these values.

Both points above proved that a cost-limited ground survey was suitable for this investigation.

6 GROUND INVESTIGATION METHODS

Correlation and interpretation of ground data over long distances and the short time available to execute the survey has led to a special set-up of the testing program. A series of tests were executed with emphasis to use the results for correlation of parameters.

6.1 Field tests

1. Visual Inspection

All bridges and surroundings were inspected by an engineering-geologist or geotechnical engineer. Emphasis was placed on the recording of geological and geomorphological features of the foundation layers. Secondly the abutments were inspected for damage possibly caused by failure of the foundations. The as-built drawings were compared with the particulars found in the field.

2. Dutch Cone Penetration Test (CPT)

A Dutch Cone Penetration test was necessary for each bridge as this was the main tool for correlating the ground layers. For larger bridges two or more CPTs were made. For a series of bridges CPTs were made available by the Indonesian State Railways but for the majority of bridges a CPT test was executed during the survey. The CPT test included the measurement of local friction so that a better indication of the ground parameters was obtained. To keep data handling errors to a minimum the measurements were recorded with a hydraulic/electronic measuring device on the top of the CPT rods and were plotted by a computer. To decrease the execution time the CPT equipment was mounted on a flat train-wagon which was loaded with a ballast of 12 ton. The advantages of this set-up were that it was possible to execute the CPT tests in a very short time which allowed these tests to be executed in-between the normal (very busy) train schedule. Also this set-up allowed the execution of the CPTs directly behind the rear of an abutment, so that soil-information over the full depth of the abutment, including the track embankments, could be recorded.

On locations where a CPT test from the track was not possible, eg. besides piers, a 2.5 ton CPT was carried out. The readings were obtained by means of hydraulic pressure gauges.

The CPT testing set-up proved to be very efficient. Approximately 325 CPTs were executed along the railway lines within a tight train schedule. Although the data obtained by CPT tests was not directly applicable to calculate drained ground parameters, it was an excellent tool for correlation purposes.

3. Boreholes

Based upon the geological information available, boreholes were drilled at regular intervals along the railway lines to obtain stratigraphic information and samples for laboratory testing. Boreholes where especially drilled at locations where changes in foundation formations were expected. Two types of drilling were executed: machine drilling with Standard Penetration Tests (SPT) up to a depth of maximum 25 m and hand drilling for obtaining samples from shallow depths and from the track embankments.

Machine drilling was executed with a rotary drilling machine. All boreholes were fully sampled. A flight auger was used in soft none-cohesive soils to obtain re-moulded samples. Shelby tube samples were taken from soft to firm cohesive soils and double and triple tube core barrels were used in firm to hard cohesive soils and in rock.

Hand drilling was executed with a flight auger in none-cohesive soils and Shelby tubes were taken in cohesive soils.

All boreholes were described according British Standard for Engineering purposes by a trained engineering geologist at the drilling location.

The borehole information was necessary for correlation and for test samples although the drilling itself was cumbersome. The quality of available equipment was often marginal and handling of borehole cores was rough. Although this later improved during the survey. Another related problem was the availability of trained engineering geologists.

6.2 Laboratory testing

The following tests were executed on samples obtained from boreholes:

1. Consolidated un-drained triaxial tests with water-pressure measurement
2. Determination of bulk, dry and wet unit weights.
3. Unconfined Compressive Strength (UCS)
4. Pocket Vane test
5. Pocket Penetrometer
6. Water content
7. Grain size analyses
8. Hydrometer analyses
9. Attenberg Limits (Liquid Limit and Plasticity Limit)
10. Specific Gravity.

The triaxial tests were executed in Bandung while the other tests were executed in field laboratories.

The consolidated un-drained triaxial tests were executed in geotechnical laboratories in Bandung because the equipment was too extensive and too sensitive to transport to the site. The results of the tests were not always very satisfactory. This was thought to be caused by sample handling, and the long distances and time involved in transporting the samples from the site to the laboratories in Bandung. Control testing was executed in The Netherlands to maintain independent control of the tests. Odd single test results were rejected if they were not confirmed by the testing in The Netherlands.

The tests executed in the site laboratory were very successful. The massive quantities of tests were executed very satisfactory.

Some of the laboratory tests were not strictly necessary for calculating the safety factors, but were executed in large quantities for correlation purposes.

7 RESULTS/INTERPRETATION

The CPT results together with the stratigraphic information and laboratory test results from the boreholes provided reasonably detailed geological profiles along the railway lines. The limestone sands/muds were distinguished from the clay formations and obviously, (weak-) rock formations were easily recognized. In many cases it was also possible to correlate boundaries within the clay formations with the available geological maps and profiles.

The strength of the limestone sands and muds and the rock formations was more than sufficient for the foundation loads and groundparameters for limestone sands and muds and for the weak rock formations proved to be no problem and will not be discussed further.

Volcanic ash was found along all railway lines. The degree of weathering and compaction of the volcanic ashes varied from none-compacted and fresh ash to fully compacted, consolidated and weathered (ash-) clay. The fully weathered volcanic (ash-) clay is geotechnically considered as a clay. The un-weathered and fresh ash was never below the foundation but at some locations the volcanic ash contributed to the active and passive soil pressures on the abutments and piers. Undisturbed sampling of volcanic ash is, in general, impossible, because the very loose structure is immediately disturbed by the sampling tools. Determination of bulk and dry densities was possible but the strength parameters had to be estimated.

Determining ground-parameters of particular clay formations in the coastal areas was difficult. Below the different approaches are described.

7.1 Bulk and Dry Density Parameters

The bulk and dry density parameters could be established from the boreholes. The variations in bulk and dry densities within the clay formations were so small that average values of bulk and dry densities could be used.

7.2 Swelling clays

Potentially swelling clays were found at a few locations. The existing bridges were only assessed at failure state and therefore the influence of the swelling clays was not evaluated. However for new bridges swelling clays will be a major design criterium.

7.3 Sensitivity of soils

Soils sensitive to earthquake vibrations were found only at a few locations and, in general, at larger depths where they had no influence on the stability of the bridges.

7.4 Phi and Cohesion Parameters

1. Searle graph
The Searle graph (Searle 1979) correlates soil type with cone resistance and local friction. For the clay formations surveyed the soil types according to Searle for cone resistance less than 1 [MPa] did not correlate at all with the borehole information. The distances between boreholes and CPTs which were compared, were small and it was reasonably certain that the same soil layers were encountered by both borehole and CPT.

PRABUMULIH - KERTAPATI

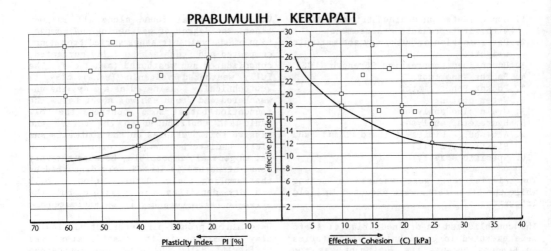

GAMBRINGAN - SURABAYA

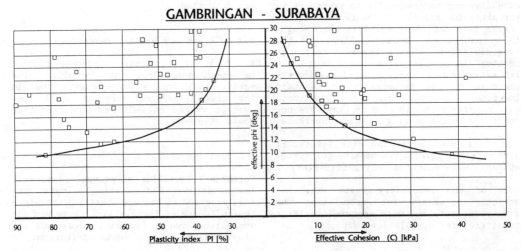

Figure 2. Examples Plasticity Index - effective phi and cohesion relation. (ch.7.4-4)

The discrepancies were most likely due to cementation between the soil grains. The Searle graph is not valid in cemented soil types.

2. Correlation between CPT cone resistance, friction and effective triaxial data

The CPT values did not correlate with the laboratory triaxial effective phi and effective cohesion values. The scattering in both sets of values was too large for a reasonable correlation.

3. Minimum effective phi and cohesion

The minimum effective phi and cohesion values measured were used in a series of calculations of the stability of the foundations. The loads on the foundations in these calculations were based on the old loading scheme. According to these calculations the majority of the bridges founded on clay worked out to be very unstable whereas the bridges in reality did not show any sign of instability. The minimum values therefore were too pessimistic for most of the bridges founded on clay.

4. Other correlations

Correlations between Unconfined Compressive Strength, pocket penetrometer, pocket vane tests and triaxial values and CPT results were all unsuccessful.

A reasonable correlation was obtained between Plasticity Index and effective phi. This relation is described in the literature (Kenney 1959, Lambe 1979). The scattering in the diagram (fig.2) shows that the relation is not optimal but accounting for the scattering by taking the lower boundary for the phi values resulted in a workable relation.

Also it was found that a plot of triaxial laboratory effective phi and cohesion for clay formations showed a relation. All phi and cohesion values of clay and ash-clay formations lie within a particular area (fig. 2). Because of the scattering in this relation the lower boundary of the range was used.

The Plasticity Index values show a good correlation with the different types of clay and ash-clay formations. To put a PI value to a particular layer was possible and through the effective phi/cohesion - PI relation it became possible to determine the ground parameters.

Along the railway lines 4 different effective phi/cohesion - PI relations were established. Safety factors calculated according to the effective phi/cohesion - PI relations under the old loading scheme, correlated very well with the state in which the bridges were at the time of the visual investigation.

8 CONCLUSIONS

Careful examination of geological information combined with Dutch Cone Penetration tests facilitates the determination of ground parameters of foundation layers. Although the Dutch Cone Penetration test might not give a strength value which can be used directly in foundation calculations, it is a very efficient and cheap test to establish different ground layers.

The relation between effective phi and cohesion, and PI values can be very useful in assessing ground-parameters. Determining Plasticity Index values is easy and cheap in contrast with relatively expensive and cumbersome triaxial tests. The authors, however, do not claim that similar relations exist in soil types different from the clay and ash-clay formations found in the coastal areas of Sumatra and Java. Further investigations will be necessary to establish the validity of effective phi/cohesion - PI relations.

ACKNOWLEDGEMENT

The authors are grateful to the Indonesian State Railways authorities (PJKA) for their support and co-operation during the fieldwork.

REFERENCES

Kenney, T.C. 1959. Discussion, Proc. ASCE, Vol. 85, No. SM3, pp. 67-79.
Lambe, T.W. & Whitman, R.V. 1979. Soil Mechanics, SI Version. 553 pp. John Wiley & Sons, New York.
Searle, I.W. 1979. The interpretation of Begemann friction jacket cone results to give soil types and design parameters in geotechnical engineering. Design parameters in geotechnical engineering. BGS, London, Vol 2, pp. 265-270.

Evaluation and microcomputer mapping on construction foundation suitability in Ningbo

Evaluation de l'aptitude de la fondation à la construction à Ningbo et cartographie par micro-ordinateur

Hu Ruilin

Institute of Hydrogeology & Engineering Geology, Chinese Academy of Geological Sciences, People's Republic of China

ABSTRACT: With the Expert Cluster and Tendency Analysis, this paper comprehensively evaluated the suitability for construction foundation in Ningbo City, and plotted A Map of Evaluation of Construction Foundation Suitability in Ningbo City (1:75,000) with the microcomputer mapping softwares(ZT) developed by the author, which set a precedent in Chinese geological mapping on microcomputer. This paper further introduces the evaluation methods and standards as well as microcomputer process, the structure of softwares and their functions.

RESUME: Grâce aux méthodes de la grappe d'expert et de l'analyse de tendance, on a synthétiquement évalué l'aptitude de la fondation à la construction de la ville de Ningbo. Avec les micrologiciels cartographiques(ZT) élaborés par lui-même, l'auteur a dressé une carte de l'évaluation de l'aptitude de la fondation à la construction en cette ville à 1/75,000, ayant fait une création dans le domaine de la cartographie géologique chinoise par micro-ordinateur. Dans cet article, les méthodes et critères de l'évaluation, le processus de cartographie par micro-ordinateur, la structure et les fonctions de logiciels ont été présentés.

1. INTRODUCTION

The construction foundation suitability means the stability and costs of formation as construction foundation. With regard to the modern construction technology level, we know that there is nearly no foundation unsuitable to construction, but we cannot easily attain a stable formation without high costs. Therefore, it is very important to chose a favorable construction site in advance.

In the past, we often rely on our qualitative observation to choose the construction site so that many results were wrong or not accurate. Hardly were geological microcomputer mapping used for foundation evaluation in China. In order to change this situation, the author were asked to quantitatively evaluate the suitability for construction foundations in Ningbo City and make a Map of Construction Foundation Suitability in Ningbo City by microcomputer as a final result.

2. METHOD AND STANDARD OF EVALUATION

2.1 Method

To analyze researched areas by some data points is the basic method of evaluation. On data points, the author adopts the Expert Cluster for evaluating its suitability which is the basis of evaluation of researched areas. The Expert Cluster can describe as follows:

$$ZF = (w1, w2, .., wi, .., wn) \times (F1, F2, .., Fi, .., Fn)^{-1}$$

where:

ZF, the individual marks of data point.

wi, the weight of evaluation factors.

Fi, the basic mark corresponding to evaluation factor(i).

n, total numbers of evaluation factors.

It means that on the basis of determination of the weights, basic marks of given evaluation factors and the grades of index, the individual marks(ZF) of the factor are first calculated, and then times them with their corresponding weights, and finally plus them with the result of total marks which shows the degree of suitability of data points.

Based on the total marks, the author adopted the Tendency Analysis to deduce the suitability of whole areas. This train of thought reflects the process from point to facet. Whole evaluation process is chiefly completed by microcomputer(Fig.1). Storing and treatment of data are carried out on the Ningbo Geotechnical Data Base built by my collaborators and me. As an independent mapping system, the author specially developed a set of softwares called as ZT which can be used for various zoning and evaluations. These softwares have combined the evaluation described above with plotting, so they could be in common use.

2.2 Standards

Considering that my current task is to evaluate the suitability of foundations of industrial and civil construction, only two kinds of concrete foundations were studied. They are Shallow Foundation and Pile Foundation. Their processes of evaluation are to first calculate respective suitabilities of them and then determine those of sites according to the combination of their foundation suitabilities.

On statistics and collective comments, the evaluation standards for two kinds of foundation were gained (Table 1, Table 2). In Table 1 and Table 2, the weights reflect the importance of evaluation factor for its suitability, which are the deductive results.

Table 1. The standard of evaluation for Shallow Foundation

Evaluation Factor	Index	Index Grade		Weight
		I	II	III
Bearing capacity of formation (KPa)	value,	<100	100-200	>200
	marks*,	10	60	100	0.26
Depth of bearing stratum(m)	value,	>3	1-3	<1
	marks,	10	60	100	0.21
Thickness of bearing stratum (m)	value,	<1.6	1.6-6.5	>6.5
	marks,	10	60	100	0.15
Coefficient of compressibility of bearing stratum (KPa)$^{-1}$	value,	>0.001	0.001-0.0005	<0.0005
	marks,	10	60	100	0.10
Coefficient of compressibility of stratum under bearing stratum (KPa)$^{-1}$	value,	>0.0017	0.0017-0.0007	<0.0007
	marks,	10	60	100	0.05
Thickness of soft clay(m)	value,	>7	2-7	<2
	marks,	10	60	100	0.14
morphologic unit	Types,	littoral	valley	plain
	marks,	10	60	100	0.09

* means basic marks.

Table 2. The standard of evaluation for Pile Foundation

Evaluation Factor	Index	Index Grade		Weight
		I	II	III
Bearing capacity of formation (KPa)	value,	<1800	1800-3200	>3200
	marks,	10	60	100	0.28
Depth of bearing stratum(m)	value,	>30	15-30	<15
	marks,	10	60	100	0.24
Thickness of bearing stratum (m)	value,	<3	3-12	>12
	marks,	10	60	100	0.18
Coefficient of compressibility of bearing stratum (KPa)$^{-1}$	value,	>0.0007	0.00025-0.0007	<0.00025
	marks,	10	60	100	0.12
Coefficient of compressibility of stratum under bearing stratum (KPa)$^{-1}$	value,	>0.0007	0.0002-0.0007	<0.0002
	marks,	10	60	100	0.07
Bearing capacity of the stratum under bearing stratum (KPa)	value,	<150	150-230	>230
	marks,	10	60	100	0.08
morphologic unit	Types,	littoral	valley	plain
	marks,	10	60	100	0.03

The basic marks embodies the quality of the foundation corresponding to the grades of index.

To two kinds of foundations, seven factors are respectively chosen on Ningbo's actual conditions. In fact, only the depth of bearing stratum absolutely incarnate the cost of construction projects.

After repeatedly counting up the total marks of all points, the dividing marks corresponding to grades of suitability were gained as follows,

1. Unsuitable foundation,
   Shallow Foundation, <50   Pile Foundation, <53
2. Suitable foundation,
   Shallow Foundation, 50-80   Pile Foundation, 53-85

3.Excellent foundation,
  Shallow Foundation, >80   Pile Foundation, >85

Table 3. Principle for determining the suitability of sites

Suitability	Combination of foundations	
	Shallow foundation	Pile foundation
Site for com-prehensive de-velopment	Excellent or Suitable	Excellent or Suitable
Site for Sha-llow founda-tion	Excellent or Suitable	Unsuitable
Site for Pile foundation	Unsuitable	Excellent or Suitable
Unsuitable site for construc-tion	Unsuitable	Unsuitable

Based on the evaluating results of the Shallow Foundations and the Pile Foundations, we can easily gain the suitability of site in the same manner as illustrated in Table 3.

3. A BRIEF INTRODUCTION OF MAPPING SOFTWARES (ZT)

3.1 Overall structure

This set of mapping softwares are chiefly composed of four parts as follows,
Part one, Operating Management System. Its function is to offer the running channels and free choices of the operation methods as well as the graphic windows for users. Besides, it could control each function modules.
Part two, Data Editing System. Used for building, editing or rearranging data files.
Part three, Numerical Analysis. Used for sorting data points and numbering mapping areas.
Part four, Mapping System. Responsible for exchanging, editing or plotting graphs.
The structure of the four parts and their basic program files are illustrated in Fig.2.
These softwares were developed on the IBM-PC/XT computer and its compatible computers. The environment of software is PC-DOS(Version 3.0 or above). The languages adopted in programs are both BASIC(V3.0) and FORTRAN(V3.14). The external device for outputting map is SPL-400 PLOTTER (Made in Japan).

3.2 Main function modules

CONTROLLING MODULE(ZT) and MENU MODULE(ZT0),As

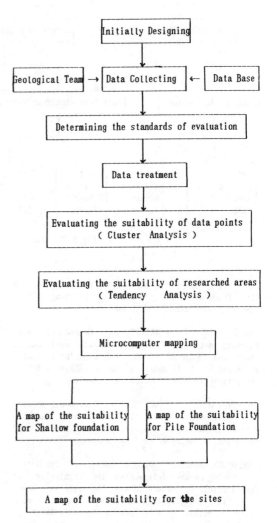

Fig. 1 Evaluating and mapping process

everyone knows, BASIC is rich in colors and good at graph on screen, while FORTRAN is quick in scientific calculation. In order to give full play to respective advantages, the author adopted appropriate language to programme for different purposes. For that the programs written by two languages are combined as a whole and easy to select running contents and forms, the author designed these two modules. The communication between the Batch File(ZT) and the Menu(ZT0) with the subjunctive files are used for this purpose.

DATA EDITING MODULE(ZT6),It adopts the conversational inputting on keyboards to build or edit contents of data files and store them in arrays. Its main functions include building a new

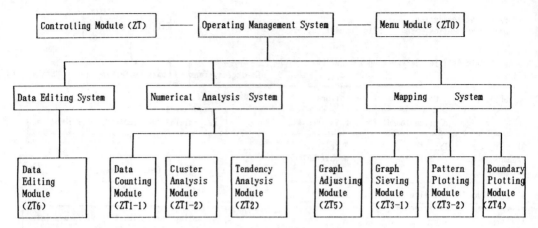

Fig. 2 The structure of mapping software

data file, correcting a old file, composing a new file from a old file, listing or printing files, executing DOS command and so on.

DATA COUNTING MODULE(ZT1-1): Method of data counting is to show the values on data distribution chart and histogram and then determine their limits according to their numerical ranges. These graphics can illustrate both on screen and plotter.

CLUSTER ANALYSIS MODULE(ZT1-2): This module is used for comprehensive classification of data, and its method is the Expert Cluster. Analysis results are the bases of zoning or sorting in areas.

TENDENCY ANALYSIS MODULE(ZT2):Regional evaluation or zoning means determining the regularity of distribution in certain ways on the basis of known data. It is a process from point to facet. Common analysis method is to deduce on logical thinking, which is often seriously related to researcher's knowledge and the precision of data, and has obvious contingency. However, quantitative Tendency Analysis could remedy these defects. Tendency Analysis is actually the direct extending from the regression analysis.

GRAPH ADJUSTING ON SCREEN(ZT5):Because of the limitations of counting analysis and the tendency analysis, calculated or evaluated results often can not exactly tally with the actual situation. It is necessary to do some of adjustment or supplement. Adjustment on screen is the most audio-visual, most convenient and quickest way, so the author developed this module to do these jobs. So long as pressing some keys, such as "C" or cursors, we are much easy to correct the dividing lines or types of some points or areas with a good result.

GRAPH SIEVING MODULE(ZT3-1):It is a common knowledge that many maps need to be plotted in several steps. For example, because plains and

mountain areas are completely different in engineering geological properties and have different standards for suitability of construction foundation, so we should first respectively draw their maps and then overlap them. On the other hands, such topography as sea, river, lake etc. often need to be separated from or drawn on map in order to treat them with some of particular devices (e.g. digitizer). It is asked that a all-round mapping software must has the function called as graph sieving.

In my software, the manner of sieving graph is to utilize its coordinates to imitate the boundary of sieved area with the result of its outline. Once the point plotted goes into outline range, plotting pen will automatically uplift and leave out it to attain the goal for graph sieving. Selected method for modelling outline is the SPLINE.

PATTERN PLOTTING MODULE(ZT3-2):The pattern plotting module, that means drawing the pattern of map, has two sets of patterns: One is block with different colors, and another is the numeral marks, which could be freely chosen. Different colors or numerals expresses different types or zones or degrees of suitability. Patterns will be plotted in the order from the bottom to the top and the interval of each scanning lines is 5 mm. The width of colored block is 5 mm, too. Now, SPL-400 plotter has only 6 pens, with which we can make up more than 30 varieties of patterns.

BOUNDARY PLOTTING MODULE(ZT4):It is used for plotting the boundaries of zones. It has the function for identifying the topographical areas. Since it adopts dense nets to draw curves, the boundary is rather smooth and beautiful.

3.2 The outstanding features of ZT

(1). It can be in common use. It can be used for great varieties of zoning or evaluating based on

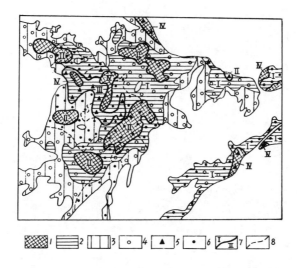

Photo 1.   A map of construction foundation suitability in Ningbo City (1:75,000)

1-unsuitable for pile foundation 2-suitable for pile foundation 3-excellent
pile foundation   4-excellent shallow foundation   5-unsuitable for shallow
foundation 6-suitable for shallow foundation 7-site's suitability and its
dividing line 8-the dividing line of foundation suitability

original data,   such as the zoning of   crustal
regional stability, engineering geological zoning,
zoning of groundwater quality and so on.
   (2). Reasonable results. To great extents, it get
rid of the interference of some stochastic events to
analysis process with a comprehensive result.
   (3). Easy to use. User can freely decide the scale
and scope of map. You can choose different running
channels or devices for outputting at will.
   (4). Popularity. Without high-grade   plotting
device, just a small-scale plotter could finish whole
mapping jobs. This situation is much suitable for the
geology trades in most of developing countries.
   (5). Graph is beautiful and easy to read. The size
of single map is 27x35 cm^2. Larger map can be plotted
in different pages, which could obtain the same
effect as a whole.

4. EVALUATION RESULTS

With the softwares, the author successfully drew up A
MAP   OF   EVALUATION OF   CONSTRUCTION   FOUNDATION

SUITABILITY IN NINGBO CITY(1:75,000)(Photo 1). At the
appraisal meeting held by the Chinese Academy of
Geological Sciences, it won widespread   acclaim.
Because original map is too big to take a clear
picture, the map on photo 1 is the simplified one
from the original.

5. CONCLUSION

With the mathematical method and   microcomputer
technology, this paper evaluated the suitability for
construction foundation in Ningbo City, and developed
a   set   of   geological   mapping   softwares   on
microcomputer. Three grades of suitability   for
construction foundation will be helpful to guide the
planning of Ningbo's construction. Although   the
mapping software(ZT) is far from perfection,   its
first application indeed set a precedent in Chinese
geological mapping on microcomputer. I am sure that
it will become more and more perfect under our
future research and with the help of   advanced
countries.

647

Anisotrophy assessment of different rocks

Estimation des anisotropies de différents rochers

J.M.Kate

Civil Engineering Department, Indian Institute of Technology, Delhi, New Delhi, India

ABSTRACT : Experimental investigation has been carried out to assess the anisotropy of rocks of different genesis. The cores from the rock blocks were drilled in three mutually perpendicular directions including one along the bedding planes. The anisotrophy of these rocks were assessed in relation to their geophysical and strength behaviour. Attempt has also been made to understand correlations between certain properties of rocks.

RESUME : l' Etude expérimentale ait fait pour éstimer l'ansisotropies des rochers d'origines divers. Les carottes des rochers etait percé en trois directions mutuellement perpendiculaire, y compris la direction de plane de couche des rochers. l' Anisotropies de rochers etait estimé en terme de ces proprietés géophysiques et ces comportements rigidités. On a essayai à comprendre les corrélations entre les divers propriétés de rochers.

1 INTRODUCTION

The knowledge about engineering behaviour of rocks is of paramount importance while designing major strcutures like high dams, deep tunnels, underground storage tanks and many others. In majority of the designs the idealized conditions of geological media viz. homogeneous, isotropic and linear elastic behaviour is assumed. However, in practice these are seldom encountered. Most of the rocks exhibit certain degree of anisotrophy in their behaviour. For design purposes it is very essential to evaluate the strengths viz. compressive, tensile and shear of rocks. Over estimation of rock strength may lead to structural disaster. In order to carry out safe structural designs it is extremely important to identify the direction along which the rock mass exhibits the lowest strength and the orientation of such weak planes with load axis. This emphasizes very clearly the need to conduct anisotropy study on rocks, which alone can provide detailed information about their strength and planes of weaknesses to be considered for safe designs. In view of this, the present laboratory study were conducted to assess an intrinsic anisotropy of some of the rocks of different genesis viz. sedimentary, igneous and metamorphic rocks.

2 OVERVIEW ON STRENGTH ANISOTROPY

The geological media always possess certain degree of anisotropy may be intrinsic, induced or combination of these. The primary anisotropy in rock is brought about by preferential orientation during crystallisation or by recrystallisation during the genetic process such as sedimentation or metamorphism. The presence of discontinuties such as bedding planes, foliations, macro and micro fissures, fault, joint and fracture planes etc. and their orientation in relation to load axis make rock to exhibit anisotropic behaviour.

Many research workers have proposed a number of empirical failure criteria for anisotropic rocks under certain basic assumptions. Mclamore and Gray (1967) conducted anisotropy studies on sedimentary rocks and proposed a failure theory by assuming that the material fails in shear and has a variable cohesion but constant value of internal friction. Based on an analogy with the nonlinear failure envelope predicted by classical Griffith theory for plane compression, Hoek and Brown (1980) developed failure criterion for both istoropic and anisotropic rocks. The nonlinear behaviour of rocks also can be predicted by the above failure criterion.

Ramamurthy (1986) suggested a failure criterion which can predict strength for anisotropic or jointed rock masses covering the entire brittle and ductile regions. The parameters qualifying and quantifying the type of anisotropy were reviewed by Singh et al (1989) and based on this, suggested a classification of anisotropy related with rock genesis. They have also proposed a relationship to estimate the compressive strength of anisotropic rocks at various orientation of planes of weakness on the basis of known strength only at three orientations.

3 EXPERIMENTAL PROGRAMME

3.1 Rocks studied

The present investigation was conducted on rocks of different origins namely sedimentary, igneous and metamorphic. The rock blocks were collected from different parts of India and are referred here by the names of places they belong. Petrographic analysis of these rocks were carried out in the laboratory. In order to understand physical anisotropy (size, shape and orientation of grains and pore spaces etc. in different directions) at micro level, scanning electron microscopic studies were conducted on all these representative rock samples. Brief description of these rocks are given below.

Sedimentary rocks

Dholpur sandstone (DS-1) and Salal dolomite (SD-2) are rocks of sedimentary origin. The former rock DS-1 is a pink coloured which exhibits perfect bedding planes. It shows coarse quartz grains with ferrugineous cementing material, micro cracks and large void spaces. The latter SD-2 looks reddish brown with bedding planes not distinctly visible. It consists of coarse tetrahedron crystals of dolomite, coarse grains with small void spaces and fine cracks between the grains.

Igneous rocks

The three igneous rocks studied here are Arpa granite (AG-3), Bodhaghat basalt (BB-4) and Onkareshwar basalt (OB-5). The rock AG-3 appears yellowish white in colour and compact. It contains mica, quartz and felspar with coarse grains and very small pore spaces. It exhibits good visible cleavage in different directions. The dark grey coloured rock BB-4 has fine grains with small voids and does not exhibit any regular shape of crystals. The another basaltic rock OB-5 is light red in colour with micaceous minerals and small pore spaces.

Metamorphic rocks

These rocks are chamera quartzitic phyllites (CP-6), Dulhasti micaceous phyllites (DP-7) and Jamrani gneiss (JG-8). The light grey coloured rock CP-6 predominantly contains quartz, kaolinite clay and mica. The flaky grains of silt and clay size with loose bonding between the grains and small open voids are the noticeable features of this rock. The rock DP-7 has a dark grey colour and predominantly contains mica. The typical scanning electron micrographs of the rock samples taken in three different directions are shown in Fig. 1. The flaky hexagonal kaolinite clay particles can be seen clearly in Fig. 1(a) whereas, the orientation of grains in other two directions as shown in Fig. 1 (b) and (c) illustrate intrusion of finer clays into void spaces. Jamrani gnciss is also a dark grey coloured rock indicating

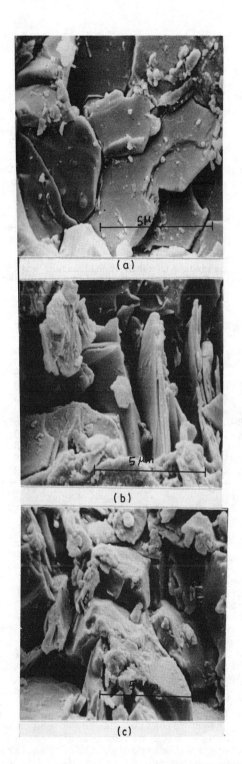

Fig.1 Micrographs of rock DP-7 in different directions

presence of quarz, mica and gnessic minerals. It shows fine grained particles with very small voids but extensive microcracks. All these three metamorphic rocks exhibit thin bedding planes.

3.2 Specimen preparation

Large blocks of rock collected from the field after blasting were cut into smaller ones and were brought to laboratory. The cores of 38 mm diameter (D) from these rock blocks were drilled in 3 mutually perpendicular directions including one across the bedding plane or plane of discontinuity with laboratory core drilling machine. All the specimens of a particular rock were drilled out from the same rock block to maintain uniformity. These core samples were cut to different lengths (L) depending upon the test requirements. The rock specimens thus prepared were conforming to tolerances as per ISRM (Brown, 1981).

The rock specimens were dried up initially in an oven at 105 $^{\circ}$C and were then cooled down to ambient temperature in a dry desiccator. For some of the tests to be conducted on fully saturated rock, the saturation was achieved by keeping them in desiccator maintaining 100% relative humidity till moisture equilibrium was reached.

3.3 Tests conducted

Tests were conducted on all these rock specimens drilled along three mutually perpendicular directions to assess their intrinsic anisotropy for strength and geophysical behaviour. Certain tests to determine their physical properties viz. density (both in dry and saturated conditions), porosity, water absorption and grain specific gravity were conducted as per ISRM suggested methods.

Strength tests

All the strength tests were conducted on completely dry rock samples. Uniaxial compressive strength (σc) and Brazilian tensile strength tests were carried out as per suggested methods of ISRM (Brown, 1981) on specimens with L/D ratios of 2:1 and 0.5:1 respectively. The tensile strength was also obtained indirectly by compressing cylindrical rock specimens between two point contacts (referred here as point loading tensile strength, σ_{tp}) suggested by Reichmuth (1962). The punching shear device (Jumikis, 1982) was specially designed and shear strengths (τ_p) of these rocks were obtained.

Geophysical tests

The electrical resistivity (ρ) and P-wave velocity (V_p) were determined in both dry and fully saturated conditions of these rocks. The electrical resistivity was determined by holding the opposite faces of cylindrical rock specimen firmly between two copper discs of same dia. and measuring the resistance (R) offered by rock with Mega-ohm-meter. From the known cross sectional area (A) of the rock specimen, the ρ is calculated by the following equation.

$$\rho = R.A/L \qquad \ldots\ldots (1)$$

The ultrasonic pulse technique (ISRM : Brown, 1981; Kate & Koundanya, 1985) was adopted to determine P-wave velocity of these rocks

4 RESULTS AND DISCUSSIONS

The three mutually perpendicular directions along which the rock cores were drilled have been referred here as X, Y and Z directions for clarity of presentation. The X-direction is normal to the bedding plane (or plane of discontinuity, Y-Z plane) and similarly Y and Z-directions (normal to X-Z and X-Y planes respectively) indicate the directions along which cores were obtained. Subsequently, the suffixes x, y and z have been used to designate properties of rock drilled along X, Y and Z directions respectively. It may also be mentioned here that each value

reported in the result is an average of atleast three determinations.

4.1 Physical properties

The physical properties of all these rocks are reported in Table 1. Amongst these, Dholpur sandstone (DS-1) is the one with lowest dry density, highest porosity and water absorption. The grain specific gravity values (not reported in table) of these rocks have been observed to range from a minimum of 2.65 for Chamera quartzitic phyllites to a maximum of 2.76 for Salal dolomite. For rest of the rocks it lies within a value of 2.69 ± 0.01.

Table 1.Physical properties of rocks

Rock	Density,kg/m^3		porosity	water
	Dry	Sat.	%	absorption %
DS-1	2450	2650	11.40	3.80
SD-2	2710	2740	2.50	0.55
AG-3	2580	2590	0.74	0.32
BB-4	2650	2680	1.84	0.48
CB-5	2640	2650	2.04	0.38
CP-6	2580	2600	0.94	0.78
DP-7	2640	2660	1.60	0.29
JG-8	2640	2660	1.20	0.68

4.2 Strength and geophysical properties

Various strengths and geophysical properties of rocks drilled along X-direction are presented in Table 2. The rock SD-2 exhibit highest strengths in uniaxial compression, Brazilian tensile and point loading tensile tests whereas, in shear strength BB-4 is strongest. The weakest amongst these seems to be DP-7 in compression and shear whereas, DS-1 in tensile strength. The lowest tensile strength observed in DS-1 appears to be in agreement with the poor cementing of grains as noticed during petrographic analysis and high porosity. Amongst the two methods of determining tensile strengths indirectly, the Brazilian test appears to provide higher values than point loading for the

Table 2 Strength and geophysical properties

Rock	σ_{c_x} MPa	σ_{tb_x} MPa	σ_{tp_x} MPa	τ_{p_x} MPa	ρ_x ohm.m		V_{p_x} $\times 10^3$ m/sec	
					$\times 10^5$ Dry	$\times 10^3$ Sat.	Dry	Sat.
DS-1	80.5	7.4	5.2	24.5	39.1	32.9	1.94	2.65
SD-2	143.5	17.1	11.5	46.3	34.5	24.5	2.43	2.93
AG-3	115.9	9.7	7.8	36.7	31.6	26.3	2.74	3.80
BB-4	141.4	14.3	9.8	48.4	29.1	21.1	2.40	2.88
OB-5	86.7	9.1	6.4	25.5	35.0	24.4	2.74	3.75
CP-6	97.4	11.4	8.5	31.8	25.5	16.9	2.00	2.70
DP-7	79.4	12.5	8.5	19.8	29.0	19.5	2.23	2.67
JG-8	137.1	15.5	10.5	43.0	42.1	24.2	3.6	3.88

same rock. The shapes of stress-strain curves obtained during uniaxial compression tests for these rocks indicated brittle failure. The typical stress-strain curves for rock OB-5 is illustrated in Fig. 2. It is interesting to note from Table 2 that, on saturation the electrical resistivity of rocks decreases exponentially and contrary to this, P-wave velocity increases considerably. The increase in V_p on

saturation has been observed to range between a maximum of 38% for rock AG-3 and a minimum of 8% for JG-8. The rocks DS-1 and JG-8 possess the lowest and highest V_p respectively in dry and fully saturated conditions. Similarly, in both the conditions the lowest value of ρ is exhibited by rock CP-6. However, the highest values of ρ have been observed for JG-8 and DS-1 in dry and saturated states respectively.

4.3 Anisotropy study

As a simplified approach to understand anisotropy, herein the properties of samples drilled along X-direction has been adopted as representatives and the properties of samples in other two directions have been expressed in terms of ratios as presented in Table 3 and Table 4 for strengths and geophysical behaviour respectively. It is interesting to note from Table 3 that for all the rocks, these ratios are more than 1 for compressive and shear strengths whereas, less than 1 for tensile strengths. This clearly indicate that the rocks drilled across the bedding plane/discontinuity are strong in compression and shear but weak in tension. Irrespective of rock origin, this has been noticed in the present study. The highest

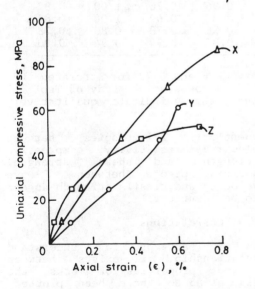

Fig.2 A typical stress-strain behaviour of OB-5 in compression test

Table 3. Anisotropy ratios for dry strengths

Rock	Uniaxial Compressive		Brazilian tensile		Point loading tensile		Punching shear	
	$\sigma_{c_x}/\sigma_{c_y}$	$\sigma_{c_x}/\sigma_{c_z}$	$\sigma_{tb_x}/\sigma_{tb_y}$	$\sigma_{tb_x}/\sigma_{tb_z}$	$\sigma_{tp_x}/\sigma_{tp_y}$	$\sigma_{tp_x}/\sigma_{tp_z}$	τ_{p_x}/τ_{p_y}	τ_{p_x}/τ_{p_z}
DS-1	1.16	1.22	0.82	0.84	0.83	0.91	1.23	1.64
SD-2	1.33	1.39	0.73	0.63	0.79	0.86	1.43	1.72
AG-3	1.14	1.11	0.84	0.83	0.81	0.74	1.30	1.47
BB-4	1.61	1.82	0.87	0:79	0.91	0.78	1.72	2.44
OB-5	1.41	1.41	0.99	0.88	0.97	0.89	1.10	1.59
CP-6	1.22	1.52	0.91	0.78	0.96	0.87	1.59	2.38
DP-7	1.16	1.56	0.70	0.54	0.74	0.68	1.06	1.82
JG-8	1.12	1.19	0.86	0.77	1.03	0.92	1.19	1.64

Table 4. Anisotropy ratios of geophysical properties

Rock	Electrical resistivity				P-Wave velocity			
	Dry		Saturated		Dry		Saturated	
	ρ_x/ρ_y	ρ_x/ρ_z	ρ_x/ρ_y	ρ_x/ρ_z	V_{p_x}/V_{p_y}	V_{p_x}/V_{p_z}	V_{p_x}/V_{p_y}	V_{p_x}/V_{p_z}
DS-1	1.20	1.35	1.57	1.57	0.83	0.78	0.82	0.76
SD-2	1.25	1.27	1.08	1.03	0.75	0.70	0.90	0.83
AG-3	1.14	1.32	1.37	1.35	0.96	0.92	0.98	0.98
BB-4	1.30	1.06	1.20	1.04	0.85	0.72	0.87	0.72
OB-5	1.56	1.22	1.22	1.09	0.96	0.76	1.00	0.87
CP-6	1.23	1.04	1.25	1.14	0.82	0.69	1.00	0.82
DP-7	1.19	1.08	1.12	1.09	0.86	0.78	0.86	0.78
JG-8	1.09	1.02	1.20	1.09	0.97	0.94	0.94	0.90

values of strength anisotropy ratio observed are 1.82 in compression for rock BB-4 and 2.44 in shear for OB-5. The lowest ratios are 0.54 and 0.68 in Brazilian and Point loading tensile tests respectively exhibited by rock DP-7 alone. The magnitudes of these ratio indicate that all these rocks fall under a category of low to medium anisotropy.

The anisotropy ratios in respect of electrical resistivity are more than 1 for both dry and saturated rocks as seen from Table 4. The highest value of this ratio under dry state is 1.56 for OB-5 and 1.57 for DS-1 in saturated condition. Unlike electrical resistivity the anisotropy ratios for P-wave velocity are less than 1 for both dry and saturated rocks. The lowest values of these ratios are 0.69 for dry CP-6 and 0.72 for saturated BB-4. A comparative study of Table 3 and Table 4 indicate qualitatively identical anisotropies between electrical resistivity, compressive strength and shear strength. Whereas, those between P-wave velocity and tensile strength appear to be identical.

4.4 Correlations

In an attempt to understand relationship, if exists, between various properties of rocks, the figures 3 to 6 have been plotted. The figures 3 and 4 illustrate variations of σ_{tp} with σ_c and

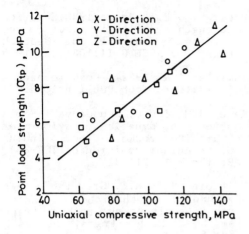

Fig.3 Variation between σ_{tp} and σ

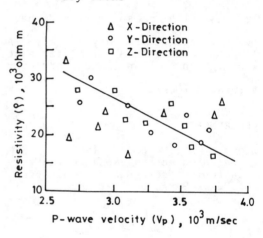

Fig.5 Variation of ρ with V_p for dry rocks

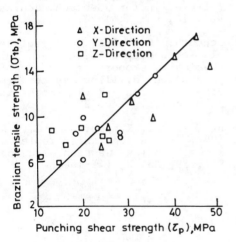

Fig.4 Relationship of σ_{tb} with τ_p

Fig.6 Plot between ρ and V_p for saturated rocks.

Brazilian tensile (σ_{tb}) with τ_p respectively. Although the points are scattered on these figures, on an average a linear increase with each other may be conceived. The variations bewteen ρ and V_p are presented in Fig. 5 and Fig. 6 respectively for dry and saturated rocks. In general, a linear variation may also be conceived between these properties but interestingly with a reverse trend. The rocks when dry exhibit increase in ρ with V_p whereas, their saturation causes decrease in ρ with increase in V_p.

5 CONCLUSIONS

The following conclusions have been summarised.

1. The strength anisotropy ratios indicate that the rocks exhibit weakness in both compression and shear along the bedding planes/planes of discontinuity. Whereas, same rocks are weak in tension across the bedding planes.

2. The anisotropies in respect of electrical resistivity, compressive strength and shear strength appear to be qualitatively identical with each other whereas, those between P-wave velocity and tensile strength seems to be identical.

3. A linear increase in σ_{tp} with σ_c, σ_{tb} with τ_p and ρ with V_p have been conceived for dry rocks. A linear decrease in ρ with increase in V_p has been conceived for saturated rocks.

REFERENCES

Brown, E.T. (ed.) 1981. ISRM suggested methods-Rock characterization, testing and monitoring. 1-221. Pergamon, Oxford.

Hoek, E. & E.T. Brown 1980. Empirical strength criteria for rock massess. Journal of Geotech. Engg. Div.ASCE,G II,106:1013-1035.

Jumikis, A.R. 1982. Rock mechanics. 1-613.Trans Tech Publications.USA.

Kate, J.M. & V.U. Koundanya 1985. A study on dynamic behaviour of Quartzite rocks. Proc. Indian Geotechnical conference (IGC-85), Roorkee, 1: 311-315.

Mclamore, R. & K.E. Gray 1967. The mechanical behaviour of anisotropic sedimentary rocks. Trans. Am. Soc. Mech. Engrs., Series B, 89: 62-76.

Ramamurthy, T. 1986. Stability of rock mass. Eighth IGS annual lecture, Indian Geotech. Journal, Vol. 16, No. 1, 1-74.

Reichmuth, D.R. 1962. correlation of force-displacement data with physical properties of rock for percussion drilling system.Proc. 5th Symp. on Rock mechanics, Minneapolis, Minn. 33-59.

Singh, J., T. Ramamurthy & G.V. Rao 1989. Strength anisotropies in Rocks. Indian Geotech. Journal, Vol. 19, No. 2, 147-166.

The zoning system for assessment of weathering states of granites
Le système de classification pour l'estimation de l'état d'altération de granites

Kehe Wei
Hehai University, Nanjing, People's Republic of China

Dianxuan Liu
North China Institute of Water Conservancy and Hydro-electric Power, Handan, People's Republic of China

ABSTRCT: The significance of qualitative assessment in weathering states(or degree) zoning is explained on the basis of site investigations of the characteristics of weathered granites in South Fujian,P.R.China. Three indices(point load strength index,longitunal wave velocity and rebound number) which can reflect the changes of weathering states well in this area are picked out by statistics on a large number of site test data. A comprehensive index (Z_f) for classification of weathered states of granites —— the comprehensive desending rate of weathering index is given on the basis of hierachical cluster analysis.Furthermore, a direct method recognizing weathering states by fuzzy pattern is given as taking Z_f as the criterion. A weathering states zoning system is built,which begins with qualitative zoning and ends as fuzzy pattern direct recognition.

RESUME: La signification de l'assertion qualificative dans la zone desagregee de granits, qui traite dans ce papier, est basée sur l'investigation sur place pour les caracthérisations désagrégées de granits dans la zone du sud de la province de Fujian de la R.P.C. Trois indices (index de soloidarité de charge de pointe, vitesse des ondes longitudinales et le nombre de rebondissement), qui peuvent refléter les changes de la désagrégation, ont été choisis par la statistique des données mesurés. Un index compréhensif (Z_f) pour la classification de la zone désagrégée de granit — le taux abaissé compréhensif de l'index de désagrégation a été donné dans la base de l'analyse du groupement hiérarchique. De plus, une méthode directe reconnaissant la zone désagrégée par une manière tatillonne a été donnée pour prendre Z_f comme le critère. Un système de zonation de la zone désagrégée a été bâti, qui prend la zonation qualificative comme le commencement et la reconnaissance directe d'une manière talilonne comme le lefinissage.

1. INTRODUCTION

Over a long period of time, a lot of studies on rock weathering classification have been carried out and many principles and methods have been proposed both domestically and abroad. But the principles and methods for accurate zoning different rock weathering sta-tes have not been established because many elements have effects on rock weathering, the nature,behavious and pattern of rock weather -ing are very complicated,it is very difficult to classify the various states of weathering with one principle,index or method usually,and there are a lot of difficulties to take sam-ples and have laboratory tests on highly-wea-thered rocks with general methods. Summarily, the principles having been established can be classified into two categories essentially,the qualitative and the quantitative.

Taking the geological features of rock weathering as the bases,qualitative zoning (or

"classification"), being simple and visual, is still widely used in engineering nowadays. But it is difficult to meet the requirements of important projects for its poor precision and unavoidable man-made fortuities.

With the development of test techniques, some studies on quantitative classification have been carried out by means of various tests since 1960's. Many quantitative asse-ssment methods,such as single-index method, multi-index method,and so on, have been pro-posed. But most of them have not become pra-ctical for following inadequacies:

(1) It is difficult for single-index method to reflect the engineering characteristics of rocks in every aspects. And it is affected by the chosen-index's susceptibility reflecting different weathering states;

(2) It is difficult to take samples or have laboratory tests for certain indices(such as uniaxial compression strength) in full range of rock weathering with multi-index methods;

(3) The weights of different indices are not
or fairly concerned in the multi-index
methods;

We have carried out a systematic study on
the weathering features and the quantitative
assessment of weathering states of Yansan
Period granites in South Fujian,P.R.China for
several years. We have proposed a "Zoning
system" for quantitative assessment of weathering states of granites on base of qualitative
assessment.

2. THE PRINCIPLES FOR QUALITATIVE
   CLASSIFICATION

Qualitative classification is the essential
geological base for quantitative classification, though it has certain defects. To overcome the man-made voluntary differences in
understanding and recognition,when making the
principles, we must begin with geological
investigation, and stress the most significant
features of each zone, and take the other
characteristics as supplementary ones, then we
will make the principles which we draw up more
practical, more accurate, and easily mastered.

We have learned from the site investigations
that the most reliable markings indicating the
degree of weathering are the growth of weather
-ed micro-joints in rocks,ie, the change of
structure feature.The degree of mineral variation,water-properties, the possibility of
being broken with hands, are also reliable
indicators. The colour and the other indices
of weathered rocks can be used as supplementary indices only. We made table 1,features
for qualitative assessment of weathering
states in granites,in terms of the above principles.

3. A ZONING SYSTEM FOR ASSESSMENT OF
   WEATHERING STATES OF GRANITES

3.1 Data of tests

We have chosen seven typical profiles of
weathered granites in South Fujian,P.R.China,
and systematically carried out a test study
on them both in laboratory and in situ for a
quantitative assessment of weathering states.
The data of point-load test ($I_{s(50)}$),longi-
tunal wave velocity test(V_p) and schmidt re-
ound test(R) are showed in table 2.

3.2 Determination of the indices for quanti-
   tative assessment of weathering states

The test data show that the changes of three
indices, $I_{s(50)}$,V_p and R all follow the

pattern that they decrease gradually with the
increasing in intensity of weathering. Therefore, we chose these three indices as the
indices for quantitative assessment of weather
-ing states of granites.   Althouth all are
granite, as we learned, rocks in different
area certainly create differences on their
absolute indices. We suggest that make a new
index,"the relative desending rate of index",
for the assessment of degrees of weathering.
It is defined as:

$$F = \frac{M_{Fresh} - M_{Weathered}}{M_{Fresh}} \times 100\% \qquad (1)$$

for a rock, there:
  F:the desending rate of the weathering
    index;
  M_{Fresh} :the average value of the index
    of the fresh rock;
  $M_{Weathered}$:the value of the index of
    the weathered rock.
The different indices are named as: the
desending rate of point-load strength index
of weathered rock($F_{Is(50)}$), the desending
rate of longitudinal wave velocity of weathered rock (F_{Vp}), and the desending rate of
rebound number of weathered rock (F_R). We
calculated the desending rate of the indices
of the seven typical profiles in table 2 and
gave the results in table 3. In table 3, we
can find that all of the three above indices
($F_{Is(50)}$,F_{Vp},F_R) increase remarkably with the
increasing in intensity of weathering, which
reflects the patterns of various weathered
stages.

3.3 The application of hierachical cluster
   analysis in weathered state zoning

Data in table 2 are gotten after the accom-
plishment of qualitative classification in
terms of the geological features in table 1.
Does the qualitaive classification reflect
the same nature as the test indices do ? We
made a hierachical cluster analysis on the
seven profiles in order to examine our conclusions.

Hierachical cluster analysis means following
rules given in advance to compare the similarity or relationship of samples (weathering
zones) in characteristics and to sort out the
samples having similar features.

With the considering of that what we study
is a problem of regional weathering classification and there are several profiles being
studied systematically, we use the average-
distance hierachical cluster analysis which
produces a good clustering effect to carry
out the cluster analysis in order to compare
and determine which state the weathering
zones of each profiles belong to.

Fig.1 is the hierachical diagram of cluster

Table 1. Features for qualitative assessment of weathering states of granite

Zone		Geological feature of each weathered zone			
		Mineral component, structure	Colour	Simple test	Core
Completely weathered* (residual overburden) (V)	Upper (homogeneous red horizon) (V-3)	Majority is kaolinite and quartz	Brick-red	Plastic--hard plastic	Monolith can be taken
	Middle (mottled horizon) (V-2)	Majority is kaolinite and quartz, coarse clast increase downwards	Brick-red, pur-plish, red with mottles and strips	Plastic--hard plastic	Monolith can be taken
	Lower(pallid horizon with sand and gravels) (V-1)	Majority is quartz and kaoli-nized feldsper, the outward appearance of the texture of parent rock is still intact	Greyish-white or brownish yellow	Slightly cohesiveness in wet state, crisp when dry; friable in hand, slaking in water	Monolith can be taken with difficu-lties
Hightly weathered (IV)		Majority is quartz and kaolinized feldspar, weath ered joints grow well in net pattern	Discol oured mostly, greyish yellow or brow-nish yellow	Specimen can be broken in hands with force; sound of friable can be heard when gripped in hand; broken clastics but not slaked in water	Core samples broken into 0.2-1.5cm breccia and coarse sand
Moderately weathered(III)		Majority is quartz and feldspar, kaoli-nized on surfaces of plagioclase, less weathered in potash feldspar with disconti-nuous weathered joints	Discolou -red sli -ghtly, brownish yellow , wea-thered plagio-clase changed into white	Specimen can not be brocken with hands; kaolinized feldspar can be carved with knife; friable, sound is not clear when hammered; not broken clastics in water	Upper, core samples are short cylinders less then 10cm, lower , more complete
Slightly weathered(II)		Majority is quatrtz and feldspar, few weathered joints, in more complete structure	With the colour of fresh rock essentia -lly, rust-ye-llow halo is present only around dark minerals	Not friable, with clear sound when hammered	Core samples are essentially complete
Fresh(I)		No visible sign of weathering of rock material, complete structure, dense and hard	No disco -loura -tion, no sign of weathe-ring	Not friable, with clear sound when hammered	Complete core samples

* Being equal to both completely weathered and residual soil together in the weathering classification standards of ISRM(1979), the upper and middle horizons in our standards matching the residual soil, the lower horizon matching the completely weathered.

analysis with the data(average value) and sample number in table 3. We can find in the hierachical diagram and the relevent sample number:
(1) if we use 0.396 for the distance coeffi-cient level, the weathered states can be divided into five classes(or "zones"), slightly weathered, lower moderately weathered, upper moderately weathered, highly weathered, lower completely weathered, which show no differences with the result of qualitative assessment.
(2) the convergence sequence of these zones shows the gradual change nature of weathering states and the change tendency of the rele-vent engineering-geological properties. Highly weathered zone and lower portion of complete-ly weathered zone have similar characteris-tics, so group them into one first(IV,V-1); both lower portion and upper portion of

Table 2. Data of point load, longitunal wave velocity and Schmidt rebound tests

Location	Rock types	Degree of weathering	$I_{S(50)}$ (MPa)			V_P (m/s)			R		
			Average value	Usual value	Number of samples	Average value	Usual value	Number of points tested	Average value	Usual value	Number of points tested
Shanhou	Grano-diorite	I	6.4411	4.8619-7.7726	27	5136.0	4900.0-5398.0	24	58.07	53-63	18
		II	5.2420	3.9574-6.5449	21	3922.4	3016.0-4886.1	18	51.25	46-56	26
		III-2	1.2389	0.4882-2.1389	30	2101.2	1433.8-2601.5	15	40.43	34-46	18
		IV-2	0.0941	0.0263-0.2181	28	1115.5	813.1-1663.0	8	16.40	14-18	24
		V-1	0.0038	0.0032-0.0046	25	394.2	312.0-485.9	10	7.68	4-11	25
Pulin		I	6.9989	5.2600-8.3787	24	5174.0	4550.0-5684.0	18	58.00	54-64	24
		II	5.7439	4.4384-6.7517	16	4451.0	3132.0-5195.0	21	49.32	43-54	34
		III-2	0.7531	0.4042-1.3850	12	2615.3	1909.0-4370.1	18	40.91	36-45	37
		IV	0.1310	0.0479-0.2488	24	983.5	617.9-1618.1	17	20.46	16-24	36
Wanyao	Migmatitic admellite	I	6.9040	5.0324-9.1098	0	5173.2	4970.6-5491.5	18	60.50	58-64	25
		II	5.7880	4.0159-7.3654	6	4195.0	3602.6-5117.3	20	57.30	52-63	25
		III-1	2.8968	2.5060-3.7509	20	3202.5	2933.9-3345.4	13	51.20	49-53	15
		III-2	1.2882	0.7134-1.7050	28	1933.6	1379.0-2443.7	7	37.29	32-45	10
		IV	0.5556	0.2015-0.9959	47	1320.1	859.8-1682.9	8	22.40	18-28	25
		V-1	0.0150	0.0069-0.0347	23	418.5	329.6-544.0	25	7.07	4-10	51
Shibakan	Monzonite granite	I	7.6500	5.9639-9.6422	37	5319.5	5210.4-5461.9	15	60.80	58-64	36
		II	5.8936	4.5642-7.7321	70	4513.9	3829.2-4958.1	33	56.20	54-60	25
		III-2	1.4618	0.6560-2.5459	46	2314.4	1334.0-3279.4	20	41.60	36-46	25
		IV	0.2761	0.0521-0.5497	26	1152.3	680.4-1795.2	22	23.10	18-28	18
		V-1	0.0170	0.0128-0.0223	19	549.7	411.5-723.4	11	7.50	6-10	16
Jinjiting		I	6.1602	5.0317-7.1001	24	5008.0	4893.1-5131.7	12	59.10	51-65	42
		II	5.2290	4.0260-6.7290	28	5072.5	4778.8-5218.6	22	56.00	50-62	105
		III-1	2.7970	1.7558-3.9373	30	3482.8	3239.7-4035.7	12	46.80	41-53	84
		III-2	1.0080	0.6171-1.9866	22	2382.5	1620.1-2891.9	13	32.58	27-40	84
		IV	0.1970	0.0389-0.5528	24	1244.9	917.1-1597.1	26	21.60	16-29	42
		V-1	0.0109	0.0090-0.0129	58	422.7	350.2-471.6	7	11.96	9-14	105
Yutingli	Biotit granite	I	7.6766	6.4013-9.1555	20	5351.2	5178.3-5527.3	4	60.90	59-64	42
		II	5.6500	4.2844-7.0337	37	4938.1	4078.6-5287.0	11	57.78	53-63	63
		III-2	1.0098	0.4914-1.6358	18	1908.0	1074.6-2503.0	11	40.30	35-46	42
		IV	0.1783	0.0719-0.3325	48	1273.0	1006.5-2024.5	6	24.03	19-30	42
		V-1	0.0354	0.0184-0.0653	36	867.2	693.6-1095.0	3	13.00	12-15	21
Lucuo		I	6.8354	4.8182-8.7371	17	5634.5	5444.7-5784.7	14	55.63	49-60	20
		II	6.1025	4.2409-7.6948	21	4827.4	4417.4-5127.0	8	48.06	41-56	39
		III-2	1.1716	0.7334-1.7298	43	2300.0	1837.5-2976.2	9	31.80	28-36	21
		IV	0.2426	0.1354-0.4475	18	1361.5	887.8-1835.3	41	20.15	14-26	42
		V-1	0.0087	0.0034-0.0178	26	433.0	386.1-512.2	7	6.80	4-10	21
Total					1069			499			1330

Note: 1. III-1 and III-2: Lower moderately weathered and upper moderately weathered
2. IV-2: upper highly weathered
3. $I_{S(50)}$ is computed with the formula $I_S=P/D_e^2$ recommended by Point Load Test Group of International Society for Rock Mechanics(ISRM, 1985)

moderately weathered zone have the similar characteristics and classify them into one secondly(III-1,III-2);then the slightly weathered zone and the (III-1,III-2) make one class(II,III-1,III-2); at the end,cluster (II,III-1,III-2) and (IV, V-1) into one class, see Fig.1.

The conclusion that we got with hierachical cluster analysis sets up a basis for us to make a comprehensive index for assessment of weathered states.

3.4 A comprehensive index for quantitative assessment of weathered states of granities

We suggest using

$$Z_f = \sum_{i=1}^{n} \rho_i F_i \quad (i=1,2,3,\cdots\cdots, n) \quad (2)$$

as the comprehensive index for quantitative assessment of weathered states of granites, where Z_f is the comprehensive desending rate of weathering(%), ρ_i is the weight of ith index and $\sum_{i=1}^{n} \rho_i =1$, F_i is the desending rate of weathering index of the ith index.
(1) Determination of "weight"
"weight" means the proportion that each index has in reflecting a feature. If an reflects weathering states more sensitively,it has a heavier weight; conversely, less sensitively, a lighter weight. See table 3,the usual change range of desending rates of weathering index in a weathered zone can reflect the changes of the features in the zone (It many involve random changes surely, but the random element is partly eliminated for our using the general range). The greater the change range is, the more sensitively the index reflects weathering

660

Table 3. Data of the desending rate of weathering index

Location	Rock types	Degree of weathering	$F_{Is(50)}$ (%)		F_{Vp} (%)		F_R (%)		Sample number
			Average value	Usual value	Average value	Usual value	Average value	Usual value	
Shanhou	Grano-diorite	II	18.6	0.0-38.5	23.6	4.9-41.3	12.0	4.0-21.0	1
		III-2	80.7	66.7-92.4	59.0	49.3-72.1	30.0	21.0-41.0	2
		IV-2	98.5	96.6-99.6	78.3	67.6-84.2	72.0	69.0-76.0	3
		V-1	99.9	99.9-99.95	92.3	90.5-93.5	87.0	81.0-93.0	4
Pulin		II	18.5	3.5-36.6	14.0	0.0-39.5	15.0	7.0-26.0	5
		III-2	89.2	80.2-94.2	49.5	32.9-63.1	29.5	22.4-38.0	6
		IV	98.1	96.4-99.3	81.0	68.9-88.1	65.0	59.0-72.0	7
Wanyao	Migmatitic admellite	II	16.2	0.0-41.8	18.9	1.1-30.4	5.0	0.0-14.0	8
		III-1	58.0	45.7-63.7	38.1	35.3-43.3	15.3	12.4-19.0	9
		III-2	81.3	75.3-89.7	62.6	52.8-73.0	38.4	25.6-47.1	10
		IV	92.0	85.6-97.1	74.5	67.5-83.4	63.0	54.0-70.0	11
		V-1	99.8	99.5-99.9	91.9	89.5-93.6	88.0	83.0-93.0	12
Shibakan	Monzonite granite	II	23.0	0.0-40.3	15.1	6.8-28.0	8.0	1.0-11.0	13
		III-2	80.9	66.7-91.4	56.5	38.4-74.9	32.0	24.0-41.0	14
		IV	96.4	92.8-99.3	78.3	66.3-87.2	62.0	54.0-70.0	15
		V-1	99.8	99.7-99.8	89.7	86.4-92.3	88.0	84.0-90.0	16
Jinjiting		II	15.1	0.0-30.6		0.0-4.6	5.0	0.0-15.0	17
		III-1	54.6	36.1-71.4	30.5	19.4-35.3	23.0	10.3-30.6	18
		III-2	83.6	67.8-90.0	52.4	42.3-67.6	45.0	32.0-55.0	19
		IV	96.8	91.0-99.4	75.1	68.1-81.7	63.0	52.0-73.0	20
		V-1	99.8	99.8-99.9	91.6	90.6-93.0	80.0	75.0-86.0	21
Yutingli	Biotite granite	II	26.4	8.7-44.2	7.7	1.2-23.8	5.0	0.0-13.0	22
		III-2	86.8	78.7-93.6	47.6	31.2-68.0	34.0	25.0-43.0	23
		IV	97.7	95.7-99.1	76.2	62.2-81.2	41.0	51.0-69.0	24
		V-1	99.5	99.1-99.8	83.8	79.5-87.0	79.0	75.0-80.0	25
Lucuo		II	10.7	0.0-38.0	14.3	9.0-21.6	14.0	0.0-26.0	26
		III-2	82.9	74.7-89.3	57.8	47.2-67.4	43.0	35.0-50.0	27
		IV	96.5	93.5-98.0	75.8	67.4-84.2	64.0	53.0-75.0	28
		V-1	99.9	99.7-99.95	92.3	90.9-93.1	88.0	82.0-93.0	29

of a same weathered zone have different usual change ranges, ie. different weights; the same index of different weathered zones has different usual change, ranges either, for example, the usual change range of $F_{Is(50)}$ in each weathered zone becomes smaller with the increasing in intensity of weathering, from nearly 40% in the slighly completely weathered zone to less than 1% in the lower completely weathered zone where $F_{Is(50)}$ is insensitive to reflect the weathering states, but F_R goes an opposite way to $F_{Is(50)}$, and F_{Vp} changes little in full range weathering states.

We define the susceplibility relflecting weathering states of an index to be the absolute value of this index's usual change range in a weathered zone. With it we calculated the weights of each index in every weathered zone(See table 4).

(2) Using Z_f (%) to assess the weathered states of granites quantitatively

We calculated all Z_f of every profile with table 4 and formula (2), and showed the

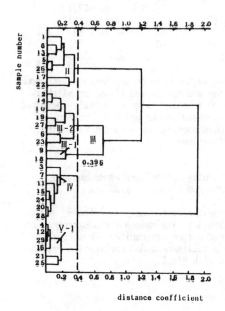

Fig. 1 the hierachical diagram of Q-mode cluster analysis with varable $F_{Is(50)}$ F_{Vp} and F_R

distance coefficient

states. Being seen in table 3, different indices

Table 4. Assessment of index's weights

weathered zone \ index	$F_{Is(50)}$	F_{Vp}	F_R
II	0.46	0.32	0.22
III-1	0.44	0.35	0.21
III-2	0.30	0.42	0.28
IV	0.25	0.35	0.40
V-1	0.05	0.35	0.62

average values of the seven profiles in table 5.We made some slight changes on table 5 and brought out Z_f's threshold values for each weathered zone in table 6 for sake of convenience in use(see table 6).

Table 5. The average values of Z_f (%) of the seven profiles

weathered zone Z_f (%)	II	III-1	III-2	IV	V-1
average	14.7	40.8	58.3	76.8	87.5
range	2.2 ~30.6	30.0 ~48.7	46.9 ~69.2	69.1 ~83.1	83.6 ~90.7

Table 6. The threshold values of Z_f (%) in every weathered zone

weathered zone Z_f (%)	II	III-1	III-2	IV	V-1
average	15	41	58	77	88
range	<30	30~50	50~70	70~85	>85

3.5 The application of fuzzy pattern direct recognition method in assessing the weathered states of granites

The degrees of weathering have a gradual change property.It is very difficult to make a precies principle to determine which one the transition section of two zones belongs to. The fuzzy pattern recognition method can be used directly to establish a pattern to describe those fuzzy characteristics because we built up the comprehensive index zoning principles already.

The direct method of fuzzy pattern recognition is "the principle of maximun membership value", which is used in single-object recognition.To weathered zones,every one represents a pattern.If we build up a membership function for each weathered zone, the problem which state a weathered zone belongs to is resolved.

For the problem of weathering classification, its domine U is within the field of real numbers. There is one, and only one point u_0 which has $\mu_A (u_0)=1$ in a regular simple fuzzy subset A. By the experience we can determine that there are a point u_1 in the left side of u_0 and a point u_2 in the right side of u_0 with $\mu_A (u_1)= \mu_A (u_2)=0$, and if $u_1 <u<u_2$,then $\mu_A (u)>0$. We assume to use linear interpolation to get the membership values for any point left, so we set $\mu_A (u)$ as:

$$\mu_A (u) = \begin{cases} [f_1 (u)]^\alpha & (u_1 \leqslant u \leqslant u_0) \\ [f_2 (u)]^\beta & (u_0 \leqslant u \leqslant u_2) \\ 0 & (else) \end{cases} \quad (3)$$

here $f_1 (u)$ and $f_2 (u)$ are linear functions with $f_1 (u_1)= f_2 (u_2)=0$, $f_1 (u_0)= =f_2 (u_0)=1$. By equation $\mu_A (u_1^*)=\mu_A (u_2^*)$

$=0.5$, we can determine the boundaries u_1^*, u_2^* for the fuzzy set in order to get exponents α and β. If we can make it sure that there are $u_1^* \in (u_1 ,u_0)$ and $u_2^* \in (u_0 ,u_2)$ by experience, then

$$\alpha =-\lg 2/\lg[f_1 (u_1^*)]$$
$$\beta =-\lg 2/\lg[f_1 (u_2^*)] \quad (4)$$

Actually, classification of weathering states is a matter to determine which category element Z_f belongs to.In terms of table 6, we got the membership functions for each weathered zone, as following:

lower moderately weathered zone

$$\mu_A (Z_f) = \begin{cases} [(Z_f -15)/26]^{1.26} & (15 \leqslant Z_f \leqslant 41) \\ [-11(Z_f -58)/17]^{0.92} & (41 < Z_f \leqslant 58) \\ 0 & (else) \end{cases} \quad (5)$$

upper moderately weathered zone

$$\mu_A (Z_f) = \begin{cases} [(Z_f -41)/17]^{1.09} & (41< Z_f \leqslant 58) \\ [-(Z_f -77)/19]^{0.69} & (58< Z_f \leqslant 77) \\ 0 & (else) \end{cases} \quad (6)$$

highly weathered zone

$$\mu_A (Z_f) = \begin{cases} [(Z_f -58)/19]^{1.51} & (58 \leqslant Z_f \leqslant 77) \\ [-(Z_f -88)/11]^{0.53} & (77< Z_f \leqslant 88) \\ 0 & (else) \end{cases} \quad (7)$$

Obviously,there is no any necessary to seek after the membership functions for slightly weathered zone and lower completely weathered zones, because we can use the functions of their neighbouring zones to determine their classifications.

3.6 A "zoning system" for assessment of weathered states of granites

Reviewing the above analysis process, we can establish a zoning system for assessment of weathered states of granites. The processes of its establishment and application are explained with Fig.2.

4 Conclusion

Our "zoning system" is carried out by site investigations and a lot of test studies on certain number typical profiles in South Fujan,P.R.China.It is practical and successful in this region, and may be helpful in other regions.

662

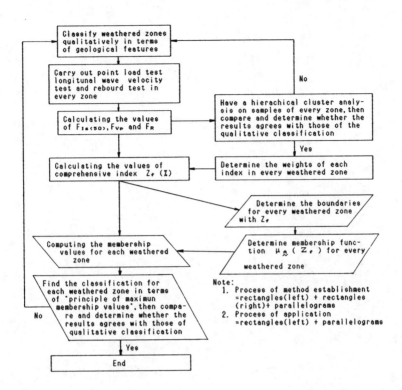

Fig.2 Flowchart of the zoning system

References

Kehe Wei, A study on application of point
    load test for quanlititative assessment
    of weathering states in granites,
    Hydrogelolgy and engineering geology,
    China geology publishing house,Feb.1982
Kaitai Fang, Hierachical cluster analysis,
    China geology publishing house,1982
Deyi Feng, Fuzzy mathematics--methods and
    applications,China Seism publishing house,
    1983
R.P.Martin, Use of index tests for
    engineering assessment of weathered rocks,
    IAEG Congress,2.1.4-pp433-450

6th International IAEG Congress / 6ème Congrès International de AIGI, © 1990 Balkema, Rotterdam. ISBN 90 6191 130 3

Engineering properties of weathered metamorphic rocks in Peninsular Malaysia

Les propriétés techniques de roches métamorphiques altérées en Malaisie

Ibrahim Komoo & Jasni Yaakub
Universiti Kebangsaan Malaysia, Malaysia

ABSTRACT: Metamorphic rocks in Peninsular Malaysia mainly consist of phyllite, schist and quartzite and occupy more than 20% of total surface area. Most of these rocks have experienced extensive chemical weathering forming a very thick regolith. This paper will discuss the nature of weathering profile, and physical characterization based on material texture and mass structure. A practical classification of the profile is here-in proposed. The basic engineering properties of selected rock materials representing each weathering grade are determined in the laboratory. The properties tested include density, porosity, friability, Brinell's hardness index, point load strength, and uniaxial compressive strength. The results indicated these properties are suitable as classification index for quantifying weathering grades.

1 INTRODUCTION

Metamorphic rocks in Peninsular Malaysia mainly consist of phyllite, schist and quartzite. These rocks which are characteristically of the greenschist facies of regional metamorphism, are associated with mesozonal granitic batholiths (Hutchison, 1973). Phyllite and occasionally quartzite, in places are interbedded with slightly metamorphosed shale and sandstone, therefore are generally called clastic metasediments. Gneiss, marble and hornfels are also quite common but their occurences as minor bodies occupy relatively very small surface area.

Due to the tropical climate, most metamorphic rocks in Peninsular Malaysia experience extensive chemical weathering. The thickness of regolith varies considerably, from few meters up to more than 100m. Several factors influence the variation in thickness, however the type of lithology, mass structure and topographic conditions play important role (Ibrahim Komoo, 1986; Ibrahim Komoo and Mogana, 1988). Bearing in mind of the more or less similar mass structure, weathering profiles on phyllitic or quartzitic rocks are generally thicker compared to schist formation.

Metamorphic rocks in Peninsular Malaysia occupy more than twenty percents of surface area and generally forming undulating to hilly topography along the flank of granitic batoliths. In recent years, considerable engineering works have been associated with these rocks, i.e. major highway projects, hydroelectric schemes and some urban developments. The engineering problems posed are mainly related to weathering condition, engineering properties and slope failure. This paper will highlight the approach to characterise these weathered metamorphic materials for engineering purposes.

2 WEATHERING

Most of the weathering classifications proposed, for example the classification refined by Dearman (1976) and lately discussed by Lee and de Freitas (1989), or as suggested by ISRM (1981), IAEG (1981) and BSI (1981) are based on works associated with granitic rocks. These classifications are generally suitable for most igneous rocks and perhaps some massive sediments and metamorphic rocks. However, field observation indicate that most of the metamorphic rocks are continuously weathered to form residual soils. Weathering profile seldom develop core boulders, and in most cases it is difficult to accurately identify the boundary between rock and soil. Some of typical charasteristic of weathering profile developed in sediments and metamorphic rocks in Peninsular Malaysia have been discussed by Ibrahim Komoo (1985) and Ibrahim Komoo and Mogana (1988).

2.1 Weathering grade

For accurate weathering classification of the rock mass one should be able to recognise several distinct changes in material characters. For practical purpose, a combination of only three main characters namely discolouration, mineral fabric (texture) and friability were used to differentiate various weathering grades. Table 1 shows how these characters are applied to define weathering grade. Please note that generally two characters are required to differentiate

between grade, but sometimes only one character is available, i.e. friability solely used to classify between soil and rock, and between grade III and II. Degree of chemical decomposition cannot be used on metamorphic rocks because they are generally fine-grained.

Table 1. Material characters used to define weathering grades of metamophic rocks.

WEATHERING		MATERIAL CHARACTERIZATION			
Term	Grade	Type of material	Discolouration	Texture	Friability by hand
Residual Soil	VI		total changes (homogenous)	completely destroyed	friable
Completely Weathered	V	soil	total changes (non-homogenous or mottled)	partially preserved	
Highly Weathered	IV b / a		completely discoloured	completely preserved	form smaller fragment
Moderately Weathered	III	rock	partially discoloured		fragment corners can be chipped
Slightly Weathered	II				fragment corners cannot be chipped
Fresh Rock	I		no change		

2.2 Weathering profile

The classification of the weathering profile or zone on metamorphic rocks is often far more complicated. This is, among other factors, due to the structural inhomogenity, and lithologic and textural variations. Classification based on scheme suggested by ISRM (1981) is frequently not suitable because the nature of metamorphic weathering profile lacks the major diagnostic features used in the classification. As mentioned earlier, both the degree of minerals decomposition and boundary between rock and soil which are two main diagnostic features in ISRM (1981) classification were not easily recognised in weathered metamorphic profile.

Practical classification of weathered metamorphic profile require recognition of both the weathering grade of the material and rock mass structure. The basis of the classification should be the material weathering grade as discussed earlier. Since the degree of inhomogenity is relatively high, one should use the most dominant weathering grade as indicative of a particular weathering zone. For example, a zone consisting of 60% grade IV, 30% grade V and 10% grade VI materials should be recognised as zone IV weathering profile. On the other hand, the rock mass structure should also be used as supplementary features to classify weathering zone. An example, zone V generally consists of the accumulation of secondary iron in the form of nodules, concretions or sheets; zone IV indicate partial preservation of mass structures, whereas in zone III and II the mass structures are completely preserved. Table 2 describes the classification scheme of rock mass weathering profile for metamorphic rocks. Figure 1 shows some typical examples of weathering profiles classified based on the above scheme.

3 ENGINEERING PROPERTIES

Several engineering properties of some metamorphic rocks have been determined to evaluate changes due to weathering processes. The properties that show correlation includes dry density, porosity, slake-durability, point load strength index, Brinell's hardness index, and uniaxial compressive strength. Rock samples representing main metamorphic rocks were collected at random from several weathering profiles. Sampling and sample preparation for weathered materials are rather difficult, therefore several sampling techniques and preparations were adopted to minimise error. Details of the procedures are discussed in Jasni

Table 2. Classification of rock mass weathering profile for metamorphic rocks

WEATHERING CLASSIFICATION		ROCK MASS DESCRIPTION
Term	Zone	
Residual Soil	VI	All rock material is converted to soil. Material fabric (texture) and mass structure are completely destroyed. The material is generally silty or clayey and shows homogenous colour.
Completely Weathered	V	All rock material is decomposed to soil. Material fabric partially preserved. The material is sandy and friable if squeeze by hand, and shows non-homogenous or mottled in colour. Mass structure mostly destroyed. This zone commonly consists of secondary iron (nodules, concretions or sheets)
Highly Weathered	IV	All rock material is in transitional stage to form soil. Material condition is either soil or rock. Material is completely discoloured but it fabric is completely preserved. Mass structure partially preserved.
Moderately Weathered	III	The rock material shows partial discolouration. Material fabric and mass structure are completely preserved. Discontinuity commonly filled by iron-rich material. Material fragment can be chipped by hand.
Slightly Weathered	II	Discolouration along discontinuity and may be part of rock material. Material fabric and mass structure are completely preserved. Material is generally weaker but fragment corners cannot be chipped by hand.
Fresh Rock	I	No visible sign of rock material weathering. Some discolouration on major discontinuity surfaces.

Yaakob (1989). Samples of fresh (grade I) phyllite and quartzite are not exposed along the profiles, therefore its engineering properties cannot be determined.

3.1 Dry density and porosity

Dry density and porosity are determined using saturation and caliper techniques as suggested by

667

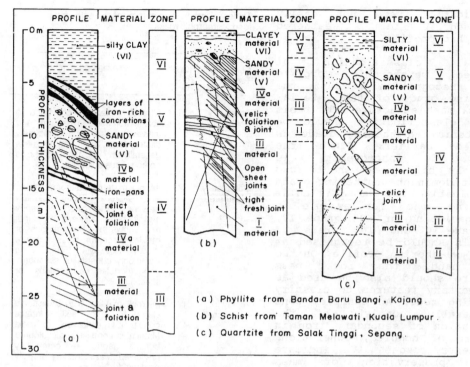

Figure 1. Some typiucal examples of weathering profiles on metamorphic rocks.

ISRM (1979a). Special care should be taken while immersing grades III and IV materials in water for porosity determination. The porosity of grade V material cannot be determined due to its high friability. Figures 2 and 3 show the relationship between weathering grade with dry density and porosity, respectively. It is interesting to note that all metamorphic rocks tested show a gradual decrease of dry density and increase in porosity with the increase in weathering grade.

3.2 Slake-durability index

Slake-durability is determined using the method suggested by ISRM (1979a). Although few other modification methods were also applied, the results show similar trends. The relationship between slaking durability index and weathering grade is shown in Figure 4. The results indicate a relatively high slake-durability

index for weathered rocks (grades III and II) and very low for completely weathered soil (grade V). However, for grade IV material the value ranges from nearly 0% to more than 90%. This parameter seem most suitable to differentiate between grade IVa (engineering rock) and IVb (engineering soils) as suggested in the weathering classification.

3.3 Brinell's hardness index

Brinell's hardness index is determined using procedure described by AIT (1981). The test was conducted using point load strength index apparatus with some modification, rather than ROHA tester as suggested in the original procedure. The modification involved the replacement of the upper cone with 10 mm steel ball indenter, and the lower cone with load transducer. Slab shaped specimens with thickness between 2.5 to 3.5 cm were used for this test. The

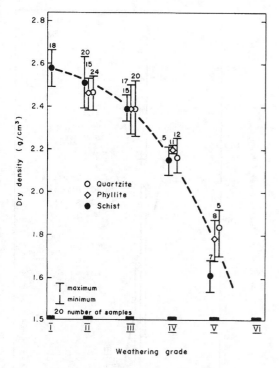

Figure 2. Relationship between dry density and weathering grades for metamorphic rocks.

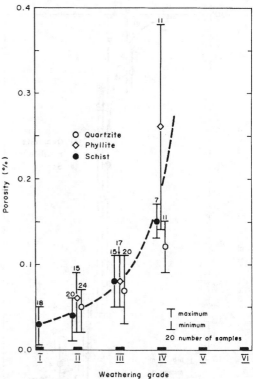

Figure 3. Relationship between porosity and weathering grades for metamorphic rocks.

results is shown in Figure 5. Note that the Brinell's hardness index is a suitable parameter to differentiate between grades III and IV materials.

3.4 Point load strength index

Point load strength index (Is50) was determined using the method suggested by ISRM (1985). Due to the relatively low strength index of the highly weathered materials, the apparatus had to be slightly modified by adding load transducer with sensitivity up to 3 newton. The AX-size core specimens with length/diameter ratio of 0.8-1.1 were used and load was applied axially. The results is shown in Figure 5. The gradual decrease of the point load strength index with weathering grade provide a valuable parameter for engineering classification of the weathering profile.

3.5 Uniaxial compressive strength

Uniaxial compresive strength was determined using the method suggested by ISRM (1979b). AX-size core specimens with length/diameter ratio about 2.0 were used for the test. The larger diameter specimens cannot be used due to a high inhomogenity of the samples. The relationship between compressive strength and weathering grade is shown in Figure 6. A relatively linear decrease of material strength with weathering grade can provide a significant classification index.

4 ENGINEERING CLASSIFICATION

The engineering properties of phyllite, schist and quartzite tested differed slightly but the changes due to the degree of weathering showed similar trend.

669

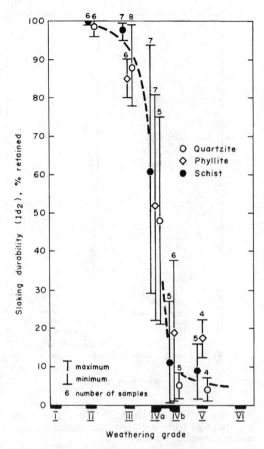

Figure 4. Relationship between slake-durability index and weathering grades for metamorphic rocks.

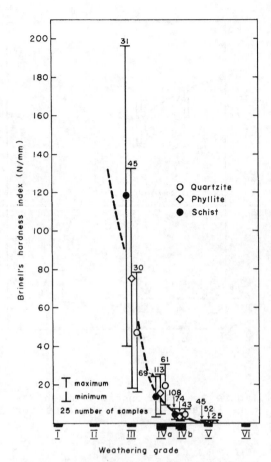

Figure 5. Relationship between Brinell's hardness index and weathering grades for metamorphic rocks.

As discussed earlier, although each engineering property of any particular metamorphic rock can be used to classify weathered material, individual parameter show some degree of overlapping and limitations to differentiate the whole spectrum of weathering. For example, while slake-durability index is best to differentiate between grades IVa and IVb materials, the Brinell's hardness index is only suitable for differentiating between grades III and IV.

For better engineering classification of the weathered materials, one should use a combination of several engineering

parameters. Table 3 shows a proposed classification of rock material weathering for metamorphic rocks based on six engineering properties which have been discussed earlier. This table is important for estimating some engineering properties of materials from any weathering zones. It can also be used by engineers to classify the weathering grade of rock samples.

5 CONCLUSION

Weathering products of metamorphic rocks in Peninsular Malaysia forms a thick weathering profile with materials gradually changing from fresh rock to residual soil.

670

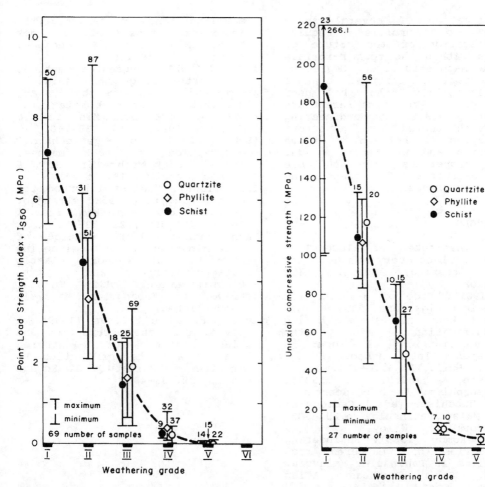

Figure 6. Relationship between point load strength and weathering grades for metamorphic rocks.

Figure 7. Relationship between uniaxial compressive strength and weathering grades for metamorphic rocks.

Table 3. Classification of rock material weathering of metamorphic rocks based on some engineering properties.

Weathering Classification		Dry Density (g/cm³)	Porosity (%)	Slake-durability, I_{d2}(%)	Brinell's Hardness Index (N/mm)	Point Load Index (MPa)	Uniaxial Compressive Strength (MPa)
Term	Grade						
Completely weathered	V	<2.00	n.d.	0 - 20	<1	<0.1	<5
Highly weathered	IV	2.00 - 2.25	> 0.12	IVb. 0 - 30 IVa. 30 - 80	IVb. 1 - 6 IVa. 6-40	0.1 - 0.5	5 -20
Moderately weathered	III	2.25 - 2.45	0.06 - 0.12	80 - 95	>40	0.5 - 3.0	20 - 80
Slightly weathered	II	2.45 - 2.55	0.02 - 0.06	95 - 100	n.d.	3.0 - 6.0	80 - 140
Fresh rock	I	> 2.55	< 0.02	n.d.	n.d.	> 6.0	> 140

Classification of several zones within a profile require identification of the nature of changes within the rock materials and the mass structures. Degree of discolouration, textural changes and friability can be used effectively to differentiate weathering grade. Six engineering properties tested show a strong correlation between these properties and weathering grade. These properties are suitable index parameters for quantifying weathering grades.

REFERENCES

Asian Institute of Technology (1981). Laboratory manual for rock testing, 256pp. AIT, Bangkok.

BSI (1981). Code of practice for site investigations. BS5930.

Dearman, W.R. (1976). Weathering classification in the characterisation of rock: A revision. Int. Assoc. Engng. Geol., Bull., 13, 123-127.

Hutchison, C.S. (1973). Metamorphism. In D.J. Gobbett & C.S. Hutchison (eds.) Geology of the Malay Peninsula, p.253-303. New York, John Wiley.

Ibrahim Komoo (1985). Engineering properties of weathered rock profiles in Peninsular Malaysia. Proc. 8th. Southeast Asian Geotech. Conf., 381-386.

Ibrahim Komoo (1986). Engineering geological aspects of clastic metasediments in the Kuala Lumpur area, Malaysia. Geol. Soc. Malaysia, Bull., 19, 597-612.

Ibrahim Komoo & Mogana, S.N. (1988). Physical characterization of weathering profile of clastic metasediments in Peninsular Malaysia. Proc. 2nd. Int. Conf. Geomech. in Tropical Soil, 37-42. Balkema, Rotterdam.

IAEG (1981). Rock and soil description for engineering geological mapping. Int. Assoc. Engng. Geol., Bull., 24, 235-274.

ISRM (1979a). Suggested methods for determining water content, porosity, density, absorption and related properties and sewlling and slake-durability index properties. Int. J. Rock Mech. Min. Sci. & Geomech. Abstr., 16, 148-156.

ISRM (1979b). Suggested methods for determining the uniaxial compressive strength and deformability of rock materials. Int. J. Rock Mech. Min. Sci. & Geomech. Abstr., 16, 135-140.

ISRM (1981). Basic geotechnical description for rock masses. Int. J. Rock Mech. Min. Sci. & Geomech. Abstr., 18, 85-110.

ISRM (1985). Suggested method for determining point load strength. Int. J. Rock Mech. Min. Sci. & Geomech. Abstr., 22, 53-60.

Jasni Yaakob (1989). Technique of determination of the engineering properties of weathered clastic metasediments and its classification. MSc Thesis, Universiti Kebangsaan Malaysia (unpublished).

Lee, C.F. & de Freitas (1989). A revision of the description and classification of weathered granite and its application to granites in Korea. Q. J. Engng. Geol., 22, 31-48.

Geotechnical uniformity of the Weichselien Loess sequence in South Limburg, the Netherlands

L'uniformité géotechnique dans la séquence du loess weichselien dans le sud du Limbourg, Pays-Bas

R.R.Kronieger

Delft University of Technology, Faculty of Mining and Petroleum Engineering, Section Engineering Geology, Delft, Netherlands

ABSTRACT: The Weichselien loess cover of the Dutch province of south Limburg is geotechnically regarded as one homogeneous layer. However in a litho-stratigraphical sense the Weichselien loess is thought to be composed of several sub-units with differences in grain structure, cementation and clay content. The question arises if these small differences are reflected in the geotechnical behaviour of the loess. To investigate the relationship between litho-stratigraphy and geotechnical behaviour, in-situ tests have been performed in a location with a known litho-stratigraphy. By use of the tor-vane test and the penetrometer test, geotechnical units were defined. In-situ shear tests and a new method of plate-loading tests were performed in each of the geotechnical units to provide additional information to aid the geotechnical characterisation of the Weichselien loess litho-stratigraphy. Initial results indicate that the geotechnical properties iof the loess are variable.

RESUME: La couche du loess weichselien dans le sud de la province néerlandaise du Limbourg est considérée d'être homogene du point de vue géotechnique. Pourtant, dans le sens lithostratigraphique on croit que le loess weichselien est composé de plusieurs éléments qui diffèrent dans la structure des grains, dans la cimentation et dans la quantité qu'il contient d'argile. On se demande si ces petites différences se traduisent dans le comportement géotechnique du loess. Pour examiner la relation entre la conduite lithostratigraphique et géotechnique on a exécuté des essais sur terrain dans un domaine dont la lithostratigraphie était connue. En utilisant le test "tor-vane" et le test pénétromètre on délimitait des unités géotechniques. Des essais de cisaillement sur terrain et un nouvelle forme d'essai de charge par plaques ont été exécutés dans chacune des unités géotechniques en vue d'obtenir plus d'information pour caractériser le loess weichselien du point de vue géotechnique. Jusqu'a maintenant les résultats obtenus démentent la question que le loess serait homogène dans le sens géotechnique du terme.

1 GEOLOGY

The major exposure of loess in the Netherlands is restricted to the south of the province of Limburg. The loess forms part of the loess "belt" which can be traced from northern France through Belgium and the Netherlands into Germany (Kuyl, 1975). The loess sequence in south Limburg, consists of three major units of Quaternary age. The Weichselien loess consisting of the yellow-brown coloured loess of the Twente formation was dated to 18.500-21.000 and 27.900 B.P. respectively (Andersen et al.,1960). The lower Saalien loess unit, the formation of Eindhoven, is locally not exposed in the province of Limburg. The undisturbed loess sequence forms a plateau incised by small rivers in an area between the cities of Maastricht, Sittard and Heerlen (fig.1). The stratigraphy of the Weichselien Loess Unit is described in detail by Kuyl & Bisschops (1969),and Mucher (1986). The unit can be subdivided in two main units generally referred to as the 'Upper Loess' and 'Middle Loess'. Characteristic structures, recognizable in the field, are formed by the carbonate boundary, an ash-layer related to the Eiffel vulcanism

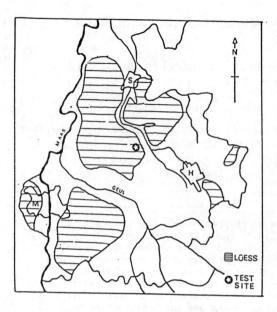

Fig.1   Loess area in south Limburg,
the Netherlands. M = Maastricht; S =
Sittard; H=Heerlen.

(Meys et al.,1983), cryoturbated layers
and paleosols.

2. GEOTECHNICAL INVESTIGATION.

The investigation was carried out in
quarry Nagelbeek, one of the major
quarries where the loess sequence is
exposed as a result of a deep excavation
to reach the underlying gravel and quartz
sand. The site for the in-situ tests was
located on the most western exposure
within the quarry, in undisturbed material
overlain by an thin veneer of redeposited
loess. The thickness of the total
sequence is in excess of 5 metres. No
other adequate exposures were available in
the loess area. In the quarry the
Weichselien loess unit forms the cover of
a Pleistocene gravel terrace of the Maas
river, the Saalien loess unit is not
exposed.

2.1 Geotechnical Units.

The loess in south Limburg shows
geotechnical behaviour which differs from
the collapsing loess found elsewhere.
High densities qualify the loess as being
almost non-collapsible (Clevenger,1956).

It fits however the criteria stated by
Lysenko (1973) for a loess in being a
yellow, highly porous, uniformly
distributed material with dominantly silt
grain sizes and macropores.

To investigate the geotechnical variation
of litho-stratigraphical units within in
the loess, a 5 metre long and 2.4 metre
high section of the quarry face was
covered with a grid with a spacing of 0.4
meter of the nodes. At each node a vane
was used to measure shear strength and
between the nodes pocket penetrometer
readings were taken to estimate
compressive strength. The readings were
mapped and contoured as z-values above the
x,y plane representing the quarry face.
Results given in fig.2 and fig.3 indicate
that the shear- and compressive strength
appear to be fairly consistent within the
horizontal direction. However a
subdivision of the Weichselien loess,
based on litho- stratigraphy
(Mucher,1986), which was used in the field
can still be recognized. The loess
sequence was therefore subdivided in
geotechnical units on basis of this
subdivision (Plasman & Kronieger,1990).
The in-situ tests were carried out for
each subdivision. The geotechnical
subdivision is given below:

Unit A - The top of the 'Upper Loess'
can be described as a friable brown-yellow
decalcified loess underlain by a brown-
yellow calcareous loess. The thickness of
this unit in the exposure is approximately
2 metre.

Unit B - Below this top unit about 2
meter yellow coloured calcareous loess is
exposed which changes gradually with depth
into a loess unit which is characterised
by cryoturbation structures and occasional
frost wedges (Kesselt paleosol). Beneath
this unit a marker horizon could be traced
throughout the exposure representing the
volcanic ash layer with several
millimetres in thickness.

Unit C - Underneath this marker horizon
a 2 meter thick yellow-brown loess
sequence forms the middle loess unit. The
laminated topunit changes with depth into
a more homogeneous 1.5 meter thick
yellow-brown loess unit.

Unit D - The lower unit of the
Weichselien is generally recognized by a
dark-brown colour, associated with an
increasing clay content above an
yellow-brown homogeneous loess unit. An
erosion surface marks the boundary with
the underlying gravel unit at the test
location.

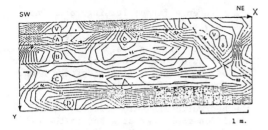

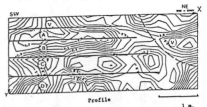

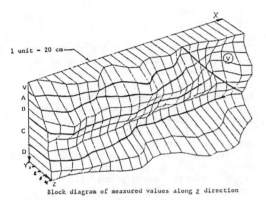

Block diagram of measured values along Z direction

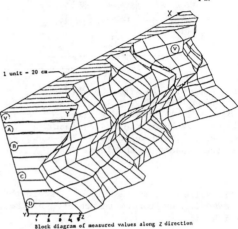

Block diagram of measured values along Z direction

Fig.2   Shear strenght as measured by use of the Tor-vane (in 10 kN/cm2 units) in a section of the Weichselien Loess.

Fig.3   Unconfined compressive strenght as measured by us of small penetrometer (in 10 kN/cm2 units) in a section of the Weichselien Loess.

Unit V - This unit consists of a red-brown friable cover of the Weichselien loess sequence. The material shows locally patches of light coloured sand and gravel pebbles.

2.2 Sampling.

From each unit a block sample was prepared by cutting a pillar from the exposed working face. This pillar was then sectioned horizontally into blocks as vertical cutting proceeded. The blocks were wrapped in plastic and transferred to Delft in tight fitting boxes from which samples were extracted for direct shearbox tests and to determine grainsize distribution and index parameters (table 1).

3. IN-SITU TESTING.

3.1 General.

In-situ testing consisted of a series shear tests and plate loading test in top and bottom unit of each subdivision of the

Table 1 . Index parameters geotechnical sub units of the Weichselien loess.

unit	dry density [kN/m3]	moisture content [%]	CaCO3 content [%]	Linear Shrinkage [%]
A	16	17	5.4	11
B	16	10	6.8	5
C	17	22	8.2	5
D	18	24	13	11
V	17	24	16	5

unit	grainsize analysis <2um [%]	2um - 63u [%]	>63um [%]	permeability (10e-7 m/s)
A	10	79	11	1.5
B	10	82	8	7.3
C	12	88	0	3
D	25	69	6	2
V	12	78	10	1.5

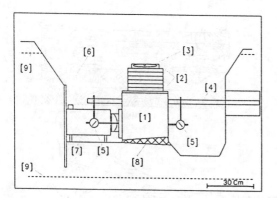

Fig.4   The in-situ shear test.
[1]=block of loess, [2]=weights,
[3]=level, [4]=reference frame,
[5]=displacement gauge, [6]=reaction
platen, [7]=hydraulic jack, [9]=boundary
of the subunit in the loess.

Weichselien and an additional test in the
redeposited unit covering the top of the
Weichselien.

3.2 In-Situ Shear Testing.

3.2.1. Preparation of the Test.

The blocks for the in-situ shear test were
prepared by cutting away material starting
in an rectangular ditch from a horizontal
plane resulting in an rectangular shaped
pillar. From this elongated pillar then
three separate cube shaped blocks of 30 cm
x 30 cm x 30 cm were prepared by further
cutting and modelling the loess thus
ensuring testing of the same unit.
   In the shear direction of each cube the
ditch was excavated about 5 cm deeper to
facilitate failure shearing. On top of
the cube of loess a square steel platen
was placed to distribute the normal load.
   Normal loading was achieved by adding
circular steel platens of 8 kg until the
desired normal load was obtained. Care
was taken not to exceed the unconfined
strength of the material underneath the
platens. For lateral loading a hydraulic
jack was used. The reaction force was
created by an large 1.5 m. x 1.0 m steel
platen dug into intact material in one of
the benches of the excavated ditch. The
centre of the jack was placed on 1/3 of
the height of the cube and pressed by
means of a spherical seating against a
square rigid steel platen.

A reference frame, leading to the non
stressed area on the bench was used to
measure displacements in the shearing
direction. Tilting due to the loading was
controlled by use of a levelling glass on
top of the sample. Tilting of the sample
happened only if the jack was placed to
high. The shear force was calculated from
the hydraulic pressure measured by a
precision manometer attached to the oil
pump. A diagram of the configuration used
is given in fig 4.

3.2.2. Testing.

The normal loads used in the test were in
the low region (11-15-19 kPa) so as to
prevent the material below the platens
from being crushed or causing premature
failure of the unsupported blocks.
Macroscopic failure started at the lower
edge of the shear platen and with
increasing load extended in shearing
direction to the other side of the cube.
The failure plane was inspected at the end
of each test and proved to be a
sub-horizontal smooth plane with some
striations parallel to the shearing
direction. The shear plane was for all
cases parallel to the stratigraphical
horizons, so that test results could be
compared (Matalucci et al.,1970). After
the failure plane reached the other side
of the cube, the residual shear strength
was calculated from the pressure needed to
facilitate further movement of the block
on the failure plane for 10 cm with the
same normal load as the direct shearing of
the block was done.

3.2.3. Discussion.

Results from three blocks combined gave
the Mohr envelope from which the angle
of internal friction and cohesion in
terms of effective stress were derived.
Moisture content of the loess at the
moment of testing was in the order of
16-20 %. The data from the in-situ tests
(table 2) compare remarkably well with the
cohesion of 12 kPa and internal angle of
friction of 32 degrees found from tests on
loess in the surrounding area's by De
Ploey (1973), Kahl (1972) and Kronieger
(1988). The residual shear stress for the
sub-units is given in table 3. The clay
containing units from unit D are marked by
a very low value of the effective residual
shear angle, the carbonate-rich units from
unit A with an high value and a high
cohesion.

Table 2.   Results of Shear Testing of Weichselien loess.
            Normal loads of 11,15 and 19 kPa.

unit	in-situ shear test		direct shear box		residual shear test	
	cohesion [kPa]	angle of internal friction [deg.]	cohesion [kPa]	angle of internal friction [deg.]	cohesion [kPa]	angle of internal friction [deg.]
A	9	32	2.5	11	1.8	5.4
B	10.6	29	1.9	18	3.6	3.2
C	6.1	30	5-7.2	4-11	2.9	3.9
D	7.8	29	-	-	2.5	0.7
V	7.6	30	5.4	7	2.7	2.9

Test results from direct shearbox tests in the lab (table 2) however were not consistent with the effective shear test data in the field. A cause for the discrepancy might be the premature collapse of the loess fabric on the loading side of the samples, as was already reported by Feda (1967). Other factors could be a different, lower moisture content at the moment of testing, as described by de Ploey (1973).

3.3.   Plate-Loading Testing

3.3.1.   General.

Minkov et al.(1979) stated that the main reason of the discrepancy between predicted and real settlement of Bulgarian loess is the value for the deformation modulus. He concluded that lab values were 2.5-3 times smaller than the in-situ found values based on plate loading tests. In order to use lab test results for settlement calculations the value of the stated multiplication factor for the Dutch loess had to be determined by plate loading tests. However since the loess sequence consists of a number of structurally different units with varying densities and porosities, it was necessary to investigate the contribution of each separate unit. The aim of the research was to investigate if the previously defined geotechnical units of the Weichselien sequence showed variations in deformation behaviour.

3.3.2.   Preparation of the Test.

In each geotechnical sub units of the loess a rectangular cavity was dug with opening of 40 cm x 25 cm and with a depth of 70 cm. Care was taken to form rounded edges and two flat parallel sidewalls. In the cavity a hydraulic jack was placed with on both sides a circular platen. In fig.5 the arrangement is shown. The

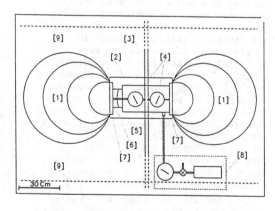

Fig.5   The plate-loading test.
[1]=sphere of influence, [2]=excavated loess, [3]=reference frame, [4]=displacement gauges, [5]=hydraulic jack, [6]=sherical seating, [7]=circular platen, [8]= pump unit with precision manometer, [9]=boundary of the subunit in the loess.

677

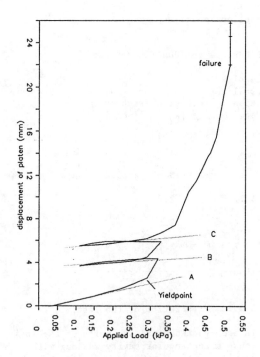

Fig.6   Typical loading diagram of the plate-loading test. A, B, C, tangent moduli of elastic deformation.

applied load was derived from the precision manometer connected onto the pressuere line leading from the pump to the hydraulic jack. The lateral displacements of the platens were separately monitored by use of dial gauges. The diameter of the platens was chosen to create a sphere of influence which remained within the loess unit to be tested. Assuming an homogeneous isotropic material within the sphere of influence, a plate diameter of 20 cm would be sufficient (Jaeger & Cook, 1979). The test was carried out with the loading direction horizontally as the loess was assumed to be homogeneous laterally. This setting had also the advantage that no reaction force had to be provided and a symmetrical geometry facilitated load calculations.

3.3.3. Testing.

Force was applied by means of increasing the pressure to the hydraulic jack until displacement measurement indicated the presence of permanent deformation by increasing displacements with identical

stress increments. Two reloading cycles were performed to measure elastic component in the elasto-plastic region. After the last loading cycle, load was increased until macroscopic failure took place, generally by wedge failure on the free surface.

By plotting the displacement of the two platens, failure of the ground was indicated when the displacement of one of the platens deviated from the symmetry axis of the elastic part of the loading or if macroscopic failure was recognized.

3.3.4. Discussion.

A typical stress-displacement curve from a loading test is shown in fig.6. In general the stress-displacement curve of the plate loading tests consisted of an elastic and an elasto-plastic part. In the elasto-plastic part the two reloading curves gave almost straight matching lines. Tangent moduli A,B and C were calculated from the apllied load-displacement curves and plotted against the depth below the surface and lithology in fig.7 and fig.8 respectively.

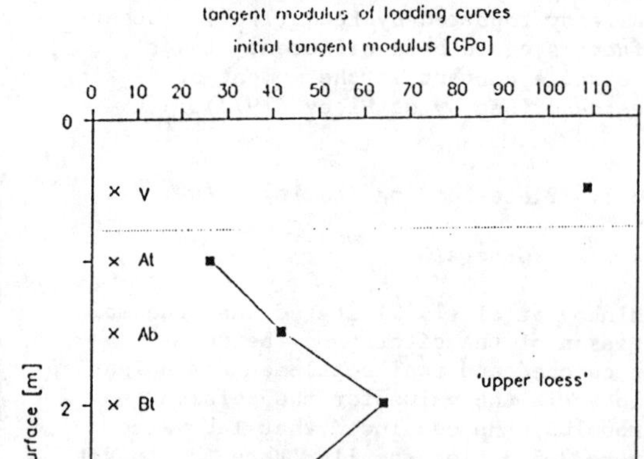

tangent modulus of loading curves

initial tangent modulus [GPa]

Fig.7   Initial tangent modulus with depth. Indicated with 'x' are the geotechnical sub units in the Loess.

678

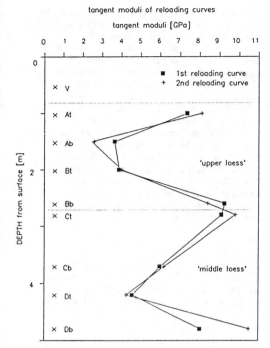

tangent moduli of reloading curves

tangent moduli [GPa]

■ 1st reloading curve
+ 2nd reloading curve

'upper loess'

'middle loess'

Fig.8   Tangent moduli of the reloading cycles with depth. Indicated with 'x' are the geotechnical sub units in the Loess.

Disturbance by cryoturbation of the cemented loess, causes the material [Bb] to have a lower tangent modulus upon initial loading than the homogeneous material [Bt,Ct,and Db]. Lower tangent moduli were also found for the decalcified [Ab] and clay containing units [Cb & Dt] when compared with moduli from the homogeneous carbonate cemented units [Bt,Ct,and Db].

Redeposited material [V] shows relative high tangent moduli. Cyclic loading could not be achieved as brittle failure took place during elastic deformation .

The tangent moduli of the loading cycles in the elasto-plastic region are about 5 times smaller. This is caused failure of the fabric of the loess, causing an increase of the density near to the platen, which was observed after removing the platen at the end of the test. Especially the undisturbed carbonate and clay cemented units show relative low values of the tangent moduli with respect to the cryoturbated, decalcified and laminated units where the carbonate or clay fabric already has

disappeared.

The elasto-plastic part of the applied load-displacement curves of each of the subunits can be described with:

$$S = k1 * exp(k2 * Q) [1]$$

where S = displacement of the platen , Q the applied load and k1,k2 constants. The correlation coefficient R^2 for constants k1 and k2 in Eq.[1] range from 0.87 to 0.99. In fig.9 the relation ship between applied load and natural logaritm of the displacement of the platens for each sub-unit is given. Lines between yieldpoint and failure reflect the e-power relationship between applied load and displacement. All sub-units have the same order for k1 and k2 however exceptions are with sub-unit [At],[Ab] and [V]. Unit [V] with a steep loading curve (large k2) leading to failure at low strenghts as does the cemented sub-unit [At]. Unit [Bb] having a intercept (low k1) with a relative low displacements related to its uncemented, cryoturbated soil structure.

The load at which failure takes place (fig.10) shows an increase with depth within the 'upper' and 'middle' loess, the cryoturbated unit [Bb] seems to have the highest strength, probably caused by a higher density. In the deepest homogeneous unit [Db] underneath this unit however an unexpected low failure strength was measured, due to the presence of clay. The yield load is about the same for the total sequence, but a slight maximum is

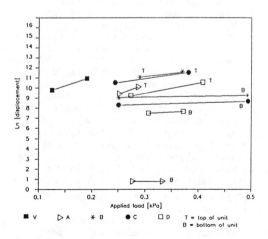

Fig.9   Applied load versus the natural logaritm of the displacement.

679

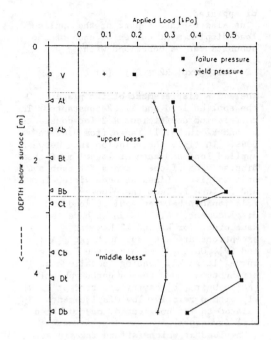

Fig. 10 Applied load versus depth.
Indicated with 'x' are the geotechnical
sub units in the Loess.

present with the carbonate cemented
homogeneous units.

4. CONCLUSION.

1. The geotechnical subdivision of the
Weichselien loess sequence based upon
lithostratigraphy, vane and pocket
penetrometer readings could be verified.
2. With the in-situ shear test, the
cohesion of 7-11 kPa also falls within the
range of values found in triaxial testing
and shows a good relationship to
cementation and clay content. However
with respect to the effective angle of
internal friction, the Weichselien shows
no differentiation. The effective angle
of friction of 29-32 degrees compares well
with the values found in triaxial testing
of Rhineland silt (Schultze & Odendahl,
1967), which indicates a lack of influence
by the cementation after cohesion is
destroyed.
3. Results from the small scale
horizontal plate loading tests appear to
give a better resolution between different
loess units. The elastic loading and
reloading cycles give a good indication of
the difference in behaviour of the

litho-stratigraphical units within the
Weichselien. Failure strengths with the
plate loading tests increase with depth
within the 'Upper Loess' and the 'Middle
Loess' of the Weichselien sequence. The
yielding strength remained, however, about
the same for all units. The tangent
moduli appeared however to be a good
criterion to discriminate the loess
grain structure.
4. The geotechnical sub-division of the
weichselien loess , based on in-situ tests
for this one location, are parallel to the
existing litho-stratigraphical boundaries.
The 'Upper Loess'- 'Middle Loess'
transition can be recognised and
cryoturbated, laminated or cemented layers
are discernible.

5. ACKNOWLEDGEMENT

The author would like to thank the
students Plasman, Molendijk, Clements,
Nijdam and Breukink for enthusiastic
assistance in the laboratory and in the
field.

REFERENCES.

Andersen,S.V.T.,De Vries,H.I.&Zagwijn,W.H.
1960. Climatic change and radiocarbon
dating in the Weichselien loess of
Denmark and the Netherlands. Geologie
en Mijnbouw. 22:38-42
Clevenger, W.A 1958. Experiences with
loess as foundation material. Transact.
Am.Soc.Civ. Engngrs. 123:121-180.
Feda,J. 1967. Stress-strain relationship
for loess soil during shear box test.
Geotech.Conf.Shear Strength Properties
of natural Soils and Rocks, Oslo,
1:187-192
Jaeger,J.C. & Cook,N.G.W, 1979.
Fundamentals of rock mechanics (3rd
ed.). 593 pp. Chapman and Hall, London.
Kahl,W. 1972. Geologie und bodemmechanik
des rheinischen Schluffs.PhD thesis
RWTH Aachen.
Kronieger, R.R. 1988. Development of
engineering-geological maps on
construction materials in south Limburg
Delft progress report 13(1/2):237-253.
Kuyl, O.S. & Bisschops,J.H. 1969 Le
Loess aux Pays-Bas.in:La Stratigraphie
des loess d'Europe (Bull. Assoc.
Franc. Etude Quat.Suppl.), 103-104
Kuyl, O.S.1975 Loess. Grondboor en
Hamer 1:2-12
Lysenko,M.P. 1973 Particle-size as basic
criteria for distinguishing between
loess and loess-like deposits. DAN SSSR
Earth Sci. Sects. 208:186-188

Matalucci,R.V.,Abdel-Hady,M. &
Shelton,J.W. 1970. Influence of grain
  orientation of loess on direct shear
  strength of a loessial soil.Engineering
  Geology 4:121-132.
Meijs,E., Mucher,H., Ouwerkerk,G.,
Romein,A. and Stoltenberg,H. 1983.
  Evidence of the presence of the
Eltviller Tuff Layer In Dutch and
  Belgian limbourg and the consequences
  for the loess stratigraphy. Eiszeit u.
  Gegenwart 33:53-56.
Minkov,M. Evtatiev,D. Alexiev,A.P. &
Donchev,P. 1977 Deformation properties
  of Bulgarian loess soils. Proc. 9th
  ICSMFE, Tokyo, 123:171-179
Mucher,H.J. 1986 Aspects of loess
  and loess-derived slope deposits: an
  experimental and micromorphological
  approach. PhD thesis Univ. of
Amsterdam.
Plasman,S.J. & Kronieger,R.R., 1990. An
  engineering-geological classification
  of the loess deposits in the
  Netherlands. 6th Int. Congr.
I.A.E.G.,
  Amsterdam (in prep).
Ploey,J.de 1973 . A soil mechanical
  approach to the erodibility of loess
  by soliufluction. Revue de Geom. Dyn.
  22:61-70.
Schultze,E. & Odendahl,R. 1967 The shear
  strength of undisturbed rhineland silts.
  Proc.Geot.Conf.,Oslo,1:239-242.

6th International IAEG Congress / 6ème Congrès International de AIGI, © 1990 Balkema, Rotterdam. ISBN 90 6191 130 3

Geological prediction in the Jundushan Tunnel
Prédiction géologique dans le tunnel de Jundushan

Liang Jinhuo & Sun Guangzhong
Institute of Geology, Chinese Academy of Sciences, Beijing, People's Republic of China

ABSTRACT: The Jundushan Tunnel, located in the northern Beijing, is a railway tunnel with 8.46 kilometers long. The tunnel passed through loess, andesite, granite, syenite veins and so on. In the andesite and granite area, collapses were occurred frequently when the tunnel was excavated in these fractured and faulted rocks. Therefore, geological prediction based on geological conditions was conducted in order to reduce the occurrence of collapses. The methods were included geometric methods, special feature method and integrated method. The accuracy of the geological prediction in the Jundushan Tunnel is over 70%.

RESUME: Le Tunnel ferroviaire de Jundushan, situé au Nord de Pékin, est 8.46 kilométres de long, en traversant différentes roches, comme loess, andésite, granite, veine de syénite, etc··· Láffaissement de terrain se produit souvent á cause des fractures et des failles, de roches. La prédiction géologique est donc nécessaire en fin de diminuer la production des affaissement de terrain. La prédiction basant sur les conditions géologiques est réalisée far application de différentes méthodes: ex. méthode géométrique, méthode d'analyse des caractéristiques anormales et méthodes d'analyse synthétique. La précision de la prédiction géologique dans le Tunnel de Jundushan est plus de 70%.

1 INTRODUCTION

Geologic conditions is a major factor controlling tunnel cost, construction speed, excavation and support methods. In many cases, geologic hazards are caused by unexpected poor geologic conditions. Because actural geologic conditions are not well known before tunnel is opened, it is increasingly required to make detail exploration and accurate geologic prediction in various stages of site investigation in order to reduce geologic hazards and tunnel cost.

Recently, a railway tunnel, the Jundushan Tunnel, was constructed in the northern Beijing. At the beginning of the tunnel construction, overbreak and rockfall occurred frequently along veins and faulted zones because of the lack of adequate geologic prediction and support. After that time, geologic prediction was concerned and was conducted by the Institute of Geology of Chinese Academy of Sciences and the Bureau of Tunnel Engineering of Railway Ministry of China in the following aspects:

1. detail surface geologic investigation for determining adverse geologic conditions, such as faults, veins, aureole and groundwater, and related geologic hazards which would probably be encountered in the tunnel opening;

2. selecting methods for the prediction of adverse geologic conditions;

3. modifying the prediction based on the analysis of correlation between surface geology and tunnel geology.

2 GEOLOGIC SETTING

The Jundushan Tunnel is 8.4 kilometers long, 10.5 meters high and 9.5 meters wide, and takes horseshoe shape. The tunnel was excavated by drilling and explosion method from five working parts: east port, west port, No.1, No.2 and No.3 inclined shafts.

The tunnel is situated in the Yanshan Orogen which has been suffered several times of tectonism. The geologic conditions in working area are very complicated. Faults and veins are main adverse geologic structures. The rocks along the tunnel route consist of andesite, coarse andesite, andesite porphyrite, granite, syenite, syenite porphyry, and lamprophyre. These rocks are strong and fractured. Andesite has been intensively fractured around the aureole of granite. In addition, there is a 600 meters long

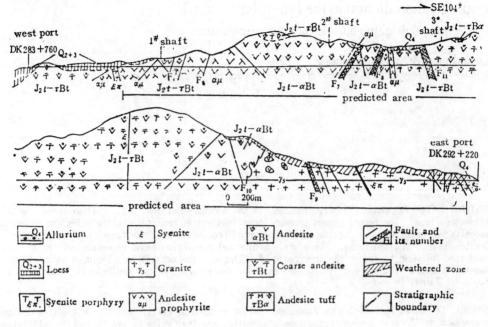

Fig. 1 Geologic section of the Jundushan Tunnel, Beijing

loess in the west port (see Fig.1). In loess area, the tunnel is covered by thin soil. The thickness of covered layer above the tunnel is 12 to 27 meters. Silt-soil layer is filled with groundwater. Groundwater table is above the tunnel crown. Construction was very difficult. In order to know the distribution of silt-soil layer and to dewater, drilling holes were densified near the tunnel. Therefore, groundwater table was under tunnel bottom, collapses were reduced, and construction were speeded up. This paper will discuss the problems of geologic prediction in rock area.

3 MAIN GEOLOGIC FACTORS AFFECTING TUNNELING ACTIVITY

Faults and veins were encountered frequently in the construction of the Jundushan Tunnel. Because they are highly fractured and rich in water, collapse and heavy groundwater inflow were easily taken place along these weak discontinuities. In addition, the aureole around granite was also intensively fractured and rich in clay and groundwater. It was also the factor causing tunnel geologic hazard (see Fig.2). Therefore, the geologic prediction, including the prediction of collapse and heavy groundwater inflow, were replaced by the prediction of faults, veins, and aureole in the Jundushan Tunnel.

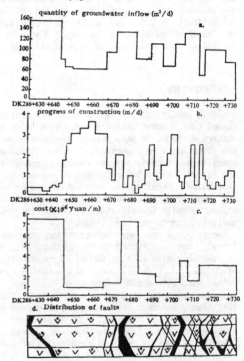

Fig. 2 Correlation of quantity of groundwater inflow, progress of construction, cost with geology from the Station DK286+630 to DK286+735 in the Jundushan Tunnel, Beijing

4 METHODS FOR TUNNEL GEOLOGICAL PREDICTION

Three kinds of geological methods were adopted for the prediction of faults, veins and aureole in the Jundushan Tunnel. Geological methods are ones based on the geologic regulations or special geologic features. The methods for the prediction of faults, vein and aureole include geometric methods, special features method and integrated method. They are basic and important methods in the geologic prediction of the Jundushan Tunnel.

4.1 Geometric methods

Many geologic structures, such as fault, vein, strata layer, aureole, are planar. They can be projected from surface to tunnel level, from pilot tunnel to main tunnel, or from working face to unexcavated part after surface geologic investigation and/or tunnel geologic mapping.

1. Surface projection method

Projection of planar structures from surface to the tunnel level is named as surface projection method. In Fig.3, the horizontal distance d between the exposed point P in the surface and the exposed point P'' in the tunnel of the fault can be expressed by Eq.(1):

$$d = h \cdot ctg\beta' \qquad (1)$$

where h is the vertical distance from P to P' in the tunnel longitudinal section;
β' is the visual dip angle of the fault in the tunnel longitudinal section.

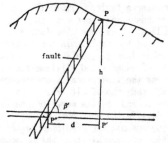

Fig. 3 Projection of fault from surface to tunnel level

The extended length l of the fault in tunnel bottom can be calculated by Eq.(2) (see Fig.4):

$$l = D \cdot ctg\alpha + t / sin\alpha \cdot sin\beta \qquad (2)$$

where D is the width of the tunnel;
t is the thickness of the fault;
α is the angle between the fault strike and the tunnel line;
β is the dip angle of the fault.

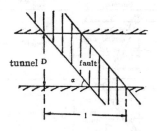

Fig. 4 Schematic diagram of the distribution of fault in tunnel bottom

EXAMPLE: The fault F7, whose attitute is SW230°∠62°, is 1.5 meters thick and exposed in the Station DK286+798 in the surface and 170 meters high above the tunnel level. The direction of the tunnel line is SE104°. The visual dip angle β' of the fault is 47°52' and

$$d = 170 \cdot ctg47°52' = 153.8m$$

$$l = 10.5 \cdot ctg36 + 1.5 / sin36° \cdot sin62° = 17.3m$$

By surface projection method, the fault F7 was predicted to be encountered from the station DK286+633.5 to DK286+651 in the tunnel level. Actually, the fault F7 distributed from the Station DK286+632 to DK286+646. The predicted and the encountered are not greatly different.

In the Jundushan Tunnel, an attempt had been made to compare the predicted faults and veins by projection with the ones encountered in the tunnel opening. Results showed that the correlation of faults in the surface with the ones in the tunnel was good. It means that surface projection method was validable for the prediction of faults in the Jundushan Tunnel. Surface projection method was also valiable for prediction of faults in the other tunnels in the world (Wahlstrom 1964). Nevertheless, its validity was lower for prediction of minor faults whose thickness is less than 0.5 meter in the Jundushan Tunnel.

2. Pilot tunnel projection method

Projection of planar structures from the pilot tunnel to the main tunnel is named as pilot tunnel projection method (see Fig.5). Because of the short projected distance, pilot tunnel projection method is more reliable than surface projection method. In the Jundushan Tunnel, the prediction of most faults by this method was successful, especially the prediction of the fault F9 whose attitute is SE120°∠71°. The tunnel was excavated with less trouble through it. The fault is 47 meters thick and was exposed in the Station DK290+765 in the pilot tunnel of the east port. It was predicted to distribute from the Station DK290+718 to DK290+773, and was encountered from the Station DK290+720 to

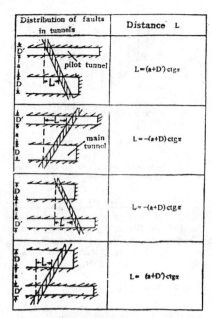

Distribution of faults in tunnels	Distance L
pilot tunnel	$L = (a + D') ctg\alpha$
main tunnel	$L = -(a + D) \cdot ctg\alpha$
	$L = -(a + D) \cdot ctg\alpha$
	$L = (a + D') \cdot ctg\alpha$

Fig. 5 Projection of faults from pilot tunnel to main tunnel

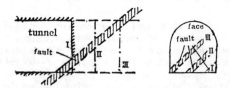

Fig. 6 Schematic diagram of the distribution of fault in tunnel longitudinal section and three working faces

DK290+770 in the main tunnel. In the Jundushan Tunnel, pilot tunnel projection method was not validable for the prediction of the faults whose thickness is less than 0.3 meter, because their distribution is not regular.

3. Working face projection method

Projection of faults and other planar structures from the tunnel face to the unexcavated part ahead of the face is named as working face projection method. Planar structures can be projected horizontally or vertically to unexcavated part with fair accuracy on the basis of their attitude, width, exposed position (see Fig.6). This method is a useful and convenient one in the tunnel construction. For example, a fault whose attitude is $NW286°\angle53°$ was encountered in the underneath part of the working face in the Station DK286+677.5 in the Jundushan Tunnel on May 16, 1986. The fault is 0.05-0.10m thick with white fault gouge. When the tunnel was continually excavated toward east, the fault was exposed in the upper part of the face and the thickness of the fault is 0.5m in the Station DK286+680 (see Fig.7). Rock is highly fractured and is rich in groundwater near the fault. The quantity of groundwater inflow was 500 tons per day along this fault. Therefore, the fault was predicted to be encountered and collapse would probably occur in the crown of the tunnel from the Station DK286+681 to DK286+684.5. In fact, collapse occurred in the crown at the Station DK286+684 and about 10m³ fragment and mud fallen.

4.2 Special feature method

There are some special features near a fault. Using these special features as the indicator of existence of a fault in the tunnel construction is named as special feature method. The special features near a fault mainly include:

1. Joint-spacing. Joint-spacing becomes smaller and smaller or joint's frequency does higher and higher with the approach to a fault (see Fig. 8).

2. Joints' attitude. Joints' attitude becomes more and more dispersed toward a fault (see Fig.9).

3. Filling materials. The filling materials in joints are thick near a fault. Moreover, many of them are mud and few are fragment.

4. Groundwater. In most cases, the fault is a good acquifer which permits relatively easy flow of groundwater through it. In the area with enough water source, groundwater inflow will rapidly increase when the tunnel is excavated through the fault (see Fig.10).

5. Strain Minerals. There are some strain minerals, such as chlorite, mica, wallonstonite, grown in or near major a fault.

In the field, not all of these special features are grown near a fault. Typically, only one or two features are apparent in the fault. Nevertheless, they are the indicator of the occurrence of the fault. In the construction of the Jundushan Tunnel, the abnormal distribution of joints' frequency and filling materials were regarded as the indicators of occurrence of the fault F3 and the fault F5, the ones of groundwater, and pyrite were for the fault F7, and so on. In the east port of the Jundushan Tunnel, the aureole was predicted by the abnormal distribution of mineral size which becomes small toward the aureole, when the tunnel was excavate from granite to andesite.

4.3 Integrated Method

Integrated method is a method synthesizing geometric methods and special feature method for the prediction of faults. It includes two cases as follows:

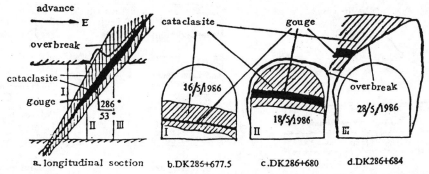

a. longitudinal section   b.DK286+677.5   c.DK286+680   d.DK286+684

Fig. 7 Distribution of fault in face and longitudinal section near the Station DK286+677.5 in the Jundushan Tunnel

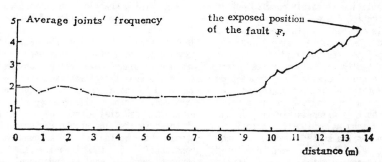

Fig. 8 Average joints' frequency near the fault F_7 in the Station DK286+632 in the Jundushan Tunnel

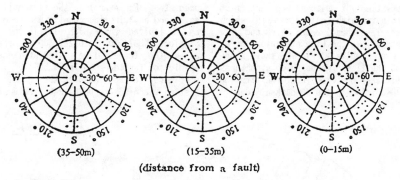

(35–50m)   (15–35m)   (0–15m)

(distance from a fault)

Fig. 9 Polar stereonet of joints' attitude in different distance from the fault near the
Station DK286+500 in the Jundushan Tunnel

1. For the prediction of faults found in the surface geologic investigation, first step is to project faults from surface to tunnel level using surface projection method; second one is to identify the indicator of faults existence; and the last one is to predict the distribution of faults in tunnel level using working face projection method and/or pilot tunnel projection method. At the same time, the linear geomorphic structures can also be regarded as a faulted zone and be predicted their distribution in the tunnel level.

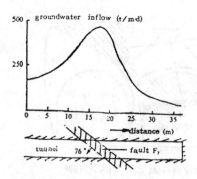

Fig. 10 Distribution of groundwater inflow
near the fault No. 7 in Jundushan tunnel

2. For the faults covered by weathered soil and alluvium, the prediction is mainly carried out in the tunnel construction stage. The procedure is that first step is to identify the indicator of fault's existence, second one is to find the exposed position of the faults, and the last one is to project the faults to the unexcavated part ahead of the face using working face projection method and/or pilot tunnel projection method.

For the prediction of faults by integrated method, a flow chart is given in Fig.11. In the Jundushan Tunnel, many major faults, especially the faults found in the surface, such as the faults F3, F5, F7, F8, F9, were predicted successfully by integrated method. Small faults were predicted after analyzing the correlation of geomorphic structures with faults encountered in tunnel excavation. Results showed that flat gully is correlative with faulted zone in which most faults are less than 1 meter thick and few are 1 to 3 meters. In the high elevation area, there are fewer faults. Fig.12 shows the correlation of the linear geomorphic structures with the faults encountered from the Station DK285+300 to DK287+100 in the Jundushan Tunnel.

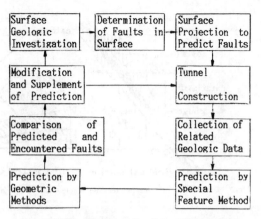

Fig.11 Flow Chart of the Integrated Method
for the Prediction of Faults

In this area, the faults F5, F7, F8, F11 are exposed in the surface. They were projected from surface to tunnel level before construction (see Table 1). When tunnel was excavated, an attention was paid to the abnormal distribution of joints' attitude, spacing, filling materials, groundwater, minerals, especially near the predicted position of these faults. The ground was supported by tube steel when the faults were encountered. In the construction of the tunnel, the correlation between surface geology and tunnel geology had been analysed. Based on the correlation, the unanticipated faulted zones were successfully predicted. For example, the flat gully between No.2 and No.3 inclined shafts had been regarded as a faulted zone which distributes from the Station DK287+030 to DK287+400 in the surface along the tunnel line. The attitude of the predicted faulted zone is SW250°∠75°. It was predicted to cut through the tunnel from the Station DK286+967 to DK287+340. Actually, many

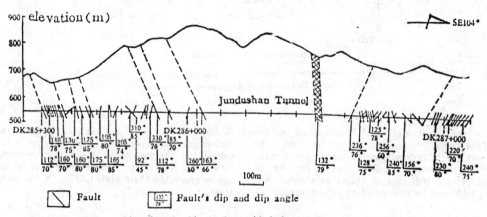

Fig. 12 Comparison of geomorphic structures with faults encountered from the Station
DK285+300 to DK287+100 in the Jundushan Tunnel, Beijing

688

Table 1 Comparision of the predicted and encountered faults F_5, F_7, F_8, F_{11} in the Jundushan Tunnel

Faults		F_5	F_7	F_8	F_{11}
predicted	attitude	SW250°∠63°	SW230°∠62°	SW238°∠75°	NE27°∠86°
predicted	thickness	1.0	1.2	1.1	3.0
predicted	distribution	DK285+315—DK285+375	DK286+600—DK286+680	DK286+915—DK286+955	DK287+370—DK287+480
encountered	attitude	SW252'∠60°	SW232°∠76°	SW230°∠68°	NE35°∠85°
encountered	thickness	1.2	1.5	1.5	2.5
encountered	distribution	DK285+355—DK285+360	DK286+632—DK286+646	DK286+935—DK287+950	DK287+380—DK287+406

faults whose attitude is SW235°∠75° were encountered from the Station DK286+960 to DK287+577 in the Jundushan Tunnel. In this area, groundwater was rich and the construction was very difficult.

In the east port of the Jundushan Tunnel, a pilot tunnel was excavated away 25 meters from the main tunnel. The pilot tunnel is on the right of the main tunnel. In this area, the pilot tunnel projection method and surface projection method were successfully used to predict the fault F9 and the aureole around granite. In addition, pilot tunnel projection was also used to predict many faults and lamprophyre veins which are covered in the surface and whose distribution is controlled by the fault F9 and aureole. Because of the successful prediction, a construction speed record of 240 meters per month was made and the most difficult part was passed through without a hicth. For example, in January 1987, when the aureole was exposed in the pilot tunnel, a great deal of fragment and mud run into the tunnel which was named Running Ground by Wahlstrom (1973). The running ground was also encountered in the construction of the main tunnel in April 1987. Running ground is a rare phenomenon in China. Running ground, which occurs in the major fault or highly fractured aureole, is a phenomenon that a large amount of fragment and mud containing enough water runs or flows toward tunnel under high or moderate water pressure. In the Jundushan Tunnel, running ground was not anticipated before the construction. But because of the successful prediction by pilot tunnel projection method, the excavation of the main tunnel through the aureole was not trouble with running ground.

5. CONCLUSIONS

In the Jundushan Tunnel, collapse and heavy goundwater inflow were main geologic hazards. Because they were caused by faults and other weak geologic structures, the object of tunnel geologic prediction was turned to the prediction

of faults, veins, aureole and goundwater, especially to faults. In tunnel geologic prediction, three kinds of geological methods were adopted which are geometric methods, special feature method and integrated method.

Surface projection method used for the prediction of faults found in the surface. But its accuracy was limited by projetcd distance. The error was within 30 meters. It was permitted in the tunnel design stage, but it seemed to be too large for the construction of the tunnel. In the construction of the Jundushan Tunnel, the effect of prediction by pilot tunnel projection method is better when the pilot tunnel is 50-100 meters beyond the ahead of the main tunnel. For prediction of faults which obliquely cut through tunnel in which $\alpha < 50°$ (see Fig.4 or Eq.(2)) by working face projection method, the effect was also desirable. Special feature method was useful for major faults whose thickness is greater than 1-2 meters.

In the geological prediction of the Jundushan Tunnel, the results showed that the integrated method had been successfully used to predict overall of 70% faults, veins and aureole.

ACKNOWLEDGEMENT

Field studies in the Jundushan Tunnel was made possible by financial suport from the Bureau of Tunnel Engineering, Minister of Railway of China. The authors wish to thank our colleagues Wu Zhiyong, Jiang Zhongrong, Tang Zhenqing, Mei Ji'an for their help in the field studies.

REFERENCES

Wahlstrom, E. E., 1964. The validity of geologic projection: a case history. Economic Geology, vol.59, pp.465-474.

Wahlstrom, E. E., 1973. Tunneling in rock. Elsevier Scientific Publishing Company, Amsterdam, 250p.

6th International IAEG Congress / 6ème Congrès International de AIGI, © 1990 Balkema, Rotterdam. ISBN 90 6191 130 3

Geological study of flow slide sensitive sediments

Etude géologique sur la sensibilité des sédiments pour la rupture par écoulement

C. E. Ligtenberg-Mak, P. V. F. S. Krajicek & C. Kuiter
Delft Geotechnics, Delft, Netherlands

Abstract: Flow slide is a large scale sediment transport that often damages the coastal protection works. The mechanism that causes the flow slide is liquefaction. A geological study was made of the occurrence and process of flow slides. This study leads to a criterion that easily traces the places where flow slides can occur. It appears that both tidal and fluvial channel fills are sensitive for liquefaction.

RESUME: Solifluxion est un mécanisme de transport de sediment sur une grande échelle qui endommage souvent les digues. Le mécanisme qui cause la solifluxion est liquéfaction du sable. Une étude geologique d'après la comportement et le phénomène de solifluxion est faite. On a pu tirer un critère de cette étude qui permet simplement de localiser les lieux ou les solifluxions pourraient se produire. Il parait que les remblements des chenaux de marée et des chenaux fluviaux peuvent être sensible à la liquéfaction.

1 INTRODUCTION

Zeeland, the southwestern part of the Netherlands consists of the estuary of the river Scheldt. A part of the area has been enclosed by a system of earth dams between the islands, but the mainstreams, the Eastern Scheldt and the Western Scheldt, have been left tidal basins. Large tidal streams generate deep channels which move dynamically and may sometimes approach the coastline quite closely. At these locations the phenomenon of flow slides threatens the dikes which protect the low lying country against the sea. At places where a flow slide may occur, the seadefences must be protected by underwater constructions.

During a flow slide liquefaction of a sand layer followed by a flow of sand occurs causing large areas to 'disappear' completely. Areas of several square kilometers have been reported in the Eastern Scheldt. Two conditions have to be fulfilled for a flow slide to occur: a thick sand layer with low density must be present and the geometry must be favourable, a low lying area must be present to which the heavy liquid can flow. The sandflow ends when a gradual equilibrium slope is generated (Lindenberg 1985)(see figure 1). The mechanism of flow slide has extensively been discussed in the soil mechanical literature (Lindenberg 1985, Kuiter 1986). From the analysis of the mechanism the density of the sand has been identified as the essential parameter which describes the sensitivity of a sediment for flow slides.

The initial mechanism of a flow slide is a deformation of the grain skeleton caused by an increase of the shear stress. In loosely packed saturated sand the pore space will decrease and the water tension will increase. When the water tension is so high that the grains loose their contact the ground will behave as a heavy liquid. This is called liquefaction. Only when it is possible for the sand to flow away a flow slide actually can develop.

A flow slide can only occur when the sand layer is loosely packed. The parameter describing the sensitivity for liquefaction is the critical density of the sand.

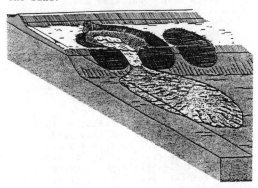

figure 1 An impression of a flow slide.

691

To locate the sand layers with a density lower than the critical density extensive field- and laboratorytests are nessecary to determine the actual density of the groundlayers as well as the critical density of the sand in question.

Since a complete mapping of the shoreline by borings is time consuming and expensive, it was decided to investigate the mechanism of flow slide from a geological point of view. The purpose of the study was to investigate the feasibility of geological characterization for location of the flow slide sensitive areas.

2 METHODS

A location at the southern border of the Eastern Scheldt, the northern side of Noord Beveland where many flow slides have occurred in the past (Wilderom 1961-1973, Rijkswaterstaat 1979) is selected to investigate the sedimentary processes in a flow slide.

Historical maps and old depth soundings are analysed to locate the places of the old flow slides and the change in morphology of the coastline. Historical maps can also give an indication where young channel fills are present in the underground. Old documents and literature is analysed to locate and date the old flow slides.

To study the groundlayering drill holes are made at the place of the old flow slide and landinwards where the underground is supposed to have the original stratification. These continuous borings give information about the stratification, composition and deposition of the sediment before and after the flow slide.

An inventory of all the flow slides known along the Eastern and Western Scheldt in Zeeland (Rijkswaterstaat 1979), is compared to the geological information of the area identifying deposits that are possibly sensitive for liquefaction according to the sedimentary processes. The correlation leads to a conclusion about the reliability of the concept for prediction of the occurrence of sand that is sensitive for liquefaction.

3 RESULTS

3.1 Geology of Zeeland

The geological study is situated in Zeeland because many of the flow slides occurring along the Eastern and Western Scheldt are described in the literature (Wilderom 1961-1973, Rijkswaterstaat 1979). For this reason the geology of Zeeland is described briefly (see figure 4)

(Rijks Geologische Dienst 1978).

At 25 meters below mean sea level lays the top of the fluvial deposits of the pleistocene Tegelen-Formation consisting of sand with gravel and some clayey, sandy silt. From 25 meters to 10 meters below mean sea level the pleistocene Twente-Formation is present. It consists of fine grained aeolean sand and medium grained fluvioperiglacial sand.

On these pleistocene formations lays the Westland-Formation, which contains all the holocene marine and perimarine deposits of the western part of The Netherlands. In the Holocene period there are two transgressive phases(Calais- and Dunkirk-Deposits) which are separated from each other by a regressive phase. They both consist of sandy tidal-channel fills that may have eroded former deposits and of clay deposited on subtidal flats. During the regressive phase a peat layer developed which is called the Hollandveen. The Dunkirk-deposits lay at the surface. All the deposits are present in the underground of Zeeland, only at the places with Dunkirk and Calais channel fills the older deposits are eroded. The channel fills are several meters to more than thirty meters thick.

3.2 Location and morphology.

The selected location for the investigation is situated along the soutern border of the Eastern Scheldt, the northern border of the island Noord Beveland (see figure 2). From this location it is known that at least twenty-one flow slides occurred between 1860 and 1940 (see figure 2). Some of those flow slides damaged the seadefences and polders were lost (Wilderom 1961-1973, Rijkswaterstaat 1979).

The sequence of the flow slides shows that the location of a new flow slide is in many cases situated between two older ones (see figure 2).

From depth soundings along the coastline of Noord Beveland shortly after the occurrence of the flow slide it appears that a more or less circular depression has appeared. This depression slowly disappeared afterwards through sedimentation of the river (figure 3).

3.3 Description of the continuous borings

The continuous borings are made at the places of the flow slides and more landinwards where the underground is supposed to have the original stratification.

In the continuous borings made at the places of the old flow slides the next stratification was found (from below to above)(see figure 4):

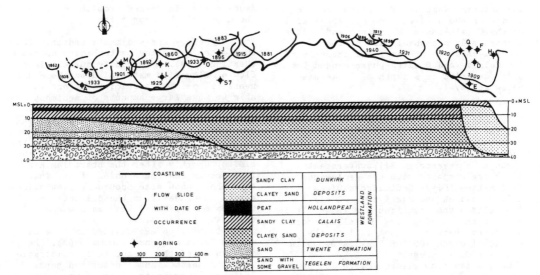

FIGURE 2 THE STRATIFICATION AND MORPHOLOGY OF THE STUDIED AREA

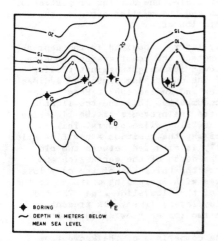

figure 3 Depthsoundings of the flow slide
of 1909

UNIT 6	CLAYEY SAND	DEPOSITED
UNIT 5	SAND AND CLAY LAYERS	AFTER THE
UNIT 4	CLAY WITH SILT LAYERS	FLOW SLIDE
UNIT 3	SAND WITH FINING UPWARD	DEPOSITED DURING AND SHORTLY AFTER
UNIT 2	STONES, PIECES OF CLAY AND PEAT	THE FLOW SLIDE
UNIT 1	SAND WITH CLAY OR SILT LAYERS	UNDISTURBED SEDIMENT

FIGURE 4   STRATIFICATION AT THE PLACES OF THE
OLD FLOW SLIDES

- unit 1, sand with thin clay or silt layers. On the top of this layer the original internal structures are disturbed,
- unit 2, stones, pieces of clay and peat; this unit is one to two meters thick,
- unit 3, sand without silt or clay, a clear fining upwards is present in this unit, one to six meters thick,
- unit 4, clay with thin silt layers, the number of silt layers increases to the top of this unit, six to twelve meters thick
- unit 5, thin clay and sand layers, the sand becoming more pronounced to the top of the unit, this unit is four to six meters thick,
- unit 6, clayey sand, one to three meters thick.

In the continuous borings made more land-inwards statifications as described in 3.1 were found. For a schematisation of the underground see figure 3. In the most western part of the area, where no flow slides occurred the following stratification is found: the deposits of the Tegelen-Formation, the Twente-Formation and the tidal flat and peat deposits of

693

the holocene Westland-Formation. In the middle of the studied area a channel fill of the Calais-phase is present and at the eastern side a channel fill of the Dunkirk-phase is present. The channels of the Calais- and Dunkirk-phase eroded the older deposits to a depth of 35 meters below sealevel.

3.4. Comparison of the flow side inventory and the geology of Zeeland.

The location of the flow slides known through an inventory (Rijkswaterstaat 1979) are compared with the geological information (Rijks Geologische Dienst 1978, Krajicek and Mak 1988). Of circa 700 flow slides the locations are known. It is obvious that most of the flowslides occur at places where a channel fill of the Calais- (circa 100) of Dunkirk-Deposits (circa 600) is present.
   Only a few flow slides (circa 10) occur at places where no holocene channel fill is present. A flow slide near Scherpenisse on Tholen on the northern border of the Eastern Scheldt is an example of one of those flow slides. At that location silty sand which could be a pleistocene channel fill was found in continuous borings. Also in the other cases a pleistocene channel fill may be present but the information on pleistocene channels available has insufficient detail to give conclusive evidence.

4. DISCUSSION

Comparison of the sequence in the continuous borings at the location of the old flow slides with the undisturbed layer sequence shows that the first unit is the undisturbed sediment. The second unit consists of dikematerial, pieces of clay and peat that tumbled down the slope during and after the flow. The third layer sedimentated immediately after the flow when the particles started to settle down from the heavy liquid. In this way the fining upward in this layer arised.
   The top of unit 3 forms the circular depression that is recorded by the depth soundings shortly after the flow slides. This depression is filled up by sedimentation.
   In a depression the current velocity is low and deposition of clay with sand layers dependent of the sediment influx is possible. When the depth of the depression decreases the current velocity increases. By higher current velocities only the coarse particles are deposited, so less clay and more sand is deposited. In this way it can be explained why the sand content of the deposit increases in the units 4, 5 and 6. The sand content in the

deposit also increases from the dike to the mouth of the depression.

The morphology and sediments resemble the turbidites deposited by turbidity currents (Reading 1978). A turbidity current is a flow of liquefied material down the continental slope. The turbidity current is able to move large amounts of sediment over several kilometers. A flow slide is smaller but the deposits and the sedimentary process are comparable. Through the turbulence the turbidity current is in the beginning erosive at the front of the turbidite. Shortly after the start laminar flow occurs and the erosion stops. That this phenomenon also occurs in flow slides can be concluded from the disturbances at the top of unit one, the basis whereover the heavy liquid flows.
   The turbidites are divided in five units of the Bouma sequence (Bouma 1962). The first unit of this sequence is comparable with the units 2 and 3 described here. That the other units of the Bouma sequence are absent can be understood by the difference in scale, the smaller proportion of the flow and the smaller depth of the sedimentation area.

The clay layers deposited in the depressions protect the coast against erosion and through its weight against flowing away of the underlying liquefied sediment (Krajicek and Kuiter 1983). When a clay layer becomes too thin and too light to compensate the pressure of the liquefied sand a new flow slide occurs. This explains why the position a new flow slide generally is situated between two old ones. The depth of the depression and therefore the thickness of the clay layer decreases with increasing distance to the centre of the flow slide area. The thinnest clay layer, if any, is present between two former flow slide positions.

From the comparison of the known flow slides with the sedimentary facies it can be concluded that flow slides occur at places where (tidal) channel fills are present. This is understandable because from literature (Reineck and Singh 1980) it is known that tidal channel fills are deposited very quickly without strong waveaction so the grain arrangement is very loose and therefore the density may be below the critical density. From the Scherpenisse situation it can be concluded that fluvial channel fills which are filled up rapidly are also sensitive to liquefaction.
   At the places where a flow slide occurs the underground exists of holocene or pleistocen channel fills. At places where no flow slides occur are no channel fills present; or the morphology of the coast is not steep enough and the sand can not flow

away; or there is no clay layer or protection work that prevent the flowing away of the sand.

5. CONCLUSIONS

It can be concluded that:
- tidal channel fills and rapidly deposited fluvial channel fills are sensitive for liquefaction. Through mapping of those sediments the potential places for flow slides are located.
- a flow slide occurs at places where sand sensitive for liquefaction is present, where the sand can flow away and where the clay layer is to thin to compensate the pressure of the sand. With these characteristics a prediction can be made for the chance of occurence of a flow slide.

This criterion for tracing the locations with sand that is sensitive for liquefaction is already used for the borders of the river Scheldt and the river Dordtse Kil in the Netherlands.(Krajicek and Kuiter 1989, Krajicek and Mak 1988)

6 ACKNOWLEDGMENTS

This study has been made possible by financial support from Rijkswaterstaat Directie Zeeland. We want to thank dr. J.K. van Deen for comments on early versions of the manuscript.

REFERENCES

Bouma,A.H.,(1962). Sedimentology of some flysch deposits: A graphic approach to facies interpretation,168 pp,Elsevier Amsterdam.

Lindenberg,J.,(1985). Verslag inventarisatie adviespraktijk zettingsvloeiingen, report Grondmechanica Delft CO-416509, 134pp.

Krajicek,P.V.F.S.,& Kuiter,C.(1983). Het ontstaan van zettingsvloeiingen, report Grondmechanica Delft CO-416096,63pp.

Krajicek,P.V.F.S.,& Kuiter,C.(1989). Dichtheidsonderzoek Dordtse Kil Fase 1, report Grondmechanica Delft CO-287781, 10pp.

Krajicek,P.V.F.S.,& Mak,C.E.(1988). Onderzoek valgevoeligheid oud-wadzand in het Ooster- en Westerscheldebekken,report Grondmechanica Delft CO-280131,10pp.

Kuiter,C.(1986). Zettingsvloeiingen bij vooroevers. Polytechnisch Tijdschrift Civiele Techniek 41,47-51.

Reading,H.G.(1978). Sedimentary Environments and Facies, 569pp.

Reineck,H.E.,& Singh,I.B.,(1980). Depositional Sedimentary Environments,549pp, New York.

Rijks Geologische Dienst,(1978). Toelichting bij de geologische kaart van Nederland 1:50.000,kaartblad Beveland,138pp, Haarlem.

Rijkswaterstaat Studiedienst Vlissingen (1979). Resultaten van het vooroeveronderzoek langs de Zeeuwse Stromen,nota 75.2,33pp.

Wilderom,M.H.,(1961-1973). Tussen Afsluitdammen en Deltadijken,Deel I-IV, Middelburg.

Dispersive soils – A constraint for embankment dams in outer Himalaya

Sols dispersés-une contrainte pour les barrages en terre dans l'Himalaya extérieur

P. L. Narula & Y. P. Sharda
Geological Survey of India, Lucknow, India

Abstract: The limitations of standard characteristics of soils for evaluating their suitability as core materials in regard to dispersivity and internal erosion have been brought out for medium height embankment structures being executed in outer Himalaya for watershed management. The data base of diagnostic tests for dispersivity has been discussed, correlated with relationships worked out by various investigators and disparities have been logically explained. Based on the dispersivity of clayshale bands in top stratigraphic levels of Upper Siwalik rocks, inferences on the environment of their deposition have been drawn.

Resume: Les limitations de caracteristique étendard de sols pour l'évaluation de leurs convenir tandis, materiaux moelle avec considerar á dispersivity et des erosion ai été discuté pour les barrages en terre des hauteur median en construction á ce moment en l' Himalaya extérieur pour management de ligne de partage des eaux. Les données basement de essais diagnostiques pour dispersivity ai été discuité, mettré en correlation entre rapports rendré par des investigateurs diverses et les disparités ai été expliqué logiquement. Par suite de dispersivity des arigile–argileux dans le haut de niveaue stratigraphique, des roches, Siwalik supérieur, inférences sur les environment de leurs déposition ai été etabli.

1 INTRODUCTION

The economical design and execution of embankment structures is, primarily, dependent on the availability of appropriate and adequate quantities of fill materials in the vicinity of a particular site. The standard test procedures, many a times, are not adequate to predict the piping performance of the impervious soils used as hearting materials resulting in serious damage by internal erosion.

1.1 The Kandi Watershed Management is an integrated scheme to develop the backward, submountainous areas, extending for a length of about 250 km. in Ropar and Hoshiarpur districts of Punjab, India, A component of this Watershed Management project envisages construction of medium height earth and rock-fill dams across a number of torrents in this tract, which carry enormous amount of water alongwith suspended and bed load materials during flash floods in the monsoon periods,

damaging vast agricultural land as well as communication routes. These storage dams have been planned with a view to moderate the flash floods and utilise the storage for irrigation purposes during lean periods through regulated releases. All the embankments are located in the poorly cemented sandrocks and clayshales exposed in the outer Himalaya. These embankments necessitated constraining the designs for the availability of fill materials in the vicinity, though many of the physical characteristics of these materials are not ideally suited. With this constraint in mind investigations of fill materials were carried out for the borrow areas which served as major input for the embankment designs.

1.2 The first author in his capacity as investigating Engg. Geologist as well as Member of the World Bank Panel of Experts and the second author as investigating Engg. Geologist have

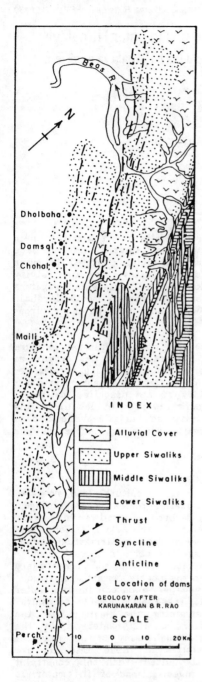

Fig.1 Geological map of Kandi watershed and the vicinity.

INDEX

Alluvial Cover

Upper Siwaliks

Middle Siwaliks

Lower Siwaliks

Thrust

Syncline

Anticline

Location of dams

GEOLOGY AFTER
KARUNAKARAN & R.RAO

SCALE

10   0   10   20Km

been associated with the assessment of the fill materials for Dholbaha dam, Damsal dam, Chohal dam, Maili dam and the Perch dam constructed or under construction in the Kandi Watershed. The test results of the impervious core materials, including the four diagnostic procedures for verifying the dispersivity of the soils, have been assessed and discussed in this paper and inferences have been drawn regarding the possible causative factors as well as the environment of deposition of the Upper Siwalik clayshale which is the source for core material.

2 GENERAL GEOENVIRONMENTAL SETTING

The Kandi watershed tract in Punjab, India is the outermost low Siwalik ranges with sub-montane undulating zone of fanglomeratic deposits which merge into the Indo Gangetic plains. In the Watershed management the outermost hilly sub-catchments are proposed to be provided with check dams for mitigation of the flood hazard causing extensive damage to cultivated land as well as providing additional irrigation facilities in the submontane plains. Out of a number of sub-catchments to be developed, the authors have taken five case studies viz. Dholbaha Dam Project, Damsal Dam Project, Chohal dam, Maili dam and Perch dam which have either been constructed or are in the advanced stages of designs (Fig.1). All the above case studies relate to earth and rock-fill dams, the heights of which vary between 25 m. and 38. 5m., with central or upstream inclined cores.

2.1 These structures are located on the southern limb of an anticline exposing an alternate sequence of poorly cemented Upper Siwalik sandrock and clayshale. The axis of this anticline almost coincides with the northern boundary of all the sub-catchments in the Kandi Watershed. The loosely cemented and poorly consolidated Quaternary sediments and the associated bad land topography precludes the construction of Masonry/concrete check dams.

2.2 There are abounding resources of impervious core materials in the form of thick exploitable clayshale bands, residual soils on the clayshale members as well as alluvial terraces.

Table 1. Physical characteristics of hearting materials of Kandi Water Shed dams in Punjab, India.

Location of Dam	Height in (m)	Quantity of Imp. material in m³	Proctor Max-Dry Density (Kg/m³)	Optimum Moisture content (Percent	Specific gravity	Effective 'C' K Pa	Angle of Internal friction (effective	Permeability (m/sec.)
DHOLBAHA	38.5	1,40,000	2050	11.55	2.66	14	30°	1×10^{-8}
PERCH	25	38,000	1808	14.5	2.645	24	23.85°	4.45×10^{-10}
DAMSAL	32	71,000	1820	15.8	2.626	12	24.5°	4.96×10^{-9}
CHOHAL	26	.-.	1820	13.5	2.66	40	27.7°	5.62×10^{-9}
MAILI	33	.75,000	1900	14.3	-	38.5	18°	3.16×10^{-9}

The conventional physical characteristics like the gradation curves of different fractions, the Atterberg's limits, the shear parameters, the coefficient of permeability, the Procter density as well as compressibility of these soils indicated their general acceptability as hearting materials. The conventional physical characteristics used as input parameters in the design along with salient features of the projects under study are given in Table-I.

2.3 It is seen from table 1 that shear strength of soils in the borrow areas for Dholbaha dam is quite good and the permeability is of the order of 10-8 m/sec. These soils are either silty sand or sandy silt with low plasticity index and these character.stics are controlled by the bedrock lithology.(Fig.2). The soils of Damsal dam borrow area are, predominantly sandy silts with plasticity index varying between 5 and 15. The effective Cohesion is 0.012 M Pa and effective shear friction angle is 24.5°. The permeability is of the order of 4×10^{-9} m/sec. The Chohal borrow area comprises soils of sandy silt, silty sand and clayey silt category with average plasticity to be around 15. The permeability and the shear characteristics are of the same order as for the Damsal project.

For Maili and Perch dam projects the impervious material was proposed to be borrowed from the Upper Siwalik

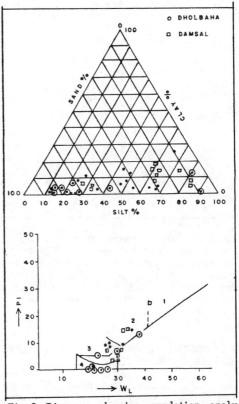

Fig.2 Diagram showing gradation analysis & relationship between PI & WL for samples from Dholbaha and Damsal borrow areas. Erodibility zones after Arylanandan (1983). Encircled samples are dispersive in nature on consensus.

clayshale bands which could be categorised as clayey silts and silty clays. The Maili clayshales are of low to medium plasticity (Fig.3),while those of the Perch dam are of Moderate to high plasticity (Fig.4).

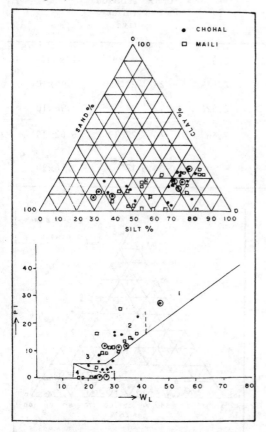

Fig.3 Diagram showing gradation analysis & relationship between PI & WL for samples from Chohal & Maili borrow areas. Erodibility zones after Arulanandan (1983). Encircled samples are dispersive in nature on consensus.

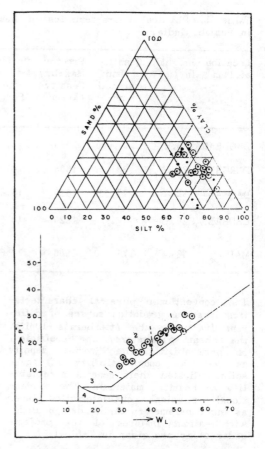

Fig.4 Diagram showing gradation analysis & relationship between PI & WL for samples from Perch borrow area. The erodibility zones after Arulanandan (1983). Encircled samples are dispersive in nature on consensus.

3 POTENTIAL DISPERSIVITY OF IMPERVIOUS HEARTING MATERIALS

Since the recognition, some 25 years ago, the hazard which dispersive soils might pose to the earth and rockfill dams, a lot of research has been done in a number of countries, demonstrating the extent of risks when such soils are used in embankment structures. With this background in view, the diagnostic tests to ascertain the dispersiveness of the soils being used as core materials in the Kandi Watershed projects were conducted. These special tests include Pin Hole Test suggested by Sherard et. al. (1976), the crumb test or aggregate coherence test of Emerson (1954), SCS dispersion test, developed by U.S. soil conservation Service and reported by Sherard, Decker & Dunnigan (1976) and chemical analysis of pore water extract, (Sherard, 1976). The results obtained are presented in the Table-2.

Table 2. Results of special tests for dispersivity.

PROJECT	Crumb Test		Pin Hole Test		SCS Test		Chemical Analysis of Pore Water				
	Samp-les of Grade III & IV. (%)	Devia-tion from consensus. (%)	Samp-les C Eros-ion >2.5% (%)	Devia-tion from consensus. (%)	Samp-les with Dispersion (%)	Devia-tion from consensus. (%)	Samp-les Dispersive (%)	Devia-tion from consensus (%)	SAR <1.7 (%)	SAR 1.7-2.9 (%)	SAR >3 (%)
DHOLBAHA	18.8	22.2	40	35	50	20	8.3	33.3	91.3	8.7	-
PERCH	42.3	57.7	61.5	27	77	27	69.3	11.5	34.6	23.1	42.3
CHOHAL	17.6	31.25	31.2	18.7	41.1	6.2	75	41.7	50	8.3	41.7
DAMSAL	0	-	6	-	16.7	-	87.5	-	6.2	31.2	62.5

3.1 The results of different tests on same samples differed, hence the final conclusions whether the sample was dispersive or not was drawn on the basis of consensus arrived at after considering the results of all the four tests. After arriving at the conclusions, the reliability analysis was carried out to determine the dependability of each test. This study indicated that in case of Dholbaha dam, the results of crumb test deviated from conslusions by 22.2%, those of pinhole test by 35% and those of S.C.S. and chemical analysis of pore water extract deviated from conclusions by 22% and 33.3% respectively (Sharda and Narula, 1987). In case of Chohal dam, results of crumb test indicated deviation from conclusion by 31.75%, those of SCS test by 6.25% and deviation of 18.7% and 41.67% was observed in case of Pin Hole and chemical analysis of pore water extract test respecti-vely, In case of Perch dam the results of crumb test deviated from conclusion by 57.67%, those of Pin Hole test by 27.0% , results of SCS test by 26.72% whereas the results of chemical analysis of pore water extract deviated by 11.54%. The reliability tests indic-ate that SCS test was found to be most dependable in case of impervious material from Dholhaha and Chohal, borrow areas where as chemical analy-sis of pore water extract was found to be most dependable in case of borr-ow area of Perch dam project. Ryker (1977) has surmised that SCStest is the most reliable diagnostic test; but the results obtained at various sites in Kandi Watershed have, how-ever, demonstrated that such a general-isation is hazardous. It has been obse-rved that majority of the chemical test results from Perch, Damsal and Chohal borrow areas plot in Zone A and Zone C, while those of Dholbaha project plot in Zone B (Fig.5).

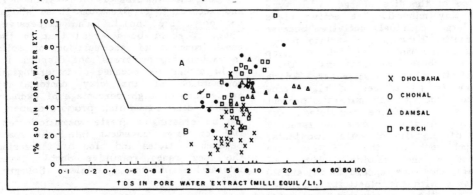

Fig.5.Results of pore water extract. Zone A: Dispersive, Zone B: Nondispersive, Zone C: Transition. X-Dholbaha samples, O-Chohal samples, Δ-Damsal samples & □-Perch samples.

4 DISCUSSIONS

The dispersivity, colloidal activity or erodibility of the soils could be attributed to the mineralogical composition of the clay fractions, the plasticity index or presence of excessive percentage of exchangeable sodium ions present in the soil in a given geoenvironment. The diagnostic tests specified for designating a particular soil to be dispersive in nature utilises these attributes. Arulanandan (1983) suggested a correlation of the erodibility of the soils relating it to the plasticity index. The zones suggested by Arulanandan have been shown in the Fig. 2 to Fig. 4, for respective borrow areas under study.

It is seen from these figures that for Dholbaha dam borrow area, majority of the samples fall in the category of high erodibility but the consensus obtained from the diagnostic tests show quite high percentage of deviations from this inference. For Chohal dam borrow area, seven samples have given dispersive characters out of which five fall in the category of high resistance to erosion. Similarly all the samples which have demonstrated dispersivity fall in the category of highly resistant to erosion. Thus the erodibility relationship suggested by Arulandandan (1983) could not be applied without caution.

The plasticity characteristics of soils have practical significance in their ability to reflect the types and amounts of clay minerals present in the fine fraction (Skempton 1953). For natural soils the plasticity index tends to increase with proportion of the clay content with a straight line relationship but the slope of the line changes with type of the clay mineral. The active clays (With large colloidal activity because of interaction between the surface forces and the water molecules) exhibit steep slope in these relationships. Thus, if plasticity characteristics are known with varying percentages of clay fraction, the type of clay minerals present in the soils could be predicted. Out of the four diagnostic tests for assessing the dispersivity of soils, two tests, the crumb test and the SCS test, are dependent on the colloidal activity of the soils and thus should have simple relationship with the plasticity characteristics or the relationship suggested by Skempton (1953).

The results of these two diagnostic tests have been plotted with relationship to the plasticity index and percentage of clay content from borrow areas of all the case studies under reference (Fig.6) and it is seen that the samples which demonstrate dispersive nature are not consistent with the colloidal activity suggested by Skempton (1953). The minimum slope of the dispersive soils by crumb test works out to be 0.625 while that of SCS test is 0.4 which according to Skempton (1953) should be inactive, thus is the inconsistency. It is possible that high colloidal activity (in relation to plasticity index) is due to presence of appreciable amounts of mica in the soils. Tubey (1961) has demonstrated that even 10% addition of fine mica in silty clay reduces the plasticity index by 33% . This could be an explanation of dispersive soils falling in the inactive zone.

The soils of borrow areas for Dholbaha as well as Damsal dam projects demonstrate average activity of 1.0 which indicates that the crumb test as well as SCS test would demonstrate dispersivity. The tests for Dholbaha dam project are consistent to this conclusion but those of Damsal borrow area are inconsistent as almost all the samples are found to be non-dispersive by crumb test, SCS test and Pin Hole Test but the chemical extract test has indicated that all the samples are of dispersive to intermediate dispersive category. The majority of samples from Perch dam borrow area, when tested by all the four diagnostic tests, were found to be dispersive in nature. As discussed earlier the colloidal activity of about 50% of the samples is less than 0.75 and according to Skempton these should not be active though found dispersive. Thus, it could be inferred that in the geoenvironment of accumulation of soils in various borrow areas, the dispersivity could either be because of mineralogical composition of the clay content and/ or because of high percentage of exchangeable sodium ions which promote dispersion by creating a greater osmotic potential for water movement into the space between particles and also by increasing the long range repulsive double layer interaction of the particles (Holmgren, 1977).

From the above discussions following conclusions could be drawn:

702

4.1 Though many investigators have been able to identify active soils on the basis of plasticity characterstics and the percentage of clay fractions, there are limitations in these relationships, particularly those soils which contain appreciable amount of fine mica and/or high percentage of exchangeable sodiumions. Thus diagnostic tests for dispersivity would be necessary for establishing their usability.

4.2 Because of the varied conditions like the mechanical, mineralogical as well as chemical which could be responsible for causing dispersivity, none of the four accredited diagnostic tests can be preferred. Many investigators Melvill and Macelar (1980), Sharda and Narula (1987) have attempted statistical dependability of various tests to assess the consistency of various test procedures for a particular site. Such a treatment of the test results could only be applicable for site specific correlations and could not be generalised.

4.3 The test results for various borrow areas have demonstrated that consensus method of assigning dispersive or non-dispersive nature may lead to hazardous complicency with disastrous results. For example the soils of borrow area of Damsal project have not indicated dispersive nature by Pin Hole Test, crumb test or by SCS test, while almost all the samples have demonstrated dispersive or intermediate dispersive nature by chemical test. If weighted averages of all the tests are taken into account, there could be a tendency of assigning non dispersive category to these borrow areas, which would be hazardous. It could thus be surmised that even if one of the diagnostic tests indicate dispersive nature, the soils should be taken as dispersive, and effective defensive measures be provided.

5 CLAY MINERALOGY/ENVIRONMENT

OF DEPOSITION

5.1The borrow areas of the case studies under discussion are located in the higher stratigraphic levels of Upper Siwalik rocks of Quaternary age. The investigations have indicated that some of the clayshale bands are highly dispersive in nature and some of them are nondispersive or less dispersive. The mineralogical composition of these clayshales

indicates that the dominant clay mineral in these formations is illite with kaolinite, montmorillonite and chlorite as other minor constitutents (Sharda and Verma, 1977). A number of embankment structures have been made using the Siwalik clayshales or the soils and are performing satisfactorily. Narula and Shome (1987) while classifying Indian soils as embankemnt materials have indicated that Siwalik clayshales, in general, are acceptable as hearting materials but for Upper Siwalik clayshales, caution of possibility of their being dispersive in nature, had also been postulated.

5.2 Skempton (1953) correlated the plasticity index of three clay minerals-kaolinite, illite and montmorillonite with present clay fraction and arrived at straight line correlations with slopes of these lines to be different for different minerals. The test results obtained from studies conducted by the authors (Fig.6) have indicated that the majority of the clay fraction percentage v/s the plasticity index fall in the zone of illite suggested by Skempton, thus strengthening the view that illites are the dominant clay minerals in the Upper Siwalik clayshales. With illites as the

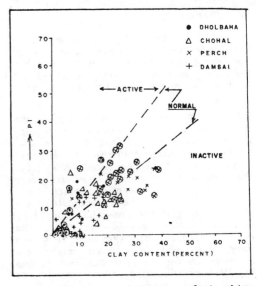

Fig.6 Diagram showing relationship between PI & clay content for samples from borrow areas of Kandi watershed projects & activity limits (after Skempton, 1953). Encircled samples are dispersive in nature on consensus.

dominant clay mineral, the shales would demonstrate normal activity but the test results studied suggest that some clayshale bands show high activity or dispersivity. This inconsistency could be attributed to high concentration of exchangeable sodium ions compared to other mobile elements in the shales as discussed earlier in this paper. The perusal of results of the pore water extract from the samples tested (Fig.5) also indicated that in the total dissolved contents, the percentage of soda is invariably more than that of potash. The survey of the literature however, indicates that in shales irrespective of their age, the potash is generally present in excess of soda except in those which contain products of abrasive action rather than the clay minerals. Some glacial clays and silts are thus characterised (Pettijohn 1969). It could thus be inferred that some of the clayshale bands of Upper Siwaliks contain glacial clays and silts deposited in fluvio-lacustrine environments and the usability of these as embankment materials should always be subjected to diagnostic dispersivity tests.

6 DEFENSIVE MEASURES FOR DESIGN OF EMBANKMENTS

In view of the dispersive nature of the impervious materials used in all the dams of medium height following defensive measures have been adopted in the design and construction of these structures:

6.1 As it has been established that the dispersivity of the soils is lithologically controlled and all the clayshale bands show different degrees of dispersivity, it was endeavoured that only those soils which do not show dispersive nature are used as hearting materials.

6.2 It has also been suggested that the filter criteria on the downstream chimney filters should be such that it is capable of retaining discrete, very fine particles which might tend to move in the case of isolated patches of dispersive soils which could get into the core material inspite of specification that such materials are not t be used.

6.3 At these dam sites, where the foundation materials are also composed of dispersive clayshales and these can not to be obviated at the foundation grades, it has been suggested that the area grouting of these clayshale bands by using 2% lime grout may be done. Alternatively the first two lifts of the impervious material used may be lime treated.

6.4 For dams where it is apprehended that some mixing of dispersive soils might take place, two layer chimney filter might be necessary with finer graded material in upstream layer so that migration of particles is effectively controlled and coarser filter to reduce the peizometric pressures.

ACKNOWLEDGEMENTS

The authors are thankful to the Director General, Geological Survey of India for permission to prepare and submit this paper. The authors are indebted to the authorities of Kandi Watershed Management, Punjab for making available the test data.

REFERENCES:

Arulanandan,K., & B.P.Edwards 1980 Erosion in Relation to filter design criteria in Earth Dams. Jour. Geotech. Engg. 109,5.

Decker, R.S., & L.P.Dunnigan, 1976. Development and use of SOS Dispersion Test. Am. Soc. for Testing Materials Symp. On Disper. Clays. Chicago. III.

Emerson, W.W. 1954. Determination of stability of soil crumbs. Jour. Soil Satu., 5:233254.

Holmgren, G.G.S.1977. Factors affecting spontaneous dispersion of soil materials as evidenced by crumb test. ASTM, STP-23: 218-239.

Melvill, A.L. & D.C.R. Mackelar. 1980. Identification and use of dispersive soils at Elands Dam, South Africa. VII Regional Confr. for Africa on Soil Mech. & Found. Eng.: 440-456.

Narula, P.L. & S.K.Shome. 1987 Geotechnical Classification of Indian Soils as Embankment materials. Proc. in Tropical Terrains, Bangi.

Pettijohn, F.J. 1969. Sedimentary Rocks. New York: Harper Bros.

Ryker, N.L. 1977. Encuntering dispersive clays on Soil Conservation service Projects in Oklahoma. ASTM, STP- 623:370-389.

Sharda, Y.P. & P.L. Narula. 1987. Geotechnical assessment of fill materials for Dholbaha Earth & Rockfill Dam, Hoshiarpur district, Punjab. Proc. All India Semi. On Earth & Rockfill Dams: 73-89.

Sharda, Y.P. & V.K. Verma. 1977. Clay Mineral studies of Murree of Siwalik Sediments around Udhampur, Jammu sub-Himalaya.Chayinca Geol. 3.1:215-219.

Sherard, J.L., R.S. Decker & N.L. Ryker. 1972. Piping in Earth dams of dispersive clay. Proc. Special Confr. on Perm. of the Earth-Supported Structures. ASCE-1: 589-679.

Sherard, J.L., L.P. Dunnigan & R.S. Decker. 1976. Identification and nature of dispersive soils. Jour. Geotech. Engg. Divn. 102. CT4:287-301.

Sherard, J.L., L.P. Dunnigan & R.S. Decker. 1976. Pinhole test for identifying dispersive soils. Jour. Geotech. Engg. Divn. 102:69-85.

Skempton, A.W. 1953. The "Colloidal activity" of clays. Proc. Int. Conf. Soil Mach. & Found. Engg. 3rd. Zurich. 1:57-62.

Tubey, L.W. 1961. A laboratory investigation to determine the effect of mica on properties of soils and stabillised soils. Brit. Road. Res. Lab. Note. 4077 (Unpublished).

Some aspects of the engineering-geological properties of swelling and slaking mudrocks

Considérations sur les propriétés des argilites gonflantes dégradables appliquées à la géologie de l'ingénieur

H.J.Olivier
Keeve Steyn Incorporated, Johannesburg, South Africa

ABSTRACT : Most mudrock materials tend to deteriorate with time. This is one of the main factors which influences the geomechanical behaviour of the rock mass after exposure to the air. However, timely preventative measures can control the deterioration. Assessment of the rock material durability is thus of considerable engineering importance.
This paper discusses the characteristics and underlying mechanisms which influence the long term behaviour of potentially swelling and slaking mudrocks. Reference is also made to a rock durability classification which has proved successful in identifying and predicting the behaviour of mudrocks of the Karoo Sequence in South Africa.

RESUME: La plupart des roches à base d'argilite ont tendance à se détériorer dans la temps. C'est là l'un des principaux facteurs influencant le comportement de la masse rocheuse une fois exposée à l'air. Des mesures préventives peuvent cependant permettre de contrôler cetter détérioration. L'ingenieur s'appliquera donc à evaluer les problèmes de durabilité d'une telle roche.
Cet article présente les caratéristiques particulières ainsl que les mécanismes qui influencent le comportement à long terme des argilites potentlellement gonflantes et dégradables. L'auteur y présente une classification des roches vis a vis de la durabilité qui s'est révélée efficace pour identifier et préviour le comportement des argilites de la série du Karoo en Afrique du Sud.

1. INTRODUCTION

The process of swelling has been considered by several authors to be the causative mechanism of disintegration of mudrocks upon exposure to the air. However it is not by any means the sole factor related to such a phenomenon.

The International Society for Rock Mechanics, Commission on Swelling Rocks (1983) gives the phenomenological definition of swelling mechanism as follows: "It is a combination of physico-chemical reaction involving water and stress relief. The physico-chemical reaction with water is usually the major contributor to swelling but it can only take place simultaneously with or following stress relief".

A fundamental study of the time-dependent behaviour of the rock material is necessary for a meaningful assessment of the geomechanical behaviour of rock masses consisting of potentially swelling mudrocks. Such a study should include observations on the weathering characteristics of the intact rock material, as well as the magnitude of the swelling and shrinkage strains which could develop after exposure (Harper et al, 1979 ; Olivier, 1979 a).

In the case of the wide-spread Karoo Sequence mudrocks of South Africa (Lower Triassic age) it was found that the tendency of the rock material to deteriorate with time was one of the main factors which influenced the geomechanical behaviour of the rock mass after exposure to the air. Other criteria, such as the occurrence of prominent near-horizontal bedding features and a low rock strength in general, were additional factors (Olivier, 1976).

This paper deals with the more important characteristics which control the engineering behaviour of such swelling and slaking mudrocks, as assessed on the scale of the rock material. Reference is also made to a rock durability classification system in predicting the swelling and slaking behaviour of mudrocks of the Karoo Sequence in South Africa.

2. RELEVANT ENGINEERING-GEOLOGICAL PROPERTIES OF KAROO MUDROCKS

It is considered that the engineering-geological properties and long-term weathering behaviour of potentially-swelling Karoo mudrocks are influenced to a varying degree by criteria such as:
1) mineralogical composition,
2) in-situ moisture content and degree of saturation,
3) anisotropical swell behaviour
4) texture (fabric) and degree of micro-fissuring,
5) influence of air-drying and moisture redistribution,
6) durability of the rock material

2.1 Mineralogical composition

A literature survey of this parameter shows that there is still much uncertainty amongst many research workers as regards the influence of mineralogy (in a quantitative sense) on the swelling behaviour of potentially-expansive mudrocks and it therefore needs further investigation (Grice et al, 1982).

A major problem appears to be the fact that it is generally difficult to distinguish on a quantitative basis, by means of X-ray diffraction or other analytical methods, between the influence of true swelling clay minerals, such as montmorillonite (Taylor and Spears, 1970), and that of stable-lattice, "non-expandible" clays such as illite. The latter mineral can also contain relative small quantities of mixed-layer (interstratified) minerals such as montmorillonite-illite (Foster, 1955).

2.2 In-situ moisture content and degree of saturation

The degree of saturation of the rock material in its in-situ condition will control the amount of moisture absorbed (or lost) after exposure of the rock in any type of excavation.

It is to be expected that in the case of excavations below the groundwater table the rock material should generally indicate a high degree of saturation, even in the case of low-permeable mudrocks. Such rock types will therefore have a low propensity to swell in a confined rock

mass state. This high degree of saturation can, however, be markedly reduced by stress relief (caused by the excavation) and also as a result of intensive air drying (Olivier, 1979 a and 1990).

Harper et al (1979) are of the opinion that the history of environmental humidity within an excavation is a principal factor which influences the development of swelling strains in the rock material.

According to Hopkins and Deen (1984) the in-situ (natural) moisture content of shales provides a strong indication of their slake-durability properties. They consider shales with a natural water content below approximately 3.5% to have high slake-durability indices while those with natural water contents of between 3.5 and 7.5% appear to have an intermediate slake-durability rating.

2.3 Anisotropical swell behaviour

The results of free swelling tests conducted on mudrock samples from the 82km long Orange-Fish Tunnel, constructed in South Africa, provided evidence of a prominent anisotropic swelling behaviour of the intact rock material, with the free swelling properties measured perpendicular to the bedding generally being several times greater than parallel to it (Olivier, 1979 a). A similar rock mass behaviour can therefore be assumed upon the ingress of moisture during and after construction, particularly in the case of thinly-bedded mudrocks. The source of moisture can be either from a humid atmosphere or can be the result of direct wetting (eg. groundwater fissures).

2.4 Texture (fabric) and degree of micro-fissuring

It has been shown (Olivier, 1979 a) that textural parameters, such as the degree of compaction of the rock particles, the extent of micro-fissuring (mainly as a result of air drying) and the presence of incipient micro-discontinuities, influence the free-swelling characteristics and weathering behaviour of Karoo mudrocks to a marked extent. These rock types consist of illite, quartz, and feldspar (in decreasing order of abundance) and only very small quantities of chlorite, hematite, calcite and swelling mixed-layer montmorillonite-illite.

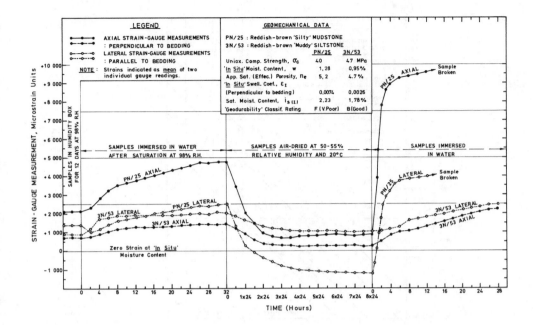

Fig. 1   Strain behaviour of typical Karoo mudrocks subjected to alternating cycles of
water absorption and intensive air drying

2.5   Influence of air-drying and moisture
redistribution

Evidence has been found of important,
anisotropic, differential swelling and
shrinkage strains developing in mudrocks
of a low durability when these are exposed
(in both underground and surface
excavations) to any of the following
criteria, either separately or in
combination:
1. A continuous variation in the
environmental humidity (Van Eeckhout,
1976; Harper et al, 1979; Olivier,
1979 a).
2. Alternating wetting and drying cycles
of variable intensities and duration
(Lemp, 1978)
3. Prolonged periods of intensive air
drying (at a relative humidity of
less than 60 percent) followed by
moisture absorption (Olivier, 1979
a).

Figure 1 illustrates the magnitude of
differential swelling and shrinkage
strains developing in typical Karoo
mudrocks when subjected to some of the
above-mentioned criteria.   Of primary
importance is the fact that although the
test samples used have approximately
similar strength and mineralogical
characteristics, they showed markedly

different durability ratings.

2.6   Durability of the rock material

The breakdown phenomena of mudrocks after
exposure appear to fall into one of the
following categories:
1. Durable rock types showing very
little breakdown.
2. Disintegrating rock types which
breakdown into hard rock fragments
of various shapes and sizes.
3. Slaking rock types which breakdown
to clay or silt size particles.
In practice there is no sharp division
between these types of behaviour.
Weathering and disintegration (or
slaking) of swell-susceptible mudrocks can
occur if they are subjected to moisture-
content changes, either as a result of
swelling-drying cycles, or intensive air-
drying. The durability of the mudrock
material is therefore of significant
importance in the time-dependent
geomechanical behaviour of such rock
masses. The phenomenon of slaking,
although often consequent upon swelling,
is complex and could involve other causes
and mechanisms than those defined for
swelling.

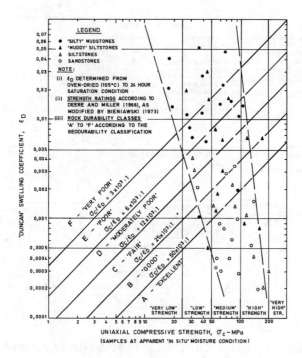

Fig.2.   "Geodurability" classification of the intact rock material
          (Orange-Fish Tunnel samples)

3.   POTENTIAL CONSTRUCTION PROBLEMS IN MUDROCKS

The presence of rock types of a low durability can lead to important construction problems in both underground and surface excavations if timely preventive measures are not taken. This has been found to apply particularly to mudrocks which generally show a fairly wide range of other engineering properties (Olivier, 1976 a).

It is considered that the tendency of the rock material to deteriorate with time is a very important factor as it dominates the time-dependent behaviour of mudrocks. With decreasing strength of the rock in the low and medium strength ranges, the mechanical properties of the rock material will also tend to play an increasingly important role in the structural behaviour of the rock mass itself (Olivier, 1990).

In the case of non-durable mudrocks, weathering after exposure can cause extensive fretting and spalling and, in extreme cases, even slaking of the rock material. Although only a small volume of the exposed rock material will be initially affected by these phenomena,

successive disintegration and breaking away of loose rock fragments will result in a fresh rock surface being exposed each time to a further cycle of weathering. Excessive overbreak and rock-mass stability problems could then be the end result. This will particularly be the case where thinly-bedded mudrocks of a low strength are exposed (Olivier, 1990).

4.   DURABILITY CLASSIFICATION OF THE INTACT ROCK MATERIAL

The engineering problems caused by swelling and slaking mudrocks have demonstrated the need for a classification system which can be relied on to identify and predict the behaviour of rocks susceptible to these phenomena.

A rock durability classification of this kind, must of course, be related to index rock properties which are relevant to the actual weathering mechanics of mudrocks.

The empirical Geodurability Classification (Olivier, 1979 b) provides such a type of engineering-geological classification.

It is based on different ranges of ratios of the uniaxial compressive strength, σ_c and the "Duncan" free swelling coefficient, $\mathcal{E}_D$ (Duncan et al, 1968) as index parameters (Fig. 2). These rock properties can be measured fairly rapidly in a field laboratory by means of simple test apparatus operated by semi-skilled labour. The extensive use of the Geodurability Classification on the Orange-Fish Tunnel project indicated such a system to be particularly relevant to compacted and weakly cemented mudrocks. This all the more so where the tendency of the rock material to deteriorate with time (and not the presence of rock discontinuities such as joints and bedding planes) dominates the time-dependent behaviour of excavated rock masses.

5.   CONCLUSIONS

The studies conducted to date to evaluate the criteria which control the engineering behaviour of swelling and slaking mudrocks revealed rock material durability to be a time-dependent factor of primary concern as regards the fretting and spalling of the rock material. It has also been shown that textural characteristics of the mudrock material, such as fabric and degree of micro-fissuring (mainly as a result of air-drying), play a much more important role in the long-term weathering and disintegration of mudrocks than mineralogical composition.

The weathering behaviour of mudrocks appear to be closely related to the following factors:
1. The partially irreversible anisotropic shrinkage of the rock material when subjected to a variable degree of air drying.
2. The partially irreversible, anisotropic expansion of the rock material when subjected to capillary action as a result of moisture absorption. The source of moisture can be either from a humid atmosphere or can be the result of direct wetting.

The combination of these factors causes important differential swelling and shrinkage phenomena in non-durable mudrocks when these are subjected to moisture redistribution. Moisture migration will also tend to open up the incipient micro-discontinuities present in the exposed rock material. The depth of influence of such moisture migration will be augmented by the presence of blasting fractures and open natural discontinuities. It is therefore important that the surface of exposed non-durable mudrocks should be sealed as soon as possible against the loss or absorption of moisture by the application of an appropriate sealing layer.

The assessment of rock durability on a routine basis during construction by means of an appropriate system, such as the Geodurability Classification, is of significant importance as it should assist in an early evaluation of the need to protect the face of mudrocks against deterioration.

The phenomenon of slaking, although often consequent upon swelling, is complex and may involve other causes and mechanisms than have been determined for swelling rocks. Further studies on slaking rocks are therefore necessary. A special Working Group of the ISRM Commission on Swelling Rock, of which the author is the co-chairman, has been formed to investigate this matter.

REFERENCES

BIENIAWSKI, Z.T.(1973).Engineering classification of jointed rock masses. Trans. S. Afr. Inst. Civ.Eng., 15(12), 335-344.

DEERE, D.U. & MILLER, R.P. (1966). Engineering classification and index properties for intact rock. Tech. Rep. Air Force Weapons Lab., New Mexico, AFNL-TR-65-116.

DUNCAN, N., DUNNE, M.H. & PETTY, S. (1968). Swelling characteristics of rocks. Water Power, May 1968, 185-192.

FOSTER, M.D. (1955). The relationship between composition and swelling in clays. Proc. 3rd Nat. Conf. Clays and Clay Minerals, Nat. Res. Counc., Washington, 205-220.

GRICE, R.H., KIM, C.S. & BROWN, G.R. (1982). Relationship of texture, composition and adsorbtion properties to the weathering of mudrocks. Geol. Surv. Canada, Current Research, Part A, Paper 82-1A, 359-367.

HARPER, T.R., APPEL, G., PENDLETON, M.W., SZYMANSKI, J.S. & TAYLOR, R.K. (1979). Swelling strain development in sedimentary rock in northern New York. Int. J. Rock Mech., Min. Sci, 16(5), 271-292.

HOPKINS, T.C. & DEEN, R.C. (1984). Identification of shales, Geotech. Testing Jnl., 7/1, 10-18.

INTERNATIONAL SOCIETY FOR ROCK MECHANICS, COMMISSION ON SWELLING ROCK. (1983). Characterization of swelling rock. Task A, Doc. (Oct. 1983), 4pp.

LEMPP, Ch. (1978). Disintegration of overconsolidated clay shales and marls (field observations and experiments). Proc. 3rd Congr. Int. Assoc. Engng. Geol., Madrid, 1, 236-241.

OLIVIER, H.J. (1976). Importance of rock durability in the engineering classification of Karoo rock masses for tunnelling, In. Z.T. Bieniawski (ed.), Exploration for Rock Engineering. 1, 137-144 Cape Town, Balkema.

OLIVIER, H.J. (1979 a). Some aspects of the influence of mineralogy and moisture redistribution on the weathering behaviour of mudrocks. Proc. 4th Congr. Int. Soc. Rock Mech, Montreux, 3, 467-474.

OLIVIER, H.J. (1979 b). A new engineering-geological rock durability classification, Engng. Geol., 14, 255-279.

OLIVIER, H.J. (1990). Assessment of the geomechanical properties of swelling and slaking mudrocks in tunnel excavations. Proc. Congr. Int. Soc. Rock Mech, Static and Dynamic Considerations in Rock Engineering, Mbabane (In the press).

TAYLOR, R.K. & SPEARS, D.A. (1970). The breakdown of British Coal Measure rocks. Int. J. Rock Mech. Min Sci., 7(5), 481 501.

VAN EECKHOUT, E.M. (1976). The mechanisms of strength reduction due to moisture in coal mine shales. Int. J. Rock, Mech. Min. Sci., 13(1), 61-67.

Physico-chemical fundamentals of soil microrheology

Les bases physico-chimiques de la microrhéologie de sols meubles

V.I.Osipov
Moscow State University, USSR

ABSTRACT: The paper describes energy heterogeneity of individual contacts between structural elements of the soils, peculiarities in distribution of effective stresses at different contact, development of intrinsic structural changes in the soils under deformation. It is stressed that the long-term strength and creep of the soil are in great measure influenced by the energy heterogeneity of the contacts. Proceeding from peculiarities in the behaviour of separate contacts, rheologie models of soils representing systems of parallel-connected element groups to simulate behaviour of the contact of specific types are suggested. Basing on the suggested model, explanation is given to the basic rheologic behaviour patterns of the soils.

RESUME: Dans la publication, sont examinés l'heterogenité enugetique des contacts individuels entre les éléments structuels des sols, les particularités de la répartition des containts effectives aux diffuents contacts, ainsi que l'évolution de la réorganisation structuelle interne lors des déformations. Il y est noté l'influence exceptionnellement impatante de l'hétérogénéité energetique des contacts sur la solidité de longue durée et sur le fluage. En partant des particularités de conduite des contacts individuels, dans la publications sont proposés des modéls rhéologiques de sols meubles representant des systèms de groupes d'éléments parallèlement relis, qui immitent le travail des typs de contacts definis. Sur la base des modéls proposés, il est donné une explication des lois fondamentale du conduite rhéologique des sols.

1 INTRODUCTION

The modern theory of linearly and non linearly-deformed bodies is based on the model of continuous and homogeneous medium. The criterium for quasi-continuity and non-uniformity of bodies serves as the relation between the dimensions of individual structural elements and the study object neglecting the non-uniform interaction of its constituent elements. The investigations conducted in the field of physico-chemical mechanics of the finely granular systems (Rebinder 1966, Osipov 1975, 1988) testify to the existance of difference in the energy of interaction between the structural elements. Necessity arises therefrom to study soils as non-uniform systems with particular care being given to the analysis of the processes taking place on the contacts of structural elements. Proceeding from the above said, attention should be given to study of the behaviour of individual contacts and basing on it an overall picture of time dependent deformation of the soil can be represented.

2 TYPE OF CONTACTS IN SOILS AND THEIR ENERGY HETEROGENEITY

All structural elements (grains and particles) composing the soils are bonded by the forces of different nature which are defined as structural bonds. The structural bonds are formed only in the points of

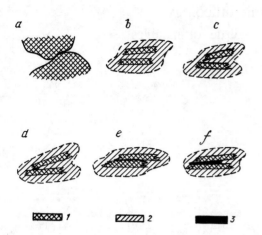

Figure 1. Type of contacts between structural elements of soils: a) interlocking; b) distant coagulation; c) close coagulation; d) transition (point); e) crystallization; f) cementation. 1 - particles, 2 - adsorbed water, 3 - new phase (cement).

particle contacts. The contacts are weakened zones through which the soil deformation and failure take place. In the first approximation the soil strength R_t can be assessed by an average value of the interparticle force of attraction at a single contacts f and a number of contacts in the unit of failure surface area χ :

$$R_t = f\chi$$

Parameter χ is the function of the particles size distribution, soil porosity and particle shape. For the particles of an isometric shape, parameter χ can be assessed using the equation (Babak 1974)

$$\chi = \frac{3}{2\pi} \cdot \frac{Z(1-n)}{(2\bar{r})^2}$$

where Z - the coordination number of particles packing defined as from $Z = 3/n$, $\bar{r}$ - average radius of particles, n - porosity of the body in unity fractions.

Value χ may vary within a considerable ranges: for instance, the sands with 40% porosity for the unit of failure surface have 9×10^2 contacts/cm², while the clay soils featuring the same porosity have 2×10^8 contacts/cm².

Alongside with the number of contacts, the strength of soils is governed by particle interactions at an individual contact (f). The structural bonds are being formed through out the geological history as a result of complex process such as consolidation, ageing, syneresis, formation of new amorphous and crystal phases from the oversaturated pore solution etc. These processes result in the formation of contacts of various types which in terms of formation, the nature of interaction forces and energy could be grouped as follows: 1) interlocking contacts, 2) coagulation contacts, 3) point contact, 4) crystallization contacts, 5) cementation contacts (Osipov 1975, 1988).

The contacts of the first type are typical for coarse clastics and un-cemented sand soils (Fig. 1a). The structural bond in such soils achieved through interlockings of the irregularities on the surface of structural elements is considerably less than the weight of the particles proper. Therefore the soils with interlocking contacts behave like granular bodies. When subjected to tensile tests they exibit extremely low strength, while under shearing force they behave according to the "dry" friction law.

The coagulation contacts develop in the water-saturated soils (clay, loam, loess, peat etc.) due to long-range molecular and in some case electrostatic and magnetic interactions. The forces acting on such contacts exceed the weight of particles and therefore the soils of coagulation contacts behave like bonded systems. Particular feature of the coagulation contacts is the presence of a thin layer of adsorbed water between the particles (Fig. 1 b, c) whose thickness is determined by a total effect of the interparticle forces of attraction and repulsion and the disjoining action of the adsorbed water film. The adsorbed water separating the particles has an altered structure and as a result of it posses specific structural-mechanical properties. Testing by various methods has revealed that it features an increased ultimate shear stress which amounted to 9.5-13 Pa as compared to 10^{-3} Pa for the free water.

Energy of the particle interaction

at the coagulation contact equal to the oppositely-directed forces of attraction and repulsion, is a complex function of the interparticle distance. In general this function is characterized by two minimums corresponding to two equilibrium values of the adsorbed water film thickness. The thickness of the equilibrium film (h_1) corresponding to the first equilibrium state totals several nanometres that of the second state totals tens and hundreds of nanometres. In this connection two varieties of the coagulation contacts are differentiated: close contact with the hydration film thickness h_1 and distant contacts with the hydration film thickness h_2. The strength of these contacts is not the same because of difference in h_1 and h_2 for the close and distant contacts. The cohesive force of the μm-size particles at the distant coagulation contact amount to 10^{-11} and -10^{-10}N, while at the close contact at reaches $10^{-9} -10^{-8}$N.

The important peculiarity of the coagulation contacts is the absence of "dry" contiguity between the particles. Therefore deformation of a system with such contacts follows the internal friction law and is governed by the properties of the absorbed water film in the contact zone.

The transition (point-type) contacts are met in the low water content silt and clay soils (including the soils with considerable porosity such as loess) as well as in water saturated soils sustained considerable gravitational consolidation. In the first case they are developed due to compression of the particles by capillary meniscus and formation of ion-electrostatic bonds (Osipov, Sokolov 1974). In the second case it is caused by breaking through of the hydrate film in separate contact areas and formation of single bonds of the chemical nature (Fig. 1, d).

Contacts formed in such a way are metastable and under certain conditions they turn into coagulation or phase (crystallization and cementation) contacts which gives the grounds to consider them as transition ones. Validity of such a definition has been confirmed also by measurements of their strength. As computation and experimental data indi-

cate the cohesion force at such a contact between two particles of micrometre sizes is by an order of magnitude higher than at the close coagulation contact and amounts to about 10^{-4}N.

From the rheological viewpoint it should be noted that at the transition contact there are areas of direct (dry) contiguity between the particles and the areas separated by the adsorbed water film. Therefore behaviour of such contacts in deformation is mostly complicated and is controlled both by properties of the particles in contact and the adsorbed water in the gap.

The strongest crystallization and cementation contacts are formed at a high degree of compaction of soils and its lithogenetic transformations. The crystallization contacts develop under high pressure and temperature causing plastic flowage of the grains in the areas of highest effective stresses (Fig. 1, e) which results in "welding" of the particles due to chemical bonds on the area A_c equal to $A_c = F/6_T$ where F - magnitude of force acting on the contact, 6_T = yield limit of mineral.

The cementation contacts are formed in the course of diagenesis and catogenesis due to precipitation of a new-phase bonding particles from the circulating solutions (Fig. 1, f).

The magnitude of interparticle cohesion force at the crystallization and cementation contacts is governed by the number valent bonds developing in the contact zone. The contact formed by at least $10^2 -10^3$ valent bonds is regarded to be the contact of the minimum strength. The area of such a contact totals $10^{-17} -10^{-16}$ m^2 which corresponds to the cohesion force $10^{-6} -10^{-7}$N. At a lesser number of valent bonds, the contact is metastable and it should fall under the transition contact category. Under external loading, the crystallization and cementation contacts behave like the material of particles proper exhibiting the properties characteristic of solid bodies.

Thus depending on the composition, formation conditions and degree of the lithification, the soils may feature a wide range of variations in the number and strength of individual contacts which control physicomechanical properties of the soils.

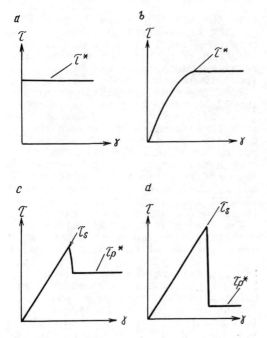

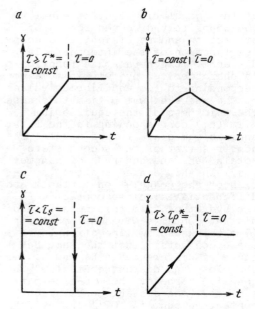

Figure 2. Deformation curves for different types of contact: a - interlocking, b - coagulation, c - transition, d - phase (crystallization or cementation).

Figure 3. Time dependent deformation curve under loading (τ = const) and unloading (τ = 0) of different types of contacts: a - interlocking, b - coagulation, c - phase and transition under loads below structural strength, d - phase under loads above structural strength.

From the rheological viewpoint it is important to distinguish the various nature of the contacts being formed and their energy heterogeneity.

3 RHEOLOGICAL PROPERTIES OF SOILS AS ADDITIVE BEHAVIOURAL CHARACTERISTICS OF SEPARATE CONTACT TYPES

Study of properties of various contact types provides physical basis for rheologic modelling of soils and explanation of their behaviour. For this purpose let us consider idealized soils with strictly definite types of contact.

Let us start with the soil formed by structural elements with interlocking contacts. Systems with such contacts behave like typical stiff-plastic bodies with plasticity developing according to the external dry friction law. Even at a low level of stress applied, the yield limit $\tau \gg \tau^*$ develops in the inter-

locking contacts and plastic displacement of the particles takes place at a constant rate which depends on the external stress (Fig. 2, a). The accumulated deformation becomes irreversible (Fig. 3, a) which is well-demonstrated by shear tests on uncemented sands and coarse debris soils.

Thus the main conditions of deformation of the interlocking contacts can be expressed as follows:

$$\dot{\gamma} = 0 \quad \text{at} \quad \tau < \tau^*$$

$$\dot{\gamma} = f(\tau, \mathbf{6}) \quad \text{at} \quad \tau \gg \tau^*$$

where τ^* - yield limit of contacts; $\dot{\gamma}$ - rate of deformation; $\mathbf{6}$ - normal load.

These conditions are satisfied by a simplest rheologic model consisting of a dry friction element (Fig. 4, a).

The deformation pattern of soils with coagulation contacts is governed by the properties of the adsorbed

716

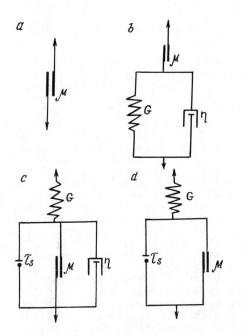

Figure 4. Rheologic models of contacts: a - interlocking, b - coagulation, c - transition, d - phase (crystallization and cementation).

water film. Depending on the acting stresses the coagulation contacts may exhibit elastic viscous and plastic properties. At stresses below the yield limit, the bodies with coagulation contacts may feature distinct elasticity and viscosity. Both properties result from a hinge rotation of the particles about the contacts. The developing elasticity is of an entropic nature (Shukin, Rebinder 1971) and it is conditioned by the orientation of particles. This causes transition of the system into a thermodynamically less apparent state and decrease of its entropy. At external unloading the system regains its thermodynamic equilibrium through return of the particles to its initial state. Rotation of structural elements is accompanied by viscous flow of adsorbed water from the gaps between particles. On the whole at $\tau < \tau^*$ the contact behaves as a viscous-elastic body exhibiting aftereffect typical for the Kelwin rheological model.

Behaviour of the model meets the following conditions

$$\tau = G\gamma + \eta\dot{\gamma}$$

where τ - stress, G - modulus of entropic elasticity, η - viscosity, γ and $\dot{\gamma}$ - deformation and rate of deformation respectively.

At stress equal or above the yield limit $\tau \geqslant \tau^*$ the soil passes into a qualitatively new state. At these stressses relative displacement of particles along the coagulation contacts and plastic flow of the soil begin (Fig. 2, b). The value tends to increase infinitely with time while at destressing a considerable permanent set remains (Fig. 3, b). It should be noted that the nature of plastic flow of the coagulation contact differs absolutely from that of the interlocking contact because it relates to the internal friction developing inside the adsorbed water film.

Proceeding from the above said, a general rheological model can be generated for coagulation contacts. It will comprise the Kelwin element and viscous element to be series-connected (Fig. 4, b).

The most complicated rheological behaviour is characteristic of the soils with transition (point-type) contacts which exhibit elastic and viscoplastic properties within certain loading ranges. The former relate to the existance of chemical or ion-electrostatic bonds at the transition contacts though on a limited area, which tend to break down irreversibly at some stress τ_s. Therefore up to τ_s the transition contact is subjected primarily to elastic strain (Fig. 2, c; Fig. 3, c). Upon breakdown of chemical bonds at $\tau \geqslant \tau_s$ the transition contact undergoes deformations as the coagulation contact with dry friction elements in the areas where chemical and ion-electrostatic bonds have been broken. Deformation goes on at stresses $\tau \geqslant \tau^*_p$ (where τ^*_p - yield limit of transition contact after break down of strong chemical and ion-electrostatic bonds) is of viscoplastic time-irreversible character.

The rheologic model of the transition contact includes elastic and plastic elements with a special element modelling brittle failure of chemical and ion-electrostatic bonds (Fig. 4, c). At stresses $\tau < \tau_s$ in such model the elastic element works while after rupture of the brittle failure element the dry friction and viscous elements come into place. The equation of state of such a model is governed by the following conditions:

$$\gamma = \tau/G \quad \text{at} \quad \tau < \tau_s$$
$$\dot{\gamma} = \frac{1}{\eta}(\tau - \tau_p^*)$$
$$\dot{\gamma} = f(\tau, \mathbf{6}) \qquad \text{at} \quad \tau \geqslant \tau_p^*$$

Rheologic properties of the soils with strong (crystallization and cementation) contacts differ tangibly from the properties of the soils considered above. Having high strength due to chemical bonds, the contact zones in such soils behave like a solid Hook's body and they undergo plastic strain practically upto the ultimate strength limit τ_s (Fig.2, d; Fig. 3, c). The value τ_s does not depend on the type and rate of loading. At $\tau \geqslant \tau_s$ the crystallization and cementation contacts tend to fail irreversibly and then they behave like the interlocking contacts, i.e. show ability for plastic flow at the stress τ_p^* which is considerably less than the ultimate strength of the structure $\tau_p^* < \tau_s$ (Fig. 2, d). The deformation of the broken contact is irrivisible and goes at a constant rate depending on the shear stress and normal load (Fig. 3, d).

Behaviour of the crystallization and cementation contacts can be simulated by the model consisting of series connected elastic and plastic elements (Fig. 4, d). Besides the model incorporates an element parallel-connected with the plastic element to simulate brittle failure at attaining the shear force τ_s. At stresses $\tau < \tau_s$ the elastic element works in the model and the deformations are reversible. At stresses $\tau \geqslant \tau_s$ the stiff element breaks down and then the dry friction elements comes into play which becomes deformed at the stress $\tau_p^* \ll \tau_s$.

Using the above model the rheologic state equation of the soils with crystallization and cementation contacts can be expressed as follows:

$$\gamma = \tau/G \qquad \text{at} \qquad \tau < \tau_s$$
$$\dot{\gamma} = f(\tau, \mathbf{6}) \quad \text{at} \qquad \tau \geqslant \tau_p^*$$

where G - shear modulus of elasticity, γ - deformation, $\dot{\gamma}$ - rate of deformation, τ_s - ultimate strength of structure, τ_p^* - yield limit of broken structure.

The considered models embrace the idealized soils because they are based on the assumption that a soil has one type of the contacts featuring the same strength. In the natural soils such conditions are extremely rare. Even in unlithified young sediment there may exist several types of contacts, such as interlocking and coagulation contacts. The latter may include both close and distant coagulation contacts varying in strength by a factor of 10^3. Diversity of the contacts increases with lithification of the soils. The greatest diversity in the types of contacts is observed in the soils of the medium degree of lithification where there may occur simultaneously interlocking, coagulation, transition and phase (crystallization and cementation) contacts. Therefore in terms of the structural bond type these soils belong to the mixed type soils (Osipov 1988).

It is understood that deformation of natural soils and their rheologic models must be more complex and reflect behaviour of each type of the contacts present in the soil. Such models represent various combination of detail models simulating behaviour of individual contacts. All detail models are parallel-connected in the general model which helps to analyze performance of each detail model when studying the soil in shear. Besides the general model can incorporate several uniform detail models featuring different mechanical parameters of the elastic, viscous and friction elements to show the strength heterogeneity of contact of the same type. Bearing this in mind rheologic models can be built for soil structures formed by one type of the contact but with different G, η and μ parameters, as well as structures formed by several types of contacts. The former comprises the coagulation, crystallization, cementation and transition

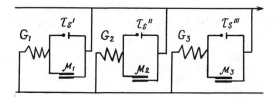

Figure 5. Rheologic model of soil with coagulation structure.

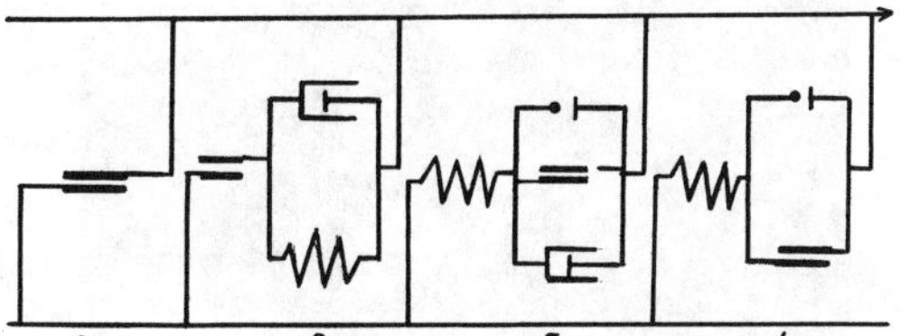

Figure 6. Rheologic model of soil with mixed structure: 1 - interlocking contact, 2 - coagulation contact, 3 - transition contact, 4 - phase contact.

structures, the latter covers the mixed-type structures.

Fig. 5 shows a rheologic model of the soil with coagulation structure. The most important feature of this model is that it behaves as an elastic-viscoplastic body with whatever large irreversible deformations within any ranges of loads and rates of deformation adopted in the soil mechanics. The second peculiarity is that the adopted model does not have a well-defined ultimate strength because of gradual involvement of the plastic elements having different μ value and restorability of the coagulation contacts. The latter permits the viscous and friction elements in the coagulation contacts to be assumed to have infinite length. Therefore the deformation curve of such a model flattens out gradually and tends to asymptote corresponding to the contact strength of a maximum value

The same concept was used to obtain the rheologic model of soils with transition and phase contacts. The major feature of the latter model is its brittle failure inspite of the presence of special elements with various value τ_s. This is accounted by the absence of viscous elements. At failure of weak contacts the load is instantaneously is transferred onto stronger contacts causing their brittle failure. Upon failure of the soil structure its strength is determined by the performance of friction elements which undergo deformation at stresses well below τ_s.

The most complex model is that of mixed structure soil (Fig. 6). It may comprise any line up of detail

models depending on the types of contact available in the soil. The total number of homogeneous detail models can be proportional to the number of contacts of the given type. The structure of such a model and its deformation behaviour can be multivariant. In each specific case it is controlled by the nature of structural bonds in the real soil.

4 CHANGES IN MICROSTRUCTURE OF SOILS AT DEFORMATIONS

Deformation of soils causes relative displacement of solid particles, changes in the microstructure of soil and its strength. At present there exists a viewpoint that rearrangement of the microstructure is one of the major factors contributing to reduction of the soil strength when it undergoes sustained deformation (Morgenstern, Tehalenko 1967; Krizek, Chanla & Edil 1977; Vyalov 1979; Sokolov, Osipov 1983).

A run of tests has been conducted by us on direct shear, triaxial and ringshear apparatus for the comprehensive study of this problem. The tests were accompanied by thorough study of microstructural alterations in the test samples using the scanning electron microscope and vacuum freeze-drying of the samples.

The findings indicate that changes in the microstructure begin under certain level of stresses corresponding to the yield limit at which relative displacement of the

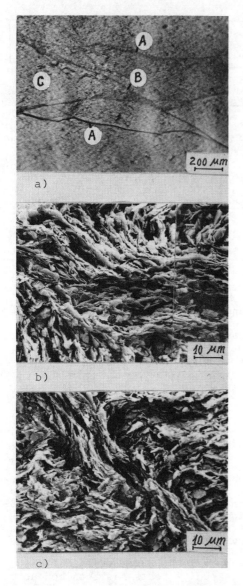

a)

b)

c)

Figure 7. Microstructural changes
in lacustrine-glacial clay in shear:
a) general view of shear zone: A -
subparallel shear planes outlining
the shear zone, B - diagonal shear
planes, C - soil blocks with undisturbed microstructure within shear
zone; b) orientation of clay particles in subparallel shear plane;
c) orientation of clay particles in
diagonal shear planes.

particles is recorded in the contact points. This process embraces
not the entire specimen but is localized in the definite zone formed
along the direction of maximum shearing stress. The resultant shear
zona has a definite width and a complex internal structure. The zone
outlines are well traced along the
shear planes confining it from top
and bottom and running subparallel
to each other (Fig. 7, a). Inside
the shear zone system of diagonal
shear planes are observed which
form the net which is typical for
volumetric deformation of granular
bodies. The net formed by shear
planes divides the entire shear zone
into separate blocks (Fig. 7, a).
The clay particles adjoining to the
shear planes have closer packing and
are well oriented along the shear
planes (Fig. 7, b, c). At the same
time microstructural changes inside
the blocks are rather inconsiderable.

The width of shear planes and degree of particle orientation within
these planes depend on the level of
normal loading. In direct shear test
it varies from 50 μm (6 = 0.03
MPa) up 30 μm (6 =0.5 MPa) while
at ring shear and triaxial tests it
stays within 5-15 μm. The total
width of the entire shear zone depends on the stress state, rate of
shearing and soil fabric. Under the
fast direct shear test (γ = 1mm/
min) the shear zone width will decrease from 16 mm to 8 mm, at the
normal load rise from 0.03 to 0.5
MPa. At the ring shear test (6 =
= 0.05 MPa, γ = 3.3 mm/min) the
width of shear zone amounted to
0.5-2.0 mm. While in the triaxial
apparatus (lateral pressure 6_3 =
= 0.05 MPa, γ = 3.3 mm/min) it
came to 0.04-0.2 mm.

At decrease of the rate of shearing microstructural alteration remain the same, but at slow shearing
(γ = 0.03 mm/min) the width of
shear zone tends to increase from
3 to 30 times as compared to quick
shearing. This phenomenon relates
to the development of viscous deformation in slow shear embracing
larger volume of the specimen as
compared to quick shear.

With the particles being somewhat oriented in the soil, the
width of shear zone depends on the
angle between the direction of
shearing and the plane of particle

orientation. Shearing at various angles (0; 22 and 45°) to the plane of particles orientation indicates that the width of shear zone decreases with reduction of the angle (Q) between the direction of shearing and the plane of orientation of the structural element. For example, at quick ring shear of lacustrine glacial clays at an angle of 90° the width of shear zone exceeded 0.5 mm, at an angle of 45° it came to 0.3 mm, at an angle of 23° - 0.08 mm, at zero angle it turned practically into a narrow shear plane upto 15 μm wide.

These data indicate that the maximum structural rearrangement takes place in shearing normal to the orientation of structural elements, while the minimum one in shearing along the direction of orientation of structural elements. It is evident that the volume of structural rearrangement of the soil is closely related to the work put in shearing. This phenomenon explains anisotropy of the mechanical strength of soils which is a well known fact in the publications. The soil strength is always higher in the direction normal to the orientation of structural elements than that parallel to the orientation.

In triaxial tests, the maximum strength was observed when the particles were oriented to the axial plane of the specimen at an angle of Q and 90° the minimum strength was observed at Q = 45°. In the latter case the orientation of particles was close to the direction of the incipient shear plane which brings on decrease in specimen strength.

Analyzing the entire process of structural changes in the soil in shear three phases can be singled out. During the first phase the specimen undergoes volumetric strain without visible alterations in the microstructure. During the second phase at further increase in stresses formation of the shear zone takes place. It is manifested by the development of subparallel shear planes which are first intermittent and then merge into the shear zone rather clearly outlines from top and bottoms. Within the shear planes, structural elements assume orientation along the shear plane. The width of shear zone being formed depends on the rate of deformation and

strength of structural bonds. During the third phase of deformation which corresponds to the off-limit stress (area of transition to residual strength), the shear zone continues to develop: there appears a network of diagonal shear planes breaking down the entire shear zone into a system of separate blocks. On completion of all microfabric alterations in the shear zone decrease in the system strength decays gradually and deformation passes onto the area of residual strength.

In conclusion it should be added that the processes of structural changes in the soils in the shear zone are closely related to dilatancy leading to changes in the quantity of contact interactions and strength of soil. During the first two phases of structural changes dilatation has a negative sign and causes some increase in the soil density. During the third phase for the majority of soils dilatancy passes from negative into positive sign bringing additional decrease in the soil strength.

5 EXPLANATIONS FOR SOME REGULARITIES IN LONG-TERM DEFORMATION OF SOILS FROM MICRORHEOLOGIC VIEWPOINT

Peculiarities of the contact interactions in the fine grained soils discussed above and microstructural changes taking place at deformation help to disclose the physical nature of some experimental regularities in long-term deformation from the microrheologic viewpoint.

Many researchers who were involved in investigation of the nature of rheologic processes focused their attention on the role of excessive pore pressure and microstructural changes in the soils. But these processes are incidental ones causing partial alteration in the contact interaction. The major factor governing the behaviour of the soils in long-term deformation is energy heterogeneity of the contacts. This factor is involved in all peculiarities of deformational behaviour of the natural soils which are controlled by the condition of their formation and degree of lithification.

The above said agrees with the general notion that the development of rheologic processes in the solid bodies is conditioned by the presence of heterogeneity in the body structures: in the crystalls it may be defects at the crystall level structure dislocation, vacancy etc., while in heterogeneity grained systems - defects at higher structural level which includes also heterogeneity in interaction of structural elements.

As was said above the attraction force of micron-size particles at a single contact may vary from 10^{-11}N (distant coagulation contact) to 10^{-2}N (phase crystallization or cementation contact). The presence of such heterogeneity in the soil structure plays a dominant role in distribution of stresses in deformation. Under quick loading the stress is transferred simultaneously onto all contacts irrespective of their strength. When taking rheologic models of soils (Fig. 5 and 6) quick transfer of stresses loads simultaneously all detail models irrespective of values G, μ and η because their viscous elements do not get deformed. Under this condition the soil fails at the maximum stress and simultaneous rupture of all contacts in the failure plane.

Another picture is being observed at slow loading of the soil. In this case gradual transfer of stresses takes place from weaker to stronger contact. As a result effective stresses get localized on a limited number of bonds whose rupture leads to the soil failure at lower stresses that at quick shear. The latter can be illustrated by the rheologic model with mixed structure. At slow loading of such a model deformation of the weakest coagulation contact is limited by more rigid transition and phase contacts which will undoubtedly lead to stress relaxation on the coagulation contact and its transfer onto stronger contacts. This mechanism is of a cascade character, embraces all stronger contacts and ultimately it localizes the total stress on a limited number of contacts which are going to break at lower stress that the soil failure at quick shear. The magnitude of relative decrease in the ultimate stress proves to be the function of two parameters: role of deformation and degree of non-uniformity of the contacts. Influence of

the latter parameter still remains unexplored but theoretically it may be assumed that the relative drop in strength of the soils at reducing the rate of deformation will be the higher, the more heterogeneous is the contact interaction.

Of great interest is investigation of influence of the pore pressure change on the long-term strength at the microlevel. Till now this matter remains controversial when testing water saturated soil in a confined system, the pore pressure in the shear zone may rise or drop depending on the magnitude and sign of dilatancy. When negative dilatancy develops there occurs rise of the pore pressure and drop of the effective stress at the contacts. This results in higher mobility of the friction elements in the rheologic model whose behaviour depends on normal stress. At the time consolidation of the soil at negative dilatancy causes decrease of water content in the shear zone which may restrain deformation of viscous elements in the system.

The positive dilatancy is accompanied by manifestation of negative pore pressure and drawing of water into the shear zone. The water content rise in the system contributes to higher mobility of the viscous elements while deformation of the friction element will be limited by development of negative porous pressure.

Such a complex mechanism of development of microrheologic processes in pore pressure changes and its relation to the dilatancy effects do not permit straightforward judgement about the character and magnitude of this factor impact on the long-term strength of the soil. It should be taken into account that changes in the dilatancy sign causes decrease or increase in the number of contacts in the shear zone which makes the whole process more complicated.

Dilatancy affecting the pore pressure change results from microstructural rearrangement of the soil in the shear zone. This gives the ground to believe that the long-term strength depends also on the processes of microstructural alterations. Aside from the compaction of soil and the pattern of structural bonds the latter is determined by the soil fabric, i.e. by the degree of orientation of the structural elements and angle Q between the direction of structural element orientation and shear plane.

Both the nature of soil strength decrease at its long-term deformation and configuration of the creep curves can be explained from the viewpoint of energy heterogeneity. For this purpose let us consider the case of attenuating creep using the rheologic model of soil (Fig.6 and 7). It accords with continued stress $\tau < \tau^*_{min}$ where τ^*_{min} - yield limit of weakest contacts. Under such condition the soil undergoes deformation at the expense of viscous and elastic elements in the weakest contacts and is of an attenuating character because unrestrained deformation of these elements is hindered by the presence of stronger contacts. Cessation of deformation is helped by the absence of stress transfer from weak onto stronger contacts because upon action of the elastic and viscous elements no stress relaxation takes place on the weak contacts for stress is being preserved on the friction elements of the same contacts.

At stresses $\tau > \tau^*_{max}$ where τ^*_{max} - yield limit of strongest contacts, the picture changes drastically. In this case there occurs gradual relaxation of stresses on the weak contacts and their transfer onto stronger contacts which undergo deformation and get relaxed with time while the stress is being transferred onto still stronger contacts. Though the outer stress remains unchanged, the magnitude of deformation tends to increase progressively with time and by its character it resembles the soil behaviour at unattenuating creep.

Between those two extreme values τ^*_{max} and τ^*_{min} there exists a range of loads under which the system gets deformed at a constant rate over a long-time. It accords with $\tau^*_{min} < \tau < \tau^*_{max}$ when the yield limit is being gradually overcome first at the weak contacts then probably at the contacts of medium strength thus helping to maintain the deformation process without attenuation at more or less constant rate over long time which meets the study case the steady-state flows. In the long run development of such a process leads to concentration of stresses on the strongest contacts and if the stress level does not exceed τ^*_{max} deformation may result in break down of the system after a contimed phase of steady-state flow.

The effects related to changes in pore pressure and microstructure of the soil in shear which cause acceleration or deceleration of the major processes. For example increase in pore pressure may result in decrease of values τ^*_{min} and τ^*_{max}. The same is true also for microstructural changes: dilatancy depending on its sign may retard or promote the creep process, while reorientation of particles along the slip surfaces increases mobility of the friction elements due to reduction of values τ^*_{min} and τ^*_{max}.

6 CONCLUSION

1. Types of contacts between structural elements of fine grained soils and their energy heterogeneity were examined. It has been shown that the particle interaction force at the contacts may vary by an order of 8-9.

2. A concept has been developed which states that the major factor governing rheologic behaviour of fine grained soils is heterogeneity of the structural elements interaction at the contacts.

3. Peculiarities of rheologic behaviour are examined and detail rheologic models are proposed for different type of contacts. It is shown that rheologic properties of the soils are additive behavioural characteristics of separate types of the contacts.

4. General rheologic models are developed basing on investigation of microrheologic processes and the physical nature and regularities of long-term strength and creep of fine grained soils have been explained.

REFERENCES

Babac, V.G. 1974. Strength of porous bodies. (In Russian). Sc.D.Thesis, Moscow University Publisher, p.20.
Vjalov, S.S. 1978. Rheologic funda-

mentals of soil mechanics. (In Russian). Visshaya Shkola Publisher.

Osipov, V.I. 1988. Structural strength of soils and physico-mechanical fundamentals of its quantification. Engineering geology today: theory, problems, practics. (In Russian). Moscow University Publisher, p. 77-91.

Osipov, V.I. & V.N. Sokolov 1974. Role of ion-electrostate forces in formation of structural bonds in clays. (In Russian). Moscow University Proceedings. Geological Series, No 6, p. 9-21.

Rebinder, P.A. 1966. Physico-chemical mechanics of finely granular structures. (In Russian). Nauka Publisher.

Sokolov, V.N. & V.I. Osipov 1983. Microstructural changes in clays in direct shear. (In Russian). Engineering geology, No 6, p.9-21.

Shukin, E.D. & P.A. Rebinder 1971. Elasticity mechanism after action of structured suspension of low concentration bentonite. (In Russian). Colloids Magazine. 3, No 3, p. 450-458.

Krizek, R.J., Chawla, K.S. & T.B. Edil 1977. Directional creep response of anisotropic clays. Geotechnique, vol. 27, No 1, p. 37-51.

Morgenstern, N.R. & J.S. Tehalenko 1967. Microscopic structures in kaolin subjected to direct shear. Geotechnique, vol. 17, No 4, p. 309-328.

Osipov, V.I. 1975. Structural bonds and the properties of clays. Bull. of the IAEG, No 12, p. 13-20.

An engineering geological classification of the Loess deposits in the Netherlands

Une classification pour la géologie de l'ingénieur des dépôts de Loess aux Pays-Bas

S.J. Plasman & R.R. Kronieger
Delft University of Technology, Faculty of Mining and Petroleum Engineering, Section Engineering Geology, Netherlands

ABSTRACT: A classification of the loess deposits in The Netherlands was made using simple tests. The conclusion was that a classification of the loess deposits is possible with the Atterberg limits and the carbonate content.

RESUME: On a cherché s'il'etait possible d'établir une classification des dépôts de loess existant aux Pays-Bas au moyen de tests simples. Le conclusion est qu'une telle classification est possible en utilisant la limite d'Atterberg et la teneur en calcaire.

1. INTRODUCTION

In the South of The Netherlands a loess deposit is found forming a dissected platform in South Limburg. This deposit can be divided in three parts.

- The Upper Loess (Twente Formation)
- The Middle Loess (Twente Formation)
- The Lower Loess (Eindhoven Formation)

Figure 1 shows the stratigraphy of these layers (Kuyl 1975). The maximum thickness of the loess deposit is about 20 meters.
As can be seen in figure 1 there are some characteristic boundaries and phenomena visible in the loess deposits.
Characteristic are:

- The carbonate boundary
- The Eltviller tuff layer
- The Kesselt paleosol
- Frost phenomena (e.g. cryoturbation)

The carbonate boundary is situated in the Upper Loess. It is generally assumed that after deposition the carbonate dissolved from the top part and precipitated in the lower part of the deposit. This happens during wet periods when the infiltration rate is bigger than the evaporation rate and bigger than the field capacity, so that a downward water movement will take place.
The Eltviller tuff layer is also situated in the Upper Loess and has a thickness of a few mm's. In this layer volcanic heavy minerals were found. The layer is related to Eiffel volcanism.
Below this layer there is a zone of humus rich layers. This zone is called the Kesselt paleosol. The frost phenomena which can be observed are cryoturbation, frost wedges and polygons.
Loess areas are distinguished by problems of erosion and collapse. In the loess deposits of South-Limburg the main problem is erosion (table 1). The velocity of the erosion process is estimated on +/- 1 mm/year (this means 15 tonnes/ha/year). This compares with the erosion rate in some parts of the USA (Mücher 1986). During heavy rainfall the loess erosion can be very high and cause many problems. The most widespread

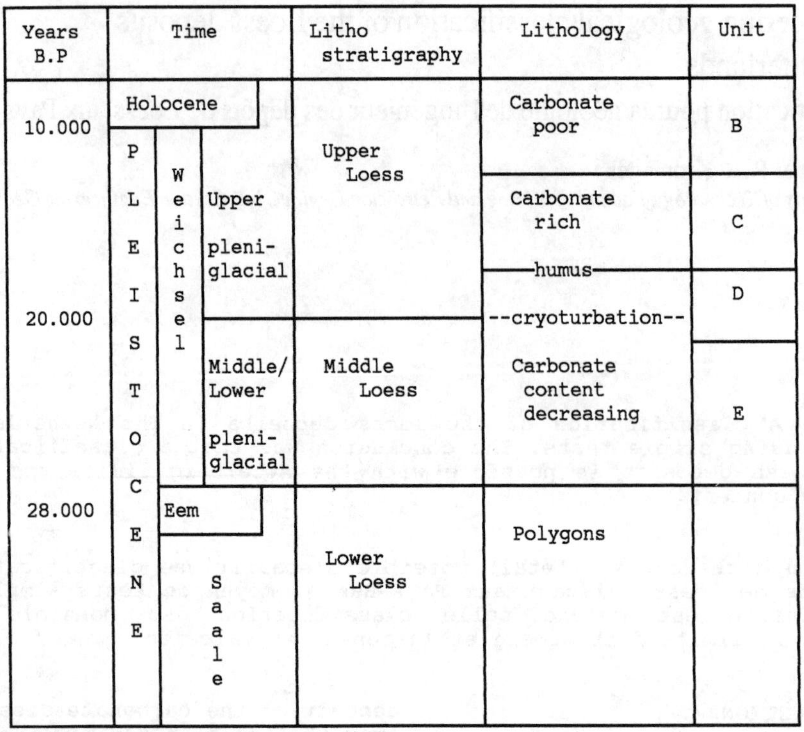

Years B.P	Time			Litho stratigraphy	Lithology	Unit
10.000	Holocene			Upper Loess	Carbonate poor	B
	P L E I S T O C E N E	W e i c h s e l	Upper pleni-glacial		Carbonate rich	C
					—humus—	D
20.000					--cryoturbation--	
			Middle/ Lower pleni-glacial	Middle Loess	Carbonate content decreasing	E
28.000		Eem		Lower Loess	Polygons	
		S a a l e				

Figure 1. Stratigraphical column of the loess in South-Limburg.

Table 1. The amount of erosion in South-Limburg (Bouten, van Eysden et al. 1985).

	ha	%
not eroded loess	10400	26
eroded loess	20100	50
colluvium derived loess	9800	24

form of erosion is sheet erosion. Another form of erosion is erosion by rill and interrill flow. On slopes rills can become very important and easily develop into gullies, if the loess is not ploughed every year, especially on slopes steeper than 10%.

The non-eroded loess can be found on the plateaus while the eroded and secondary loess can be found on the many valley slopes. The secondary loess is also situated in the dry valleys.

Saturation of loess at low loading may lead to collapse and slumping. In view of this collapsing behaviour loess can be regarded as a subsident or metastable soil. This type of soils is deposited in such a way that it exists in a loose unconsolidated structure at a relative low density. At normal moisture contents it possesses high apparent strength but is susceptible to a large reduction in void ratio (collapse) upon wetting. The cohesion and angle of internal friction decrease. This is not typical of the loess in South-Limburg.

2. OBJECTIVE OF THE INVESTIGATION

The aims of the investigation now underway are :

1. to distinguish differences between loess units.
2. to find a possibility to map the units using simple tests.
3. to find out if there is a relation between distinguished units and there behaviour.

The work here reported deals with investigation point 1 and 2.
For the fieldwork a 76 km2 area was chosen as shown in figure 2. This area is part of the loess plateau of Beek. In this area 42 samples were taken from 16 locations. Because of the difficulties in identifying loess units on geological maps the characteristic boundaries, which were discussed in the intro-duction, were taken as guidance for the identification of the loess units.

Before the fieldwork started the general idea was that the clay content might be varying in the different loess units. Unknown was how much the clay-content varied and if it was characteristic for a specific loess unit.
The following tests were done:

- Grain size analysis (BS 1377:1975, Test 7(B and D)
- Atterberg limits (BS 1377:1975, Test 2(A) and 3)
- Linear shrinkage (BS 1377:1975, Test 5)
- Specific gravity (Pycnometer)
- Proctor test (BS 1377:1975, Test 12)
- XRD analysis
- SEM analysis
- Organic content (BS 1377:1975, Test 8)

3. GEOTECHNICAL PROPERTIES

The grain size distribution of loess in general shows a dominant silt fraction between 2-63um and has usually less than 10% sand.
In South-Limburg the silt content is mostly 70% or more (Kuyl 1975 and Mücher 1973).

In table 2 all the determined geotechnical properties of the South-Limburg loess are shown. When a comparison is made between the South-Limburg loess and loess from China, Britain and Canada (Derbyshire and Mellors 1988, Sweeney and Smalley 1988) the conclusion is that the soils have more or less the same characteristics, but the loess in South-Limburg seems better sorted than the other loesses.

4. CLASSIFICATION

A distinction between five different units was made.

1. Secondary and/or eroded deposits. (Unit A)
2. Carbonate poor Upper Loess. (Unit B)
3. Carbonate rich Upper Loess. (Unit C)
4. Carbonate poor/rich Middle Loess disturbed by cryoturbation. (Unit D)

Table 2. Geotechnical properties of the loess in South-Limburg.

UNIT	DENSITY					NMC		ATTERBERG LIMITS						GRAIN SIZE				SH.STR.		PROCTOR		ORG.CONT.	CARB.CONT.
	jb KN/m3	jd KN/m3	Sp.gr	Xd	Poros. %	nmc %	Xd	LL %	Xd	PL %	Xd	PI %	Xd	Sort.	Xd	Skew.	Xd	p	c Kpa	dens. Mg/m3	mc %	organic cont. %	carbonatic cont. %
A						20.7	1.2	32.9	1.2	21.7	0.9	11.8	1.1										0
B	19-20	16-17	2.58	0.02	34-38	18.2	4.1	30.6	1.6	21.9	1.3	9.0	0.9	2.05	0.27	0.58	0.09	35	23	1.82	14	1.5	0 - 1
C	19-20	16-17	2.61	0.02	35-39	17.9	3.7	28.3	0.8	22.5	0.7	5.9	1.1	1.76	0.15	0.70	0.17	29	19	1.78	15	1.6	24
D	20-21	17-18	2.62	0.02	31-35	20.1	2.2	29.3	0.5	20.8	1.2	8.4	1.2	1.98	0.08	0.66	0.04	30	17	1.78	15	2.1	0 - 18
E	19-20	16-17	2.61	0.01	35-39	19.2	2.3	27.7	0.9	22.5	0.5	5.4	0.5	1.59	--	0.84	--			1.84	14	1.1	16

Table 3. The Atterberg limits.

Atterberg Limits

Sample	nmc (%)	LL (%)	PL (%)	PI (%)	PI-LS (%)
3A	22.10	33.1	21.0	12	13
3A/C	21.43	32.2	22.0	10	12
3A/C	18.45	32.2	22.9	9	6
6A/B	21.31	35.0	22.9	12	14
14A	19.80	33.5	20.8	13	13
14A/B	21.05	31.1	20.7	10	8
1B	21.66	30.8	22.5	8	10
2B	24.87	31.9	23.0	9	15
4B	16.23	27.7	22.8	5	5
5B	14.84	31.7	22.4	9	12
7B	22.20	30.1	20.6	9	13
8B	17.74	33.7	23.8	10	11
9B	17.47	30.4	22.2	8	9
10B	20.42	29.4	18.8	11	10
11B	13.56	31.0	22.9	8	11
12B	17.55	29.8	21.8	8	9
13B	22.77	31.5	21.3	10	12
15B	14.82	29.3	20.6	9	10
16B	18.52	30.8	21.5	9	10
1C	18.75	27.6	21.8	6	10
1C/D	13.51	29.1	22.9	6	6
2C	23.32	28.2	22.7	6	5
4C	10.04	26.9	23.5	3	4
5C	10.38	29.9	24.0	6	7
6C	23.93	29.2	22.0	7	9
7C	20.42	27.6	22.5	5	6
8C	15.61	29.0	22.7	6	7
8C	14.26	30.0	22.5	7	7
9C	20.81	27.6	22.8	5	6
10C	15.00	27.8	21.7	6	8
11C	19.29	28.0	24.2	4	5
12C	14.18	28.4	22.2	6	8
14C	18.87	28.1	21.7	6	7
16C	16.21	27.5	22.6	5	6
2D	21.56	29.1	20.1	9	7
7D	22.70	30.2	20.1	10	13
7D/E	19.94	28.8	21.1	8	10
8D	15.94	29.7	22.8	7	9
9D	20.25	28.9	19.6	9	9
11D	20.61	28.6	21.5	7	10
13D	18.62	27.6	22.0	6	9
2E	21.67	27.9	23.2	5	5
2E	16.63	26.5	21.4	5	6

5. Carbonate rich Middle Loess (Unit E)

(figure 1)

Most of the 42 samples were taken with a hand-auger. The maximum depth which could be reached with this device was 5.5 meters. Samples were also taken from two sand and gravel pits which were situated in the fieldwork area. With table 2 and table 3 an evaluation of the results can be made.

The results show that the main difference between loess A - E is the clay content and the carbonate content. The Scanning Electron Microscope (SEM) analysis and the X-Ray Diffraction (XRD) results confirm the importance of the clay minerals. The carbonate content is 0% for loess A, 0-1% for loess B, 24% for loess C, 0-18% for loess D and 16% for loess E. In the diffractograms two carbonate minerals are found, calcite and dolomite. These minerals were found in the following proportions.

	Calcite (vol.%)	Dolomite (vol.%)
B	1%	0%
C	20%	4%
D	13%	5%
E	13%	3%

The fact that there is much more calcite, which dissolves better in water than dolomite is surprising. The ratio calcite-dolomite for the loess in South-Limburg is in general 3 : 1.

The conclusions drawn from the SEM analysis are that there are swelling clays present in the Limburg loess. In the carbonate rich loess carbonate can be found as a coating or encrustation on the grains. And finally it was concluded that weathering and leaching processes are active in the loess. From the XRD analysis the following observations were made.

1. The clay minerals that can be detected are: Muscovite, Illite, Montmorrilonite and Chlorite.
2. According to the peak intensity loess C contains less clay minerals as the other loesses.
3. Loess B has a higher Muscovite content as the other loesses.
4. The feldspar content is increasing going from loess C to E.
5. The carbonate content is decreasing going from loess C to E.

The Atterberg limits could be useful as a test for classification of the loess units.

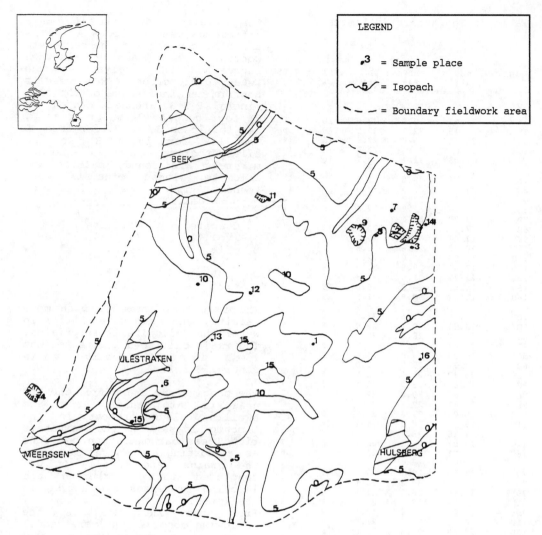

Figure 2. The fieldwork area.

The results of the Atterberg limits (table 3) point out that there is a relation between the plasticity index (PI) and a specific loess unit. The carbonate poor Upper Loess (B) has a relative high plasticity while the carbonate rich Upper Loess (C) has a low plasticity. The upper part of the Middle Loess has again a high plasticity in the cryoturbation zone (D) and a low plasticity in the lower part (E). The most fluctuations are in the PI values of the carbonate poor Upper Loess. This is caused by different degree in weathering which causes different clay contents. The fluctuations of the other lower lying units is much less.

The PI seems also useful as a mapping parameter (figure 3). The horizons which were followed in the field give the same PI values. For instance the carbonate rich Upper Loess (C) has a PI of 5-6% for all the samples. So a characteristic unit identified in the field can also be identified on its PI value in the laboratory. Very high PI values of samples from the carbonate poor Upper Loess (A) seem to point out secondary/eroded deposits. Comparing their position with the

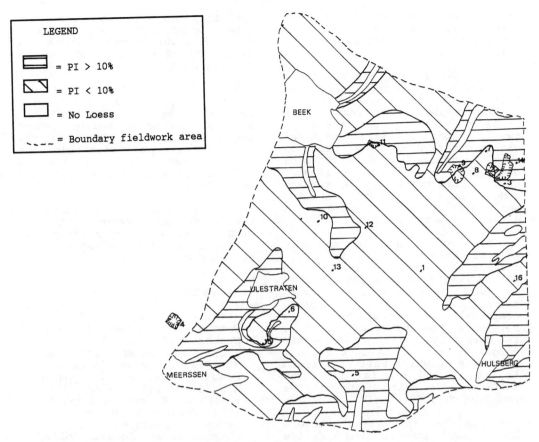

Figure 3. The PI used as mapping parameter.

soil map of the area confirmed that their position is in an area of secondary deposits.

During mapping the conclusion was that the boundary between the loess units A and B was at a PI of 10%.

From table 3 a comparison can be made between PI values determined from the linear shrinkage (PI-LS) and the "normal" PI. In general the PI-LS is a bit higher than the PI but the relation between them for the different units is the same.

Because of these results the conclusion can be drawn that the Atterberg limits are an useful tool for characterising the loess in South-Limburg.

5. DISCUSSION

Table 2 shows that the density of loess D is somewhat higher than the density of the other loesses. This can be explained by the fact that the cryoturbation process densified the loess. Field CPT tests also give higher values for loess D.

The lower sorting of loess D agrees with the higher amount of clay.

Table 2 also shows that the carbonate content combined with the Atterberg limits can be used to make a loess classification.

For a good classification of the loess units the carbonate content and the Atterberg limits are used combined with some stratigraphical information.

a. When the PI of a loess sample is higher than 10% and the carbonate content is 0% the sample is from unit A

b. When the PI of a loess sample

731

is lower than 10% and higher than 7.5% and the carbonate content is higher than 0% the sample is from unit D.

c. When the PI of a loess sample is lower than 10% and higher than 7.5% and the carbonate content is 0% the sample can be from unit B or D. To distinguish these two some stratigraphical characteristics should be used. In unit D cryoturbation structures and humus layers can be found. Unit B lies often at the surface and is about 2.5 - 3 meter thick While unit D is found on 4.5 - 5 meters below the surface.

d. When the PI of a loess sample is lower than 7.5% and the carbonate content is higher than 0% the sample is from unit C or E. To make a distinction between these two also some stratigraphical information is necessary. The colour of unit C is very light yellow while the colour of unit E is dark Yellow. The depth on which unit C is found is between 1 - 5 meter below the surface while unit E can only be found below 5.5 meter.

The fact that unit A and B have a high clay content and a low carbonate content explains the erosion problem of the loess. The carbonate dissolves during rainfall from the upper loess units. The moist conditions of these upper units accelerates the weathering of some minerals (e.g. feldspars) into clay minerals. As the result of this the bonds between the loess grains loosen. Because of this the loess along the slopes becomes unstable during rainfall (swelling clays). On this basis it seems justified to say that loess units with high PI values are more sensitive for erosion processes than units with low PI values.

6. CONCLUSIONS

The conclusions are:

1. On basis of the plasticity index (PI) and the clay content three engineering geological different types of loess can be

distinguished on the first 5.5 meter. Loess B and D who have a PI of 8 - 10%, loess C and E who have a PI of 4 - 6%, and loess A who has a PI of 12 - 13%.

2. The plasticity index (PI) can be used to map the different loess types. It was found that in an area of secondary and/or eroded deposits, the PI values for the first 2 meters of loess below the surface were higher than 10%. In undisturbed areas the PI was less than 10%.

3. With the Atterberg limits and the carbonate content combined with some stratigraphical information a loess classification can be made.

REFERENCES

Bouten, W., Eysden van, G., Imeson, A.C., Kwaad, F.J.P., Mücher, H.J. and Tiktak A. 1985. Ontstaan en erosie van de lössleemgronden in Zuid-Limburg. KNAG Geogr. Tijdschr. XIX. 3:192-208

Derbyshire, E. and Mellors, T.W. 1988. Geological and geotechnical characteristics of some loess and loessic Soils from China and Brittain: a comparison. Engineering Geology. 25:135-175.

Goedemoed, S., Plasman, S.J. & Bik, M.L. 1987. Report fieldwork South-Limburg: The quarry Bruls. TU Delft, Faculty of Mining and Petroleum Engineering, Section Engineering Geology.

Kuyl, O.S. 1975. Löss. Grondboor en Hamer. 1:2-12.

Mücher, H.J. 1973. Enkele aspecten van de loss en zijn noordelijke begrenzing, in het bijzonder in Belgisch en Nederlands Limburg en in het daaraangrenzende gebied in Duitsland. KNAG Geogr. Tijdschr. VII. 4:259-276.

Mücher, H.J. 1986. Aspects of loess and loess-derived slope deposits: an experimental and micromorphological approach. KNAG/Fys. Geogr. Bodemk. Lab. Universiteit van Amsterdam.

Sweeny, S.J. and Smalley, I.J. 1988. Occurence and Geotechnical properties of Loess in Canada. Engineering Geology. 25:123-134

6th International IAEG Congress / 6ème Congrès International de AIGI, © 1990 Balkema, Rotterdam. ISBN 90 6191 130 3

Regional engineering geological evaluation of weak zones
Evaluation géotechnique régionale des zones affaiblies

M.Šamalíková
Department of Geotechnics Brno Technical University, Brno, Czechoslovakia

ABSTRACT: In this article the summarization of the information obtained until present about the weak zones of some parts of the Bohemian Massif in Moravia is presented. In individual regions weak zones have been described from the point of view of the genesis, petrology, engineering geological properties and hydrogeology. From the regional point of view the weak zones of crystalline complex, Proterozoic and Paleozoic sediments, Cretaceous sediments as well as neovolcanic rocks are classified. According to the genesis the tectonic zones, zones of hydrothermal alteration, zones of metasomatic decomposition, zones of dissolution and weak intercalations of subhorizontal layered rocks are described.

RESUME: L´article fait un résumé les connaissances du territoire du Massif de Bohémia et des environs des constructions significatines de génie en Moravie. Dans les régions individuelles on classifie les zones affaiblies du point de vue de la geneze, de pétrologie, des propriétes géotechnique et de la hydrogéologie. Du point de vue régional on distingue des zones affaiblies du cristallin, des sédiments paléozoiques, crétacés et des néovulcanites. D´aprés genése on décrit des zones d´origin tectonique, des zones de désaggrégation des zones d´altération hydrothermale, de désintégration metasomatique et intercales affaiblies des sédiments déposés horizontalement.

1 DEFINITIONS

Weak minerals

Mostly argillitized minerals (kaolinized feldspars and feldspars with smectites or chloritized biotite) or mineral grains affected by mechanical deformatin with microcracs, or grains partly dissolved and corroded.

Weak rocks

Heavy fractured or weathered rocks to even disintegrated ones, soft to slurry, originally strong rocks which differ from their corresponding undisturbed strong rocks by mechanical, physical, hydraulic and deformation properties. Most soft mylonites, phylonites and soft rocks of metasomatism and hydrothermal alteration belong to them too. However, evaporites, coal and fresh mudstones or marlites do not rank among the weak rocks.

Weak rock massif

Rock unit faulted due to tectonic activity or other effects and loosened rock enviroment with numerous pinched discontinuities or with fillings of clay character weakened by a mineral filling.

Clay residuum

Nontransported accumulation of sheet silicates formed in situ, incoherent or only slightly coherent, having direct spatial relation to the parent rock (Konta 1972).

Microcracks

Cracks which originate in the region of ultimate strength by a brittle deformation of the rock mass. Their origin is accompanied by the growth of inelastic volume deformation. Their density may be the criterion for brittle deformation at the breaking strength.

Fractures

Are discontinuities which are divided into tight fractures and proper fractures which may be either with a filling or without this or with incrustations on planes of separation. According to the size these are divided into small, large and fracture zones.

Argilitization

Process of weathering during which clay minerals originate. This is the most frequent cause of weakening the rocks having substantial content of feldspars such as granitoides, gneisses and migmatites.

2 THE PROCESSES OF WEAKENING THE ROCK ENVIRONMENT

The weakening of the rock environment occurs due to the effects of endogenous and exogenous geological processes. The character of weakening may be purely mechanical and purely chemical an also the combination of both. All constituents of the rock environment, i.e. mineral grains, rocks as rock material and the complete rock massif may be weakened. The schematic representation of processes leading to the weakening of the rock environment is shown in Fig.1.
The mechanical weakening of rock debris as rock material is basically given by the degree of reducing the concentration both by mechanical fracturing and disintegration and also by goffering which is usually in connection with original texture. The mechanical weakening of rock massifs is shown by the existence of tectonic zones, by pressure faulted and crushed parts and also as a total reducing of concentration

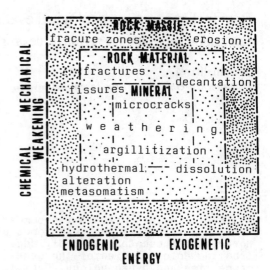

Fig.1 Schematic presentation of the processes resulting in weakening the rock environment

according to the discontinuities.
The chemical weakening is connected with petrological changes. This refers above all to argillitization as a result of hydrolysis, hydrothermal alteration and metasomatism. Weakening may also occur due to the effect of cold water in the surface parts of litosphere (oxidation). An important part in these processes is played by the ph of the rock environment.

3 REGIONAL EVALUATION OF THE WEAK ZONES OF THE BOHEMIAN MASSIF IN MORAVIA

This evaluation is based on a long-term engineering geological evaluation of weak zones in the vicinity of important engineering, underground and water-management structures in the territories of Bohemia and Moravia in Czechoslovakia. From the regional point of view the following engineering geological regions are distinguished in the above territories:
- The region of crystalline complex
- The region of Paleozoic sediments
- The region of Cretaceous sediments
- The region of neovolcanic rocks.
  From the West Carpathians the

territory of our special interest is
affected by sediments of Neogene
which are not presented in this
paper.

3.1 Weak zones of the crystalline complex

With the problem of the weakening
of the rock material of crystalline
rocks the author was concerned in
her previous papers (Šamalíková
1983, 1985).

The rocks of the crystalline
complex represent the main source of
aggregates for structural purposes
and for the building stone. Above
all granitoids and gneisses are
used, to a lesser extent migmatites,
amphibolites and granulites.

The rocks with substantial content
of feldspars, i.e. granitoids,
gneisses and migmatites may be
divided into three categories due to
their quality. The decisive
mechanical properties are simple
compresssion strength, volume weight
and frost resistance which comply
with the criteria set by Czechoslo-
vak standards. As results from Fig.2
these categories may be distinguished from the weakening of the
main mineral grains already, which
are feldspars. The degree of argillitization of feldspars corresponds
to the degree of weakening of rock
lumps.

Another type of weakening is
limonitization spreading from
fractures up to feldspar microfissures, which also corresponds to
a certain degree of weakening.

Argillitization of feldspars
appears in these rocks in two ways.
Partly by kaolinization, which
causes the loss of strength and thus
the impropriety of theses rocks
aggregates, which, however, mean the
exclusion of rocks as foundation
soil remaining constant as to the
volume. Partly by argillitization
during the formation of volume
unstable clay residua, especially
the minerals of the smectites group.
In such a case, which is signalized
by shiftinf the line d_{001} on the
X-ray diffraction analysis up to
2 nm, a significant weakening of the
rock environment occurs - not only
as a material but also as a foundation soil. The rock becomes not
only less strong but also sqeezing

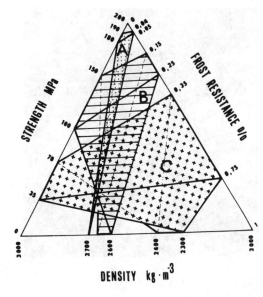

Fig. 2 The diagram of the quality of
the rocks with substantial part of
feldspars
A - Very good quality
B - Conditional quality
C - Unsuitable quality

and unstable and may cause the
beginning of displacement of rock
blocks on slopes.

The argillitization of biotite
occurs most frequently in the form
of chloritization. Chlorite may also
originate due to the effect of tectonic pressure on fault traces.
Besides diseptechlotite, the mixture
of chlorite and montmorillonite, so-
called false swelling chlorite was
found very often (Fig.3).

Weak zones of tectonic origin
represents one of the most frequent
cause of the weakening of rock
massifs of crystalline complexes.
They include large fault zones,
crushed zones and individual faults
as well. The filling of these zones
is composed of swelling polymineral
tectonic clays.

The typical example is the territory of the tectonic contact of Mora-
vicum and Svratka crystalline complex
in the vicinity of the construction
of water supply piping. This weak
zone was predicted according to the
engineering geological and geophysical survey before the construction

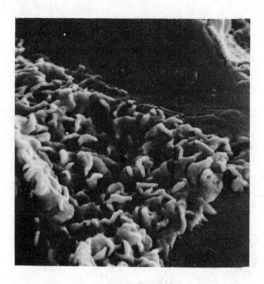

Fig.3 The character of swelling chlorite, SEM, enl. 2300 x

Fig.4 The deformation of the gallery in the swelling phyllonite

started (Šamalíková, Hašek 1981).

During the tunnelling of the exploration gallery the water seepage and a small fall occured.

After five years break of tunnelling for economic reasons considerable deformation of the gallery profile and the bracing appeared.

The extreme deformation originated in the part formed by silvergray to light green phyllonite, which was greasy when touching (Fig.4). According to the resultes of the X-ray analysis it consists of quartz, muscovite, sericite and admixtures of swelling illite. The moisture contant is 15 %, w_P 55,2 %, w_L 31 %.

On the direct contact of the phyllonite and strong gneiss a several cm thick graphite zone was formed (Fig.5).

The above weak zone is approximately 50 m thick and represents the eastern part of the contact of both above mentioned regional units. In its western part two more similar weak positions were formed. Also along these a deformation of the gallery occured during tunnelling.

The weak zones of hydrothermal origin of crystalline complex was investigated mainly in granitoids. This concerns, in most cases, the so-called epidotized filling of

Fig.5 The contact of strong gneiss and swelling phyllonite with graphite zone

faults and fissures. These fillings are enriched by clay residua, quartz and calcite. They are green up to white green of thickness being up to several centimeters.

An example is presented in Fig.6 from Brno massif of Brunovistulicum.

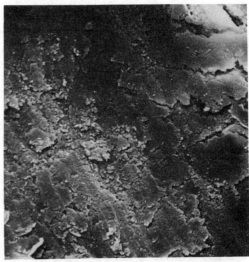

Fig.6 The character of weak filling of hydrothermal origin with clay mineral and calcite, SEM, enl. 2300 x

Fig.7 The character of tectonic clay from Culmian sediments with nontronite, SEM, enl. 2300 x

This weak zones represented the main causes of instability and rock fall in this region.

3.2 Weak zones of paleozoic sediments

In our region of interest clastic and carbonate complexes occur.

Clastic Devonian psephites and psammites offer mostly good and reliable rock environment. Weakening mostly occurs along small lutaceous layers laid between rougher grained psammites and psephites. The main constituents of those layers are kaolinite, quartz, hematite and admixture of muscovite. Even when kaolinite is a volume-stable clay mineral the weakening of those strata occur due to the weather effects. Thus overlying rocks break off due to frost weathering and cause unexpected collapses of rock walls in the built-in area.

Sediments of Culm also represent reliable foundation soils. Weakening occurs only along tectonic lines with filling of tectonic clay. Their character is shown in Fig.7. It is the clay whose main clay mineral is swelling nontronite. During construction these zones should be given a special consideration from the very beginning of the engineering geological investigation.

In Devonian carbonate complex the most important factor of the weakening is the degree of the carst phenomena.

Other weak zones within the rock environment originate mostly because of the tectonics. Some weak zones are enriched with graphite, as shown in Fig.8.

3.3 Weak zones of Cretaceous sediments

Cretaceous sediments appear in the region of the Bohemian massif as the first platform unit which lay discordantly and horizontally on older rock complexes. Petrologically this refers to the alteration of quartz thick-bedded sandstones with marlites to marles and arenaceous marl. Almost horizontal position causes that weakening occurs either parallel to the clayey interlayers or by rock fall along vertical fissures.

A part of cretaceous sediments in the vicinity of Svitavy in Moravia and also the territory of Bohemian-Saxon Switzerland may be taken as an example.

It was found out that the rock

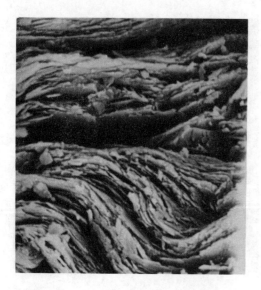

Fig.8 The deformation of Devonian
limestone with graphite, SEM, enl.
2300 x

3.4 Weak zones of neovolcanic rocks

The neovolcanic rocks represent a
relatively strong rock environment.
The weak zones are connected with
the tectonic or with the disintegration due to some secondary minerals
formed as a result of postvolcanic
activities. Their location within
the rock massif is unregullar.

CONCLUSION

From the regional point of view the
veak zones of crystalline complex,
Paleozoic and Cretaceous sediments
and neovolcanic rocks have been
classified according to the genesis,
mineral content and mechanical
behaviour. The tectonic zones, zones
of weathering, zones of hydrothermal
alteration and decomposition and
weak intercalations of horizontal
layered rocks have been described.

REFERENCES

Konta, J. 1973. Kvantitativní systém
    reziduálních hornin, sedimentů a
    vulkanoplastických usazenin.
    Praha: Universita Karlova.
Šamalíková, M. 1983. Scanning elec-
    tron microscopy examples of clay

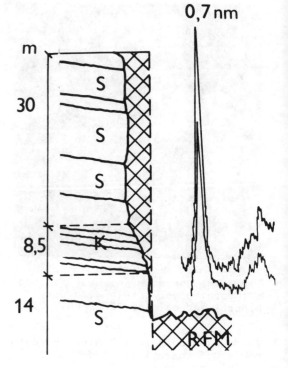

Fig.9 Weak kaolinite layers (K),
sandstone (S), rock fall material
(RFM), the line d_{001} of kaolinite
from Cretaceous typical rock wall

residua from crystalline rocks.
    Bull. IAEG 28:91-102.
Šamalíková, M. 1985. Characteristics
    of discontinuities and their in-
    fluence on the mechanical behavi-
    our of the rock mass. Bull. IAEG
    31:111-121.
Šamalíková, M. 1986. The importance
    and use of the SEM for the quality
    forecast of aggregates for rock-
    fill dams. Proc. 5th Congr. IAEG
    4.5.5-1461-1467.
Šamalíková, M. et V.Hašek 1981.
    Engineering-geological character-
    istics of weak zones on tectonic
    contacts in some metamorphic rocks
    in Czechoslovakia. Proc. Int. Symp.
    Weak rock. 429-434.
Šamalíková, M. et J.Klengel et J.
    Richter 1989. Ingenieurgeologische
    Untersuchungen zur Standfestigkeit
    von Felsböschungen im Sandstein
    der Sächsischen Schweiz. Brno: VUT
    5:84-91.

The fractal analysis of strength of intercalated clay layer in dam foundation
Etude d'analyse fractionnelle de la résistance de la couche d'argile intercalée dans la fondation d'un barrage

Shufang Xiao
Engineering Geomechanic Laboratory of Chinese Academy of Science, People's Republic of China

Renjiu Pang
Changchun College of Geology, Changchun, People's Republic of China

Minxun Ding
Xian Institute of Highway, Xian, People's Republic of China

ABSTRACT: The method used to determine the shear strength of intercalated clay layer stoday is macromechanical test in situ, which costs a lot of money. Fractal as a new quantitative method is applied to describe the intercalation microstructure which is of "Chaos" behaviour, i.e. disorderly outwardly but self-similar and multiscale inwardly. The grain exponent dg as a fractal index is established. The effect of the structure of different scale can be determined and the shear strengeh of intercalated clay layer can be evaluated by means of the correlation analysis between the dg of different range of grain and the coefficient of friction(f).

RESUME: On utilisait les experiences macromécaniques sur les lieux pour déterminer la résistance delacouche d'aigile intercalée contre le cisaillement; Cette methode s'avere ties cher et demande beaucoup d'argent. Fraction considerée comme une nouvelle methodequantitative, est appliquée pour décrire la microstructure d'intercalation; qui, avecune structurede chaos c'ad extérieurement désordonnée main présente a l'intérieur plusieurs couches secondairesayant une certaine propre ressemblance. On prend une grainede dimension dgpour représenter l'index de fraction. Les effects desstructures des differents niveaux sur larésistance peuvent être déterminés et les forces de cisaillement entre la couche d'argile intercalee peuvent etre evaluees par l'analyse de correlation entre les dg des différents types de graines et par le moyen du coefficient de friction(f). Par la même analyse on peut predire la resistance de la couche intercalee contre le cisaillement.

1 INTRODUCTION

Dimension is a important characteristics of geometry objects. In Euclidean space, the integral dimensions express the point, line, cube. However, in the last more than decade, the geometry objects with fractal dimension have made the scholars pay more attention. The concepts of fractal and fractal dimension have entered many field, such as engineering, physiology and economics. The purpose of this paper is to use the principle of fractal to find a new mathematical model which expressed fabric characteristics of intercalated clay layer and to provide a new method of strength prediction by establishing the relationship between the fractal dimension and strength parameter.

2 SOME CONCEPTS OF FRACTAL GEOMETRY

2.1 Fractal and fractal dimension

In nature, there are many objects with "chaos" behaviour which is disorderly outwardly but self-similar and multiscale inwardly, and the self-similarity is

Take the famous Cantor Set as an example.(Fig.1)

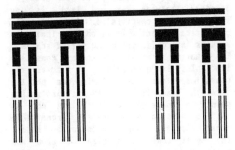

Fig.1 The Cantor Set

For this few and scatted point set, its dimension can be calculated by follow procedure. For a geometry object with d dimension, the length increased by l times in every independent direction yields N former objects. The relationship among three values is as follows:

$$l^d = N \qquad (1)$$

It can be proved that Eq.(1) is suitable for all general geometry objects.
Taking the log of equation(1) yields:

$$d = \ln N / \ln l \qquad (2)$$

Where d must not be integer and the extensively difined dimention is called factal dimention expressed by D

$$D_0 = \ln N / \ln l \qquad (3)$$

The fractal dimension of Cantor Point Set can be calculated by Eq.(3)

$$D_0 = \ln 2 / \ln 3 = 0.6309$$

Plotting N and l in the double-log coordinate, the slope of this plot is D_0.
The Cantor Point set is regular geometry object, in which the unified change times of length can be used. However the fact that fractal is not limited by regular object leads to the concept of "non-scale".

2.2 Characteristic scale and non-scale

Objects have their own characteristic length which needs to be measured by fair characteristic scale, inch used to measure the Great wall or clay grain is too short for former and too long for latter. The characteristic scale is the key in establishing and solving mathematical model and quantitatively describing natural phenomina. Whether are there any problems without characteristic scale? Without charateristic scale means that many scales from small to large must be considered simultaeously, i.e. "non-scale". How long is the coast of Great Britain put forward by Mandelbrot is a example of non-scale.
Boundary between sea and land is an irregular, complicated, multiscale geometrical object. It is obvious that the results are different using varied scales.
The shorter ruler is, the longer coastline is. This is "non-scale". So far as the coastline, the lower limit can be the straigth lines linking several protruding points of outer edge of the island and it has no meaning using scales larger than these straight lines. On the other hand, the least scale is the size of atom and molecule. It has no meaning using smaller scale using the least scale we can get the

upper limit of the length of coastline. Between these two limits, there is a non-scale section in which the length obviously is not a good characteristic value to decribe the coastline, and the fractal dimension will be the more suitable concept.

2.3 The Basic Definition of Fractal

For the Chaos object with the structure of self-similarity and non-scale, its dimensions are also suitable for Eq.(2). Returning to the example of coastline, suppose the distant between two protruding points is to be measured (J.R.Carr, 1989). The length is calculated as:

$$L = Ny + b \qquad (4)$$

When y is the ruler length, N is the number of rulers, and b is the remainder. This equation can also be expressed as:

$$L = [N+(b/y)] y^D \qquad (5)$$

If coastline is straigt line, D=1. Whereas the coastline is irregular curve, D is the fractal dimension. (Mandebrot, 1982). Taking the log of equation and normalizing L=1, yields:

$$-D = \log(y) = \log(N + b/y) \qquad (6)$$

$$-D = \log(N + b/y) / \log(y) \qquad (7)$$

For any different y, there iscorresponding different N and b. Plotting (N+b/y) and y in the double-log coordinate. the slope is the fractal dimension D. So the basic concept of fractal can be expressed as:

$$Ni = C/r^D \qquad (8)$$

Where Ni is the number of objects with characteristic scale r_i, D is the fractal dimension, C is the proportional costant.
For regular fractal objects, the fractal dimension are constant (as Cantor Set). For irregular fractal objects, the curve in the double-log coordinate often present broken line with different fractal dimensions, according to the number of the broken line n the structure can be divided into n levels, the structure of each levels shows self-similarity in this section of scale. It can be found from above that the function of fractal geometry is to use the fractal dimension to describe the objects with Chaos behaviours, and to distingrish and quantitatively describe the structure with varied scales. Therefore, by analysing of correlation between the fractal dimension of structure of different scale and

macromechanical parameters, we can judge the relative extent of the effect of the structure of varied scale on mechanical porperties, analysis the failure mechanism and predict the strength.

3 FRACTAL CHARACTERISTICS OF FABRIC OF INTERCALATED CLAY LAYER

The fabric of intercalated clay layer is a structural system with multi-compositions. It is composed of grains and pores with various shape and size. These grains and pores are called units of structure. The fabric of intercalated clay layer is the sum of these units. Whether this fabric has the fractal characteristics. A great number of tests shows that the fractal characteristics of structure of intercalated clay layer can be explained from following three structure factors, i. e. grain distribution, pore distribution and the extent of fracture roughness under scanning electron microscope.

3. 1 The fractal characteristics of grain distribution of intercalated clay layer and quantitative describe

The range of the size of grain unit is very large, from gravel grain to clay grain and colloid in succession. The distribution of various size of grains is very scatted and fragmented outwardly. However, a number of test data show that the curves of grain size distribution of various intercalated clay layers present broken lines in the double-log coordinate as shown in Fig. 2.

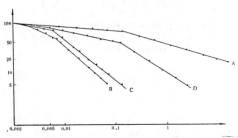

Fig 2. The curves of grain distribution of four type of intercalated clay layers in double-log coordinate

Each straight line in the broken lines can be expressed as

$$G = C_0 / r_i^a \qquad (9)$$

When r_i is the grain size, G is the cumulative content of grain larger than r_i, Co is the proportional coefficiat dg is grain exponential, i. e. the slope of

straight line in fig 2. It is easy to prove that the Eq. 9 can be changed into Eq. 8. Therefore, the grain distribution of intercalated clay layer has fractal characteristics. Different straight lines correspond with different fractal dimension D. For convenience, the grain exponential can be applied directly as a parameter to express the fractal characteristics of structure of intercalated clay layer. Grain sizes corresponding to turning points of the brokin lines are boundary values between structures with different fractal characteristics. The statistics of present data indicate that the double-log curves of grain distribution present two or three broken lines. It means that the structure can be divided into two or three scales according to their fractal characteristics which can be expressed respectively by dg_1, dg_2, dg_3.

3. 2 The fractal characteristics of pore distribution of intercalated clay layer and quantitative describe

The curve of pore distribution which is similar to that of grain distribution can be obtained by plotting the pore size r_i and corrsponding Xr, the cumulative content of pore larger than r_i in the double-log coordinate as shown in Fig. 3 :

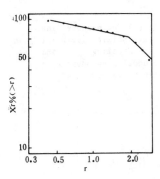

Fig. 3 The differential size distribution curve for pores

The curve presents two straight line segments and has obvious turning point is 1. 9 m which approximately is equal to the boundary value between the pore inside the microaggregates and pore between the microaggregates. The slope of the straight line is defined as pore exponential (dp_1, dp_2) which express respectively the fractal characteristics of different scale of microstructure.

3.3 The fractal characteristivs of the fructure roughness for intercalated clay layer under scanning electron microscope

Usually, the photograph of scanning electron microscope can only be used to describe the microstructure qualitatively. The curve in Fig.4 a secondary-reflection scanning curve of fracture of the sample. It represents the graphy of the fracture and also express the microstructure.

Fig. 4 The photograph of scanning electron microscope of intercalated clay layer, the white curve among the photo is the secondary-reflection scanning curve.

Using different scale (l) to measure the length of the curve can obtain different number of rulers. Plotting l and corresponding N in the double-log coordinate yields the relationship curve as shown in Fig.5:

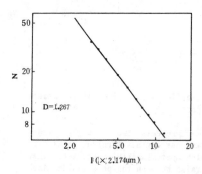

D=1.267

l(×2.174um)

Fig.5 The relationship curve between scale(l) and number of rulers(N)

The slope of the curve is the fractal dimension D of the fracture of sample D=1-2. The rougher fracture is, the larger D is.

4 THE APPLICATION OF FRACTAL TO ANALYSIS AND PREDICTION OF STRENGTH

Numbers of tests show that the distritution curves of grain size of intercalated clay layer can be classified into four types as shown in Fig.2.

4.1 Type A

The content of gravel grains is larger than 10%. The cruve presents two or three broken lines. The grain exponential and coefficient of friction f are listed in table 1

Table 1 The value of f and dg for 12 groups of samples

sample No.	−dg1	dg2	f
37	0.105	0.105	0.46
25	0.123	0.213	0.44
29	0.123	0.09	0.35
4	0.360	0.140	0.36
27	0.230	0.305	0.28
28	0.123	0.213	0.46
32	0.141	0.344	0.3
35	0.212	0.445	0.25
31	0.212	0.404	0.26
30	0.212	0.445	0.34
26	0.577	0.404	0.34
5	0.267	0.624	0.14

The relationship between absolute value of grain exponential dg1, dg2 and frictional coefficient f is shown in Fig.6

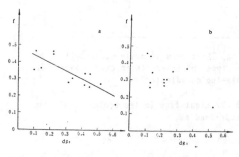

Fig.6 The relationship between f and dg
a. f − dg2   b. f − dg1

Fig. 6 shows that the correlation between f and dg_t is more obvious than that between f and dg_i. dg_i expresses the structural feature of fine grains smaller than sand size. dg_t represents that of coase grains. The obvious corretion between dg_t and f means that the structure between dg_t and f the structure of coarse grains plays a leading role in f value. The effect of structure of fine grain on strength is comparatively small due to less content. Follwing equation can be yielded from Fig.6:

$$f = 0.48-0.445dg_t$$

4.2 Type B

The content of clay grains is larger than 20 % and without gravel grain. The curve presents three or two broken lines. The grain size corresponding two turning points are $r_{21}=0.002mm$, $r_{22}=0.005 - 0.05mm$.
The relationship between the absolute value of dg_i, dg_t and f is shown in Fig.7. The Fig.7(b) shows that the f is not correlative obviously with dg_i. It means that while the content of clay grains is larger than 20%, the silts and grains are surrounded by clay.

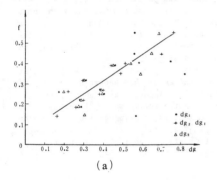

(a)

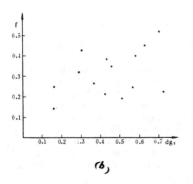

(b)

Fig. 7 The relationship between dg_i, dg_t, dg_i and f

and shear is underway in the clay. Therefore the structures of silt and sand do not exist a obvious influesce on f. The correlation between dg_t and f is more obviour than the correlation between dg_i and f. But Fig.7(a) shows that the most obvious correlations is that between f and the combination of dg_i and dg_t($2/3dg_i + 1/3dg_t$). The correlation equation is shown as follow:

$$f = 0.06+0.625(2/3dg_i+1/3dg_t)$$

The dg_i and dg_t express the fractal characteristics of microstructure which is composed of clay microaggregates and clay grains, and relate to the clay mineral composition. It is concluded that the microstructure and the composition of clay mineral play a key role in the strength of this type. Some intercalated clay layers in the foundations of Xiaolangdi Dam, Datenxia Dam and Wuqiaxi Dam belong to this type.

4.3 Type C

It is composed of mainly silt grains. The content of clay grains is less than 20%. The intercalated clay layers in the foundation of Zanfan Dam, Qinshan Dam belong to this type. The curve of grain distribution presents two broken lines and dg_t, dg_i. The grain size of turning point is $0.005 - 0.01$ mm. The correlation between dg_i, dg_t and f are shown in Fig.8 separetely. It is found that the correlation whether between f and dg_i or f and dg_t are less obvious than the correlation between f and ($2/3dg_i+1/3dg_t$). It means that the friction of silt grains, the structural bond of clay microaggregates and clay grains play simultaneously a important role in the shear strength.

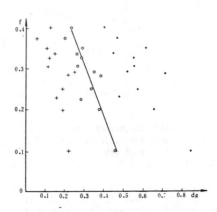

Fig.8 The relationship between dg_i, dg_t and f + $dg_i \sim f$. $dg_t \sim f$ ○ dg_i, $dg_t \sim f$

The following equation can be yielded:
$$f= 0.4-2.47(2/3dg_i+dg_r)$$

4.4 Type D

The feature of this type is between that of type A and type B. The silt and sand grains are not encircled completely by clay. The effect of friction, various microstructure bond act simutaneously in shear strength.

5. THE FRACTAL DIMENSION OF FEACTURE AND SHEAR STRENGTH

The fractal dimension of a fracture under scanning electron microscope is very useful index which can be used to describe quatitatively the feature of microstructure, such as the size and the shape of grains, pores and the arrangment of them. So the fractal dimension of the fracture roughness should be related to the shear strength. Some test results of intercalated clay layer in the foundation of Tongjieji Dam are shown in Fig. 9. It seems that the fractal Dimension D and the friction coefficient f exist the positive correlation. Conversely, the D and the cohesion C represent the negative correlation.

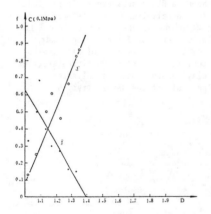

Fig. 9 The relationship between the fractal Dimension of roughness of fracture and The parameters of shear strength.
(1) $D \sim f$ (2) $D \sim C$

The relationship shown in Fig. 9 is an initial result. It must be comfirmed further using a great number of test data.

CONCLUSION

1. The fractal geometry is a disipline of mathematics, which can be used to describe quatitatively the objects with Chaos behaviour.
2. The fabric or structure of intercalated clay layer is a dissipative system with Chaos behaviour, i.e. disorderly outwardly but similar and multiscale inwardly.
3. The fractal characteristic of intercalated clay layer is represented by three aspects: The fractal characteristecs of grain distribution, pore distrubution and fracture roughness under scanning electron microscope. The fractal characteristic of three aspects can be expressed by three index, i.e. the grain exponental dg, the pore exponental dp, and fractal dimension D of fracture roughness respectively.
4. It has been comfirmed that there is a correlation between the macromechanical parameters f, c and the index of fractal dimension D, dg, dp.
5. A new way of prediction of strength of intercalated can be established by means of measuring the fractal index of the disturbed sample and analysing the correlation between these index and parameters of strength. It is obvious that the new way is much more cheaper than the macromechanical test in situ, and because of its simplicity, it can be performed in a large number for an engineering project.
6. The fractal geometry, as a very useful mathematical means, has become a subject of considerable theoretical and practical interest in the world, we try to apply it to the field of engineering geolog. With the developing of study, the fractal geometry will indicate increasingly its function in solving fuzzy, undifinied and confuse problems.

REFERENCE

Jeder, Jens, 1988. Fractals. Plenum Press, New York
James R. Carr, 1989. Pelationship between the Fractal Dimension and Joint Roughness Coefficient. Bulletin of the Association of Engineering Geologess. Vol XXVI. No. 2, 1989
Belin Hao, 1986. Fractal and Fractal dimension. Science, vol. 38, No. 1
Mandebrot B.B, 1977 Fractals, W.H.Freeman
Mandelbrot B.B, etal. 1984 Fractal character of fracture surfaces of metals. Nature Vol.308 19 Appril.
Pietronero L. , Tosatti E.(editors) 1986 Fractals in Physics, Elsevier Science Publishers B.V.

Geotechnical classification and determination of the rock materials properties used for the embankment of Wadaslintang Dam, Central Java, Indonesia

Classification et détermination géotechnique des propriétés des matières rocheuses utilisées au remblai du barrage de Wadaslintang, Java centrale, Indonésie

Soedibjo
Ministry of Public Works, Directorate General of Water Resources Development, South Kedu Multipurposes Project, Gombong, Central Java, Indonesia

ABSTRACT : Among the most important engineering geological aspect of the dam construction are related to the quality of the construction materials.
For the purpose of identification and classification, geotechnical properties of the rock materials which will be used for the dam construction were ditermined either in the laboratory or in the field.
Selected geotechnical properties of the construction materials have been plotted against each other in order to be able to estimate the value of one property from the value of another property.
The interrelationships are used for the selection of cheap and simple field - testing methods of the purpose of quality control during dam emplacement.

RESUME : Parmi le plus important à l´aspect en éngineering géologique dans la construction de barrage a une relation à la qualité des materiaux utilisées.A propos d´identification et classification, les propiétés géotechnique de roche matière qui seront utilisées dans la construction de barrage sont déterminées soit en laboratoire soit sur le terrain. Les propiétés géotechnique choisis des materiaux de construction ont été tracées point par point. l´un pour l´autre, afin que pouvoir à éstimer leur propiétés. Cettes relation mutuelles ont été utilisées pour selectioner des methodes simple qui moin cher à l´éssai sur terrain pour controller des qualités pendant l´émplacement de barrage

1 INTRUDUCTION

Any classification of rock to be used as material for a rock fill type dam should be based on specific geotechnical properties.
The rock materials used for the dam embankments should be adequate in durability, strength and density.
For the purpose of quality control during the construction stage of the dam project a number of regular measurement of the geotechnical properties of the rock are needed.
All these measurement are often costly and time consuming.
So, the suitability of the various methods for identifaction and classification to be used during quality control of these materials will be discussed in the following sections.
The methods for identification and classification will have to be :
1.1 Rapid and simple involving a minimum of sample preparation.
1.2 Relevant to the essential dam material properties.
1.3 Capable of discriminating between grades of engineering suitability

All datas for these purposes were collected during the contruction of Wadaslintang Dam in Central Java, Indonesia.
The rock material consists of volcanic breccia.

2 GEOTECHNICAL PROPERTIES OF THE ROCK MATERIALS.

2.1 Basic Properties.

Basic properties include the fundamental characteristics of the materials and are used mainly for identifacation and corre - lation. Some are used in engineering calculation.These properties consist of porosity, void ratio, moisture content, density, durability and sonic velocity.

745

2.2 Engineering propertis.

These properties consist of hydraulic and mecahnical properties. Hydraulic properties are expressed in terms of permeability. It is measured for the possibility of flow of fluids through the rock material.

Rupture strength and deformation characteristics are mechanical properties, the magnitude of which can be measured by static or dynamic methods.

The engineering properties include shear strength, compressive strength and angle of friction .

2.3 Index properties

Index properties define certain physical characteristics and are used basically for classifications, but also for correlations with engineering properties.

Some index properties with should be considered for dam materials are point load strength index and Schmidt hammer rebound number

3 METHODS OF MEASUREMENT

For the Wadaslintang dam study, rock materials were tested in the laboratory for the most part and occasionally also in the field. Rock cores which were extracted from some boulders in the quarry, were tested for basic, engineering and index properties in the laboratory.

An index property of the rock materials was also measured in the field with the Schmidt hammer .

The result of the measurement of geotecn - nical properties for rock materials are shown in Table 1.

4 PROPERTIES OF THE ROCK MATERIALS.

4.1 Description of the weathering grades

The geotechnical properties of most rock are influenced by t he effects of weathering .

For the quality control during the dam con struction a description of the weathering grades in term of degree of change from the fresh rock material is needed follow - ing either a qualitative or a semi - quantitative scheme .

Dearman (1976) has established a descrip - tive scheme of weathering grades of rock material on the basis of the change associated with mechanical and chemical weathe ring (see Table 2).

From the excavated rocks in the dam, spill-way and quarry sites, samples were select-ed which showed distinct changes of fresh rock in terms of discolouration, chemical decomposition and desintegration.

Four grades of weathering were recognized on this basis and the rock blocks were classified as follows :

Table 1 Result of the measurement of geotechnical properties of rock materials.

Sample no.	Natural water content W %	Poro-sity n %	Absorp-tion A %	Saturated density Ds gr/cm^3	Dry density Dd gr/cm^3	Slake dur-ability index Id2 %	Sonic wave velocity V m/sec	Elastic modulus Static Es MPax10^3	Elastic modulus Dynamic Ed MPax10^3	Point load strength index Is(50) Pls MPa	Schmidt hammer rebound value Shv	Unconfined compressive strength Ucs MPa
1	6.79	26.84	14.34	2.14	1.87					0.3	16	6.9
2	7.36	19.73	8.78	2.44	2.25		2,350	8.3	12.5	2.6	46	63.2
3	6.59	20.78	10.14	2.26	2.05		1,100		2.75	0.78	24	19.2
4	4.87	11.82	5.27	2.36	2.24		2,750	6.51	16.3	2.20	41	52.9
5	9.53	28.42	14.03	2.31	2.03		1,608	1.40	4.99	1.08	24	21.9
6	2.94	10.22	4.20	2.53	2.43	97.58	3,984	8.35	27.6	2.60	50	64.1
7	10.3	30.20	15.38	2.27	1.96	72.52	1,078		2.78	0.74	26	18.3
8	4.99	27.68	14.20	2.24	1.94	62.54	902		1.68	0.71	23	18.3
9	5.35	17.84	8.23	2.34	2.17	93.26	1,935	5.65	5.92	1.65	39	32.6
10	3.69	18.82	8.35	2.44	2.17	95.08	2,525	6.23	11.4	1.9	44	44
11	2.36	7.48	3.07	2.51	2.43		4,154		31.8	4	56	87.7
12	3.21	19.20	8.64	2.41	2.22	81.49	1,909	5.61	7.54	1.69	38	39.4
13	3.78	18.99	8.46	2.44	2.24	96.67	2,571	5.22	13.8	2.10	40	43.1
14	7.71	23.44	11.28	2.31	2.08	90.11	1,605	1.07	5.93	1.10	24	26.5
15	6.55	16.97	8.68	2.12	1.95		2,092	4.94	8.03	1.65	38	33.5

Table 2 : Description of weathering grades
of the rock material (after Dearman,1976).

Term	Description
Fresh	No visible sign of weathering of the rock material.
Discoloured	The colour of the original fresh rock material is changed and is evidence for weathering. The degree of change from the original colour should be indicated. If the colour change is confined to particular mineral constituents, this should be mentioned
Decomposed	The rock is weathered to the condition of a soil in which the original material fabric is still intact, but some or all of the mineral grains are decomposed.
Desintegrated	The rock is weathered to the condition of a soil in which the original material fabric is still intact. The rock is friable, but the mineral are not decomposed.

Fresh rock.

The rock shows no changes attributable to weathering. The rock is grey to dark grey in colour and the Schmidt hammer rebound value is 50 or more.

Sligthly weathered

The rock is stained brownish grey on the exposed surface and the joint surface only. There is no penetration of weathering Deeper in to the rock. The Schmidt hammer rebound follows are in the range of 40 to 50.

Moderately weathered.

Discolouration is penetrated inward from joint or fracture surface. The discoloured rock is grey to brownish grey in colour. A lot of fragments in the rock are still grey coloured. The Schmidt hammer rebound values may very from 30 to 40.

Highly weathered.

Discolouration occurs within the matrix as well as within the fragments. The outer rim of rock is yellowish brown in colour and the inner part is brownish grey . The Schmidt hammer rebound follows are between 20 and 30 .

Resume of the geotechnical properties for rock material in the ranges of fresh to highly weathered are shown in Table 3.

Table 3 : Resumes of the geotechnical properties of rock materials in the ranges of fresh to highly weathered .

Rock Description	Specific gravity Sg (gr/cm^3)	Poro sity n (%)	Natural water content w (%)	Dry density Pd (gr/cm^3)	Absorp tion A (%)	Slake durability index Id2 (%)	Ultra sonic velocity $Vx10^2$ (m/sec)	Point load strength Pls (Mpa)	Schmidt hammer rebound value Shv	Uniaxial compressive strength Ucs (Mpa)	Elastic modulus Static $Es \times 10^3$ (Mpa)	Elastic modulus Dynamic $Ed \times 10^3$ (Mpa)
1.Fresh. No changes attributable to waethering, grey to dark grey in colour.	2.63	10.22	2.94	2.43	4	97.58	39.85	2.6	50	64.1	8.35	27.6
2.Slightly weathered. The rock stained brownish grey on the exposed surface and the joint surface.	2.51	15.42	3.74	2.21	8.5	95.88	26.6	2.15	43	53.15	66.76	15.05
3.Moderately weathered. Discolouration has penetrated inward from the joint or fracture surfaces. The discoloured rock is grey to brownish grey in colour.	2.41	18.52	5.95	62.06	8.5	87.38	19.22	1.67	38.5	33.05	65.3	6.98
4.Highly weathered. Discolouration occurs within the matrix and the fragments. The outer rim of rock is yellowish brown and the inner part is brownish grey.	2.20	28.94	7.65	1.95	14.5	67.53	9.9	0.74	24.5	18.75	1.08	2.23

4.2 Strength test.

The strength of the rock material is very important because it enters into many en - gineering problems. For example, the strength of concrete depends on the strength of the rock aggregate and also the stability of the rockfill will be in - fluenced by the strength of the rock mate- rials. The adequate strength should be a- ble to resist the compactive effort,other- wise the equipment will crush the materi- als into a smaller gradation.

The strength was determined either in the field by the Schmidt hammer or in the la - boratory by point load and uniaxial com - pressive strength testing equipments. Some samples, representative for the range of - fresh to weathered rock material from the quarry site, from excavated rocks in the spillway and from the dam site were collec- ted for these strength tests.

Irregular lumps and cylindrical test spe - cimen were used for the point load and the uniaxial compressive strength test respec- tively .

The results of these tests on rocks at va- rious degrees of weathering show that the point load strength is in the range bet- ween 0.7 and 4 MPa,while the uniaxial com- pressive strength is in the range between 18 and 90 MPa.

4.3 Durability test.

The durability expresses the ability of the rock materials to resist the gradation oy mechanical or chemical agents. It is a factor controlling the stability of the materials which are used as concrete aggre gate and as rockfill.

A test method for this purposes is the slake durability test,information of which gives an indication of breakdown expected to occure during stock-pilling, handling, transpoting and mixing.

The results of the test show that the rock materials in fresh,slightly weathered, mo- derately weathered and highly weathered conditions have the slake durability index (Id2), average values of 98 % , 96 %, 87 % and 68 % respectively.

4.4 Elastic modulus test

The elastic modulus (E) of rock materials will influence the strength of concrete if the rock is used as concrete aggregate. Generally the elastic modulus of the mate- rials will be considered in the design calculation of the concrete structure.

This E modulus as can be carried out with static or dynamic methods.

The result of these tests that the rock rock materials in the study area have a static modulus between 1.1×10 and 8.4×10 MPa.

The dynamic elastic modulus is between 2.2×10 and 27.6×10 MPa.

4.5 Ultrasonic wave velocity test

The velocity of ultrasonic waves (V) through a rock sample can be correlated with the unconfined compressive strength of the rock.

The rock materials which have a high com - pressive strength with also show a high velocity .

The result of test show that ultrasonic wave velocity of the rock materials in the ranges of fresh to highly weathe- red are between 2,750m /Sec and 900 m / Sec.

4.6 Density test.

The density is needed for identification , correlation and engineering analysis. Ge - nerally speaking, the rock materials with a higher density have better geotechnical properties than those with a lower density Using high density rock materials for dam embankment, it will be possible to incre- ase the steepness of the embankment slopes So, it will reduce the volume of the em - bankments and decrease the construction cost

Rock material of high density is usually more satisfactory for concrete agre - gates with respect to soundness and strength .

The results of the density test in the ra- nges of fresh to highly weathered rock ma- terials are between 2.12 and 2.53 g/cm^3 for the saturated condition, and between 1.95 and 2.43 g/cm^3 for the dry condition. The dry density is influenced by amount of cement or matrix material occupying the pores (Bell, 1978).

Due to the fact that the matrix of volcanis breccia is less resistant to we- athering, the density of the rock material tends to decrease with increasing degree of weathering.

4.7 Natural water content, absorption and porosity tests.

The natural water content and absorption

are very important to be known for concrete aggregates because the strength of concrete is influenced by the water - cement ratio.The porosity influences the strength as well as the absorption behaviour of the rock materials.

The results of natural water content determination in the ranges of fresh to highly weathered conditions give values between 2,36 % and 7.71 % absorption values are between 4 % and 15 % , porosities are between 10.22 % and 28.94 % .

According to the USBR (1956) , an absorption value of much over 1 % indicates that the aggregate is of poor quality, but this property alone is not a cause for rejection because the other properties of the materials and the required strength of the concrete have to be considered as well . Considering also the other properties, for example slake durability index, rock materials with a high absorption characteristic (A > 9 %) should be rejected as concrete aggregate.

4.8 Interrelationships between the various geotechnical properties.

Selected basic, index and engineering properties of the rock in the study area have been plotted against each other in order to be able to predict and estimate one property to another property. These interrelationships are investigated to be able to determine the most reliable cheap and simple field testing methods for quality control.

The result of the correlations are as follows :

1. Point load strength (Pls) versus Schmidt hammer rebound value (Shv).

The relationship is best expressed by exponential function. Point load strength and Schmidt hammer rebound value decrease gradually with the change from fresh to highly weathered condition. The relationship is shown in Fig.1.

2. Point load strength (Pls) versus absorption (A).

This relationship is linear with a very high correlation coefficient . The point load strength decreases gradually with an increase in absorption (see Fig.2) . The increase in absorption is the result of an increase in effective porosity. Due to the fact that chemical weathering is intense in the study area, the rock with a high

porosity will be weathered more easily and causes the decrease in strength.

3. Point load strength (Pls) versus sonic wave velocity (V).

The relation between these two parameters is linear and linear regresion gives a very high correlation . Point load strength decreases gradually with the decrease in sonic velocity (see Fig.3).

4. Point load strength (Pls) versus dry density (Dd)

The relationship is linear and its correlation coefficient is very high. Point load strength decrease sharply with the change of dry density (see Fig.4).

5. Point load strength (Pls) versus uniaxial compressive strength (Ucs).

The relationship between these two parameters has been studied by Broch & Franklin (1972), Dearman (1974) and Bieniawski (1975). According to these authors,uniaxial compressive strength is in the range of 16 to 24 times the point load strength. In the study area the relationship is linear, and linear regresion gives a very high correlation coefficient and a multiplication factor of approximately 22.5 (see Fig.5).

6. Point load strength (Pls) versus elastic modulus (E)

The relationship between point load strength and static elastic modulus (Es)is linear with a high correlation coefficient (see Fig.6). The correlation with dynamic elastic modulus gives an equation of an exponential function (see Fig.7).The point load strength decreases gradually with decreasing static or dynamic elastic modulus. Both properties decrease with increased weathering.

7. Schmidt hammer rebound value (Shv) versus uniaxial compressive strength (Ucs)

This relationship is linear with a high correlation coefficient (see Fig.8). Uniaxial compressive strength decreases sharply with decreasing Schmidt hammer rebound value. Both properties decrease with increased weathering.

8. Schmidt hammer rebound value (Shv) versus elastic modulus (E) .

The relationship between the Schmidt hàmmer rebound value and the static elastic modulus (Es) is linear with a high corre -lation coefficient (see Fig.9). The relationship between the Schmidt hammer rebound value and dynamic elastic modulus - (Ed)is an exponential function with a correlation coefficient of only 0.86 (see - Fig.10). The Schmidt hammer rebound value decreases rather sharply with a decreasing static or dynamic elastic modulus . Both properties decrease with increased weathering.

9. Schmidt hammer rebound value (Shv) versus ultrasonic wave velocity (V).

Both properties are influenced by discontinuities in the rock materials, porosity, weathering, grain size of the constituents, mineral composition , compactness of the grains and cementation. This relationship is linear with a high correlation coefficient. The Schmidt hammer rebound value increases gradually with increasing sonic velocity (see Fig.11).

10. Schmidt hammer rebound value (Shv) versus absorption (A).

The strength indication derived by the test hammer is influenced by the porosity of the rock material, which also affects the absorption property of the rock. This relationship is linear with a high corre -lation coefficient (see Fig.12). The Schmidt hammer rebound value decreases gradually with increasing absorption.

11. Absorption (A) versus uniaxial compressive strength (Ucs).

This relationship is a logarithmic function with a high correlation coefficient (see Fig.13). The uniaxial compressive strength decreases with a linear relationship,when the absorption values are increasing over approximately 7 %. When the absorption values are decreasing less than 7 %, the uniaxial compressive strength increases very gradually .

5 SELECTED INDEX TEST

The test with fulfill the requirements of the index test for the classification and characterization of rock material, as defined in the introduction, are the Schmidt hammer,the point load strength and the absorption tests.
The Schmidt hammer test has a strict pro -cedure to be followed during application to provide reasonable accurate results. This mainly concerns the smooth and fresh testing location (the operation procedure of the Soiltest Inc.CT-320,1976). For the quality control of volcanic breccia in the quarry site, complying that procedure will be time consuming . Also the rock materials which are produced for rockfill have mostly small size which will not be suitable for the Schmidt hammer test. Due to these reasons it is proposed here that the applicability of the Schmidt hammer test in determining the quantitative weathering index should be limited to the rock used as rip-rap.
The point load strength test has been studied by several authors (Broch & Franklin, 1972), Bieniawski, 1975) and the testing was done by a standardized method (ISRM, - 1973). The result of these studies show that the test is reliably applicable over a wider range of strengths than the Schmidt hammer test. For the quality con -trol of rock materials in the quarry site the point load strength test will be sui -table, because irregular lump samples can be obtained more easily for the purposes of this test.
The absorption test gives a rapid approximation of the strength of rock materials. It has the advantage that field determination can be made if a weighing balance and an oven are available. The test with also indicate wether or not the materials under test are affected by slaking.
The three tests outlined above can be regarded as true field tests, because they do not require any special preparation of the test specimen and can be performed in a simple procedure.

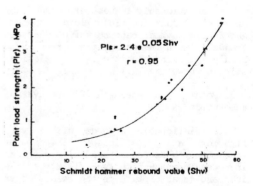

Fig.1 Correlation between Pls ãnd Shv

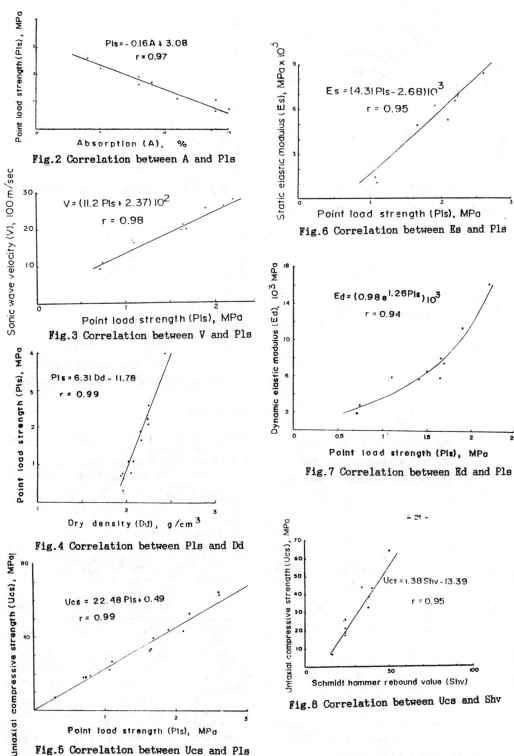

Fig.2 Correlation between A and Pls

Pls = -0.16 A + 3.08
r = 0.97

Fig.3 Correlation between V and Pls

$V = (11.2 \, Pls + 2.37) 10^2$
r = 0.98

Fig.4 Correlation between Pls and Dd

Pls = 6.31 Dd - 11.78
r = 0.99

Fig.5 Correlation between Ucs and Pls

Ucs = 22.48 Pls + 0.49
r = 0.99

Fig.6 Correlation between Es and Pls

$Es = (4.31 \, Pls - 2.68) 10^3$
r = 0.95

Fig.7 Correlation between Ed and Pls

$Ed = (0.98 \, e^{1.26 Pls}) 10^3$
r = 0.94

Fig.8 Correlation between Ucs and Shv

Ucs = 1.38 Shv - 13.39
r = 0.95

751

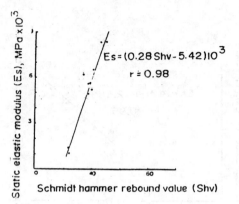

Fig.9 Correlation between Es and Shv

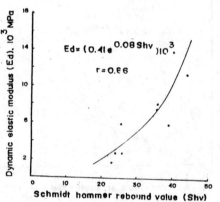

Fig.10 Correlation between Ed and Shv

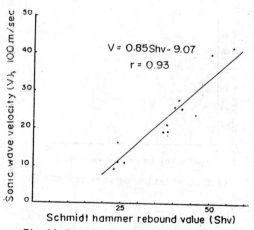

Fig.11 Correlation between V and Shv

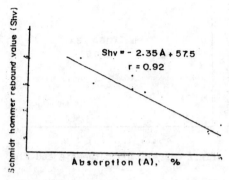

Fig.12 Correlation between Shv and A

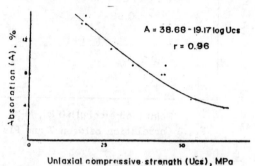

Fig.13 Correlation between A and Ucs

REFERENCES

ASTM 1977. Annual book for ASTM standards. part 19, Easton, Md.USA.

Bell,F.G. 1983. Engineering properties of soil and rocks. 2nd edition, Butterworks & Co., London.

Bieniawski,Z.T. 1975. The point load test in geotechnical practise. Eng. Geol. 9, pp.1-11, Elsevier Scientific Publishing Company, Amsterdam.

Dearman, W.R. 1976. Weathering classisfication in the classification of rock. A revision, Bull. Int. Ass. Eng. Geol., Krefeld. No. 13, PP. 123 - 127.

Fookes, P.G, W.R. Dearman & J.A. Franklin, 1971. Some engineering aspects of rock weathering with field examples from Dart moor and elsewhere. Q.J. Eng. Geol.Vol. 4, No.3, pp.139-185.

Irfan, T.Y. & W.R. Dearman 1978. Engineering classification and index properties of a weathered granite. Bull. Int. Ass. Eng. Geol. Krefeld, No.17,pp.79-80.

Rengers, N. 1981. Principles of engineering geology. Lecture note, ITC, The Netherlands.

Roberts, A. 1977. Geotechnology, an introductory text for students and engineeers Pergamon Press Ltd, Oxford.

6th International IAEG Congress / 6ème Congrès International de AIGI, © 1990 Balkema, Rotterdam. ISBN 90 6191 130 3

Engineering-geological classification of clay microstructures
Classification des microstructures des roches argileuses en géologie de l'ingénieur

V.N. Sokolov
Moscow State University, Moscow, USSR

ABSTRACT:The paper delivers a clay microstructure classification accounting morphometric (size and shape of structure elements), geometric
(structure elements orientation), and energy (structural bonds) characteristics which displays the interrelation between the structure and
properties of clays. The classification distinguishes classes and subclasses on the basis of integral characteristics - dispersion and anisotropy coefficient (degree of structural elements orientation). Groups
are identified according to energy characteristic - value of strength
and character of deformation. There were identified three microstructure classes - finedispersed, medium dispersed and coarsedispersed;
three subclasses - with low, medium and high degree of structural elements orientation; three groups - coagulative, mixed and crystallized-
cementated. The combination of the classes, subclasses and groups provides 12 microstructure types. Each type is characterized by a certain
set of morphometric, geometric and energy indexes which enables to predict the strength properties and character of deformation in clays.

RESUME: Nous proposons dans cet article une classification des microstructures des argiles qui tient compte des caracteristiques morphometriques (la grandeur et la forme des elements structuraux), géometriques
(l'orientation des élements structuraux) et énergetiques (les liaisons
structurales), et qui reflète la relation entre la structure et les
propriétés des argiles. Cette classification distingue des classes et
subclasses à la base des charactéristiques integrales - la dispersion
et le coefficient d'anisotropie (le degré d'orientation des elements de
la structure). Les groupes se distinguent suivant une caractéristique
energetique - une valeur de la resistance et un caracter de la deformation. Les trois classes de microstructures ont été identifiées comme
ayant la dispersion fine, moyenne et grosse; les trois subclasses ayant
le degré de l'orientation des elements de la structure bas, moyen et
grand; les trois groupes - coagulant, mixte et celui cristallisé-ci-
menté. La combinaison des classes, subclasses et groupes donne les 12
types de la microstructure. Chaque type est caracterisé par le certain
nombre d'index morphometriques géometriques et energétiques suffisant
pour pronostiquer la résistance mechanique et le caracter de la deformation des argiles.

The classifications now being in
use do not take into account the
entire structural characteristics
of clays. Complex investigation of
clay structure gives a reliable
scientific foundation enabling to
predict the strength and deformation behaviour of clays as well as
the changes in physical and physico-mechanical properties under the
impact of various factors. Therefore to make a structure classification accounting morphometric,
geometric and energy characteristics of clays and displaying the
interrelation between their structure and properties is a task of
major significance at present time.

To develope engineering-geological microstructure classification of clays there was made a quantitative evaluation of both morphometric and geometric characteristics on microlevel using SEM-microcomputer system and also structure energy characteristics of a large number of samples diverging in age, genesis and the degree of lithification. The following quantitative structure parameters have been obtained: porosity from SEM images n_i, total square S_i and pores perimeter, the average radius of particles, a number of contacts between them, the degree of structure elements orientation according to the method of intensity gradient signal Ag. The integral structure characteristics such as anisotropy A and dispersion D index taking into account the structure elements sizes and their orientation have been calculated on Fourie spectrum SEM images. The technique of quantitative microstructure analysis on SEM images was described in full details by Sergeev et al.(1985). Morphometric characteristics were taken from a series of SEM images (magnification range is from 500 to 32000 times). To get reliable information SEM analysed the samples being of homogeneous structure (Sokolov et al., 1988).

The standard methods were used to analyse mineral, granulometric and microaggregate composition; soil density, strength and deformation properties of samples, totally 22 parameters.

To find the interrelation between quantitative characteristics obtained while analysing SEM-images and the data of composition and properties of clay samples investigated we apply factor analysis (Jöreskog et al., 1976). The corresponding factor model in coordinates III and I of factor axis has been obtained which gives an appropriate structure of interrelations between morphometric, geometric and energy microstructure characteristics as well as their properties in a selection of 101 samples under investigation (fig. I).

In correspondence with the maximum loading of variable D on III factor axis within the total agregation of points there formed three

point clouds stretched along the subhorisontal direction, their boundaries being marked with dashline.

Mutually orthogonal location of dispertion and orientation variables resulted in a division of the whole sample selection into three isolated point clouds, stretched along the vertical. Their boundaries are marked with dash-stipple lines.

In correspondence with the high loading I factor axis is affected with the following variables: compressive strength R_{cl}, maximum shearing strength τ_{maxl}, shearing strength anisotropy τ_{maxl}/τ_{maxll}, ratio of maximum shearing strength to residual strength τ_{maxl}/τ_{minl} as well as the variables correlated with them density ρ_s and total pore square S_i. Their directed action devides the selection analysed into three vertical point clouds (a,d,g;b,e,h; c,f,i). Their boundaries are shown with dash-stipple-stipple lines.

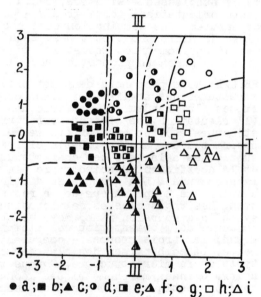

● a;■ b;▲ c;o d;▢ e;△ f;o g;▢ h;△ i

Fig.I. Factor model displaying the quantitative data distribution of clays microstructure characteristics, composition and some properties in coordinates III and I of factor axis. Specific signs show the microstructure types a-i.

Factor analysis gave an advantage to identify the objective laws displaying the interrelations between basic characteristics in clay structures, being of various age, genesis, composition and degree of lithification. The method used allowed to make an original engineering-geological microstructure classification. The classification is of parallel type involving three basic items.

Microstructure classes has been distinguished on the basis of morphometric characteristics (sizes and shapes of particles and pores) which is quantitatively characterized by integral parameter of dispersion D. This parameter is calculated through the ratio of a number of pores and particles sizing more than $10\mu m$ to the number of pores and particles being less than $5um$ (Sergeev et al.,1985). In plastic clays there exists a high degree of correlation between D and plasticity index I_p, the correlation coefficient being $R_{D-I_p}=-0,73$. The value of D and plasticity index I determine the identification of three microstructure classes (Table).

The first class - finedispersed microstructure, index **A**. The typical clays with such microstructure are fat and medium clays (clay fraction content (c.f.c.) more than 30%) and clay shales. The mixed-layered minerals and illite predominate in mineral composition of clay fraction. The age varies from Q_{IV} up to **Pg**.

The second class - mediumdispersed microstructure, index B. This class mainly involves medium and lean clays, loams, argillites of various age and genesis. The presence of a large number of silt graines (fraction $10\mu m$) is rizing in them which affects the decreasing of plastisity index in clays still keeping their mixed-layered-illite mineral composition.

The third class - coarsedispersed microstructure, index C. For this class the dispersion D is of the largest values while I_p is of the lowest ones. Clays of coarsedispersed microstructure contain a large number of microaggregates, silt and sand grains of sizes increasing $10\mu m$. Illite predominates in mineral composition of the clay fraction mixed-layered minerals can be found as well along with kaolinite and chlorite. This class involves light clays, loams, argillites and aleurolites. Genesis and age is indicated in table.

Microstructure subclasses are distinguished according to geometric characteristics. Their values are expressed quantitatively by anizotropy coefficient A or Ag.The dependence between these two values is very high.

Microstructure subclass with low degree of structural elements orientation (loworiented), index I. It involves mainly marine and lacustrine loams of Holocene age. All the microstructures belonging to this subclass are isotropic,the circular form of the rose of orientation and Fourie-spectrum proves the fact.

Microstructure subclass with medium degree of structural elements orientation (mediumoriented), index II. It involves marine, glacial,lacustrine, lacustrine-glacial and alluvial deposits of various age with medium degree of lithification. The process of consolidation occuring in these clays results in growing of structural elements degree of orientation. The ellipsoid form of the rose of orientation and Fourie spectrum proves this.

Microstructure subclass with high degree of structural elements orientation (highoriented), index III. It involves medium and high lithified clays. For the microstructures of this kind typical are high values of anisotropy coefficients and very stretched elliptical form of the rose of orientation and Fourie spectrum. The microstructures belonging to the subclasses II and III are anisotropic.

Microstructure groups are identified according to energy characteristics that is to the predominating energy type of individual contacts between structural elements (structural bounds). The quantitative indexes characterize inderectly the availiability of a certain type of energetic contacts between structural elements, they are: one axial compression strength along the direction perpendicular to bedding $R_{c\perp}$; the ratio of relative elastic strain ε_e to the total deformation ε_t, the ratio of maximum shearing strength $\tau_{max\perp}$ to the residual one $\tau_{min\perp}$.

The calculations showed the strength of individual contacts P_l varies from $2 \cdot 10^9$ N in coagulative up to $1,0 \cdot 10^5$ N in phase ones (Sokolov, 1985). For clays with coagulative microstructure $\varepsilon_e/\varepsilon_t$ does not exceed 0,15 and $\tau_{max\perp}/\tau_{min\perp}$ –1,70. As the number of transitional and phase contacts in clays increases (mixed and crystallization-cementation microstructures) $\varepsilon_e/\varepsilon_t$ reaches 1,00 and $\tau_{max\perp}/\tau_{min\perp}$ goes up to 65.0.

Taking into consideration such values as compression strength, $\varepsilon_e/\varepsilon_t$ and $\tau_{max\perp}/\tau_{min\perp}$ and the character of deformation behaviour there have been identified three groups of microstructures.

Coagulative microstructure, index –a. This group distinguishes from the others with a small compression strength, elasticoviscous character of deformation with viscousplastic destruction. Close coagulative contacts predominate in such clays. Clays have high (up to 3%) content of organic matter of different indistructibility, high porosity n and high water content exceeding the liquid limit. They display vividly thixotropic properties. Both in undisturbed and disturbed conditions they do not swell or swell insignificantly.

Mixed microstructure, index b. The types of contacts between structure elements belonging to this group are coagulative as well as transitive and phase. This group involves clays with medium degree of lithification, plastic consistency, porosity up to 61%, water content not exceeding the liquid limit. The distinctive features of this group are: rather high compression strength, the deformation is elasticoviscous or with fragilplastic or fragil distruction.They are reversible being affected by water and swell considerably. The swelling in disturbed condition is higher than in nondisturbed one.

Crystallized – cementated microstructure, index c. This group involves clays with high and more seldom medium degree of lithification, where phase contacts of crystallization and cementation nature predominate between structure elements. This type of microstruc-

ture is characterized by a high compression strength, elasticoviscous or elastic deformations with fragile distruction. The consistency of clays forming this group is hard or semihard, the porosity and water content decreases, swelling is insignificant, but increases in disturbed condition.

Various combinations of classes, subclasses and groups result in identification of 12 microstructure types (table). The coagulative microstructure group involves three types, the group of mixed microstructures involves six types, and the group of crystallized-cementated microstructures involves three types.

Index of a type is a combination of class, subclass and group indeces. For example: finedispersed, loworiented, coagulative microstructure has index –A-I-a. Every type of microstructure is characterized by particular limits of content and property variations as well as quantitative values of morphometric and geometric characteristics. The measuments are given in the table. Figures 2-5 show SEM-pictures of the most typical clay samples belonging to different microstructure types. The most common morphometric feature of clay microstructures being found in most types is the presence of microaggregates – structure elements of anisometric shape having complicated construction. The length varies from 5 to $15\mu m$. They are composed of ultramicroaggregates – associations of clay particles, coming into contact with each other by basis surfaces. Ultramicroaggregates are of anisometric shape, their length generally does not exceed $1\mu m$ and thickness $0,3\mu m$. Such specific structure elements composition results in the formation of four categories of pores: interparticle ultramicropores, an indicator of clays (equivalent diameter $d_{ip} \approx 0,06\mu m$); fine and small interultramicroaggregate micropores ($d_{iu_1} \approx 0,2 - 0,4\mu m$; $d_{iu_2} \approx 2,5\mu m$); small and large intermicroaggregate micropores ($d_{im_1} \approx 3,9\mu m$, $d_{im_2} \approx 10 - 30\mu m$).

The transition from finedispersed to coarsedispersed microstructures displays the growth of evarage equivalent diameter in inter-

ultramicroaggregate and intermicroaggregate pores as well as the
availiability of intergraine and
intermicroaggregategraine pores.
The major part of pore space (85-
97%) is occupied by intermicroaggregate pores, but the multitude
of interparticle and interultramicroaggregate pores occupy only 3-
15% of the total porosity.

The transition from loworiented
to highoriented microstructure
groups displays the growth of anisotropy in strength and deformation properties.

A number of types which do not
involve clays are marked with dashes in the table. Nevertheless
applying the classification elaborated for other lithological types
like carbonates, cherts, sandstones and others the examples can be
found for any of 27 identified microstructure types. Therefore cemented sandstone of loose construction belongs to coarse dispersed loworiented crystallized-cementated microstructure (C-I-c) and
carbonate silt to finedispersed,
mediumoriented coagulative microstructure (A-II-a).

Thus we can state that the microstructure classification elaborated is of universal character. Its
major advantage is to furnish a
researcher with a reliable information enabling to forecast the
strength and deformation behaviour
of clays belonging to various microstructure types.

REFERENCES

Jöreskog, K.G., Klovan, T.E., Reyment, R.A. (1976). Geological
factor analysis. 312 pp. Elsevier Sci. Pub.Com. Amsterdam. Oxford. New York.
Sergeev, J.M., Osipov, V.I., Sokolov, V.N. (1985). Quantitative
analysis of soil structure with
the microcomputer system. Bulletin IAEG 31, 131-136.
Sokolov, V.N. (1985). Physico-chemical aspects of clays mechanical behaviour. Eng.Geol.4, 28-41
(in Russian).
Sokolov, V.N., Rumjantseva, N.A.,
Kovbasa S.I. (1988). Some spesi-
fications of solid structure
quantitative analysis on SEM
images. Izv. Ac.Sci. of the USSR,
ser. Physics, 52, 7, 1350-1353
(in Russian).

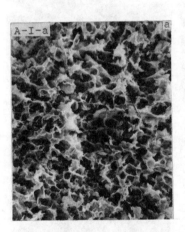  

Fig.2. Microstructure loworiented, coagulative: a - finedispered (marine silt Q_{IV}); b - mediumdispersed (marine clay Q_{IV}); c - coarsedispersed (marine clay Q_{III})

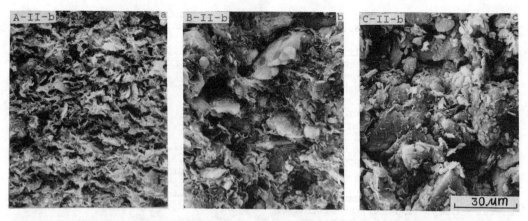

Fig.3. Microstructure mediumoriented, mixed: a – finedispersed (marine clay Q_{III}); b – mediumdispersed (alluvial clay Q_{III}); c – coarsedispersed (glacial loam Q_{II})

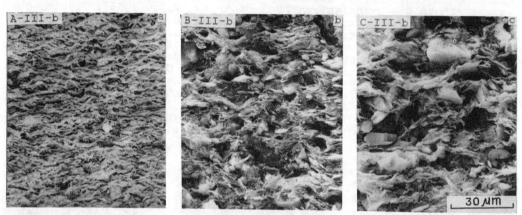

Fig.4. Microstructure highoriented mixed: a – finedispersed (varved clay Q_{III}); b – mediumdispersed (marine clay J_3); c – coarsedispersed (marine clay J_3)

Fig.5. Microstructure highoriented, crystallized-cemented: a – finedispersed (marine clay N_l); b – mediumdispersed (marine clay Θ); c – coarsedispersed (aleurolite PR_3)

Table. Clay microstructures classification

Class	Microstructure subclass					
	I. Loworiented A = 0,93-1.07; Ag = 0-7,0%			II. Mediumoriented A = 1,08-1,50; Ag = 7,1-22,0%		
	Microstructure group			Microstructure group		
	a. Coagulative $R_{c\perp}< 0,05$MPa $P_1 =2,1\cdot10^9-5,3\cdot10^{-8}$ N $\mathcal{E}_e/\mathcal{E}_t=0,00-0,15$ $\tau_{max\perp}/\tau_{min\perp} = 1,02-1,70$	b	c	a	b. Mixed $R_{c\perp}=0,05-0,5$MPa $P_1=4,2\cdot10^8-1,5\cdot10^{-6}$ N $\mathcal{E}_e/\mathcal{E}_t=0,16-0,52$ $\tau_{max\perp}/\tau_{min\perp}= 1,71-4,30$	c
A. Finedispered D =0,50-0,80; $I_p>32$	A-I-a Marine and lake silts, Q_{IV}; n=61-81%; w=52-161%; c.f.c. 30%; $R_{c\perp}=0,0016-0,038$MPa; $\tau_{max\perp}=0,0012-0,017$ MPa; $E \leqslant 0,20$MPa; $n_i=52-63\%$; $S_i=12250-14630\mu m^2$ $d_{ip}=0,06\mu m$; $d_{iu}=0,22\mu m$; $d_{im}=5-8\mu m$	−	−	−	A-II-b Marine lacustrine glacial fat and medium clays, $Q_{IV}-J_3$; n=46-61%; W=42-72%; c.f.c > 30%; $R_{c\perp}=0,06-0,32$MPa; $\tau_{max}=0,05-0,15$MPa; E=0,8-5,7MPa; $n_i=46-58\%$; $S_i=11024-13412\mu m^2$; $d_{ip}=0,08\mu m$; $d_{iu}=0,42\mu m$; d=8,4um	
B. Mediumdispersed D =0,81-1,10; $I_p=20-32$	B-I-a Marine, medium and lean clays, $Q_{IV}-Q_{III}$; n =44-60%; W =30-80%; c.f.c.=28-46%; $R_{c\perp}=0,007-0,03$MPa; E =0,1-1,4MPa; $n_i=36-42\%$; $S_i=8541-9864\mu m^2$; $d_{ip}=0,08\mu m$; $d_{iu}=0,24\mu m$; $d_{im1}=9,0\mu m$; $d_{im2}=30,0\mu m$	−	−	−	B-II-b Marine glacial, alluvial and lacustrine, medium and lean clays, loams, $Q_{IV}-J_3$; n =33-53%; W =28-44%; c.f.c =13-38%; $R_{c\perp}=0,1-0,32$MPa; $\tau_{max\perp}=0,04-0,22$MPa; E =2,0-9,0MPa; $n_i=30-47\%$; $S_i=7051-11021\mu m^2$; $d_{ip}=0,06\mu m$; $d_{iu}=0,22\mu m$; $d_{im1}=8,0\mu m$; $d_{im2}=20,0\mu m$;	−
C. Coarsedispersed D =1,11-2,40; $I_p=3-20$	C-I-a Marine lean clays, loams, $Q_{IV}-Q_{III}$; n=35-45%; W =32-52%; c.f.c. =6-28%; $R_{c\perp}=0,01-0,028$MPa; $\tau_{max\perp}=0,004-0,02$MPa; E =0,08-0,45MPa; $n_i=32-40\%$; $S_i=7767-9486\mu m^2$; $d_{ip}=0,06\mu m$; $d_{iu1}=0,2-0,6\mu m$; $d_{iu2}=2-5\mu m$; $d_{im}=8-30\mu m$	−	−	−	C-II-b Marine, glacial, lacustrine-glacial alluvial medium and lean clays, loams, $Q_{IV}-Q_{II}$; =27-46%; W=13-38%; c.f.c.=15-33%; $R_{c\perp}=0,1-0,45$MPa; $\tau_{max\perp}=0,07-0,18$MPa; E=5,1-10,8MPa; $n_i=25-41\%$; $S_i=6580-9668\mu m^2$; $d_{iu1}=0,36\mu m$; $d_{iu2}=4,5\mu m$; $d_{im}=19,0\mu m$	−

Microstructure subclass	
III. Highoriented. A =1,51–3,00; Ag =22,1–78,0%	
Microstructure group	

a.	b. Mixed $R_{c\perp}$ =0,05–0,5MPa P_1 =4,2·10^{-8}–1,5·10^{-6} N $\varepsilon_e/\varepsilon_t$ =0,16–0,52 $\tau_{max\perp}/\tau_{min\perp}$ =1,71–4,30	c. Crystallized – cemented $R_{c\perp}$ > 0,5MPa; P_1 =7,1·10^{-8} –1,0·10^{-5} N $\varepsilon_e/\varepsilon_t$ =0,52–1,00; $\tau_{max\perp}/\tau_{min\perp}$=4,31–65,0
—	A–III–b Lacustrine-glacial fat and medium clays, Q_{IV} –Q_{III}; n=44–60%; W=45–55%; c.f.c. > 30%; $R_{c\perp}$ =0,1–0,4MPa; $\tau_{max\perp}$ =0,09–0,14MPa; E =0,4–1,0MPa; n_i =44–48%; S_i =10064–11226μm; d_{ip} =0,06μm; d_{iu} =0,22μm; d_{im} =2,8μm	A–III–c Marine fat clays; Q_{III} –Pg_3; n=47–56%; W=31–39%; c.f.c.>30%; $R_{c\perp}$ = 0,7–4,0MPa; $\tau_{max\perp}$=0,2–1,5MPa; E = 20,0–30,0MPa; n_i =43–50%; S_i =9985–12074μm; d_{ip} =0,06μm; d_{iu} =0,22μm; d_{im1}=3,0μm; d_{im2}=10,0μm
—	B–III–b Marine medium and lean clays, Q_{II} –C_1; n =41–54%; W=27–63%; c.f.c.= 36–51%; $R_{c\perp}$ =0,1–0,5MPa; $\tau_{max\perp}$=0,03–0,14MPa; E =19,0–50,0MPa; S_i =8434–11396μm; d_{ip} =0,06μm; d_{iu} =0,24μm; d_{im1}=5,0μm; d_{im2}=13,0μm; n_i = 36–49%	B–III–c Marine medium and lean clays, clay shales, argillites; N_1 –C_1; n=5–51%; W=3–49%; c.f.c.=30–55%; $R_{c\perp}$ =0,6–91,0MPa; $\tau_{max\perp}$ =0,4–12,5 MPa; E =15,0–160,0MPa; n_i =4–48%; S_i =942–11307μm; d_{ip} =0,06μm (in clay shales and argillites are absent); d_{iu} =0,24μm; d_{im1}=3,0μm; d_{im2} =12,0μm
—	C–III–b Marine, alluvial, lacustrine-glacial medium and lean clays, loams, Q_{III} –C_2; n =36–40%; W=20–36%; c.f.c.=25–34%; $R_{c\perp}$ =0,12–0,33MPa; $\tau_{max\perp}$ =0,10–0,18MPa; E =5,0–12,8MPa; n_i =32–36%; S_i = 7380–8590μm^2; d_{ip} =0,06μm; d_{iu1}= 0,28μm; d_{iu2}=2,4μm; d_{im} =16,0μm	C–III–c Marine medium and lean clays, argillites, aleurolites, J_3 –PR_3; n=22–30%; W=7–24%; c.f.c.=15–35%; $R_{c\perp}$ =1,4–9,2MPa; $\tau_{max\perp}$ =0,8–1,72 MPa; E =8,9–117,0MPa; n_i =20–35%; S_i =4638–8200μm; d_{ip} =0,06μm (are sharply reduced in argillites and aleurolites); d_{iu} =0,22μm; d_{im1}=3,0μm; d_{im2} =10,0μm

The classification of rocks for underground structures and results of engineering – geological observations

La classification des roches pour les constructions souterraines et les résultats de l'observation de la géologie de l'ingénieur

O. Tesař

The Project Institute of Transport and Engineering Structures, Prague, Czechoslovakia

ABSTRACT: The rock quality in the present classification system is expressed by a coresponding number of QTS classification points. These are algorithms at from the uniaxial strength, average distance of discontinuity planes, degree of rock mass loosening, groundwater pressure and inflow rate, quality direction and dip of discontinuity planes with respect to the driving direction. The number of classification points can be determined at any survey stage, is unambignous simple and accurate. Based on the number of the QTS classification points and the cross-section of the tunnel, the working to be assessed is classified into technological groups which include rock stability, disintegration method and driving method data, and information on its temporary and permanent support.

RESUME: La qualité de la roche est dans la classification exprimée à l'side du nombre des points de classification QTS (Qualité, Texture, Structure). Ceux-ci sont définis par l'algorithme de la résistence à la compresion simple de la roche, de la distance moyenne des surfaces de contact des fissures, du degré du foisonnement du massif de la roche, de la valeur et de la poussée de la nappe souterraine, de la qualité, du sens et de l'inclinaison des surfaces des fissures en égard au sens de l'abattage. Le nombre des points de classification peut être déterminé au cours de toutes les étapes de la prospection géologique d'une manière univoque, simple et précise. Le nombre des points de classification QTS et la dimension de profil transversal du tunnel definissent le classement de l'abattage dans un groupe technologique qui contient les données sul la stabilité de la roche, sur la technologie du détachement et de l'abattage et la détermination du soutènement provisoire sinsi que du revêtement définitif.

1 PROPOSED CLASSIFICATION OF ROCKS FOR UNDERGROUND STRUCTURES

The newly elaborated rock classification method for tunnel structures represents an effort to assess the rock quality as best as possible and to classify the rock at any stage of the engineering-geological survey. Although originally designed for underground structures built in the region of Prague, its field of application is likely to cover all rocks. (Tesař 1979).

The principal parameters A, B and C are based on the most important structural and textural properties of the rock (TS).

The influence and importance of separate structual and textural properties of the rock are expressed by a basic number of classification points (ST) according to the following equation:

$$TS = 10 \log \sigma_D + 26.2 \log d + 6.2 \cdot \log D + 61.4$$

where TS = basic number of classification points

σ_D = uniaxial compression strength of rock fragments (MPa)

d = average distance of discontinuity planes (m)

D = depth of the classified rock under the soil deposits (m) or opening of discontinuity planes (mm)

The parameter A is determined by the equation:

$$TS_A = 10 \log \sigma_D - 3$$

If the uniaxial compression strength of a rock fragment is lower than 2 MPa, the rock is not classified and is assumed to be soil.

In the preliminary stage of the survey, the uniaxial compression strength of rock fragments is likely to have to be determined in an orientative manner, according to the petrological description of the rock and appropriate tables (Záruba, Mencl, 1957), if no archival results of laboratory tests are available. In the foll ing survey stages when rock samples are available, its value is determined using laboratory tests performed with cube-, prism- or cylinder-shaped test specimens. Differences introduced due to varying shapes of test specimens should not influence the number of classification points in an essential manner. If a pilot adit is driven, it is recommendable to supplement the laboratory tests by an in situ index measurement, such as by Schmidt hammer, to find out possible fluctuations in the rock strength.

The parameter B is determined according to the average distance of discontinuity planes (d).

The appropriate equation is as follows :

$$TS_B = 26.2 \log d + 52.4$$

The average distance of discontinuity planes can be measured directly in situ, regardless to which system they fall into. The measurement is performed over certain fixed length at least in two mutually perpendicular directions. As far as drillholes are concerned, it is possible to take into account the average size of cores and its relation to the real degree of fracturing of the rock mass (Tesař 1981). If the distance between discontinuity planes is greater than 4.0 m, the parameter B assumes the maximum value of 68 points.

The number of classification points in the parameter C depends on the degree of loosening of the rock mass as shown by the opening of discontinuity planes. As drillholes cannot provide this information, the parameter C can be estimated on the basic of the depth of the assessed rock under the base of deposits (soils), as this criterion reflects influences of physical weathering. The depth is determined from an geological section. If the depth exceeds 30 m, the parameter C assumes the maximum value of 21 points. If loosened rock is observed in a tunnel stope, the parameter C is determined according to the opening of discontinuity planes, regardless the fact that the loosening may not be due to weathering processes.

In initial stages, when only drilling works are available, the parameter C is determined as follows :

$$TS_C = 6.2 \log D + 12$$

The designed underground working having been opened, the parameter C is assessed according to the following table :

Crack opening (mm)	Number of classific. points TS_C
5	14
3 - 5	16
1 - 3	18
1	19
closed	21

The sum of the three parameters gives a total number of classification points (TS) representing the basic quality of rock mass.

The basic number of classification points must be reduced in the following instances :

Reduction :

α = if main discontinuity planes show a dip from 30 to 80°, if their dip discords with the direction of driving or if they are parallel to the tunnel axis and dipping from 30 to 90°,

β = if the dip of discontinuity planes is infavourable (see α) and, at the same time, discontinuity planes are even, smooth or filled with clay, and continuous over a distance greater than half the tunnel diameter,

γ = if there is a groundwater flow without hydrostatic pressure through the rock,

δ = if there is a pressurized groundwater flow through the rock or if finer material is being washed out.

The reduction is performed according to the following table :

Number of classif. points (TS)	Reduction of classif. points			
	α	β	γ	δ
< 30	0	0	4.0	12
30	0	0	3.5	10.5
35	1	1.5	3	10
40	1.5	2.5	3	9.5
45	2.5	3.5	3	8
50	3.5	5	2.5	8
55	4	6	2.5	7.5
60	5	7.5	2	7
65	6	9	2	6
70	6.5	10	2	5.5

75	7.5	11	1.5	5
80	8.5	12.5	1.5	4
85	9.0	14	1.0	3.5
90	10.0	15	1.0	3.0
95	11.0	16	1.0	2.5
100	11.5	17.5	0.5	2.0

The number of classification points reduced by α, β, γ and δ (QTS) indicates the rock quality as related to driving works proper. The reduction coefficients can be exactly determined only when a pioneer or pilot adit is driven, or when geological documentation of the tunnel itself is made. In simpler geological (tectonic and hydrogeological) conditions, it is possible to reduce the basic number of classification points (TS) as early as during the drilling or orientation survey.

For the purpose of classification it is possible to use equations given above or tables (Tesař 1979). For an estimate of the number of classification points, one may use the chart given in Table 1 (Tesař 1983). Each subsequent survey stage results in a more precise number of classification points.

When analyzing relations between the number of classification points and results of rock mechanics in situ tests, a significant correlation was found between the basic number of classification points (TS) and the moduli of elasticity and deformability (Young's moduli) (Tesař 1981b). The observed relationship is defined by the following equations:

$$\log E_0 = \frac{TS - 7.9}{20.1} \quad \text{for deformability modulus (MPa)}$$

$$\log E = \frac{TS + 6.6}{23.0} \quad \text{for elasticity modulus (MPa)}$$

The relationship has been verified for various geological conditions, thus being able to be used in practice for quick estimates of deformability moduli. Its use may significantly simplify and improve the quality of design works and construction processes. The relationship has been used also in mathematical and physical modelling, where either the number of classification points can be determined according to values of the moduli or vice versa.

2 ROCK MASS RESPONSES TO TUNNEL CONSTRUCTION

Detailed engineering-geological documentation of adit- and tunnel-driving using various technologies was used to find out and statistically process a relationship between the number of classification points and the overbreak size. The overbreak size can be easily observed in the course of driving and can be used for assessing the degree of stability of the tunnel stope and the advance of driving works. The statistical processing, which is probably most important for practical purposes, has resulted in curves of probability of occurrences of overbreaks having certain dimensions (0.5, 1.0 and 2.0 m) for various tunneling methods in relation to the number of classification points. (Tesař, Kameníček 1981). These results represent a base for technological categorization which directly determines the suitability of various driving methods, type and number of supporting structures etc.

A comparison of the results of mathematical modelling and the results of driving advance assessments showed that the risk of a collapse (sinking) was imminent when the rock mass quality corresponded to approximately 2.0 m loosening of the rock as determined by mathematical modelling. First overbreaks are encountered when the degree of rock loosening is approximately 1.0 m. The tunnel cross-section is essentially unimportant in this respect. The degree of loosening is a function of rock quality and stope dimensions, but may be influenced by technological aspects (rigidity and activation time of supporting structures etc.) to a considerable extent. (Tesař, Kameníček 1981). The obtained data on the degree of rock loosening can be used to determine the load on supporting structures and to assess the tunnel face stability.

3 PRACTICAL APPLICATIONS

The results of driving advance monitoring and theoretical calculations represent a basis for the selection of particular driving method according to the rock quality.

A similar approach was used to determine the possibilities and conditions of employing a TVM (DEMAG) 24-27 and RS 24-27 driving machines. For both circular cross-sections (2.67 and 3.50 m in diameter), technological groups were set forth, which reflect the necessity of e-recting a temporary support. The realations between costs and advance rates of the driving machines suggest that it is necessary to consider carefully the percentage of low-quality rocks in the proposed section of a tunnel, because it influences the construction economy. As an indication of rock disintegrability by driving mechanisms, the rock fragment uniaxial compression strength was used, which in combination of QTS classification points (representing the rock stability) determines such sections, where a driving machine can be made use of in an optimum manner.

A complex issue of driving three-aisle subway stations was solved using a detailed analysis of the driving technology as related to the rock quality. Values of classification points and their percentual distribution along the station s longitudinal axis were determined, which enable the following basic combination of station tunnelling:
- full-face driving of all three tunnels
- full-face driving of two tunnels, while the remaining one is part-face driven
- either part-face or full-face driving of all three tunnels, following after the rock quality has been improved

The results presented in the theoretical and practical parts of the work serve as a basis for instituting technological groups, into which parts of the stope are included according to the number of classification points and tunnel diameter (see Table 2). The importance of these technological groups in the selection of an appropriate driving method is illustrated in Table 2. (Tesař 1983).

The individual technological groups can be briefly characterized as follows:
Ia) Driving conditions are very good. Tunnels are driven in high-quality rocks (Classes A and B) (see Table 1), if the breaking width does not exceed 7.5 m. Under certain circumstances, this group may also include Class C rocks (adit driving). The rock is stable irrespective of the underground working diameter. The purpose of using shotcrete or rock protective coating is just to eliminate atmospheric influences and dessication, as there are no stope stabilization requirements. If the overburden overlying Class A rocks is too thick or if there are tectonic stresses, so-called rock peeling may be encountered. In this case, it is advisable to disintegrate the rock

by blasting, the objective being to break it in the circumferential zone of the stope. Class A rocks are hard to break, thus representing an equivalent of intrusives.
Ib) Similar to Group Ia, this group also includes Classes A, B and C rocks, which are stable irrespective of the tunnel diameter, with only rarely occurring overbreaks. The rock is secured in the maximum possible distance from the face, using shotcrete or steel wire meshing with short bolts, mainly when the diameter is large and the ratio of principal stresses infavourable, or in tension zones. As far as Class A rocks in deep--seated tunnels are concerned, the application of blasting and costs are identical to those of Group Ia. In other instances, these rocks fall within the first category of drivability.
II) This technological group mainly includes rocks of good and poor quality (Classes B and C); however, even very poor quality rocks (Class D) may occur, predominantly in smalldiameter tunnels and adits (< 7 m). Driving conditions are worsened in this group. The rock is characterized as temporarily stable. In adits and tunnels of any diameter, it is necessary to proceed by short attacks and to minimize the unsupported section of a tunnel. As far as diameters over 7 m are concerned, it is advisable to use part-face operation, providing this approach is economic and improves working safety. Machine disintegration can be used to advantage in this group. The stope is secured after each attack, either permanently (precast concrete rings), or temporarily (steel mesh, anchors and shotcrete or rolled sections).
III) This group is prevalently represented by very poor quality rocks (Class D). Class C rocks occur only in tunnels whose diameter exceeds 6 m. Driving conditions in this group are infavourable. The rock is unstable, even if only short attacks are employed. Adits and small--diameter tunnels can be driven using a full-face method, in short attacks not over 0.5 m, with temporary stabilization of the tunnel face. It is also possible to make use of advance spilling. If the tunnel diameter is large, it is necessary to use a part-face approach and to stabilize the face, or possibly even the overburden. Shotcrete may be combined with anchored steel centering. This group is characterized by increased time requirements as far as closing up of the invert is concerned. Part-at-

tack machine disintegration can be advantageously employed in Class C rocks in tunnels having a larger diameter.
IV) This group includes very poor quality rocks (Class D) in tunnels whose diameter exceeds 5 m. In these tunnels, the rock is quite unstable, very squeezing, its properties identical or compatable to those of cohesionless soils. For this reason, it is necessary to erekt a robust temporary supporting of part-face breaks and stope face. Driving works in these circumstances cannot dispense with advance spilling or shielding. Blasting is not necessary. Permanent support is of a heavyduty, invert type.

The submitted technological classification is applicable for rocks whose minimum number of QTSclassification points is 30. Below this limit, the rock possesses the charakcter of soil, with all resulting consequences as far as tunnelling is concerned.

An individual evaluation of the rela tion between the rock quality and adit or tunnel driving technology is necessary in the following cases :
- when the breaking width exceeds 12 m,
- when the cross-section of an underground working is atypical, i.e. when the difference between the width and height of the stope is higher than one half of the smaller dimension,
- when driving in rocks whose properties have been improved artificially (draining, strengthening, grouting, anchoring etc.). In these instances, the rock is evaluated according to its improved properties, i.e. not according to its natural (initial) ones,
- when driving is performed using mechanisms characterized by loading and subsequent unloading of rings, resulting in disturbances of low-strength rocks to such an extent that temporary supporting of a tunnel/adit is necessary,
- when several tunnels are driven close to one another and the thickness of pillars between the tunnels does not guarantee a sufficient load-bearing capacity,
- when the rock has been disturbed by an improperly located or insufficiently secured exploratory or technological adit,
- when there are single- or double - level tunnel or adit intersections,
- when the overburden thickness is less than twice the breaking width.

The technological groups not only illustrate the influence of an underground

working diameter upon construction requirements, but also include data on how safeguarding the tunnel interferes with the driving works proper. This is very important for cost analyses. The classification has been worked out to use it for all driving methods, including mechanized driving and NATM.

The classification and categorization have been published in this form (Tesař 1983a) and included in a Branch Standard "Design and Construction of Railway Tunnels", as well as in "Regulations of Engineering-Geological Survey for Underground Workings Driven by Tunnelling Mechanisms"

CONCLUSION

Works aimed at verifying the relations between the rock quality and driving technology will continue, both in the theoretic al and practical fields.

The use of the results improves the quality of engineering-geological survey, and provides designers and contractors with such data, which enable them to design and build an underground working safely and economically.

Practical applications of the QTS rock classification are included in the "Regulations of Survey Works for Adits Driven by Full-Face Tunnelling Mechanisms" and in the "Recommendations Concerning Engineering-Geological and Geotechnical Survey of Higway Tunnel Sites in the Slovak Socialist Republic". Consultants and commenting organizations have recommended this classification to be included in drafts of related standards.

All data necessary for the classification are included in the "Engineering-Geological and Geotechnical Databank of Prague".

REFERNCES

Tesař, O. 1979. Klasifikace skalních hornin a její využití pro ražení podzemních staveb v Praze. Inženýrské stavby č.8. roč. 28
Tesař, O. 1981. Využití klasifikace skalních hornin pro určení přetvárných modulů. Inženýrské stavby č.6.
Tesař, O. & Kameníček, I. 1981. Možnosti praktického využití indexu QTS pro návrh technologie ražení štol a tunelů. Zpravodaj Metro roč.XII. č.2.
Tesař, O. 1983. Určení a význam technologických skupin pro projektování a ražení podzemních staveb ve skalních a poloskalních horninách.Inž.stavby č.3.

TABLE 1

ALIGNMENT CHART OF THE DETERMINATION OF CLASSIFICATION POINTS NUMBER QTS (TS)

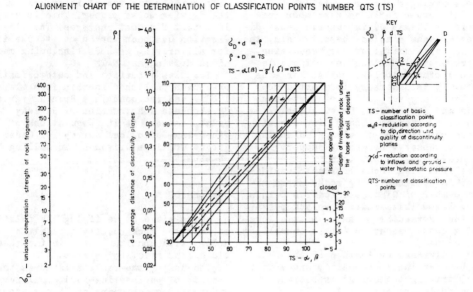

Table 2

DETERMINATION OF TECHNOLOGICAL GROUPS AND DRIVING TECHNOLOGY IN DEPENDENCE
ON THE NUMBER OF CLASSIFICATION POINTS AND STOPE DIMENSIONS

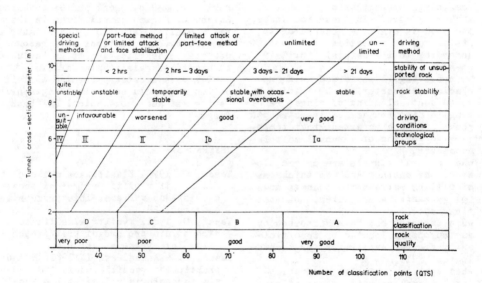

6th International IAEG Congress / 6ème Congrès International de AIGI, © 1990 Balkema, Rotterdam. ISBN 90 6191 130 3

Use of an expert system in site investigation
Utilisation d'un système expert dans l'investigation des sites

D.G.Toll & P.B.Attewell
School of Engineering and Applied Science, University of Durham, Durham, UK

ABSTRACT: The basic components of a knowledge based (expert) system for site investigation are geological/geotechnical and structural interaction. Some examples of the former are suggested. Two further components - financial and contractual/legal - are added because the design and conduct of a site investigation, and its reporting, cannot be isolated from these related issues. An interpretation of the state of the ground by interpolation between local and limited pieces of evidence (usually boreholes) provides a severe test for knowledge based systems. It is important that the knowledge base and the site specific geological/geotechnical database permit interrogation and interpretation at different levels of complexity, ranging from broad classifications, through detailed qualitative descriptions to quantitative analysis. One example of this scheme is suggested.

RESUMÉ: Les elements fondamentaux d'un systeme de connaissances basées dans l'investigation des sites sont geologiques/geotechniques et les interactions structurelles. Quelques examples sur le premier cas sont proposés. Deux autres elements - financier et contractuel/legal - sont ajoutés, car l'étude et la conduite de l'investigation des sites, et ces compte-rendus, ne peuvent pas être isolés de ces deux issues. Une interpretation de l'etat du sol par interpolation entre les pieces d'evidence locales et limitées (habituellement vides des forures) fournit un test severe pour les systemes de connaissances basées. Il est important que chacune des bases de connaissances et des specifigues bases de données geologiques/geotechniques du site permettent l'interrogation et l'interpretation à different degrés de complexité, variant des classifications generales aux descriptions qualitatives et analyses quantitatives detaillées. Un example à ce sujet est proposé.

INTRODUCTION

Geological materials *in situ* are usually highly variable both spatially and in terms of their geotechnical properties. Precise, but very local, information on the materials and their properties is derived from ground investigation boreholes and/or trial pits. Design within the ground beyond the boundaries of these holes or pits requires the interpolative/extrapolative injection of further information derived from the site investigation desk study and ground surface observations, together with the experienced decision making of professional engineers and geologists.

There are two important matters to note. First, professional decisions are based on different types of knowledge. In some cases, codified knowledge may exist in the form of standards and codes of practice. However, in many situations, particularly related to ground engineering problems, codified knowledge will not exist or will be insufficient. Established knowledge - that provided by text books and other published literature - will then be needed. In the absence of these two sources of information, the professional will fall back on his/her own personal knowledge. It may even be the case that this personal knowledge will conflict with the codified or established knowledge base, and will be used in preference.

The size of the established knowledge bank that can be accessed grows with time as technology advances. Codification is also progressing and, with increasing risks of litigation, professionals rely more and more on codes and standards as justification for their decisions. An important part of personal knowledge can therefore be the skill of knowing where and what to access. The ability to order the information in a manner that will generate an optimum solution grows with experience and of course in the specific context of ground engineering the type, volume and depth of knowledge to be accessed does depend to some extent on the nature of the construction (road, bridge, tunnel, and so on).

Second, any technical decisions in civil engineering cannot ultimately be isolated from their cost and contractual ramifications. A technically sensible answer may be forthcoming from purely technical considerations. However, a small omission in the knowledge base could lead to contractual problems and more expense being incurred than had been originally envisaged.

Man-made decisions stemming from a distillation of personal and other knowledge may also be affected by personal extraneous factors that should strictly not bear on the outcome. The ultimate aim of an 'expert system' is to detach the data base from the person for the primary decision-making process. By interactive consultation with the system, subjective decisions will be highlighted and alternative solutions can be considered.

Tunnelling is an example of a civil engineering project which is dominated by geotechnical considerations. In this paper the concepts needed for developing a knowledge based system for site investigation are based around the tunnelling problem.

DEFINITIONS

Knowledge based systems are computer programs which contain large amounts of varied knowledge that are brought to bear on (accessed for) a particular job. *Expert systems* or *intelligent knowledge based systems (IKBS)* are computer programs that embody the knowledge and capability which allow them to operate at an expert's level. Because knowledge is frequently expressed in the form of rules they may be termed *rule based systems*. Heuristics are 'hunches', 'rules of thumb', guesses, or beliefs. A heuristic rule might be 'if A then maybe B or possibly C', different weightings being applied to each possibility. A *fuzzy set* is a set which allows for 'graded' membership (one which does not require sharply-defined all-or-no membership). *Fuzzy logic* is a method of reasoning using *fuzzy set* theory and is adopted as a means of dealing with

uncertainty. The present interest is with IKBS. However, reasoning with uncertainty will be an essential part of an IKBS for site investigation.

KNOWLEDGE DEFINITION

In contrast to a system which merely acts as a computerised check list (the user simply being asked a series of questions such as 'Have you thought of...?', or 'What about...?') the need is for an intelligent system capable of interacting with the engineer at his level of experience and knowledge. The system would not, in fact, complete the decision-making progress for the engineer but would question him about which geotechnical routes to adopt, would point out where oversights appeared to have been made, would identify problem areas (which would promote different decision paths) which appeared not to have been addressed in sufficient detail (or not at all), and, if required, would offer suggestions about the best of several procedures to follow. In order to achieve such a high level of system objectives the fundamental requirement is for an extensive, refined and detailed data base to be accessed by software written from a precise understanding of what needs to be known.

Toll (1989) has identified the fundamental ground-structure interaction problem in defining two types of knowledge that a geotechnical expert system requires: **Geotechnical** and **Structural**. It may also be argued that the engineering geologist, geotechnical engineer and project engineer need to be aware of two further knowledge bases: **Financial** and **Legal/Contractual**. Estimates of soil and rock quantities and the billing of the work for a contract depend upon geological and geotechnical inferences based on incomplete evidence. If those estimates are incorrect then there may be contractual claims for extra payment and/or extension of time for completion. All these types of knowledge may be defined at different levels of complexity. Some typical levels of geotechnical information with their knowledge definitions are given in Table 1.

Degree of confidence in the outcome of an assessment is usually deemed to increase from the top of the Response column to the bottom since each succeeding element of the Input Information includes the integration of the previous elements. Rules of thumb and qualitative empiricism offer qualitative 'overview' conclusions which may be suitable for outline planning purposes. Semi-quantitative and

Table 1. Levels of Geotechnical Knowledge.

Input Information	Response
Outline design brief	Rules of thumb
Geological maps, plans and other descriptive input from the desk study	Qualitative empiricism
Field information on soil/rock types	Semi-quantitative empiricism
In situ field test data and laboratory classification tests	Quantitative empiricism
Material properties (laboratory measurements)	Simplified theory
Interpreted fundamental properties	Theory

Quantitative empiricism are able to provide responses that are rooted in precedent for similar engineering and geological conditions and which trigger decisions on a project's engineering viability. Simplified Theory and Theory provide a quantitative result that is usually essential for project design and contract letting.

An oversimplified tunnelling example is given in Table 2 which illustrates these subdivisions of information and confidence.The example addresses only two aspects of a quite complicated problem at the purely geotechnical level. Whereas in the case of soil tunnelling, excavation in the dry is unlikely to create any problems, contract billing for excavation in rock, and for which a financial knowledge base is needed, does require the firm definition of strength (usually the unconfined compressive strength) and failure to get this right can lead to disputes and claims (legal/contractual knowledge base) for additional payment. (For example, rock unconfined compressive strength in excess of about 50MPa usually implies uneconomic cutting performance with a road heading machine.) Support problems in soil are not confined to the tunnel proper but lead to ground surface settlement and potential structural damage. Specific questions with respect to the former, and the guidance rules detailed by Attewell et al. (1986) on the latter, need to be incorporated in any knowledge based system related to tunnelling.

In a problem of ground-structure interaction much more information may be available about one of the elements in the design than about the other. For example, a soil may have been assessed at the liquid and plastic limit level only whereas the analytical framework for defining the response of concrete support may be well known. If the strength of the soil can be inferred from its limit values then the problem as a whole can be addressed at the simplified theory level. The degree of confidence in the conclusion will then be influenced by the confidence in the relationship between limit values and strength. If strength cannot be inferred, the problem must then be tackled at the semi-quantitative empirical level.

INFORMATION CATEGORIES

A fundamental requirement of any knowledge based system for interpretation and analysis in geotechnical engineering is the provision of site specific information. Backing this up, and separate from it, will be relevant information stemming from early experiences which can be accessed for assessment against the current information suite if the ground conditions can be reasonably matched.

An information structure capable of handling geological and interpreted fundamental properties, as well as any factually observed data, is shown in Figure 1 after Toll (1989). Information structuring is referenced both to its sources (field investigation, laboratory testing, data interpretation) and to the level within the knowledge definition scheme. The components of the information are:

Profile: The sequence of ground conditions referenced to depth and to reduced level at a specific position within a site.

Geological Horizon: A specified bed of stratified rock or soil within a geological sequence.

Table 2. An example illustrating different levels of information.

Response	Excavation	Support
Rule of thumb	If tunnelling is in soil	
	then excavation problems are possible	then support problems are possible
Qualitative empiricism	If tunnelling is in alluvium and an open-face shield is used	
	then excavation problems are likely	then support problems may lead to settlement
Semi-quantitative empiricism	If tunnelling is in water-bearing gravel at sub-artesian water pressure and an open-face shield is used	
	then excavation and soil disposal problems may arise	then compressed air support pressure is likely to be needed
Quantitative empiricism	If tunnelling is in medium-dense gravel and the water table is three metres above the tunnel soffit	
	then excavation will be easy and soil disposal problems may not arise	then compressed air support pressure must match the water pressure at tunnel axis
Simplified theory	If open-shield tunnelling and the soil has unit weight γ and h is the depth to tunnel and q is the surface surcharge pressure	
	then excavation and soil disposal problems may not arise	then stability ratio is $(\gamma h + q - \sigma_i)$ and the internal support pressure σ_i must be adjusted to keep it less than about 4

Layer: A specific rock or soil type within a geological horizon.

Samples: Discrete volumes of material removed from the ground at known locations and horizons during field investigations.

In situ tests: Tests that are performed in the ground during a field investigation.

Classification tests: Simple laboratory tests which categorise and geotechnically identify rock and soil types.

Material Properties: Geotechnical measurements and definition of rock and soil material in a manner suitable for engineering design.

Fundamental Behaviour: Parameters of material behaviour interpreted within a fundamental framework.

Groundwater: Observations of groundwater levels during and after the field investigation.

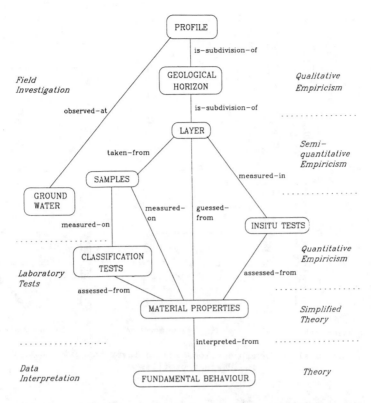

Figure 1 - A Geotechnical data structure

INTERPRETATION OF GROUND CONDITIONS

An expert needs to exercise his skill to interpolate between the discrete bits of geological and geotechnical evidence and build up a three-dimensional visualisation of its geological and engineering character. At its simplest, the operation involves linear matching between stratigraphical horizons identified within boreholes or trial pits, but there are numerous occasions where such simplistic interpolation may not be valid and resort to expert appraisal is needed. A bed of rock may show a thinner section in one borehole than in an adjacent hole. This might be due to local erosion or to interruption caused be a fault. Several scenarios of this type (some shown diagrammatically in Figure 2) need to be foreseen and incorporated within a KBS. Again, changes in strata horizons between adjacent boreholes without significant changes in the strata thicknesses in the boreholes could be attributable to conformable dipping of the beds, to local folding (which would not be apparent without very careful examination of rock

cores and, in the context of a KBS, would be misinterpreted without sufficient information), or to faulting (see Figure 3). Dumbleton and West (1974) give some useful examples of possible misinterpretations.

Between-hole/pit 'anomalies' of the above form can carry particular cost implications. Estimates of quantities for excavation usually rely heavily on direct interpolation. If, for example, a tunnel is to be driven through mixed soil/rock ground and the effect of such folding or unforeseen faulting is to project more rock into the tunnel face than had been expected at the design and billing stage, there will be claims for extra payment and perhaps for an upward re-rating of unit price based on the greater volumes of rock requiring to be excavated. (The latter would not usually be conceded because the equipment for dealing with rock excavation would be on site anyway, and because the unit start-up cost would then be spread over a greater volume of rock any re-rating should really be downward in price.) There are also special problems at the interface between glacial clays and Coal Measures rock and in

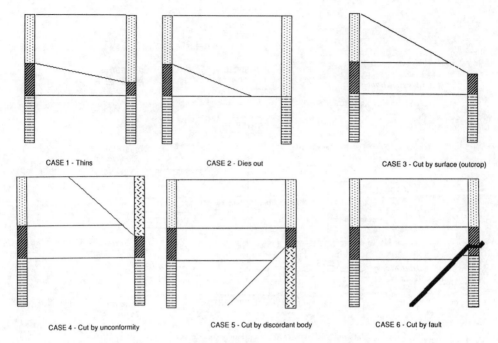

CASE 1 - Thins   CASE 2 - Dies out   CASE 3 - Cut by surface (outcrop)

CASE 4 - Cut by unconformity   CASE 5 - Cut by discordant body   CASE 6 - Cut by fault

Figure 2 - Some possible interpretations when a layer has different thickness
in adjacent profiles

the glacial clays themselves. The KBS therefore needs to consider different possible interpretations and to be aware of the financial implications of each possibility.

Contractually, rock will usually be defined in terms of its excavatability (and perhaps its ease or otherwise of support) which, in semi-quantitative empirical terms, would be described by its weathering grade or by a perception of its strength from core examination. This latter, if expressed on the borehole log in the site investigation report would be presented in semi-quantitative terms of 'strong', 'moderately strong', and so on which will then correlate with quantitative values of unconfined compressive strength (50-100MPa, 12.5-50MPa...). The contractual implications of making such interpretations must be included.

There are further 'anomalous' situations that need to be accommodated in any expert system for triggering the awareness of the contract designers. One example relates to the presence of boulders in till (boulder

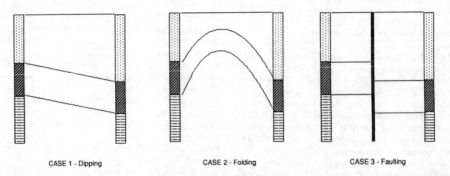

CASE 1 - Dipping   CASE 2 - Folding   CASE 3 - Faulting

Figure 3 - Some possible interpretations when a layer has different levels in
adjacent profiles

772

clay). The primary purpose of boreholes is to probe and define the presence of stratified rock and of groundwater. They will not adequately define the presence and spatial density of cobbles and boulders. The contractual difficulties related to tunnel construction in such clay must be included in the legal knowledge base. Also, the financial knowledge base should incorporate a bill item for boulder rock. In Britain this would generally be based on the recommendations in the Civil Engineering Standard Method of Measurement (Institution of Civil Engineers, 1985) by which individual boulders of volume less than $0.25m^3$ (equivalent to a diameter of 780mm if a spherical shape is assumed) need not be measured for payment, irrespective of the tunnel cross-section size. The implication is that boulders and cobbles of size less than $0.25m^3$ are technically clay for the purposes of the contract and would be paid as such for their excavation. Even boulders below this critical size may present difficult handling problems in a small diameter tunnel, and even more so when they extend beyond the payment line, cannot easily be removed as a body, and have to be cut off at the tunnel extrados. Notwithstanding the provisions of CESMM, a well-designed contract will usually attempt to relate critical boulder size to tunnel size. It can therefore be seen how an interpretation of ground conditions requires an input of structural, legal and financial knowledge if it is to promote an underlying awareness of particular statements in a site investigation report.

CONCLUSIONS

A site investigation should not be conducted, and the results of that investigation should not be presented, without an awareness of the wider contractual implications of the information. Four classes of information for building into a site investigation expert system have been identified:

Geological/geotechnical: attempting to interpret the nature of the ground that will affect and be affected by the construction, and being able to envisage and characterise alternative interpretations.

Structural interaction: inference rules (empiricism) or theoretical relations that define the effects of the construction in the ground.

Financial: design of the works, estimation of quantities, and billing of the contract on the basis of the ground information and the interaction of the ground with the structure.

Legal/Contractural: understanding the effects of departures from the strict provisions of the works contract, particularly unforeseen ground conditions.

A suitable knowledge based system needs to be able to respond at one level to generalised perceptions and at a higher level to confident quantitative data. It must be suitable for use by an engineer or geologist who understands the concepts of ground engineering, but may not be an expert in the particular area of application. It should also be capable of providing objective advice to an expert.

REFERENCES

Attewell, P.B., Yeates, J. and Selby, A.R. 1986. Soil movements induced by tunnelling and their effects on pipelines and structures. Blackie, Glasgow, 325p.

Dumbleton, M.J. and West, G. 1974. Guidance on planning, directing and reporting site investigations. TRRL Report LR625, Transport and Road Research Laboratory, Crowthorne, Berkshire, England.

Toll, D.G. 1989. Representing the ground. Proc. NATO Advanced Study Institute: Optimisation and Decision Support Systems in Civil Engineering, Heriot-Watt University, vol. II, June 1989.

6th International IAEG Congress / 6ème Congrès International de AIGI, © 1990 Balkema, Rotterdam. ISBN 90 6191 130 3

Another approach to discontinuity shear strength assessments, based on investigations of bore cores

Une autre approximation des estimations de la force de coup d'incohérences basée sur des investigations de carottes de forage

Martin Th. van Staveren

Instituut Geotechniek Nederland b.v., Netherlands

ABSTRACT: The roughness of discontinuity surfaces has an important influence on the discontinuity shear strength. Several conventional methods of quantitative roughness determination have disadvantages when applied to the relatively small discontinuity surfaces available in cores.
The presented new roughness measuring method and classification system is especially developed for measurements within cores. The measuring method is simple and quickly applied. The system has been used on a site investigation project in The Netherlands.
A correlation was made between shear strength characteristics, obtained by direct shear tests, and 6 roughness parameters of the developed system. Also the joint roughness coefficient (JRC) and the rock material properties of the samples were determined.
The correlated 6 roughness parameters seem to identify the differences in shear strength characteristics better than the mentioned JRC. The presented roughness classification system is believed to improve discontinuity shear strength assesments, based on core investigations.

RESUME: La rudesse des surfaces incohérentes a une influence importante sur la force de coup d'incohérences.
Plusieurs méthodes conventionnelles de détermination quantitative de rudesse ont des désavantages quand elles sont appliquées à des surfaces incohérentes relativement petites qui se présentent dans des noeuds de forage. La nouvelle méthode présentée de mesurage d'incohérences et le système de classification ont été développés spécialement pour des mesurages dans des noeuds. La méthode de mesurage est facile et rapide à exécuter. Le système de classification a été appliqué sur un projet d'investigation d'un terrain à batir en Hollande.
On a fait une correlation entre les caractéristiques de la force de coup qu'on a obtenu grâce aux épreuves directes, et 6 paramètres de rudesse du système développé de classification. On a déterminé les coefficients incohérentes de rudesse (JRC = joint roughness coefficient) et les qualités du matériel des échantillons de pierre aussi.
On constate que les 6 paramètres en correlation expriment mieux les différences en force de coup que la méthode susdite (JRC).
Il semble que le système de classification présenté, basé sur des descriptions de noeuds de forage, améliore l'estimation de caractéristiques de force de coup.

1 INTRODUCTION

1.1 Theory

Rock core drilling is a wellknown method of geotechnical data acquisition. It is often necessary in order to build up a three dimensional geotechnical model, on which the design of an engineering project should be based.

In this geotechnical model the discontinuity shear strength is an important factor. the discontinuity shear strength can be characterised by the roughness of discontinuity surfaces occuring in cores.

Unfortunately, several conventional methods of quantitative roughness determination have disadvantages when applied to the discontinuity surface areas available in cores (Van-Staveren 1987). These methods are based on visual classification only, or rather complicated and time consuming. Some methods are not quite suitable for relatively small discontinuity surface areas.

1.2 Practice

The actual reason of interest of discontinuity roughness in bore cores was a feasibility study for an underground pump accumulation system in southern Limburg in The Netherlands. The study was based principally on the data from one vertical deep borehole. The core logging procedure of 1200 m of cores included roughness measurements and classification of each mechanically separated discontinuity encountered. Therefore a roughness measuring method and classification system especially for discontinuities within cores was developed.

2 THE PROPOSED ROUGHNESS MEASURING METHOD AND CLASSIFICATION SYSTEM

2.1 The roughness measuring method in theory

The developed roughness measuring method is based on the principles described by Patton (1966). Patton proposed a distinction between waviness (large scale roughness) and uneveness (small scale roughness). Waviness is presented by the length of undulation LU of the large scale roughness on a surface. LU is defined as the average distance between the large scale peaks and throughs on a discontinuity surface. This average distance is considered parallel to the discontinuity plane. The discontinuity plane is defined as the average direction of the tangents touching the large scale peaks and throughs. LU is expressed by formula (1) in figure 1.
    Uneveness is considered as the amplitude of the small scale roughness. The depth of asperities DA represents the amplitude. DA is defined as the average distance between the tangents touching the small scale peaks and throughs caused by the asperities. DA is expressed by formula (2) in figure 1.

2.2 The roughness classification system in theory

Based on the presented discontinuity roughness measuring method, a discontinuity roughness classification system has been developed. This system is presented in table 1. Values of LU and eventually supplementary requirements determine the discontinuity roughness expressed in a roughness parameter. A number of 17 roughness parameters are distinguished, and presented in table 1.

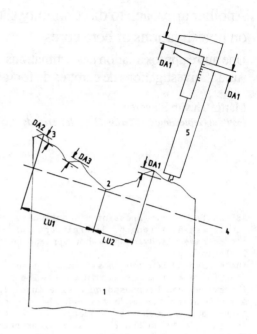

$$LU = \frac{LU_1 + LU_2 + \ldots + LU_n}{N} \quad (1)$$

$$DA = \frac{DA_1 + DA_2 + \ldots + DA_n}{N} \quad (2)$$

LU = length of undulation

DA = depth of asperities

1 = piece of core

2 = through on discontinuity surface

3 = peak on discontinuity surface

4 = proposed discontinuity surface

5 = Vernier calliper

Figure 1 : The proposed roughness measuring method

2.3 The presented method and system in practice

Values of LU and DA were determined by mea-

TABLE 1 : The proposed roughness classification system

ROUGHNESS PARAMETER	D (mm)	L (mm)	SUPPLEMENTARY REQUIREMENTS
polished	0	0-30	no asperities visible or palpable; discontinuity surface shows lustre.
polished & slickensided	>0	0-30	asperities are linear, parallel and polished ridges
slickensided	>0	0-30	asperities are linear, parallel ridges
smooth	0-0.5	0-30	
smooth & listric	0-0.5	0-30	asperities are polished, discontinuity surface shows lustre.
stepped	>0.5	0-30	asperities are stepshaped.
rough	0.5-2	0-30	
rough & listric	0.5-2	0-30	asperities are polished, discontinuity surface shows lustre.
very rough	>2	0-30	
very rough & listric	>2	0-30	asperities are polished, discontinuity surface shows lustre.
undulating		>30	
undulating & polished	0	>30	see polished
undulating & slickensided	>0	>30	see slickensided
undulating & smooth	0-0.5	>30	
undulating & stepped	>0.5	>30	see stepped
undulating & rough	0.5-2	>30	
undulating & very rough	>2	>30	

-surements with use of a Vernier calliper. For each discontinuity encountered, some measurements of LU and DA were made. The average and ultimate values of LU and DA were considered. Based on these values the discontinuity was classified according to table 1.

The proposed measuring method and classification system have been proven to be simple and quickly applied. It was possible to measure on the surfaces of all types of mechanically separated discontinuities.

3 TESTING PROCEDURES

3.1 General comments

In general discontinuity roughness will be determined to obtain an impression about the shear strength characteristics of the discontinuities. This implies that the accuracy of the proposed roughness measuring method and classification system have to be investigated. Therefore, a correlation with actual shear strength characteristics is necessary (Price 1985).

3.2 Sampling of discontinuities

To execute the correlation, 31 samples of mechanically seperated discontinuities were selected. These samples consisted of mudstone and limestone, which were the most common rock types occuring within the mentioned borehole.

The samples involved 6 roughness parameters: polished, slickensided, smooth, rough, stepped and very rough. Because of the test equipment available, discontinuities with large scale, undulating roughness could not be tested and were not considered. The surfaces of the samples fitted close together without infilling material. Except for the slickensided surfaces, it was presumed that no previous movements had occured on these surfaces.

Of course, the roughness of each sampled discontinuity was measured and classified according to the previously presented new method and system.

3.3 Determination of the joint roughness coefficient

The joint roughness coefficient JRC is a parameter to characterize the discontinuity surface roughness. The JRC varies between 0 and 20 from the smoothest to the roughest surfaces. The JRC can be estimated by visual comparison of profiles of the discontinuity surfaces with 10 typical roughness profiles (Barton and Choubey 1977).

Roughness profiling is done by using a contour gauge (Stimpson 1982). For each sample the JRC value was estimated.

3.4 Direct shear testing

Direct shear testing involves the determination of the peak shear strength of discontinuities. Under an applied constant normal force, a shear force is increased until failure starts to occur, usually after a small shear displacement. The normal force is applied perpendicular to the discontinuity surface. The shear force is applied along the discontinuity plane.

Each sample was tested in direct shear. The shear forces had been applied in those directions which were expected to result into the lowest peak shear strengths. The testing was undertaken in a portable shear box, manufactured by Robinson International. The samples had to be cast in concrete. Testing procedures followed the suggested methods of The International Society of Rock Mechanics (Brown 1982).

3.5 Determination of rock material properties

Rock material properties will influence direct shear test results (Crawford and Curran 1981). Therefore, rock material properties had to be determined of the tested discontinuity samples.

Testing included determination of the material unconfined shear stress, Brasilian tensile strength, ultrasonic velocity and unconfined compressive strength. The procedures laid down by the American Society for Testing and Materials and the International Society for Rock Mechanics were followed.

4 CRITICAL CONSIDERATIONS

4.1 Sampling of test specimen

The number of tests is rather limited in comparison with the number of roughness parameters tested. Only 2 rocktypes were sampled. Often the weakness of the mudstone made sampling of appropriate test specimen impossible. Therefore, the mudstone test results will represent the stronger mudstone only.

Table 2 : Rock material properties of the discontinuity samples

rock type	unconfined shear strength (MPa)	Brazilian tensile strength (MPa)	ultrasonic velocity (m/s)	unconfined compressive strength (MPa)
mudstone	11,6 (6,4)*	7,1 (3,5)	4074 (320)	57,8 (-)
limestone	31,9 (10,4)	16,6 (3,4)	5922 (321)	266 (100)

* The standard deviation is presented between brackets.

4.2 The joint roughness coefficient

It should be emphasized that the JRC is estimated by visual comparison of roughness
profiles. This can be considered as a rather
subjective procedure. Tse and Cruden (1979)
pointed out that fairly small errors in
these estimations would result in serious
errors when estimating peak shear strength
based on JRC.

4.3 Direct shear testing

With regard to direct shear testing, a lot
of factors will influence the test results.
They include the relatively small surface
areas of the discontinuity samples tested
(Bandis et al. 1981) and the influence of
small shear displacements when applying more
tests on the same sample (Rengers 1971).
Furthermore the choice of the test parameters is rather critical (Crawford and
Curran 1981). Also the stiffness of the test
equipment (Jager and Cook 1969) and the
concrete quality in which test specimen is
encapsultated (Van Staveren 1987) might
influence the direct shear test results.
   By direct shear testing usually the angle
of shearing resistence and the apparent
cohesion is determined. However, the direct
shear test results indicated that the mentioned factors influence in particular the
apparent cohesion. The apparent cohesion is
often considered as the shear strength without a normal load. To minimize the influence of the test results on the proposed
correlation with values of JRC and roughness
parameters as much as possible, the apparent
cohesion is left out of consideration in
the following.

5 TEST RESULTS

5.1 Rock material properties

Average values of the test results regarding the rock material properties are presented in table 2.

5.2 JRC versus the angle of shearing resis-
tance

A correlation between the JRC and the
angle of shearing resistance is presented
in the graphs of figure 2. If more than 1
sample got an equal JRC-value, the average
value for the angle of shearing resistence
is plotted. The spread of those values contributing to the average values is indicated by the verticals plotted on the
average values.

5.3 Roughness parameters versus the angle of
shearing resistance

The graphs in figure 3 represent a correlation between the roughness parameters and
the angles of shearing resistance.
Again average angles of shearing resistance
have been plot-ted when more samples got
equal roughness parameters. The spread of
the average values is plotted by the verticals in figure 3.

6 DISCUSSION OF TEST RESULTS

6.1 Rock material properties

The standard deviations in table 2 show
that the rock material properties varied
widely for both mudstone and limestone

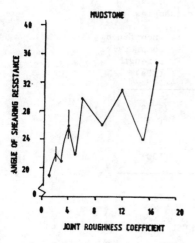

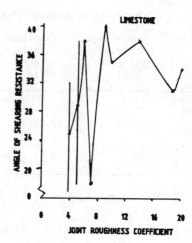

Figure 2 : Joint roughness coefficient versus the angle of shearing resistance

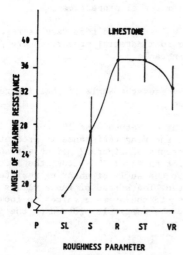

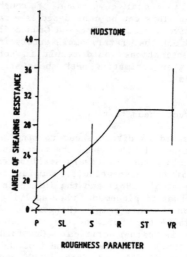

Roughness parameters : p = polished   r = rough
                       sl= slickensided st = stepped
                       s = smooth   vr = very rough

Figure 3 : Roughness parameters versus the angle of shearing resistance

samples. Both rock types have to be considered as inhomogeneous.

The rock material properties of limestone samples are higher than the properties of mudstone samples. This implies that limestone can be considered as stronger. Therefore test results are presented in the graphs in figure 2 and 3 separately for mudstone and limestone.

The graphs show that in general equal JRC values and roughness parameters in mudstone are related to lower angles of shearing resistance than in limestone. Because due to the lower rock material

properties, asperities in mudstone will yield at lower shear stress levels than asperities in limestone.

6.2 JRC versus the angle for shearing resistance

One would expect that increasing values of JRC, thus increasing discontinuity surface roughness, would be related towards increasing angles of shearing resistance.

But according to figure 2, increasing JRC-values might also be related to de-

creasing angles of shearing resistance.
And rather small differences in JRC-
values seem to be related to considerable
differences in shear strength.

This can be seen for mudstone as well as
limestone and shows the consequence of subjective visual roughness classification.

For limestone samples in particular it can
be seen that one particular value of JRC
might cover quite a wide range of angles of
shearing resistance. The lower angles of
shearing resistance for limestone samples
with JRC-values higher than 8 are unexpected. Unfortunately only a minority of the
tested samples had a JRC-value higher than
8. This makes it even more difficult to explain the mentioned decrease. Possibly it
is caused by one or more of the restric-
ting factors of the direct shear test.
With regard to mudstone samples the decrease in angles of shearing resistance for
higher JRC-values is less clear. Therefore
it might also be influenced by the greater
rock material strength of limestone, possibly in combination with the mentioned
restrictions.

More detailed research might contribute
to an explanation.

6.3 Roughness parameters versus the angle of shearing resistance

The graphs in figure 3 show that most of
the roughness parameters are related to different average angles of shearing resistance. Thus the roughness parameters seem
to represent different shear strength characteristics indeed. Increasing roughness
coincides with increasing angles of shearing resistance, at least until the parameter rough. In partilar for limestone
samples, the roughness parameter very rough
is related to decreasing angles of shearing
resistance. This effect is also mentioned
for the JRC and thus does not seem to depend on the presented new roughness measuring method and classication system.

Like the JRC, the roughness parameters
also cover a rather wide range of angles of
shearing resistance, in particular for the
weaker mudstone samples.

7 CONCLUSIONS

7 1 Critical considerations

It will be complicated to eliminate all
of the disturbing factors in the direct
shear test. Thus it will be difficult to
obtain real reliable data to control invented roughness methods and classification
sytems.

7.2 Rock material properties

Discontinuities with lower rock material
properties tend to have lower shear
strength characteristics. It seems therefore not possible to make reliable assesments of discontinuity shear strength,
based on discontinuity roughness measurements and classification only.

7.3 Joint roughness coefficient

Higher values of JRC do not automatically result in higher shear strength characteristics. Discontinuity shear strength
assesments based on JRC-values only seem
rather doubtful.

7.4 Roughness parameters

The proposed new method of discontinuity roughness measurement and classification seems to improve the reliability of discontinuity shear strength
assessments, based on core investigations. The roughness parameters seem to
identify differences in shear strength
characteristics better than the JRC. The
method is easily and simply applied in
practice.

More extensive and accurate investigations will give more insight in its practical value.

8 REFERENCES

Bandis, S. Lumsden, A.C., Barton, N.R.
(1981). Experimental studies of scale effects on the shear behaviour of rock joints.
International Journal of Rock Mechanics
and Mining Sciences and Geomechanics Abstracts 18, 1-21.

Barton, N., Choubey, V. (1977). The
shear strength of rock joints in theory
and practice. Rock Mechanics 10, 1-54.

Brown, E.T. (1981). Rock Characteri-
sation Testing and Monitoring; I.S.R.M.
Suggested Methods 5-52 Pergamon Press,
Oxford, England.

Crawford, A.M., Curran J.H. (1981). The
influence of shear velocity of the frictional
resistance of rock discontinuities. International Journal of Rock Mechanics and Mining
Sciences and Geomechanics Abstracts 18,
505-515.

Hoek, E., Bray J.W. (1977). Rock slope engineering 91-94. Institution of Mining and Metallurgy, London, UK.

Jaeger, J.C., Cook, N.G.W. (1969). Fundamentals of rock mechanics, 1s 167 Methuen, London.

Patton, F.D. (1966). Multiple modes of shear failure in rock. Proceedings of the International Symposium on Rock Mechanics, 509-513. Lisbon

Price, D.G. (1985). General Engineering geology Q38, part 2, 137-160. Delft University of Technology, Delft.

Staveren, M.Th. van (1987). Discontinuity studies for the Opac project, 87 pp. Delft University of Technology, Delft.

Stimpson, B. (1982). A rapid field method for recording joint roughness profiles. International Journal of Rock Mechanics and Mining Sciences and Geomechanics Abstracts 19, 345-346.

Tse, R. Cruden, D.M. (1977). Estimating joint roughness coefficients (JRC). International Journal of Rock Mechanics and Mining Sciences and Geomechanics Abstracts 16, 339-362.

6th International IAEG Congress / 6ème Congrès International de AIGI, © 1990 Balkema, Rotterdam. ISBN 90 6191 130 3

Classification of Dutch peats

Classification des tourbes néerlandaises

A.A.M.Venmans & E.J.den Haan
Delft Geotechnics, Delft, Netherlands

ABSTRACT: New specific design methods for constructions on peat are being developed. As a base for use of these methods the suitability of classification systems and testing procedures existing outside the Netherlands have been studied. It is concluded that modification of certain procedures is required because of the different average composition of Dutch peats. A brief description of the modified testing procedures is given. In addition, a new method to determine the magnitude of the anisotropy of the fiber orientation from anisotropic shrinkage is presented.

RESUME: Des méthodes de calcul nouvelles et spécifiques pour constructions sur tourbe sont en course de développement. Comme base pour l'utilisation de cettes méthodes, la propriété des systèmes de classification et des procédures d'essais existantes hors de Néerlande a été étudiée. On a trouvé que la modification des certaines procédures est nécessaire à cause de la composition différente des tourbes Néerlandaises. Une description brève des procédures modifiées est donnée. Aussi, un essai nouveau pour la détermination de la magnitude de l'anisotropie de l'orientation des filaments est présenté, en utilisant l'anisotropie du rétrécissement.

1 INTRODUCTION

The subsoil of a large part of the western Netherlands consists of soft holocene peat and clay deposits on firm pleistocene sands. These materials were deposited in the lower Rhine/Meuse delta behind a coastal barrier in a brackish to fresh water environment (figure 1). The thickness of the layers decreases from 15 m in the south-west and 5 m in the north-east of the area. Also the western Netherlands also is the most densily populated and built-up part of the country. The present widening of roads and railways and the strengthening of the existing river and polder embankments require optimization of design for minimal space occupation and minimal damage to existing constructions. The main problems are related to large horizontal and vertical deformations and slope stability. The need for optimization puts ever increasing demands on the quality of constitutive models of soil behaviour used for calculations.

At present, the modelling of the mechanical behaviour of clay is sufficiently advanced for most practical problems. For convenience, peat is usually considered as a special type of clay, allowing the use of the same models, however ignoring some major differences between clay and peat:

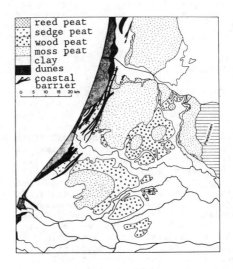

reed peat
sedge peat
wood peat
moss peat
clay
dunes
coastal barrier

0 5 10 15 20 km

Figure 1: Maximal distribution of peat deposits in the western Netherlands in 1200 AD

- much greater compressibility caused by the open structure of plant remains
- strongly anisotropic shear behaviour caused by interaction between the matrix of orientated organic fibers and intermediate material
- great susceptibility to weathering
- greater variability of peat deposits, reflected in geotechnical properties

In order to bring the modelling of the mechanical behaviour of peat to a level, compatible with that of clay, the Dutch Technical Committee on Embankments (TAW) of the Dutch Ministry of Public Works commissioned Delft Geotechnics in 1987 to start a 5-year research programme. The main stages in this programme are:

- identification of knowledge gaps
- improvement of basic engineering methods for one-dimensional consolidation and slip surface analysis
- definition of a constitutive model of peat for application in Finite Element codes
- definition of recommendations regarding the balance between the natural variability of peat deposits, schematization of the subsoil stratigraphy and accuracy of calculation methods and results

In an early stage of the research programme it was apparent that the geotechnical classification of peat required further research. The purposes of this further study were:

- description of peat in terms related to the geotechnical behaviour for mapping of geotechnical units
- definition of reliable and cost-effective procedures for the determination of classification parameters

2 PEAT FORMATION IN THE WESTERN NETHERLANDS

Peat formation in the western part of the Netherlands begins approximately 4000 BP with the warping of the area behind the coastal barrier, allowing the formation of mineral-rich reed peat on the tidal flats (figure 1). The plants grow in shallow brackish water with a rich supply of anorganic nutrients; conditions are eutrophic. Following the uplift of the ground level by plant growth, sedge peat is deposited in a fresh water environment. The quantity of anorganic nutrients is limited because water circulation is impeded by plant growth; conditions are eutrophic to mesotrophic. As the zone of plant growth further rises, the plants become more dependent on nutrient-deficient rain water. In oligotrophic conditions moss peat is deposited, containing little anorganic material. The usual transition stage with mesotrophic wood peat is not found in this succession.

In all stages, peat formation in the western Netherlands is strongly influenced by the presence of several rivers (Oude Rijn, Oude IJssel) depositing anorganic material from the backland. These rivers are part of the Rhine/Meuse delta. The lowlying areas behind the natural levees are a good habitat for shore forests. Because these areas are frequently flooded, conditions are eutrophic and the wood peat deposited has a large ash content. At a greater distance from the rivers, the nutrient supply is again dominated by rain water. Wood peat of mesotrophic origin is found along the shores of creeks originating in the peat area, such as the Amstel and the Holendrecht.

Around 1200 AD a large part of the western Netherlands was covered with layers of moss peat, reaching thicknesses up to 6 meters. Nowadays, moss peat is only found in polder embankments, which are left over from exploitation of the peat for fuel. Moss peat was considered very suitable because of its low ash content; in some places also the sedge peat was excavated. The thickness of the peat deposits remaining has been further reduced by drainage, powered initially by the famous Dutch windmills, causing compaction and weathering. On the average, the present thickness of the remaining sedge and wood peat is 5 meters.

An analysis was made of the botanical composition of approximately 500 samples taken at random from polder embankments in the western Netherlands. The results are given in figure 2; the typical values for the ash content given have been determined on 50 samples. In figure 2, similar data are given for peats from the north-eastern part of the U.S.A. (Nichols & Boelter, 1984) and Finland (Karesniemi, 1972).

Compared to conditions in North America and Scandinavia, where large areas are covered with mesotrophic and oligotrophic peats, eutrophic and mesotrophic peats are

botanical composition of main constituent	percentage of samples		
	Nether-lands	U.S.A.	Finland
reed, sedge	60	44	61
wood	25	9	7
moss	15	46	32
ash content[1]			
< 10 %	15	76	78
10 % - 20 %	60	18	11
> 20 %	35	6	11

Figure 2: Typical botanical composition and ash content of peats from several countries

784

dominating in the western Netherlands. Also the average ash content of Dutch peats is considerably higher as a result of the specific deltaic setting of the western Netherlands.

Resuming, both the natural depositional setting and the large scale of human post-depositional activities in the Netherlands are responsible for the difference.

3 CLASSIFICATION OF PEAT

An excellent discussion of the relationship between depositional processes and basic geotechnical properties is given by Hobbs (1986). It is concluded that basic geotechnical engineering methods may well be improved by making distinction between different types of peat. Allthough most engineers are aware that a further specification may have practical use, they lack the means to express and quantify the differences.

From the viewpoint of common engineering problems a classification of peat and organic soils should therefore contain the following information:

1. Mapping:
   - main description
   - botanical origin of main constituent
   - water and ash content
   - content of fine and coarse fibers
   - content of woody parts
   Determination of these parameters should be possible in the field by simple visual inspection
2. Compression behaviour:
   - water and ash content
   - specific gravity
   - bulk density
   Many authors (summarized in Hobbs, 1986) give relationships between the initial void ratio or initial water content and compression index C_c, implying a non-linear relation between void ratio or water content and the logarithm of stress. For some time a method based on a similar relationship, known as Fokkens method, is in use in the Netherlands; as part of the present research programme, this method has been improved by den Haan (1989b).
3. Shear behaviour:
   - type and degree of weathering of organic fibers
   - quantity and length of organic fibers
   - main orientation of fibers; anisotropy
   Since the shear strength of peat is strongly influenced by the organic fibers, great attention should be given to express this property in classification. Also anisotropy must be determined, because a pronounced horizontal fiber orientation, induced by compaction, is usually present.

Depending on the purpose of study the classification can be more or less elaborate. The recommended extent of a geotechnical classification is summarized in figure 3.

parameter	field	laboratory	
	survey	com-pression	shear strength
main description	X	X	X
botanical composition	X	X	X
degree of decomposition	X	X	X
ash content	X/V	X	X
water content	X/V	X	X
bulk density	–	X	X
specific gravity	–	(X)[1]	(X)[1]
content of fine and coarse fibers	X/V	X/V	X
content of woody parts	X	X	X
fiber content	–	–	X
anisotropic tensile strength	X	X	X
anisotropic shrinkage	–	–	X

X   = recommended
X/V = recommended, visual field determination
–   = not necessary
[1] = optional

Figure 3: Recommended minimal extent of a classification

4 DESCRIPTION OF CLASSIFICATION TESTS

The classification tests described below are summarized from the report "Geotechnical classification of peat and organic soils" (Venmans, 1990), written for the Technical Committee on Embankments TAW.

4.1 Main descriptive term

The descriptive term "peat" has several meanings; a usual way to define peat is based on the ash content, i.e. the content of anorganic matter. The exact type of anorganic particles is of minor interest.

Recently, a new Dutch standard NEN 5104 for the description of soils has been published. According to this standard the main descriptive term is given primarily by the position of a material in the organic matter-clay-silt+sand classification

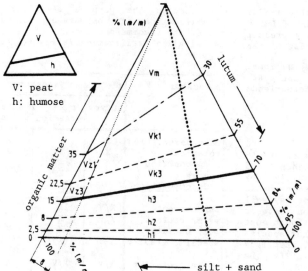

Figure 4: Classification triangle for peat and organic soil according to the Dutch standard NEN 5104

•••••••• = scope of natural deposits in the Netherlands

field	additional term	
	peat	humose soil
Vm	–	
Vk1	slightly clayey	
Vk3	strongly clayey	
Vz1	slightly sandy	
Vz3	strongly sandy	
h1		slightly
h2		moderately
h3		strongly

triangle (figure 4). The division of the triangle is dictated by field experience: if the anorganic fraction contains more clay, the soil should contain more organic matter to be classified visually as equally humose. The triangle suits well agricultural purposes; however, it still is a subject of research if it also reflects the geotechnical behaviour of peat with maximum effect. For reasons of standarization use of the triangle for geotechnical purposes is supported at the moment.

On the other hand it is felt that the internationally accepted definition of peat as material with a loss on ignition exceeding 80 % (Landva, 1983) is too strict for Dutch peats. From figure 2 it is apparent that many true Dutch peats would be excluded, whereas this definition applies very well to foreign peats.

4.2 Botanical composition

The botanical composition of the constituting plant remains is an important classification parameter, as it contains information on the depositional setting of the peat.
A first subdivision in geotechnical units will often be based on botanical composition. Furthermore, the mechanical behaviour of the fibers is related to their botanical origin.

For classification purposes, distinction is made between the following species: aquatic herbs, reeds, sedge, wood (oak, alder, birch, willow), mosses (spaghnum, hypnum sp.) cottongrass, heather. If recognizable, also the part of the plant should be mentioned, e.g. roots, stems.

4.3 Degree of decomposition

The strength of the individual fibers will be an important factor in the shear strength of peat. In the process of decomposition of organic fibers cellulose and lignin, which give the fiber its strength, are reduced to amorphous gelly-like humic acids. Though the strength of fibers in peat cannot be measured easily in a direct way, the progress of breakdown of the fibers can be estimated by a simple test. In this test, described by von Post in 1922 (Stanek & Silc, 1977), a disturbed sample of peat is squeezed in the hand. The degree of decomposition is expressed on a scale ranging from 1 to 10, depending on the substance squeezed out between the fingers and the remaining solids.

The test appears to be most succesful for foreign peats with ash contents below 10 % average. The presence of greater quantities of anorganic matter in Dutch peats often obstructs an accurate determination of the condition of the plant remains. To overcome this difficulty, other tests are a subject of research. The pyrophosphate solubility test, decribed by Stanek & Silc (1977) seems most promissing.

4.4 Ash content

For classifiction purposes, dry ashing in a furnace is recommended. As a compromise between undesired desintegration of anorganic matter and incomplete oxydation of the organic fraction, charring at 550°C for 5 hours is suggested as a standard. The ash content is calculated from the loss on ignition according to:

$$a = 1.04 \cdot (100 - N) \qquad (1)$$

in which:
  a = ash content (%)
  N = loss on ignition (%)
The factor 1.04 accounts for partial desintegration of anorganic material.

4.5 Water content

The water content is usually expressed as the mass of water in a sample, related to the mass of the sample after drying in an oven. It is recommended to use a drying temperature of 105°C during 24 hours. Some oxydation of organic material may occur, but the error is sufficiently small for classification purposes. Drying at lower temperatures (60°C, 85°C) will not remove all free water at atmospheric pressure. If desired, more accurate results can be obtained by drying at 85°C at reduced air pressure.

4.6 Bulk density

The bulk density is determined best by measuring the mass of a regular volume of material, e.g. a sampling tube. The bulk density of irregularly shaped samples can be determined by the mercury displacement method. In this method the volume of the sample is calculated from the mass of mercury spilled after immersion of the sample in a container, filled to the rim with mercury.

4.7 Specific gravity

For classification purposes, the specific gravity can be determined with sufficient accuracy from correlation with the loss on ignition, according to the following equation:

$$G = \frac{3.60}{1.34 \cdot \dfrac{N}{100} + 1.31} \qquad (2)$$

in which:
  G = specific gravity (-)
  N = loss on ignition (%)
The equation holds for Dutch peats, implicitly assuming a specific gravity of 1.36 for pure organic matter and 2.65 for pure anorganic material and partial des-

integration of anorganic material. The accuracy of the calculated value is within 4% as compared to values measured by the pycnometer method, described in BS-1377. section 2.6.

4.8 Wood content

The wood content is determined visually by estimation of the area percentage of wood on a representative horizontal section through a sample.

4.9 Fiber content

Special attention has been paid to the determination of the fiber content and the fiber size. Nichols and Boelter (1984) describe a method for wet sieving analysis in which the sample is rubbed gently between the fingers before sieving. By rubbing the sample large but partially decomposed fibers, which will not contribute to the strength of the material in a degree proportional to their size, are broken into smaller pieces. For this reason the "rubbed" fiber size distribution is considered a better classification datum than the "unrubbed" fiber size distribution.
  The openings of the sieves used are 0.106 mm, 0.150 m, 0.250 mm, 0.500 mm, 1.00 mm and 2.00 mm. The Rubbed Fiber Content is defined as the mass percentage coarser than 0.150 mm.
  The method described by Nichols & Boelter has been modified in the following respects:
- before sieving the sample is soaked for 24 hours in a solution of sodium-pyrophosphate to loosen the particles and dissolve humic acids
- the ash content of the fractions retained on the sieves is determined to obtain the sieve curve of the organic material solely; again, this modification is imposed by the high ash content of the Dutch peats as compared to foreign peats

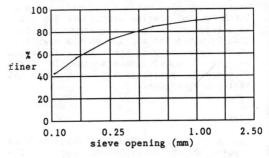

Figure 5: Sieve curve of a slightly weathered reed peat

An example of the fiber size distribution of a slightly weathered reed peat is given in figure 5.

In the test described, the fiber diameter is decisive for the fiber size. Thus, the fiber size distribution obtained in the test is in fact a fiber diameter distribution. Because fiber length is an equally important parameter, it is recommended that after drying and weighting of the sieve fractions, the ratio of fiber length and fiber diameter is determined visually.

4.10 Anisotropic tensile strength

The anisotropic tensile strength can easily be estimated in the field by manually testing small samples in horizontal and vertical direction. This test gives an indication about the magnitude of anisotropy of the fiber orientation, assuming a horizontal preferential orientation.

4.11 Anisotropic shrinkage

Den Haan (1989a) describes a new method for a more quantitative determination of the magnitude of the anisotropy of the fiber orientation. The test is based on the fact that organic fibers tend to shrink more in a direction perpendicularly to the long axis than along this axis.
However, anisotropic shrinkage on a macroscopic scale will only show if a preferential fiber orientation is present. Assuming a eventual preferential orientation to be horizontal, the magnitude of the anisotropy can be determined by measuring the difference between shrinkage of a peat sample in horizontal and vertical direction.

4 CONCLUSIONS

The average composition of Dutch peats is different from that of foreign peats due to the specific deltaic depositional setting and large scale human activity in the Netherlands. For optimal classification of Dutch peats, testing procedures described in foreign literature require modification. Valuable information on the magnitude of the anisotropy of the fiber orientation can be obtained from measurements of the differential shrinkage in horizontal and vertical direction.

REFERENCES

Den Haan, E.J. 1989a. Shrinkage test on peat. Delft Geotechnics report CO-290341/39 (in Dutch)
Den Haan, E.J. 1989b. Compressibilty and other properties of peat in a characteristic Dutch peat profile. Delft Geotechnics report CO-305862/3 (in Dutch)
Hobbs, N.B. 1986. Mire morphology and the properties and behaviour of some British and foreign peats. Quart. Jnl. Engng. Geol., vol. 19, pp. 7-80
Karesniemi, K. 1972. Dependence of humification on certain properties of peat. Proc. 4th Int. Peat Congr., Otaniemi, vol. 2, pp. 273-282
Landva, A.O. e.a. 1983. Geotechnical classification of peats and organic soils. In: Testing of Peats and organic soils, ASTM STP 820, P.M. Jarrett, Ed., ASTM, pp. 37-51
Nichols, D.S. & Boelter, D.H. 1984. Fiber size distribution, bulk density, and ash content of peats in Minnesota, Wisconsin, and Michigan. Soil Sci. Soc. Am. 48:1320-1328
Stanek, W. & Silc, T. 1977. Comparison of four methods for determination of the degree of peat humification (decomposition) with emphasis on the von Post method. Can. Jnl. Soil Sci. 57:109-117
Venmans, A.A.M. 1990. Classification of peat and organic soils. Delft Geotechnics report CO-305863/7 (in Dutch)

A numerical simulation study of core disking

Recherche sur la simulation numérique de carotte discoïde

S.T.Wang & R.Q.Huang
Chengdu College of Geology, Sichuan, People's Republic of China

ABSTRACT: Rock core disking has been generally accepted as a special rock mechanic phenomena in high geo-stress areas since 1970s, but its mechanism and stress conditions of occurrence has not yet been clarified. In this paper, on summarization of geological conditions of rock core disking at Laxiwa damsity, a FEM numerical model is presented to simulate the process of core rupture during drilling. The results of simulation indicate that different stress conditions lead to different occurring mechanism of rock core disking. Three basic types may be summed up, e.g., simple-tensioned completely, tensioned in the surface part and sheared in the internal part and tensioned in the surface part only. Furthermore, a series of criteria describing the stress conditions of different core disking mechanism are established according to the simulation results, and their correctness is confirmed through surface texture studies of core disking using a scanning electronic microscope. Obviously, it is quite convinent for us to use these criteria to estimate geo-stress magnitude in projected site where show high geo-stress and rock core disking phenomena.

RESUME: Depuis les années soixante-dix vingtième siècle, la carotte discoïde a été acceptée généralement comme un phénomène de roche mecanique dans les zones à haut tension géologique, mais leur mécanisme et condition de tension n'ont pas été clarifiées. Dans cet article, sur la base de condition géologique de la carotte discoïde en laxiwa barrage, un FEM modèle numérique est présenté pour simuler le processus de rupture de carotte pendant forage. Le résultat de la simulation indique que les conditions des tensions défférentes conduisent à la production de la carotte discoïde à mécanisme defférente. Trois types des mécanismes sont peut-être résumés, c'est-a-dire, simple tension complète, la tension en la partie superficiele et le cisaillement dans la partie interne et la tension en la parite superficiele seulement. En outre, d'après les résultats de la simulation, les conditions de la tension de mécanisme différente de carotte discoïde sont expliquées, et leur corrections sont confirmées par la recherche de microscope balayage électronique sur les textures superfécie les de carotte discoïde. Il est convaincu que on peut utiliser les critères pour estimer la magnitude de; a tension géologique dans le chantier du barrage projecté où les phénomènes de haute tension géologique et la carotte discoïde sont développés.

1 INTRODUCTION

A phenomenon that a core is ruptured into a series of disks during drilling is referred to as core disking. It was discovered in 1960s, but has been generally accepted as a special rock mechanic phenomenon in high geostress areas only since 1970s. Problems related to the core disking have been studied by many researchers, but its mechanism and stress conditions of occurrence have not yet been clarified. Core disking phenomena were discovered at many damsites in China. Based on geological investigations and numerical simulations, problems related to the core disking at Laxiwa Damsite are studied in detail in this paper.

2 GEOMECHANICAL CONDITION AND CORE DISKING CHARACTERISTICS IN LAXIWA DAMSITE

As a general rule, rock core disking in Laxiwa Damsite occurred in such a geomechanical condition (Table 1) in which the rock masses have a higher capacity to storage the strain energy due to its

Table 1 Geomechanical condition in which core disking occurs at Laxiwa Damsite

Type of rock masses	Structure of rock mass	Modulus of elasticity of rock mass (Mpa)	Range of stress magnitude in rock mass (Mpa)
Granite	Intact RQD 95%	57000	30-58
Metamorphic limestone	Ditto	60000	30-45

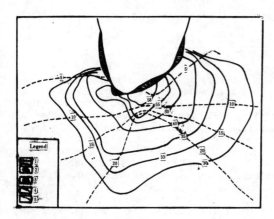

Fig.1 Extension of high stressed rock masses beneath the valley bottom at Laxiwa Damsite
1-fault; 2-area ruptured by tension 3-area ruptured by shearing; 4-contour of σ_1 ; 5-contour of σ_3

intact structure and higher elasticity and they are high stressed in the geostress field.

The following features were observed in the development of core disking at Laxiwa Damsite, namely:

(1) Core disking was large in number and extension. It was observed in all ten bore holes passing through the river bed and distributed in the range of high stressed rock masses from 2.5m to 300m beneath the bed rock surface (Fig. 1 and 3), but the densest distribution was observed in the depth from 70m to 130m.

(2) Rock disks were small in thickness and large in density. For example, 959 rock disks were observed in the depth interval from 138m to 158m in bore hole No.4 and their average thickness was 0.7-0.8 cm (Fig. 3).

(3) In the aspect of form, there were three types of disks: with slightly concave surface (Fig. 4), with even surface (Fig. 5) and, in small numbers, with uneven surface.

(4) The surfaces of core disks were fresh, and in most of the cases they were rough in their peripheral parts and rather smooth with slickensides along the same direction in their interior parts.

(5) Some cores apparently fractured at the surface but not yet ruptured into individual disks and in most of the cases, such a rock core can easily be separated into individual disks even by hands (Fig. 6).

3 NUMERICAL SIMULATION ANALYSIS OF CORE DISKING, RELATED TO THE MECHANISM AND STRESS CONDITIONS OF ITS GENERATION

3.1 Mechanism analysis

The stress state of a rock core at its base during drilling in a high stressed area is very complex. In order to analyze this kind of problems, a two-dimensional FEM numerical model is established (Fig. 7).

It is demonstrated by the numerical analysis that there exists a tensile stress concentration area in the peripheral part of a core base (Fig. 8), the tensile stress is maximum at the surface of the core, decreasing rapidly towards its interior, and in a certain distance (about 0.8m) from the surface the stress becomes compressive instead of the tensile one (Fig. 9). Further nonlinear analyses show that the mechanism of core disking is different under different environmental stress conditions, for the magnitudes of tensile stress in its interior part of a core depend directly on the environmental stress level (Fig. 9). Thus the following three basic types of core disking can be distinguished:

(1) Simple-tensioned completely. When the environmental stress is high enough, for example σ_x =110Mpa, a core can completely be ruptured into individual disks by simple-tension due to stress redistribution during the drilling.

(2) Tensioned in the surficial part and sheared in the interior part. When the

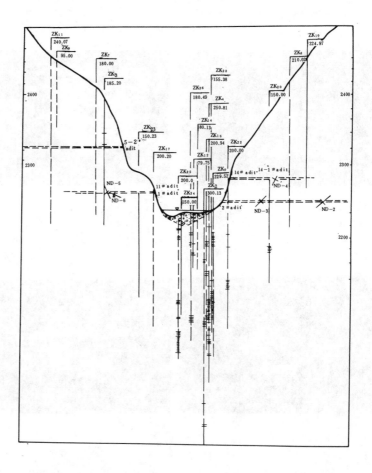

Fig. 2 Projection showing the distribution of core disking phenomena at Laxiwa Damsite

Fig. 3 Core disking phenomenon in bore hole No.4

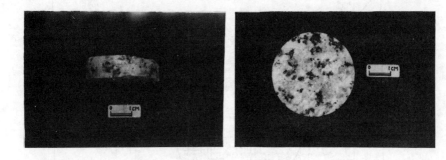

Fig. 4 Rock disk with slightly concave surface

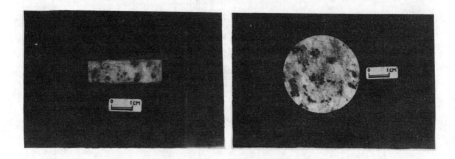

Fig. 5 Rock disk with even surface

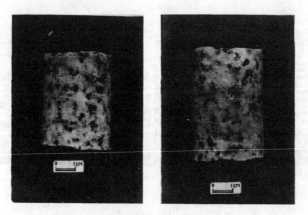

Fig. 6 A core apparently fractured at the surface, but not yet
ruptured into individual disks

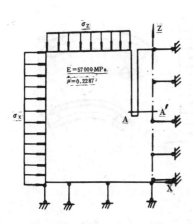

Fig. 7 A two-dimensional FEM numerical model for the analysis of core disking

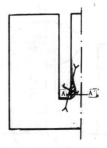

Fig. 8 Tensile stress distribution in a core at its base section A-A'

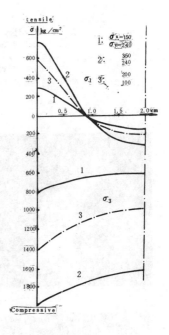

Fig. 9 Stress state in the core at its base section A-A'

environmental stress is lower to some extent than that in the case above-mentioned a core may be ruptured into individual disks by this mechanism.

(3) Tensioned in the surficial part only. When the environmental stress is rather low, a core may be fractured mainly by tension in its surficial part only.

Actually, the stress state in a rock core at its base section during a drilling process can be analyzed as an axial symmetry problem in three dimensions. It is shown by the analysis using such a model (Fig. 10) that the result of the analysis is in well agreement with that obtained by the two-dimensional model.

3.2 stress condition analysis

3.2.1 Relationship between tensile stress concentration in rock core surface at its base and environmental stress

It is necessary first to clarify this relationship because the fracturing process of a core is proceed from its surface and the tensile stress concentration at the core surface in its base section plays a very important role in the development of core disking. Through FEM numerical calculations using the model illustrated in Fig. 7 and regression analyses, the following equations were derived:

$$\sigma_{stx} = -7.682 + 1.562\sigma_x + 0.148\sigma_z \ldots\ldots(1)$$
$$\sigma_{sty} = -7.682 + 1.562\sigma_y + 0.148\sigma_z \ldots\ldots(2)$$
$$(r=0.999)$$

where σ_x --Maximum horizontal environmental stress

σ_y --Minimum horizontal environmental stress

σ_{stx} , σ_{sty} --Tensile stress concentration in the core surface at its base along σ_x and σ_y respectively.

It can be seen from above equations that the maximum horizontal environmental stress plays a predominant role in the core disking process.

3.2.2 Stress conditions of core disking for different mechanism.

1) Tensioned in the surficial part of a core only. The tensile strength (σ_t) of

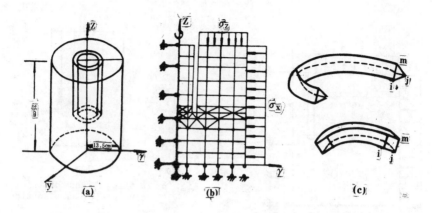

Fig. 10 A three-dimensional FEM numerical model for the analysis of core disking
   a-model view
   b-element division
   c-form of element

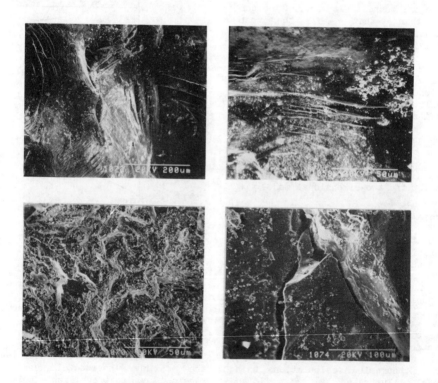

Fig. 14 Surface textures of specimens No. 2 and 3

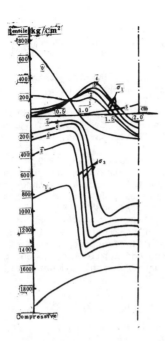

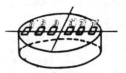

Fig.12 Sampling position at the surface of a core disk

Fig.13 Surface texture of specimen No.1

Fig.11 Change of the stress in section A-A' during nonlinear analyses by successively eliminating tensile stress when σ_{stx} =74Mpa

the fresh granite at the damsite was estimated at 9 Mpa according to the experiments. So, the stress conditions of fracturing for this kind of mechanism are derived by substituting it into equations (1) and (2):

$$\sigma_x \geqslant 10.68-0.095\,\sigma_z \ldots \ldots (3a)$$
$$\sigma_y \geqslant 10.68-0.095\,\sigma_z \ldots \ldots (3b)$$

It can be seen from Fig. 1 that both equations (3a) and (3b) are tenable in the depth interval less than 300m for the damsite, so this kind of core fracturing could be observed during a drilling process in the area.

2) Simple-tensioned completely

The change of the stress in section A-A' during nonlinear analyses by successively eliminating tensile stress when σ_{stx} =74Mpa is demonstrated in Fig. 11. It is shown by the further analysis that a rock core can completely be ruptured by simple-tension only if $\sigma_{stx} \geqslant$ 110Mpa, so according to equation (2) the stress conditions of core disking for this kind of mechanism can be expressed as:

$$\sigma_x \geqslant 75.35-0.095\,\sigma_z \ldots \ldots (4a)$$
$$\sigma_y \geqslant 75.35-0.095\,\sigma_z \ldots \ldots (4b)$$

Actually, for Laxiwa damsite only equation (4a) is tenable within a local rock mass at the right side of the valley bottom,

and only there rock core disking by simple-tension along σ_x direction could be observed.

3) Tensioned in the surficial part and sheared in the interior part.

Nonlinear analysis by successively eliminating tensile stress shows that when σ_{stx} =45Mpa a core can be ruptured into individual disks by tension at its surficial part of some 0.8cm deep and shearing in its interior part. According to this, the stress conditions of core disking for this mechanism may be expressed as:

$$\sigma_x \geqslant 33.75-0.095\,\sigma_z \ldots \ldots (6a)$$
$$\sigma_y \geqslant 33.75-0.095\,\sigma_z \ldots \ldots (6b)$$

only equation (6a) is tenable in the condition of Laxiwa damsite, indicating that core disking of this kind of mechanism may occur only along σ_x direction.

It is demonstrated by further analyses that when σ_{stx} =32Mpa, or $\sigma_x\,(\sigma_y)\geqslant$25.4-0.095 σ_z, a core may be fractured by tension and shearing only in its external part of 0.5cm deep, but not ruptured completely into individual disks.

According to the real environmental stress condition in Laxiwa Damsite, most of core diskings should be ones tensioned in its surficial part and sheared in its interior part, and it is shown by the observation that such is just the case.

795

4 SURFACE TEXTURE STUDY OF CORE DISKING

In order to examine the correctness of the numerical analyses and to find out a useful method to distinguish mechanism of core disking, a surface texture study of core disks using a scanning electronic microscope was undertaken. As shown in Fig. 12, specimens were cut out from a core disk along its surficial hollow, which was taken from the depth of 135m in bore hole No.22, where σ_1 =33Mpa, σ_3 =21.6Mpa.

The surface textures of specimens No.1 and 1' cut out from the periphery part of the core are characterized by the uneven and keen-edged sections (Fig. 13), indicating a tension fracture.

The surface textures of specimens No.2,3 and 2',3' cut out from the interior part of the core are characterized by the even sections with microslickensides, parallel to the surficial hollow, and microsteps, perpendicular to it, indicating a shearing fracture (Fig. 14). So, the surface texture study shows that the core was ruptured by tension at its surficial part and shearing in its interior part. This conclusion is in well agreement with that derived from above-mentioned analyses according to the stress state in which the rock core was situated in the nature.

5 CONCLUSION

Through above-mentioned researches, the following conclusions can be made:

(1) A series of criteria describing the stress conditions of core disking for different mechanism are established according to the numerical analyses, and their correctness is confirmed through surface texture studies of core disks.

(2) The surface texture study is a useful measure to distinguish core disking mechanism.

(3) It is quite convenient for us to use above-mentioned methods to estimate the geostress level in a projected site where show high geostress and core disking phenomema.

REFERENCE

R.Q. Hang, 1988, Engineering geological studies of high rock slopes in complicated rock mechanics environmental conditions, Ph.D. Thesis, Chengdu College of Geology

Miller, C.H. and Skinner, E.H.,1980, "The natural of fracturing and stress distribution in quartze around the 1128m level of the Crescent Mine, Idaho", Eng. Geol. Vol.16, pp321-338.

Brunner, F.K. and Scheidegger, A.E., 1973 "Exfoliation", Rock Mechanics, Vol.5 N. pp43-62.

Twidal, C.R., "On the origin of sheet jointing", Rock Mechanics, Vo15 (1973), pp163-187.

Lajati, E.Z., 1977, "A mechanistic view of some aspects of joints of jointing in rocks", Tectonophysics, Vol.38, No.3-4, pp327-338.

M.Fridman and J.M.Logan, 1970, "Influence of residual strain on the orientation of experimental fractures in three quartzose sagstones", J. Geophysical Research, Vol.75, No.2, pp387-405.

F.L. Hou, Y.R. Jia, 1984, Residual stress states of rock cores in drill holes and analysis of fracture stress of cores, Research and Application of Geo-stress, Earthquake Publishing House, Beijing.

6th International IAEG Congress / 6ème Congrès International de AIGI, © 1990 Balkema, Rotterdam. ISBN 90 6191 130 3

Simulation tridimensionnelle de la blocométrie naturelle de massifs rocheux
Three-dimensional simulation of natural rock mass granulometry

Xu Jixian & R.Cojean
CGI, Ecole des Mines de Paris, France

RESUME: Cet article présente un modèle général de simulation à 3D de la blocométrie naturelle des massifs rocheux. Il est constitué de quatre parties: 1) analyse statistique et simulation à 3D des discontinuités. 2) analyse de la connectivité des discontinuités simulées. 3) identification des blocs discrets intersectés par des discontinuités connectées. 4) Caractérisation de la blocométrie de massifs rocheux. Trois types de représentation (distribution de taille, distribution de taille pondérée, distribution de l'orientation des blocs) sont utilisés. Les résultats d'un exemple réel sont donnés dans les sections correspondantes de l'article.

ABSTRACT: This paper presents a general model for simulating 3D natural rock mass granulometry. It consists of four parts: 1) Statistic analysis and 3D simulation of rock mass fractures. 2) Connectivity study of simulated fractures. 3) Identification of 3D distinct blocks intersected by the fractures. 4) Characterization of rock mass granulometry. Three representative methods, such as size distribution, weighted size distribution and orientation distribution of blocks, are used. The results of a real example are given in the correspondent sections of the paper.

INTRODUCTION

Les développements récents dans l'analyse de la stabilité et dans l'évaluation de la fragmentation des massifs rocheux par explosif exigent une évaluation initiale de la blocométrie naturelle de ceux-ci. Depuis le travail original de Cundall (1972), la méthode des éléments discrets s'est développée pour étudier le comportement mécanique des milieux fissurés (Cundall,1988). Cette méthode considère un massif rocheux comme un système de blocs discrets intersectés par des discontinuités, et donc elle exige un système de blocs pré-défini. Les propriétés de fragmentation d'un massif rocheux par explosif sont contrôlées pour partie par sa blocométrie initiale (distributions de taille et d'orientation des blocs). Un modèle de la blocométrie naturelle nous permet de faire l'étude comparative entre celle-ci et la granulométrie du tas abattu pour optimiser l'utilisation de l'énergie explosive.

Certains programmes de simulation des discontinuités à 3D ont été développés pour l'analyse d'écoulements souterrains. Long et al (1985) ont réalisé un modèle de

système de discontinuités à 3D. Andersson et Dverstorp (1987) ont développé un modèle de simulation conditionnelle de réseaux de discontinuités discrètes. Mais l'application de systèmes de discontinuités simulées à l'analyse mécanique ou à l'évaluation de la blocométrie d'un massif rocheux rencontre la difficulté d'identification des blocs discrets intersectés par des discontinuités.

Lin et al (1987) ont proposé une méthode géométrique pour identifier le système de blocs. Dans cette méthode, un bloc (polyèdre) est considéré comme un complexe orienté dont quelques propriétés topologiques peuvent être utilisées comme critères pour identifier les blocs. Cette méthode peut considérer, pour la première fois, les blocs convexes ou concaves.

Dans cet article, nous présentons la méthode utilisée dans le programme SIMBLOC de simulation tridimensionnelle de la blocométrie naturelle des massifs rocheux. A partir de la configuration des discontinuités simulées, une étude de la connectivité des discontinuités est faite d'abord pour éliminer les discontinuités non-connectées et fournir un modèle géométrique pour l'analyse d'écoulements de fluides. Des propriétés topologiques d'un polyèdre sont utilisées pour identifier un bloc discret. Trois méthodes de caractérisation de la blocométrie d'un massif rocheux sont présentées: distribution de taille, distribution de taille pondéré et distribution de l'orientation des blocs.

1 SIMULATION A 3D DES DISCONTINUITES

La méthode de simulation présentée dans cet article est une méthode généralement utilisée par d'autres auteurs comme Baecher et Lanney (1978), Long et al (1985), Andersson et Dverstorp (1987). Une discontinuité est modélisée comme un disque de rayon r avec une orientation α et un pendage β. Ces paramètres géométriques sont aléatoires et suivent certaines lois de distribution pour chaque famille de discontinuités.

L'estimation des paramètres à 3D comme l'extension et la densité des discontinuités se base sur des relations probabilistes appliquées aux paramètres mesurables à 2D comme l'extension et la densité des traces des discontinuités sur un affleurement. Un programme statistique GEOSTAT est développé spécialement pour ce faire (Xu, 1987). Dans ce programme, la méthode de la classification automatique est utilisée pour déterminer les familles directionnelles des discontinuités, et pour chaque famille, certains paramètres statistiques comme les moyennes et les écarts-types de l'extension, de l'orientation, du pendage et de la densité sont calculés simultanément. Les résultats d'un exemple réel de 183 discontinuités mesurées sur la paroi d'un tunnel de reconnaissance dans un massif de granite sont montrés dans le tableau 1. La figure 1 est la projection stéréographique des discontinuités sur laquelle on peut visualiser approximativement trois familles.

Dès que les paramètres statistiques de la géométrie des discontinuités sont obtenus, la simulation peut être réalisée d'une manière standard suivant la loi de distribution de chaque paramètre. Dans le programme, le domaine de simulation est dilaté par rapport au domaine réel suivant la valeur de l'extension des discontinuités pour générer un marge tampon et diminuer ainsi l'influence de marge (Xu,1987; Rouleau et Gale, 1987). Un exemple de configuration de discontinuités simulées dans un talus suivant les données des cinq familles du tableau 1 est montré dans la figure 2.

Tableau 1 Résultats d'analyse statistique des données de discontinuités

num fam	nombr dicon	E extens	E densit	E orient	E pendag	S.D. orient	S.D. pendag	disper param
5								
1	55	2.37	.226	341.2	81.9	20.4	9.2	12.3
2	89	3.15	.379	74.4	46.3	17.9	15.7	16.2
3	12	3.47	.057	251.2	45.7	25.0	11.2	16.2
4	15	2.10	.056	287.0	80.5	6.5	8.2	45.1
5	12	2.06	.056	240.7	88.5	14.8	7.8	22.4
4								
1	55	2.37	.226	341.2	81.9	20.4	9.2	12.3
2	89	3.15	.379	74.4	46.3	17.9	15.7	16.2
3	24	2.93	.106	245.8	67.2	21.1	20.6	7.7
4	15	2.10	.056	287.0	80.5	6.5	8.2	45.1
3								
1	55	2.37	.226	341.2	81.9	20.4	9.2	12.3
2	89	3.15	.379	74.4	46.3	17.9	15.7	16.2
3	39	2.47	.163	262.1	72.5	26.4	18.4	6.7
2								
1	55	2.37	.226	341.2	81.9	20.4	9.2	12.3
2	128	2.03	.612	77.4	61.2	21.1	21.2	5.0
1								
1	183	2.01	.856	97.1	72.6	43.6	23.7	2.7

E: moyenne; S.D.: écart-type

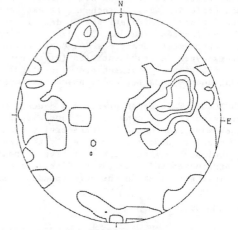

Fig 1 Projection stéréographique des discontinuités

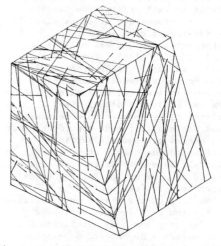

Fig 2 Configuration des discontinuités simulées

2 ETUDE DE LA CONNECTIVITE DES DISCONTINUITES

Une discontinuité peut constituer un cheminement d'écoulement ou une facette d'un bloc à condition quelle soit connectée avec le réseau global des discontinuités. Pour un problème d'écoulement, elle doit intersecter au moins deux autres discontinuités ou les limites du domaine. Pour un problème de blocométrie, elle doit intersecter au moins trois autres discontinuités ou les limites du domaine et les traces des intersections doivent se connecter mutuellement sur la discontinuité. Donc l'étude de la connectivité se base sur l'identification des intersections entre les discontinuités ou les limites du domaines.

L'intersection entre le disque d'une discontinuité i et le disque d'une discontinuité j peut être identifiée par les trois étapes suivantes: 1) déterminer la ligne d'intersection entre deux plans contenant les disques i et j; 2) calculer la distance entre la ligne d'intersection et les centres des disques i et j. Si la condition de l'extension n'est pas satisfaite, les deux discontinuités ne s'intersectent pas; 3) déterminer la partie commune d'intersection dès deux discontinuités. Si celles-ci ne sont pas décalées, il existe une partie commune et les deux discontinuités s'intersectent. Les limites du domaine de simulation peuvent être interprétées comme des polygones convexes, leurs intersections avec des discontinuités peut être identifiées par une méthode semblable à celle présentée ci-dessus. Quand cette procédure est appliquée à toutes les discontinuités, toutes les intersections auront été identifiées.

L'étude de la connectivité pour l'écoulement de fluides conduit à éliminer une discontinuité possédant moins que deux intersections. Pour le cas où une discontinuité n'intersecte aucune autre discontinuité, l'élimination peut être réalisée directement. Mais pour le cas où une discontinuité intersecte une autre, l'élimination de la discontinuité doit être réalisée ainsi que l'élimination de l'intersection correspondante et de la connexion avec la discontinuité intersectée. Le résultat de cette étude pour le système de discontinuités de la figure 2 est montré dans la figure 3.

L'étude de la connectivité pour la blocométrie de massifs rocheux consiste en une étude des traces des intersections sur chaque discontinuité et est réalisée dans un système local de coordonnées installé sur chaque discontinuité. La procédure est semblable à celle expliquée ci-dessus si une discontinuité est considérée comme un segment d'intersection et une trace d'intersection comme un point d'intersection. Une discontinuité sur laquelle il n'existe pas un réseau de traces d'intersection doit être éliminée en accompagnant l'élimination des intersections et des connexions avec les discontinuités intersectées. Le résultat de cette étude pour le système de discontinuités de la figure 2 est montré dans la figure 4.

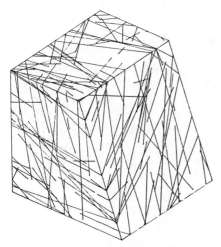

Fig 3   Réseau des discontinuités pour l'écoulement de fluides

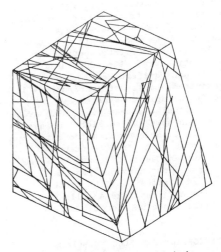

Fig 4   Réseau des discontinuités pour la blocométrie

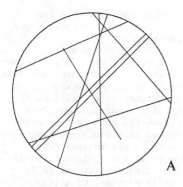
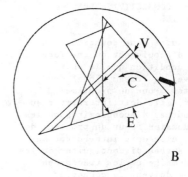

Fig 5 Sommets, arêtes et cycles sur une discontinuité. A: Traces des intersections;
       B: Cycles élémentaires orientés.

3 IDENTIFICATION DES SOMMETS, DES ARETES ET DES FACETTES

L'identification globale des sommets, des arêtes et des facettes est réalisée premièrement dans un système local de coordonnées installé sur chaque discontinuité. La figure 5-A montre une discontinuité circulaire et les traces des intersections avec d'autres discontinuités: un sommet V est un point d'intersection entre deux traces et donc est un point d'intersection entre trois discontinuités, une arête est un segment entre deux sommets et donc est un segment sur la ligne d'intersection entre deux discontinuités, une facette est définie comme un cycle élémentaire orienté dont la frontière est constituée d'arêtes orientées (cf figure 5-B).

Du fait qu'un sommet global est une intersection entre trois discontinuités, il peut apparaître dans différentes discontinuités. De même, du fait qu'une arête globale est une intersection entre deux discontinuités, elle peut aussi apparaître dans différentes discontinuités. La transformation vers le système global doit considérer ce phénomène, le même sommet ou la même arête doit avoir un même numéro global unique quand il est transformé de différents systèmes locaux. Une facette étant présente sur une seule discontinuité, le problème précédent n'existe pas. La méthode d'identification d'un cycle élémentaire est identique à celle d'identification d'un bloc bidimensionnel.

4 IDENTIFICATION DES BLOCS DISCRETS

L'identification d'un bloc à 3D commence par une facette (représentée par un cycle orienté) et se déroule en trois étapes: 1) orientation des facettes candidates; 2) choix de la facette la plus à gauche comme nouvelle facette du bloc; 3) test de la fermeture du bloc. Une facette intérieure peut constituer la facette commune à deux blocs voisins correspondant à deux normales opposées (cf figure 6). Par contre, une facette sur la surface du domaine ne correspond qu'à un bloc réel, l'autre est un bloc virtuel constitué par toutes les facettes sur la surface. La différence entre eux est que le bloc virtuel a un volume négatif et ne sera pas compté dans la blocométrie. Dans le programme, nous ne distinguons pas une facette intérieure d'une facette sur la surface du domaine afin de ne pas perdre la généralité de la méthode.

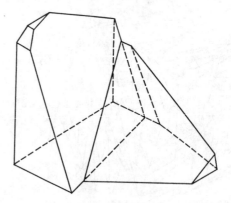

Fig 6 Deux blocs voisins correspondant à
       une facette intérieure.

Quand une facette est orientée, nous pouvons définir la normale à cette facette simplement en calculant le moment du cycle par rapport à un point quelconque de référence sur le plan de la facette. Dans le processus d'identification, un cycle est orienté pour que la normale à la facette soit toujours vers l'extérieur du bloc. Ceci peut être réalisé en considérant que, pour un bloc donné, si toutes les facettes sont orientées de sorte que leurs moments sont dirigés vers l'extérieur du bloc, alors l'orientation de l'arête commune à deux facettes voisines y est opposée. Suivant ce critère, nous pouvons orienter une facette candidate en vérifiant l'orientation de ses arêtes par rapport à l'orientation des arêtes des facettes adjacentes identifiées.

Si toutes les facettes candidates sont orientées, nous choisissons la facette la plus à gauche par rapport aux facettes identifiées précédemment comme une nouvelle facette du bloc.

Du fait qu'une arête a une orientation opposée sur deux facettes voisines, la somme de toutes les arêtes sur toutes les facettes d'un bloc sera nulle. Du point de vue topologique, cela signifie que la frontière de la surface d'un bloc est nulle. Si nous définissons un ensemble dynamique des arêtes pendant le processus d'identification d'un bloc comme l'ensemble des arêtes des facettes identifiées, nous pouvons l'utiliser pour tester la fermeture du bloc, c'est à dire, l'achèvement d'identification du bloc. Si l'ensemble est nul, le bloc est identifié. Le processus doit être arrêté et nous commençons l'identification du bloc suivant.

La surface engendrée par un cycle orienté peut être calculée en évaluant le moment du cycle et est égale à la moitié de la norme du moment. Le volume V d'un bloc de n facettes peut être calculé par les surfaces des facettes et les normales extérieures aux facettes, et s'écrit

$$V = (1/3) \sum_{i=1}^{n} (\vec{r}_{0i} \cdot \vec{n}_i) \, S_i$$

où $\vec{r}_{0i}$ est un point de référence sur la facette i, $\vec{n}_i$ est la normale extérieure à la facette i et S_i est la surface de la facette i

La méthode présentée ci-dessus pour identifier un bloc ne fait pas l'hypothèse de la convexité du bloc. Un bloc concave peut être identifié d'une même manière qu'un bloc convexe, mais nous avons

supposé que les blocs sont homomorphiques en terme de topologie. Cela sera généralement satisfait si une étude de la connectivité des discontinuités et des traces des intersections sur chaque discontinuité est faite et si les réseaux locaux de discontinuités (non connectés au réseau principal) sont éliminés.

5 BLOCOMETRIE DE MASSIFS ROCHEUX

La taille d'un bloc peut être représentée par son volume ou tout simplement par son diamètre équivalent. Nous utilisons ici la racine cubique du volume pour représenter le diamètre équivalent d'un bloc. La distribution de taille (diamètre équivalent) F(x) montre certaines propriétés de la structure d'un massif rocheux. Par exemple, on peut mesurer le pourcentage de nombre de blocs dans un intervalle de taille. Cette distribution et l'histogramme de fréquence du système de blocs de la figure 4 sont montrés dans la figure 7.

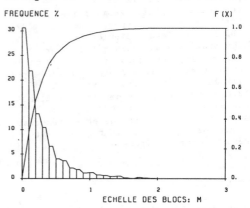

Fig 7   Distribution de taille des blocs

L'inconvénient de cette représentation est qu'elle n'insiste que sur l'importance en nombre, et ne montre pas clairement l'importance de la taille elle-même. Dans un massif rocheux, l'intersection entre des discontinuités aléatoires peut générer un très grand nombre de blocs de petite taille, le volume de ces petits blocs pouvant être très faibles. Leur influence sur la stabilité ou la fragmentation d'un massif rocheux ne sera pas aussi importante qu'on pourrait le penser au vu de leur pourcentage en nombre. Ainsi nous utilisons une autre méthode de représentation de la blocométrie, appelée distribution de taille pondérée G(x) définie comme le pourcentage du volume des blocs de diamètre inférieur à une valeur

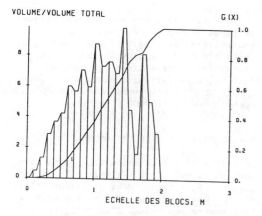

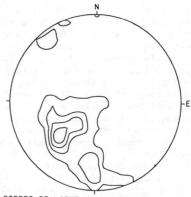

STEREO DE L'AXE MAX. DES BLOCS

Fig 8 Distribution de taille pondéré

Fig 9   Projection stéréographique de l'orientation des blocs

x, sur le volume total des blocs du système. Cette distribution et l'histogramme de fréquence du système des blocs de la figure 4 sont présentés dans la figure 8. On peut voir la différence importante avec la figure 7.

Une autre propriété géométrique importante des blocs pour l'analyse de la stabilité et l'évaluation de la fragmentation est la distribution de l'orientation des blocs. L'orientation d'un bloc est définie dans cet article par celle de l'axe principal maximum. La distribution sphérique de l'orientation des blocs peut être présentée par la projection stéréographique et le résultat du système de blocs de la figure 4 est présenté dans la figure 9.

CONCLUSION

La méthode présentée dans l'article pour identifier un bloc discret intersecté par des discontinuités utilise les méthodes simples de la géométrie, et ne se base pas sur l'hypothèse de la convexité des blocs. Les blocs concaves peuvent être identifiés de la même façon que les blocs convexes. Les études de la connectivité des discontinuités nous permettent de fournir les modèles géométriques nécessaires à l'analyse des écoulements, à l'analyse mécanique ou à l'analyse couplée hydromécanique.

BIBLIOGRAPHIE

Andersson, J. and B. Dverstorp 1987 Conditional simulation of fluid flow in three dimensional networks of discrete fractures. Water Resources Research, Vol.23, No.10, 1876–1886.

Baecher G. B. 1983 Statistical analysis of rock mass fracturing. Mathe. Geol. vol.15(2), 329–348.

Cundall P. A. 1971 A computer model for stimulating progressive large scale movements in blocky rock systems. Proc. Int. Symp. Rock Fracture, ISRM, Nancy.

Cundall P. A. etc 1988 Formulation of a three dimensional distinct element model, Int. J. Rock. Mech. Min. Sci. & Geomech. Abstr. Vol.25, 107–125.

Hudson J. A. and S. D. Priest 1979 Discontinuities and rock mass geometry. Int. J. Rock. Mech. Min. Sci. & Geomech. Abstr. Vol.16. 339–362.

Hudson J. A. and S. D. Priest 1983 Discontinuity frequency in rock masses. Int. J. Rock. Mech. Min. Sci. & Geomech. Abstr. Vol.20. 73–89.

Lin D., C. Fairhurst & A. M. Starfield 1987 Geometrical identification of three dimensional rock block systems using topological techniques. Int. J. Rock. Mech. Min. Sci. & Geomech. Abstr. Vol.24. 331–338

Warburton P. M. 1980 A stereological interpretation of joint trace data. Int. J. Rock. Mech. Min. Sci. & Geomech. Abstr. Vol.17. 181–190

Warburton P. M. 1980 Stereological interpretation of joint trace data: Influence of joint shape and implications for geological surveys. Int. J. Rock. Mech. Min. Sci. & Geomech. Abstr. Vol.17. 305–316.

Xu Jixian 1987 Déscription statistique et simulation des discontinuités de massifs rocheux. Mémoire de DEA, CGI, Ecole des Mines de Paris.

Application of surface spline function to engineering geology
Application de fonction 'spline' de surface à la géologie de l'ingénieur

Zhou Chuangbing
Wuhan University of Hydraulic & Electric Engineering, Wuhan, People's Republic of China

ABSTRACT: In this paper, SSF(surface spline function) is introduced according to the feature of engineering geology. The mathematical model of SSF and its algorithm are presented. The main application of SSF is to analyse the form of geological discontinuities such as fault, weak interlayer, slip surface and interface between different weathered zones. It is also used for the anti-calculation of the strength parameters of a slip surface. Three illustrative examples are given here. It is shown that SSF, as an effective means processing geological data, has the advantage of being flexible, accurate and distinct for a practical use.

RESUME: Selon la caractéristique de la géologie de l'ingénieur, la fonction "spline" de surface se introduit dans cette exposé. Le modèle mathématique et l'algorithme sont proposés. Les principaux applications sont à analyser les formes des discontinuités géologiques comme la faille, l'intercalation faible, l'interface entre plusieurs zones altérées. Elle est encore utile pour calculer le forme et les parametres mécaniques de la surface de glissement. Trois examples illustratives sont présentés. On a trouvé que la fonction "spline" de surface est un moyen excellent, car elle est maniable, précise et distincte en pratique.

1 INTRODUCTION

It is often required to evaluate the form of a geological plane in engineering geology. A conventional way is to estimate it approximately using boreholes data. Since recent years, trend surface analysis has been applied to engineering geology, and certain reasonable results achieved. The method of trend surface analysis, however, poses certain major difficulties for the user, at least in the selection of the order of a surface, and the collection of the sufficient data used for statistical analysis. It is therefore the objective demand for the assessment of engineering geological problem to establish an approach to the form analysis. The proposed method should be instituted on the following basis:

It should make full use of the practical data which have been examined based on geological condition.

It should transform a geological problem to a relative mathematical one rigorously.

The solution derived from the model should be applied to geological prediction.

This results in the use of SSF, which is proved to satisfy the principles mentioned above. One-dimensional spline function was originally proposed by I.J. Schoenbeg (1946), and two-dimensional one was developed late. Up to 1960, Brikhoff & Garabedien extended initially the theory of third-order spline to higher dimensions, and ever since that time, Boor & Ahlberg et al. have made comprehensive researches. However, the great majority of present splines are merely suitable for a regular region, whilst the study of two-dimensional interpolation problem for an irregular region is not still perfect. Because of the fact that the layout of boreholes is controlled by such factors as geology, hydrology and landform, it is generally impossible to form a regular netty arrangement of boreholes. In addition, according to the second-order or third-order interpolation, the partial derivative at boundary is required, and difficult to be realised. For this reason, SSF is introduced here, and its advantage is that the coordinates of points might not be arranged in a regular way, and only natural boundary condition is adopted instead of any information relating to partial derivatives. It should be pointed out that geological planes not only have complicated forms, but also are usually discontinuous due to tectonic interference. In geological sense, because a discontinuous plane is fitted to a smooth one, SSF may distort the original appearence of the plane. However, the anomaly often occurs just in the distorted place. As a result, various geological phenomena can be recognised by means of certain mathematical treatments.

2 MATHEMATICAL MODEL OF SSF

Suppose that P is a set of known points relating to a certain plane, and the coordinates of a known point are x_i, y_i and z_i, then P can be expressed as

$$P = [(x_i, y_i, z_i)] \qquad (1)$$

Suppose that function $Z(x,y)$ controlled by set P within the known boundary represents the form of the plane, obviously, at the known points there are:

$$Z_i = Z(x_i, y_i) \qquad (2)$$

The function $Z(x,y)$ is constituted by using the elastic deformation equation of an infinite flatten plate, and can be written as:

$$Z(x,y) = a_0 + a_1 x + a_2 y + \sum_{i=1}^{n} t_i \, \eta_i^2 \ln(\eta_i^2 + e)$$

$$\eta^2 = (x - x_i)^2 + (y - y_i)^2 \qquad (3)$$

Where a_0, a_1, a_2 and t_i are unknown coefficients needed determining, e is a parameter adjusting the curvature of the plane generally, a small value of e, such as 10^{-4} to 10^{-5}, is needed when the plane is greatly curved, on the contrary, a relative large value of e, such as 10^{-1} to 10^{-2}, is expected for a flatten plane. There are n+3 unknown coefficients in formula (3) which can be determined by the matrix equation below:

$$A \cdot X = B \qquad (4)$$

where $X = (t_1, t_2, \cdots, t_n, a_0, a_1, a_2)^T$, and $B = (Z_1, Z_2, \cdots, Z_n, 0, 0, 0)$

Matrix A is a symmetry one in which all diagonal elements are equal to zero. Hence, Householder transformation or principal element method should be adopted to resolve the above matrix equation.

To differentiate $Z(x,y)$ with respect to x and y, the expressions of the first-order and the second-order partial derivatives are as follows:

$$\frac{\partial Z(x,y)}{\partial x} = a_1 + 2 \cdot \sum_{i=1}^{n} t_i \left[\eta^2 + e \right] + \ln(\eta^2 + e)] \cdot (x - x_i) \qquad (5)$$

$$\frac{\partial Z(x,y)}{\partial y} = a_2 + 2 \cdot \sum_{i=1}^{n} t_i \left[\eta^2/(\eta^2 + e) + \ln(\eta^2 + e)] \cdot (y - y_i) \right. \qquad (6)$$

$$\frac{\partial^2 Z(x,y)}{\partial x^2} = 2 \cdot \sum_{i=1}^{n} t_i \left[\eta^2/(\eta^2 + e) + \ln(\eta^2 + e) + 2 \cdot (x - x_i)^2 (e/(\eta^2 + e)^2 + 1/(\eta^2 + e)) \right] \qquad (7)$$

$$\frac{\partial^2 Z(x,y)}{\partial y^2} = 2 \cdot \sum_{i=1}^{n} t_i \left[\eta^2/\eta^2 + e) + \ln(\eta^2 + e) + 2 \cdot (y - y_i)^2 (e/(\eta^2 + e)^2 + 1/(\eta^2 + e)) \right] \qquad (8)$$

The apparent dip angle can be calculated from the first-order partial derivatives:

$$\alpha_x = \operatorname{arctg}\left[\frac{\partial Z(x,y)}{\partial x}\right]$$
$$\alpha_y = \operatorname{arctg}\left[\frac{\partial Z(x,y)}{\partial y}\right] \qquad (9)$$

and the variance of the apparent dip angle from the second-order partial derivatives:

$$\Delta\alpha_x = \operatorname{arctg}\left(\frac{\partial^2 Z(x,y)}{\partial x^2} \cdot \Delta x + \frac{\partial Z(x,y)}{\partial x}\right) - \operatorname{arctg}\left(\frac{\partial Z(x,y)}{\partial z}\right)$$

$$\Delta\alpha_y = \operatorname{arctg}\left(\frac{\partial^2 Z(x,y)}{\partial y^2} \cdot \Delta y + \frac{\partial Z(x,y)}{\partial y}\right) - \operatorname{arctg}\left(\frac{\partial Z(x,y)}{\partial y}\right) \qquad (10)$$

Based on the gradient concept, the real dip angle of a plane can be susquently calculated:

$$\alpha = \operatorname{arctg}\left(\sqrt{(\frac{\partial Z(x,y)}{\partial x})^2 + (\frac{\partial Z(x,y)}{\partial y})^2}\right) \qquad (11)$$

The direction of dip can be then defined as:

when $\dfrac{\partial Z(x,y)}{\partial y} < 0$

$$S_t = \begin{cases} \operatorname{arctgs} & \text{when } \dfrac{\partial Z(x,y)}{\partial x} < 0 \\ 360 - \operatorname{arctgs} & \text{when } \dfrac{\partial Z(x,y)}{\partial x} > 0 \end{cases} \qquad (12)$$

when $\dfrac{\partial Z(x,y)}{\partial y} > 0$

$$S_t = \begin{cases} 180 - \operatorname{arctgs} & \text{when } \dfrac{\partial Z(x,y)}{\partial x} < 0 \\ 180 + \operatorname{arctgs} & \text{when } \dfrac{\partial Z(x,y)}{\partial x} > 0 \end{cases} \qquad (13)$$

where $s = \dfrac{\partial Z(x,y)}{\partial x} / \dfrac{\partial Z(x,y)}{\partial y}$

On the basis of the above calculation, the graphical presentation of a plane can be easily produced by means of CAD, and the anomaly can be clearly shown in the place where the occurrence of the plane changes considerably.

Because of the ignorence of the uncertainty of geological parameters, the above model still has limitations. For this reson, a statistical model is necessary to describe the uncertainity of geological parameters. The matrix B in formula(4) can be then broken into as follows:

$$B_1 = (Z_1, Z_2, \cdots, Z_i, Z_k)^T$$

$$B_2 = (Z_{k+1}, Z_{k+2}, \cdots, Z_n, 0, 0, 0)^T \qquad (14)$$

Where matrix B_2 is considered to be deterministic, while B_1 to be statistic. Obviously, the function $Z(x,y)$ and the matrix X in formula(4) are correspondingly statistic, moreover $Z(x,y)$ is a linear function with respect to matrix B through the process of matrix A.

Suppose Z_i (i=1,2,$\cdots$,k) tends to Gauss distribution,then pdf can be expressed as :

$$f(x) = \frac{1}{\sqrt{2\pi}\,\sigma_i} \exp[-\frac{(x-\mu_i)^2}{2\sigma_i^2}] \qquad (15)$$

Where μ_i is mean value of Z_i, σ_i is mean square derivation of Z_i. According to the property of linear function of indepentent Gaussian random variables. $Z(x,y)$ also tends to Gauss distribution with the distribution parameters as:

$$\mu = Z(x,y) \Big|_{Z_i = \mu_i}$$
$$\sigma = \sqrt{\sum_{i=1}^{k} \sigma_i} \qquad (16)$$

Therefore, the varying interval of a plane can be predicted, and the probability of $Z(x,y)$ smaller than the presupposed value Z_e is

$$P_e = \Phi(\frac{Z_e - \mu}{\sigma}) \qquad (17)$$

While the probability of $Z(x,y)$ larger than the presupposed value Z_e is

$$P_f = 1 - \Phi(\frac{Z_e - \mu}{\sigma}) \qquad (18)$$

Where $\Phi(\cdot)$ is the probability distribution function of Gauss distribution.

3 APPLICATION TO ENGINEERING GEOLOGY

In order to provide scientific basis for reasonable engineering assessment, the application of SSF is mainly to investigate the spatial distribution and the form of any large-scale geological discontinuity.

3.1 Approach to buried fault

The emphasis of approach to buried fault is to deal with two kinds of problems which are frequently encountered in practice, they are:

1. Although the fault has been identified, further investigations are still required to inspect its extent and its form feature as detailed as possible, and to analyse its influence on engineering structure;

2. According to combinatorial tectonic analysis as well as geomorphological anomaly analysis, the deep-seated fault is supposed to be covered, but it has to be confirmed.

For the first case, a certain number of boreholes are generally required. From the data obtained and with the use of SSF, the spatial distribution of the fault can be displayed in terms of contour line map and occurrence variance map as well. With those maps profiles in different directions can be plotted in light of the engineering demands. Those profiles are meaningful for the layout of engineering structures and the choice of remidical schemes.

For the second case, at least two marker beds have to be selected. The contour line map and occurrence variance map of the marker beds are subsquently produced depending on available data. Through comparative analysis of the anomaly which occurred in cross profiles and longitudinal profiles, it can be discriminated that whether the fault exists or not. If the fault is approved, the age estimation and the mechanical property analysis of the fault are required. An illustrative example is shown as follows:

At the beginning of the preliminary feasibility study of E Zhou power plant, it was felt quite trouble to find out the tectonic framework, because the ground surface was extensively covered by residual soils. A large number of boreholes were then designed in different places, and a large amount data were achieved, from which it would appear that the geology of the region consists largely of sedimentary rocks and volcanic rocks, which were seriously folded during Cretaceous period. Inspite large-scale fault had not been found out, it was suggested that faults in northeast direction would exist according to the comprehensive analysis of landform and geomorphological anomaly. In order to confirm the suggestion, SSF was adopted in the investigation. 37 boreholes were carefully examined, and the data collected from these were analysed statistically to reveal the geological and hydrological factors known to contribute to tectonics. Two formations that were verified to be stable both in extension and in thickness were selected as marker beds; one is volcanic rocks (J_3') of upper Jurassic, and the another is DAYE limestone (T_1) of lower Triassic. In the study, a standard statistical program entitled SPLINE was used to handle the data. The techniques were employed range from simple graphical presentation to more complicated statistical routines (e.g. discriminant analysis). All those work is conveniently fulfilled with the help of CAD. It was noted that a certain quantity of profiles in different directions were necessary to make the interpretation sucessful and reliable. Hence, 8 profiles parpendicular and parallel to the anomaly respectively were established with CAD in the study. In addition, the contour line maps and occurrence variance maps of two marker beds were obtained. Fig. 1 shows only the contour line map of the first marker bed together with the geological interpretation.

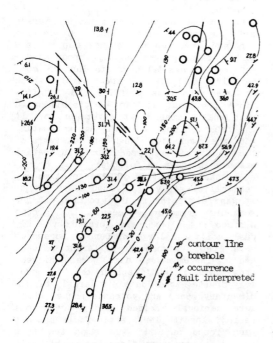

Fig.1 contour line & occurrence of 1st marker bed

As shown in Fig.1, the strata dip northwestward on the whole, but there are local domings and depressions that may be tectonic indication. It is interpreted that there are two sets of deep-seated faults developed; one set, striking northeast, is of a tensional shear fault, the another set, striking northwest, is of a shear slip fault.

According to the geological horizon where faults extended, fault formed during Cretaceous period. The above recognition had been eventually confirmed by seismic prospecting.

3.2 Approach to weak interlayer

It is well known that weak interlayer widely developed within rock foundations is an object of many geological and geomechanical investigations owing to its slightly dipping and lower strength. The spatial form of the interlayer and the combinatory relationship with other tectonic elements can be distinctly shown by using SSF. Based on the contour line map obtained, Profiles in the direction of engineering force can be clearly displayed

As a result, graphical presentation relating to the relief variance can be generated after comprehensive analysis in all profiles. The achievements usually provide reference to the determination of the shear strength of the interlayer. A few case studies had been undertaken in the dam construction of China, and satisfactory results obtained.

3.3 Approach to back analysis

It is necessary to examine the distribution regularity of slip surface and its mechanical parameters in the stability assessment and in carrying out any preventive works found necessary. It is certain that drillings and mechanical tests can be noticably reduced, the economic benifit can be largely improved so long as the method of SSF combined with optimization technique is properly employed. It is called as back analysis of slip surface form and mechanical parameters, which is instituted on the following basis:

1.Along the known boundary or potential boundary that can be identified if there were signs of incipient failure, coordinates of points should be measured as much as possible such as N1.

2. Boreholes or trenchings should be arranged purposefully in the slope, and be designed to reveal the potential slip

surface. The number of boreholes are expected appropriate, not necessaryly more but enough, such as N_2.

3. Two or more representive profiles should be selected, and be sliced in light of the structural characteristics of the slope. Using conventional equilibrium analysis, the factor of safety can be calculated if the extension and the mechanical parameters of the slip surface are known which, however, are not available before calculation, and can be anti-calculated by means of optimization technique.

Suppose that N_3 points indicating slip surface except for N_2 boreholes have been obtained. Therefore, the set P referred to the distribution of slip surface are set up. It is obvious to see that if increasing in the capacity of set P is demanded, then increasing in N_1 or N_3 is relatively feasible compared with the increasing in N_2 which frequently costs a lot.

On the basis of the above consideration, a practical slope has been studied. The geology of the slope consists of LONGTAN feldspar sandstone of upper Permian. Rock mass has medium strength, but are weathered to considerable depth. 18 boundary points, 7 boreholes, two profiles as shown in Fig.2 were selected, and Box method adopted in the back analysis.

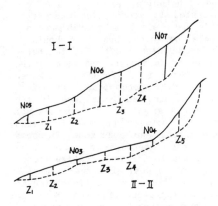

Fig.2 calculational profiles for back analysis.

In accordance with the slope displacement data, the factor of safety was reckoned to be close to unity. With reference to test data and with a view to the unhomogeneity of rock mass, the constraint condition of strength parameters were assumed as:

$$0 < C < 200 KPa \qquad (19)$$
$$15° < \varphi' < 35°$$

Where C is cohesion intercrpt in terms of effective stress, φ' is the angle of internal friction in terms of effective stress.

The results obtained are listed in Table 1.

Table 1　Results of back analysis

Profiles	Z_1*	Z_2	Z_3	Z_4	Z_5	C	φ'
I - I	8.58	12.76	19.88	27.25	/	136	28.0
II - II	7.21	9.90	19.60	21.62	35.88	59	31.3

* Z_i (i=1,2,···,5) is the elevation of the slip surface as shown in Fig.2, its unit is metre, C is cohesion intercept, φ' is is internal friction angle.

The contour line map and the occurrence of the slip surface are shown in Fig.3, which can explain why the strength parameters gained by back analysis are slightly larger than that of tests.

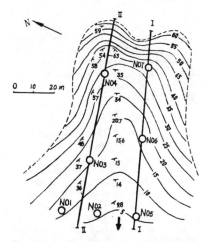

1 contour line, 2 boundary line
3 occurrence of slip surface
4 borehole & its code
5 profile & its code
6 sliding direction

Fig.3 contour line and occurrence of the slip surface

807

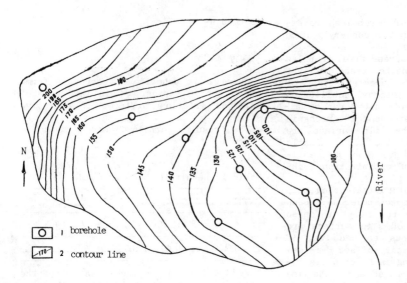

Fig.4 contour line of the top of heavily weathered zone in BANLIN slope

For the analysis of sliding direction, 277 points at different parts of the slip surface were reconsidered, and those points were broken into 5 groups based on dip direction. The sliding direction can be calculated in such a way that the mean dip direction of every group is used as vector direction, the number of points as vector magnitude, then composite analysis of vectors is taken. The direction of composite vector in the study is 250.2° which is quite consistent with factural sliding direction. Moreover, the volume of slide mass had been estimated by numerical integration.

3.4 Approach to interface between different weathered zones

It is known that rocks are generally weathered to considerable degree at ground surface, but the weathering intensity is gradually reduced with the increase in depth, except certain anomalous zones probably due to tectonic interference or infiltrating water. Such routines as drilling, trenching and acoustic sounding are required to study the weathering feature, from which a large amount of data can be collected. The relationship between rock weathering and geological, hydrological factors could be clearly recognised with SSF and relative techniques.

The case study aimed at the bank slope stability was carried out during the pre-liminary investigation of LECHANG power station. The steepness of the slope is about 30 to 40 degree, and ring-like ditches in geomorphic expression is clear. The slope consists of phyllitic micaceous quartzose sandstone intercalated with siltyslate of Cambriarn. The formation extends with varying occurrence, and was fractured with faults.

As suggested by geomorphic interpretation, this bank slope is probably an ancient rockslide which, however, should be examined by further studies. It is therefore considered to analyse the form of the interface between different weathered zones with SSF. 8 boreholes, 4 trenches, 11 outcrop points had been used in the analysis. Fig.4 shows the contour line of the interface between heavily weathered zone and entirely weathered zone. It seems clear that the interface is quite irregular, unfitted to the land configuration, especially it has weathered deep trenches on it. It is exceedingly enlightening to note the analogy between the form of the interface and the slip surface. Further surveyings were then carried out, and a layer of red clay adjacent to the top of heavily weathered zone, marked with sliding trace, 0.2 to 0.4 metre thick, wavelike extending, was discovered. The age of the red clay was determined to be about 1,600 years by means of heat release technique. Therefore, it was confirmed that the bank slope is an ancient rockslide.

4 CONCLUSIONS

On the practical application, following conclusions can be achieved:

Inspite of the fact that the application of SSF to engineering geology is still at the tentative stage, the use of SSF is possibly the most ambitious attempt carried out to process complex geological data and to establish quantitative models. It is because that the use of SSF can present various geological index and diagrams mainly depending on boreholes data without any specific care. This can raise the efficiency of borehole data utilization, it can also overcome the limitation on ground investigation, moreover, it can account for deep-seated structure and other adverse geological phenomena. SSF, as an effective means processing data, can be applied to any places where the form analysis is required, and expected to have extensive applications.

The combination of SSF with optimization technique is a unique development in the assessment of slope stability. Not only the form and the mechanical parameters of the slip surface for a failed slope can be quantitatively evaluated, but also the potential slip surface for a deformational slope can be predicted under three-dimensional condition.

It is an important consideration that the analysis of diagrams obtained with SSF should be referred to the firsthand materials, and the anomaly should be reasonably interpreted according to the specific condition. Furthermore, the interpretation should be explained in terms of geological process.

REFERENCES

Ahlberg, J.H., Nilson, E.N. & Walsh,J.L., (1967) The theory of splines and their applications, New York.
Lyche, T. & Schamaker, L.L., (1975) Local spline approximation methods. J.Approx. Theory vol. 15, p. 294~325.
Schoenberg, I.J., (1973) Cardinal spline interpolation, soc. for Indust. and Appl. Math.
Zhou, C.B. & Pen, Y.H. (1988) Back analysis of slip surface form and strength parameters, Proc. 3th national engin. geo. conference, p.421~429. (in Chinese)